MEDICAL CYTOGENETICS

MEDICAL CYTOGENETICS

edited by

Hon Fong L. Mark

Brown University School of Medicine
Rhode Island Department of Health
Providence, and
KRAM Corporation
Barrington, Rhode Island

MARCEL DEKKER, INC. NEW YORK • BASEL

Financial support from the following organizations for reproduction of the color art is gratefully acknowledged: Applied Spectral Imaging, Applied Imaging, Inc., Carl Zeiss, Inc., Chroma Technology Corporation, Oncor, Inc., and Vysis, Inc.

ISBN: 0-8247-1999-9

This book is printed on acid-free paper.

Headquarters
Marcel Dekker, Inc.
270 Madison Avenue, New York, NY 10016
tel: 212-696-9000; fax: 212-685-4540

Eastern Hemisphere Distribution
Marcel Dekker AG
Hutgasse 4, Postfach 812, CH-4001 Basel, Switzerland
tel: 41-61-261-8482; fax: 41-61-261-8896

World Wide Web
http://www.dekker.com

The publisher offers discounts on this book when ordered in bulk quantities. For more information, write to Special Sales/Professional Marketing at the headquarters address above.

Current printing (last digit):
10 9 8 7 6 5 4 3 2 1

PRINTED IN THE UNITED STATES OF AMERICA

To my family, without whom all worldly achievements are meaningless; and to the memory of my sons, Lancelot and Vincent, who died of a rare genetic disease contracted from parents born worlds apart and brought together by destiny.

Foreword

In this age of electronic communication and rapid advancements in genetic knowledge and technology, one might wonder about the usefulness of a textbook in any field of genetics. However, many of the basic tenets of genetics have stood the test of time. Understanding and applying of the newest advances requires a sound appreciation of these basic tenets. This book, edited by Hon Fong L. Mark, with chapters by recognized experts in the field, provides a thorough and clear starting point for students new to cytogenetics. At the same time, it provides complete coverage of the many current applications of cytogenetics in medicine. These include a variety of culturing and staining techniques as well as new methods of in situ hybridization ranging from simple chromosome identification with centromeric α-satellite probes to the newest spectral array of the entire karyotype.

The format is well suited to allow the reader to grasp basic concepts. There are helpful glossaries to ensure clear understanding of the factual material and a current bibliography for those who want to read the original sources of the material provided.

It has been within the last few decades that the correct number of chromosomes in humans has been appreciated, that we have seen that sexual dimorphism in mammals is caused by dimorphisms of sex chromosomes, and that chromosomal analysis has been shown to have potential as a powerful tool in molecular understanding of many forms of cancer. With the advent of the polymerase chain reaction and other powerful tools for the study of human diseases at the molecular level, some felt that these advancements signaled the demise of cytogenetics as a major player in the study of human genetic disorders. Rather, cytogenetics has continued to advance in technology and knowledge at a pace comparable to molecular genetics, and the two have become wedded in a marriage in which each partner is increasingly difficult to identify individually. *Medical Cytogenetics* provides an excellent resource to bring new students and old dogs into fascinating fields of cytogenetics.

Thaddeus E. Kelly
Professor of Pediatrics
Director, Division of Medical Cytogenetics
University of Virginia Health Sciences Center
Charlottesville, Virginia

Preface

While preparing for the certification examination for the American Board of Medical Genetics in Clinical Cytogenetics a number of years ago, my colleagues and I observed that, while there are numerous reference books and technical texts on cytogenetics, books giving an overview of medical genetics and clinical cytogenetics are at best few and at worst nonexistent. There is a unique niche for such a text. Thus, this book provides an overview of the field of cytogenetics in medicine, with an emphasis on the practice of clinical cytogenetics in a variety of settings, such as a department of pathology or pediatrics in an academic medical center, a government-sponsored laboratory, or a commercial enterprise.

The targeted audience for this book consists of physicians and scientists in training or in practice, medical and advanced undergraduate and graduate students interested in medical genetics, cytogeneticists studying for the certifying examination in Clinical Cytogenetics offered every three years by the American Board of Medical Genetics (ABMG), and board-certified medical geneticists desiring a review of the field of cytogenetics for recertification* and other purposes. It is for this audience that great pains have been taken to compile a large number of practice questions following many chapters, together with answers to these questions in a separate manual. For the same reason, a glossary has also been compiled for each chapter. It is hoped that other medical practitioners, such as pathologists, family medicine physicians, internists, neurologists, occupational medicine physicians, hematologists, medical oncologists, pediatricians, endocrinologists, obstetricians and gynecologists, maternal and fetal medicine physicians, urologists, public health personnel, and other specialists will find this publication to be a useful reference on cytogenetics as well.

*The Editor was a member of the seven-member national Recertification Committee of the ABMG.

It is not the goal of this book to be encyclopedic. Rather, we aim to convey the excitement of this rapidly evolving field in as simple and comprehensible a fashion as possible. We strive to make this intrinsically interesting topic even more interesting by offering bountiful examples and illustrations.

This book contains 20 chapters. The first chapter discusses the discipline of medical genetics and the place of cytogenetics in the bigger scheme of things. It is only logical that the former President of the American College of Medical Genetics, Dr. Reed Pyeritz, be invited as a co-author. Chapters 2 to 6 provide basic cytogenetic information on the human genome and the techniques for gathering and interpreting that information. The Tharapels, co-authors of Chapter 3, were responsible for the organization of the Memphis Conference, which made a significant contribution to the clarification of human cytogenetic nomenclature. Dr. Sandra Wolman is a pathologist and cytogeneticist with many years of experience and is well qualified to write with me Chapter 4, on culture and harvest techniques. (Sandy also authored Chapter 15, on breast cancer, which is one of her many areas of expertise.) Chapter 6 provides a thorough introduction to the technique of FISH. Drs. Blancato and Haddad have extensive research experience using this molecular cytogenetic technique.

Chapter 7, by Dr. Bruce Korf and Dr. Nancy Schneider, discusses the applications of cytogenetics in medicine. Nancy was formerly Chair of the College of American Pathologists (CAP)/American College of Medical Genetics (ACMG) Cytogenetics Resource Committee. Bruce was a recent Chair of the American College of Medical Genetics Program Committee, on which I also served, and is one of the most respected geneticists in the New England region.

Following this chapter, more specialized topics in cytogenetics are discussed. Whereas many chapters in this book deal with important and essential topics that can be found in other cytogenetics texts, other chapters—such as those on chromosomal instability and fragile sites, breast cancer, male infertility and transfusion medicine—are unique. Neither Dr. Allen Lamb nor Dr. Wayne Miller, co-authors of Chapter 8 on prenatal diagnosis, needs introduction; both are well known in the cytogenetics community. Dr. Dorothy Warburton, author of Chapter 9, on the cytogenetics of reproductive wastage, is a recognized authority in the field. Dr. Mark Sigman, my colleague and friend at Rhode Island Hospital and co-author of Chapter 10, on the cytogenetics of male infertility, is an expert on this subject. Chapter 11 was written by Drs. John Anastasi and Diane Roulston. Dr. Anastasi is a hematopathologist and was one of the first pathologists to pioneer the use of FISH in hematopoietic disorders. Dr. Roulston, a valued colleague, is a director of the Cancer Cytogenetics Laboratory at the University of Chicago, working with Drs. Michelle Le Beau and Janet Rowley.

Dr. Susana Raimondi has published extensively on the cytogenetics of lymphoid malignancies and therefore was a most logical person to write Chapter 12. Dr. Avery Sandberg is the Editor-in-Chief of *Cancer Genetics and Cytogenetics* (CGC), on whose editorial board I have the honor of serving. Dr. Zhong Chen is the Associate Editor of CGC. These co-authors of Chapter 13 need little introduction. My colleagues and friends Drs. Herman Wyandt, Vijay Tonk, and Roger Lebo have many years of experience in the field of cytogenetics and have done an outstanding job writing Chapter 14. Chapter 16 focuses on the cytogenetics of transfusion medicine, a topic not usually covered in a cytogenetics text. Dr. Carolyn Young is the Medical Director at the Rhode Island Blood Center and is solely responsible for the transfusion medicine aspect of the chapter.

Whereas the goal of Chapter 17 is to summarize the possible applications of FISH for the average clinical cytogenetics laboratory of the 1990s, Chapter 18 (by Gabriela Green et al.) provides a glimpse of what may become routine in the average clinical cytogenetics laboratory of the future. The laboratory of co-author Dr. Thomas Ried is one of the top laboratories in the world and actively conducts cutting-edge research on molecular cytogenetic technologies. It was Thomas's group that first pioneered the technique of spectral karyotyping (SKY), which has recently generated so much excitement in the cytogenetics community.

Many cytogenetics texts omit a discussion of specimen preparation and other preanalytic variables that might affect the quality of cytogenetic testing. Thus, Chapter 19 was commissioned for this purpose, together with Chapter 20 on quality control and quality assurance issues. Dr. Nancy Schneider was responsible for suggesting the former and I am pleased that my colleague and friend, Dr. Jila Khorsand, a very experienced pathologist, agreed to take on the project. My colleague and friend Dr. Gerald Hoeltge is a well-known pathologist and cytogeneticist from the Cleveland Clinic. Few are more qualified than Gerry to write about setting standards in clinical laboratories. He is the Inter-Regional Commissioner for the Laboratory Accreditation Program of the College of American Pathologists.

In summary, I am indeed fortunate to have so many commitments from such a large number of distinguished chapter authors, many of whom are considered top experts in their fields. I have worked very hard to co-ordinate the chapters so that topics do not significantly overlap. However, a certain degree of redundancy was deemed valuable for emphasis.

Before closing, I wish to thank Dr. Michele Sinoway and Sandra Beberman of Marcel Dekker, Inc., who invited me to be the editor of this volume. Lia Pelosi and Elizabeth Curione, as production editors, and other staff members, also assisted during the production of this book. These highly competent individuals extended themselves above and beyond their duties to ensure the successful completion of this project. For their assistance and support I am deeply grateful.

I am grateful to the contributors who, in addition to completing their own chapters, served freely as reviewers and readers for other chapters.

The following colleagues helped read and critically review the chapters of this book: Dr. Zhong Chen, Dr. Janet Cowan, Dr. Lewis Glasser, Dr. Gerry Hoeltge, Dr. Abby Maizel, Dr. Yvonne Mark, Mr. David Mello, Ms. Anne Richardson, Dr. Mary Sandstrom, Dr. Nancy Schneider, Dr. Dorothy Warburton, Dr. Lee Wells, and Dr. Sandy Wolman. Many of these colleagues reviewed more than one chapter. Many chapters were reviewed by colleagues of the chapter authors, and I wish to thank them as well. Dr. Roger Mark read each chapter of the book at least five times.

My students, especially Ms. Jennifer Gilbert and Ms. Sandy Chang, were especially good at detecting errors that various readers and reviewers missed.

It is customary for book editors and principal authors to state that they are to blame for any remaining errors. Let me not depart from that tradition: I hereby claim responsibility for any undetected stubborn errors, big or small, that are still lurking among these pages, despite multiple rounds of careful double-checking.

Last, but not least, I would like to thank the corporate sponsors who made it possible to publish many of the figures in color. Vysis and Oncor (now Vantana), two companies at the forefront of FISH research, provided funding for the figures of fluorescent in situ hybridization. Carl Zeiss, Inc., and Applied Imaging, Inc., whose instruments adorn many clinical cytogenetics laboratories all over the country, were also generous contributors. Chroma Technologies and Applied Spectral Imaging were sponsors for the beautiful SKY and CGH pictures in the chapter by Dr. Green, Dr. Ried and their colleagues at the National Institute of Human Genome Research (formerly the National Center for Human Genome Research), National Institutes of Health.

No book is complete without a Foreword. I wish to thank its author, Dr. Thaddeus Kelly, a past President of the American Board of Medical Genetics, who agreed to take time out of his very busy schedule to write it. In addition to the Foreword, this book also has an Afterword. Dr. Avery Sandberg, a cancer geneticist known to cytogeneticists all over the world, graciously agreed to accept this writing assignment.

Publishing a book is in many ways like having a child, and in my lifetime I have had the privilege of bringing forth into this world several children. Few joys can be compared with that of seeing one's baby come to life. It has taken more than nine months and much effort to complete this book. Thus, I have awaited its birth with excitement, hope and anticipation, and a prayer that the book's readers will be enthusiastic and that its critics will be kind.

Hon Fong L. Mark

Contents

Contributors

John Anastasi, M.D. Department of Pathology, The University of Chicago Medical Center, Chicago, Illinois

Jan K. Blancato, Ph.D., F.A.C.M.G. Institute for Molecular and Human Genetics, Georgetown University Medical Center, Washington, D.C.

Zhong Chen, M.D. Department of Cytogenetics and Pediatrics, University of Utah School of Medicine, Salt Lake City, Utah

Gabriela Adelt Green, M.D. Genome Technology Branch, National Human Genome Research Institute, National Institutes of Health, Bethesda, Maryland

Bassem R. Haddad, M.D. Department of Obstetrics and Gynecology and Institute for Molecular and Human Genetics, Georgetown University Medical Center, Washington, D.C.

Kerstin Heselmeyer-Haddad, Ph.D. Division of Clinical Sciences, Department of Genetics, National Cancer Institute, National Institutes of Health, Bethesda, Maryland

Gerald A. Hoeltge, M.D. Department of Clinical Pathology, The Cleveland Clinic Foundation, Cleveland, Ohio

Jila Khorsand, M.D. Department of Pathology, Boston University, Boston, Massachusetts, and Department of Hematology Laboratory, Roger Williams Medical Center, Providence, Rhode Island

Bruce R. Korf, M.D., Ph.D. Partners Center for Human Genetics, Harvard Medical School and Partners HealthCare, Boston, Massachusetts

Alan N. Lamb, Ph.D., F.A.C.M.G. Department of Cytogenetics, Genzyme Genetics, Santa Fe, New Mexico

Roger V. Lebo, Ph.D., F.A.C.M.G. Department of Pediatrics, Center for Human Genetics, Boston University School of Medicine, Boston, Massachusetts

Hon Fong L. Mark, Ph.D., F.A.C.M.G. Brown University School of Medicine and Rhode Island Department of Health, Providence, and KRAM Corporation, Barrington, Rhode Island

Yvonne Mark, M.D. Department of Family Medicine, Memorial Hospital of Rhode Island, Pawtucket, and Brown University School of Medicine, Providence, Rhode Island

Wayne A. Miller, M.D. Perinatal Diagnosis Center, Inc., Lexington, Massachusetts

Hesed M. Padilla-Nash, Ph.D. Department of Genetics, National Cancer Institute, National Institutes of Health, Bethesda, Maryland

Reed E. Pyeritz, M.D., Ph.D. Center for Medical Genetics, Allegheny General Hospital, Pittsburgh, Pennsylvania

Susana C. Raimondi, Ph.D. Department of Pathology and Laboratory Medicine, St. Jude Children's Research Hospital, Memphis, Tennessee

Thomas Ried, M.D. Department of Genetics, National Cancer Institute, National Institutes of Health, Bethesda, Maryland

Diane Roulston, Ph.D. Division of Hematology/Oncology, Department of Medicine, The University of Chicago Medical Center, Chicago, Illinois

Avery A. Sandberg, M.D., D.Sc. Department of DNA Diagnostics, St. Joseph's Hospital and Medical Center, Phoenix, Arizona

Nancy R. Schneider, M.D., Ph.D. Department of Pathology, University of Texas Southwestern Medical Center, Dallas, Texas

Evelin Schröck, M.D. Genome Technology Branch, National Human Genome Research Institute, National Institutes of Health, Bethesda, Maryland

Mark Sigman, M.D. Division of Urology, Department of Surgery, Brown University School of Medicine, Providence, Rhode Island

Avirachan T. Tharapel Ph.D. Department of Pediatrics and Pathology, The University of Tennessee, Memphis, Tennessee

Sugandhi A. Tharapel, Ph.D., F.A.C.M.G. Veterans Affairs Medical Center and Department of Pediatrics and Pathology, The University of Tennessee, Memphis, Tennessee

Vijay S. Tonk, Ph.D. Department of Pediatrics, Texas Tech University Health Science Center, Lubbock, Texas

Tim Veldman, B.S. Genome Technology Branch, National Human Genome Research Institute, National Institutes of Health, Bethesda, Maryland

Dorothy Warburton, Ph.D. Departments of Genetics and Development and Department of Pediatrics, College of Physicians and Surgeons, Columbia University, New York, New York

Sandra R. Wolman, M.D. Department of Pathology, Uniformed Services University of the Health Sciences, Bethesda, Maryland

Herman E. Wyandt, Ph.D., F.A.C.M.G. Departments of Cytogenetics and Pathology, Center for Human Genetics, Boston University School of Medicine, Boston, Massachusetts

Carolyn Te Young, M.D. Department of Pathology and Laboratory Medicine, Rhode Island Blood Center and Brown University, Providence, Rhode Island

MEDICAL CYTOGENETICS

1
Cytogenetics as a Specialty of Medical Genetics

Hon Fong L. Mark
Brown University School of Medicine and Rhode Island Department of Health, Providence, and KRAM Corporation, Barrington, Rhode Island

Reed E. Pyeritz
Center for Medical Genetics, Allegheny General Hospital, Pittsburgh, Pennsylvania

I. INTRODUCTION

Cytogenetics is a relatively young science. The correct number of chromosomes in humans was established less than 50 years ago (1). The current explosion of knowledge in the field of human and medical genetics is astonishing, with discoveries being made monthly and sometimes weekly. Progress has accelerated in the last few years, as a result of the advent of the Human Genome Project, whose goals include mapping all of the estimated 50,000 to 100,000 human genes and sequencing the estimated 3 billion nucleotides in a haploid set of chromosomes. In this chapter we trace the history of cytogenetics in a nonsystematic, somewhat nostalgic fashion.

II. CYTOGENETICS PRIOR TO THE 1960s

Cytogenetics, simply defined, is the study of chromosomes. Modern genetics began in 1900 with the rediscovery of the studies of the monk Gregor Johann Mendel (1822–1884), who formulated laws of inheritance by experimenting with the garden pea. Mendel hypothesized that the basis of heredity was particulate, not a blending phenomenon as most people of his era thought. He suggested in 1865 that each characteristic of the pea plant, such as height, was caused by the effect of an "element," and that each cell contained a pair of elements for each trait. These elements are what modern-day geneticists refer to as alleles.

Mendel's principle of segregation predicts that the segregation of two alleles on the same chromosome is random, with a 50% chance of each allele being inherited. His principle of independent assortment predicts that the assortment of two different alleles on two different chromosomes is independent. Mendel also discovered the principle of dominance (versus recessiveness) in gene expression. A summary of the life and work of Mendel, who twice failed a licensing examination for a teaching credential and was never recognized for his scientific contribution to the field in his lifetime, was described by Dewald (2) and Orel (3). Dewald told the story of how, in the absence of peers to appreciate and further confirm his results and because of his frustrations with the hawkweeds, Mendel eventually abandoned his study of inheritance with plants during the later years of his career. What is also particularly noteworthy is the fact that as Mendel advanced up the administrative ladder to dean of the monastery, he also became entangled in political struggles that greatly hampered his ability to focus on research—a rather familiar story for modern-day scientists/administrators.

The significance of Mendel's contribution was not recognized until 1900 when, within a two-month period, Hugo de Vries of Amsterdam, Holland, Carl Correns of Tübingen, Germany, and Erich von Tschermak of Vienna, Austria, rediscovered Mendel's work during literature searches in preparation for publication of their own research. From that point onward, the pattern of nuclear inheritance has been called "Mendelian."

Walter Flemming, professor of anatomy at Kiel, first described chromosomes in 1876 and published pictures of them in 1882 (4). The term "chromosome," derived from the Greek word for "colored body," was coined by Waldeyer in 1888 (5). The chromosomal basis of heredity was discovered by William Sutton when he was still a graduate student in genetics (6). In 1909, Wilhelm Johannsen, a Danish biologist, first coined the word "genes," derived from the Greek word for giving birth.

In 1914, Theodor Boveri proposed the somatic mutation theory of cancer, which first suggested the clonal origin of neoplasms (7). Boveri proposed that cancer developed from a single cell that acquires a genetic alteration. However, Boveri's hypothesis could neither be confirmed nor refuted because the tools for testing his hypothesis were not available at that time. With advances in the techniques for analyzing chromosomes, support of Boveri's hypothesis began to appear in the literature in the 1960s (8).

Evidence that DNA is the genetic material was provided by Oswald Avery and colleagues in 1944 (9). The elucidation of the structure of DNA as a double helix by Watson and Crick occurred in 1953. At that time, the number of human chromosomes was thought to be 48. In 1956, Tjio and Levan (1) exploited advances in tissue culture, the use of colchicine to arrest cells in metaphase, hypotonicity to disperse the chromosomes and to enhance the quality of the cell preparation for study. Working with cultures of embryonic fibroblasts, they first identified the correct number of chromosomes to be 46. This finding was confirmed in the same year by Ford and Hamerton (10) using direct preparations of human testicular material.

By 1956 it was also known that a sexual dimorphism existed in the interphase nuclei of humans. A dense sex chromatin body is present in many cells of females, but not in normal males. In some conditions, notably Klinefelter syndrome and Turner syndrome, the phenotypic sex is often at variance with the number of chromatin bodies (11). However, based on the common belief that sex determination in humans was identical to that in *Drosophila*, the chromatin body (also known as the Barr body) was assumed to consist of the material of both X chromosomes. Chromatin-positive males and chromatin-negative females were incorrectly considered to be examples of true sex reversal, with males and females expected to have a 46,XX and a 46,XY constitution, respectively (12). However, in 1959 Jacobs and Strong (13) found that individuals with Klinefelter syndrome had an extra X chromosome in addition to the X and Y chromosomes which normal males possess. In 1959 Ford and colleagues found that Turner syndrome females were missing one of the two X chromosomes which normal females possess (14). Jacobs and her co-workers also provided the first evidence of the 47,XXX female, which, following the terminology for *Drosophila melanogaster*, was called, unfortunately, the "super-female." These studies made it possible for researchers to conclude that the genes on human Y chromosome was male-determining and to deduce the existence of a testis-determining factor (TDF). Within a very short interval of time, the chromosomal bases for Patau syndrome (trisomy D) and Edward syndrome (trisomy 18) were also identified. Another landmark discovery in cytogenetics during this era was the demonstration of an extra G-group chromosome in Down syndrome by Lejeune in 1959 (15). This historic finding of trisomy 21 by Lejeune was described by Hecht (16) in a short article after Lejeune's death. Lejeune employed a used microscope that had cost the equivalent of $7, had a defective ratchet, and needed an aluminum cigarette wrapper wedged into the scope's ratchet to keep the objective lens from crashing into the slide.

Despite the numerous advances cited above, the discipline of medical genetics did not yet exist as such nor as a recognized area of specialization within the practice of medicine in the United States. As Mark et al. (1995)

(17) pointed out, genetics was still "an academic, 'ivory tower' topic far removed from in-patient wards and out-patient clinics.''

III. CYTOGENETICS AND MEDICAL GENETICS IN THE 1960s

Several areas began to converge into the specialty of medical genetics in the early 1960s. A number of technical advances made this process possible.

In 1960, Peter Nowell, a pathologist teaching at the University of Pennsylvania at the time, discovered that phytohemagglutinin (PHA), a crude powder extracted from the common navy bean, *Phaseolus vulgaris*, and used to agglutinate red blood cells, also served as a mitogen for lymphocytes (18). The same year, Moorhead and his colleagues (19) from Philadelphia described the technique for culturing peripheral blood lymphocytes that still serves as the basis for chromosome studies on peripheral blood and bone marrow all over the world. A picture of Dr. Moorhead with his granddaughter, Annie Rose, more than 30 years after this fundamental discovery, can be found in Chapter 4.

A simple and reliable method for obtaining metaphase spreads from peripheral blood lymphocytes for chromosome analyses led in the 1960s to the rapid development of cytogenetic laboratories at academic medical centers. The first clinical laboratory geneticists in this country were mainly physicians, along with a few Ph.D.s trained in plant (primarily maize) cytogenetics. The cytogenetics laboratory at Rhode Island Hospital in Providence, Rhode Island, for example, was established in 1963 by a pediatrician, Dr. Paul H. LaMarche, who before coming to Rhode Island had trained at the Wistar Institute in Philadelphia under Dr. Moorhead but also received training from Dr. Nowell and Dr. Hungerford. About this time, conferences and focused courses among those with common interests in cytogenetics emerged, including one held at Rhode Island Hospital in the mid-1960s that counted as participants Dr. T. C. Hsu (who perfected the protocol for hypotonic treatment), Dr. Paul LaMarche, Dr. William Mellman, Dr. James Patton, Dr. Herbert Lubs, and Dr. Frank Ruddle (20).

In 1960, Nowell and Hungerford reported the first consistent chromosomal abnormality associated with a single cancer type, chronic myelogenous (or myeloid) leukemia (CML) (21). The marker chromosome was given the name Philadelphia chromosome (Ph), in honor of the city where it was discovered. This was one of the first pieces of evidence in humans supporting the chromosome theory of cancer. Unfortunately, few other consistent chromosomal changes could be discerned at the time, since the analyses were limited by the state of technology in the pre-banding era. Only with the advent of the various banding techniques developed in the 1970s could many more specific chromosomal rearrangements be identified.

Also in 1960, the "Bar Harbor Course" at the Jackson Laboratory in Bar Harbor, Maine, was initiated as the "Short Course in Medical Genetics" where, under the leadership of Dr. Victor A. McKusick and his colleagues, many medical geneticists, including the authors, have participated as students, faculty, or both. The course was supported exclusively by the National Foundation—March of Dimes through 1984. Although March of Dimes support has continued, the prominent support in recent years has come from the National Institute for Child Health and Human Development of the National Institutes of Health and from the Lucille P. Markey Charitable Trust (22).

The Denver conference for standardization of human cytogenetic nomenclature, at which autosomes were arranged in seven groups (A to G) based on size and position of centromere, was also held in 1960 (23).

Mary Lyon's hypothesis of X chromosome inactivation was reported in 1961 (24) and 1962 (25). It was proposed as a mechanism to explain dosage compensation. In 1966, the first edition of McKusick's *Mendelian Inheritance in Man* (26) was released, with 1487 entries. The 12th edition of this document is currently in preparation, and a continuously updated electronic version is available through the Internet.

The same year, the Chicago Conference on the standardization of nomenclature in human cytogenetics was held (27). The designations *p* and *q* were adopted for chromosome arms.

The first banding technique, developed in 1968, was Q-banding using quinacrine fluorescence (28,29). This was followed by Giemsa banding (30,31), C-banding (32), and R-banding (33).

The development of human cytogenetics permitted the first assignment of a hereditary phenotype to an autosome, with the linkage of the Duffy blood group locus to a benign polymorphism (heteromorphism) of the size of centromeric heterochromatin of chromosome 1 (34).

Another notable cytogenetic finding in this decade was the first description of the marker X chromosome in 1969 (35). This subsequently became known as the fragile X chromosome, associated with mental retardation, and was the subject of numerous reports in the modern cytogenetic literature (see, for example, Ref. 36).

It was also in the 1960s that Jacobs and co-workers found an excess of XYY individuals in mental-penal settings (37,38). The suggestion that the additional sex chromosome might contribute to the antisocial behavior of such men created a great deal of controversy and led to a reconsideration of both what constituted appropriate research in this general area and how it should be structured (39,40).

The use of cross-species somatic cell hybrids for the assignment of genes was described by Weiss and Green (41) during the same period.

The application of increasingly sophisticated biochemical analysis to inborn errors of metabolism began to suggest a clinically relevant means for management. The field of biochemical genetics owed its early development to Guthrie's development of a bacterial inhibition assay for newborn screening of PKU. The ease and economy of this test served as a basis for a recommendation to implement newborn screening programs in every state of the United States.

Concomitant with advances in laboratory technology, quality assurance and quality control (QA/QC) issues in clinical pathology laboratories became of increasing concern. Recognizing the need for regulation, the College of American Pathologists (CAP) began to develop standards for accreditation that would define parameters for physical resources and environment, organizational structure, personnel, safety, QA/QC and record keeping. The CAP also began to develop a "checklist" so that laboratory inspections could be standardized. These issues are discussed in Chapter 20 (Hoeltge).

IV. CYTOGENETICS AND MEDICAL GENETICS IN THE 1970s

In the early 1970s, medical genetics began to expand as a medical discipline in part as a result of expanding opportunities for fellowship training, notably with Dr. Victor McKusick in Baltimore, Maryland, and with Dr. Arno Motulsky in Seattle, Washington.

In the 1970s, a variety of techniques emerged: C-banding of centromeric heterochromatin (32), Giemsa-banding (30,31), and reverse or R-banding (33).

The Paris conference was held in 1971 (42) to standardize the nomenclature used in human cytogenetics, and banding numerology was introduced.

Through banding, Rowley (43) was able to identify the Philadelphia chromosome as a reciprocal translocation between chromosomes 9 and 22. It is now known that this translocation, in which the *c-abl* oncogene at 9q34 is fused to the breakpoint cluster region (*bcr*) locus on chromosome 22, results in a hybrid gene, which produces a fusion protein with increased tyrosine kinase activity that predisposes to myeloid cell transformation.

Recognizing the need to organize and share experiences, the Association of Cytogenetic Technologists (ACT) was born in 1974, although it was much later (in 1981) that a certification process was launched. Over the years, this organization has served the needs of cytogenetic technicians, technologists, supervisors, managers and laboratory directors. The ACT produces a fine journal, various educational materials, and an international directory for cytogenetic laboratories.

A protocol for preparing prometaphase spreads and expanded chromosomes was described by Yunis in 1976 (44). This high-resolution technique has been especially useful for the detection of microdeletions.

It was also in the 1970s that Jacobs and his colleagues introduced the first method for conducting direct chromosome analysis of human spermatozoa by allowing them to fertilize hamster eggs (45). The chromosomes of the male pronucleus were subsequently analyzed. This method has been utilized quite extensively for the study of infertility (see Chapter 10, by Sigman and Mark, in this volume).

Concomitant with the advances in cytogenetics, Southern blotting, used in the analysis of restriction fragments, was introduced in 1975 (46). Soon after, in 1977, methods of sequencing DNA were described (47,48).

V. CYTOGENETICS AND MEDICAL GENETICS IN THE 1980s

Among the many scientific developments in the 1980s, three notable ones related to the development of the specialty of medical genetics. First and foremost, the American Board of Medical Genetics (ABMG) was formed in 1980 under the initial sponsorship of the American Society of Human Genetics (ASHG). The first certifying examination was given in 1982. Second, commercial laboratories, especially those involved with cytogenetics, began to appear and flourish. Third, important advances in molecular genetics touched all areas of biology. Restriction fragment length polymorphism (RFLP) (49) was recognized as an important tool to map the human genome. It was also during the 1980s that the polymerase chain reaction (PCR) was born in the redwood forests of California (17). The introduction of PCR has made it possible to amplify small quantities of DNA (50). During the same period, Jeffreys et al. (51,52) established the protocol for DNA fingerprinting, which was later used extensively in forensic studies.

The 1980s also saw the application of recently developed molecular diagnostic techniques for cytogenetics, from the creation of DNA probes for the identification and mapping of oncogenes to the widespread use of fluorescent probes for chromosome staining (FISH).

The Alliance of Genetic Support Groups was established in 1986. The Council of Regional Networks for Genetic Services (CORN) and the Maternal and Child Health Branch of the Department of Health and Human Services played an increasingly important role in the development of medical genetics related to public health. The SPRANS program (Special Projects of Regional and National Significance) of the Maternal and Child Health Branch became an occasional source of funding for pilot projects and applied research in clinical and clinical laboratory genetics (53).

VI. CYTOGENETICS AND MEDICAL GENETICS IN THE 1990s

With the arrival of the 1990s came a steady succession of disease genes identified using a positional cloning approach, the discovery of the triplet-repeat diseases, gene therapy, and mitochondrial DNA diseases. In addition, there was the resurrection of anticipation, autosomal imprinting, and imprinting diseases in humans. The genetics of common cancers has been greatly elucidated. The study of tumor suppressor genes is increasing our understanding of the process of carcinogenesis. Microdeletion syndromes became common words in cytogenetics laboratories. New methodologies for prenatal screening and diagnosis emerged. Concepts such as uniparental disomy became a challenge to conventional thinking. The Human Genome Project, launched in 1990, is making excellent progress. New and powerful techniques of molecular cytogenetics are being developed, including comparative genomic hybridization (CGH) and spectral karyotyping (SKY) (54).

With the arrival of the 1990s also came the recognition of the American Board of Medical Genetics (ABMG) as the 24th primary specialty board of the American Board of Medical Specialties (ABMS). The professional organizations representing medical geneticists have been restructured into a tripartite structure characteristic of most medical specialties. The ABMG is the accreditation body for training programs and the certification board for Ph.D. medical geneticists and laboratory geneticists. It assumed the responsibility for quality control by developing and administering examinations to evaluate the competence of individuals and by certifying training programs. The first of the certifying examinations was given in 1982. A Residency Review Committee (RRC) in Medical Genetics now accredits training programs for M.D. clinical geneticists. The American College of Medical Genetics (ACMG), founded in 1991, is the clinical standards and practice arm of the medical genetics community. The American Society of Human Genetics (ASHG) remains the scientific research and education arm of the field. The American Board of Genetic Counseling was established in 1993 for certifying masters-level genetic counselors, who were previously certified by the ABMG. The Council of Medical Genetics Organizations (COMGO) was formed in 1992 as an umbrella group to foster communication among all of the organizations mentioned, as well as with other groups involved in medical genetics. It is a consortium of professional genetics societies, certifying boards, research societies, genetics consumer groups and related organizations. Rowley (55) listed the following organizations as members of COMGO: Alliance of Genetic Support Groups, American Board of Genetic Counseling, American Board of Medical Genetics, American College of Medical Genetics, American Society of Human Genetics, Association of Cytogenetic Technologists (now called Association of Genetic Technologists), Association of Professors of Medical and Human Genetics, Canadian Association of Genetic Counselors, Canadian College of Medical Genetics, Council of Regional Genetics Networks, International Society of Nurses in Genetics, National Society of Genetic Counselors, Royal College of Physicians and Surgeons and Society of Craniofacial Genetics.

The certificates issued each year by specialty by the ABMG are listed in Table 1. A list of AMBG-certified professionals by specialty and by state can also be obtained by consulting the Official American Board of Medical Specialties Directory of Board Certified Medical Specialists. Figure 1 is a pie chart depicting the distribution of certified medical genetics personnel.

In 1974 and again in 1978, the American Board of Medical Specialties and its member boards endorsed the policy of recertification. One of the conditions for the acceptance of the American Board of Medical Genetics as a board member was that the ABMG agreed to adopt recertification from the onset and to issue time-limited certificates to Diplomates certified in 1993 and thereafter. Therefore, the American Board of Medical Genetics has appointed an ad-hoc committee which is currently formulating a proposal and developing a recertification process for its diplomates.

In 1995 the ACMG became a full member of the Council of Medical Specialty Societies (CMSS). It received a seat in the American Medical Association's House of Delegates in 1996.

Table 1 ABMG Certificates by Examination: Six Examination Cycles Are Shown

Medical genetics specialty	1982	1984	1987	1990	1993	1996	Total certificates
M.D. Clinical	283	128	113	134	136	123	917
Ph.D. Medical	56	30	27	12	12	11	148
Clinical Biochemical	57	26	25	0	29	22	159
Clinical Cytogenetic	123	79	103	61	61	66	493
Clinical Molecular	0	0	0	0	144	82	226
Genetic Counselor	167	144	179	143	182	258	1073
	686	407	447	350	564	562	3016

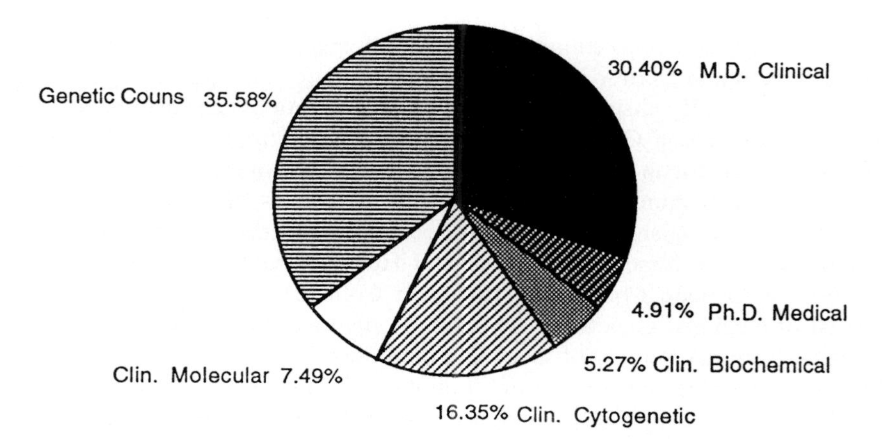

Figure 1 Distribution of certified medical genetics professionals. M.D. Clinical are M.D. clinical geneticists. Ph.D. Medical are Ph.D. medical geneticists. Clin. Biochemical are clinical biochemical geneticists. Clin. Cytogenetic are clinical cytogeneticists. Clin. Molecular are clinical molecular geneticists. Genetic Couns are genetic counselors.

Box 1. The American Society of Human Genetics

The American Society of Human Genetics (ASHG) was founded in 1948. According to the records, 60 people were in attendance at the first meeting of the ASHG, held on September 11–13, 1948, in Washington, D.C. (56). The ASHG is the primary professional membership organization for human geneticists in North America and holds its annual meeting each year in a major U.S. or Canadian city in the fall. The mission of the ASHG is to provide leadership in research, education, and service in human genetics, by providing an interactive forum for sharing research findings affecting individuals and families with inherited conditions. The society is governed by a board of directors consisting of 16 members, composed of the president, the president-elect, secretary, treasurer, editor, the two most recent past presidents, and nine other members of the society elected for a three-year term, with three positions designated for election each year. In addition to the *American Journal of Human Genetics*, the ASHG also publishes programs of its scientific meetings, including abstracts of the papers to be presented. Other publications of the ASHG include the *Guide to North American Graduate and Postgraduate Training Programs in Human Genetics*, *Solving the Puzzle—Careers in Genetics*, and the *Membership Directory*, published in conjunction with the Genetics Society of America, the American Board of Medical Genetics, the American Board of Genetic Counseling, and the American College of Medical Genetics. The business of the ASHG is conducted through its various standing and ad-hoc committees, chairs and members of which are appointed by the president. The ASHG also has a Student Awards Program and issues Awards for Excellence in Human Genetics Education. In 1961, the William Allan Award was established in memory of William Allan (1881–1943), one of the first American physicians to conduct extensive research in human genetics. The Allan Award is presented annually to recognize outstanding contributions and continued productivity in the field of human genetics through research and/or teaching.

Box 2. The American College of Medical Genetics

The American College of Medical Genetics (ACMG) is the clinical practice arm of the genetics tripartite structure. It is a national organization for biochemical geneticists, clinical geneticists, clinical cytogeneticists, Ph.D. medical geneticists, molecular geneticists, as well as genetic counselors and other health-care professionals committed to the practice of medical genetics.

In 1990, the ASHG appointed a committee to establish the ACMG. The ACMG was incorporated in 1991 to provide national representation for providers of genetic services and their patients with genetic disorders, and full representation to and advocacy for the emerging specialty of medical genetics in medical service, certification, and the associated regulatory organizations. The ACMG advocated the ABMS's recognition of medical genetics as a primary specialty in 1991, and in turn supported the formation of the ABGC in 1993. In 1995, the ACMG became a full member of the Council of Medical Specialty Societies, and received full Delegate status in the American Medical Association House of Delegates in 1996.

The mission of the ACMG is to provide education, resources, and a voice for the medical genetics profession. In order to make genetic services available to the public, the ACMG promotes the development and implementation of methods of diagnosis, treatment, and prevention of genetic diseases. The ACMG sponsors continuing education programs for geneticists, other heath-care providers, and the public. Membership in the ACMG offers a collegial forum in which to address issues, raise concerns, and discuss matters related to medical genetics services. Postgraduate education and other collective efforts are made available to benefit the professional development of those involved in the clinical practice of medical genetics. Through its AMA House of Delegates seat, the ACMG is involved in policy-making decisions and other issues of medical practice at the local and national levels. Important in this area is the development of CPT codes and policy statements concerning genetic practices.

Much of the work of the ACMG is performed through committees, which are open to all members. From time to time, position statements on relevant issues in genetics as well as guidelines on clinical and laboratory practices are released. The college is increasingly recognized by organized medicine, public health officers, and federal and state governments as an official authority on issues regarding genetic services. Through its national activities and liaison with other organizations, the college has heightened the awareness of the medical as well as lay communities with regard to the field of genetics. In addition, the ACMG holds an annual meeting, in conjunction with the March of Dimes—Birth Defects Foundation. The educational activities of the ACMG emphasize issues of clinical importance and focus on the professional development of medical geneticists. In May 1996 the ACMG also co-sponsored the Genetics Review Course with Baylor College of Medicine. Beginning in 1998, the college publishes an official journal, Genetics in Medicine. This new journal will be a forum for original research in all aspects of medical genetics, including cytogenetics, position papers and policy statements of the college, and correspondence regarding issues of importance to members.

As of October 1996, the American College of Medical Genetics (ACMG) reported its membership at approximately 1029, with 851 Fellows, 68% of whom are MDs. At present, 56% of the ACMG members are M.D. Fellows as reported by the ACMG Membership Committee (J. Phillips, Chair). The first president of the ACMG was Dr. David Rimoin (1993–1996). The second president of the ACMG was Dr. Reed Pyeritz (1997–1998).

VII. CONCLUDING COMMENTS

At the same time as the structure of medical genetics is evolving, the delivery of health care in the United States is undergoing a painful transformation.

The arrival of managed care has led to numerous consolidations, downsizings, and organizational restructurings, especially of academic medical centers, throughout the United States. Against this background, medical genetics is seeking to secure a viable and effective place in the practice of medicine. Medical geneticists must carve out and identify a generally accepted role based on the genetic information that they alone have the training to access, interpret, and deliver. The evolution of managed health care requires new and effective strategies to deliver clinical and laboratory services with an emphasis on adaptability. Strategies may include the development of alliances among related laboratories, subcontracting with large, multiservice laboratories, and the development of formalized marketing plans (57). By necessity, the cytogenetics laboratory of tomorrow will be one with greatly increased efficiency and cost consciousness, or it will cease to exist in our current, highly competitive marketplace.

With the often newsworthy advances of the Human Genome Project, medical geneticists are increasingly brought under close scrutiny by their peers, the public, medical regulators, and legislators. Currently, clinical laboratories in the United States that provide information to referring physicians are certified under the Clinical Laboratory Improvement Amendments of 1988 (CLIA 88). Cytogenetic laboratories must be registered with the Health Care Financing Administration (HCFA). The ACMG has assumed responsibility for developing standards of practice for the medical genetics community. The document entitled *Standards and Guidelines*: *Clinical Genetics Laboratories* is an evolving document that reflects the consensus of participating ACMG members. In addition, most, if not all, cytogenetics laboratories are inspected by the College of American Pathologists' Laboratory Accreditation Program and participate in the Inter-laboratory Comparison (Surveys) Program. Issues related to setting standards in the clinical cytogenetics laboratory are discussed in Chapter 20 (Hoeltge).

The 1994 International Cytogenetic Laboratory Directory lists 320 laboratories in the United States that provide genetic laboratory services (17). The 1997 International Cytogenetic Laboratory Directory lists 285 laboratories in the United States, 35 from Canada, and 163 from 41 other countries (58). Between January 1, 1995, and December 31, 1995, a total of 559,232 cytogenetic procedures were performed in U.S. cytogenetic laboratories. Cytogenetics as a specialty of medical genetics has come of age.

ACKNOWLEDGMENTS

We thank Dr. Yvonne Mark for computer graphics assistance, and Ms. Anne Richardson, Dr. Paul LaMarche, Dr. Yvonne Mark, and Dr. Roger Mark for reading the manuscript. The assistance of Dr. Paul Moorhead and Dr. T. C. Hsu, who provided us with valuable historical information, was instrumental in making this chapter both informative and interesting. As always, the support of Dr. Roger Mark and the staff of the Lifespan Academic Medical Center Cytogenetics Laboratory at Rhode Island Hospital is also acknowledged.

REFERENCES

1. Tjio JH, Levan A. The chromosome number of man. Hereditas 1956; 42:1–16.
2. Dewald GW. Gregor Johann Mendel and the beginning of genetics. Appl Cytogenet 1996; 22:1–5. Reprinted from Mayo Clin Proc 1977; 52:513–518.
3. Orel V. Gregor Mendel: The First Geneticist. New York: Oxford University Press, 1996.
4. McKusick VA. The growth and development of human genetics as a clinical discipline. Am J Hum Genet 1975; 27:261–273.
5. Waldeyer W. Über Karyokinese und ihre Beziehung zu den Befruchtungsvorgängen. Arch Mikrosk Anat 1888; 32:1.
6. McKusick VA. Walter S. Sutton and the physical basis of Mendelism. Bull Hist Med 1960; 34:487–497.
7. Sirica AE, ed. The Pathobiology of Neoplasia. New York: Plenum Press, 1989.
8. Mark HFL. Bone marrow cytogenetics and hematologic malignancies. In: Dale KS, Kaplan BJ, eds. The Cytogenetic Symposia. Burbank, CA: The Association of Cytogenetic Technologists, 1994.
9. Avery OT, MacLoed CM, McCarty M. Studies on the chemical nature of the substance inducing transformation of *Pneumonococcus* types. J Exp Med 1944; 79:137–158.
10. Ford CE, Hamerton JL. The chromosomes of man. Nature 1956; 178:1010–1023.
11. Moore KL, Barr ML. Smears from the oral mucosa in the detection of chromosomal sex. Lancet 1955; ii:57.
12. Jacobs PA. The William Allan Memorial Award address: human population cytogenetics: the first twenty-five years. Am J Hum Genet 1982; 34:689–698.
13. Jacobs PA, Strong JA. A case of human intersexuality having a possible XXY sex determining mechanism. Nature 1959; 183:302–303.
14. Ford CE, Jones KW, Polani PE, de Almeida JC, Briggs JH. A sex-chromosome anomaly in a case of gonadal dysgenesis (Turner syndrome). Lancet 1959; 1:711–713.
15. Lejeune J, Gautier M, Turpin MR. Etude des chromosomes somatiques de neut enfants mongoliens. C R Acad Sci (III) 1959; 248:1721–1722.
16. Hecht, F. Jerome Lejeune (1926–94): in memoriam. Am J Hum Genet 1994; 55:201–202.

17. Mark, HFL, Kelly T, Watson MS, Hoeltge J, Beauregard L. Current issues of personnel and laboratory practices in genetic testing. J Med Genet 1995; 32: 780–786.
18. Nowell PC. Phytohemagglutinin: an initiator of mitoses in cultures of normal human leukocytes. Cancer Res 1960; 20:462.
19. Moorhead PS, Nowell PC, Mellman WS, Battips DM, Hungerford DA. Chromosome preparations of leukocytes cultured from human peripheral blood. Exp Cell Res 1960; 20:613–616.
20. Hsu TC. Human and Mammalian Cytogenetics—An Historical Perspective. New York: Springer-Verlag, 1979.
21. Nowell PC, Hungerford DA. A minute chromosome in human chronic myelocytic leukemia (CML). Science 1960; 132:1497.
22. Guethlein LA. The Bar Harbor Course: a 30-year veteran in the teaching of human genetics. Am J Hum Genet 1990; 46:192–206.
23. Denver Conference. A proposed standard system of nomenclature of human mitotic chromosomes. Lancet 1960; i:1063–1065.
24. Lyons MF. Gene action in the X-chromosome of the mouse (*Mus musculus* L.). Nature 1961; 190:372–373.
25. Lyons MF. Sex chromatin and gene action in the mammalian X-chromosome. Am J Hum Genet 1962; 14:135–148.
26. McKusick VA. Mendelian Inheritance in Man: Catalogs of Autosomal Dominant, Autosomal Recessive and X-Linked Phenotypes. Baltimore: Johns Hopkins University Press, 1966.
27. Chicago Conference. Standardization in Human Cytogenetics. Birth Defects Original Article Series 2. New York: National Foundation—March of Dimes, 1966:1–21.
28. Caspersson T, Farber S, Foley GE, et al. Chemical differentiation along metaphase chromosomes. Exp Cell Res 1968; 49:219–222.
29. Caspersson T, Zech L, Johansson C, Modest EJ. Identification of human chromosomes by DNA-binding fluorescent agents. Chromosoma 1970; 30:215–227.
30. Seabright M. A rapid banding technique for human chromosomes. Lancet 1971; 2:971–972.
31. Sumner AT, Evans HJ, Buckland RA. A new technique for distinguishing between human chromosomes. Nature New Biol 1971; 232:31–32.
32. Arrighi FE, Hsu TC. Localization of heterochromatin in human chromosomes. Cytogenetics 1971; 10:81–86.
33. Dutrillaux B, Lejeune, J. New techniques in the study of human chromosomes: methods and applications. Adv Hum Genet 1975; 5:119–156.
34. Donahue RP, Bias WB, Renwick JH, McKusick VA. Probable assignment of the Duffy blood group locus to chromosome 1 in man. Proc Natl Acad Sci USA 1968; 61:949–955.
35. Lubs HA. A marker X chromosome. Am J Hum Genet 1969; 21:231–244.
36. Mark HFL, Bier JB, Scola P. The frequency of chromosomal abnormalities in patients referred for fragile X analysis: a five-and-a-half year study. Ann Clin Lab Sci 1996; 26:323–328.

37. Jacobs PA, Brunton M, Melville MM, Brittain RP, McClemont WF. Aggressive behaviour, mental subnormality and the XYY male. Nature 1965; 208:1351–1352.
38. Hook EB. Rates of XYY genotype in penal and mental settings. Lancet 1975; i:98.
39. Pyeritz RE, Beckwith J, Miller L. XYY disclosure condemned (letter). N Engl J Med 1975; 293:508.
40. Pyeritz RE. Potential benefits of XYY research. In: The XYY Controversy: Researching Violence and Genetics. Hastings-on-Hudson, NY: Institute of Society, Ethics and the Life Sciences, 1980: 11–15.
41. Weiss MC, Green H. Human-mouse hybrid cell lines containing partial complements of human chromosomes and functioning human genes. Proc Natl Acad Sci USA 1967; 58:1104–1111.
42. Paris Conference. Standardization in Human Cytogenetics. Birth Defects Original Article Series 8:1–46. New York: The National Foundation—March of Dimes, 1971.
43. Rowley JD. A new consistent chromosomal abnormality in chronic myelogenous leukemia identified by quinacrine fluorescent and Giemsa staining. Nature 1973; 234:290–293.
44. Yunis JJ. High resolution of human chromosomes. Science 1976; 191:1268–1270.
45. Rudak E, Jacobs PA, Yanagimachi R. Direct analysis of the chromosome constitution of human spermatozoa. Nature 1978; 274:911–913.
46. Southern EM. Detection of specific sequences among DNA fragments separated by gel electrophoresis. J Mol Biol 1975; 98:502–517.
47. Maxam AM, Gilbert W. A new method for sequencing DNA. Proc Natl Acad Sci USA 1977; 74:560–564.
48. Sanger F, Nicklen S, Coulson AR. DNA sequencing with chain-terminating inhibitors. Proc Natl Acad Sci USA 1977; 98:502–517.
49. Botstein D, White RL, Skdnick M, Davis R. Construction of a genetic linkage map in man using restriction fragment length polymorphisms. Am J Hum Genet 1980; 32:314–331.
50. Saiki RK, Sharf S, Faloona F, et al. Enzymatic amplification of b-globin genomic sequences and restriction site analysis for diagnosis of sickle cell anemia. Science 1985; 230:1350–1355.
51. Jeffreys AJ, Wilson V, Thein SL. Hypervariable minisatellite regions in human DNA. Nature 1985a; 314:67–74.
52. Jeffreys AJ, Wilson V, Thein SL. Individual-specific ''fingerprints'' of human DNA. Nature 1985b; 316:76–79.
53. Cohen MM. 1994 ASHG presidential address: Who are we? Where are we going? Anticipating the 21st century. Am J Hum Genet 1995; 56:1–10.
54. Epstein, CJ. 1996 ASHG presidential address: towards the 21st century. Am J Hum Genet 1997; 60:1–9.
55. Rowley JD. 1993 American Society of Human Genetics presidential address: Can we meet the challenge? Am J Hum Genet 1994; 54:403–413.
56. Neel JV. Our twenty-fifth. Am J Hum Genet 1974; 26:136–144.

57. Mark, HFL, Prence E, Beauregard LJ, Greenstein R. Managed care and genetic testing laboratories. Cytobios 1996; 88:43–52.
58. Knutsen, T., ed. International Cytogenetic Laboratory Directory. 1996–1997 Edition. Lenexa, KA: Association of Cytogenetic Technologists.

APPENDIX: GLOSSARY OF TERMS

ABMG
American Board of Medical Genetics. The 24th primary specialty board of the American Board of Medical Specialties (ABMS)

ACT
Association of Cytogenetic Technologists. A member of the Council of Medical Genetics Organizations (COMGO), now known as the Association of Genetic Technologists (AGT).

AGT
Association of Genetic Technologists. A member of the Council of Medical Genetics Organizations (COMGO), formerly known as the Association of Cytogenetic Technologists (ACT).

ASHG
American Society of Human Genetics. Founded in 1948, the ASHG is the primary professional membership organization for human geneticists in North America.

CAP
College of American Pathologists. The organization that develops standards for accreditation of clinical pathology laboratories, defines parameters for physical resources and environment, organizational structure, personnel, safety, QA/QC, and record keeping.

C-banding
A technique that produces specific staining of constitutive heterochromatin, giving bands that are found mostly in the centromere region. Prominent C-bands are found on human chromosomes 1, 9, 16, and the long arm of the Y chromosome.

centromere
The central location of the chromosome to which spindle fibers attach, also called the primary constriction.

CGH
Comparative genomic hybridization. A molecular cytogenetic method for analyzing a tumor for genetic changes, such as gain or loss of chromatin material.

CLIA 88
Clinical Laboratory Improvement Amendments of 1988.

COMGO

Council of Medical Genetics Organizations. An organization founded in 1992 as an umbrella group to foster communication among the American College of Medical Genetics, the American Society of Human Genetics, the American Board of Medical Specialties, as well as other groups involved with medical genetics.

CORN

Council of Regional Networks for Genetic Services. One of the organizations that played an important role in the development of medical genetics related to public health during the eighties. As the name implies, the membership of CORN is comprised of the regional genetics networks: GENES (Genetics Network of the Empire State), GLaRGG (Great Lakes Regional Genetics Group), GPRGSN (Great Plains Regional Genetics Services Network), MARHGN (Mid-Atlantic Regional Human Genetics Network), MSRGSN (Mountain States Regional Genetics Services Network), NERGG (New England Regional Genetics Group), PacNoRGG (Pacific Northwest Regional Genetics Group), SERGG (Southeastern Regional Genetics Group), and TEXGENE (Texas Genetics Network).

cytogenetics

The study of chromosomes.

dosage compensation

The phenomenon in humans and other mammals in which a balance in the number of sex-linked alleles in both sexes is achieved by the random inactivation of all but one of the female X chromosomes.

Edwards syndrome

Trisomy 18; some common characteristics of those affected with the syndrome include "faunlike" ears, rockerbottom feet, a small jaw, a narrow pelvis, and death within a few weeks of birth.

heterochromatin

Chromatin that is tightly coiled and devoid of genes, usually stained positive with C-banding, which can be constitutive or facultative.

HRB

High-resolution banding. The special technique that provides enhanced visualization of chromosome bands by elongating and arresting the chromosomes during prophase and prometaphase.

Klinefelter syndrome

XXY; a condition that occurs 1 in 1000 male live births. Some characteristics of those affected include gynecomastia, slight mental retardation, tall stature, underdeveloped testes, and sterility.

Lyon Hypothesis

The hypothesis, conceived by Mary Lyon, that states that one of the X chromosomes in somatic cells of mammalian females becomes randomly inactivated during early development. The Lyon hypothesis thus provides a mechanism whereby dosage compensation can be achieved.

marker chromosome
A structurally abnormal chromosome that cannot be identified with currently available technology.

Mendelian
The mode of inheritance as predicted by the principles established by Gregor Mendel.

mitogen
A substance that stimulates cells to enter mitosis.

neoplasm
A new, abnormal growth of tissue that divides uncontrollably and continuously.

p
Short arm of the chromosome.

Patau syndrome
Trisomy 13; some common characteristics of those affected include a hairlip, a tiny deformed head, and death within a few days of birth.

PCR
Polymerase chain reaction. A method for making multiple copies of a given DNA sequence.

Ph
Philadelphia chromosome. The first marker chromosome to be identified with a single cancer type—chronic myelogenous leukemia (CML)—and was then named in honor of the city in which it was discovered.

PKU
Phenylketonuria. A human disease that is inherited recessively, through simple Mendelian genetics, which is the result of the inability of the body to convert phenylalanine into tyrosine, thus allowing for the accumulation of phenylalanine in the body that leads to mental retardation.

polymorphic variant
Normal variations of chromosome morphology, especially of prominent C-band regions.

positional cloning
A method of cloning genes, formerly known as "reverse genetics."

q
Long arm of the chromosome.

Q-banding
A technique that utilizes quinacrine mustard and other related compounds to stain chromosomes in order to induce differentially stained areas along each chromosome, observable with a fluorescent microscope.

R-banding

A technique that produces bands on human chromosomes that are nearly the reverse of what are seen in Q- and G-banding. This type of banding is useful in the visualization of bands that show light staining in Q- and G-banding.

RFLP

Restriction fragment length polymorphism. Markers of known DNA variation that are used for chromosome mapping.

sex chromatin body

A Barr body, or an inactivated X chromosome in the somatic cells of female mammals.

SKY

Spectral karyotyping. The labeling of chromosomes using a spectrum of different colors, one of the newest molecular cytogenetic techniques.

Southern blotting

A molecular biologic technique which involves gel electrophoresis of DNA fragments and their transfer onto a nitrocellulose membrane for autoradiography.

TDF

Testis-determining factor. The factor that determines maleness (i.e., secondary sex characteristics) in early fetal development.

Turner syndrome

A female with a single X chromosome; such cases occur 1 in 10,000 female live births. Some characteristics of those affected include short stature, poor breast development, underdeveloped reproductive organs, characteristic facial features, a constricted aorta, and sterility.

2

The Normal Human Chromosome Complement

Hon Fong L. Mark
Brown University School of Medicine and Rhode Island Department of Health, Providence, and KRAM Corporation, Barrington, Rhode Island

Yvonne Mark
Memorial Hospital of Rhode Island, Pautucket, and Brown University School of Medicine, Providence, Rhode Island

I. THE STRUCTURE OF CHROMOSOMES

Cytogenetics is the study of chromosomes. What are chromosomes? A simplified definition of a chromosome is that it is an organelle composed of DNA and proteins. The structure of the chromosome seems to remain rather constant from organism to organism. The most widely accepted chromosome structure is shown in Fig. 1.

For many years, geneticists debated whether the chromosome is composed of one strand (unineme model) or multiple strands (polyneme model) of DNA (1). It is now known that each cytogenetically visible metaphase chromosome contains a single molecule of DNA organized into several orders of packaging. The genetic material of the cell is tightly bound to histones, which are basic proteins, and other acidic nonhistone proteins. There are five major types of histones, designated as H1, H2A, H2B, H3, and H4. In the nucleus, DNA is arranged into structures called nucleosomes, as depicted in Fig. 1. Each nucleosome consists of a DNA strand wrapped around a histone core that is connected to other histone molecules (2). Specifically, approximately 140 base pairs of DNA are wound in two turns around a block of eight (octamer) core histones (H2A, H2B, H3, and H4) to produce one nucleosome (3). A short segment of spacer DNA, associated with a single H1 molecule, links one nucleosome to the next, thereby forming a

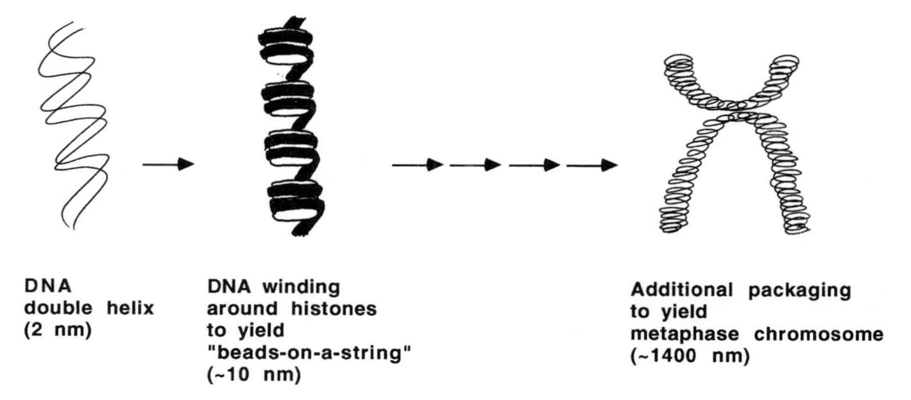

Figure 1 A simplified diagram of the structure of a chromosome.

10-nm chromatin fiber comprised of a linear chain of closely packed nucleosomes. Further supercoiling and condensation produce the higher levels of organization seen in Fig. 1.

Chemical modifications are responsible for the binding and release of DNA/histone complexes. Of particular importance is the chemical reaction called acetylation, in which simple chemical groups known as acetyls are added to histones. Upon acetylation, the modified proteins then hold less tightly to the nearby DNA, opening the way to activating gene expression. Four distinct nuclear histone acetylating enzymes and five more enzymes that undo the reaction by removing acetyls from histones had been identified as of the beginning of 1997, but the field is literally exploding.

Chromosomal material has been classified into two types: euchromatin and heterochromatin. Euchromatin is universally perceived as gene-bearing chromatin that makes up the bulk of the chromosome. In contrast, heterochromatin is usually considered to be variable in amount and mostly devoid of genes. Heterochromatin is, in turn, grouped into two types: constitutive and facultative. Constitutive heterochromatin is heterochromatin that is found near the centromeres of chromosomes, in the satellites and short arms of the acrocentric (D- and G-group) chromosomes, and on the long arm of the Y chromosome. These areas contain highly repetitive DNA sequences and are the areas stained by C-bands (4) (Fig. 2). Facultative heterochromatin is chromatin that is in a transcriptionally inactive state. The mammalian inactivated X chromosome, which is manifested as the Barr body (5), is an example (Fig. 3).

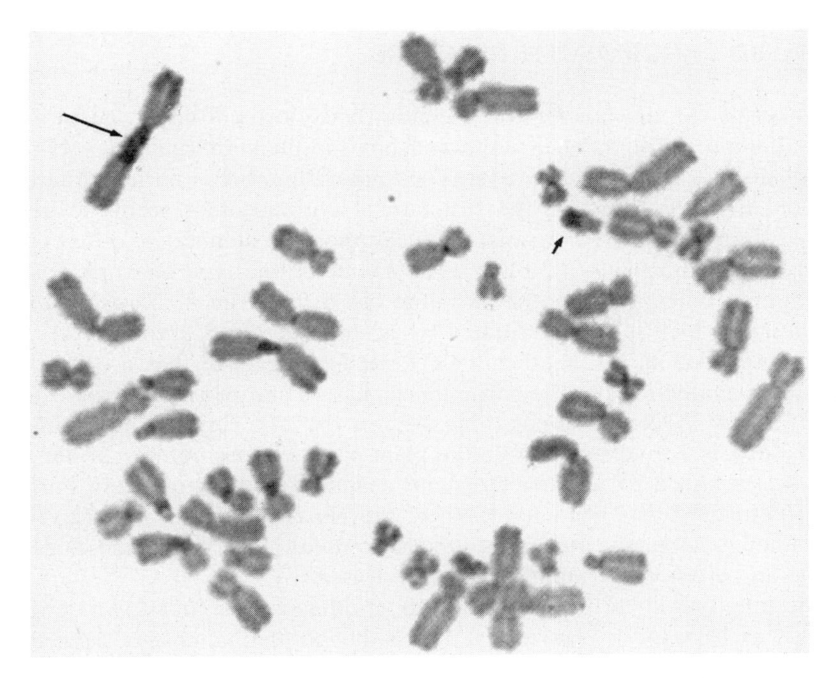

Figure 2 Constitutive heterochromatin found in areas stained by C-bands. The large arrow points to chromosome 1 C-band. The small arrow points to the distal long arm of the Y chromosome.

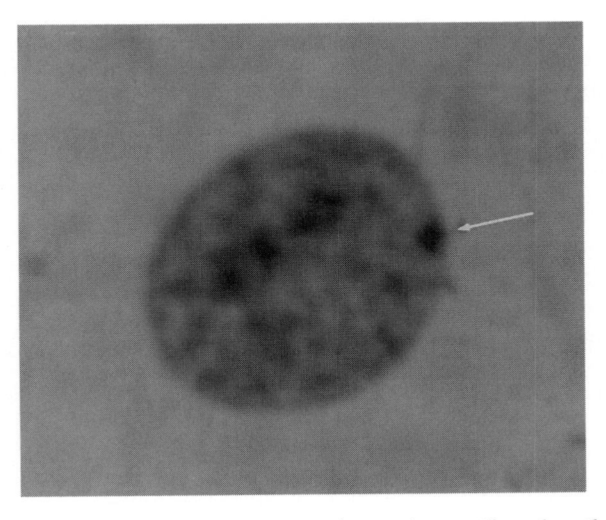

Figure 3 Facultative heterochromatin manifested as the Barr body.

II. DIPLOID CHROMOSOME NUMBERS

The number of chromosomes in the somatic (body) cells of an organism is its diploid number. The number of chromosomes in the germ (gametic) cells of an organism is its haploid number. The haploid number is normally half the diploid number. Although the structure of chromosomes seems to be conserved from organism to organism, the chromosome numbers of different organisms differ drastically (Table 1). For example, one type of worm has only 2 chromosomes, whereas the mosquito has 6. The fruit fly, *Drosophila melanogaster*, which is widely studied by geneticists, has 8 chromosomes, whereas corn (*Zea mays*), another widely studied organism, has a diploid chromosome number of 20. The house mouse has 40 chromosomes, whereas humans possess 46 chromosomes (Figs. 4a and 4b). The chromosome number of the dog is 78, whereas a fernlike plant has a diploid number of 480, only to be surpassed by another fern with a diploid number of 1260. Fortunately, cytogeneticists need not analyze the chromosomes of ferns routinely, although in many tumor specimens complex aneuploidy and huge variations in chromosome number are often found.

The true diploid chromosome number of humans was not known until as recently as 1956 (6,7). Prior to that, it was thought that humans possessed 48 chromosomes. Most individuals with Down syndrome (8) possess 47 chromosomes, with an extra copy of chromosome 21. Most individuals with Turner syndrome possess 45 chromosomes, with a missing X chromosome (9). On the other hand, most individuals with Klinefelter syndrome have an extra X chromosome (10). These and other examples of sex chromosomal abnormalities are discussed in Chapter 7 (Korf and Schneider).

Each of the two sexes contains 22 pairs of autosomal chromosomes, or autosomes. Each sex also possesses two sex chromosomes: XX in females and XY in males. Prior to the banding era (11), the only chromosomes that could be identified unequivocally were chromosomes 1, 2, 3, 16, and the Y chromosome. To identify the X chromosome, investigators sometimes used the time-consuming and labor-intensive technique of autoradiography.

Half of the chromosomes in an individual are normally derived from the mother, and half of the chromosomes are derived from the father. According to Mendel's law of segregation, one homolog of each chromosome goes randomly (with equal probability) to each daughter cell. According to Mendel's law of independent assortment, different chromosome pairs assort independently, whether they are derived maternally or paternally. The diploid number in germ cells is halved during the process of meiosis to produce haploid gametes (sperm and eggs). During fertilization, gametes combine to form a zygote. Thus, the diploid number is once again reconstituted.

Table 1 Diploid Number of Chromosomes in Selected Organisms

Common name	Genus-species	Diploid number (2*N*)
Worm	*Ascaris lumbricoides*	2
Small composite plant	*Haplopappus*	4
Mosquito	*Culex pipiens*	6
Indian muntjac	*Muntiacus muntjak*	6 in female, 7 in male
Fruit fly	*Drosophila melanogaster*	8
Roundworm	*Caenorhabditis elegans*	12
House fly	*Musca domestica*	12
Broad bean	*Vicia faba*	12
Tasmanian rat kangaroo	*Potorous tridactylus*	12 in female, 13 in male
Slime mold	*Dictyostelium discoidium*	14
Pink bread mold	*Neurospora crassa*	14
Evening primrose	*Oenothera biennis*	14
Garden pea	*Pisum sativum*	14
Black bread mold	*Aspergillus nidulans*	16
Garden onion	*Allium cepa*	16
Bee	*Apis mellifera*	16
Snapdragon	*Antirhinum majus*	16
Creeping vole	*Microtus oregoni*	17 in female, 18 in male
Corn	*Zea mays*	20
Chinese hamster	*Cricetulus griseus*	22
Cricket	*Gryllus domesticus*	22
Jimson weed	*Datura stramonium*	24
Tomato	*Lycopersicon esculentum*	24
Rice	*Oryza sativa*	24
Grasshopper	*Melanoplus differentialis*	24
Lily	*Lilium longiflorum*	24
Oak tree	*Quercus alba*	24
Mushroom	*Agaricus campestris*	24
Frog	*Rana pipiens*	26
Hydra	*Hydra vulgaris*	32
Alligator	*Alligator mississippiensis*	32
Black bread mold	*Rhizipus nigricans*	32
Yeast	*Saccharomyces cerevisiae*	34
Spider monkey	*Ateles paniscus*	34
Green algae	*Chlamydomonas reinhardi*	36
Earthworm	*Lumbricus terrestis*	36
Cat	*Felis domesticus*	38
House mouse	*Mus musculus*	40
Rhesus monkey	*Macaca mulatta*	42
Wheat	*Triticum aestivum*	42
Rat	*Rattus norvegicus*	42
Human	*Homo sapiens*	46

Table 1 Continued

Common name	Genus-species	Diploid number (2*N*)
Tobacco	*Nicotiana tabacum*	48
Chimpanzee	*Pan troglodytes*	48
Potato	*Solanum tuberosum*	48
European field vole	*Microtus agrestis*	50
Amoeba	*Amoeba protens*	50
Cotton	*Gossypium hirsutum*	52
Snail	*Helix pomatia*	54
Silkworm	*Bombyx mori*	56
Cattle	*Bos taurus*	60
Horse	*Equus caballus*	64
Dog	*Canis familiaris*	78
Chicken	*Gallus domesticus*	78
Pigeon	*Columba livia*	80
Water fly	*Nymphaea alba*	160
Crayfish	*Cambarus clarkii*	200
Adder's tongue fern	*Ophioglossum vulgatum*	480
Fern like plant	*Ophioglossum reticulatum*	1260

References: Altman and Dittmer, 1972 (34); Gardner, 1971 (35); Hsu, 1979 (11); Klug and Cummings, 1994 (36); Wurtster and Benirschke, 1970 (37).

III. THE CELL CYCLE

The cell cycle (Fig. 5) consists of a dividing phase, mitosis, and a nondividing phase, interphase. Through the process of mitosis, a cell divides to yield two genetically identical daughter cells. Somatic (body) cells undergo mitosis only, but gonadal cells undergo both mitosis and meiosis. These cells undergo mitosis to increase their numbers before the final stages of development, in which meiosis occurs.

A. Mitosis

The diploid number (2*N*) is maintained from cell division to cell division in somatic cells through the process of mitosis (Fig. 6). Duplicated chromosome homologs are distributed equally to each pole and are subsequently included in two daughter cells. It has been hypothesized that tension is the key parameter which determines if cells will get the correct number and types of chromosomes (12). Failure of such cell cycle checkpoint mechanisms can lead to both birth defects (e.g., trisomy 21 in Down syndrome) and cancer (e.g., certain trisomies in leukemia).

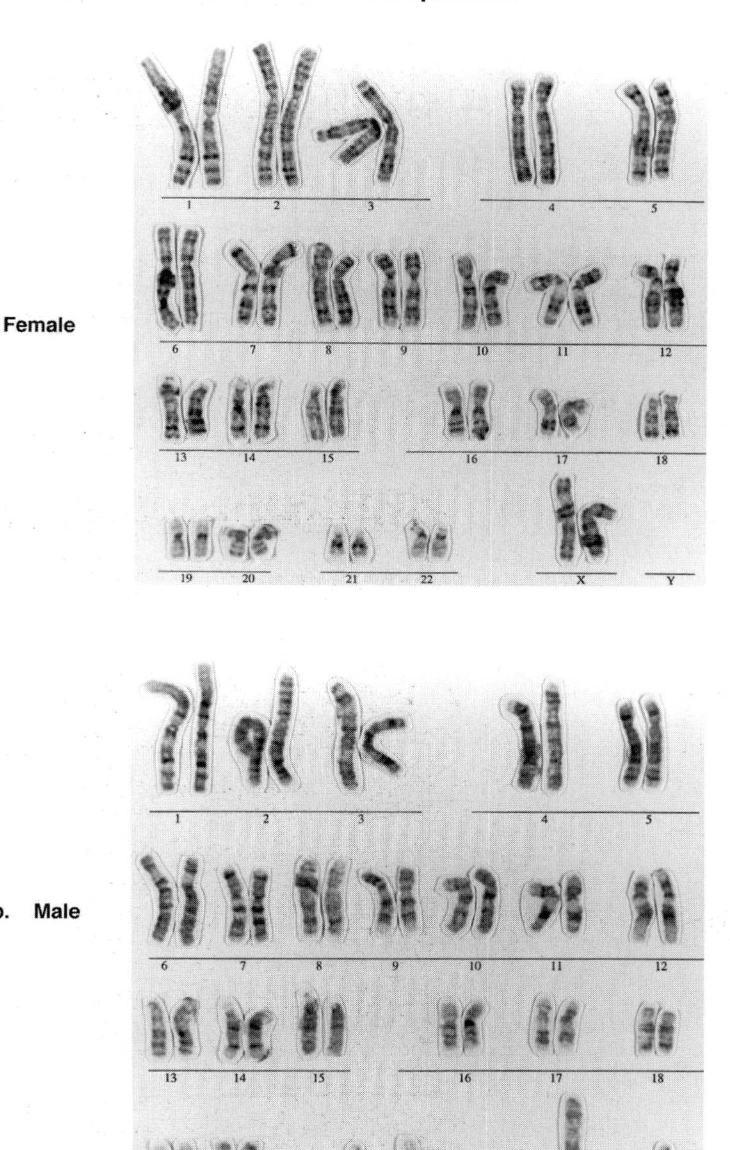

Figure 4 (a) The normal human female chromosome complement. (b) The normal human male chromosome complement.

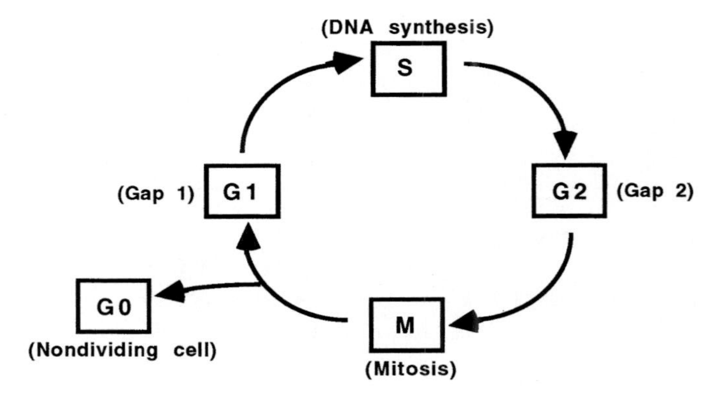

Figure 5 The cell cycle.

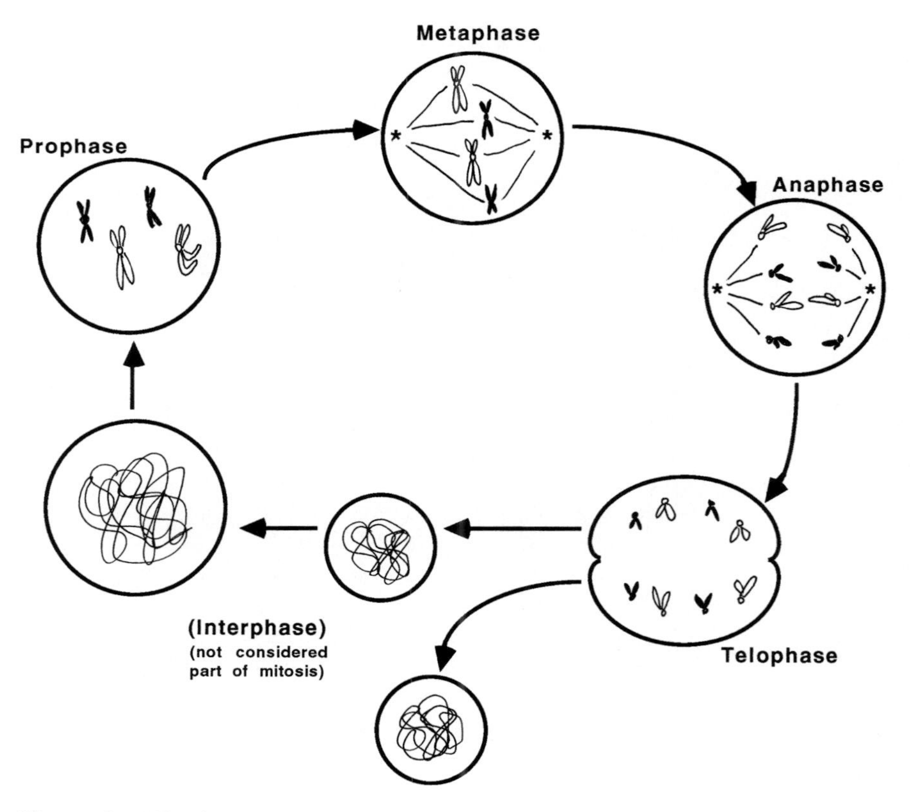

Figure 6 Mitosis.

Mitosis is divided into four stages: prophase, metaphase, anaphase, and telophase. At the conclusion of interphase, mitotic prophase begins. During prophase, chromosomes condense and become visible under brightfield microscopy. Each chromosome consists of two parallel strands (called sister chromatids), held together by the centromere. In addition, the nuclear membrane disappears. Mitotic metaphase is the stage during which the chromosomes contract completely and move to the center of the cell. Spindle fibers radiating from each of two centrioles located at opposite poles of the cell become attached to the kinetochores of the chromosomes at the centromeres. In mitotic anaphase, the centromere of each chromosome has divided and the sister chromatids begin to separate and move to opposite poles of the cell. In mitotic telophase, the daughter chromosomes reach the poles of the cell and begin to decondense. The cytoplasm divides, and the nuclear membranes reform. Upon the completion of telophase, interphase begins. The process of mitosis is depicted in Fig. 6.

Errors can occur both in meiosis and mitosis. Both numerical and structural chromosome abnormalities can occur. These abnormalities, which can involve autosomes as well as sex chromosomes, will be discussed further elsewhere in this volume.

B. Interphase

Interphase is comprised of the part of the cell cycle between mitoses. It is the stage during which the cell has either completed its mitotic division but has not yet undergone DNA synthesis (gap 1 or G1), is actively synthesizing DNA (S phase) or is preparing for cell division (gap 2 or G2). The majority of the time, the cultured cell remains in interphase. Cells in vivo remain in a noncycling stage termed G0, during which the cells remain in a temporarily suspended state until they are stimulated to rejoin the cell cycle. In a typical growing cell culture, the G1, S, and G2 phases together last a total of 16–24 h, although there is great variation among cell types. The replication time of cells in culture varies depending on the source of the tissue. It has been found that chromosomes appear to occupy clearly defined, although not individually distinguishable, domains within the interphase nucleus (13–15). Unfortunately, individual chromosomes are visible and amenable to cytogenetic analysis only during the brief period of mitosis, which lasts only 30 min to 2 h.

During the G1 phase prior to the synthesis phase, the chromosome consists of one chromatid (a single double-helical molecule of DNA). In contrast, during the G2 phase after the synthesis phase, the chromosome consists of two sister chromatids (two identical DNA molecules). This is an important fact to remember when one observes "double" fluorescent signals in G2 interphase nuclei by interphase fluorescent in-situ hybridization (FISH) (refer to Chapter 6, by Blancato and Haddad, and Chapter 17, by Mark, in this volume).

C. Meiosis

A simplistic view of meiosis is that it is two consecutive cell divisions without an intervening interphase and cycle of DNA replication (16). During meiosis, homologous chromosomes pair, cross over, and exchange genetic materials. The first division of meiosis (meiosis I) begins with prophase. Prophase I is usually subdivided into leptotene, zygotene, pachytene, diplotene, and diakinesis. During leptotene, the chromosomes appear under the light microscope as thin threads, and sister chromatids cannot be distinguished. Zygotene involves the process of synapsis. Homologous chromosomes pair, and the synaptonemal complex, a specialized structure that lies between chromatin fibers, is formed. During pachytene, the tightness of the synapses cause homologous chromosomes to appear as bivalents. Recombination takes place via the process of crossing over. In diplotene, the pairs of homologous chromosomes begin to separate at all sites except at points of crossovers, which are seen cytologically as chiasmata (singular, chiasma). Chromosomes reach maximal condensation during diakinesis. The next stage in meiosis I is metaphase I, where the nuclear membrane disappears and the bivalents line up on a plane in the center of the cell. Spindle fibers, which radiate from centrioles at opposite poles of the cell, become connected to the centromeres of the chromosomes. In anaphase I, the homologous chromosomes that comprise each bivalent separate from each other and move to opposite poles of the cell. During telophase I, the two sets of chromosomes reach opposite poles of the cell, and the cytoplasm divides. Meiosis I is a reductional division in which homologs separate, and the diploid chromosome number (of 46 duplicated chromosomes) is reduced to the haploid number (of 23 duplicated chromosomes). Centromeric division does not occur during meiosis I.

After a brief interphase, and without DNA synthesis, the second meiotic division begins. Meiosis II is similar to mitosis in that the chromosomes, each composed of two sister chromatids attached at the centromere, line up at the equatorial plate. The centromeres divide and the sister chromatids separate with the help of spindle fibers that radiate from centrioles at opposite poles. Meiosis II is thus an equational division in which the centromeres of the 23 chromosomes divide and each sister chromatid segregates into a daughter cell.

D. Gametogenesis

Gametogenesis is the process that leads to the formation of ova or sperm. Expansion of germ cells in both sexes occurs through mitosis in the gonads. Meiosis in the male (spermatogenesis) begins in the seminiferous tubules of the testes at the time of puberty and continues throughout life. Mitotic germ cell precursors called spermatogonia produce primary spermatocytes, each of which undergoes meiosis I to produce secondary spermatocytes. Each secondary spermatocyte undergoes meiosis II to form two spermatids, producing a total of four spermatids, which mature into sperm through spermiogenesis. The production of mature sperm from a spermatogonium takes about 75 days (17). Spermatogenesis, also discussed in Chapter 10 (Sigman and Mark), is depicted in Fig. 7a.

Meiosis in the female (oogenesis) begins during fetal life. All primary oocytes are formed before birth. Primary oocytes become suspended in dic-

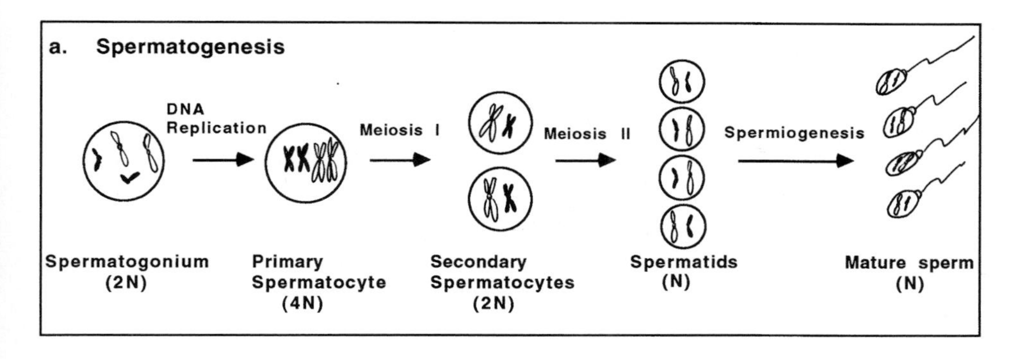

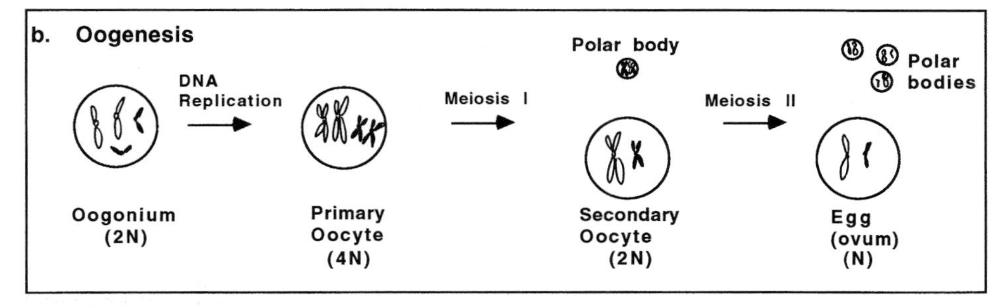

Figure 7 (a) Meiosis in the male. (b) Meiosis in the female.

tyotene, a type of diplotene stage of prophase I, until the female reaches sexual maturity, at menarche. During ovulation, individual ovarian follicles mature and release their primary oocytes from dictyotene. This prolonged arrest of female gametes in prophase I may account for the increased risk of nondisjunction associated with advanced maternal age (AMA), discussed elsewhere in this volume. Mitotic division of germ cell precursors in females is restricted to the first few months after conception. Oogonia produce primary oocytes, which initiate meiosis I. Meiosis I in females results in unequal division of the cytoplasm, with most of the nutrient-rich cytoplasm allocated to the secondary oocyte and relatively little distributed to the first polar body. Meiosis II also results in an unequal division of the cytoplasm, with most dispersed to the ovum and relatively little given to the second polar body. The first polar body may or may not divide to yield two other polar bodies. The single ovum functions as a gamete, but the polar bodies are normally nonfunctional. Oogenesis is depicted in Fig. 7b.

IV. SEX DETERMINATION

The sex chromosomes are sometimes called gonosomes, as opposed to nonsex chromosomes, which are called autosomes.

In humans, the female is the homogametic sex, which possesses two X chromosomes (Fig. 4a), whereas the male is the heterogametic sex, which possesses an X chromosome and a Y chromosome (Fig. 4b). This is in contrast to some other organisms, such as birds, in which the female is the heterogametic sex (which possesses ZW chromosomes) and the male is the homogametic sex (which possesses ZZ chromosomes).

Genes on the human Y chromosome determine the sex of the individual (Table 2). In contrast, in other organisms such as fruit flies, sex is de-

Table 2 Sex Determination in Human versus Fly

Sex chromosome composition	Human	Fly
XX	Female	Female
XY	Male	Male
XXY	Male	Female
X	Female	Male
XXXY	Male	Female
XXYY	Male	Female

termined by the ratio of the number of sex chromosomes to the number of autosomes.

In mammals, one of the two X chromosomes in normal females is randomly inactivated during embryonic development through the formation of facultative heterochromatin (18–21). This inactivated X chromosome can be distinguished cytologically in certain interphase nuclei (e.g., in buccal mucosal cells) as the darkly staining Barr body (5) or the sex-chromatin body (Fig. 3). X chromosome inactivation is also accompanied by a delay in DNA replication during the S phase. There is usually a random inactivation of either the paternally derived X chromosome or the maternally derived X chromosome. Once inactivation occurs, it is maintained through subsequent cell divisions. Reactivation of the inactive X takes place only in germ cells.

V. THE GTG-BANDED HUMAN KARYOTYPE

The normal human chromosome complement is shown in Fig. 4a (female) and Fig. 4b (male). This orderly display of an organism's chromosomes by shape, size (starting with the largest chromosome), and banding pattern is called a karyotype. Present karyotypes are GTG-banded karyotypes, derived from metaphases on slides that are treated with trypsin and stained with Giemsa, a Romanowsky stain, to produce a series of distinctive transverse bands spaced along the entire chromosome complement (the G-banding pattern). Banding is used to identify individual chromosomes unambiguously. It is especially critical for distinguishing chromosomes of similar sizes and shapes. The different kinds of banding and the various methods employed to induce banding are discussed in detail in the chapter by Mark (Chapter 5) in this volume.

VI. CHROMOSOME LANDMARKS

The centromere of the chromosome is also called the primary constriction (22,23). It is that part of the chromosome that is involved in spindle fiber attachment and chromosome movements toward opposite poles during cell division. The kinetochore is a trilaminar structure located in the centromere which mediates spindle fiber attachment. The centromere also serves to bind together the two sister chromatids of mitotic metaphase chromosomes until they separate in anaphase.

In humans, as in most other animals and plants, centromeres tend to be surrounded by a variable amount and distribution of heterochromatin, which leads to chromosome polymorphic variants (discussed below). Heterochromatin is usually present in the largest amounts on human chromosomes 1, 9, 15, and 16 and also the Y chromosome. This material, which is made up of highly repetitive DNA sequences that are not transcribed, is condensed throughout interphase and replicates late in the S period.

Each end of the chromosome is called a telomere. Based on his radiation studies in *Drosophila* many years ago, H. J. Muller postulated that telomeres cap the ends of chromosomes to prevent chromosomes from sticking to each other (24,25). Chromosomes without telomeres are susceptible to breakage, fusion, formation of dicentric chromosomes, and anaphase bridges (26), and subsequent cell death. Telomeres are composed of the sequence (TTAGGG), which is present in multiple copies at the terminal ends of chromosomes as well as interstitially. Interstitial telomeric sequences have been hypothesized to be sites of nonrandom chromosome breakage, fragile sites, viral integration sites, and hot spots of recombination.

The nucleolus organizer region (NOR) is the site of transcriptionally active, tandemly repetitive, ribosomal RNA genes. The NOR is visible as a secondary constriction using bright-field microscopy. NORs can also be visualized by ammonical silver staining. The size of these regions varies a great deal but is generally correlated with the amount of DNA at the site (27). Since rRNA genes are organized in long, tandemly repetitive arrays, it is not surprising that they show the same variability in copy number as satellite DNA and other sequences that are similarly organized. AgNOR staining is discussed further in the chapter by Mark (Chapter 5) in this volume.

The centromere and the telomere are chromosome landmarks which have been used to divide the chromosome into regions. Certain very prominent bands can also serve as chromosome landmarks.

Each region of the chromosome is divided into bands. Each band is further divided into subbands. The numbering of regions, bands, and subbands is such that the most proximal to the centromere is numbered 1, and the most distal has the highest number. These and other nomenclature-related concepts, as illustrated in Fig. 8, are further explored in Chapter 3.

VII. SHAPES OF CHROMOSOMES

The shapes of chromosomes are illustrated in Fig. 9. Metacentric chromosomes have centromeres that lie near the centers of the chromosomes. Submetacentric chromosomes contain centromeres in off-center locations on the chromosomes. The short arm of the chromosome is called *p* (for petit) arm, whereas the long arm is called the *q* arm (Fig. 8). Acrocentric chromosomes have very short *p* arms. These chromosomes usually end in structures called

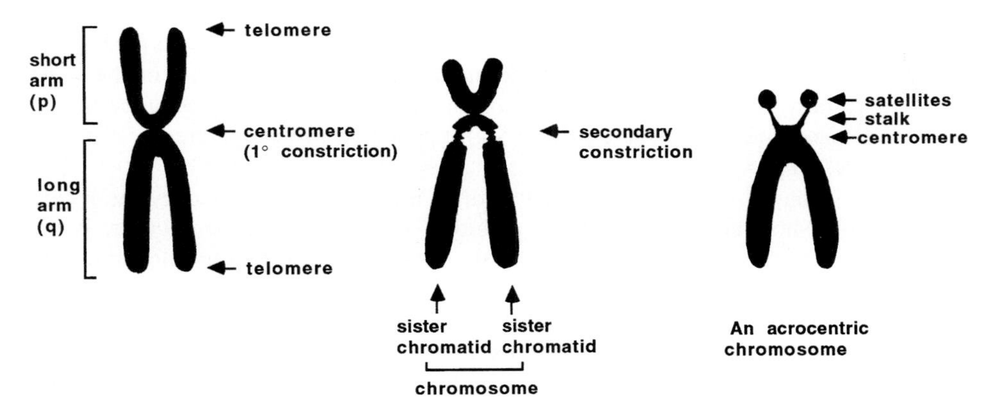

Figure 8 Chromosome landmarks.

satellites, which are connected to the main structure of the chromosome by stalks. Telocentrics are chromosomes that lack short arms. These chromosomes do not exist in humans.

Chromosomes are grouped by size and position of the centromere. For historic reasons, however, chromosome 21 is shorter than chromosome 22. The human Y chromosome is similar in size and shape to chromosomes 21 and 22, but it lacks satellites and has consistent, albeit subtle, characteristic features such as morphology of the long arm.

VIII. OTHER CHROMOSOME ATTRIBUTES

The modal chromosome number is the number of chromosomes per cell most often encountered in the metaphases scored from a sample. The modal chromosome number in a normal human individual is the diploid number of 46. Due to artifacts of the technique, chromosomes are sometimes lost

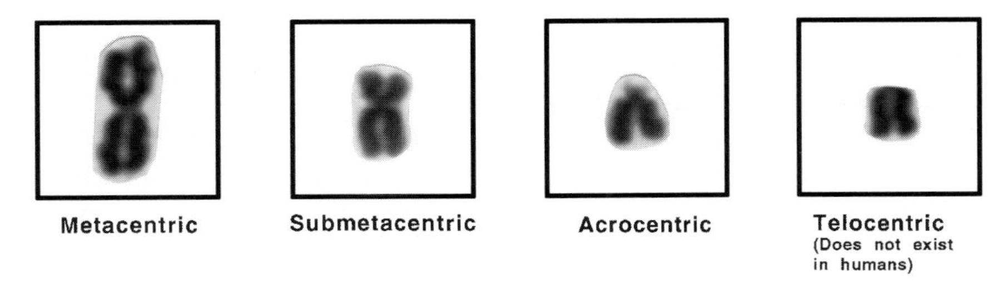

Figure 9 Shapes of chromosomes.

on the slide. This is termed random loss. An International System for Human Cytogenetic Nomenclature (28) defines random loss as the loss of the same chromosome in no more than two metaphase cells. Loss of a particular chromosome is considered to be significant if it is encountered in three or more cells.

The arm ratio, q/p, is the length of the long arm divided by the length of the short arm. The centromere index, expressed as a percentage, is the length of the short arm divided by the total length of the chromosome, $p/(p + q)$, (see, for example, Ref. 29). The relative length, expressed as a percentage, is the length of that individual chromosome divided by the total haploid lengths of all the chromosomes in the human genome. Alternative approaches to estimating relative lengths have been proposed in the literature (see, for example, Refs. 30 and 31).

IX. CYTOGENETIC NOMENCLATURE

As mentioned above, the term karyotype refers to a display of the chromosomes of a representative cell of an organism in a systematic fashion. By convention, the short arm of each chromosome is placed on top. An ideogram is a diagrammatic karyotype based on chromosome measurements from a large number of cells. An ideogram of the human genome can be found in Chapter 3.

A constitutional karyotype is the karyotype of the entire individual. An acquired karyotype is the karyotype acquired by somatic cells postzygotically. Acquired chromosomal abnormalities are often found in cancer.

Mosaicism occurs when two or more cell lines develop in the same individual. A chimera is the result of a rare event in which an embryo develops from what are essentially two separate fertilization events. Often these will be detected as a mixed population of male and female cells. The new technique of fluorescent in-situ hybridization, discussed in Chapter 6 (Blancato and Haddad), and also in Chapter 17 (Mark), is especially useful in studying these two phenomena.

X. COMMENTS ON KARYOTYPING

Prior to the banding era, chromosomes from nonbanded (or solid-stained) metaphase cells were karyotyped based on size and position of the centromere. Such a nonbanded karyotype is shown in Fig. 10. During the prebanding era, other features, such as prominent satellites and secondary constrictions, were often utilized as well.

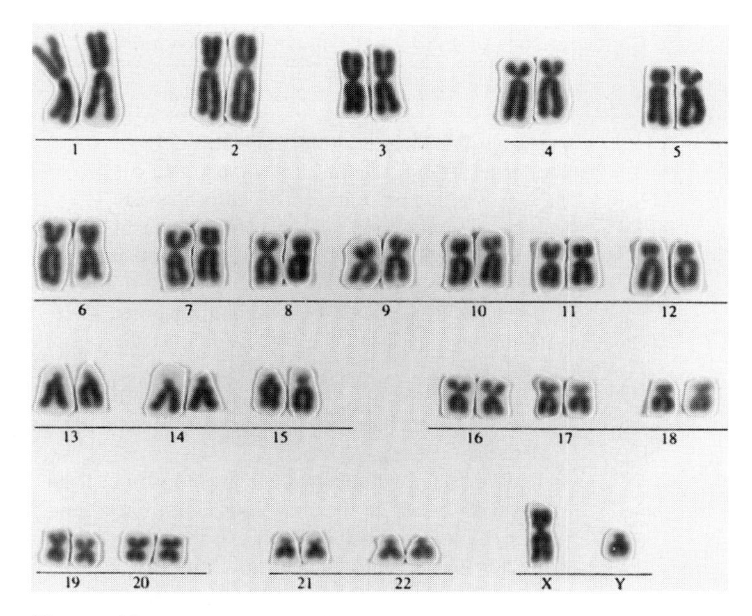

Figure 10 Nonbanded male karyotype.

With the advent of banding in the late 1960s and early 1970s, each chromosome in the human genome could be identified unambiguously. Based on the unique chromosome bands, cytogeneticists all over the world were able to identify abnormalities in the structure as well as the number of chromosomes.

Over the years many students and residents have rotated in the first author's laboratory. The first author has observed that the uniform initial reaction of trainees, when they are handed the first metaphase for karyotyping, is that the task is nearly impossible to perform: "One chromosome looks just like another." As with playing the piano, proficiency in manual karyotyping needs diligence and practice. Table 3 gives a tested protocol for identifying each and every chromosome on a GTG-banded metaphase.

Prior to the purchase of an automated computer karyotyping system, many students and trainees in this laboratory experienced hours of enjoyment developing chromosome pictures in the dark room (with the radio on, of course). With the availability of instantaneous karyotypes, our darkroom has become a virtually deserted place. But the automated karyotyping system is neither entirely automated nor perfect. It is usually easier to perform semimanual karyotyping via instructions of the technologists than to let the

Table 3 Morphologic Characteristics of Individual Human Chromosomes

Chromosome	Group	Morphologic characteristics
1	A	The largest metacentric chromosome.
2	A	The largest submetacentric chromosome.
3	A	The second largest metacentric chromosome
4	B	A large submetacentric chromosome. Its arm ratio is approximately 4:1. It has a small dark band below the centromere.
5	B	It is the same size and shape as chromosome 4. It has a large dark band in the middle of the long arm.
6	C	The largest of the C group. It has a submedian centromere and a large light area in the middle of the short arm.
7	C	A medium-sized submetacentric chromosome. It has two dark bands in the long arm and a dark band near the tip of the short arm.
8	C	A medium-sized submetacentric chromosome. It is the least distinguishable of the C group. It has a dark band near the end of the long arm.
9	C	It looks similar to chromosome 7, but is smaller and has less distinct bands. A secondary constriction can sometimes be seen near the centromere on the long arm of chromosome 9.
10	C	It has three dark bands in the long arm. The band near the centromere is much darker than the other two.
11	C	It has a large dark band in the middle of the long arm.
12	C	It has a large dark band in the middle of the long arm. The short arm is noticeably shorter than the short arm of chromosome 11.
13	D	An acrocentric chromosome with a dark lower half.
14	D	An acrocentric chromosome. It has two dark bands: one near the centromere and the other near the end of the long arm.
15	D	An acrocentric chromosome. It is the lightest stained of the D group. It has a small dark band in the middle of the long arm.
16	E	The most metacentric of the E group chromosomes. It is the same size as the D group. It has a dark band at the centromere.
17	E	It is the same size as chromosome 16, but is more submetacentric. It has a large light area in the long arm and a dark band at the end.

Table 3 Continued

Chromosome	Group	Morphologic characteristics
18	E	It is slightly smaller than chromosome 17. It is mostly dark. It has an occasional noticeably dark band near the centromere.
19	F	A small metacentric chromosome with a dark band at the centromere.
20	F	A small metacentric chromosome that is grayish.
21	G	A small acrocentric chromosome with a dark band near the centromere.
22	G	A small acrocentric chromosome that is grayish.
X	C	It is almost as large as chromosome 6. It has two distinguishing dark bands: one in the middle of the short arm and the other in the long arm.
Y	G	Although usually the same size as chromosomes 21 and 22, it does not possess satellites and is morphologically distinct.

computer do the job entirely by itself and then to correct its many mistakes. However, the technology may continue to improve in the future.

XI. NORMAL VARIATIONS

Normal variations are called polymorphic variants, polymorphisms, or heteromorphic variants. Some of the uses of these cytogenetically distinguishable normal variants include the following: use as markers for population studies, determination of the origin of nondisjunction, paternity determination, detection of maternal contamination in amniotic fluid cultures, establishment of the origin of ovarian teratomas, determination of the zygosity of twins, determination of the origin of certain deleted or translocated chromosomes. Normal variants can also be used in follow-up studies in bone marrow transplantation (BMT) (32), although the Y chromosome may be a better marker in sex-mismatched BMTs.

Examples of polymorphisms include the large dark heterochromatic block below the centromere in the long arm of chromosome 1 that varies in size from individual to individual but is constant from cell to cell within an individual. Note that this heterochromatic variant led to the localization of the first autosomal locus (Duffy blood group locus) on human chromosome 1 (33). Another human chromosomal polymorphism is the pericentric inversion of chromosome 9. A series of these inversions is shown in Fig. 11.

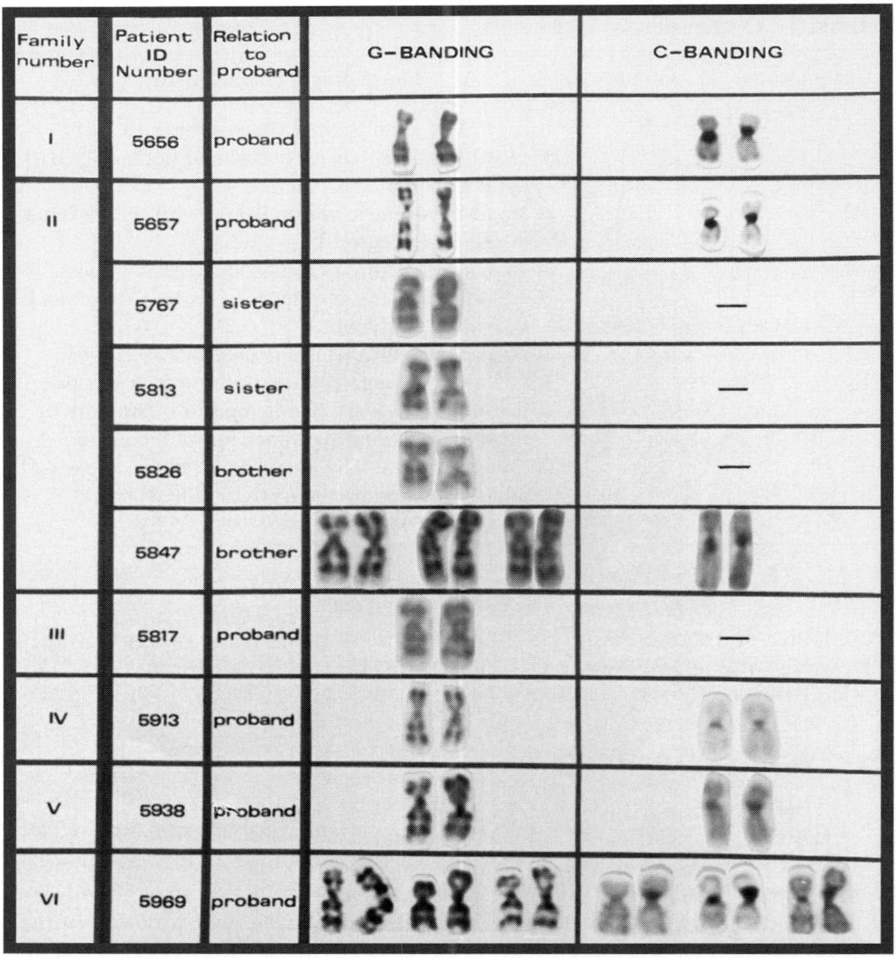

Family number	Patient ID Number	Relation to proband	G-BANDING	C-BANDING
I	5656	proband		
II	5657	proband		
	5767	sister		—
	5813	sister		—
	5826	brother		—
	5847	brother		
III	5817	proband		—
IV	5913	proband		
V	5938	proband		
VI	5969	proband		

Figure 11 Pericentric inversion of human chromosome 9.

XII. EVOLVING TECHNOLOGIES FOR STUDYING THE HUMAN GENOME

It is difficult to believe that less than 45 years ago the correct number of chromosomes in the normal human genome was not even established. Just 30 years ago, many of the chromosomes in the human genome could not be identified unequivocally. The discovery of banding for human chromosome identification was truly a significant event that opened a new era for cytogenetics. The establishment of a practical protocol for FISH for the average cytogenetic laboratory was another significant milestone (refer to Chapter 6, by Blancato and Haddad, and to Chapter 17, by Mark, in this volume). Other evolving technologies, such as comparative genomic hybridization (CGH) and spectral karyotyping (SKY), discussed elsewhere in this volume, have further contributed toward the study of the human genome.

ACKNOWLEDGMENTS

We thank Dr. Roger Mark and Ms. Anne Richardson for reading the manuscript and Ms. Jennifer Gilbert for help with the literature search. The support of the staff of the Lifespan Academic Medical Center Cytogenetics Laboratory at Rhode Island Hospital is also acknowledged.

REFERENCES

1. Taylor JH, Woods PS, Hughes WL. The organization and duplication of chromosomes as revealed by autoradiographic studies using tritium labeled thymidine. Proc Natl Acad Sci USA 1957; 43:122–127.
2. Pennisi E. Opening the way to gene activity. Science 1997; 275:155–157.
3. McGhee JD, Nickol JM, Felsenfeld G, Rau DC. Higher order structure of chromatin orientation of nucleosomes within the 30 nm chromatin solenoid is independent of species and spacer length. Cell 1983; 33:831–841.
4. Arrighi FE, Hsu TC. Localization of heterochromatin in human chromosomes. Cytogenetics 1971; 10:81–86.
5. Barr ML, Bertram EG. A morphological distinction between neurons of the male and female, and the behaviour of the nucleolar satellite during accelerated nucleoprotein synthesis. Nature 1949; 163:676–677.
6. Tijo JH, Levan A. The chromosome number of man. Hereditas 1956; 42:1–16.
7. Ford CE, Hamerton JL. The chromosomes of man. Nature 1956; 178:1020–1023.
8. Lejeune J, Gautier M, Turpin R. Etude des chromosomes somatiques de neuf enfants mongoliens. Compt Rend 1959; 248:1721–1722.
9. Ford CE, Jones KW, Polani PE, et al. A sex-chromosome anomaly in a case of gonadal dysgenesis (Turner's syndrome). Lancet 1959; i:711–713.
10. Jacobs PA, Strong JA. A case of human intersexuality having a possible XXY sex-determining mechanism. Nature 1959; 183:302–303.
11. Hsu TC. Human and Mammalian Cytogenetics. New York: Springer-Verlag, 1979.

12. Nicklas RB. How cells get the right chromosomes. Science 1997; 275:632–637.
13. Haaf T, Schmid M. Chromosome topology in mammalian interphase nuclei. Exp Cell Res 1991; 192:325–332.
14. Manuelidis L. Individual interphase chromosome domains revealed by *in situ* hybridization. Hum Genet 1985; 71:288–293.
15. Manuelidis L. A view of interphase chromosomes. Science 1990; 250:1533–1540.
16. Wolstenholme J. An introduction to human chromosomes and their analysis. In Rooney DE, Czepulkowski BH, eds. Human Cytogenetics: A Practical Approach, Volume I. Oxford, UK: IRL Press, 1992.
17. Friedman JM, Dill FJ, Hayden MR, McGillivray BC. The National Medical Series for Independent Study: Genetics. Baltimore: Williams & Wilkins, 1992.
18. Lyon MF. Gene action in the X-chromosome of the mouse (*Mus musculus L.*) Nature 1961; 190:372–373.
19. Russel LB. Genetics of mammalian sex chromosomes. Science 1961; 133:1795–1803.
20. Lyon MF. The quest for the X-inactivation centre. Trends Genet 1991; 7:69–70.
21. Lyon MF. Some milestones in the history of X-chromosome inactivation. Annu Rev Genet 1992; 26:17–28.
22. Earnshaw WC. Mitotic chromosome structure. BioEssays 1988; 9:147–150.
23. Russel PJ. Genetics, 3rd ed. New York: Harper Collins, 1992.
24. Muller HJ. The remaking of chromosomes. The Collecting Net 1938; 13:181–190.
25. Muller HJ, Herskowitz IH. Concerning the healing of chromosome ends produced by breakage in *Drosophila melanogaster*. Am Naturalist 1954; 88:177–208.
26. McClintock B. The stability of broken ends of chromosomes in *Zea mays*. Genetics 1941; 26:234–282.
27. Miller OJ. Structure and organization of mammalian chromosomes: normal and abnormal. Birth Defects: Original Article Series 1987; 23:19–63.
28. Mitelman F. An International System for Human Cytogenetic Nomenclature (1995). Basel, Switzerland: S. Karger, 1995.
29. Mark HFL, Mark R, Pan T, Mark Y. Centromere index derivation by a novel and convenient approach. Ann Clin Lab Sci 1993; 23:267–274.
30. Mark HFL, Wyandt HE, Pan T, et al. A cytogenetic and morphometric study of homologous chromosomes using a new approach. Genome 1993; 36:1003–1006.
31. Mark HFL, Parmenter M, Campbell W, et al. A novel, convenient, and inexpensive approach for deriving ISCN (1985) length: validation by a morphometric study of 100 karyotyped metaphase cells. Cytogenet Cell Genet 1993; 62:13–18.
32. Richardson A. Analysis of cytogenetic abnormalities. Chapter 11: Chromosome analysis. In Barch MJ, Knutsen T, and Spurbeck JL, eds. The AGT Cytoge-

netics Laboratory Manual, 3rd ed. Philadelphia and New York: Lippincott-Raven, 1997.
33. Donohue RP, Bias WB, Renwick JH, McKusick VA. Probable assignment of the Duffy blood group locus to chromosome 1 in man. Proc Natl Acad Sci USA 1968; 61:949.
34. Altman PL, Dittmer DS. Biology Data Book. Bethesda, MD: Federation of American Societies for Experimental Biology, 1972.
35. Gardner AL. Karyotypes of two rodents from Peru, with a description of the highest diploid number recorded for a mammal. Experientia 1971; 27:1088–1089.
36. Klug WS, Cummings MR. Concepts of Genetics, 4th ed. Englewood Cliffs, NJ: Prentice Hall, 1994.
37. Wurster DH, Benirschke K. Indian muntjac, *Muntiacus muntjac*: a deer with a low diploid number. Science 1970; 168:1364–1366.

APPENDIX: GLOSSARY OF TERMS

acquired karyotype
The karyotype acquired by somatic cells postzygotically.

aneuploidy
The state of having an abnormal unbalanced set of chromosomes.

arm ratio
The length of the long arm divided by the length of the short arm.

autosomes
Chromosomes 1 to 22; chromosomes other than the sex chromosomes.

banding
The staining of a chromosome using various methods to produce a pattern that distinguishes it from the other chromosomes.

Barr body
An inactivated X chromosome that upon Papanicolau staining is manifested as a darkly stained mass in the somatic cells of the mammalian female.

centromere
The central location of the chromosome to which spindle fibers attach, also called the primary constriction.

centromere index
Expressed as a percentage, this number is calculated by taking the length of the short arm divided by the total length of the chromosome.

chiasma
The cytologically identifiable point of crossover between members of a homologous pair of chromosomes during meiosis.

chimera
An organism or tissue that is composed of two or more genetically distinct subpopulation of cells, derived from two different zygotes.

chromosome
Molecular structure composed of DNA and proteins.

constitutional karyotype
The karyotype of the entire individual which began in the zygote.

constitutive heterochromatin
Heterochromatin found near the centromeres of chromosomes.

cytogenetics
The study of chromosomes.

dictyotene
A period in the diplotene stage of prophase I of female meiosis in which primary oocytes become suspended from further development until menarche.

diploid number
The number of chromosomes in the somatic cells of an organism.

euchromatin
Chromatin that is universally perceived as gene-bearing and which also stains normally.

euploidy
The state of having a normal balanced set of chromosomes.

facultative heterochromatin
Chromatin that is transcriptionally inactive, such as the inactivated X chromosome in a mammalian female which is manifested as a Barr body.

gametogenesis
The process that leads to the formation of ova or sperm.

germ cells
Cells of the reproductive system that carry the genetic information to be bequeathed to offspring.

gonosomes
Another name for sex chromosomes.

haploid number
The number of chromosomes in the egg or sperm cells of an organism.

heterochromatin
Chromatin that is tightly coiled and devoid of genes, usually stained positive with C-banding, which can be constitutive or facultative.

histones
Basic proteins that, together with DNA, comprise the nucleosomes in eukaryotic chromosomes.

homologous chromosomes
More or less identical chromosomes comprising the two members of each pair of chromosomes.

ideogram
A diagrammatic karyotype based on chromosome measurements from a large number of cells.

interphase
The stage of the cell cycle between mitoses, composed of the synthesis phase, G_1 and G_2, in which the cells are actively functioning and synthesizing DNA.

kinetochores
Trilaminar structures located in the centromeres which mediate spindle fiber attachment.

Klinefelter syndrome
XXY; a condition that occurs 1 in 1000 male live births. Some characteristics of those affected include gynecomastia, slight mental retardation, tall stature, underdeveloped testes, and sterility.

menarche
The first menstruation for a female adolescent, signaling the onset of puberty.

metacentric chromosome
A chromosome with its centromere located at the center, equidistant from each arm.

mosaic
Having more than one kind of genetically distinct cell types that are derived from the same zygote.

NOR
Nucleolus organizer region. The site of transcriptionally active, tandemly repetitive, ribosomal RNA genes usually found in the secondary constriction of the chromosome.

nucleosome
The fundamental packaging unit of DNA, composed of eight histones strung together by two DNA coils.

oogenesis
Female meiosis which leads to the development of ova or haploid eggs.

polyneme model
A proposed model of the chromosome which suggests that its structure is composed of multiple DNA molecules.

primary oocytes
Female germ cell precursors that remain suspended in the dictyotene stage of diplotene in prophase I until the female reaches menarche.

primary spermatocytes
The male germ cells that arise directly from the development of spermatogonia and through further division give rise to secondary spermatocytes.

relative length
Expressed as a percentage, this number is derived by taking the length of an individual chromosome divided by the total haploid lengths of all the chromosomes in the human genome.

secondary oocytes
The female germ cell precursors that arise directly from the nuclear division of primary oocytes through meiosis I.

secondary spermatocytes
The male germ cell precursors that arise directly from the nuclear division of primary spermatocytes through meiosis I.

sister chromatids
Identical structures that are held together at the centromere after replication, which together form a chromosome. Each sister chromatid will become a new chromosome after centromeric division.

somatic cells
Cells of the body.

spermatogenesis
The male meiotic process leading to the formation of sperm.

spermatogonia
Male germ cell precursors.

spermiogenesis
The differentiation process in male development whereby spermatids become mature sperm.

synapsis
The close pairing of homologous pairs of chromosomes during meiosis in preparation for crossing over.

synaptonemal complex
A complex structure that brings the homologs together during the prophase stage of meiosis.

telocentric chromosome
A chromosome with a total lack of a short arm and whose centromere is located at one end. Telocentric chromosomes are not found in humans.

telomere
TTAGGG sequences which usually appear at the ends of a chromosome and which also serve as caps. Interstitial telomeric sequences can also be present in a chromosome.

Turner syndrome
A female with a single X chromosome; such cases occur 1 in 10,000 female live births. Some characteristics of those affected include short stature, poor breast development, underdeveloped reproductive organs, characteristic facial features, a constricted aorta, and sterility.

unineme model
A proposed model of the chromosome which suggests that its structure is composed of a single DNA molecule.

3
Human Cytogenetic Nomenclature

Sugandhi A. Tharapel
Veterans Affairs Medical Center and The University of Tennessee, Memphis, Tennessee
Avirachan T. Tharapel
The University of Tennessee, Memphis, Tennessee

I. INTRODUCTION

The history of human cytogenetics began with the discovery of the correct diploid chromosome number (46) in man (1,2). This discovery was not only of monumental significance in the understanding of chromosomal variations and aberrations, it formed the backbone of present-day human genetics, providing answers to human problems such as behavior, reproduction, aging, and an array of genetically determined conditions. It has been estimated that one of every 100 newborn will have a chromosome rearrangement, and nearly one-third of all pediatric hospital admissions are because of a genetic disorder. The launching of an international project to map the entire human genome by the year 2005 is a testimonial to the role that human genetics has played in the understanding and prevention of human diseases.

The first chromosomal syndrome in humans was recognized in 1959 with the demonstration of the presence of three copies of a G-group chromosome (designated as 21) in a Down syndrome patient (3). Soon thereafter began the understanding of the chromosomal etiology of several severe malformation syndromes such as Patau syndrome (4) and Edward syndrome (5), as well as some sex chromosomal aberrations leading to phenotypically benign syndromes such as Turner syndrome and triple X syndrome (6,7).

As this new knowledge began to unfold, the number of people working in this field grew quickly. It became apparent that there was a need for a uniform system to report chromosome variations and anomalies to facilitate communication among investigators. It was with this background that, in 1960, fourteen investigators (representing each of the fourteen laboratories that had published the human karyotype up to that time) and three consultants, led by Charles E. Ford, met in Denver, Colorado, and proposed guidelines for the description of normal and abnormal human chromosomes. This historic document was published as the "Denver Conference (1960): A proposed standard system of nomenclature of human mitotic chromosomes" (8).

While the basic principles adopted at the Denver Conference have prevailed to date, new technologies and ever-increasing knowledge in human cytogenetics necessitated periodic revisions and updates of this original document. In Table 1, we have listed the major conferences that took place since 1960 leading up to the Memphis conference of 1994 (8,9,11–18).

Each conference was unique and contributed significantly to the modification and clarification of the nomenclature. Of particular significance is the Paris conference, which took place with the backdrop of an unusual technological advancement in the field of human cytogenetics. In 1968, Caspersson and colleagues, working with plant chromosomes, discovered that when chromosomes were stained with a fluorescent dye (quinacrine mustard) and observed under a fluorescent microscope, they revealed a bright- and dull-appearing pattern all along their lengths. The pattern was unique to each chromosome pair (10). This phenomenon was found to be true for human chromosomes as well. It was later observed that these unique banding patterns seen on each of the chromosomes can be obtained by simply treating chromosomes with a proteolytic enzyme, such as trypsin. The Paris Con-

Table 1 Major Human Nomenclature Conferences

Conference/document		Year of publication
Denver Conference		1960
London Conference		1963
Chicago Conference		1966
Paris Conference		1971
Paris Conference (Supplement)		1975
Stockholm 1977	ISCN	1978
Paris 1980	ISCN	1981
	ISCN	1985
Cancer Supplement	ISCN	1991
Memphis 1994	ISCN	1995

ference (1971) provided the first banded karyotype and ideogram to enable an interpretation of structural rearrangements (12).

Human chromosome nomenclature reached a new level in 1976 when, for the first time, a seven-member International Standing Committee on Human Cytogenetic Nomenclature was elected at the 5th International Congress of Human Genetics held in Mexico City, to update the nomenclature based on developments since the Paris Conference. The nomenclature document was given the name "An International System for Human Cytogenetic Nomenclature 1978" or ISCN (1978) (14). All future documents on nomenclature were identified with the same name followed by the year of publication. The development of high-resolution chromosome banding and prometaphase banding necessitated further modifications to the nomenclature, which were published as ISCN (1985) following the 6th International Congress of Human Genetics held in Jerusalem (15,16). For nearly 10 years this document was in wide use for both constitutional and acquired chromosome nomenclature. However, to meet the needs of the ever-increasing information on cancer, in 1991, a cancer supplement was published which was called "ISCN (1991): Guidelines for Cancer Cytogenetics, Supplement to An International System for Human Cytogenetic Nomenclature" (17).

The ISCN (1995), the most current document, was the result of developments on a number of fronts (18). First, the need for a comprehensive single nomenclature to describe both constitutional and acquired chromosome variations was obvious. A number of major developments took place in the 1980s on the technological front as well. For example, the 1980s saw the birth of a new discipline within cytogenetics called molecular cytogenetics, as a result of increased application of DNA technology in cytogenetic laboratories. It brought the two powerful technologies together and bridged the gap between the chromosome band and the DNA sequence. Every human chromosome is known to contain a large number of repeat sequences of DNA in the pericentromeric region. The number of repeats and the subtle differences in the nucleotides between chromosomes enabled scientists to make chromosome-specific pericentromeric probes (alpha-satellite probes). The probes were labeled enzymatically and then hybridized on metaphase or interphase nuclei to delineate abnormalities that were not amenable to cytogenetic protocol. DNA sequences have also been isolated, sequenced, and cloned from single-copy gene sequences which potentially can identify any part of a chromosome region at a resolution of 2000 base pairs of a DNA segment. A wide range of molecular techniques are now used routinely in laboratories, and the most common one is called the in situ hybridization. It became obvious that the nomenclature needed another update to keep up with the technical developments on several fronts. Furthermore, the inconsistencies or ambiguities between the ISCN 1985 and the cancer supplement

(1991) became apparent. In October 1994, the International Standing Committee of ISCN and several consultants met in Memphis, Tennessee, at the invitation of one of the standing committee members (Avirachan T. Tharapel). The proceedings of this meeting resulted in ISCN (1995) (18). This document has incorporated a new section of nomenclature for in-situ hybridization and has combined the ISCN 1985 and 1991 cancer supplement into a single document. More important, this document has provided a uniform system of nomenclature for the description of both constitutional and acquired chromosome changes. In this chapter, we will discuss the guidelines and general principles of ISCN (1995) with examples whenever possible. However, for a detailed understanding of ISCN (1995), the reader is advised to refer to the original document.

II. HUMAN CHROMOSOMES

Of the 46 chromosomes in a normal human somatic cell, 44 are autosomes and 2 are sex chromosomes (XX in a female and XY in a male). When chromosomes are stained with a chromatin stain, such as Giemsa or Wright's stain, they do not produce bands. Nevertheless, they can be classified into 7 groups based on the size and centromere position. The letters A–G were assigned to the groups at the London conference (1963) (19).

Group A (1–3)	Large metacentric or near metacentric chromosomes distinguishable from one another by the size and the position of the centromere
Group B (4–5)	Largest of the submetacentric chromosomes
Group C (6–12 and X)	Medium-sized submetacentric chromosomes (most difficult to distinguish from one another)
Group D (13–15)	Larger acrocentric chromosomes with usually polymorphic satellites on the short arms
Group E (16–18)	Shorter metacentric or submetacentric chromosomes
Group F (19–20)	Smallest of the metacentric chromosomes
Group G (21–22 and Y)	Smaller acrocentric chromosomes usually with polymorphic satellites; Y chromosome can be distinguished from the rest of the G group chromosomes by its lack of satellites and usually darkly stained heterochromatic long arm

A. Chromosome Banding and Identification

Unequivocal identifications of individual chromosomes and chromosome regions became possible with the technical developments of the late 1960s.

When chromosome preparations are treated with dilute solutions of proteolytic enzymes (trypsin, pepsin, etc.) or salt solutions (2XSSC) and treated with a chromatin stain, such as Giemsa, there appeared alternating dark and lightly stained demarcations called *bands* which can be seen along the length of each chromosome. The banding patterns produced are unique to each chromosome pair, thus enabling the identification of individual chromosomes as well as their regions. Methods commonly used to produce these discriminative banding patterns include Giemsa or G-banding, quinacrine mustard or Q-banding, reverse or R-banding, and constitutive heterochromatin or C-banding, each with its own attributes. In the United States and Canada the most frequently used methods for routine cytogenetic analysis are the G- and Q-bands (Figs. 1a and 1b). R-banding and C-banding are occasionally used to delineate specific abnormalities. Table 2 lists abbreviations commonly used to identify the type of banding technique.

B. Chromosome Regions and Band Designations

The chromosomal details revealed by banding techniques necessitated introduction of additional terminology and modifications of certain existing ones. This task was accomplished by a standing committee appointed at the 4th International Congress of Human Genetics in Paris. The recommendations of the committee was published as "Paris Conference (1971): Standardization in Human Cytogenetics" (12). Through a diagrammatic representation of banding patterns, the document elucidated the typical band morphology for each chromosome, which came to be known as an *ideogram*. The Paris Conference (1971) introduced a numbering system helpful in designating specific bands and regions while describing a structural abnormality. A partial list of recommended symbols and abbreviations appears in Table 3.

The most recent ideogram of human chromosomes depicting the regions and bands at three different resolutions as revealed by the G-banded method is shown in Fig. 2. The centromere *cen* divides a chromosome into a short *p* arm and a long *q* arm. Each chromosome arm is divided into *regions*. This division is based on certain *landmarks* present on each chromosome. By definition, a landmark is a consistent and distinct morphologic area of a chromosome that aids in the identification of a chromosome. A *region* is an area that lies between two landmarks. The regions are numbered in increasing order starting from the centromere and moving toward the telomere on both arms. The two regions immediately adjacent to the centromere are designated as 1, the next distal as 2, and so on. The regions are divided into *bands* and the bands into *subbands*. A band is that part of a chromosome which is distinctly different from the adjacent area by virtue of being lighter or darker in staining intensity. Sequential numbering of

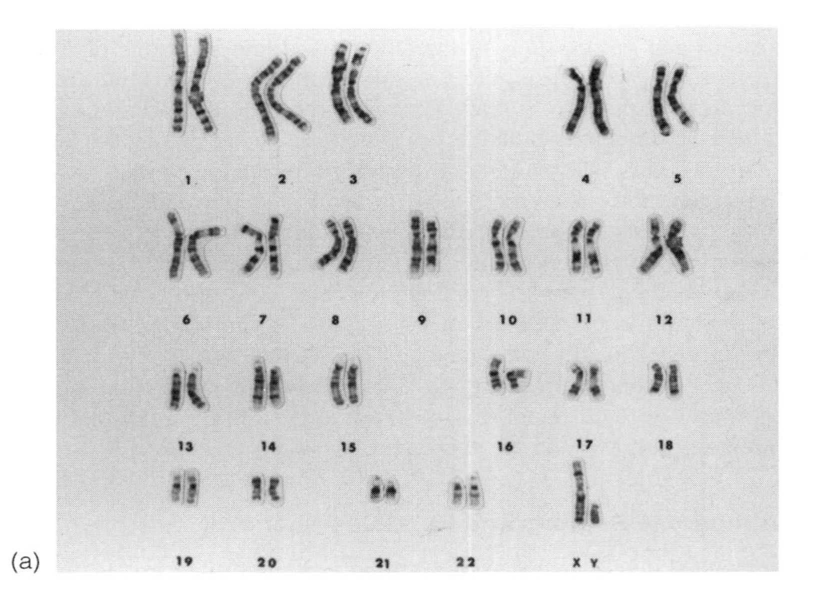

(a)

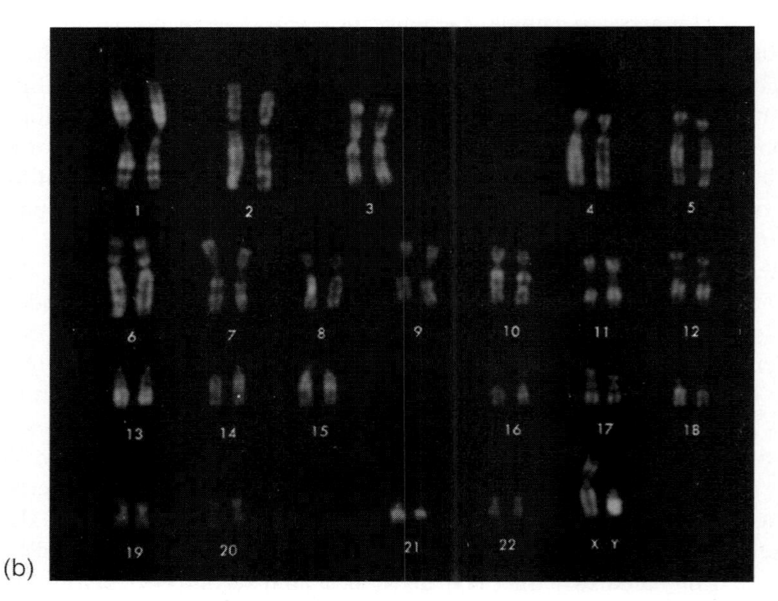

(b)

Figure 1 (a) G-banded karyotype of a normal male 46,XY at an approximate resolution of 500 bands. (b) Q-banded karyotype of a normal male 46,XY at an approximate resolution of 400 bands.

Table 2 Frequently Used Banding Methods and Their Abbreviations

Banding method	Abbreviation
Q-bands	Q
Q-bands by fluorescence microscopy and quinaqerine	QFQ
G-bands	G
G-bands by trypsin and Giemsa	GTG
C-bands	C
C-bands by barium hydroxide and Giemsa	CBG
R-bands	R
R-bands by fluorescence microscopy and acridine orange	RFA
R-bands by bromodeoxyuridine (BrdU) and Giemsa	RBG
Telomere bands or T-bands	T

chromosome arms and bands helps to make the designation of specific bands easy. For example, the terminal band on the long arm of chromosome 2 can be written as 2q37, for chromosome 2, long arm, region 3, band 7.

For descriptive purposes, the centromere is divided arbitrarily into two parts. The region between the middle of the centromere and the first band on the short arm (cen→p11 or p11.1) is designated *p10*. Similarly, the region between the middle of the centromere and the first band on the long arm (cen→q11 or q11.1) is designated *q10*. The designations p10 and q10 allow us to describe isochromosomes, whole-arm translocations, and Robertsonian translocations more accurately.

C. Karyotype Descriptions

The nomenclature for a karyotype follows certain basic rules. When designating a karyotype, the first item specified is the total number of chromosomes, including the sex chromosomes present in that cell, followed by a comma (,) and the sex chromosomes, in that order. Thus a normal female karyotype is written as 46,XX and a normal male karyotype as 46,XY. The format is a continuous string of characters, without a space between characters. A chromosome abnormality, when present, follows the sex chromosome designation along with an abbreviation or symbol denoting the abnormality (see Table 3). With a series of hypothetical cases, we will further illustrate the use of ISCN (1995) in the following pages.

Table 3 Selected List of Symbols and Abbreviations Used in Karyotype Designations[a]

Abbreviation	Explanation
add	Additional material, origin unknown
(→ or →)	Arrow from or to
[ ] square brackets	Number of cells in each clone
cen	Centromere
chi	Chimera
single colon (:)	Break
double colon (::)	Break and reunion
comma (,)	Separates chromosome number from sex chromosome and also separates chromosome abnormalities
del	Deletion
de novo	A chromosome abnormality which has not been inherited
der	Derivative chromosome
dic	Dicentric
dmin	Double minute
dup	Duplication
fis	Fission
fra	Fragile site
h	Heterochromatin
i	Isochromosome
inv	Inversion
ins	Insertion
mar	Marker chromosome
mat	Maternal origin
minus sign (−)	Designates loss
mos	Mosaic
multiplication sign (×)	Multiple copies
p	Short arm
pat	Paternal origin
Ph	Philadelphia chromosome
plus sign (+)	Gain
q	Long arm
question mark (?)	Questionable identification
r	Ring chromosome
rcp	Reciprocal
rec	Recombinant chromosome
rob	Robertsonian translocation
s	Satellite
/ slant line	Separates cell lines and clones
semicolon (;)	Separates altered chromosomes and breakpoints in structural rearrangements involving more than one chromosome
stk	Satellite stalk
t	Translocation
upd	Uniparental disomy

[a]For a complete listing of symbols and abbreviations refer to ISCN 1995.

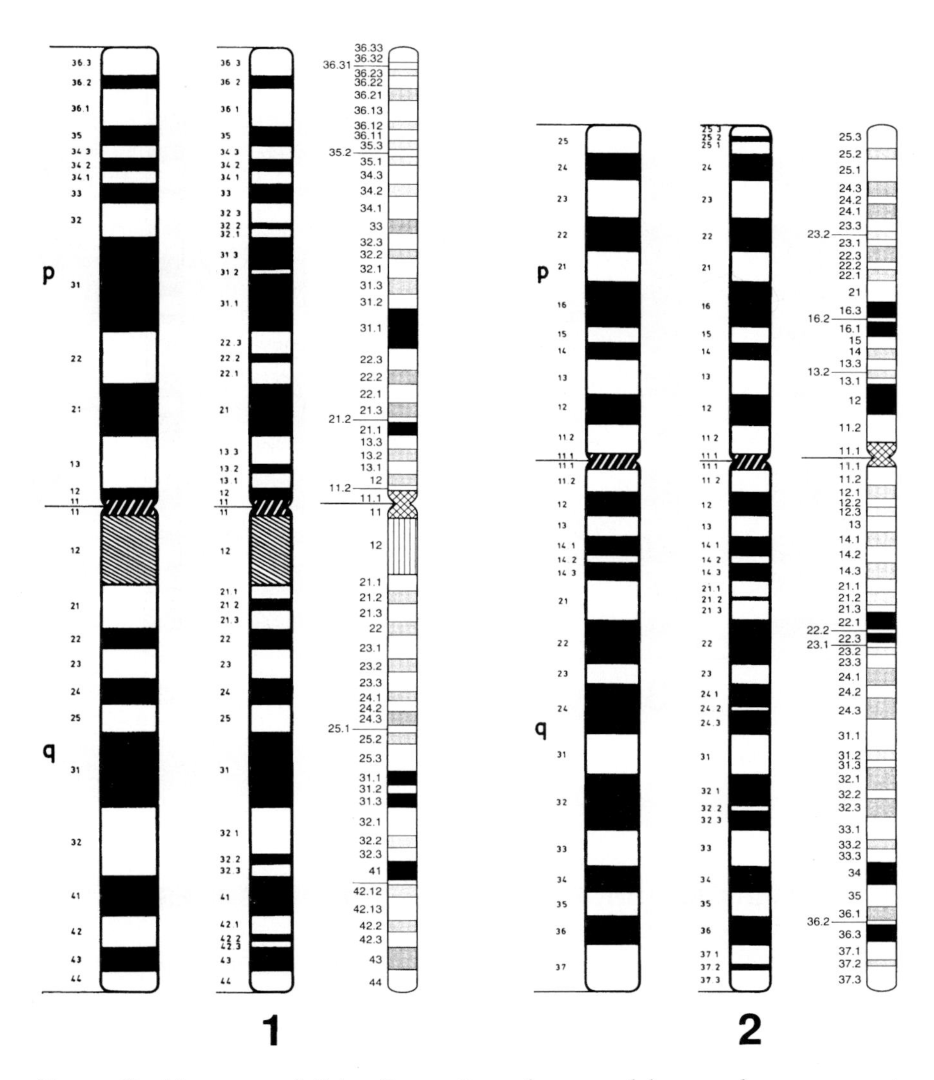

Figure 2 Ideograms of G-banding patterns for normal human chromosomes at three different resolutions. The left, center, and right at 450, 550, and 850 band levels, respectively. (Reproduced with permission from S. Karger, ISCN 1995).

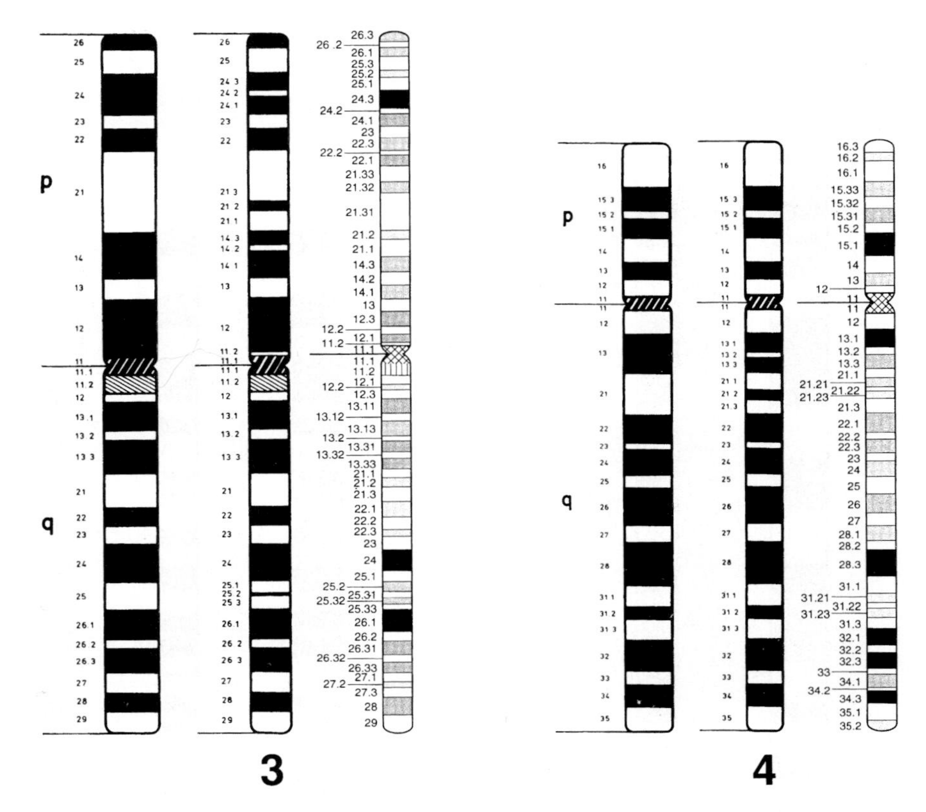

Figure 2 Continued

III. NUMERICAL ABNORMALITIES OF CHROMOSOMES

All numerical abnormalities (aneuploidies) in a karyotype are presented in increasing order of the chromosome number except for the sex chromosomes. Among sex chromosomes, X will precede Y. In describing aneuploidy involving more than one chromosome, the lower number chromosome will be placed first. As a general principle, aneuploidies are written by using the symbols plus (+) and minus (−). However, there are some subtle but distinct differences in the nomenclature for aneuploidies involving autosomes and sex chromosomes as well as for constitutional and acquired aneuploidies. The following examples will help clarify the differences.

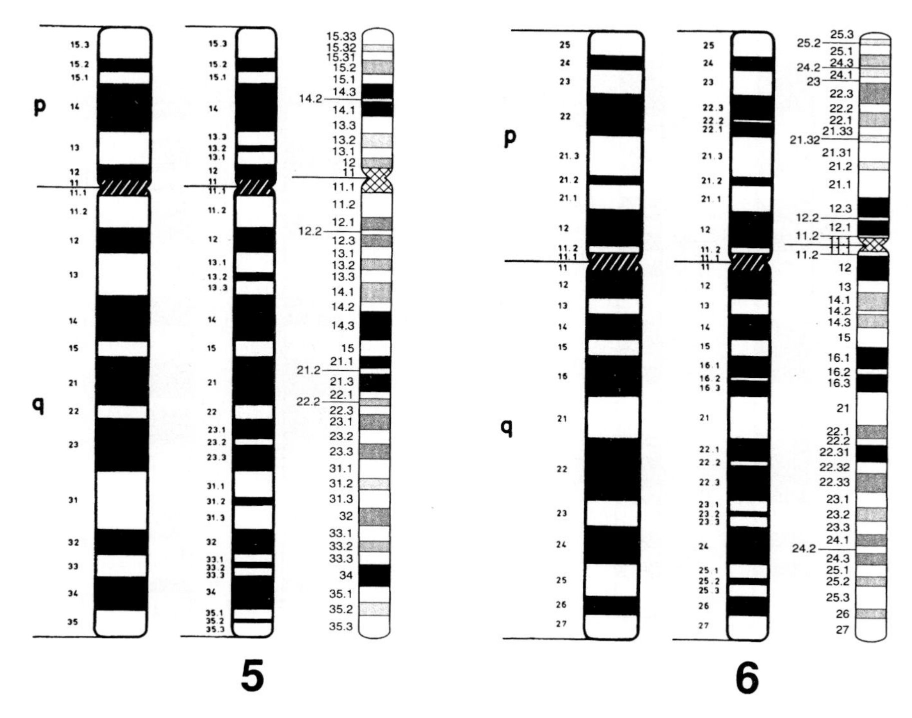

Figure 2 Continued

A. Sex Chromosome Aneuploidies

Sex chromosome aneuploidies can be constitutional (congenital) or acquired. ISCN (1995) has provided special ways to distinguish between the two. Shown below are some examples of constitutional and acquired sex chromosome aneuploidies.

1. Constitutional Sex Chromosome Aneuploidies

45,X	X monosomy as seen in Turner syndrome
47,XXY	Typical karyotype seen in Klinefelter syndrome
47,XXX	A female with three X chromosomes
48,XXYY	Variant of Klinefelter syndrome with two X and two Y chromosomes

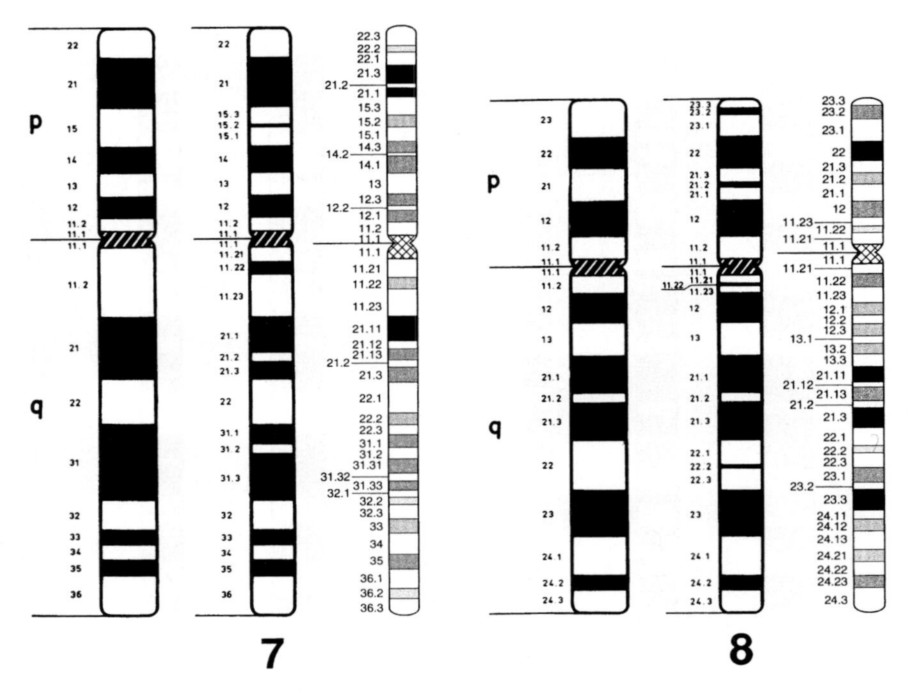

Figure 2 Continued

2. Acquired Sex Chromosome Aneuploidies

Acquired sex chromosome aneuploidies are presented by using the symbols plus (+) or minus (−). When presenting a case with both constitutional and acquired sex chromosome anomalies, the letter *c* (for constitutional) is placed after the sex chromosome complement which was seen constitutionally. However, it is not necessary to use *c* if the constitutional sex chromosome complement is normal. The following examples show both scenarios.

45,X,−X

A normal female with two X chromosomes and with the loss of one of the X chromosomes in her tumor cells.

47,XX,+X

A normal female with two X chromosomes and gain of an extra X chromosome in her tumor cells.

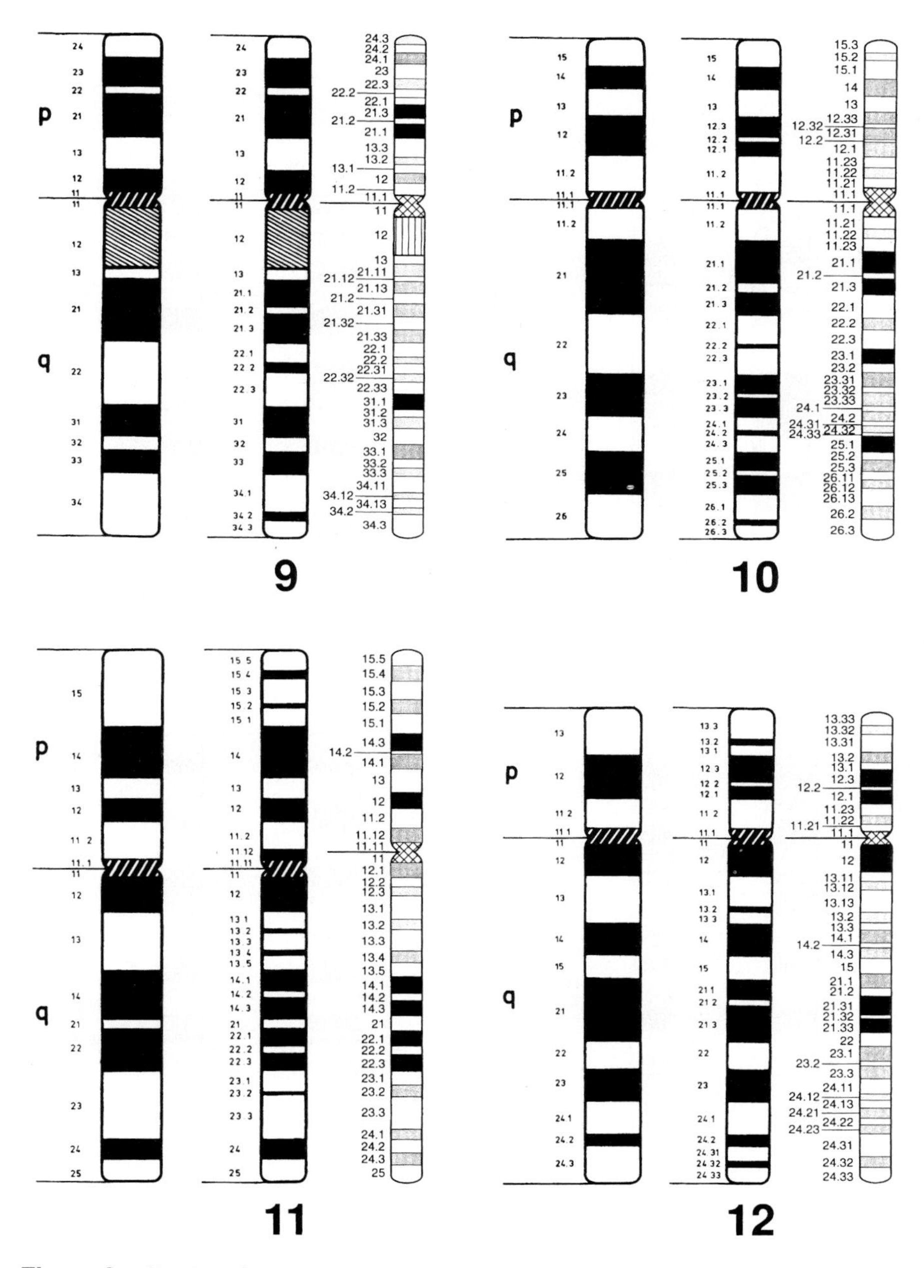

Figure 2 Continued

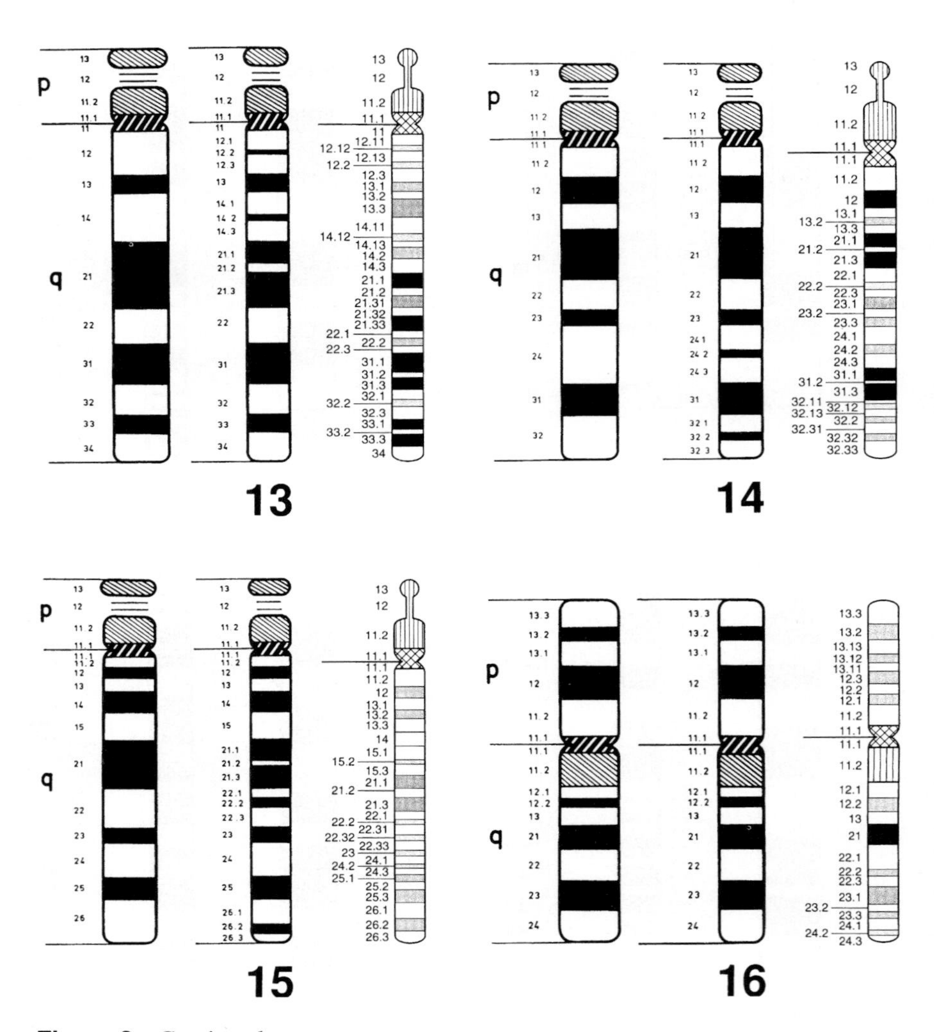

Figure 2 Continued

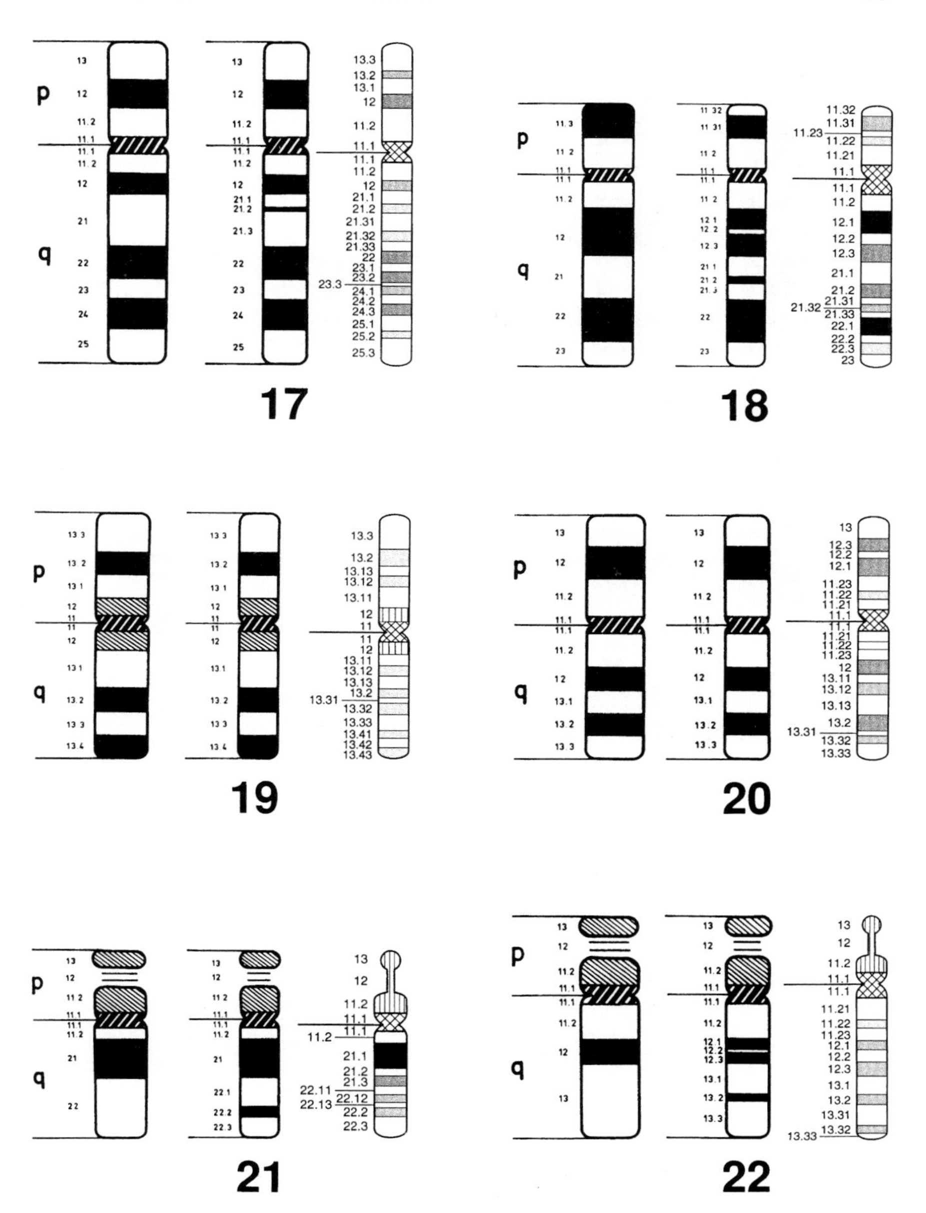

Figure 2 Continued

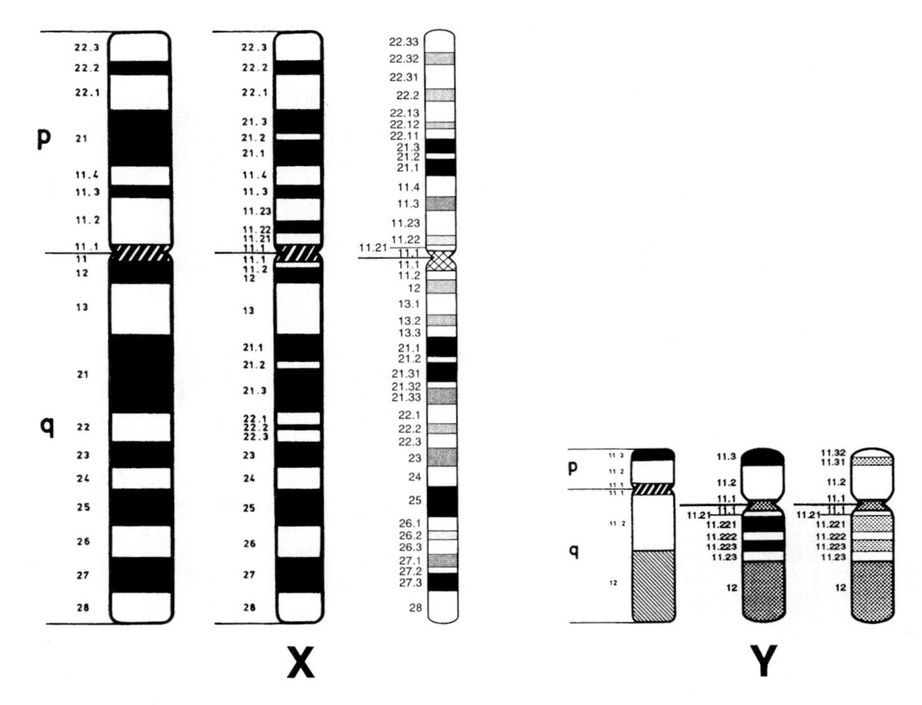

Figure 2 Continued

45,X,−Y

A normal male with XY chromosomes and loss of the Y chromosome in his tumor cells.

48,XY,+X,+Y

A normal male with an acquired X and Y chromosomes in his tumor cells.

48,XXYc,+X

A patient with Klinefelter syndrome with an acquired X chromosome in his tumor cells. The small letter *c* is placed next to XXY to show that constitutional sex chromosome complement of the patient is XXY and not XXXY.

46,Xc,+X

A Turner syndrome patient (45,X) with gain of an X chromosome in her tumor cells.

B. Autosome Aneuploidies

The autosome aneuploidies, both constitutional and acquired, are described by simply placing a + or − sign before the chromosome in question. When both constitutional and acquired aneuploidies are seen together, place letter *c* after the chromosome which is involved in constitutional aneuploidy.

47,XY,+18	Male with trisomy 18
48,XX,+18,+21	Female with both trisomy 18 and trisomy 21
45,XY,−21	Male with monosomy 21
46,XY,+21c,−21	Male trisomy 21 patient with loss of a chromosome 21 in his tumor cells
48,XX,+8,+21c	Female trisomy 21 patient with gain of a chromosome 8 in her tumor cells

C. Mosaics and Chimeras

An individual with two or more cell types differing in chromosome number or structure is either a mosaic or a chimera. If the two cell types in a specific instance originated from a single zygote, the individual is a mosaic (*mos*). If the cell types originated from two or more zygotes, the individual is a chimera (*chi*). When designating mosaic or chimeric karyotypes, a slant line (/) is used to separate the cell types. The actual number of cells detected in each clone is given within square brackets [ ]. Usages such as percentage and ratios should be avoided in the karyotype, but may be used in the text. When two or more cell lines are present and the cell lines are unrelated, the largest clone is recorded first, the next largest second, and so on. When two or more related abnormal cell lines are seen, they are written in order of increasing complexity. In all situations, the normal cell line is always given last.

mos 45,X[4]/46,XX[16]

A mosaic with two cell lines. An analysis of 20 cells showed that this individual has 4 cells with 45,X and 16 cells with 46,XX karyotypes.

mos 45,X[5]/47,XYY[5]/46,XY[10]

A mosaic with three cell lines including normal. Note that the normal cell line is given last.

mos 47,XX,+13[15]/46,XX[5]

A mosaic with trisomy 13 and normal cell lines. In a chimera where the two cell lines are normal 46,XX or 46,XY and they are represented in equal proportion, any one of them may be listed first. If one cell line is larger than the other, the larger clone is listed first.

chi 46,XX[10]/46,XY[10]

A chimera with female and male cells in equal number.

chi 47,XX,+21[15]/46,XY[5]

A chimera with female and male cell lines. The female cell line shows trisomy 21, while the male cell line is normal.

chi 69,XXX[20]/46,XY[5]

A chimera with triploid-diploid cell lines. The triploid line is female, while the diploid line is male.

D. Uniparental Disomy (upd)

In upd, which is also a numerical abnormality, both members of a specific chromosome pair are inherited from the same parent. Examples include some patients with Angelman and Prader-Willi syndromes.

46,XY,upd(15)pat

Male karyotype with uniparental disomy for paternally derived chromosome 15.

46,XY,upd(22)pat[10]/47,XY,+22[6]

Mosaic male karyotype with one cell line with upd for a paternally derived 22 and the other with trisomy 22. Here, both cell lines are abnormal and therefore the largest is recorded first.

46,XX,upd pat

Paternal upd for all 23 pairs of chromosomes, as may be seen in complete hydatidiform moles.

46,XY,upd pat

A complete hydatidiform mole with XY sex chromosomes (very rare). All 46 chromosomes are paternally derived.

46,XX,upd mat

Maternal upd as may be seen in ovarian teratoma. All 46 chromosomes are maternally derived.

IV. STRUCTURAL CHROMOSOME ABNORMALITIES

This class of abnormalities include several subclasses and will be discussed under separate headings. As stated previously, all abnormalities are presented in increasing order of chromosome number. However, when the X and Y chromosomes are involved in structural rearrangements they are listed first, with X always before Y. When designating an abnormality which is limited to a single chromosome, the abbreviation of the abnormality is given and then the number of the chromosome is written within parentheses, such as, del(2), ins(4), dup(5), and r(X). If two or more chromosomes are involved in a rearrangement, such as in a translocation, a semicolon (;) separates each chromosome. Some examples are t(3;4), t(2;3;4), or t(15;17). The chromosome arm and the breakpoint for the chromosomes involved are separated by a semicolon within a second set of parentheses. The chromosome number that is entered first is usually the one with the smallest number, unless a sex chromosome is involved. In such an event the sex chromosome is always designated first, e.g., t(X;1) or t(Y;15). If in the same cell a specific chromosome is involved in both a numerical and a structural rearrangement, the numerical abnormality is designated first, e.g., +13,der(13;14). The chromosome nomenclature may be written in two different ways, namely, the short form and the long form. The long form allows better understanding of the karyotype by way of suggesting the chromosome region present in a karyotype and designating the bands where the break and the reunion occur. In rare situations, particularly in structural rearrangements, the long form may be the only way to show where and how the rearrangement occurred. However, most practicing clinical cytogeneticists use the short form to report their results since it is concise and brief. Cytogenetic reports must contain description of the rearrangements in words, particularly in reporting abnormal karyotypes. This is an essential element of the laboratory report because the consumers of the report, most often, are not scientists. Following are a few examples of using short and long forms of nomenclature.

A. Additional Material, Origin Unknown

When additional chromosome material is attached to a chromosome, usually its origin is not identifiable through conventional banding methods. This is especially true if the abnormality is subtle and originates de novo. The abbreviation *add* is then used to record the rearrangement:

46,XX,add(17)(p13)	Short form
46,XX,add(17)(?::p13→qter)	Long form

Additional material of unknown origin is attached to chromosome 17 band 17p13. Here the type of rearrangement that resulted in the abnormality is also unknown.

46,XX,add(9)(q22)	Short form
46,XX,add(9)(pter→q22::?)	Long form

Additional material of unknown origin attached to band 9q22. It is assumed that the chromosome region 9q22→qter is deleted and replaced by the added material.

B. Deletions

A deletion is an aberration in which a part of a chromosome is lost. Deletions can be either terminal or interstitial.

1. Terminal Deletions

46,XY,del(1)(q32)	Short form
46,XY,del(1)(pter→q32:)	Long form

A karyotype with terminal deletion of chromosome 1. A single colon (:) indicates a break at 1q32 and deletion of the region distal to it. The remaining chromosome consisting of regions 1pter to 1q32 is present in the cell. (The terminal regions of the short and long arms of a chromosome are designated by *pter* and *qter*, respectively).

2. Interstitial Deletions

46,XY,del(1)(p21p32)	Short form
46,XY,del(1)(pter→p21::p32→qter)	Long form

A break and reunion represented by a double colon (::) occurred at bands 1p32 and 1p21. The segment between the two breakpoints is deleted.

C. Derivative and Recombinant Chromosomes

1. Derivative Chromosomes

A structurally rearranged chromosome generated by events involving two or more chromosomes or due to multiple events within a single chromosome is a derivative chromosome. Thus, each unbalanced product of a translocation event is a derivative chromosome and is designated by the abbreviation *der*. The identity of the centromere in a derivative chromosome determines its chromosome number. For example, der(3) indicates that this derivative chromosome has the chromosome 3 centromere.

46,XY,der(3)t(3;6)(p21;q23)

The derivative chromosome 3 in this example resulted from an adjacent 1 segregation of a balanced translocation between the short arm of a chromosome 3 at band p21 and the long arm of a chromosome 6 at band q23. The modal number of 46 chromosomes in this example indicates that there are two normal chromosomes 6. The der(3) replaces one normal chromosome 3. This unbalanced karyotype shows loss (monosomy) of region 3p21→pter and gain (trisomy) of 6q23→qter.

45,XY,der(3)t(3;6)(p21;q23),−6

The der(3) is same as in the above example and replaces one of the normal chromosome 3. However, there is only one normal chromosome 6 in the cell, the result of a 3:1 segregation. Note that the chromosome number in a cell is critical in understanding the nomenclature. Also note that the −6 is placed at the end of the karyotype. This unbalanced karyotype represents monosomy for the 3p21 to the *pter* region of chromosome 3 and monosomy for the 6pter to the 6q23 region of chromosome 6.

47,XY,+der(3)t(3;6)(p21;q23)mat

The der(3) is the same as in the first example. As a result of 3:1 segregation, the mother contributed a normal 3 and the derivative 3. The father contributed a normal 3 as well. The patient is therefore trisomic for both 3p21→qter and 6q23→qter.

---

2. Recombinant Chromosomes

A recombinant chromosome is also a structurally rearranged chromosome. It arises from meiotic crossing over between two homologous chromosomes in a heterozygote: one a structurally rearranged chromosome and the other a normal homolog. Recombinant chromosomes commonly arise due to crossing over within the limits of the inversion in an inversion heterozygote.

46,XY,inv(3)(p21q32)

A pericentric inversion of a chromosome 3 is designated above. During meiosis, crossing over within the inverted segment could result in two recombinant chromosomes.

46,XY,rec(3)dup(3p)inv(3)(p21q32)

Duplication from 3pter to 3p21 and deletion from 3q32 to 3qter.

46,XY,rec(3)dup(3q)inv(3)(p21q23)

Duplication of 3q23 to qter and deletion from 3p21 to 3pter.

D. Insertions

An insertion is a structural rearrangement wherein a part of a chromosome is inserted into a new place on a chromosome. Insertions can be within a chromosome or between two chromosomes, and can be direct or inverted.

1. Insertion Within a Chromosome

46,XX,ins(3)(p21q27q32)

This represents a direct insertion within a chromosome. The long arm segments between band 3q27 and 3q32 have been inserted into the short arm of the same chromosome at band 3p21.

---

2. Insertion Involving Two Chromosomes

46,XX,ins(4;9)(q31;q12q13)

The long arm segments between bands 9q12 and 9q13 have been inserted into the long arm of chromosome 4 at band q31. The recipient chromosome is always specified first, regardless of the chromosome number.

E. Inversions

A chromosomal aberration in which a portion (segment) of a chromosome is reversed in orientation. Inversions are of two types. Paracentric inversions involve only one arm of a chromosome, while pericentric inversions involve both arms of a chromosome.

1. Paracentric Inversion

46,XY,inv(3)(q21q27)

A break and a reunion occurred at bands 3q21 and 3q27 (long arm) of a chromosome 3. The interlying segment was reattached with its bands in an inverted sequence.

2. Pericentric Inversion

46,XY,inv(2)(p21q31)

A break and a reunion occurred at bands 2p21 in the short arm and 2q31 in the long arm of a chromosome 2. The interlying segment was reattached with its bands in an inverted sequence.

F. Isochromosomes

An abnormal chromosome with duplication of one of the arms (as a result of a misdivision of the centromere) resulting in a metacentric chromosome with identical gene sequences on both arms is referred to as an isochromosome. In the karyotype the isochromosome is abbreviated as *i*. When the nature of the centromere is not known (monocentric or dicentric), the breakpoint in an isochromosome is assigned to the arbitrary centromeric bands p10 and q10 depending upon whether the isochromosome consists of the short arm or the long arm. The use of p10 and q10 will be further demonstrated under Robertsonian translocations and whole arm translocations.

46,XX,i(18)(p10)

An isochromosome for the short arm of a chromosome 18. The breakpoint is assigned to p10.

46,XX,i(18)(q10)

An isochromosome for the long arm of a chromosome 18. The breakpoint is assigned to q10.

1. Isodicentric Chromosomes

Unlike the isochromosome, the isodicentric chromosome contains two copies of the same centromeres. One of the two centromeres is usually inactive. Designated by *idic*, the breakpoints in isodicentric chromosomes are usually on the band adjacent to the centromere on the opposite arm.

46,XX,idic(18)(q11.2)

An isodicentric chromosome for the entire short arm of chromosome 18 as well as for the long arm region between the centromere and the band 18q11.2.

G. Marker Chromosomes

Marker chromosomes *mar* are structurally abnormal chromosomes of which no part can be identified. If any part of such a marker is identifiable it is no longer a marker but a derivative chromosome. The presence of a *mar* in a karyotype is always recorded by a plus (+) sign.

47,XY,+mar

A male karyotype with a marker chromosome.

48,XY,+2mar

A male karyotype with two marker chromosomes.

48,XY,t(5;12)(q13;p12),+21,+mar

A male karyotype with a t(5;12), an extra chromosome 21 and a marker chromosome.

H. Ring Chromosomes

A structurally abnormal chromosome as a result of two breaks, one on the short arm and one on the long arm. The broken ends are attached to form a ring configuration. The net result is deletions of the terminal ends of both arms.

46,X,r(X)

A female karyotype with only one normal X chromosome and a ring X chromosome with no information on breakpoints.

46,X,r(X)(p22q24)

A female karyotype with one normal X chromosome and a ring X chromosome with breakage and reunion at bands Xp22 and Xq24.

I. Translocations

The interchange or transfer of chromosomal segments between two chromosomes is defined as a translocation.

1. Reciprocal Translocations

If the translocation involves a mutual exchange of segments between two chromosomes, it is a reciprocal translocation. To describe a reciprocal translocation the abbreviation *rcp* or the letter *t* can be used. The latter is more often used. In translocations involving two chromosomes, the autosome with the lowest number is specified first. If an X or Y chromosome is involved in the translocation, the X or Y is specified first, in preference to the autosomes. If the translocation involves three or more chromosomes, the same rule applies. However, in such rearrangements the first chromosome will be the one with the lowest number (or sex chromosome), the second chromosome specified will be the one that received the segment from the first, and so on.

46,XX,t(7;10)(q22;q24)

Breakage and reunion occurred at bands 7q22 and 10q24. The segments distal to these bands were interchanged. The translocation event has not altered the total DNA content of this cell. Therefore, the translocation is cytogenetically balanced.

46,X,t(X;1)(p21;q32)

Breakage and reunion occurred at bands Xp21 and 1q32. The segments distal to these bands were interchanged. The translocation is balanced. As per the general rules, the X chromosome is specified first.

46,X,t(Y;15)(q11.23;q21.2)

Breakage and reunion occurred at subbands Yq11.23 and 15q21.2. The segments distal to these subbands were interchanged. This translocation is cytogenetically balanced. Here again, the sex chromosome is specified first.

46,XY,t(9;22)(q34;q11.2)

Breakage and reunion occurred at bands 9q34 and 22q11.2. The segments distal to these bands have been interchanged. This represents the typical translocation resulting in the Philadelphia (Ph) chromosome.

46,XX,t(1;7;4)(q32;p15;q21)

This is an example of a translocation involving three chromosomes. The segment on chromosome 1 distal to 1q32 has been translocated onto chromosome 7 at band 7p15, the segment on chromosome 7 distal to 7p15 has been translocated onto band 4q21, and the segment on chromosome 4 distal to 4q21 has been translocated onto chromosome 1 at 1q32. The translocation is cytogenetically balanced.

The general principles also apply to designating translocations involving more than three chromosomes.

2. Whole-Arm Translocations

Whole-arm translocation is also a type of reciprocal translocation in which the entire arms of two nonacrocentric chromosomes are interchanged. Such rearrangements are described by assigning the breakpoints to the arbitrary centromeric region designated as p10 and q10. The breakpoint p10 is assigned to the chromosome with the lower number of the two chromosomes involved or the sex chromosome. Consequently, the second chromosome will have a breakpoint at q10. This assignment is particularly useful in describing whole-arm translocations in which the nature of the centromere is not known.

46,XX,t(3;8)(p10;q10)

A balanced whole-arm translocation between chromosome 3 and chromosome 8. In this example the short arm of chromosome 3 and the long arm of chromosome 8 have been fused. Reciprocally, the long arm of chromosome 3 has fused with the short arm of chromosome 8. However, it need not be written in the karyotype. The modal number 46 is also indicative of the presence of the reciprocal product. Which abnormal chromosome carries the centromere of 3 or 8 is not known. The karyotype is apparently balanced.

46,XX,t(3;8)(p10;p10)

Balanced whole-arm translocation in which the short arms of chromosomes 3 and 8 and the long arms of chromosome 3 and 8 have been fused.

45,X,der(X;3)(p10;q10)

A derivative chromosome consisting of the short arm of X and the long arm of 3. The reciprocal product consisting of the long arm of X and the short arm of 3 is missing. Note: the total chromosome number is 45, suggesting the loss of the reciprocal product. The net result is monosomy for the entire long arm of X and the entire short arm of 3.

47,XX,+der(X;3)(p10;q10)

This karyotype has an extra derivative chromosome consisting of the short arm of X and the long arm of 3 (same as in the previous example). Also present are two normal X chromosomes and two normal chromosomes 3. The net result is trisomy for the entire short arm of X and the entire long arm of 3.

---

3. Robertsonian Translocations

Robertsonian translocations originate through centric fusion of the long arms of acrocentric chromosomes consisting of pairs 13, 14, 15, 21, and 22. Since Robertsonian translocations are also whole-arm translocations, they can be described adequately using the same nomenclature approach as for whole-arm translocations. In a Robertsonian translocation, the short arms of the chromosomes involved are lost. In the translocation heterozygote the loss of short arms is not known to be causally related to an abnormal phenotype. However, in order to maintain uniformity in the nomenclature, a Robertsonian translocation product is considered to be a "derivative chromosome" and therefore the symbol *der* is used. For historical reasons, the abbreviation *rob* may also be used.

45,XX,der(13;14)(q10;q10)

A balanced Robertsonian translocation occurred between the long arm of a chromosome 13 and the long arm of a 14. As a result, the chromosome number was reduced to 45. The origin of the centromere nature is unknown. Breakage and reunion have occurred at bands 13q10 and 14q10. This derivative chromosome has replaced one chromosome 13 and one chromosome 14. There is no need to indicate the missing chromosomes because the chromosome number is reduced to 45. The karyotype now contains one normal 13, one normal 14, and the der(13;14). The short arms of the 13 and 14 are lost, which is not associated with an adverse clinical outcome.

46,XX,+13,der(13;14)(q10;q10)

A derivative chromosome consisting of the long arms of a 13 and a 14, same as in the above example. However, in this karyotype there are two normal 13 and one normal 14. The additional 13 is shown by the designation +13. In this example a 13 is involved in both numerical and structural abnormalities. In such instances the numerical abnormality is designated before the structural abnormality.

V. NEOPLASM, CLONES, AND CLONAL EVOLUTION

All principles dictating the nomenclature are identical to both constitutional and acquired changes. However, some unusual situations and circumstances apply uniquely to acquired chromosome changes. We will define some of the terminology used in the description of acquired changes. We will also give some examples of complex chromosome rearrangements not usually seen in constitutional karyotypes.

A. Clones

A clone is defined as a cell population derived from a single progenitor. In acquired chromosome changes, a clone is defined as two cells with the same structural aberration. In numerical aberrations, a trisomic clone must have at least two cells with the same extra chromosome; for a monosomy clone, a minimum of three cells must show the loss of the same chromosome. A clone is not necessarily homogeneous, since subclones may develop during evolution.

1. Mainline

The mainline is the most frequent chromosome complement of a tumor cell population. It is a quantitative term and does not indicate the sequence of origin, such as primary versus secondary.

46,XX,t(9;22)(q34;q11.2)[3]/47,XX,+8,t(9;22)(q34;q11.2)[17]

In the above karyotype, the clone with 47 chromosomes is the mainline, even though it is likely that the basic or the primary clone was the one with 46 chromosomes.

---

2. Stemline and Sideline

The stemline (*sl*) is the most basic clone of a tumor cell population. All additional subclones are termed sidelines (*sdl*).

46,XX,t(9;22)(q34;q11.2)[3]/47,XX,+8,t(9;22)(q34;q11.2)[17]/48,XX,+8,
t(9;22)(q34;q11.2),+19[12]

In the above example, the clone with 46 chromosomes is the stemline and the clones with 47 and 48 chromosomes are the sidelines. The clone with 47 chromosomes is the mainline since the largest number of cells show that karyotype.

B. Clonal Evolution-Related Clones

Cytogenetically related clones are presented in the order of increasing complexity as far as possible, regardless of the size of the clone. This means that the stemline is presented first, followed by the subclones in the order of increasing complexity.

46,XX,t(8;21)(q22;q22)[2]/45,X−X,t(8;21)(q22;q22)[19]/46,X,−X,+8,t(8;21)(q22;q22)[5]

In tumors with related clones such as the one above, the term *idem* may be used followed by the additional changes. The *idem* can replace only the stemline, which is usually given first.

Example: 46,XX,t(8;21)(q22;q22)[12]/45,−X,idem[19]/46,−X,+8,idem[5]

C. Unrelated Clones

Clones with completely unrelated aberrations are presented according to their size, the largest first, then the second largest, and so on. The normal diploid clone is always presented last.

46,XX,t(8;21)(q22;q22)[10]/47,XX,+8[6]/45,XX,i(8)(q10)[4]

If a tumor contains both related and unrelated clones, the related clones are written first in the order of increasing complexity, followed by the unrelated clones in order of decreasing size.

Numerous other more complex structural karyotypic changes may be encountered. These cases are rather rare and are not addressed here. However, one can easily extrapolate from the basic principles eluded to earlier and throughout this chapter and arrive at the proper nomenclature.

VI. MOLECULAR CYTOGENETICS—NOMENCLATURE FOR IN-SITU HYBRIDIZATION

Recent advances in human cytogenetics include the development and application of in situ hybridization *ish* protocols to incorporate and bind fluorochrome-labeled, cloned DNA and RNA sequences to cytological preparations. The technique allows for the localization of specific genes and DNA segments onto specific chromosomes, ordering the position and orientation of adjacent genes along a specific chromosome, and identification of microduplication or microdeletion of loci that are beyond the resolution of conventional cytogenetics. Molecular cytogenetics also involves the use of *ish* to identify whole chromosomes or chromosome regions (translocations, deletions, duplications, etc.) using appropriate DNA probes on interphase nuclei. For these reasons a nomenclature to designate various *ish* applications was introduced in ISCN 1995. The symbols and abbreviations used in *ish* nomenclature are listed in Table 4.

A. Prophase or Metaphase Chromosome in Situ Hybridization

Even though fluorescence microscopy is used mostly to view the in situ hybridization signals, the abbreviation FISH is *not* used in karyotype descriptions. Nomenclature for *ish* results may be presented in two ways: with and without cytogenetic results. If a chromosome analysis was done prior to *ish*, the karyotype is written first following the conventional rules. A period (.) is then placed to record the end of cytogenetics findings. This is

Table 4 Selected List of Symbols and Abbreviations Used for in Situ Hybridization *ish* Nomenclature[a]

Abbreviation	Explanation
Minus sign (−)	Absent on a specific chromosome
Plus sign (+)	Present on a specific chromosome
Double plus sign (++)	Duplication on a specific chromosome
Multiplication sign (×)	Precedes the number of signals
Period (.)	Separates cytogenetic results from ish results
con	Connected or adjacently placed signals
ish	Refers to in situ hybridization; when used by itself, ish refers to hybridization on metaphase chromosomes
nuc ish	Nuclear or interphase in situ hybridization
pcp	Partial-chromosome paint
sep	Separated signals
wcp	Whole-chromosome paint

[a]For a complete listing of symbols and abbreviations refer to ISCN (1995).

then followed by the *ish* results. If a standard cytogenetic analysis was not done and only *ish* studies were done, the *ish* results are presented directly.

When presenting an abnormal *ish* result, the abbreviation for that specific abnormality is recorded in order of chromosome number by the region tested, followed by the probe name. All entries are within separate sets of parentheses, similar to the nomenclature for a cytogenetic abnormality [e.g., *ish* del(4)(p16p16)(D4S96−)]. Whenever possible, Genome Data Base (GDB) designations for loci are used. The locus designation must be given in uppercase or capital letters only. When a GDB designation is not available, the commercial probe name can be used. If more than one probe from the same chromosome is used, they are listed in the order, pter→qter and the probe designation is separated by a comma (,). If probes for two different chromosomes are used, they are separated by a semicolon (;), just as in the cytogenetic translocation breakpoint designation.

46,XX.ish 22q11.2(D22S75x2)

This example illustrates the basic rules in describing an *ish* karyotype when both chromosome and *ish* results are normal. The test was performed on prophase/metaphase chromosomes. The probe D22S75 detects about 80% of the deletions leading to the DiGeorge and velo-cardio-facial syndrome. The cytogenetics result is given, followed by a period (.), followed by the abbreviation *ish*, and a single space, followed by the chromosome number and region. Note that when the locus tested is normal on *ish*, the chromosome number as well as the region tested are written without parentheses. The GDB locus designation and the number of signals seen are placed within parentheses (D22S75x2).

Given below are a series of examples to illustrate *ish* karyotype designations using patients suspected of different syndromes. The breakpoints and probes used in these examples are for illustration purposes only and may not reflect real situations.

1. Patients with Possible DiGeorge Syndrome

46,XX.ish del(22)(q11.2q11.2)(D22S75−)

A normal karyotype on cytogenetics. Hybridization with a D22S75 probe for the critical region for DiGeorge syndrome showed a deletion of the locus on one chromosome, represented by the minus (−) sign. The probe D22S75 is localized to the 22q11.2 region. To emphasize the fact that the deletion is within this region, the region in question should be repeated (q11.2q11.2) when the deletion is identified.

46,XX,del(22)(q11.2q11.2).ish del(22)(q11.2q11.2)(D22S75−)

A karyotype in which a deletion was identified on chromosome analysis and confirmed by *ish* using a probe for locus D22S75.

2. Patients with Possible Prader-Willi/Angelman Syndrome

46,XX.ish 15q11.2q13(D15S10x2,SNRPNx2)

Normal chromosome on cytogenetics, but suspected of the Prader-Willi/Angelman deletion. In situ hybridization using probes D15S10 and SNRPN both mapped to the 15q11.2→q13 region, showing two copies each of the two probes. Result suggests that both loci are present.

46,XX,del(15)(q11.2q13).ish del(15)(q11.2q11.2)(D15S10−,SNRPN−)

Deletion of 15q11.2→q13 was detected and was confirmed on *ish*. The probes used were D15S10 and SNRPN. Both were absent on one chromosome 15, which confirms the cytogenetics results. The deletion is denoted by the minus (−) sign after the probe name.

3. Patients with Possible Williams Syndrome

46,XX.ish 7q11.23(ELNx2)

A patient with Williams syndrome and normal chromosomes. *ish* with a probe for Elastin (ELN) locus showed hybridization at band 7q11.23 region on both chromosomes 7. Therefore, there is no deletion.

46,XX.ish del(7)(q11.23q11.23)(ELN−)

A patient with Williams syndrome and normal cytogenetics results. *ish* with a probe for ELN locus showed deletion of the locus on one chromosome 7.

46,XX.ish del(7)(q11.23q11.23)(ELN−x2)

A patient with Williams syndrome and normal cytogenetics results. *ish* with a probe for ELN locus showed deletion on both chromosome 7s.

4. Charcot-Marie-Tooth Syndrome

46,XX.ish 17p11.2(CMT1Ax2)

A patient with Charcot-Marie-Tooth syndrome. *ish* with a probe for CMT1A locus showed normal hybridization on both chromosome 17s and thus no deletion or duplication of the locus.

46,XX.ish del(17)(p11.2p11.2)(CMT1A++)

A patient with Charcot-Marie-Tooth syndrome. *ish* with a probe for CMT1A locus showed duplication of the locus on one chromosome. Note the use of the ++ sign to indicate that the two signals were seen on the same chromosome, which indicates that the probe locus is duplicated on one chromosome, versus ×2, indicating two signals, one each on the homologs, suggesting a normal result.

5. Chromosome Abnormalities Identified by Whole Chromosome Paint

46,XX,add(20)(p13).ish dup(5)(p13p15.3)(wcp5+)

In this patient a chromosome 20 has an extra material attached at band p13. By using whole-chromosome painting for 5, the extra chromosome segment was identified as a duplication of 5. By comparing the band morphology on Giemsa-banded preparations, the duplicated region was identified as 5p13→p15.3.

46,X,r(X).ish r(X)(p22.3q13.2)(wcpX+,DXS1140+,DXZ1+,XIST+,DXZ4−)

A ring X identified on G-banding was further defined by *ish* using a whole-chromosome paint for X, and probes localized to Xp22.3 (DXS1140), Xcen (DXZ1), Xq13.2 (XIST), and Xq24 (DXZ4). The last probe DXZ4 was found missing, documenting the deletion of Xq24 and regions beyond that.

46,X,r(?).ish r(X)(DYZ3−,wcpX+)

A small ring chromosome was identified on G-banding. Hybridization with DYZ3 showed the ring was not derived from the Y. Follow-up hybridization with a whole-chromosome paint for the X showed the ring to have originated from the X.

47,XX,+mar[10]/46,XX[12].ish der(12)(wcp12+,D12Z1+)[10]

An extra marker chromosome in a mosaic state identified by *ish* as a der(12) using a whole-chromosome paint probe for 12 and a probe D12Z1 for the centromeric region of chromosome 12.

46,XX.ish inv(16)(p13.1q22)(pcp16q sp)

Normal female karyotype on routine G-banding was found to have a pericentric inversion of chromosome 16 on *ish*. When a partial chromosome paint was used for band 16q22, the signal was found to be split (*sp*), showing one signal on the long arm and the other on the short arm.

6. Identification of Cryptic Translocations by *ish*

46,XX,?t(4;7)(p16;q36).ish (wcp7+,D7S427+,D4S96−;wcp4+,D4S96+, D7S427−)

A cryptic translocation between the short arm of a chromosome 4 and the long arm of a chromosome 7 was suspected and was confirmed by *ish* using a combination of unique sequence probes as well as whole-chromosome painting probes. The probes were chosen to cover the regions in question: D7S427 localized to 7qter, D4S96 localized to 4pter, whole-chromosome paint for both chromosomes 4 and 7. The der(4) was wcp7+, D75247+, and D4S96−. The der(7), on the other hand, was wcp4+, D4S96+, and D75247−. This confirms a cryptic reciprocal translocation involving the distal short arm of chromosome 4 and the distal long arm of chromosome 7.

B. Interphase or Nuclear in Situ Hybridization *nuc ish*

In situ hybridization can be performed on interphase nuclei to obtain information on the copy number of a given probe. Thus, it can be used as a screening method for the rapid detection of chromosome copy number. Interphase FISH results are presented using the abbreviation *nuc ish* rather than just *ish* as used in metaphase in situ hybridization. Usually in *nuc ish*, the conventional karyotype information may not be available. Therefore, the following examples deal only with interphase or nuclear hybridization results and the nomenclature. However, in situations where a chromosome study is done prior to an interphase analysis, it could be given using similar principles as used for metaphase *ish*. Interphase *ish* analysis can be performed to verify the presence or absence of a unique sequence or single-copy gene sequence (by the number of signals detected), as well as for the detection of certain known translocation events, such as t(9;22) in CML.

1. Designation of Number of Signals

When designating interphase FISH results, the abbreviation *nuc ish* is followed by a space, the chromosome band to which the probe is mapped, followed (in parentheses) by the GDB locus designation, a multiplication sign, and the number of signals detected.

nuc ish Xcen(DXZ1×1)

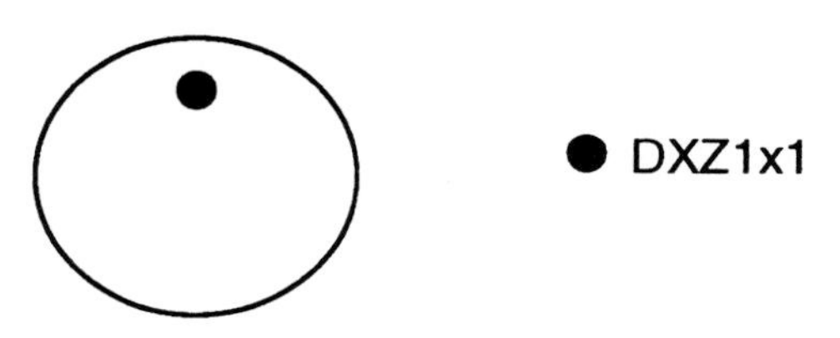

One copy of the DXZ1 locus was detected. The presence of one X chromosome centromere is implied. No other information is available. It is unclear whether the cell came from a normal male or a patient with Turner syndrome.

nuc ish Xcen(DXZ1×2)

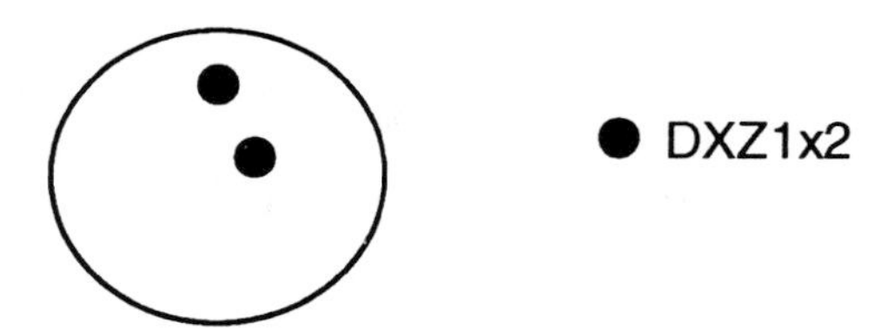

Two copies of the DXZ1 locus were detected. The presence of two X chromosome centromeres is implied.

nuc ish Xcen(DXZ1×3)

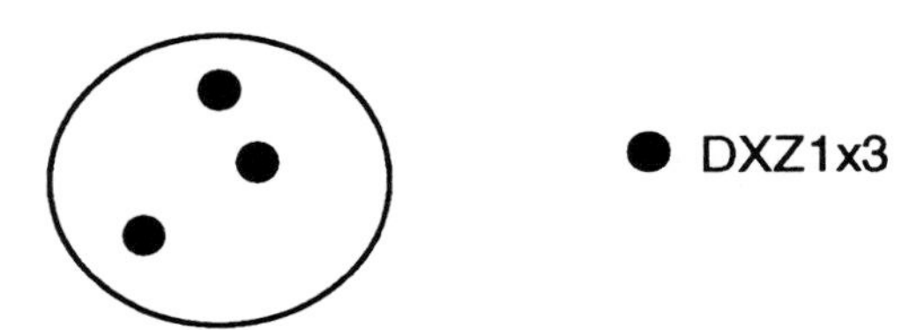

Three copies of the DXZ1 locus were detected. The presence of three X chromosome centromeres is implied.

nuc ish 13q14(Rb1×1)

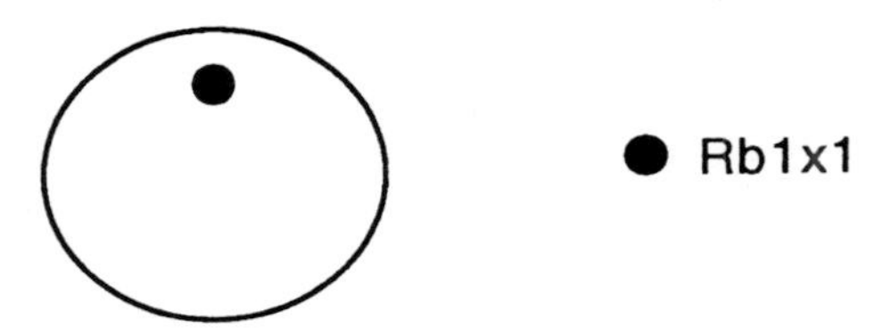

One copy of the Rb1 locus was detected. The deletion of Rb1 gene from one chromosome 13 is implied.

nuc ish Yp11.2(DYZ3×2)

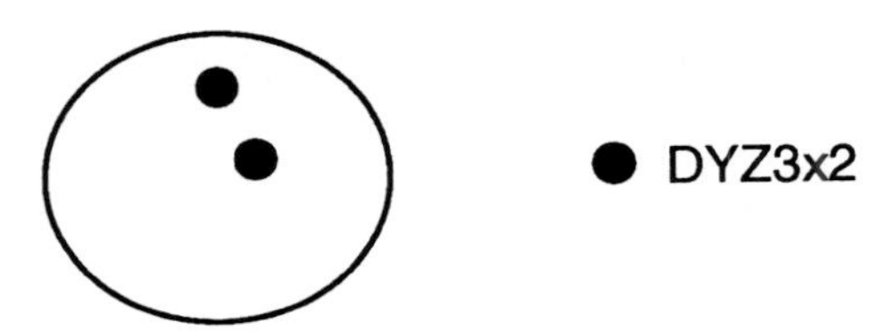

Two copies of the DYZ3 locus were detected, which implies that an extra copy of the locus was present in this cell. It is unclear if the extra copy is due to the presence of two Y chromosomes or due to other rearrangements

nuc ish 4cen(D4Z1×2),4p16.3(D4S96×1)

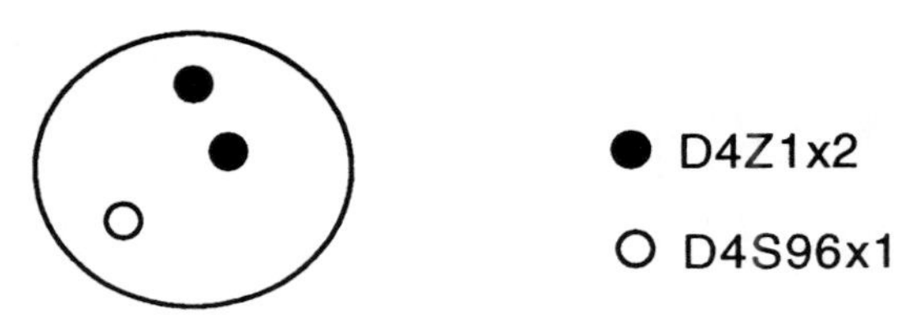

Two copies of the D4Z1 locus and one copy of the D4S96 locus were detected. The deletion of the D4S96 locus from one chromosome 4 is implied.

nuc ish Xp22.3(STS×2),13q14(Rb1×3)

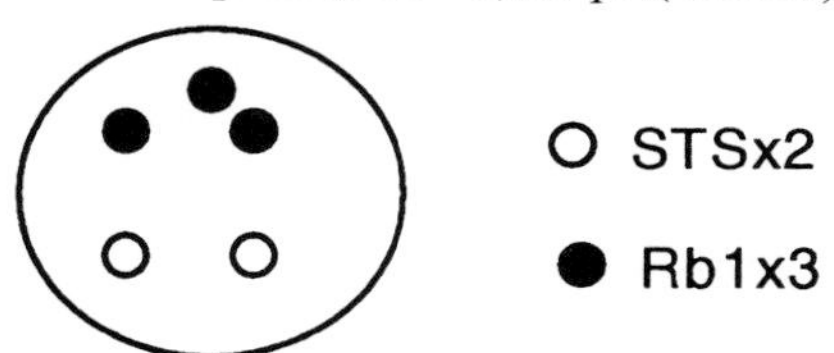

Two copies of the STS locus and three copies of the Rb1 locus are present. The presence of two X chromosomes and three chromosome 13s is implied.

---

2. Designation of Relative Position of the Signals

Under normal conditions, if probes from two chromosomes are tested simultaneously, the signals are expected to appear separated. However, chromosome rearrangements can alter the normal status and can bring the signals together, such as the one produced by the fusion of BCR and the ABL loci.

---

nuc ish 9q34(ABL×2),22q11.2(BCR×2)

BCRx2
ABLx2

Two ABL and two BCR loci are seen and are well separated. Normal chromosome 9s and 22s are implied.

nuc ish 9q34(ABL×2),22q11.2(BCR×2)(ABL con BCR×1)

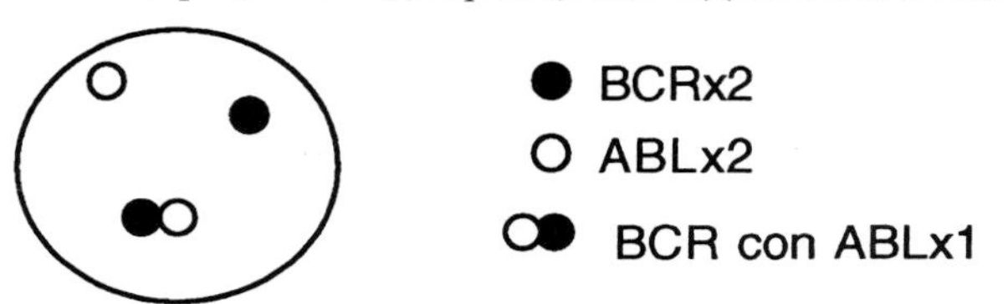

Two ABL and two BCR loci seen. However, one ABL and one BCR are connected and juxtaposed on one chromosome. A t(9;22) resulting in a Philadelphia chromosome is implied.

nuc ish 9q34(ABL×3),22q11.2(BCR×3)(ABL con BCR×2)

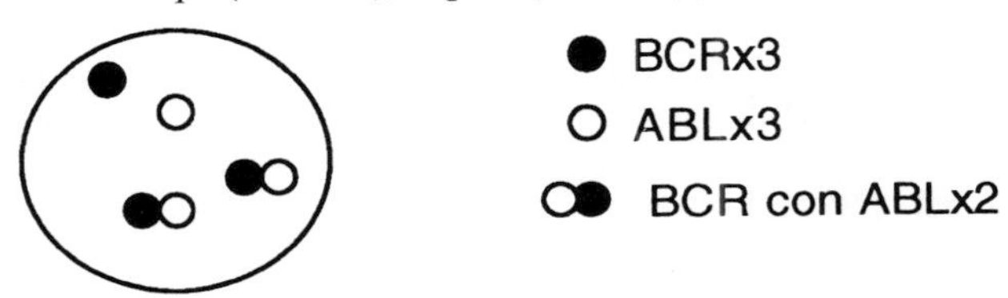

Three ABL and three BCR loci are seen. However, the BCR/ABL signals are connected and juxtaposed twice. The presence of two der(22) or Philadelphia chromosomes is implied.

nuc ish Xp22.3(STS×2),Xp22.3(KAL×2)

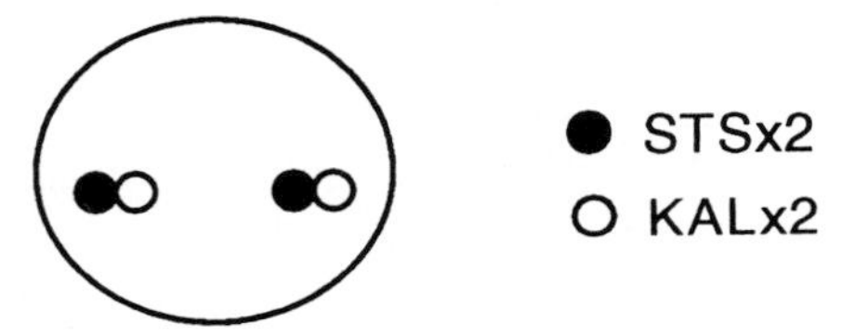

STS and KAL are two loci on Xp22.3 that are adjacent to each other. Under normal circumstances they will appear side by side.

nuc ish Xp22.3(STS×2),Xp22.3(KAL×2)(STS sep KAL×1)

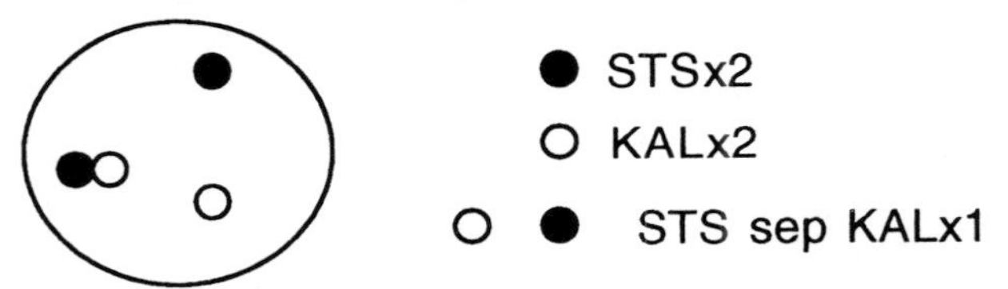

In this last example, one STS and one KAL loci are separated. The implication is that some type of rearrangement involving that area of the X chromosome has occurred, and has separated the two sequences that are usually found together in normal cells. No information can be obtained on the type of rearrangement.

REFERENCES

1. Tjio JH, Levan A. The chromosome number in man. Hereditas 1956;42:1–16.
2. Ford CE, Hamerton JL. The chromosomes of man. Nature 1956;178:1010–1023.

3. Lejeune J. Le mongolism. Primer example d'aberration autosomique humaine. Ann Genet Sem Hop 1959;11:116–119.
4. Patau KA, Smith DW, Therman EM, Inhorn SL, Wagner HP. Multiple congenital anomaly caused by an extra chromosome. Lancet 1960;1:790–793.
5. Edwards JH, Harden DG, Cameron AH, Wolff OH. A new trisomic syndrome. Lancet 1960;1:787–790.
6. Ford CE, Jones KW, Polani PE, De Almeida JC, Briggs JH. A sex chromosome anomaly in a case of gonadal dysgenesis (Turner's syndrome). Lancet 1959;1: 711–713.
7. Jacobs PA, Baikie AG, Court-Brown WM, Mac Gregor TN, Maclean N, Harnden DG. Evidence for existence of the human "super female." Lancet 1959; 11:423–425.
8. Denver Conference (1960): A proposed standard system of nomenclature of human mitotic chromosomes. Lancet 1:1063–1065; also reprinted in Chicago Conference, pp. 12–15.
9. London Conference on the Normal Human Karyotype. Cytogenetics 1963;2: 264–268.
10. Caspersson T, Farber S, Foley GD, et al. Chemical differentiation along metaphase chromosomes. Exp Cell Res 1968;49:219–222.
11. Chicago Conference (1966): Standardization in Human Cytogenetics. Birth Defects: Original Article Series, 2:(2). New York: The National Foundation, 1966.
12. Paris Conference (1971): Standardization in Human Cytogenetics. Birth Defects: Original Article Series, Vol. 8, No. 7. New York: The National Foundation, 1972; also in Cytogenetics 1972;11:313–362.
13. Paris Conference (1971), Supplement (1975): Standardization in Human Cytogenetics. Birth Defects: Original Article Series, Vol. XI, No. 9. New York: The National Foundation, 1975; also in Cytogenet Cell Genet 1975;15:201–238.
14. ISCN (1978): An International System for Human Cytogenetic Nomenclature. Birth Defects: Original Article Series, Vol. 14, No. 8. New York: The National Foundation, 1978; also in Cytogent Cell Genet 1978;21:309–404.
15. ISCN (1981): An International System for Human Cytogenetic Nomenclature. Birth Defects Original Article Series, Vol. 17, No. 5. New York: March of Dimes Birth Defects Foundation, 1981; also in Cytogenet Cell Genet 1981;31: 1–32.
16. ISCN (1985): An International System for Human Cytogenetic Nomenclature, Harnden DG, Klinger HP, eds.; published in collaboration with Cytogenet Cell Genet (Basel: S. Karger, 1985); also in Birth Defects: Original Article Series, Vol. 21, No. 1. New York: March of Dimes Birth Defects Foundation, 1985.
17. ISCN (1991): Guidelines for Cancer Cytogenetics, Supplement to An International System for Human Cytogenetic Nomenclature, Mitelman F, ed. Basel: S. Karger, 1991.
18. ISCN (1995): An International System for Human Cytogenetic Nomenclature, Mitelman F, ed. Basel: S. Karger, 1995.

APPENDIX: GLOSSARY OF TERMS

band
Alternating light and dark areas on a chromosome.

centromere
Primary constriction along the chromosome, a region at which the sister chromatids are held together.

chimera
Two or more cell lines in an individual derived from different zygotes.

clone
Cell population derived from a single progenitor.

derivative chromosome
Structurally rearranged chromosome.

diploid number
Number of chromosomes in a somatic cell in a given species.

disomy
Two homologous chromosomes of a given pair.

FISH
Fluorescent in situ hybridization.

ideogram
Diagrammatic representation of the banding pattern of a karyotype.

insertion
Presence of a chromosome segment in a new area on a chromosome.

isochromosome
Structurally abnormal chromosome with identical gene sequence on both arms; isochromosome could be for the short or the long arm.

inversion
Altered orientation of loci due to break and rearrangement.

ish
In situ hybridization.

karyotype
Arrangement of chromosomes according to their morphology and banding pattern as per the nomenclature.

landmark
Discriminating region of a chromosome from its adjacent region.

mainline
Clone that has the maximum number of cells with identical karyotype.

mosaic
Individual with two or more cell lines as part of the karyotype.

nuc ish
Nuclear in situ hybridization.

recombinant chromosome
Rearranged chromosome as a result of meiotic events.

region
Part of a chromosome that lies between two landmarks.

subband
Area within a band.

sideline
Any deviating subclone in addition to the stemline.

stemline
The most fundamental clone of a tumor cell population.

telomere
Ends of the chromosomes.

translocation
Exchange of chromosome segments between homologous and nonhomologous chromosomes.

triploid
Cell with three copies of each chromosome.

uniparental disomy
Two homologs of a pair inherited from a single parent.

4
Culture and Harvest of Tissues for Chromosome Analysis

Hon Fong L. Mark
Brown University School of Medicine and Rhode Island Department of Health, Providence, and KRAM Corporation, Barrington, Rhode Island

Sandra R. Wolman
Uniformed Services University of the Health Sciences, Bethesda, Maryland

I. INTRODUCTION

Conventional cytogenetic analysis is based on the examination of dividing cells. For some sources, such as peripheral blood, stimulation of the nucleated lymphoid cells is necessary to induce division, or mitosis. The stimulated culture then yields dividing cells with analyzable chromosomes that can be arrested in metaphase. For sources such as bone marrow with a higher fraction of cells which divide spontaneously, both unstimulated cultures (direct preparation) and short-term cultures with stimulants are employed.

Cells can be cultured for chromosome analysis for shorter or longer periods. Short-term culture usually refers to culture of 1–4 days. Short-term cultures are usually carried out for peripheral blood, bone marrow, amniotic fluid, and chorionic tissue derived from chorionic villus sampling (CVS). Direct preparations from these and other solid tissues are also often attempted. Long-term culture may be maintained for several weeks or even months. Fibroblast cultures are the best-known examples of long-term cultures. Other examples are solid-tumor cell lines and lymphoblastoid cell lines derived from immortalization of B lymphocytes in cultures, using Epstein-Barr virus (EBV) for cell transformation. They will be discussed further below.

II. TISSUES FOR CYTOGENETIC STUDY

The most common source of cells for cytogenetic analysis is the lymphocytes found in peripheral blood. A peripheral blood study is usually used to define or rule out a constitutional chromosomal abnormality in an individual. A constitutional chromosomal abnormality is one that an individual is born with and is presumably present in the zygote. An acquired chromosomal abnormality is a chromosomal abnormality acquired postzygotically (e.g., by mechanisms such as chromosomal nondisjunction, which is discussed elsewhere in this volume). Acquired chromosomal abnormalities are usually present in different proportions distributed in the somatic (body) tissues of an individual; such an individual is mosaic for the abnormality. Acquired chromosomal abnormalities are also characteristic of tumor cell populations and are localized to the tumor cells.

Bone marrow is the tissue of choice for the cytogenetic study of most hematologic disorders (1); see also Chapters 11 and 12 in this volume (by Anastasi and Roulson, and by Raimondi, respectively), although peripheral blood can be used in cases where large numbers of nucleated cells (blasts) are present in the circulation, such as in chronic myeloid leukemia (CML), chronic lymphocytic leukemia (CLL), and in acute lymphoblastic leukemia (ALL) with high white-cell counts (2). Cells from the bone marrow divide spontaneously. Thus, the cytogenetic analysis of bone marrow cells may reflect more closely the situation in vivo. The culture and harvest of bone marrow will be discussed further below. When a hematologic disorder is suspected and the specimen of bone marrow appears to be inadequate, peripheral blood is often obtained to supplement the results. In addition, a peripheral blood sample is often obtained together with the bone marrow in order to rule out a constitutional chromosomal abnormality in the patient (when all bone marrow cells sampled show the same abnormality).

Another common source of tissue for chromosome analysis is amniocytes, obtained from amniotic fluid by the procedure of amniocentesis. A volume of fluid from the amniotic sac is withdrawn through a needle, usually introduced transabdominally. Amniotic fluid samples are used for the prenatal diagnosis of genetic conditions.

Chorionic villi from chorionic villus sampling (CVS) also serve as a valuable source of tissue for prenatal diagnosis that can be obtained earlier in pregnancy than an amniotic fluid sample. The amniocentesis and CVS procedures, as well as percutaneous umbilical blood sampling (PUBS), are discussed in greater detail in Chapter 8 (Lamb and Miller) in this volume.

Fibroblasts are another excellent source of tissue for chromosome analysis (3). A biopsy of the skin may be needed to provide an independent confirmation of a suspected chromosomal abnormality or to make the diagnosis when peripheral blood studies are not informative. Skin fibroblast cytogenetics represents the "gold standard" for ruling out mosaicism in an individual. Other tissues that are sampled frequently include gonadal tissues, usually because sex chromosomal abnormalities are suspected (4,5).

Theoretically, any tissue with cells capable of undergoing division can be sampled and cultured, given the correct culture medium and optimal culture conditions. In practice, the above-mentioned are the most frequently sampled tissues for chromosome analysis because of accessibility, adaptability to growth in culture, and relevance to the clinical question addressed.

One other source of human chromosome studies should be mentioned. With the advent of the Human Genome Project, many laboratories have utilized the availability of somatic cell and radiation hybrids to isolate discrete segments of the human genome for study (see, for example, Ref. 6).

III. PERIPHERAL BLOOD CULTURE AND HARVEST: AN OVERVIEW

The most convenient and readily available tissue for cytogenetic study is peripheral blood. The optimization of the procedure for obtaining chromosome spreads from peripheral blood culture by Moorhead et al. (7) and Nowell (8) represents a significant milestone in the history of cytogenetics.

The procedure calls for a small volume (anywhere from 1 to 10 ml) of blood to be taken, generally from the arm by venipuncture (or, in the case of small infants, from a finger or heel stick), and placed into a heparinized (green top, sodium heparin) tube. Sodium heparin is the anticoagulant of choice, whereas lithium heparin, EDTA, and sodium citrate produce less than optimal results. For transport, blood should be kept at room temperature, with care taken to avoid extremes in temperature. Drops of whole blood (microprocedure) are cultured in medium containing a mitogen. An alternative macroprocedure can be used and involves drops of the leukocyte-enriched portion of the blood. For adults, 0.5–0.8 ml of whole blood is desirable for a 10-ml culture. Peripheral blood cultures are placed in a 37°C incubator for 72 h. Initiation of duplicate cultures from each specimen should be performed as soon as possible.

The agent (mitogen) most often used to induce mitosis in lymphocyte culture is phytohemagglutinin (PHA), which stimulates thymus-derived small lymphocytes (T cells) to enter cell division. However, other mitogens, such as cytochalasin B, Epstein-Barr virus (EBV), protein A, and lipopolysaccharides (LPS) are used to stimulate B cells (bursa or bone marrow-derived cells). Interleukin-2 (IL-2) and concanavalin act on both B cells and T cells. For tissues such as bone marrow or fibroblasts which divide spontaneously, an unstimulated culture may yield adequate results.

Phytohemagglutinin, originally available as a crude powder extracted from the common navy bean, *Phaseolus vulgaris*, was shown to have a mitogenic effect by Nowell in 1960 (8). It is the most widely used mitotic stimulant. Phytohemagglutinin (M form) is supplied in the lyophilized state. It can be reconstituted in sterile distilled water and stored frozen.

Because the culture time is relatively short, the choice of a medium for routine peripheral blood culture is not critical (9,10), as long as the essential ingredients are present. For example, Eagle's Minimum Essential Medium (MEM), TC 199, RPMI 1640 and RPMI 1603, Basal Metabolic Medium, MEM Alpha, and Dulbecco's Modified Eagles Medium are all adequate to support cell growth and yield satisfactory numbers of mitoses. Our laboratory uses Gibco 199 and Gibco RPMI 1640 whole-blood medium for peripheral blood cultures, supplemented with 15% fetal bovine serum, penicillin, streptomycin, and PHA. In addition, various additives are used for special studies, such as high-resolution banding, discussed in Chapter 5 in this volume (Mark), and chromosome fragile sites, discussed in Chapter 14 (Wyandt, Tonk, and Lebo).

Numerous protocols for routine peripheral blood culture and harvest have been reported. We will describe one for illustrative purposes, but it is important for the reader to keep in mind that many variations are possible.

IV. PERIPHERAL BLOOD CULTURE AND HARVEST

The protocol for the culture and harvest of peripheral blood was first optimized by Moorhead et al. (7) (Fig. 1). This method has withstood the test of time and remains essentially unchanged. It consists of the following steps.

A. Culture in the Presence of a Mitogen

In normal individuals the myelocytic cells in the circulation are nondividing. The success of short-term blood cultures depends on the addition of cellular growth factors which will induce resting T (thymus-dependent) lymphocytes to divide. T lymphocytes are involved with cell-mediated immunity, whereas B lymphocytes are involved with humoral immunity and are not mitotically responsive to PHA.

Figure 1 Dr. Paul Moorhead, shown here with granddaughter Annie Rose, first optimized the protocol for lymphocyte culture and harvest. (Photograph courtesy of Dr. Moorhead.)

B. Addition of a Mitosis-Arresting Agent

An arresting agent usually works by disrupting spindle fiber formation, thus causing an arrest of progression through mitosis and an accumulation of cells at the metaphase stage. The first such substance described was colchicine (11), an alkaloid from the bulb of the Mediterranean plant, the autumn crocus, *Colchicum* (12). Colchicine interferes with the formation of the mitotic spindle and arrests the cell at metaphase, thus increasing the number of mitotic cells available for chromosome analysis. The longer a culture is exposed to colchicine, the greater is the potential number of arrested metaphases. However, this increase in yield is offset by the risk of marked condensation and contraction of the chromosomes upon prolonged exposure to colchicine.

A commonly used arresting agent is the commercially available synthetic analog of colchicine, deacetylmethyl colchicine (Colcemid). Other compounds, such as vinblastine sulfate (Velban), are also mitotic-arresting agents.

C. Hypotonic Treatment to Swell Cells and Disperse Chromosomes

The use of hypotonic solutions to improve chromosome morphology in cytogenetic preparations was discovered serendipitously by Hsu (12) (Fig. 2) when a laboratory technician accidentally washed cells in a solution with the incorrect tonicity (a hypotonic solution) prior to cell fixation for chromosome analysis. In a hypotonic solution, the amount of solute in the extracellular solution is less concentrated than the amount of solute that is intracellular. When cells are exposed to a hypotonic solution, they take up water across the semipermeable cell membrane to reach an equilibrium state. The time of exposure to the hypotonic treatment is critical, as is the concentration, temperature, and type of hypotonic solution employed. Overtreatment in a hypotonic solution leads to overspreading with rupture of the cell membrane and random loss of chromosomes. Undertreatment, on the other hand, leads to inadequate spreading and an inability to distinguish individual chromosomes because of overlaps.

Many dilute physiologic substances, including water, saline, potassium chloride, sodium citrate, dilute balanced salt solution, dilute medium, and dilute serum, can be used to swell cells. One very commonly used hypotonic solution is 0.075 M potassium chloride (KCl). Based on over 30 years of experience in this laboratory, the use of KCl as a hypotonic solution is very effective in swelling the cells while maintaining good chromosome morphology. Experience has shown, however, that the cell suspension must be well mixed before, during, and after KCl treatment. It is important to admix the first drops slowly to avoid cell rupture. It is also necessary to use a range of KCl treatment times, usually from 20 to 35 min, to ensure a sufficient yield of optimally spread chromosomes. In this laboratory, 5 ml of KCl is added to approximately 2 ml of cell suspension. Given that the treatment time is varied, there is almost never a need to change the volume of KCl used when working with peripheral blood.

D. Fixation of Cells

Fixation presumably dehydrates the cells and enhances chromosome morphology and its ability to take up stain. As Dale and Reid (10) described,

Figure 2 Dr. T.C. Hsu, who first discovered the use of the hypotonic solution in his laboratory. (Photograph courtesy of Dr. Hsu.)

most cytogenetic fixatives are composed of a combination of methanol and acetic acid. Alcohol removes water from the cell proteins and denatures them, causing cell shrinkage and hardening. Acetic acid precipitates nucleic acids and dissolves nuclear histones, thereby enhancing the morphology of the chromosomes. Acetic acid is an important component of the fixation process because it prevents excessive shrinkage of cells by alcohol by rapidly penetrating and swelling the cells. A commonly used fixative is a modified Carnoy's solution which consists of three parts of methanol to one part of glacial acetic acid. There are two important points to remember pertaining to the fixation process. First, fixatives should be prepared fresh prior to use. Second, the cell suspension should be thoroughly mixed before, during, and after the addition of fixatives.

E. Slide Preparation

Clean slides should be used, although the preferred methods of slide cleaning differ from laboratory to laboratory. Some laboratories also see no need to clean commercially precleaned slides. Another critical factor in slide preparation is the use of freshly prepared fixative in the last step prior to the dropping of cells onto slides.

F. A Typical Peripheral Blood Culture and Harvest Protocol

In a typical procedure, four tubes of frozen whole-blood medium, previously tested for sterility, are thawed in preparation for inoculation. Labels with patient accession number, tube number, and date due for harvesting are prepared and double-checked. Tubes containing 8 ml of complete medium are inoculated with 0.5–0.8 ml of blood. The cultures are incubated at 37°C for 72 h. Whenever handling peripheral blood specimens, universal precautions should be followed meticulously.

On the day of harvest, approximately 60–90 min prior to the commencement of harvest, one to two drops (from a tuberculin needle) of a 10-mg/ml Colcemid stock solution is added to each culture for a final concentration of 0.03–0.06 μg/ml. The tubes are reincubated until commencement of harvest.

The tubes of blood culture are centrifuged for 5 min at 800 rpm. The supernatant is suctioned off with a Pasteur pipette except for approximately 1 ml. The pellet of cells is resuspended by gentle agitation. Five milliliters of a previously prepared working solution of 0.075 M KCl and 1.6 ml of 1/10,000 units/ml of sodium heparin is added to the resuspended pellet. A range of exposure times for hypotonic treatment is usually employed.

While the cells are being exposed to the hypotonic KCl solution, a fresh solution of fixative is prepared by adding 30 ml of absolute methanol to 10 ml of glacial acetic acid. At the end of the hypotonic treatment, fixative is added to each tube for a total cell suspension volume of 10 ml. After gentle agitation to mix the cells in the fixative, the tubes are centrifuged for 5 min at 800 rpm.

Tubes are then decanted to 1 ml, the cells are resuspended, and 5 ml of fresh fixative is added. The tubes should then sit for 10–30 min. Tubes are then centrifuged. This step is repeated three to four more times until the suspension no longer appears brownish. Several changes of fixative are necessary to remove extracellular debris and water. After the final fixation step, the cells are centrifuged and most of the supernatant is removed. The pellet is resuspended and inspected.

To make a slide, a proper concentration of cells is needed. This is usually judged by the appearance of the suspension, which should be cloudy. If cell growth is good and the suspension is very cloudy, the cell suspension is diluted by adding fixative drop by drop to obtain a lighter concentration of cells. If cell growth is meager, enough fixative is added to make only a single slide (1 or 2 drops). On average, approximately 0.5 ml of fixative is usually added.

Traditionally, two techniques for slide preparation have been described: the flame-drying technique and the air-drying technique. To make a slide by the flame-drying technique, a clean glass slide is dipped into double-distilled water. Two to three drops of the suspension of fixed cells are placed on the wet surface of the slide. The slide is passed through a low flame which will ignite the fixative, causing the cells to spread on the surface of the slide. The slide is immediately turned on its edge and the surface fluid is drained onto absorbent paper. Air-dried slides, which are preferred by many laboratories, are not subjected to flaming. If there are remaining cells after preparation of the needed number of slides, they may be stored in fixative for an indefinite amount of time in the freezer. Upon retrieval from storage, the cell pellet should be rinsed with fresh fixative at least once prior to use for making slides.

Slides must be thoroughly cleaned to ensure an optimal spreading of chromosomes. Many laboratories use clean slides soaked in distilled water for slide making. A slide warmer/dryer is often used to dry slides.

The quality of the metaphase spreads from a particular harvest can be evaluated immediately by examining the cells under a phase-contrast microscope. Alternatively, if there are many slides to spare, one slide can be stained with Giemsa and examined under regular bright-field microscopy to assess the mitotic index as well as the quality of the mitotic spreads.

For breakage analyses, variations on these culture procedures are used, the details of which are discussed in Chapter 14 (Wyandt, Tonk, and Lebo) in this volume. For certain conditions, use of an alternative tissue is advised. For example, the chromosome aberration in Pallister-Killian syndrome is preferentially expressed in fibroblast cultures and therefore the latter should be established in addition to lymphocyte cultures.

V. FIBROBLAST CULTURE AND HARVEST

Fibroblast cultures of the skin tissue (Fig. 3) are self-terminating cultures which are characterized by a limited number of cell generations. Two common reasons for fibroblast culture come to mind. One is to rule out mosaicism or chimerism. Another is in the case of a suspected abnormality that is tissue-specific in expression so that it may not be seen in peripheral blood.

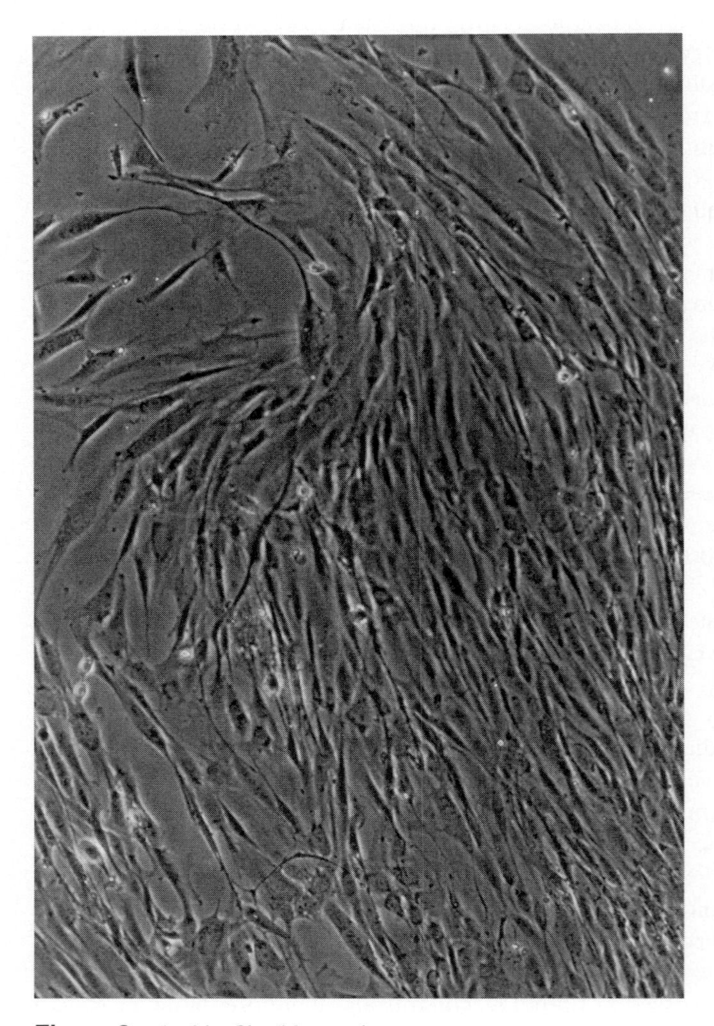

Figure 3 A skin fibroblast culture.

For example, the Pallister-Killian syndrome (isochromosome 12p or tetrasomy 12p) is preferentially expressed in fibroblasts and is rarely seen in blood cells. Dale and Reid (10) cited other uses for long-term cultures, including the determination of normal karyotype for comparison with abnormal blood karyotype, karyotyping of autopsy material, experimental studies, and fibroblasts appropriate for biochemical analysis.

The same sterile techniques observed for tissue culture of tumors, to be discussed further below, should be observed for tissue culture of skin. Separate hoods and incubators should be maintained for cell lines and malignant tissues to prevent cross-contamination. Other standard safety precautions should be observed.

To set up a fibroblast culture, a skin biopsy is performed under sterile conditions. A minimum of 0.5 cm^2 of tissue is required. Immediate delivery of the specimen to the laboratory is desirable. Specimen collection and other quality control/quality assurance issues are discussed in Chapter 19 (Khorsand).

To perform tissue culture, previously sterilized scissors and tweezers are used. The specimen is cut into two pieces. One half of the specimen is placed in a 10-ml centrifuge tube with 0.8% collagenase and 10 ml of medium and incubated overnight. The other half is cut into small pieces. Five or six fine pieces of tissue are put into each tissue culture flask with one drop of medium on each piece and with medium sprayed on the top and sides of the flasks. Many types of medium for tissue culture are available commercially. These include RPMI 1640, MEM, TC 199, and McCoy's 5A as well as other tissue culture media. The flask is capped loosely. Upon proper identification and labeling with the date, the flask is placed in the CO_2 incubator.

Our laboratory employs collagenase treatment overnight to disperse the cells, as an additional means of ensuring a successful outcome. Trypsin and pronase may be preferred by other laboratories. On the morning following initial incubation of tissue in collagenase, the supernatant is poured off and the tissue is disaggregated by gentle mechanical dissociation. The disaggregated cells are distributed into flasks and allowed to adhere. The flasks, with caps loosened, are identified and dated and placed in the 37°C incubator. After 5–7 days, growth can be checked by phase-contrast microscopy.

Once the culture is established, the medium is changed twice weekly. When feeding continuous cell cultures, one-half to two-thirds of the medium should be replaced by fresh medium. Upon confluency, cells are split into two or more subcultures by first loosening the cells from the bottom of the flask by either scraping or by the use of trypsin or pronase. Our laboratory uses 0.25% trypsin (Gibco BRL #15050-065).

The optimal time of harvest will depend on the cell type. In general, cells in the flask can be harvested when they reach 50–75% confluency (10). Prior to harvest, 60 μl of Colcemid are added to each 25-cm^2 tissue culture flask (Falcon #3013) to yield a concentration of 0.06 μg/ml. The flask is incubated at 37°C for 90 min, 3 h, or overnight. The medium with the Colcemid is suctioned off from the flask and saved in centrifuge tubes. The flask is rinsed with a small amount (approximately 1 ml) of trypsin and the trypsin is added to the centrifuge tubes with the medium mentioned above. The flask is then trypsinized with a small amount (approximately 1.5 ml) of trypsin for 5–10 min until the cells detach. Ten milliliters of medium per flask are then added to stop the action of the trypsin, and the contents of the flask are then transferred to centrifuge tubes. The tubes are centrifuged for 5–10 min. The supernatant is removed by aspiration. The pellet is resuspended. Five milliliters of 0.075 M KCl is added to each tube. After approximately 13 min in hypotonic solution, the cells are fixed in Carnoy's fixative and washed several times before slides are made. Thus, the harvesting procedure for tissue culture above is a simple modification of that for the peripheral blood protocol (7). Many variations exist.

VI. CULTURE AND HARVEST OF SOMATIC CELL AND RADIATION HYBRIDS

Somatic cell hybrids (Fig. 4a) containing one or more human chromosomes in a rodent background have proven to be extremely useful in the mapping of genes to human chromosomes (6). They have also been used as sources of DNA for the cloning of DNA sequences from specific chromosomes. Radiation hybrids (Fig. 4b) containing chromosome fragments also have demonstrated value in the physical mapping of the human genome and the region-specific cloning of DNA sequences. The utility of both types of hybrids depends on the initial identification and periodic verification of the human chromosome content. Conventional cytogenetics is a powerful tool which has traditionally been used in the characterization of somatic cell hybrids. Fluorescent in situ hybridization (FISH) has been used recently as a more sensitive technique for the detection of small fragments of additional chromosomal material which have gone undetected either by cytogenetic techniques or by DNA hybridization with probes from mapped human chromosomes. FISH can be important in the analysis of hybrids which contain subchromosomal fragments, such as those generated by radiation, that are too small to be detected using chromosome banding. Various applications of FISH will be explored elsewhere in this volume, in Chapter 17 (Mark).

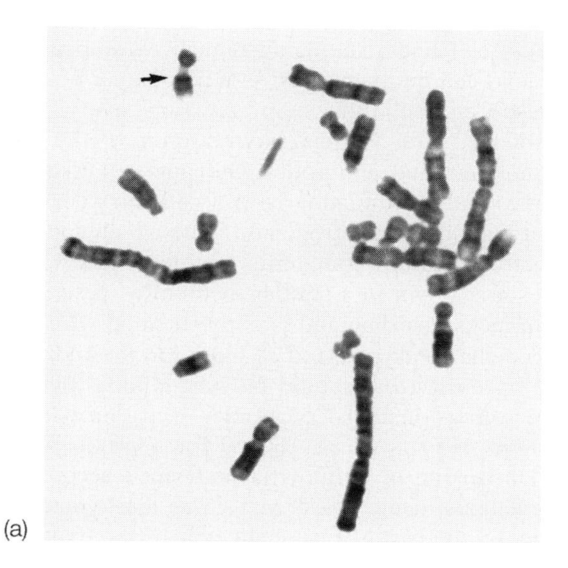

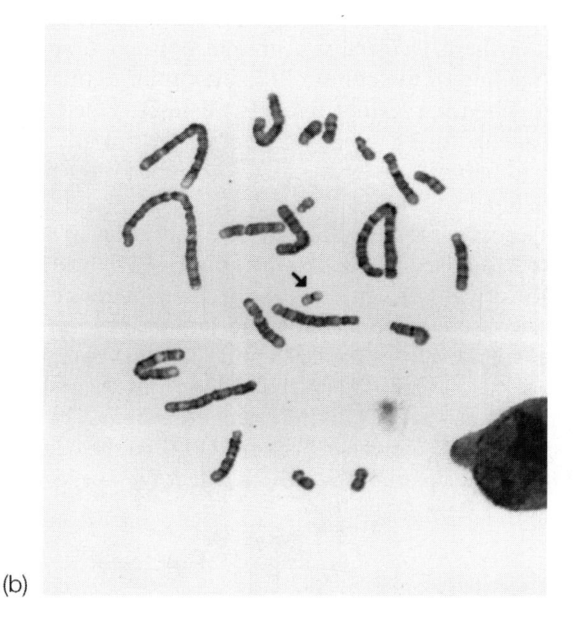

Figure 4 (a) Somatic cell hybrid containing human chromosome 9 in Chinese hamster background. (Reprinted with permission from *Applied Cytogenetics*.) (b) Somatic cell hybrid containing human chromosome 19 in Chinese hamster background. (Reprinted with permission from *Applied Cytogenetics*.)

Monochromosomal hybrid cell lines containing human chromosomes in a Chinese hamster background can be maintained in a monolayer culture using conventional techniques. Cells subdivide approximately every 5–7 days in DME medium supplemented with 5% fetal bovine serum (FBS), 5% calf serum (CS), proline, glutamine, and antibiotics. Somatic cell hybrids containing human chromosomes and the radiation hybrid cell lines derived from them contain an inserted histidinol dehydrogenase gene which confers the ability to grow in a selective medium containing histidinol instead of histidine (13). These cell lines are grown in a similar medium as described above except that the medium lacks histidine and is supplemented with 0.5 mM histidinol. For harvest, the cells are trypsinized 18 h prior to the addition of Colcemid, and replated at a density that should be subconfluent by the time of harvest. Thirty-two microliters of stock Colcemid at 10 μg/ml (Gibco Life Technologies, Grand Island, NY) is usually added for approximately 90 min prior to harvesting. Harvesting of cultures is performed according to conventional cytogenetic methods, using 0.075 M KCl as the hypotonic agent, and three parts methanol to one part glacial acetic acid as the fixative. After repeated rinsing and centrifugation, the cell suspensions are dropped onto glass slides, which are either air-dried overnight or flamed briefly.

For each hybrid cell line there is usually a unique optimal time for harvesting after the splitting/feeding of the culture. It is important to monitor the percentage of confluency when harvesting hybrid cell lines. Our experience is that a schedule of subculturing of a 90% confluent culture at a 1:2 to 1:4 dilution, 18 h prior to harvest, gives adequate metaphases for most human/hamster monochromosomal hybrids.

Although the harvesting procedures for somatic and radiation hybrids are simple modifications of the standard protocols, the culture of hybrid cell lines is by no means straightforward. Even seemingly trivial parameters, such as the kind of plasticware used, were found to be important in maximizing the yield of good metaphases from each somatic cell and radiation hybrid cell line. For example, in our laboratory the Falcon polypropylene 15-ml tubes (Becton Dickenson Labware, Franklin Lake, New Jersey) yield better harvests with better spreads and a cleaner background than the Falcon polystyrene 15-ml tubes. Many other parameters which affect the successful growth of cells in culture operate here as well.

VII. BONE MARROW CYTOGENETIC STUDIES: GENERAL CONSIDERATIONS

A. Direct Harvest versus Culture

Bone marrow chromosome studies are important in the diagnosis and management of patients with leukemias. They are also helpful in the newborn period, when rapid diagnosis of constitutional aberrations may be needed. Direct preparation is the preferred method in many laboratories because of speed and cost considerations, but cultured bone marrow samples often yield improved chromosome preparations. Most laboratories use short-term cultures (18–72 h) routinely to obtain better-quality metaphase cells and higher mitotic indices. In some cases, a better reflection of the true karyotypic picture in vivo is obtained from direct preparations. However, in some leukemias, for example, in acute promyelocytic leukemia, the t(15;17) abnormality may not be observed in direct marrow harvest but is only observed after 1–3 days of cultivation. Neither approach is ideal for detection of tumor metaphases in all cases. Therefore, it is recommended that more than one method of harvest be used routinely. The selection of preferred methods should be determined in each laboratory based on experience, time pressures, and the type and volume of diagnostic work encountered.

B. Stimulation and Synchronization

One effective means to increase mitotic activity in cultures of myeloid and undifferentiated leukemias is the addition of conditioned medium (CM) to short-term cultures to act as a stimulant to cell division. CM can be derived from many sources, including a human bladder carcinoma cell line (5637) that generates colony-stimulating factors supportive of growth of myeloid stem cells. It results in a pronounced enhancement of mitotic activity and chromosome morphology even in samples in which conventional cultures fail to yield mitoses. (We use 10% CM in 48–72 h cultures, in parallel with conventional 24–48 h unsupplemented culture.) Moreover, in some cases, minor chromosomally aberrant subclones, which are not found in the conventional cultures, are detected in cultures supplemented with CM. On the other hand, in some cases of myelodysplastic syndrome, the CM-containing cultures appear to favor normal cell growth (the percentage of abnormal cells either decreases or disappears entirely). Therefore, it is important that samples always be cultured in parallel without supplement, especially when the diagnosis is uncertain.

Synchronization by methotrexate (MTX) to increase chromosome elongation and permit better resolution of chromosomal banding patterns was first applied to bone marrow harvest by Hagemeijer et al. (14) and further modified by Yunis (15). The technique is labor-intensive and requires precise timing (especially the duration of thymidine incubation). It involves increasing the proportion of late prophase and early metaphase cells by synchronizing the cell cycle with MTX block, followed by thymidine release. Ethidium bromide is an alternative method to obtain elongated chromosomes. Although synchronized cultures are reliable for myeloid leukemias, synchronization may interfere with tumor cell growth in acute lymphocytic leukemia (ALL), resulting in a low mitotic index or overgrowth of normal cells.

C. Lymphoid Proliferations

Cytogenetic studies of acute lymphocytic leukemias (ALL) are technically more difficult than those for myeloid leukemias. The chromosomes often are resistant to spreading and morphology is poor, with indistinct bands. Several improvements in cell culturing, harvesting, and staining procedures can reduce these problems and lead to better identification of chromosome abnormalities. Culture periods for ALL cases should be very brief (6–24 h) because longer-term cultures (48 h or more) tend to favor the growth of cells with a normal karyotype. Chronic lymphoproliferative disorders (LPD) often proliferate slowly and there is accumulation of relatively mature lymphocytes in the bone marrow and blood.

Approximately 90% of all LPD are of B-cell origin. The identification of chromosome abnormalities in chronic LPD has lagged because of low spontaneous proliferation rates, inconsistent responses to mitogens, and competitive proliferation of residual normal T and B lymphocytes.

The recognition of mitogens that primarily stimulate B cells has increased our ability to diagnose and detect chromosome aberrations in some of these malignancies. The optimal B-cell mitogens are LPS, EBV, protein A, and TPA (4-tetradecanol phorbolic acid or 12-O-tetradecanoylphorbol-13-acetate or phorbol ester). The remaining 10% of LPD are T-cell disorders; PHA and pokeweed mitogens alone or in combination are alternative mitogens for T-cell LPD. The optimal culture times for LPD differ from those for myeloid diseases, with maximum stimulation seen usually after 4–5 days in culture. Many of the poorly mitotic forms of LPD can also be studied effectively with FISH probes.

VIII. PRINCIPAL STEPS IN BONE MARROW HARVEST

The general precautions in labeling, handling, and slide preparation that were described for blood also apply to the harvesting of bone marrow for chromosome studies.

A. Culture Conditions

Samples of no more than 1–2 ml in volume should be received in a sterile heparinized container. If the sample volume is larger, it may have been diluted significantly with peripheral blood. The cells should be washed with Hank's solution or another basal salt solution and then cultured in 10 ml of complete bone marrow culture medium (BMCM); we have used a medium consisting of 80% RPMI 1640 and 20% FBS, supplemented with 1% penicillin, 1% streptomycin, and 1% L-glutamine. The optimal cell density for a bone marrow culture is approximately 10^6 cells/ml. Cultures containing substantially greater numbers of cells are more likely to fail or to yield poor-quality metaphases. We incubate cultures in T-25 flasks at 37°C, in a 5% CO_2 atmosphere, with the cap of the flask loosened. Bone marrow cultures appear to grow better if the surface area is large, so flasks should be placed horizontal during the period of culture. The time in culture and other additives will depend on the presumptive diagnosis as discussed above.

Harvests are routinely set up for direct (6–18 h), short-term culture (48 h), and a 72-h culture with CM.

B. Chromosome Harvest

Colcemid is added at a final concentration of 0.05 μg/ml and the culture is returned to the incubator for 45 min. The cells are then centrifuged, exposed to a hypotonic solution, and fixed with several changes of fixative, similar to the procedures described above for peripheral blood, but with a few modifications, as given below.

The cells are transferred into a 15-ml centrifuge tube and spun at 1000 rpm for 10 min. After the supernatant is discarded, the cell pellet is slowly and gently resuspended. A prewarmed hypotonic solution of 0.075 M KCl is added drop-wise at first to a final volume of 7 ml. The hypotonic period is a brief incubation for 12 min in a water bath at 37°C. After centrifuging the sample and discarding the supernatant, the cell pellet is resuspended as before. Then a few drops of freshly made fixative are added slowly, allowing the fixative to run down the sides of the tube. Fixative is added until the total volume becomes 10 ml, and the tube is kept on ice for 1 h. The fixative is replaced at least twice by repeating the centrifugation and resuspension steps. Finally, the pellet is resuspended in a small volume (15–20 drops) of fixative depending on the amount of sediment. The cell suspension should be turbid, but not milky.

The final step is slide preparation. The suspension is mixed thoroughly with a Pasteur pipette and the same pipette is then used to drop 2–3 drops of the cell suspension onto a wet ethanol-cleaned slide. To maximize spreading of the chromosomes, the cell suspension is dropped from a height of 2–3 feet onto a 70% ethanol-cleaned slide placed at a 30° angle. The slide is allowed to dry spontaneously at room temperature, after which it is checked under phase-contrast microscopy to determine metaphase quality.

Only 2–3 drops of the final cell suspension should be sufficient for each prepared slide. Appropriate staining and banding methods are discussed in Chapter 5 (Mark).

C. Methotrexate Synchronization

Cultures are set up at a lower density than for routine harvest. We prepare 3–5 $\times$ 10^6 nucleated bone marrow cells for inoculation into 10 ml BMCM at 37°C after washing the cells twice in Hank's solution. After 3–5 h in culture, methotrexate is added to a final concentration of 10^{-7} M to block cells entering metaphase. The timing is critical: incubation of the culture should be at 37°C for 17 h. Cells are released from the methotrexate block by two 10-ml washes with Hank's solution and incubated again at 37°C with 10^{-5} M thymidine for 6 h. During the last 10 min of incubation, the cells are exposed to 0.05 μg/ml of Colcemid at 37°C and then harvested according to the method described above. After hypotonic treatment, the cells are fixed two to three times in freshly made fixative and stored at 4°C overnight. The next day the fixative is changed twice again. Similar to the standard bone marrow preparation, 2–3 drops per slide should be sufficient to produce excellent spreading.

IX. SOLID TISSUES AND TUMORS—GENERAL CONSIDERATIONS

A. Selection in Culture

Relatively few tissues are characterized by levels of spontaneous cell division that permit direct chromosome preparation, without culture in vitro. Tissues other than blood and bone marrow must generally be cultured for longer periods (days to weeks) in order to build up a population of dividing cells. Because a period of growth in culture is needed, it is important to keep in mind that growth may result in cells that do not necessarily represent the original tissue or tumor cell population. The possibility of outgrowth of selected, nonrepresentative subpopulations is greater for tumors than for normal tissues, because tumor cell populations are assumed to be less stable and more intrinsically heterogeneous than those originating from normal tissues. Factors that influence selection in culture include the source of the tumor material, the methods of tissue transport and disaggregation, the method of primary culture, and the type of culture medium. These factors may eliminate cells with abnormal karyotypes and preserve cells with normal karyotypes.

B. Validation

Because of the potential for selection, attempts should be made to validate the results of chromosome studies on cultured cells. Some confirmation that the harvests are representative of the original tissue can be obtained from cellular morphology in culture. Additional validation may be obtained from observations of how specialized structures or growth patterns in culture have formed (e.g., the secretory domes characteristic of glandular epithelium), from histochemical evidence of antigenic similarities to the cells of origin, or by demonstration of specialized cell products. However, one of the hallmarks of a tumor is loss of differentiated features, increasing the difficulties of validation. DNA content or ploidy evaluation can also provide important information for comparison of cell populations in vivo and in vitro (16).

More direct evidence of chromosomal similarities can now be obtained using in situ hybridization techniques to examine interphase cells from the original sample. If the same profile of chromosome changes can be illustrated in both original and cultured cells, then the latter studies are likely to be representative. For example, we have used centromere- and site-specific chromosome probes to confirm cytogenetic observations: in one study we showed similar types and extent of aneuploidy in cultured cells from renal tumors and in disaggregated cells from the original tissue blocks (17); another study compared chromosome changes in squamous cell tumors in culture with tissue sections of the original tumor (18). Concurrent use of other markers (e.g., histochemical, molecular, or cytometric) will be helpful in assessing the significance of cytogenetic results.

X. GROWTH OF SOLID TISSUES

A. Culture Vessels

Plastic flasks are suitable for the cultivation of most cell types, but an unfortunate disadvantage of flask culture is that during subculture to expand the population or during harvesting for chromosome analysis the cells must be dislodged, leading to some inevitable cell loss. When tissue samples are very small, it is preferable to culture the cells in smaller vessels. Chamber slide and coverslip cultures have smaller surfaces and yield metaphase cells suitable for harvest in shorter times than if samples are grown out in flasks. Harvests in situ are generally faster (i.e., within 3–7 days at a very low cell density) and have the advantage of minimum cell loss. Moreover, this method is very important in solid tumors for demonstrating cytogenetic heterogeneity and clonality.

B. Disaggregation

Solid tissues, including solid tumors, usually require disaggregation by mincing and enzyme treatment so that single cells can attach and then grow as a monolayer attached to the vessel surface. Loss, possibly highly selective, is likely to occur during these steps. Success in obtaining metaphases is dependent on the tumor type, as well as on technical factors such as specimen transport. Mechanical dissociation, such as mincing and forcing the tissues through needles or wire mesh, usually results in a reduction of cell viability. The introduction of enzymatic tissue digestion has been an important advance in solid tissue culture and metaphase analysis. Enzymatic digestion reduces the extent of mechanical processing and increases the yield of viable suspended cells. The most satisfactory enzyme for tissue disaggregation at present is collagenase. Trypsin, formerly the most widely used enzymatic method, has been associated with cell lysis. Different disaggregation protocols have been described (19). Evidence that loss of aneuploid populations may result from disaggregation procedures, appears particularly true for enzymatic digestion (20).

C. Attachment and Medium

Proliferation and growth of most solid tumors in culture depends on the adherence of cells to a substrate. A major attachment factor, fibronectin, is present in serum. Additional supplementation of the medium is usually needed to promote the growth of epithelial cells, especially when a serum-free or low-serum medium is used to deter fibroblast growth. Cholera toxin, glutamine, hydrocortisone, epidermal growth factor, various hormonal extracts, and a mixture of insulin/transferrin and selenium are some of the additives that have been used to support epithelial growth. Many other growth factors, hormones, vitamins, and detoxification reagents have been used to promote the attachment or growth of specialized cell types. Flasks may be coated with collagen or extracellular matrix (ECM) to enhance attachment. The use of feeder layers (of viable or irradiated cells), three-dimensional growth in agar, or growth on floating collagen gels are other possibilities. Some tumor cells can be cultured in a simple standard medium (e.g., RPMI 1640, Eagle's MEM, Ham's F12, or McCoy's 5A), but other cell types may require a complex, carefully defined medium with multiple supplements. Generally, 15–20% FBS, 1–2% glutamine, and 1% penicillin/streptomycin are added to the basic medium. For longer-term culture, antibiotics to protect against fungal contamination are also helpful. However, because antifungal agents are often detrimental to growth, it is a better to delay their use until cultures are well established.

The methods described below were developed for the culture and cytogenetic analysis of human cells. Most of these methods are applicable with minor modifications to other mammalian cells from the same tissue sources. In general, chromosome preparation from cell culture consists of the same steps that applied to blood and marrow: mitotic arrest, hypotonic treatment, and fixation.

XI. SOLID TISSUE SAMPLE PROCESSING

A. Initial Plating

When tissue is received, areas of hemorrhage, infarction, or contamination should be dissected from the sample before processing. Vital dye exclusion is sometimes useful to assess cell preservation. The sample should be finely minced and/or treated with an enzyme to achieve a suspension of single cells and small clumps. These cells should be plated at final concentrations of approximately $1-2 \times 10^6$ cells/5 ml of medium in T-25 tissue culture flasks (or $5-10 \times 10^4$ cells/ml in coverslip culture). Cultures are incubated at 37°C in a humidified 5% CO_2 chamber. It is important to leave cultures undisturbed for 1–2 days to permit attachment. Then the culture medium should be exchanged either partially or completely. If floating cells are present, they should be transferred to another flask to examine for the growth of cells in suspension.

B. Growth

Cultures should be inspected daily with an inverted microscope for cellular attachment, flattening, and proliferative activity. Most cell types show rounding and increased refractility when they are entering or in mitosis. Usually the best time to harvest is when the surface of the culture vessel is slightly less than 75% confluent. Cultures should also be monitored for fibroblast cell growth and should be harvested for chromosome analysis before these cells take over the culture. One approach that is useful when attempting to study epithelial tumors or normal cells is to decrease the proportion of normal fibroblasts by differential trypsinization: after attachment of the primary culture, a brief (approximately 1-min) exposure and wash with trypsin will remove many fibroblasts because they are more sensitive than epithelial cells to the enzyme. Because of cellular heterogeneity and the likelihood of asynchronously dividing populations within a tumor, multiple harvests after differing time periods is highly recommended.

If colonies of cells with differing morphology are observed within the same culture, the different cell types should be documented by photography.

Chamber slide cultures are particularly useful for the demonstration of cytogenetic heterogeneity and clonality. After an individual colony is photographed, the same colony can be localized after chromosome preparation, and correlation of cytogenetic results with cell and colony morphology is feasible.

If growth is maintained for long periods and over many subcultures, the probability of evolution in culture increases. Thus, the importance of obtaining chromosomal harvests as soon as possible after culture initiation should be emphasized. If growth persists for months or longer, as is the case for immortalized lines, harvests at regular intervals are recommended to monitor the course of chromosomal evolution.

It cannot be emphasized too strongly that different tissues and tumors will respond differently to the same conditions of culture. Careful observation and incorporation of many variations on nutrients, stimulants, culture times, and conditions will increase the probability of success. The following are intended as guidelines rather than rules of procedure.

C. Chromosome Harvest

1. Preparation from Monolayer Flask

Determining the appropriate harvest time is important. First harvest should be performed when the culture is subconfluent and cells are actively proliferating. If multiple flasks have been set up, the first harvest should be performed as soon as possible, and the remaining cultures should be harvested over several consecutive days.

Colcemid is added to a final concentration of 0.05 μg/ml for 1–3 h depending on the relative mitotic activity of the culture, which can often be estimated by the number of refractile, rounded cells. The medium is collected and added to the cells removed from the flask by trypsinization. During the period of enzyme treatment, the flask should be checked under an inverted microscope for cell detachment. The hypotonic and fixation steps are described above for bone marrow, although volumes and incubation times will vary according to the cell size and properties of the cell membranes.

2. In Situ Harvest

Cultures should be harvested when colonies are still small and have actively dividing cells, usually at the periphery of individual colonies. When nearly confluent, the cell cultures are less likely to show good chromosome spreading. After exposure to colcemid, the medium is removed and a hypotonic solution (we prefer 0.75% Na citrate) is added for 30 min. Fixative is then added to the hypotonic solution for a few minutes, followed by several changes of fixative. After complete drying, the coverslip should be slide-mounted (cell side up) for ease of handling, checked by phase microscopy for mitotic index and chromosome spreading, and stained by the selected banding method.

XII. LYMPHOBLASTOID CELL CULTURES

A. Immortalization and EBV-Treated Cell Lines

Fibroblast cultures and even many tumor-derived cultures have a limited life span in vitro, although occasionally tumor cells and very rarely cells of nonmalignant origin will give rise to permanent, immortal cell lines. Cell lines are cell populations which can propagate essentially indefinitely, under adequate culture conditions. Immortalization of diploid human cells can be achieved infrequently by chemical carcinogens and oncogene transfection. Several tumor viruses, most notably SV40, facilitate the immortalization of human cells. However, most transformations with SV40 also result in extensive and unstable chromosome alterations. The immortalized fibroblastic and epithelial lines resulting from SV40, adenovirus, or papillomavirus transformation are dependent on inactivation of tumor suppressor gene (e.g., p53, Rb) products, and are adherent to the culture vessel. In contrast, immortalization of T or B lymphocytes depends on the cultivation of cells in suspension and provides relatively homogeneous populations of cells for many types of studies. Long-term T-cell lines are generated through stimulation of cell division by either specific or nonspecific antigen exposure. These cell lines have been valuable tools for the examination of immune functions and lymphoid maturation. The immortalization of B lymphocytes to lymphoblastoid cell lines via Epstein-Barr virus (EBV) infection yields cultures of genetically stable populations with indefinite cell proliferation. Such lines have been critical tools in the linkage analyses that have permitted regional localization, and the eventual cloning of genes such as BRCA 1. These lines have also permitted the development of in vitro models for the study of diseases characterized by premature aging, in which case the life span of cultured fibroblasts may be too brief to permit detailed investigations.

Several EBV genes are probably essential to continuing infection and immortalization; EBNA-2 is clearly necessary, although its biochemical activity remains poorly understood. Infection with EBV permits the immortalization of normal diploid B lymphocytes and results in lines that are initially polyclonal. Eventually, after prolonged cultivation, one or a few clones predominate. They are capable of secreting all types of immunoglobulins, and are probably maintained because the cells secrete a variety of autocrine and paracrine growth factors. Although cell lines may be genetically stable for long periods, it is not uncommon for these cells to become tetraploid.

Because many individuals are immune to EBV, it is important to remove T cells before attempting to immortalize with EBV (21). The original protocols by Sugden and Mark (22) and Nilsson and Klein (23) are important resources for investigators who plan to develop and utilize B-lymphoid cell lines. Biosafety precautions are especially important because of the risk of human disease resulting from EBV, even though the immortalized cells generally do not produce infectious virus.

B. Generation of Lymphoblastoid Lines

1. Supernatant fluid from a virus-producing cell line previously tested for its potency is added to the cells to be immortalized. The source of EBV-containing fluid should have a minimum of 10^3 infectious particles (transforming units)/ml and should be stored in a deep freeze in complete medium (e.g., RPMI-10).
2. The target lymphocyte population can be prepared from any source of lymphocytes (e.g., samples of lymph node, spleen, tonsil, or bone marrow), but the usual source is whole peripheral blood. Ten to 15 ml whole blood is sufficient.
3. The whole peripheral blood is diluted 1:2 in phosphate-buffered saline and 12 ml of the diluted sample is placed in a large conical centrifuge tube. An equal volume of Ficoll-Hypaque is added to the bottom of the tube. An alternative method to obtain B lymphocytes is the use of a hemolysis buffer followed by centrifugation. After centrifugation the cells at the interface of the two fluid layers are transferred to another tube, washed, and recentrifuged several times. They are then resuspended in a small volume of complete medium and counted.

Approximately 10 million cells are incubated at 37°C with 10^{5-6} units of infectious virus for 2 h in a final volume of 5 ml of complete medium. After 2 h, another 5 ml of complete medium containing cyclosporin A is added to inactivate any remaining T cells, and the mixture is transferred to a culture flask for prolonged incubation. With culture growth, the medium should become acidic. After 2 weeks, the culture should begin to contain clumped enlarged "hairy" cells with thin cytoplasmic projections. After 3 weeks the culture should be split and re-fed.

C. Maintenance and Chromosomal Harvest

Subculture of cells in suspension is achieved by centrifugation, discarding the spent medium, and resuspending cells in fresh medium. Viability can be monitored by dye exclusion (Trypan Blue). Briefly, cells are incubated with a 0.04% solution of the dye in PBS for 3–5 min. An aliquot is examined microscopically and the presence or absence of dye in 200 cells is recorded. The percent viability is derived by dividing the number of dye-excluding cells by the total number of dye-containing and dye-excluding cells. Approximately 10^5cells/ml is an appropriate starting density for subculture. A 1:2 split may require subculture in 3–4 days, and splitting at a 1:3 ratio permits subculture on a weekly basis. Chromosome harvests are best done within 24–48 h after subculture at a 1:2 or 1:3 split ratio. Colcemid is added and the harvest is performed according to procedures described in Section VIIIB.

XIII. CONCLUDING REMARKS

Conventional cytogenetics is a highly informative technique based on the analysis of chromosomes from dividing cells. The field blossomed with the first description of lymphocyte culture by Moorhead (7) and has developed many clinical applications as well as contributions to a wide range of research problems. Notably, cytogenetic studies of bone marrow have been immensely useful in the diagnosis and classification of leukemias and lymphomas. In this chapter, we have attempted to give an overview of the methodology and challenges associated with the culturing and harvesting of various tissues for conventional cytogenetics.

Contrary to some predictions, the advent of recombinant DNA technology did not diminish the importance of conventional cytogenetics. Rather, the availability of molecular probes has propelled the field of cytogenetics into the next century with powerful adjunct techniques, such as FISH and the recent FISH-based techniques of CGH and SKY, all of which are discussed elsewhere in this volume.

In summary, with continuing improvement of techniques and reagents for the culture and harvest of tissues, we anticipate that cytogenetic studies of solid tumors will contribute to improved diagnosis and better understanding of these lesions and that their assessment will become part of standard practice in clinical cytogenetics laboratories in the forthcoming years.

ACKNOWLEDGMENTS

We thank Ms. Anne Richardson, Dr. Yvonne Mark, and Dr. Roger Mark for reading the manuscript. The support of the staff, students, and trainees of the Lifespan Academic Medical Center Cytogenetics Laboratory at Rhode Island Hospital is also acknowledged.

REFERENCES

1. Mark HFL. Bone marrow cytogenetics and hematologic malignancies. In: Kaplan BJ, Dale KS, eds. The Cytogenetic Symposia. Burbank, CA: The Association of Cytogenetic Technologists, 1994.
2. Rooney DE, Czepulkowski BH. Human Cytogenetics: Essential Data. West Sussex, UK: John Wiley & Sons, published in association with BIOS Scientific, 1994.
3. Martin GM. Human skin fibroblast. In: Kruse PF Jr, Patterson MK Jr, eds. Tissue Culture: Methods and Applications. New York: Academic Press, 1973.
4. Mark HFL, Meyers-Seifer C, Seifer D, DeMoranville BM, Jackson IMD. Assessment of sex chromosome composition using FISH as an adjunct to GTG-banding. Ann Clin Lab Sci 1995; 25:402–408.
5. Mark HFL, Bayleran JK, Meyers DB, Meyers-Seifer CH. Combined cytogenetic and molecular approach for diagnosing delayed development. Clin Ped 1996; 35:62–66.
6. Mark HFL, Santoro KG, Jackson CL. Characterization of chromosomes 9 and 19 somatic cell and radiation hybrids: a comparison of approaches. Appl Cytogenet 1992; 18:149–156.
7. Moorhead PS, Nowell PC, Mellman WJ, et al. Chromosome preparations of leukocytes cultured from human peripheral blood. Exp Cell Res 1960; 20:613–616.
8. Nowell PC. Phytohemagglutinin—an initiator of mitosis in cultures of normal human leukocytes. Cancer Res 1960; 20:462–466.
9. Zackai EH, Mellman NJ. Human peripheral blood leukocyte culture. In: Yunis JJ, ed. Human Chromosome Methodology. New York: Academic Press, 1974: 95–125.
10. Dale KS, Reid SD. Peripheral blood cell culture and harvest. In: Kaplan BJ, Dale KS, eds. The Cytogenetic Symposia. Burbank, CA: The Association of Cytogenetic Technologists, 1994: 4-1–4-5.
11. Ford EHR. Human Chromosomes. New York: Academic Press, 1973.
12. Hsu TC. Human and Mammalian Cytogenetics: An Historical Perspective. New York: Springer-Verlag, 1979.
13. Hartman SC, Mulligan RC. Two dominant-acting selectable markers for gene transfer studies on mammalian cells. Proc Natl Acad Sci USA 1988; 85:8047–8051.

14. Hagemeijer A, Smit EM, Bootma D. Improved identification of chromosomes of leukemic cells in methotrexate-treated cultures. Cytogenet Cell Genet 1979; 23:208–212.
15. Yunis JJ. New chromosome techniques in the study of human neoplasia. Hum Pathol 1981; 12:540–549.
16. Mohamed AN, Zalupski MM, Ryan JR, Koppitch F, Balcerzak S, Kempf R, Wolman SR. Cytogenetic aberrations and DNA ploidy in soft tissue sarcoma: a Southwest Oncology group study. Cancer Genet Cytogenet. In press.
17. Wolman SR, Waldman FM, Balazs M. Complementarity of interphase and metaphase chromosome analysis in human renal tumors. Genes, Chromosomes, & Cancer 1993; 6:17–23.
18. Worsham MJ, Wolman SR, Carey TE, Zarbo RJ, Benninger MS, Van Dyke DL. Common clonal origin of synchronous primary head and neck squamous cell carcinomas: analysis by tumor karyotypes and fluorescence in situ hybridization. Hum Pathol 1995; 26:251–261.
19. Trent JM, Crickard K, Gibas Z, Goodacre A, Pathak S, Sandberg A, Thompson F, Whang-Peng J, Wolman SR. Methodologic advances in the cytogenetic analysis of human solid tumors. Cancer Genet Cytogenet 1986; 19:57–66.
20. Costa A, Silvestrini R, Del Bino G, Motta R. Implications of disaggregation procedures on biological representation of human solid tumors. Cell Tissue Kinet 1987; 20:171–180.
21. Tosato G, Pike SE, Koski I, Blaese RM. Selective inhibition of immunoregulatory cell functions by Cyclosporin A. J Immunol 1982; 128:1986–1991.
22. Sugden B, Mark W. Clonal transformation of adult human leukocytes by Epstein-Barr virus. J Virol 1979; 23:503–508.
23. Nilsson K, Klein G. Phenotypic and cytogenetic characteristics of human B lymphoid cell lines and their relevance for the etiology of Burkitt's lymphoma. Adv Cancer Res 1982; 37:319–380.

APPENDIX: GLOSSARY OF TERMS

acquired chromosomal abnormality

A chromosomal aberration that is acquired postzygotically by mechanisms such as chromosomal nondisjunction.

ALL

Acute lymphoblastic leukemia. The most prevalent type of acute leukemia in children, which results in symptoms of headaches, pyrexia, rigors, and inflammation of the cervical nodes.

AML

Acute myelogenous leukemia. A type of acute leukemia in which there is an accumulation of leukemic cells in the bone marrow, which end up circulating into the bloodstream and infiltrating other tissues such as the lymph nodes, the liver, the spleen, the skin, other viscera, and the central nervous system.

amniocentesis
A prenatal diagnostic procedure which involves removal of amniotic fluid for cytogenetic and/or biochemical analysis.

B lymphocytes
Cellular precursors of antibody-forming plasma cells which express immunoglobulins on their surface.

chimerism
The state of having tissues which are composed of two or more genetically distinct cell types, derived from different zygotes.

CLL
Chronic lymphocytic leukemia. A chronic lymphoproliferative disorder which is common in the U.S. and Europe and occurs almost exclusively in middle-aged and elderly patients.

CM
Conditioned medium. Medium that has already been used to grow cells in culture, usually added to fresh medium to culture new cell lines for the purpose of increasing mitotic activity in short-term cultures of myeloid and undifferentiated leukemias.

CML
Chronic myeloid leukemia. A condition characterized by excessive production of granulocytes that is caused by a defect in the proliferation of bone marrow stem cells.

colchicine
An alkaloid from the bulb of the Mediterranean plant, the autumn crocus, *Colchicum*, which interferes with the formation of the mitotic spindle and arrests the cell at metaphase, thereby increasing the number of mitotic cells available for chromosome analysis.

collagenase
An enzyme used to disaggregate cells.

constitutional chromosomal abnormality
An inherited chromosomal aberration that was already present in the zygote.

CVS
Chorionic villus sampling. A type of prenatal diagnostic procedure which involves removal of a sample of chorionic villi cells for cytogenetic analysis.

disaggregation
The separation of clustered cells from each other.

EBV
Epstein-Barr virus. A human DNA tumor virus, herpes-like in nature, that leads to several human tumors—Burkitt's lymphoma and nasopharyngeal carcinoma—as well as mononucleosis.

fibronectin
A major attachment factor, present in serum, that serves as a substrate to which culture-grown solid tumor cells attach.

germ cells
Cells of the reproductive system that carry the genetic information to be bequeathed to offspring.

hypotonic solution
A very dilute salt solution used to treat cells before fixation to cause them to absorb water from their surroundings so that chromosome spreading as well as morphology are improved. Some commonly used hypotonic solutions for swelling cells are water, saline, potassium chloride, sodium citrate, dilute balanced salt solution, dilute medium, and dilute serum.

immortal cells
Cells that divide continuously without inhibition.

interleukins
Glycoproteins secreted by many different types of leukocytes with stimulatory effects on other white blood cells.

lymphoblastoid cells
Transformed cells resulting from infection of lymphocytes, usually by the Epstein-Barr virus, which grow in suspension cultures.

lymphoid cells
Lymphocyte-producing cells of the lymph glands that are nongranular.

metaphase
The stage of cell division in which chromosomes become aligned along the equatorial plate, perpendicular to spindle fibers.

mitogen
A substance that stimulates cells to enter mitosis.

mitosis
The process whereby daughter cells are produced which are genetically identical to the parent cells.

monochromosomal hybrid
A somatic hybrid cell containing one human chromosome, usually in a rodent background.

mosaicism
The state of having tissues that are composed of two or more genetically distinct cell types, derived from the same zygote.

myeloid cells
A subpopulation of cells that resemble or are derived from the bone marrow.

Pallister-Killian syndrome
Isochromosome 12p or tetrasomy 12p; some characteristics of this condition are short limbs and diaphragmatic defects.

PHA
Phytohemagglutinin. The most common mitogen used to induce mitosis in lymphocyte culture, which stimulates thymus-derived, small lymphocytes to begin cell division.

pronase
An enzyme used for tissue disaggregation.

PUBS
Percutaneous umbilical blood sampling. A type of prenatal diagnostic procedure performed to obtain fetal blood for analysis, allowing for rapid fetal karyotyping and usually to confirm amniotic fluid results.

sodium heparin
An anticoagulant commonly used to coat collection vessels of blood and bone marrow specimens for cytogenetics.

somatic cell hybrids
Cells formed by a fusion between cells from two different species (e.g., human–mouse, human–hamster, or mouse–hamster combinations) which are utilized extensively by researchers participating in the Human Genome Project.

somatic cells
Cells of the body.

stimulated culture
A cell culture that has a mitogen added to it.

T-lymphocytes
Immunologically competent cells that are derived from the lymphoid cells that migrate from the bone marrow to the thymus and are involved in cell-mediated immunity.

Trypan blue
A blue dye used to stain cells on a slide for better visualization when the cells are counted. Trypan blue dye exclusion is used as a test for cell viability.

trypsin
An enzyme used for tissue disaggregation.

unstimulated culture
A cell culture that does not contain any mitogens.

5

Staining and Banding Techniques for Conventional Cytogenetic Studies

Hon Fong L. Mark
Brown University School of Medicine and Rhode Island Department of Health, Providence, and KRAM Corporation, Barrington, Rhode Island

I. INTRODUCTION

What the clinical cytogeneticist refers to as "banding" is the staining method by which dark and light differential staining is induced along the lengths of the chromosomes. According to the Paris Conference (1971) (1), a chromosome band is part of a chromosome that can be distinguished from adjacent segments by appearing darker or lighter by one or more techniques. By this definition, chromosomes consist of a continuous series of dark and light bands, with no interbands. Three levels of resolution were standardized, namely, 400, 550, and 850 bands per haploid karyotype. Further descriptions of the universally adopted cytogenetic nomenclature can be found in *An International System for Human Cytogenetic Nomenclature* (1995) (2) and also in Chapter 3 in this volume.

Prior to the advent of banding in the late 1960s and early 1970s, chromosomes were grouped and classified based on their size, shape, the position of their centromeres, and the gross chromosome morphology, as discussed in Chapter 2 (Mark and Mark) in this volume. Unequivocal identification of each unbanded chromosome of the human genome was not possible. The development of Q-banding in 1970 (1,3,4) ushered in an era where many protocols for banding flourished, leading to an unprecedented number of discoveries of numerical and structural chromosomal abnormalities. More descriptive terms were coined, such as the QFQ technique for Q-banding and the GTG technique for G-banding. These three-letter codes were described in the Paris Conference (1971) Supplement (1975) (5). The first letter describes the kind of banding. The second letter stands for the technique or agent used to induce banding. The third letter signifies the kind of stain employed. For example, GTG stands for G-banding by trypsin using Giemsa as a stain; QFQ stands for Q-banding by fluorescence and quinacrine derivatives; CBG stands for C-banding by barium hydroxide and Giemsa; RFA stands for R-banding by fluorescence and acridine orange; and RHG stands for R-banding by heating using Giemsa (6). There are occasional exceptions, however. One example is NOR staining, which will be described later.

Banding nomenclature has been thoroughly discussed in Chapter 3 in this volume, and also in ISCN (1995) (2).

II. SOLID STAINING

Solid staining is sometimes called conventional staining which produces nonbanded preparations (Fig. 1). This method is largely obsolete since the introduction of banding methods in the late 1960s and early 1970s. However, for selected applications it may still be utilized. The scoring of chromosome and chromatid breaks and gaps, for example, can be facilitated by solid staining (see, for example, Refs. 7 and 8). Satellites, secondary constrictions, dicentrics, ring chromosomes, double minutes, and fragile sites can be better visualized by solid staining. Chromosome morphometry can also be better performed with nonbanded chromosomes (9).

The stains employed are usually one of the Romanovsky-type dyes such as Giemsa, Leishman's, or Wright's stains. A typical solid-stained cell is shown in Fig. 1.

III. G-BANDING

Giemsa banding (G-banding) is the most widely used technique for the routine analysis of mammalian chromosomes in cytogenetics laboratories. With this method a series of dark- and light-stained regions along the length of the chromosome are produced that are grossly similar to Q-bands discussed below. A number of protocols are used to induce G-bands. Most commonly used are the treatment of slides with a protease such as trypsin or the incubation of the slides in hot saline-citrate (10,11) after appropriate slide aging. The chromosome banding patterns resulting from this treatment are distinct and quite consistent from experiment to experiment, leading some

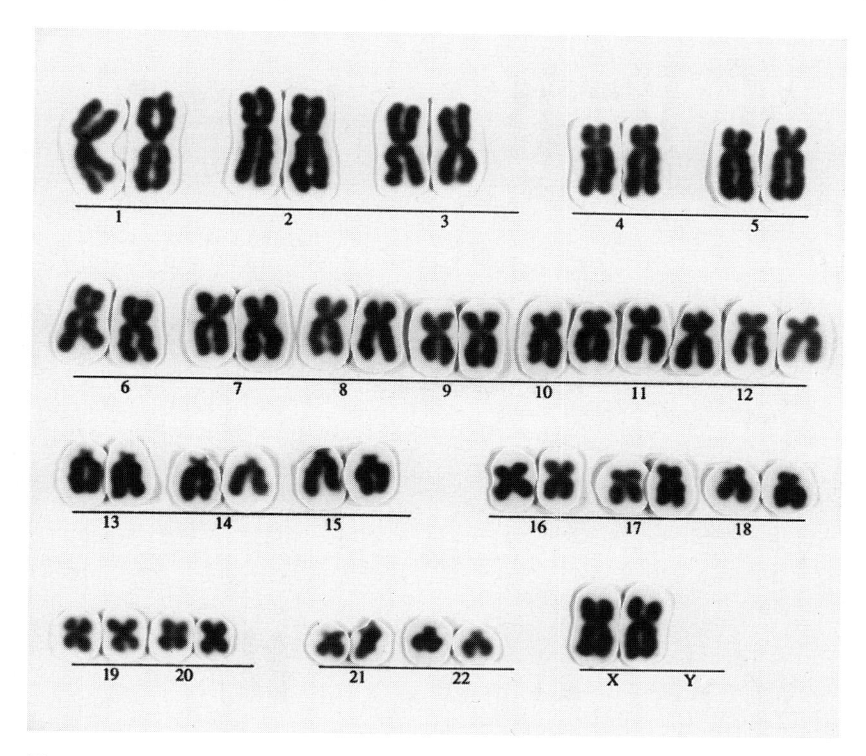

Figure 1 A nonbanded human karyotype produced by solid staining.

to postulate that G-bands reflect the inherent genetic composition and underlying structure of the chromosome. Dark G-bands have been found to correlate with pachytene chromomeres in meiosis. Generally, they replicate their DNA late in S-phase, contain DNA rich in adenosine and thymidine (AT) base pairs, appear to contain relatively few active genes, and may differ from light bands in terms of protein composition. It has been hypothesized that the differential extraction of protein during fixation and banding pretreatments from different regions of the chromosome may be important in the mechanism by which G-bands are obtained (see discussion by Rooney and Czepulkowski, Ref. 12).

It is not appropriate to go into the details of the various banding protocols in a book of this nature. The reader is referred to Refs. 13 and 14 for an encyclopedic collection of banding protocols. It should be noted, however, that although it has been approximately 30 years since chromosome banding was first discovered, the optimal banding conditions for each prep-

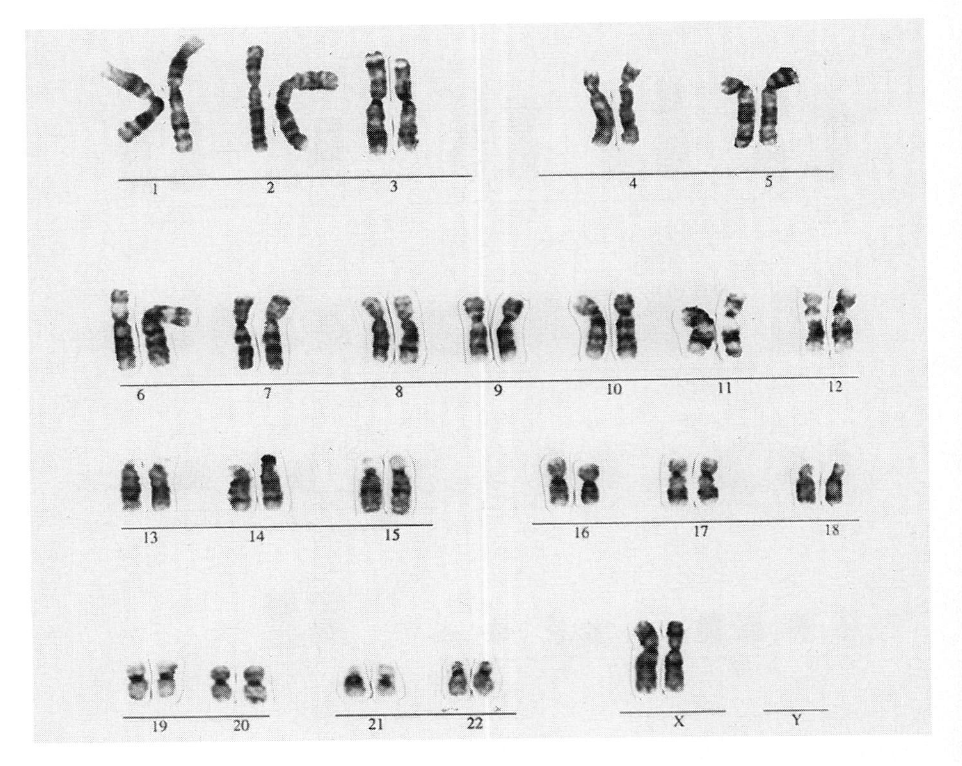

Figure 2 A G-banded karyotype.

aration are still often determined by trial and error, for cytogenetic technology is still more of an art than a science. A typical G-banded cell is shown in Fig. 2. The morphologic characteristics of individual chromosomes in the normal human chromosome complement can be found in Table 3 in Chapter 2 (Mark and Mark). All bands mentioned in the table refer to G-bands.

IV. HIGH-RESOLUTION BANDING

Although early G-banding was instrumental in delineating many structural chromosomal abnormalities, the level of its resolution is relatively low, with 400–450 bands per haploid genome, depending on the nature of the specimen. To obtain a higher level of resolution, at a band level of 550–850 or even greater, longer chromosomes at an earlier stage of mitosis, such as prophase or prometaphase, are needed. Protocols utilized to obtain such high-resolution G-banding were first described by Yunis and Sanchez (15) and Yunis (16). High-resolution banding protocols involve cell synchronization by blocking cell division with methotrexate followed by subsequent release of the mitotic block and the use of lower levels of exposure to blocking agents, such as Colcemid or Velban. High-resolution banding, which is a modified G-banding technique, is most useful for the delineation of structural chromosomal rearrangements, such as microdeletions and microduplications. Examples of these include deletions and cryptic translocations involving 17p13.3 in Miller-Dieker syndrome, deletions of chromosome 15q12 in Prader-Willi and Angelman syndromes, 22q11.2 deletions in DiGeorge syndrome and velocardiofacial syndrome, deletions of 17p11.2 in Smith-Magenis syndrome, deletions of 7q11.23 in Williams syndrome, deletions of 4p16.3 in Wolf-Hirschhorn syndrome, deletions of 5p15.2 in cri-du-chat syndrome, deletions of Xp22.3 in steroid sulfatase deficiency/Kallman syndrome, and microduplication at 17p11.2 which leads to Charcot-Marie-Tooth disease. However, even with high-resolution analysis, some of these deletions and duplications can often be missed. For the detection of subtle structural rearrangements, fluorescent in situ hybridization (FISH) with molecular probes is often a more sensitive technique than G-banding and high-resolution banding analysis (17). A typical high-resolution banded karyotype is shown in Fig. 3.

V. Q-BANDING

Quinacrine banding (Q-banding), using either quinacrine mustard or quinacrine dihydrochloride, was the first banding method reported (3). Prior to Q-banding, the only chromosomes in the human genome that could be identified unequivocally by size and morphology alone were chromosomes 1, 2, 3, 16, and the Y chromosome.

Quinacrine dihydrochloride (quinacrine, atebrin) is an acridine dye which binds to DNA by intercalation or by external ionic binding. Upon treatment the chromosomes exhibit a series of bright and dull fluorescent bands. The distal long arm of the Y chromosome is particularly bright with Q-banding. The Q-banding pattern resembles that of G-banding, with Q-bright bands corresponding to G-dark bands. Notable exceptions are centromeric regions of chromosomes 1, 9, 16, and the acrocentric satellite regions.

As in G-banding, it has been postulated that the pattern of banding reflects the underlying chromosome structure and genetic composition of the chromatin. It has been hypothesized that regions which fluoresce brightly are rich in adenosine and thymidine (AT) base pairs (18), whereas the gua-

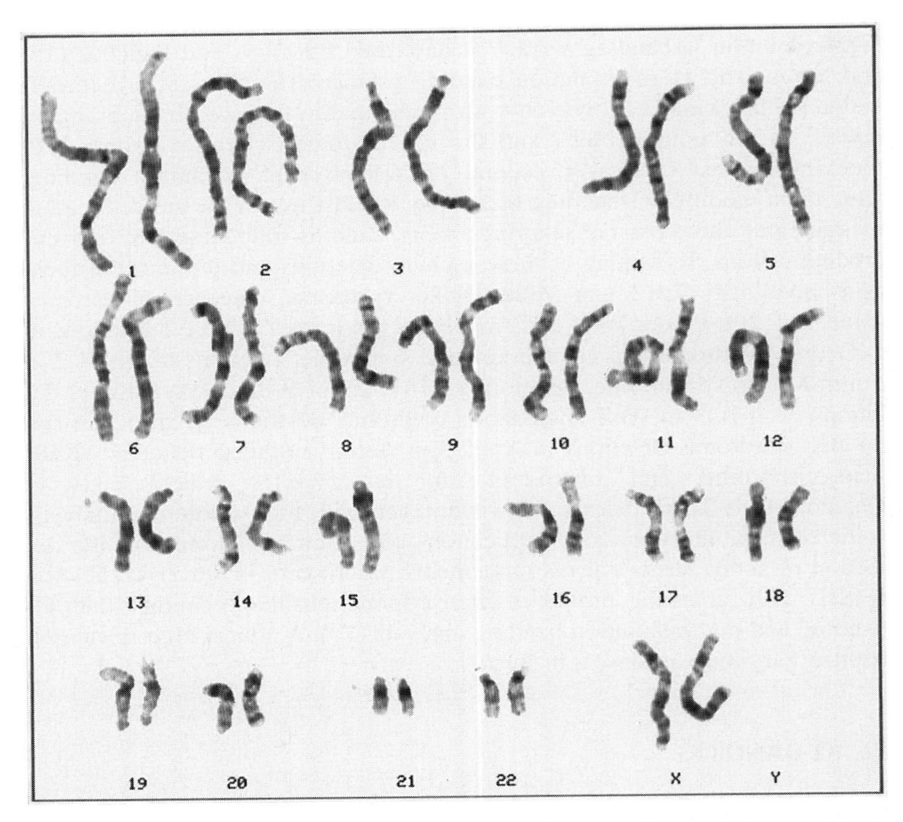

Figure 3 A high-resolution banded karyotype. (Courtesy of Dr. Bai-Lin Wu and Dr. Bruce Korf, Harvard Medical School.)

nine and cytosine (GC)-rich areas of the chromosomes tend to exhibit duller fluorescence (19,20).

One of the most useful features of this method of banding is the identification of the human Y chromosome (Fig. 4). With Q-banding, the distal long arm (Yq12) of a normal Y chromosome appears very bright and is easily distinguishable. Thus, Q-banding is especially useful for identifying sex chromosome variants, such as XYY, shown in Fig. 5. Prior to FISH and the advent of interphase cytogenetics using FISH, the fluorescent Y-body (or Y-chromatin) was often used to examine Y-chromosome copy number in interphase cells (21). The fluorescent Y-body is illustrated in Fig. 6.

Although simple, quick, and reliable, Q-banding has two serious drawbacks. One is its requirement for fluorescent microscopy. The other is the

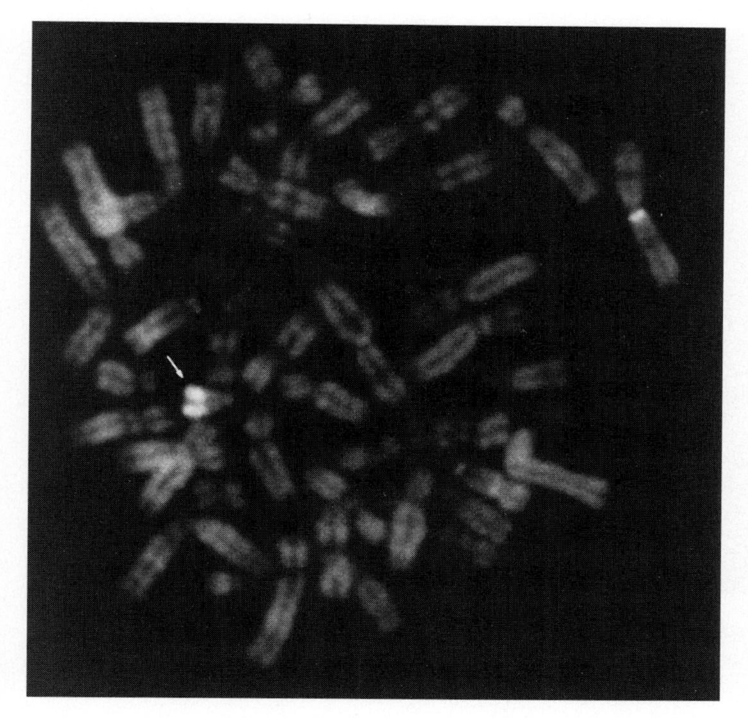

Figure 4 The human Y chromosome (arrow) illustrated by Q-banding.

rapid quenching of the fluorescence, making it less than an ideal method for microscopic analysis, especially compared to methods such as G-banding, which produces permanent preparations for microscopic examination. For the reasons above, most cytogeneticists today no longer perform Q-banding on a routine basis.

VI. CONSTITUTIVE HETEROCHROMATIN BANDING (C-BANDING)

Constitutive heterochromatin, first described by Heitz (22), represents approximately 20% of the human genome. It is the structural chromosomal material seen as dark-staining material in interphase as well as during mitosis (12). Constitutive heterochromatin is distinguished from facultative heterochromatin, which is represented by the inactivated X chromosome in mammalian females (23) and which manifests itself as the Barr body or sex

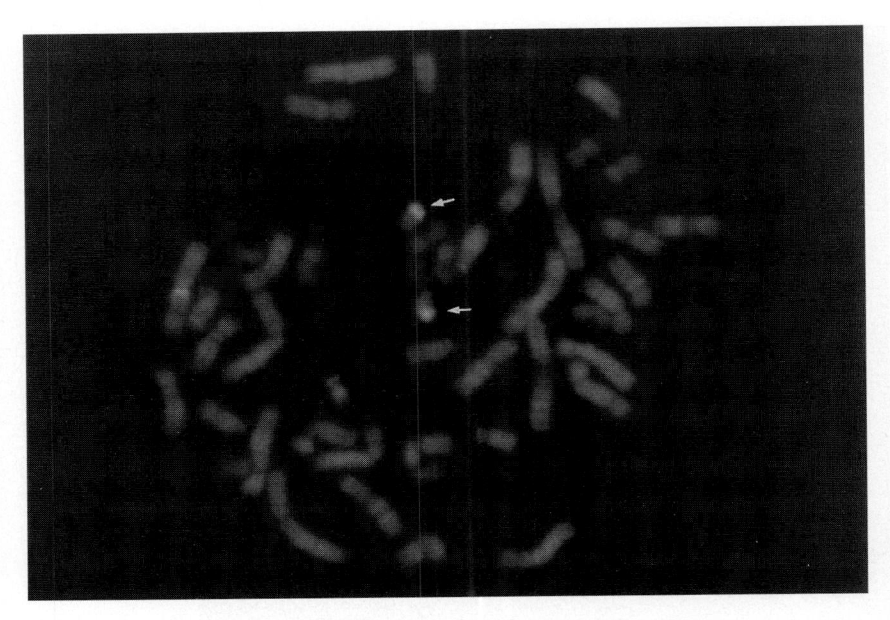

Figure 5 Metaphase cell of an XYY individual, QFQ technique.

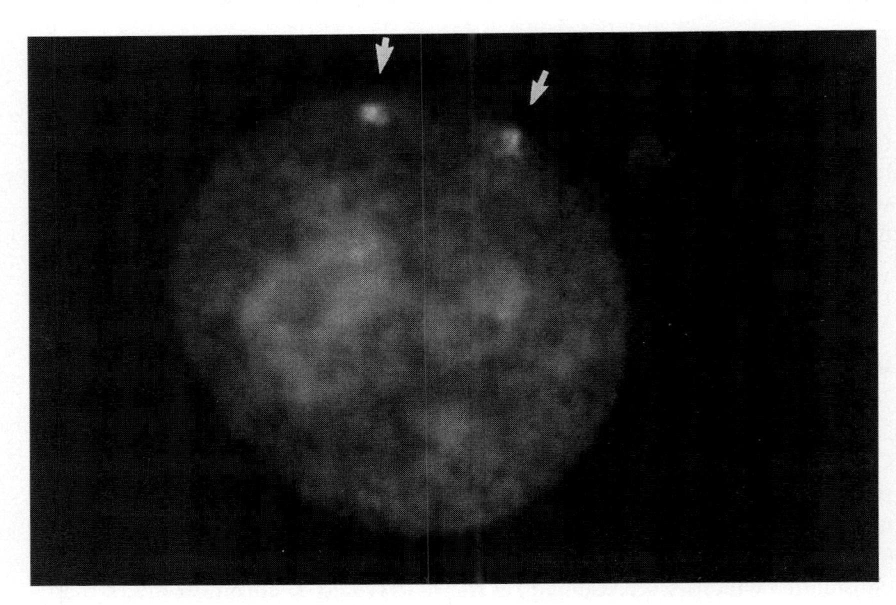

Figure 6 Fluorescent Y-body technique, XYY individual.

chromatin (24). Whereas constitutive heterochromatin can be visualized with the C-banding technique, facultative heterochromatin cannot.

C-banding stains constitutive heterochromatin located mostly at the primary constriction (centromeric) regions of human chromosomes and is most prominent at the secondary constrictions of chromosomes 1, 9, 16, and the distal long arm of the Y chromosome (Fig. 7).

Thus, human chromosomes that have prominent C-bands are chromosomes 1, 9, 16, and the distal long arm of the Y chromosome. C-banding therefore can be used sometimes to characterize karyotypic abnormalities involving these chromosomes (Fig. 8). C-dark bands are postulated to contain satellite DNA and are ideal for studying heteromorphic variants in the human genome, such as variations in the lengths and positions of the secondary constriction regions of chromosomes 1, 9, 16, and distal Yq. Figure 9 shows a C-banded karyotype of a female with a pericentric inversion of a chromosome 9, which altered the position of the centromere and the secondary constriction region of one of the chromosome 9 homologs. This particular photograph is special because it was from a case which we studied during the early days of banding, more than 20 years ago.

C-banding protocols involve treatment with acid, alkali, and hot saline and subsequent staining with Giemsa. A popular protocol uses a saturated solution of barium hydroxide (25), whereas the original method (26) used sodium hydroxide as the denaturing agent. The formation of C-bands is

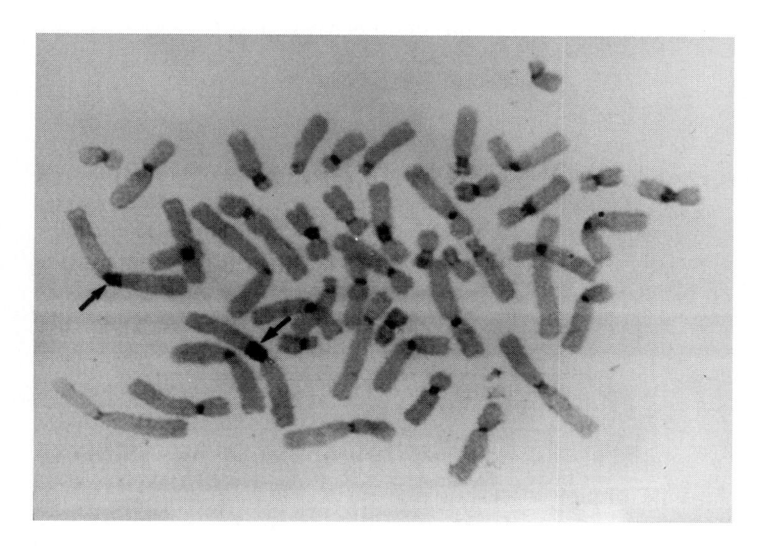

Figure 7 A C-banded metaphase. Arrows point to C-bands on chromosome 1.

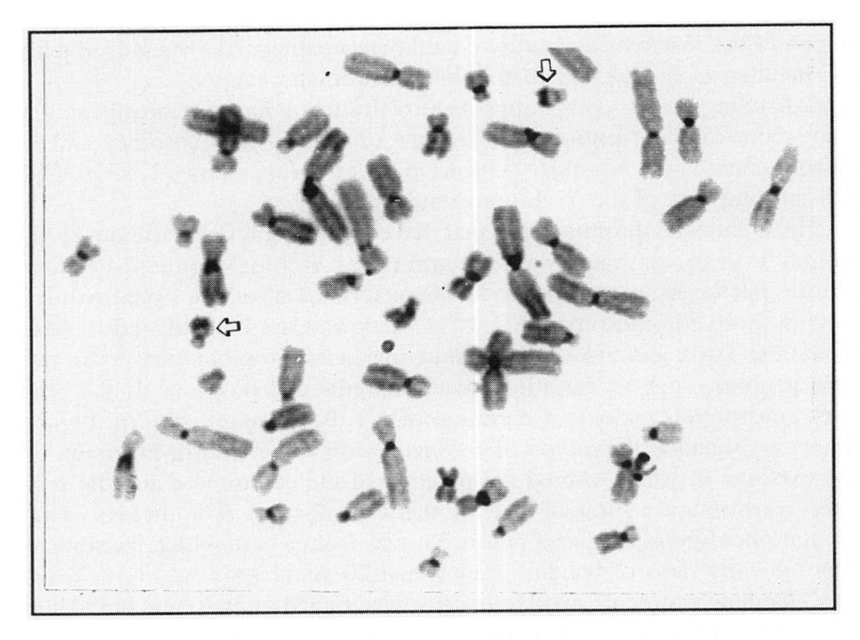

Figure 8 Metaphase cell of an XYY individual, CBG technique.

postulated to be due to preferential destruction of DNA in non-C-banded regions of the chromosomes (27,28), but the mechanism of C-banding is not at all clear.

VII. REVERSE OR R-BANDING

R-banding (29) is so called because bands that are pale by G-banding often appear darkly stained by R-banding, and vice versa. In general, R-banding is the reverse pattern to that obtained with either Q- or G-banding. R-bands replicate their DNA early and are said to contain active genes.

R-banding can be induced by incubation in a saline solution at a high temperature followed by Giemsa staining, in which case metaphases are analyzed by conventional bright-field microscopy. This is the so-called RHG technique, or R-banding by heat using Giemsa. It can also be induced by staining with acridine orange after hot phosphate buffer treatment, which results in bands being green and red in color, in which case fluorescent microscopy is required. The latter is called the RFA technique (30), or R-

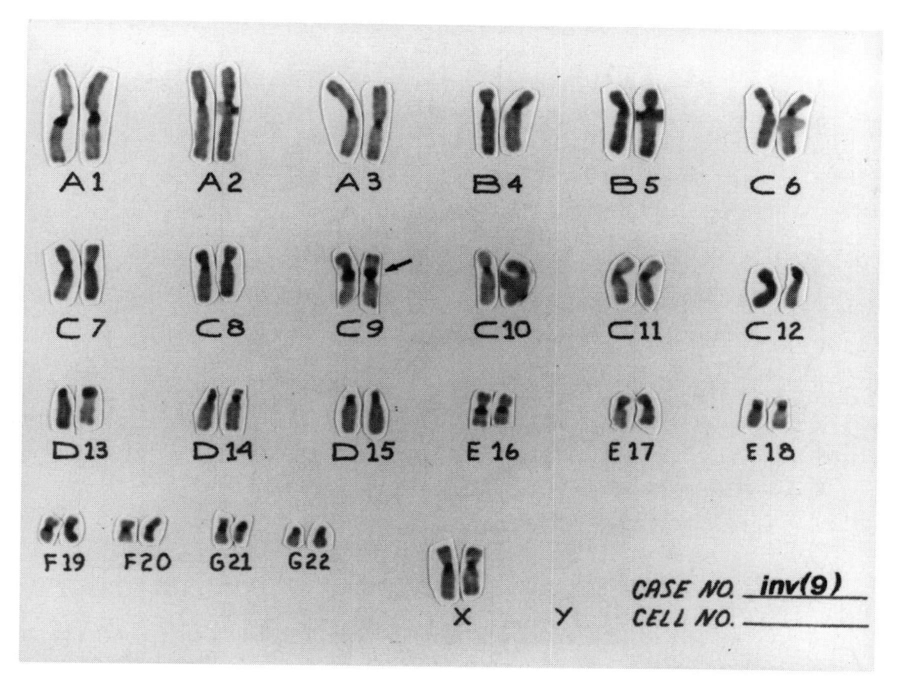

Figure 9 A C-banded karyotype. Arrow points to a pericentric inversion of chromosome 9.

banding by fluorescence using acridine orange. Other variations of the protocols used to induce R-banding are also possible.

Although the exact mechanisms of chromosome banding remain unclear, the fact that R-bands fluoresce yellow-green and G- and Q-bands fluoresce orange-red is consistent with the hypothesis that R-bands are GC-rich and are more resistant to denaturation than AT-rich regions, since acridine orange fluoresces yellow-green when bound to double-stranded DNA and orange-red when bound to single-stranded DNA. The main advantage of using R-banding over G-banding is that the telomeric regions of certain chromosomes which stain faintly using Q- and G-banding can be better visualized using R-banding. This method of banding is useful in cases where a structural chromosomal rearrangement near the terminal region of a chromosome is suspected.

An R-banded karyotype illustrating a Robertsonian translocation between the long arms of two chromosome 21s is shown in Fig. 10. This photograph, derived from a case which we studied during the early days of R-banding, is approximately 25 years old.

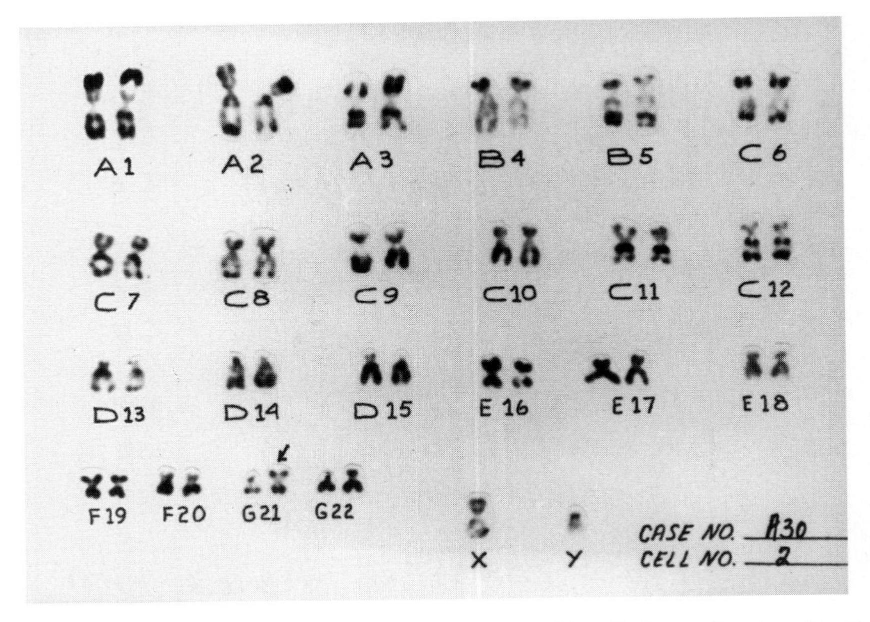

Figure 10 An R-banded karyotype of a male with a Robertsonian translocation between the long arms of two chromosome 21s.

VIII. NUCLEOLAR ORGANIZER REGIONS (NOR) STAINING

NOR staining uses silver stain, which stains active nucleolar (or nucleolus) organizer regions on acrocentric chromosomes. The nucleolar organizer regions of mammalian chromosomes are known to contain the genes for 18S and 28S rRNA. Regions in which the genes are thought to be actively transcribed can be selectively stained using silver nitrate (31–33). The stain is thought to selectively identify a protein adjacent to the nucleolar organizer region rather than the nucleolar organizer regions themselves. For human chromosomes, the genes for 18S and 28S rRNA are located on the short arms of the acrocentric chromosomes, which are chromosomes 13, 14, 15, 21, and 22. With silver staining, NOR bands appear as one or more dotlike structures of varying size located on the stalks (not the satellites) of the acrocentric chromosomes. Heritable polymorphic variations are often observed using NOR staining with silver nitrate. Many feel that silver staining plays a crucial role in discriminating small bisatellited marker chromosomes from other markers of similar appearance. In addition, this method has the advantage of allowing an additional banding technique to be carried out on the same slide after the silver staining has been completed.

Based on the results of experiments performed using RNase, DNase, and digestion with proteolytic enzymes, it is now known that the NOR-specific proteins are the target chromosomal components for silver staining (34). Subsequent studies have also shown that only the active NORs are impregnated by silver.

IX. DIFFERENTIAL REPLICATION STAINING

Based on studies using the technique of autoradiography with tritiated thymidine, it has been known that different parts of the human chromosome replicate at different times. The study of differential replication can be enhanced by the introduction of methods which detect the incorporation of 5-bromodeoxyuridine (BrdU), a thymidine analog, into chromosomes. Besides being used for the study of differential replication, such as in cases of sex chromosomal abnormalities, BrdU is also used for studies of sister chromatid exchanges such as in the diagnosis of diseases characterized by chromosomal instability, such as in Bloom syndrome, discussed in Chapter 14 (Wyandt, Tonk and Lebo) in this volume.

X. DA-DAPI STAINING

When chromosomes are stained with 4′,6-diamidino-2-phenylindole (DAPI) as the primary stain, followed by counterstaining with a nonfluorescent antibiotic, distamycin A (DA), a specific subset of chromosomes is highlighted (35). This is a useful method for studying heterochromatic regions of chromosomes 1, 9, 16, and the distal long arm of the Y chromosome as well as the proximal short arm of chromosome 15, which stain brightly while all other regions of the human genome appear dull. Thus, the distamycin-DAPI method can be used to identify abnormalities or variants of these chromosomes, especially chromosome 15, which often appear in the form of a small satellited marker. The limitation of this method is similar to other fluorescence-based methods. The fluoresent banding pattern tends to quench quickly, making rapid photographing a necessity.

XI. TELOMERE STAINING

Telomere staining (36) can be considered a special case of R-banding in which a more destructive treatment results in diminished staining except at terminal bands. T-bands were first reported by Dutrillaux (37). Modified methods to induce both C- and T-bands (CT-bands) were subsequently described by Scheres, and by Chamla and Ruffie (38,39); CT-bands stain centromeres, telomeres, satellites, and heterochromatic regions. This is not a routine technique employed by the average clinical cytogenetics laboratory, but it did play a role in identifying structural chromosomal rearrangements involving the most distal regions of the chromosomes in the past, prior to the advent of FISH and the availability of telomere probes.

XII. GIEMSA-11 STAINING

Although its mechanism of action is uncertain, staining chromosomes with Giemsa at pH 11 is a useful technique for the study of human–rodent somatic and radiation hybrids (40). With Giemsa-11 staining, human chromosomes stain blue, with the centromeric regions of chromosomes (notably chromosome 9) staining magenta. Rodent chromosomes stain uniformly magenta. However, the Giemsa-11 technique is not as sensitive as FISH (40).

XIII. KINETOCHORE STAINING

Kinetochore, or centromere dots (Cd) staining (41), as the name implies, is used to stain the kinetochore, which is the point of attachment of a chromosome to the mitotic spindle, located at the centromeric constriction of the chromosome. The pairs of dots identified by this technique may represent the kinetochore itself or they may represent the chromatin associated with this structure (12). Cd staining has been used to study centromere structure and function. It is useful in distinguishing an active centromere from an inactive one, because only the active centromere is stained positive (42,43). Inactive centromeres, also referred to as latent or suppressed centromeres, are not stained. An interesting observation by Nakogome et al. (44) is that chromosomes in older women tend to lose their Cd-bands which represent functional centromeres. This could be the causative mechanism for nondisjunction and the production of chromosome aneuploidy during mitosis and meiosis.

Thus, Cd staining can be useful in resolving active centromeres in Robertsonian translocations, dicentrics, pseudodicentrics, ring chromosomes, and other abnormal marker chromosomes (45).

XIV. BARR BODY TECHNIQUE

The Barr body technique is used to identify the Barr body (24), which is sometimes called the sex chromatin or the X-chromatin. The Barr body represents the inactivated X chromosome, which remains condensed throughout interphase in the somatic cells of the mammalian female. Theoretically, the number of Barr bodies in any given cell equals the number of inactivated X chromosomes, i.e., the total number of X chromosomes in the cell minus one. Various applications of this technique, usually performed on buccal smears, have been described by Mark et al. (46). As was previously pointed out, this has been considered an obsolete technique for routine clinical applications because of variability in results. However, its use has been resurrected since the advent of FISH, which can be performed on cytological specimens (see Chapter 17 on FISH (Mark)).

XV. OTHER MISCELLANEOUS TECHNIQUES

A multitude of other banding and staining techniques induced by a variety of agents, including antibiotics, antibodies, and restriction enzymes, have been described in detail by Verma and Babu (14). For example, in restriction endonuclease banding, the treatment of slides with restriction enzymes prior to staining with Giemsa generally produces a modified C-banding pattern depending upon the choice of enzymes. A banding pattern similar to G-banding can be obtained with HaeIII (12,14). This particular technique is useful in occasional instances where one of the more common techniques is not informative and newer molecular technologies such as FISH are not available.

Most of the above staining and banding techniques depend on bright-field microscopy, discussed thoroughly in many other texts (for example, Refs. 12, 13, and 47), the details of which are beyond the scope of this volume. Q-banding and fluorescent R-banding, on the other hand, need instrumentation for fluorescent microscopy, discussed in Chapter 6 on FISH (Blancato and Haddad).

In the United States, G-banding is the method of choice for many cytogenetics laboratories doing diagnostic testing. C-banding is used occasionally to study centromere position, dicentric chromosomes, and Y-chromosome variants. Q-banding is rarely used nowadays, although in the past it was used more often for routine analysis as well as for the study of heteromorphisms of the centromeres, shorts arms of the acrocentric chromosomes, and the distal long arm of the Y chromosome. The application of solid staining has been restricted as described above. NOR (silver) staining was used to analyze short-arm variants of the acrocentric chromosomes and for the identification of small markers. With the advent of FISH and other FISH-based techniques, as discussed elsewhere in this volume, the importance of some of the older staining and banding techniques has been somewhat diminished. However, conventional cytogenetics using G-banding remains a highly informative test in clinical pathology and will most likely continue to serve as the preferred standard of reference.

ACKNOWLEDGMENTS

The author would like to thank the former staff of the Lifespan Academic Medical Center Cytogenetics Laboratory at Rhode Island Hospital for their technical help, and Ms. Anne Richardson for critically reviewing the manuscript. The continued support of Dr. Roger Mark is also acknowledged.

REFERENCES

1. Paris Conference (1971): Standardization in Human Cytogenetics. Birth Defects: Original Article Series, VIII(7). New York: The National Foundation, 1972; 11:313–362.
2. Mitelman, F, ed. ISCN (1995): An International System for Human Cytogenetic Nomenclature (1995). Published in collaboration with Cytogenet Cell Genetics, S. Karger, Basel, 1995.
3. Caspersson T, Farber S, Foley GD, Kudynoski J, Modest EJ, Simmons E, Wagh U, Zech L. Chemical differentiation along metaphase chromosomes. Exp Cell Res 1968; 49:219–222.
4. Caspersson T, Zech L, Johansson C, Modest EJ. Identification of human chromosomes by DNA-binding fluorescent agents. Chromosoma 1970; 30:215–227.
5. Paris Conference, Supplement: Standardization in Human Cytogenetics. Birth Defects. Original Article Series, 1975, Vol. 11, No. 9. New York: The National Foundation, 1975; also Cytogenet Cell Genet 1975; 15:201–238.
6. Harnden DG, Klinger HP. ISCN: An International System for Human Cytogenetics Nomenclature (1985). Basel: S Karger, 1985.
7. Mark HFL, Naram R, Pham T, Shah K, Cousens LP, Wiersch C, Airall E, Samy M, Zolnierz K, Mark R, Santoro K, Beauregard L, LaMarche PH. A practical cytogenetic protocol for in vitro cytotoxicity and genotoxicity testing. Ann Clin Lab Sci 1994; 24:387–395.
8. Mark HFL, Naram R, Singer JT, Rich RW, Bastan W, Beauregard LJ, LaMarche PH. Cytotoxicity and genotoxicity of wood drying condensate from southern yellow pine: an in vitro study. Mutation Res 1995; 342:191–196.

9. Priest JH. Banding procedures and special stains. In Kaplan BJ, Dale KS, eds. The Cytogenetics Symposia. Burbank, CA: The Association of Cytogenetic Technologists, 1994.
10. Seabright M. A rapid banding technique for human chromosomes. Lancet 1971; 2:971–972.
11. Sumner AT, Evan HJ, Buckland RA. A new technique for distinguishing between human chromosomes. Nature 1971; 232:31–32.
12. Rooney DE, Czepulkowski, BH, eds. Human Cytogenetics. A Practical Approach. Volume I. Constitutional Analysis. 2nd ed. New York: Oxford University Press, 1992.
13. Barch MJ, Knutsen T, Spurbeck JL, eds. The AGT Cytogenetics Laboratory Manual. 3rd ed. Philadelphia: Lippincott-Raven, 1997.
14. Verma RS, Babu A. Human Chromosomes. Principles and Techniques. 2nd ed. New York: McGraw-Hill, 1995.
15. Yunis JJ, Sanchez O. The G-banded prophase chromosomes of man. Humangenetik 1975; 27:167–172.
16. Yunis JJ. High resolution of human chromosomes. Science 1976; 191:1268–1270.
17. Mark HFL, Jenkins R, Miller WA. Current applications of molecular cytogenetic technologies. Ann Clin Lab Sci 1997; 27:47–56.
18. Ellison JR, Barr HJ. Quinacrine fluorescence of specific chromosome regions: late replication and high AT content in *Samoaia leonensis*. Chromosoma 1972; 36:375–390.
19. Comings DE, Kavacs BW, Avelino E, Harris DC. Mechanisms of chromosome banding. V. Quinacrine banding. Chromosoma 1975; 50:115–145.
20. Comings DE. Mechanisms of chromosome banding and implications for chromosome structure. Annu Rev Genet 1978; 12:25–46.
21. Pearson PL, Bobrow M, Vosa CG. Technique for identifying Y chromosomes in human interphase nuclei. Nature 1970; 226:78–79.
22. Heitz E. Das heterochromatin der moose. I. Pringsheims Jb. wiss. Botonay 1928; 69:762–818.
23. Lyon MF. Sex chromatin and gene action in the mammalian X-chromosome. Am J Hum Genet 1962; 14:135–148.
24. Barr ML, Bertram EG. A morphological distinction between neurons of the male and female, and the behavior of the nucleolar satellite during accelerated nucleoprotein synthesis. Nature 1949; 163:676–677.
25. Sumner AT. A simple technique for demonstrating centromeric heterochromatin. Exp Cell Res 1972; 75:304–306.
26. Arrighi FE, Hsu TE. Localization of heterochromatin in human chromosomes. Cytogenetics 1971; 10:81–86.
27. Comings DE, Avelino E, Okada T, Wyandt HE. The mechanism of C- and G-banding of chromosomes. Exp Cell Res 1973; 77:469–493.
28. Pathak S, Arrighi FE. Loss of DNA following C-banding procedures. Cytogenet Cell Genet 1973; 12:414–422.
29. Dutrillaux B, Lejeune J. Sur une novelle technique d'analyse du caryotype humain. C R Acad Sci (D) (Paris) 1971; 272:2638–2640.

30. Bobrow M, Madan K. The effects of various banding procedures on human chromosomes studied with acridine orange. Cytogenet Cell Genet 1973; 12: 145–156.
31. Howell WM, Denton TE, Diamond JR. Differential staining of the satellite regions of human acrocentric chromosomes. Experientia 1975; 31:260–262.
32. Goodpasture C, Bloom SE. Visualization of nucleolar organizer regions. III. Mammalian chromosomes using silver staining. Chromosoma 1975; 53:37–50.
33. Bloom SE, Goodpasture C. An improved technique for selective silver staining of nucleolar organizer regions in human chromosomes. Hum Genet 1976; 34: 199–206.
34. Goodpasture C, Bloom SE, Hsu TC, Arrighi FE. Human nucleolus organizers: the satellites or the stalk? Am J Hum Genet 1976; 28:559–566.
35. Schweizer D, Ambros P, Andrle M. Modification of DAPI banding on human chromosomes by prestaining with a DNA-binding oligopeptide antibiotic, distamycin A. Exp Cell Res 1978; 111:327–332.
36. Bobrow M, Madan K, Pearson, PL. Staining of some specific regions of human chromosomes, particularly the secondary constriction of No. 9. Nature New Biol 1972; 238:122–124.
37. Dutrillaux B. Nouveau systeme de marquage chromosomique: les bandes T. Chromosoma 1973; 41:395–402.
38. Scheres JMJC. CT banding of human chromosomes. Description of the banding technique and some of its modifications. Hum Genet 1976; 31:293–307.
39. Chamla Y, Ruffie M. Production of C and T bands in human mitotic chromosomes after heat treatment. Hum Genet 1976; 34:213–216.
40. Mark HFL, Santoro K, Jackson C. A comparison of approaches for the cytogenetic characterization of somatic cell and radiation hybrids. Appl Cytogenet 1992; 18:149–156.
41. Eiberg H. New selective Giemsa technique for human chromosomes. Cd staining. Nature 1974; 248:55.
42. Nakagome Y, Teramura F, Kataoka K, Hosono F. Mental retardation, malformation and partial 7p monosomy 45,XX,tdic(7;15)(p21;p11). Clin Genet 1976; 9:621–634.
43. Daniel A. Single Cd band in dicentric translocation with one suppressed centromere. Hum Genet 1979; 48:85–92.
44. Nakagome Y, Abe T, Misawa S, Takeshita T, Iinuna K. The "loss" of centromeres from chromosomes of aged women. Am J Hum Genet 1984; 36:398–404.
45. Gustashaw KM. Chromosome stains. In Barch MJ, Knutsen T, Spurbeck JL, eds. The AGT Cytogenetics Laboratory Manual. 3rd ed. Philadelphia: Lippincott-Raven, 1997.
46. Mark HFL, Mills D, Kim E, Santoro K. Quddus M, Lathrop JC. Fluorescent *in situ* hybridization assessment of chromosome copy number in buccal mucosal cells. Cytobios 1996; 87:117–126.
47. Kaplan BJ, Dale KS, eds. The Cytogenetics Symposia. Burbank, CA: The Association of Cytogenetic Technologists, 1994.

APPENDIX: GLOSSARY OF TERMS

banding

The staining method by which dark and light differential staining is induced along the lengths of the chromosomes.

Barr body

The Barr body is sometimes called the sex chromatin or the X-chromatin. The Barr body represents the inactivated X chromosome, which remains condensed throughout interphase in the somatic cells of the mammalian female. Theoretically, the number of Barr bodies in any given cell equals the number of inactivated X chromosomes, i.e., the total number of X chromosomes in the cell minus one.

BrdU

5-Bromodeoxyuridine. It is a thymidine analog which is readily incorporated into chromosomes. BrdU is used in differential replication banding.

CBG banding

C-banding by barium hydroxide and Giemsa.

Cd staining

Kinetochore staining. As the name implies, Cd staining is used to stain the kinetochore, which is the point of attachment of a chromosome to the mitotic spindle, located at the centromeric constriction of the chromosome. The pairs of dots identified by this technique may represent the kinetochore itself or may represent the chromatin associated with this structure. Cd staining has been used to study centromere structure and function. It is useful in distinguishing an active centromere from an inactive one, because only the active centromere is stained positive. Inactive centromeres, also referred to as latent or suppressed centromeres, are not stained.

chromomeres

Foci of chromatin condensation separated by more extended regions.

chromosome band

A part of a chromosome that can be distinguished from adjacent segments by appearing darker or lighter by one or more techniques. By this definition, chromosomes consist of a continuous series of dark and light bands, with no interbands.

constitutive heterochromatin

Constitutive heterochromatin is located mostly at the primary constriction (centromeric) regions of human chromosomes and is most prominent at the secondary constrictions of chromosomes 1, 9, 16, and the distal long arm of the Y chromosome. It is composed mainly of satellite DNA and stains dark by the C-banding technique. Constitutive heterochromatin is to be distinguished from facultative heterochromatin, represented by the inactive X in the mammalian female.

DAPI

4′,6-Diamidino-2-phenylindole. It is used as a counterstain in fluorescent in situ hybridization but is also used together with distamycin A (DA) in DA-DAPI staining for identifying abnormalities of certain chromosomes, such as chromosome 15.

GTG banding

G-banding by trypsin using Giemsa stain. GTG-banding is the most widely used technique for the routine analysis of mammalian chromosomes in cytogenetics laboratories.

ISCN

An International System for Human Cytogenetic Nomenclature. The reader is referred to Chapter 3 in this volume for a thorough discussion of this topic.

NOR staining

Nucleolar (or nucleolus) organizer regions staining. NOR staining uses silver stain, which stains active nucleolar organizer regions on acrocentric chromosomes.

QFQ banding

Q-banding by fluorescence and quinacrine derivatives.

RFA banding

R-banding by fluorescence and acridine orange.

RHG banding

R-banding by heating using Giemsa.

Robertsonian translocation

A structural chromosomal rearrangement resulting from centric fusion of two acrocentric chromosomes.

6
Fluorescent In Situ Hybridization (FISH): Principles and Methodology

Jan K. Blancato and Bassem R. Haddad
Georgetown University Medical Center, Washington, D.C.

I. INTRODUCTION

The cytogenetics laboratory has witnessed major advances in diagnostic and prognostic capability in the past decade with the use of molecular cytogenetics. The technique of fluroscence in situ hybridization (FISH), a hybrid of cytogenetics and molecular biology, has increased the resolution and application of cytogenetics (1). The availability of quality-controlled DNA probes from commercial sources has expedited the clinical use and acceptance of these tests. FISH is a technique that allows the detection of DNA sequences on metaphase chromosomes and interphase nuclei from a broad range of cells and tissues. In situ hybridization techniques use DNA probes which can hybridize to entire chromosomes or to single unique sequence genes and serve as a powerful adjunct to classical cytogenetic tests. The applications of FISH include aneuploidy detection, translocation and structural breakpoint analysis, microdeletion detection, and gene mapping, to be elaborated further in another chapter in this book. In this chapter we focus on the principles and methodology of FISH.

II. BASIC STEPS OF A FISH EXPERIMENT

The steps of a FISH experiment are similar to those of a Southern blot hybridization but are performed in situ; i.e., the DNA is studied in its original place in the cell or on the chromosome rather than extracted and run in a gel. The DNA and surrounding material are generally fixed in Carnoy's fixative (3:1 mixture of methanol and glacial acetic acid). The specimen is then treated with heat and formamide to denature the double-stranded DNA so that it can be rendered single-stranded. The target DNA is made available for binding to a DNA probe with a complementary sequence that is similarly denatured and made single-stranded. The probe and target DNA then hybridize or anneal to each other in a duplex, based on complementary base pairing. The probe DNA is tagged with a hapten such as biotin or digoxigenin or is labeled directly with a fluorescent dye. Following the hybridization, detection of the hapten can be accomplished through the use of an antibody tagged with a fluorescent dye. Subsequent fluorescent microscopy then allows clear visualization of the hybridized probe on the target material. The steps of a typical FISH experiment are shown in Fig. 1.

In situ hybridization (ISH) with radionuceotide-labeled probes has been used by a select number of laboratories for over 20 years (2). Autoradiography, however, requires long periods of exposure that are not acceptable for clinical applications. In the mid-1980s the concept of biotin-labeled DNA probes was put into practice and fluorescence detection of alpha-satellite sequences became practical for the cytogenetics laboratory. Fluorescence has the advantage of providing dual or multiple target detection through the use of different fluorescent dyes in the same experiment. The

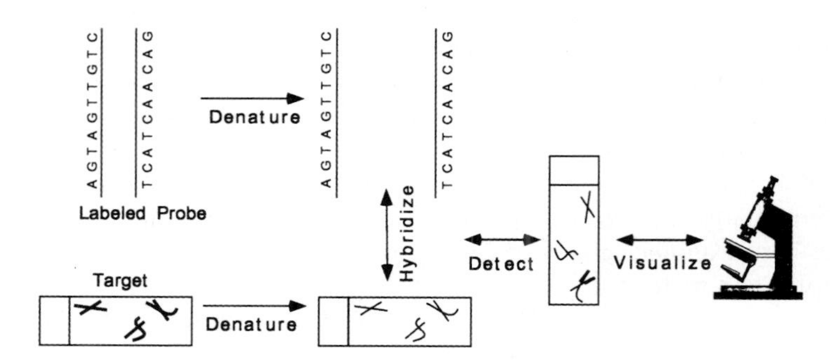

Figure 1 Diagram of the steps of a FISH experiment. The double-stranded DNA on the microscope slide is denatured (made single-stranded) and hybridized to a labeled probe, which has also been denatured. Nonhybridized probe is washed way, and the region of study is visualized with fluorescence microscopy.

signal intensity with fluorescence is greater than with in situ hybridization performed with immunochemical stains such as horseradish peroxidase. Versatile labeling systems, streamlined protocols, and vastly improved fluorescence microscopes and cameras have enabled most cytogenetics laboratories and some pathology laboratories to perform FISH tests as a part of their repertoire.

This chapter focuses on FISH with DNA probes, which involves copy number analysis and the investigation of structural rearrangements. In situ hybridization experiments which investigate aspects of RNA which involve gene expression almost exclusively require the use of polymerase chain reaction techniques or the increased sensitivity of radionucleotides. A review of these techniques can be found for various applications (3–6).

III. FISH PROBES

Three major categories of DNA sequences are used as probes in FISH studies: alpha-satellite DNA or centromeric probes, whole chromosome probes, and unique sequence probes. Satellite sequences are repetitive DNA sequences in the genome which do not code for a gene product and are polymorphic. Different individuals have variations in the number of copies of these DNA sequences. The alpha-satellite DNA is a 171-base-pair DNA monomer which is repeated *N* times as a tandem repeat. That block of tandem-repeated DNA is then copied *N* times in a higher-order repeat at the centromere of each chromosome (7,8). The majority of this DNA is identical in all human chromosomes, but 2–3% of the DNA is variable to the degree that centromeres of each individual chromosome can be distinguished and probes to those chromosomes can be produced (9). Color Plate 6.2 shows a FISH experiment in which metaphase chromosomes are hybridized with a probe to the common human centromeric sequence. The exception is the shared homology between chromosomes 13 and 21 and between chromosomes 14 and 22, which are not distinguishable through centromeric FISH studies. Other satellite DNAs include the beta-satellite, which is a 68-bp repeat which repeats in the same fashion as the alpha-satellite DNA and is found at the tip of the short arm of the acrocentric chromosomes 13, 14, 15, 21, and 22 (10). The classical satellite I DNA is an AATGG repeat found on chromosomes 1, 9, 16, and the Y chromosome (11). The telomeric repeat found at the ends of the short (p) and long (q) arms of all chromosomes is conserved over species and is composed of a TTAGGG sequence which repeats *N* times (12).

Repeat sequence probes are useful for counting specific chromosomes as in aneuploidy determinations and can be used on both interphase and metaphase cell preparations. Simultaneous visualization can be accomplished with multiple separate satellite probes through the use of different colored labels or label mixing. These probes are robust because the targets are large and repeated many times. This enables the hybridization to take place rapidly, allowing for a very large signal. Color Plate 6.3 shows an interphase bone marrow smear hybridized with chromosome 8 alpha-satellite probe.

Whole chromosome probes (WCP) are composed of numerous unique and repetitive sequences from an entire chromosome (13). They can be isolated through somatic cell hybrids, flow sorting of the specific chromosome, or microdissection of specific chromosomes with PCR amplification of the DNA (14,15). Whole chromosome probes are also called Paint™ probes because of the painted appearance of the metaphase chromosomes when hybridized. Color Plate 6.4 shows a metaphase chromosome spread hybridized with a chromosome 9-specific whole chromosome probe illustrating the staining of both chromosome 9 homologs. These probes are designed for use on metaphase chromosome preparations. Their use in interphase cells results in a splotchy, undefined fluorescence because the interphase chromatin to which they hybridize is decondensed as opposed to the compact, condensed state of metaphase. They are not useful in interphase studies and are best used for the identification of specific chromosomes in metaphases with structural abnormalities. Color Plate 6.5 shows the use of whole chromosome probes for chromosomes 3 and 9 on a breast cancer cell line which has characteristic rearrangements between the two chromosomes.

Whole chromosome probes from all the chromosomes are available for use in studies of markers and complex rearrangements involving multiple chromosomes. Chromosome arm-specific probes are DNA probes made from the p and q arms of the chromosomes and can be used for directed studies. For example, if an investigator is studying a tumor-suppressor gene on chromosome 4p, then a 4p probe may be a good research tool for loss of heterozygosity testing.

Unique sequence probes have target regions that are not repeated in the genome and may code for a gene. The various sorts of FISH unique sequence probes currently used in clinical cytogenetic laboratories are aimed at identifying chromosome microdeletions, oncogenes such as *n-myc*, *c-myc*, and *Her-2/neu*, and unique sequences in subtelomeric regions. Color Plate 6.6 shows hybridization of a fluorescein-labeled unique sequence probe to the q arm of chromosome 21.

Telomeric and subtelomeric probes can be produced from the repetitive ends of chromosomes of the region proximal to the telomere and contain unique sequences that are specific for the chromosome end. Actually, these sequences are highly conserved in eukaryotes. Unique sequences in close proximity to the ends of chromosomes are used for studies of cryptic translocations and for gene mapping (16) and ordering loci on a specific chromosome.

IV. PROBE LABELS

Probes used for in situ hybridization are best visualized with fluorescence detection methods, although some success has been demonstrated with light microscopic detection methods using horseradish peroxidase and other immunocytochemical reagents. Fluorochromes or fluorophores such as fluorescein, rhodamine, or coumarin are stains which can attach themselves to organic matter and are used to visualize the target DNA. Fluorochromes are capable of absorbing and emitting light. Fluorochromes used in FISH studies are able to absorb the excited light used in fluorescence microscopy.

In general, only alpha-satellite probes can be easily detected without the use of fluorescent labels. Unique sequence in situ hybridization probes cannot be easily resolved when using light microscopy. There are cases in which light microscopic detection becomes desirable, such as when autofluorescent material is being analyzed. Certain specimens, tissue types, and plastic embedding reagents or coverslips which have fluorescent capacity may need to be analyzed by FISH. The internal fluorescence of the specimen interferes with the discernment of the label. Some pathology specimens or research studies in which adjunct experiments involve the use of antibodies detected with light microscopy are best coupled with light in situ hybridization. Some laboratories use centromeric probes exclusively and adhere to the light ISH systems, because updating the microscope system is not possible due to budget limitations.

Two types of nonradioactive labeling methods are available: direct and indirect methods. In the direct method the reporter molecules, usually a fluorescent dye, is bound directly to the nucleic acid probe. The labeled probe and target can be visualized directly with fluorescence microscopy. Nucleic acids can be labeled nonradioactively in different ways. Fluorescent tags can be incorporated directly into nucleic acids using polymerase enzymes and labeled nucleoside triphosphates.

Biotin- and digoxigenin-tagged DNA probes are the most popular ones used both by research and clinical laboratories for indirect labeling. Biotin is a commonly used reporter molecule because it binds to avidin or streptavidin with high affinity. Streptavidin is used for the detection of biotin-labeled probes. Fluorochrome antibodies toward digoxigenin are available for the detection of this hapten. Labeling with enzymes which can be detected through a reaction process, such as alkaline phosphatase or immunoperoxidase, is a good method because signals can be amplified. One advantage of the indirect method is that amplification of the fluorescent signal can be accomplished easily with an additional round of antibody incubation. Antibodies directed specifically at the fluorophore, such as antifluorescein antibody labeled with fluorescein, can increase signal intensity without providing additional background fluorescence.

Dual-color and triple-color labeling and detection schemes have gained wider usage because of the flexibility in the use of labeling reagents and fluorescence filters. Multiple distinguishable colors, usually two or three, can be used concurrently to study the relationships among various target regions of interest (17).

Visualization of the surrounding DNA or nuclear material is accomplished through the use of counterstains. This allows the microscopist to scan for cells under low power and have a frame of reference to analyze the DNA sequences of interest. The common counterstains used in FISH studies are propidium iodide and diamidino-2-phenylindole (DAPI). Both are DNA intercalators and fluoresce under similar wavelengths as the commonly used fluorochromes, fluorescein, Texas Red, rhodamine, Spectrum Orange, and Spectrum Green. When working primarily with a yellow/green dye such as fluorescein, one can use the orange propidium iodide dye as a counterstain.

When using a red fluorochrome, such as Texas Red or rhodamine or in a dual-label study which uses red and green/yellow dyes, the blue DAPI counterstain is best. This is because it is difficult to see the red dye on an orange background. The DAPI counterstain gives the additional advantage of allowing a DAPI reverse banding pattern of the chromosomes, which helps in their identification. When performing studies with fluorochromes, an antibleaching chemical is commonly used to preserve the signal during storage, microscopy, and photography. Fading of fluorochromes upon excitation is a photochemical process. Light-induced damage in the presence of oxygen and non-oxygen radical generation are sources of photochemical fluorochrome destruction. Mounting media containing diphenylene diamine or other agents act as radical scavengers and antioxidants, which alleviate quenching without altering experimental results.

A. Minimizing Background Signal

Background is influenced by factors such as probe size, hybridization and wash stringency, and signal detection. One type of specific background is cross-hybridization, which refers to the hybridization of a particular probe to similar or related sequences. An example of cross-hybridization may be seen when using an alpha-satellite probe. The alpha-satellite sequence is present at the centromere of all human chromosomes and although chromosome-specific sequences have been identified, these sequences are similar to each other. In order to obtain a chromosome-specific signal, the stringency of the hybridization and wash must be carefully controlled. High-stringency washes after hybridization have lower salt concentrations, such as 0.25× Saline Sodium Citrate. A lower stringency wash, coupled with washes with a higher salt concentration of 2.0× SSC, may result in cross-hybridization at one or more additional chromosomes pairs.

Another example of cross hybridization may occur when using whole chromosome probes, or whole arm probes. In addition to the unique sequences, these probes contain repetitive sequences shared by all chromosomes. These sites are blocked by mixing the probes in blocking DNA composed of repetitive sequences and allowing them to prehybridize before applying to the specimen. Failure to perform such a blocking step may lead to a dotted labeling on all the chromosomes in the complement.

Nonspecific background may be found on the cells and areas surrounding the cells, possibly caused by a poorly prepared probe, nonspecific binding of the detection reagents, or amplification of the probe signal. In order to minimize background from the probe it is important to evaluate the size of the probe on a sizing gel after labeling. The size should be between 200 and 440 bp. Nonspecific binding of the detection reagents used with indirect labeling systems may be minimized by decreasing the incubation time. Amplification of the probe signal may also increase the apparent background, and use of such steps should be evaluated.

V. SPECIMENS ANALYZED BY FISH

FISH can be performed on most specimens because DNA is a very stable molecule. Cases in which DNA is degraded, such as in formalin-fixed, paraffin-embedded tissues from spontaneous abortions or stillbirths, are among the most challenging specimens. Theoretically, any cell with a nucleus can be examined using FISH techniques. To date, FISH has been performed on specimens ranging from archival pathology specimens to bladder washed epithelial cells and squamous cells from buccal smears (18). Applications of FISH in archival material and other specimens is discussed further in Chapter 17 by Mark.

A. Interphase Samples

1. Blood, Blood Marrow, and Buccal Smears

Blood and bone marrow can be applied directly in a thin smear on the surface of a silanized slide for rapid FISH analysis. Lysis of red blood cells and fixation of cells can be accomplished after the smear is made. This analysis is usually performed in situations where a quick result in necessary. The morphology of the cells in the smears may be suboptimal or unsuitable for unique sequence probe use and alpha-satellite probes are generally used in such cases.

Similar to blood smears, buccal cell scrapes are also smeared on silanized slides for use in FISH with alpha-satellite probes. These cell scrapes are comprised of squamous epithelium. Such a technique is useful for sex determination using X and Y centromeric probes on cells of newborns with ambiguous genitalia and in cases of suspected mosaicism for an aneuploidy. Color Plate 6.7 shows hybridization of buccal smear epithelial cells with dual-label X and Y chromosome-specific alpha-satellite probes. Results are confirmed through standard chromosomal analysis.

2. Cytospin Preparations

A sample of cells can be deposited onto a silanized slide with a cytocentrifuge and subsequently prepared for FISH experiments. A cytocentrifuge is a tabletop instrument that assists in the concentration of small amounts of cells in a solution by spinning the solution and delivering the cells onto a circumscribed region of a microscope slide.

This sort of preparation is helpful when the cell type for analysis or the probe is limited. It can be used for a variety of samples, such as pleural effusions, fine needle aspirates, and cell suspensions from culture. Cytospin preparations have been used successfully on cells obtained from bladder washes, or midstream first morning urine specimens (19) following a series of preliminary centrifugations steps. FISH on bladder wash specimens has been shown to be a successful noninvasive method to monitor bladder cancer patients for aneuploidy. Color Plate 6.8 shows a photo of cells obtained from a bladder wash that have been deposited onto a slide with a cytocentrifuge. These cells were hybridized with an alpha-satellite probe for the chromosome 8 centromere.

3. Touch-Print Preparations

A touch preparation is prepared by cutting a fresh, unfixed tissue specimen to expose a fresh surface. The tissue is touched directly to the surface of a silanized microscope slide in the area where hybridization will take place. The specimen can then be fixed in Carnoy's fixative and digested with enzyme prior to hybridization, if necessary. The strength of this approach is that it can provide a single-cell-depth representation of a tissue while avoiding the extra steps in processing and difficulties in scoring results that are encountered in formalin-fixed, paraffin-embedded tissues. The weakness of the approach is that a serial section is not necessarily available to identify the specifics of the tissue, such as whether the cell morphology is normal or abnormal.

4. Formalin-Fixed Paraffin-Embedded Tissue Sample Preparations

Tissues fixed in formalin or paraformaldehyde are appropriate for FISH analysis. The first step in addressing tissues that are paraffin-embedded is the thickness of the section. A 4-μm section is optimal for FISH studies. Some laboratories have had success with thicker sections and the use of confocal microscopy. It is best to cut sections and float them onto a microscope slide coated with trimethoxysilane to avoid sample loss through the many experimental steps. Most investigators air-dry these preparations and bake them overnight at 65°C before storage.

The paraffin is removed from the tissue section with successive washes in xylene. Dehydration of the tissue is accomplished in a wash of 100% ethanol. Most tissues will require pretreatment with proteinase K, pepsin, or a chaotropic solution before denaturation. The concentration and duration of the treatment depends on the tissue type. Some tissues, such as placenta, kidney, and adrenal gland, are more resistant to digestion than others, such as lung tissue require limited pretreatment. These conditions must be determined for each specimen block and even may vary for similar specimens. Conditions such as duration and temperature of fixation and age of specimen affect these variables and require careful monitoring of the pretreatment conditions. It is noted that tumor cells require less rigorous digestion than normal cells within the same tissue. When evaluating digestion, focus on the cell type and the area of the slide of most interest.

Pretreatment and digestion can be evaluated before hybridization by staining the sample with propidium iodide and viewing with fluorescence microscopy using an FITC/propidium iodide filter to see signs of underdigestion, appropriate digestion, or overdigestion. In general, overdigested tissue shows loss of nuclear borders. It will be difficult to see where one nucleus begins and another ends. If greatly overdigested, the nuclei may appear "ghostly" or be lost altogether. Such a slide should be discarded because the DNA will not be suitable for hybridization.

Underdigested tissues show persistent green autofluorescence and poor propidium iodide staining. Additional proteinase digestion is required if this result is seen. The amount of additional time and concentration of enzyme used is dependent on the extent of underdigestion. Appropriately digested tissues which are ready for hybridization show very little or no background fluorescence, nuclei that are evenly and brightly stained with propidium iodide, and clearly defined nuclear borders. Following digestion, slides are dehydrated in a graded ethanol series.

VI. CONTROLS AND ANALYSIS

Guidelines for the clinical diagnostic use of FISH have been developed for some indications, and quality assurance recommendations continue to evolve (20). This technology is approved by the U.S. Food and Drug Administration for a limited number of probes. The probe sets that are not FDA-approved are labeled by the manufacturer as "for *research or investigational use*." It is recommended that laboratory workers dedicate time for formal training and practice runs when setting up to begin FISH or when using FISH in a new application.

There are numerous sources of variation in FISH data, particularly when interlaboratory comparisons are attempted. This variation may stem from differences in sample preparation, probes used, intrasample heterogeneity, hybridization protocols, counting criteria within and between observers, fluorescence microscopes, and filters (21). Scoring criteria for interphase cell analysis is important, particularly in tissue sections in which signals may be seen in different focal planes. In general, these criteria include what size and type and cells should be scored, what size and type of signals should be scored, and what samples to discard because of cross-hybridization and/or autofluorescence. Criteria should be developed for all systems in which the laboratory is active, including translocation probe systems in interphase and metaphase, alpha-satellite probes in tissue sections, and unique sequence probes in tissue sections for monitoring amplifications.

Control probes and cells should be used when performing FISH studies. Control probes can be run to assure that successful hybridization has been accomplished. Control specimens provide essential information about the success of the experiment. For interphase studies, a control probe can be hybridized and both the control and test probe can be counted with 100 or more cells. The number of cells needed to score for a particular experiment depends on the sensitivity and specificity required for the test. Statistical measures can quantify true positives and true negatives relative to another test for the same endpoint, such as karyotype analysis. These numbers can be derived from pilot studies, which should include hybridization efficiency, percentage of cells discarded, and normal cutoffs derived from the results of normal specimens. A consultation with a statistician is helpful when establishing new test systems (22).

Reporting the results of FISH studies should include the number of test and control cells scored and the number of signals of the test and control probes. A narrative analysis of the result should accompany the appropriate nomenclature as given in *An International System for Human Cytogenetic Nomenclature* (23), which is also discussed elsewhere in this book. Specific information about the probe, such as manufacturer and lot number, is also useful in reports.

A. Gene Amplification Analysis

FISH analysis for other unique sequence probes is targeted at genes which, when amplified, provide information on the diagnosis and prognosis of specific cancers. These amplifications take the form of double minutes and homogenously staining regions (HSRs), which are cytological expressions of gene amplifications seen in human and animal tumors. Double minutes are extrachromosomal amplicons, while HSRs are incorporated into chromosomes. To date the genes which are most commonly analyzed for amplifications include *HER-2/neu*, *n-myc*, and *c/myc*. The *HER-2/neu* amplification is a prognostic indicator in node-negative breast cancer (24). The *n-myc* gene is amplified in neuroblastoma and *c-myc* is sometimes amplified in breast carcinoma and ovarian tumors. Color Plate 6.9 illustrates *n-myc* amplification in a paraffin-embedded section from a neuroblastoma tumor through the use of FISH. In general, these probes are used on interphase cells in touch preparations or paraffin-embedded tissues. Amplifications of *c-myc* are visualized as double minutes in the FISH study shown in Color Plate 6.10 using a metaphase preparation from a breast tumor cell line. In these cases FISH is used to enhance and correlate with data from other biological marker studies.

B. Translocation Analysis

Dual-colored probes are composed of unique sequences known to be involved in fusion gene events which have been associated with leukemias and lymphomas. Detection of these translocations, used as both diagnostic and prognostic indicators, is useful in patient follow-up studies as well. In general, the two genes involved in a translocation event, located on different chromosomes, can be easily visualized in metaphase and interphase studies. The power of FISH in translocation detection is that interphase analysis can be performed, thus allowing for many more scorable cells, providing answers on questions of chimerism (25).

The translocation probes are labeled with fluorochromes of different colors, resulting in a fusion color when the translocation brings those chromosomal segments into close proximity to each other. Normal controls and fusion-positive controls should be used for these studies, as a small percentage of interphase cells show fusion signals because they are in close proximity in the interphase nucleus and are not associated with a translocation event. For this reason and because of the large-sized probes that are sometimes used, some translocation probe systems have a false positive rate of 10% or more. Most laboratories have strict guidelines for scoring translocation probe sets which require analysis of many interphase cells.

The most common analyses performed with translocation probes are for the *bcr/abl* translocation which occurs in the blood cells of all patients with chronic myelogenous leukemia (CML) (26). Color Plate 6.11 demonstrates fusion of the *bcrabl* fusion gene in an interphase cell. Translocation systems are also used for the t(15;17) to visualize the promyelocytic leukemia (PML)/retinoic acid receptor alpha (RARA) translocation associated with acute promyelocytic leukemia (APL) (27). More complex FISH systems can also be designed for the detection of split signals in which a single-color-labeled DNA probe shows a large signal in the normal chromosome and two smaller signals when mutations such as deletions or translocations take place in disease states. This is the general design of a multiple myelogenous leukemia (MLL) probe set which is associated with different mutations that disrupt a gene at 11q23 (28).

Bone marrow transplant follow-up studies in cases with opposite-sex donor–recipient pairs can be accomplished using FISH probes to alpha-satellite sequences in the X and Y chromosomes. Results of studies of dual-color detection on the same cell can be used to determine the proportions of male and female cells in bone marrow or peripheral blood samples. Samples for evaluation can be scored at timed intervals following bone marrow transplantation. Low levels of chimerism can be detected using this methodology because of the high sensitivity of the test and the ease with which a large number of cells can be scored.

C. Solid Tumor Analysis

FISH studies in tumors have generally used the approach of examination of serial sections. The sections for FISH are customarily 4 μm in thickness. Adjacent slides are used for tumor grading, evaluation of the cells used in the FISH slides as normal or malignant. Some labs use a section for ploidy analysis by flow sorting to compare to the FISH studies.

Chromosomal abnormalities are common in solid tumors and hematologic malignancies. The determination of recurrent abnormalities and consistent rearrangements is important in the study of tumor progression. Diagnostic and prognostic information may also be provided by these specific abnormalities. Through combining current methodologies, most chromosomal abnormalities can be discerned in the target tissue. FISH for analysis of specific oncogene activity is a semiquantitative assessment method.

A description of the early events in tumor progression from benign to more malignant involves genetic events such as loss of heterozygosity and deletion of tumor supressor genes and specific amplification of oncogenes. For example, prostate specimens from 25 cases were studied with FISH to evaluate the relationship between high-grade prostatic intraepithelial neoplasia, prostatic carcinoma, and metastases. The study was designed to evaluate the number of copies of the oncogene *c-myc* on band 8q24 relative to the number of signals from the chromosome 8 centromere, an internal control. Substantial amplification of *c-myc* relative to the chromosome 8 centromere was strongly correlated with increasing cancer nuclear grade and immunohistochemical evidence of c-myc protein overexpression (30).

The *HER-2/neu* gene amplification can be determined by FISH as noted and has been validated with archival breast tumor tissue previously characterized for gene amplification by Southern hybridization, Western immunoblot, or immunohistochemistry. The FISH studies have high sensitivity and specificity relative to the other methods, and have some distinct advantages over Southern blotting. *HER-2/neu* amplification is associated with poor prognosis in node-negative breast cancer and is a better predictor of poor clinical outcomes than tumor size (31). Color Plate 6.12 shows amplification of the *HER-2 neu* gene in a tissue section of a breast tumor. Other abnormalities in archival breast tissue have been reported such as trisomy 8, 17, and *c-myc* amplifications (32).

D. Confined Placental Mosaicism

Confined placental mosaicism (CPM) is a phenomenon described in infants born with unexplained intrauterine growth retardation. Unlike mosaicism, which is characterized by the presence of two or more karyotypically different cell lines within both the fetus and placenta, confined CPM represents tissue-specific chromosomal mosaicism affecting only the placenta (34). This situation has been observed in cases where chorionic villus sampling (CVS) karyotypes show mosaicism but follow-up amniocentesis or fetal blood samplings show normal diploid results.

In order to study these cases, karyotypic analysis of the placenta has been pursued to study placentas from fetuses with unexplained intrauterine growth retardation (IUGR). Use of FISH to study trisomies of 2, 3, 7, 8, 9, 16, and 22 on these tissues has been accomplished (35). It is thought that the localized chromosomal abnormalities are responsible for the compromised development of the placenta in these cases. Samplings of the placenta are taken at various sites as these patterns of abnormal cells can be localized in portions of the tissue.

E. Marker Chromosome Analysis

Markers are a heterogenous group of chromosomes which cannot be definitively characterized by conventional cytogenetic techniques. Markers can be small ringlike structures or large complex rearrangements witnessed in tumor cells. They are also called extra structurally abnormal chromosomes (ESACs) or supernumerary chromosomes. The use of FISH has made it possible to characterize many of these markers. Small ring-like markers found in newborn and prenatal studies can be characterized through the use of FISH for acrocentric sequences of the satellited chromosomes, the centromere of chromosome 15, and other centromeric and whole chromosome probes (35). Applications of FISH for marker identification are further explored elsewhere in this book by Mark.

VII. THE FLUORESCENCE MICROSCOPE

A very important part of FISH is the visualization and recording of the results with a fluorescence microscope and camera system. An optimally functioning microscope is required for FISH analysis. Two types of fluorescence microscopes are available, which differ in the way the light contacts the specimen. Transmitted light microscopes illuminate from the bottom of the specimen, whereas incident or epi-illuminated microscopes reflect light onto the specimen from above. Of the two types, incident illumination is better for FISH analysis.

Of the various bulbs available, a high-pressure mercury lamp is optimal for fluorescence detection. These lamps are generally available at 50 or 100 W. The 100-W bulb is preferable for use with unique sequence probes and in situations where smaller signals are expected. Microscope bulb age and alignment can also affect the apparent strength of the signal. If the microscope is not properly aligned or the bulb exceeds the number of operational hours, the nuclei may appear dark and the signal may appear weak.

Objectives that are manufactured specifically for fluorescence microscopy are optimal because they are made of low-fluorescing glass and have few lens corrections. The laboratory should have a low-power objective (10×, 20×, or 16× oil) for scanning slides for interphase nuclei or metaphase spreads. The high-power objectives such as a 63× oil and 100× oil are used for the visualization of the fluorescence signal and for photography. A 40× oil or dry objective may also be useful, depending on the specimen and the preferences of the operators. Nonfluorescing oil should be used with the oil immersion objectives. Caution should be taken to avoid mixing immersion oils from different microscope vendors, as they are not miscible and will create a cloudy image on a slide if combined.

The optimal filters allow certain wavelengths of light through while blocking others. Filters are designed individually for specific fluorophores and should be chosen with that consideration. The counterstain filter should be used to scan the slide for cells or spreads, whereas the FISH probe filter allows visualization of both the counterstained chromosomes and the probe signal. An effort should be made to subject the probe signal to as little excitation light as possible to minimize quenching of the fluorescence. When visualizing FITC signals with propidium iodide counterstain, one should use a propidium iodide filter to scan, and a FITC/PI filter to visualize both the probe and the counterstain. This filter enables visualization of FITC while allowing PI to ''bleed through.'' If the experiment requires dual-label FISH probes such as FITC and Texas Red, and a DAPI counterstain, then a triple bandpass filter can be used to visualize the probe signals and counterstain simultaneously.

FISH experiments require that the fluorescence microscopy be conducted in a dark room. This has been defined by some microscopists as ''so dark that you cannot see your own hand in front of your face.'' This darkening can be accomplished by taping the door or moving the microscope to a specially prepared darkroom. The darkness is particularly important when using cosmid or other probes that provide low-intensity signals.

Many fluorescence microscopes are equipped with an automatic camera device. The optimal camera settings will depend on the type of microscope and should be determined empirically. High-speed color film such as Kodak Ektachrome (ASA 400 color slide film) perform well. When film is exposed for long periods of time, as with FISH, the sensitivity gradually becomes lower than the labeled value, and a longer exposure time than indicated is necessary. This phenomenon is known as reciprocity failure and will vary with film type. Most cameras have correctional settings for reciprocity failure. The user will need to experiment with these settings to assure the best outcome. Some applications may call for the use of black-and-white film if the fluorescence signal is strong. In this case, ASA 100 speed film should be used for the best resolution.

The number of digital imaging systems commercially available is increasing rapidly. Their use in both the clinical and research laboratory, in conjunction with sensitive cameras such as CCD (charge-coupled device) cameras, allows the scientist to take advantage of several potential applications of multicolor fluorescence in situ hybridization. In addition to the benefits of automation, these systems offer the advantage of higher sensitivity, image analysis, and signal quantitation, and allow the use of a wider variety of fluorochromes, especially those in the infrared spectrum.

VIII. RECENTLY DEVELOPED FISH TECHNOLOGIES

A number of cytogenetics research laboratories have been working on FISH-based experimental procedures that would be useful in answering some basic genetics questions. Although these procedures are not currently available in all clinical cytogenetic laboratories, the technology is rapidly becoming available in specialized centers. Some of these procedures are particularly useful in patient samples in which the location of the gene or chromosome involved is not known. In standard clinical FISH studies, the FISH probe is chosen because of phenotypic or genotypic clues provided by either the patient or the karyotype. One or more of these tests may prove useful when specific direction is not available, such as in the case of an uncharacterized solid tumor sample or blood sample with an ESAC.

A. Comparative Genomic Hybridization (CGH)

Conventional karyotype analysis is the ideal way to diagnose chromosomal imbalances. However, it requires cell culture and metaphase cell preparations. Recently, a novel molecular cytogenetic technique, termed comparative genomic hybridization (CGH), has been developed (36). It uses a reverse hybridization strategy in which the genetic material, such as tumor tissue DNA, is labeled as a probe and hybridized to normal chromosome preparations. This technique permits the rapid screening of chromosomal imbalances (gains or losses) within the test genome without the need for metaphase preparations from these samples. Only DNA is needed. A reference or control DNA isolated from an individual with a normal karyotype and a test DNA (e.g., from a tumor) are differentially labeled with two different reporter molecules (e.g., biotin-dUTP for the tumor genome and digoxigenin-dUTP for the control genome). The two DNA probes are then hybridized simultaneously in the presence of excess Cot-1 DNA to a metaphase preparation of normal male chromosomes. Subsequently, the probes are detected with two different fluorochromes (e.g., FITC or green fluorescence for the tumor and rhodamine or red fluorescence for the control). The ratio of the intensity of each fluorochrome is then measured along each chromosome and compared with the normal control, usually through the use of a specialized digital image analysis system with appropriate software. For each chromosome, an average ratio profile is generated using fluorescence ratio data from 8–10 different metaphase spreads. Chromosomal regions that show increased or decreased fluorescence intensity of the test genome relative to the normal control are then demonstrated to reflect gains or losses of these regions, respectively. Thus, a ratio of 1:1 reflects equal copy numbers in the test and control, whereas a 3:2 ratio reflects trisomy and a 1:2 ratio reflects a monosomy. Regions of gene amplifications such as double minutes or homogeneously staining regions (HSR) can be readily detected. Color Plate 6.13 presents the steps of a comparative genomic hybridization experiment.

Comparative genome hybridization (CGH) has been instrumental in elucidating chromosomal regions of interest in a number of solid tumors, in which little data were previously available (36,37). Color Plate 6.14 demonstrates the use of CGH for the evaluation of a breast tumor. Color Plate 6.14a shows a representative ratio image and Color Plate 6.14b shows a ratio profile. The fact that actual chromosome preparations from the study tissue are not needed is a major advantage of CGH in tumor studies. This overcomes the technical difficulty of preparing chromosomes from tumor samples and allows the characterization of complex chromosomal rearrangements often present in tumors. In addition, retrospective studies on archived samples can be conducted using DNA extracted from formalin-fixed material.

B. MicroFISH

MicroFISH, or FISH following microdissection of probe DNA, is another form of reverse in situ hybridization. Microdissection of DNA from cytogenetic preparations can be performed in order to obtain DNA to produce a FISH probe to a whole chromosome, an arm of a chromosome, a specific band, or marker. In MicroFISH studies, the DNA of interest is scraped from a number of metaphase spreads with a glass needle from a micromanipulator while visualizing the process with microscopy (38). The scraped DNA is then amplified to sufficient quantity with universal polymerase chain reaction (PCR) techniques using a degenerate oligonucleotide primer system. The DNA probe can then be labeled and used as a probe in a FISH experiment. If the microdissected probe is derived from an unknown region (e.g., an HSR, double minute, or other marker), then the probe will bind to its chromosome region of origin when hybridized to a normal male metaphase chromosome spread. For example, if the DNA from a microdissected double minute is the *N-myc* gene, then the MicroFISH probe will hybridize to chromosome 2 in the p24 band, the region of the genome where that gene maps. Counterstaining of the target metaphase with DAPI provides a banding pattern of cohybridization with a known FISH probe and can assist in defining the chromosomal region of hybridization of the microdissected probe.

MicroFISH with PCR amplification has been shown to be an effective method for FISH probe production in general. In the area of identification of genome localization of amplicons (e.g., double minutes, HSRs, and markers), this technique has been shown to be powerful (39,40). To compare CGH with MicroFISH, one should recall that MicroFISH requires metaphase spreads whereas CGH does not. However, MicroFISH has a higher sensitivity around areas of the genome with a large number of repeats, such as around the centromere.

C. PRINS

The primed in situ labeling (PRINS) technique is a method which allows for the direct detection of the chromosome, based on annealing of specific oligonucleotide primers and primer extension by Taq polymerase (41). This method is a derivative of an in situ PCR experiment in which the chromosomal sequences of interest from a sample preparation on a glass microscope slide are amplified by a PCR reaction. The reaction incorporates fluorochrome-labeled oligonucleotides so that the signal can be viewed directly with fluorescence microscopy.

PRINS experiments can be completed rapidly and have been used most effectively for alpha-satellite sequences in both interphase and metaphase cells. It is possible with PRINS to distinguish between the alpha-satellite sequences on chromosomes 13 and 21 because of the specificity of the primers used in the PCR reaction, while standard FISH with alpha-satellite probes does not. In fact, large target probes on the q arms of chromosomes 13 and 21 are recommended for these aneuploidies, and both require overnight hybridizations (42). PRINS is a relatively simple, inexpensive technique compared to the other two techniques presented here. It is similar to FISH in that the probe or primer of interest must be identified by the investigator. Only one sequence can be amplified or probed at one time because of the constraints of the PCR reaction, but sequential reactions of the same specimen are possible.

D. FICTION

An approach that combines the use of FISH and cell surface markers to characterize cells is FICTION, or fluorescence immunophenotyping and interphase cytogenetics, as a tool for the investigation of neoplasms (43). Characterization of cytogenetically defined tumor cells within cytospin preparations and frozen sections can answer specific questions about the origins of genetic changes. Such immunophenotyping has also been performed on paraffin-embedded tissues that have been digested with trypsin or lymphoma tissue after preparing the tissue with a microwave pretreatment in citrate buffer and subsequent mild trypsin digestion (44). As discussed previously, standard-tissue FISH treatment includes proteinase digestion steps. These pretreatments may degrade the antigenic determinants recognized by an antibody.

IX. SUMMARY

As the field of molecular cytogenetics continues to evolve, clinical and research laboratories are adopting the useful techniques described here. Data derived from the use of FISH-based studies have been useful in increasing the sensitivity of chromosomal analysis. The strategy of combining these methods to resolve interesting questions is successful and rewarding to the classical cytogeneticist. The applications of FISH for the clinical cytogenetics laboratory are explored further in Chapter 17.

ACKNOWLEDGMENTS

Many thanks to Susie Airhart, Mary Williams, and Roman Giraldez for their many contributions to this work.

REFERENCES

1. Pinkel D, Gray J, Trask B, van den Engh G, Fuscoe J, van Dekken. Cytogenetic analysis by in situ hybridization with fluorescently labeled nucleic acid probes. Cold Spring Harbor Symp Quant Biol 1986; 51:151–157.
2. Gall J, Pardue ML. Molecular hybridization of radioactive DNA to the DNA of cytological preparations. Proc Natl Acad Sci USA 1969; 63:378.
3. Angerer L, Angerer R. In situ hybridization to cellular RNA with radiolabeled probes. In: Wilkenson DG, ed. In situ Hybridization: A Practical Approach. Oxford, UK: IRL Press. 1993:15–32.
4. Hockfield S, Carlson S, Evans C, Levitt P, Silberstein L. Selected Methods for Antibody and Nucleic Acid Probes. Cold Spring Harbor, NY: Cold Spring Harbor Press, 1993.
5. Nuovo GJ. PCR in Situ Hybridization. New York: Raven Press, 1994.
6. Isaacson S, Asher D, Gajdusek C, Gibbs C. Detection of RNA viruses in archival brain tissue by in situ RT-PCR amplification and labeled probe hybridization. Cell Vision 1994; 1:25–28.
7. Jabs EW, Wolf SF, Migeon BR. Characterization of a cloned DNA sequence that is present at centromeres of all human autosomes and the X chromosome

and shows polymorphic variation. Proc Natl Acad Sci USA 1984; 81:4882–4888.
8. Waye JS, Willard HF. Chromosome specific alpha satellite DNA: nucleotide sequence analysis of the 2.0 kilobase repeat from the human X chromosome. Nucleic Acids Res 1985; 12:2731–2734.
9. Aleixandre C, Miller D, Mitchell A, et al. p82H identifies sequences at every human centromere. Hum Genet 1987; 77:46–50.
10. Waye JS, Willard H. Human beta satellite DNA: genomic organization and sequence definition of a class of highly repetitive tandem DNA. Proc Natl Acad Sci USA 1989; 86:6250–6254.
11. Nakahori Y, Mitani K, Yamada M, Nakagome Y. A human Y chromosome specific repeated DNA family (DYZ1) consists of a tandem array of pentanucleotides. Nucleic Acids Res 1986; 14(19):7569–7580.
12. Moyzis RK. The human telomere. Sci Am 1991; 265:48–55.
13. Cremer T, Lichter P, Borden J, Ward DC, Manuelidis L. Detection of chromosome aberrations in metaphase and interphase tumor cells by in situ hybridization using chromosome-specific library probes. Hum Genet 1988; 80: 235–246.
14. Lichter P, Ledbetter SA, Ledbetter DH, Ward DC. Fluorescence in situ hybridization with ALU and L1 polymerase chain reaction probes for rapid characterization of human chromosomes in hybrid cell lines. Proc Natl Acad Sci USA 1990; 85:9138–9142.
15. Guan XY, Meltzer P, Trent J. Rapid generation of whole chromosome painting probes (WCPs) by chromosome microdissection. Genomics 1994; 22:101–107.
16. National Institutes of Health and Institute of Molecular Medicine Collaboration. A complete set of human telomeric probes and their clinical application. Nat Genet 1996; 14:86–90.
17. Mundy CR, Cunningham MW, Read CA. Nucleic Acid Labeling and Detection in Essentials of Molecular Biology, A Practical Approach, Vol 11, 1991:57–82.
18. Cajulis RS, Frias-Hidvegi D, Yu GH, Eggena S. Detection of numerical chromosomal abnormalities by fluorescence in situ hybridization of interphase cell nuclei with chromosome-specific probes on archival cytologic samples. Diagn Cytopathol 1996; 14(2):178–181.
19. Cajulis RS, Haines GK, Frias-Hidvegi D, McVary K, Bacus J. Cytology, flow cytometry, image analysis, and interphase cytogenetics, by fluorescence in situ hybridization in the diagnosis of transitional cell carcinoma in bladder washes: a comparative study. Diagn Cytopathol 1995; 13(3):214–223.
20. American College of Medical Genetics. Standard and Guidelines: Clinical Genetics Laboratories. American College of Medical Genetics, 1996; Metaphase fluorescence in situ hybridization, p: 23–1; Interphase, p: 23–6.
21. Moore DH, Epstein L, Reeder J, Wheeles L, Waldman F. Interlaboratory variability in fluorescence in situ hybridization analysis. NCI Bladder Tumor Marker Network, Cytometry 1996; 25(2):125–132.
22. Schad C, Dewald G. Building a new clinical test for in situ hybridization. Appl Cytogenet 1995; 21(1):1–4.

23. Mitelman F, ed. ISCN 1995, An International System for Human Cytogenetic Nomenclature. Basel: Karger, 1995: 94–104.
24. Slamon D, Clark G, Wong S, Levin W, Ullrich A, McGuire W. Human breast cancer: correlation of relapse and survival with amplification of HER-2/neu oncogene. Science 1987; 244:707–712.
25. Tchuk DC, Westbrook CA, Andreef M, et al. Detection of bcr-abl fusion in chronic myelogenous leukemia by in situ hybridization. Science 1990; 250(4980):559–562.
26. Dewald G, Schad C, Cristensen, et al. The application of fluorescence in situ hybridization to detect Mbcr/abl fusion in variant Ph chromosomes in CML and ALL. Cancer Genet Cytogenet 1995; 71:7–14.
27. Schad C, Hanson C, Pairatta E, Casper J, Jalal S, Dewald G. Efficacy of fluorescence in situ hybridization for detecting PML/RARA gene fusion in treated and untreated acute promyelocytic leukemia. Mayo Clin Proc 1994; 69: 1047–1053.
28. Martinez-Climent JA, Thirman MJ, Espinosa R 3rd, LeBeau MM, Rowley JD. Detection of 11q23/MLL rearrangements in infant leukemias with fluorescence in situ hybridization and molecular analysis. Leukemia (England) 1995; 9(8): 1299–1304.
29. Wolman S. Application of fluorescent in situ hybridization (FISH) to genetic analysis of human tumors. In Rosen, Fechner, eds. Pathology Annuals. Stamford CT: Appleton and Lange, 1995; 227–243.
30. Jenkins RB, Qian J, Lieber MM, Bostwick DG. Detection of c-myc oncogene amplification and chromosomal anomalies in metastatic prostatic carcinoma by fluorescent in situ hybridization. Cancer Res 1997; 57:524–531.
31. Press MF, Bernstein L, Thomas P, et al. HER2 neu gen amplification characterized by fluorescence in situ hybridization: poor prognosis in node-negative breast carcinoma. J Clin Oncol 1997; 15(8):2894–2904.
32. Afify A, Bland KL, Mark HF. Fluorescence in situ hybridization assessment of chromosome 8 copy number in breast cancer. Breast Canc Res Treatment 1996; 38(2):201–208.
33. Kalousek DK, Vekemas M. Confined placental mosaicism. J Med Genet 1996; 33(7):529–533.
34. Lomax BL, Kalousek DK, Kuchinka BD, Barrett IJ, Harrison KJ, Safavi H. The utilization of interphase cytogenetic analysis for the detection of mosaicism. Hum Genet 1994; 93(3):243–247.
35. Lennow E, Anneren G, Bui T-H, Berggren E, Asadi E, Nordenskjold M. Characterization of supernumerary ring marker chromosomes by fluorescence in situ hybridization. Am J Hum Genet 1993; 53:433–442.
36. Kallioniemi A, Kallioniemi OP, Suder D, et al. Comparative genome hybridization for molecular cytogenetic analysis of solid tumors. Science 1992; 258: 818–821.
37. Kallioniemi A, Kallioniemi OP, Piper J, et al. Detection and mapping of amplified DNA sequences in breast cancer by comparative genome hybridization. Proc Natl Acad Sci USA 1994; 91(6):2156–2160.

38. Su Y, Trent J, Guan XY, Meltzer P. Direct isolation of genes encoded within a homogenously staining region by chromosome microdissection. Proc Natl Acad Sci USA 1994; 91:9121–9125.
39. Muller-Navia J, Nebel A, Schleiermacher E. Complete and precise characterization of marker chromosomes by application of microdissection in prenatal diagnosis. Hum Genet 1995; 96(6):661–667.
40. Guan XY, Meltzer P, Dalton W, Trent J. Identification of cryptic sites of DNA sequence amplification in human breast cancer by chromosome microdissection. Nat Genet 1994; 8:155–160.
41. Koch J, Hindkjaer J, Mogensen J, Kolvaa S, Bolound L. An improved method for chromosome specific labeling of alpha satellite DNA in situ using denatured double-stranded DNA probes as primers in a primed in situ labeling (PRINS) procedure. GATA 1991; 8(6):171–178.
42. Pellestor F, Girardet A, Andreo B, Charlieu J. A polymorphic alpha satellite sequence specific for human chromosome 13 detected by oligonucleotide primed in situ labeling (PRINS). Hum Genet 1994; 94:346–3348.
43. Weber-Matthiesen K, Pressi P, Schlegelberger B, Grote W. Combined immunophenotyping and interphase cytogenetics on cryostat sections by the new FICTION method, 7:4:646–649.
44. Nolte M, Werner M, vonWasielewski R, Nietgen G, Wildens L, Georgii A. Detection of numerical karyotype changes in the giant cells of Hodgkin's lymphomas by a combination of FISH and immunohistochemistry applied to paraffin section. Histochem Cell Biol 1996; 105:401–404.

APPENDIX: GLOSSARY OF TERMS

alpha-satellite DNA
Class of repetitive DNAs located at the chromosomal centromere.

aneuploidy
In human genetics, refers to the loss or gain of whole chromosome(s).

chimerism
An individual composed of cells derived from different zygotes; in human genetics, especially with reference to blood group chimerism, in which dizygotic twins exchange hematopoietic stem cells in utero and continue to form blood cells of both types (distinguished from mosaicism, in which the two genetically different cell lines arise after fertilization).

chorionic villus sampling (CVS)
Procedure used for prenatal diagnosis at 8–10 weeks' gestation. Fetal tissue for analysis is withdrawn from the villus areas of the chorion, either transcervically or transabdominally, under ultrasonographic guidance.

classical satellite I DNA
Class of repetitive DNAs located on chromosomes 1, 9, 16, and the Y chromosomes.

confined placental mosaicism (CPM)
Tissue-specific chromosomal mosaicism.

degenerate oligonucleotide primer system
A PCR technique using degenerated oligonucleotides as primers, thus allowing for the sequence-independent, universal amplification of genomic DNA. Its main application is degeneration of significant probe quantities from limited amount of original template.

digoxigenin
Clinical derived from digitalis plant used for indirect probe labeling affecting placental tissues in a localized manner.

direct methods
Procedure in which a DNA probe is labeled with a detection molecule such as a fluorophore.

double minutes
Extrachromosomal DNA of amplification genes seen in some tumor cells and drug-resistant cells.

extra structurally abnormal chromosomes (ESACs)
Marker chromosome of unknown origin.

flow sorting
Analysis of biological material by detection of the light-absorbing or fluorescing properties of cells or subcellular fractions (i.e., chromosomes) passing in a narrow stream through a laser beam. An absorbance or fluorescence profile of the sample is produced. Automated sorting devices, used to fractionate samples, sort successive droplets of the analyzed stream into different fractions depending on the fluorescence emitted by each droplet.

fluorochromes
Various fluorescent dyes used in biological staining used to produce fluorescence in samples.

homogeneously staining regions (HSR)
Medial G-band-positive gene amplification incorporated into chromosomes from tumor cells and drug-resistance cells.

hybridization
Process of joining two complementary strands of DNA or one each of DNA and RNA to form a double-stranded molecule.

indirect labeling methods
Procedure in which a DNA probe labeled with a hapten such as biotin or digoxigenin. It is necessary to use a detection reagent to detect hybridization when a probe is labeled this way.

microdeletion
Chromosomal deletion too small to be seen using a microscope such as in standard cytogenetic preparations.

microFISH
Reverse in situ hybridization using a microdissection DNA segment versus a probe.

PCR amplification
Method for amplifying a DNA base sequence using a heat-stable polymerase and two 20-base primers, one complementary to the (+) strand at one end of the sequence to be amplified and the other complementary to the (−) strand at the other end. Because the newly synthesized DNA strands can subsequently serve as additional templates for the same primer sequences, successive rounds of primer annealing, strand elongation, and dissociation produce rapid and highly specific amplification of the desired sequence. PCR can also be used to detect the existence of the defined sequence in a DNA sample.

telomeric repeat
Repeated DNA at ends of eukaryotic chromosomes.

unique sequence probes
Probes for specific genes or loci found twice in a diploid cell.

whole chromosome probes
Collection of unique sequence probes covering an entire chromosome.

7
Cytogenetics in Medicine

Bruce R. Korf
Partners Center for Human Genetics, Harvard Medical School and Partners HealthCare, Boston, Massachusetts
Nancy R. Schneider
University of Texas Southwestern Medical Center, Dallas, Texas

I. INTRODUCTION

Techniques for analysis of chromosome number and structure in humans were developed in the same decade as the elucidation of the structure of DNA. Almost immediately upon the development of routine methods for laboratory cytogenetic analysis, a set of syndromes was identified as being due to chromosomal abnormalities. As the technology has gradually improved, there has been a corresponding refinement in the ability to detect increasingly subtle changes in chromosome structure that are clinically important. Most recently, cytogenetic methodology has merged with DNA technology through fluorescent in situ hybridization (FISH) to permit detection of chromosomal deletions too small to be visualized through the conventional light microscope.

Chromosomal analysis is now integrated into the routine practice of medicine. The identification of a chromosomal abnormality may provide a diagnosis, prognostic information, and knowledge of genetic recurrence risks for a family. This chapter will review the major medical indications for performing chromosomal analysis using standard staining, FISH, or special methods such as sister chromatid exchange testing. It will also explore some of the major counseling issues that arise when specific chromosomal abnormalities are found. The focus will be on constitutional chromosomal changes identified after birth. Topics of prenatal diagnosis and somatic genetic change in cancer are covered elsewhere in this volume.

II. APPROACHES AND INDICATIONS FOR CYTOGENETIC ANALYSIS

Cytogenetic analysis is most effectively performed if the cytogeneticist is aware of the indications for testing and can customize the study to the clinical question. Although a standard level of scrutiny is expected for all routine studies, in many instances special studies or focus on specific chromosomal regions may be clinically appropriate. Cytogeneticists are expected to be familiar with clinical disorders due to chromosomal abnormalities, and must demonstrate this knowledge in the qualifying examination administered by the American Board of Medical Genetics. This section will consider the common indications for routine analysis as well as other means of analysis, including microdeletion analysis by FISH and other special studies.

A. Routine Chromosomal Analysis

"Routine" chromosomal analysis consists of counting chromosomes in 20 cells and karyotyping two cells (1). These karyotypes are banded at the 400 band level or better. Most laboratories use G-banding for routine analysis, although Q- or R-banding may be used in some laboratories. The most common indications for postnatal analysis are the occurrence of a recognized chromosomal syndrome, the occurrence of multiple congenital anomalies or developmental problems suggestive of a syndrome, a history of recurrent miscarriage, or a family history suggestive of a possible chromosomal abnormality.

B. Recognized Chromosomal Syndromes

The most common disorders due to chromosomal abnormalities were recognized in the early days of cytogenetic analysis. Most are due to numerical abnormalities and result in characteristic syndromes that can be identified clinically. Some structural chromosomal abnormalities give rise to sufficiently characteristic phenotypes to permit clinical identification, but most produce more variable changes in gene dosage for small chromosome regions and are more difficult to recognize phenotypically. Overall it is estimated that the incidence of chromosomal abnormalities in live births is about 0.5% (2,3). This includes both aneuploidies and structural rearrangements, but is based largely on data collected before high-resolution molecular cytogenetic studies were possible. The actual rate is therefore likely to be higher. The incidence of chromosome abnormalities among spontaneous abortions is also substantially higher (4,5).

Complete monosomy for all autosomes and most complete trisomy syndromes are lethal in the early embryonic period, resulting in miscarriage. The only viable monosomy is monosomy X, and the only viable complete (i.e., nonmosaic) trisomies involve chromosomes 13, 18, 21, X, and Y. Mosaic trisomies for chromosomes 8, 9, and 20 may also be viable, and there are rare examples of low-level mosaicism for other trisomies. Major clinical features for the common aneuploidy syndromes are summarized in Table 1.

Down syndrome was the first human disorder recognized to be due to trisomy; the chromosome involved was considered to be number 21 (although subsequent studies have shown it to be the smallest autosome and thus it should have been identified as number 22). Phenotypic features include hypotonia at birth, characteristic facies (Fig. 1a) (brachycephaly, flat facial profile, upslanted palpebral fissures, epicanthal folds, small ears), excessive nuchal skin, brachydactyly, and hypotonia. There is delay in motor and intellectual development, and there may be congenital anomalies of the heart and gastrointestinal system. Children with Down syndrome are sus-

Table 1 Major Chromosome Aneuploidy Syndromes Compatible with Live Birth

Syndrome	Major features	Chromosomal basis
Down syndrome	Characteristic facies, hypotonia, developmental delay, heart defects	Trisomy 21 (may be translocation or mosaic)
Patau syndrome	CNS malformation (especially holoprosencephaly), congenital anomalies	Trisomy 13 (may be translocation or mosaic)
Edwards syndrome	CNS anomalies, heart defects, congenital anomalies	Trisomy 18 (may be mosaic)
Trisomy 8	Dysmorphic features, developmental delay	Mosaic trisomy 8
Turner syndrome	Gonadal dysgenesis, short stature	45,X
Klinefelter syndrome	Infertility, tall stature, learning disabilities	XXY
Triple X	Tall stature, learning disabilities	XXX
XYY	Tall stature, learning disabilities	XYY

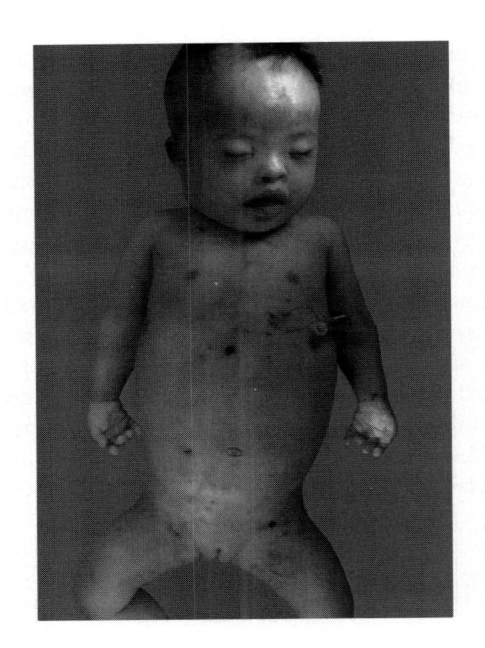

(a)

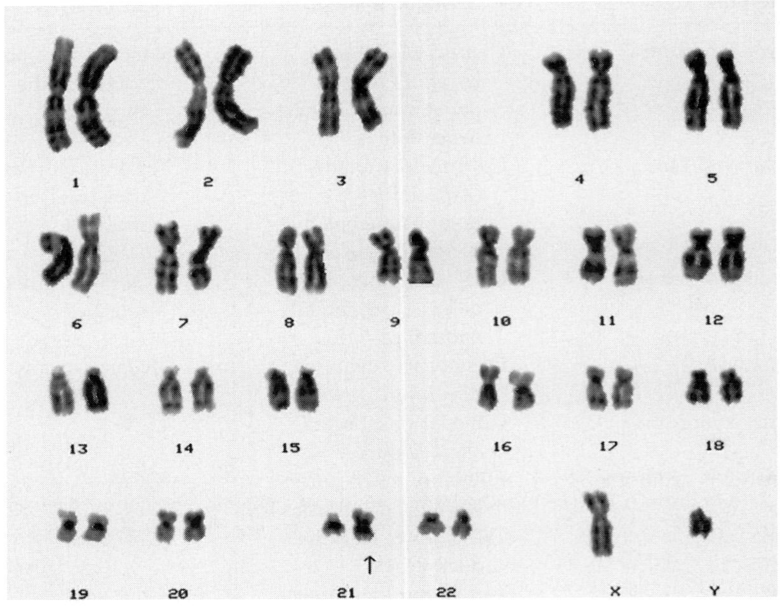

(b)

ceptible to respiratory infections and have an increased risk of leukemia (6). Adults tend to develop an Alzheimer-like dementia with characteristic Alzheimer neuropathology at an early age (third decade or later) (7). Free trisomy 21 accounts for approximately 95% of instances of Down syndrome (8). This is assumed to arise by nondisjunction, most often in the first meiotic division in the mother (9,10). Mosaic trisomy 21 occurs in approximately 1% of individuals with Down syndrome. Some may have milder features, but this cannot be counted on and the proportion of trisomic cells can vary in different tissues. The remaining 4% have trisomy 21 due to translocation, usually between chromosomes 21 and 14, but sometimes involving 21 and another acrocentric chromosome, including 21–21 translocations (Fig. 1b).

The other two viable autosomal trisomies, trisomy 13 (Fig. 2) and trisomy 18 (Fig. 3), are both lethal in the newborn period in most instances (11,12). Children with either disorder tend to have multiple congenital anomalies, including life-threatening cardiac malformations. The relatively few survivors invariably have severe cognitive impairment.

Monosomy X results in Turner syndrome (Fig. 4), characterized by a female phenotype with short stature, ovarian failure, and lack of secondary sex characteristics (13). Most fetuses with monosomy X miscarry; some girls with monosomy X have edema at birth, and there is an increased incidence of aortic outflow anomalies or coarctation of the aorta as well as renal anomalies. A high proportion of girls with Turner syndrome have mosaicism for a structurally rearranged X or Y chromosome. Those with a rearranged Y may have some degree of masculinization ("mixed gonadal dysgenesis"), and the gonads may be at risk for developing malignant tumors (gonadoblastomas) (14).

The XXY karyotype results in Klinefelter syndrome, wherein males have underdeveloped testes and infertility, and may have tall stature, hyperextensible joints, and may have learning disabilities. Females with the XXX karyotype may have learning disabilities but are fertile (15). Either males or females with additional X chromosomes tend to have more severe developmental impairment with increasing numbers of X chromosomes. Males

←

Figure 1 (a) Postmortem photograph of a 7-month-old infant with trisomy 21 (Down syndrome). Note characteristic features of upslanted palpebral fissures, epicanthal folds, flat nasal bridge, open mouth secondary to large tongue, abnormal ear, single transverse palmar crease, and midthoracic healing surgical incision secondary to attempted repair of congenital heart defect. (Courtesy of Dr. Charles F. Timmons, University of Texas Southwestern Medical Center at Dallas.) (b) Karyotype demonstrating trisomy 21 due to an isochromosome for the long arm of chromosome 21.

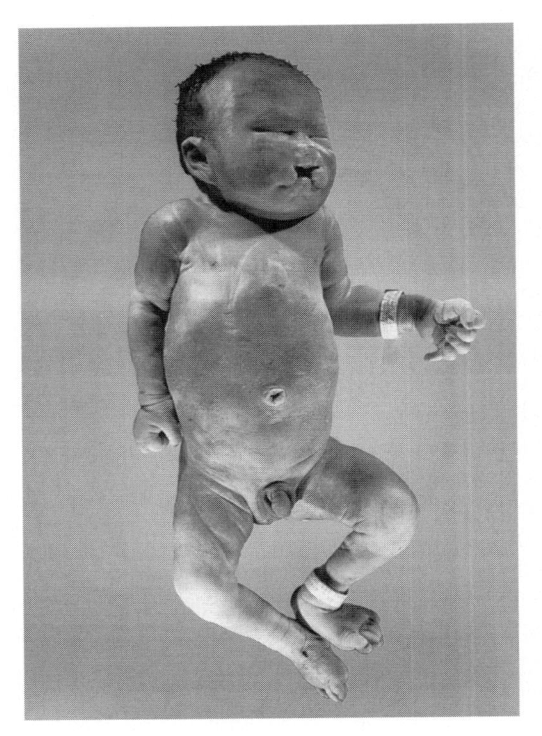

Figure 2 Postmortem photograph of a term neonate with trisomy 13. Note characteristic features of microcephaly (with holoprosencephaly), bilateral cleft lip (and palate), malformed nose and ear, and polydactyly. The cause of death was complex congenital heart defects. (Courtesy of Dr. Charles F. Timmons, University of Texas Southwestern Medical Center at Dallas.)

with an XYY karyotype tend to have tall stature and may have learning disabilities and behavioral problems, but generally are fertile and do not have specific health problems (16).

A wide variety of types of chromosome abnormalities may lead to partial deletion or duplication of chromosomal material (Fig. 5). Isochromosomes consist of a duplication of either long arms or short arms, forming a symmetrical chromosome. This is usually attributed to "misdivision" of the centromere. Duplications or deletions may also occur for parts of chromosome arms. Syndromes resulting from partial deletion or duplication of segments of chromosomes are more variable, depending in part on the size of the abnormal chromosome segment. Some of the more common syn-

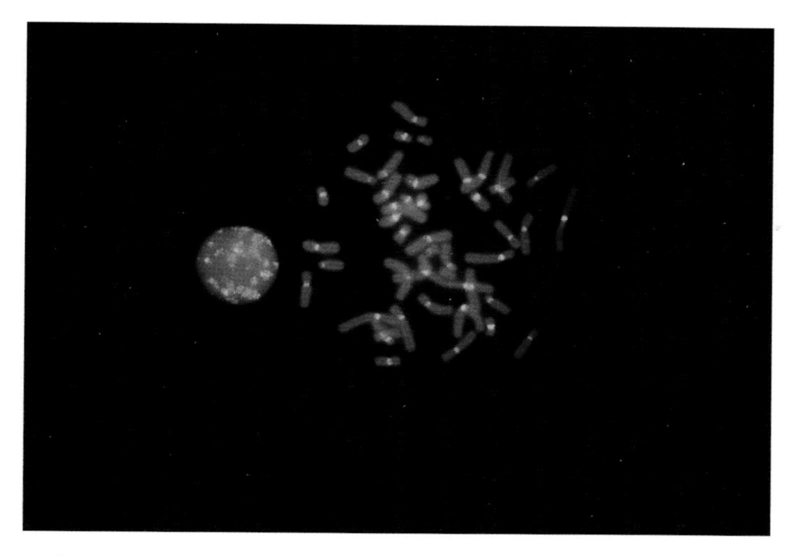

Figure 6.2 A human lymphocyte metaphase chromosome preparation shows hybridization of the common centromeric sequences at all human centromeres with a fluorescein-labeled probe. The chromosomes are counterstained with propidium iodide.

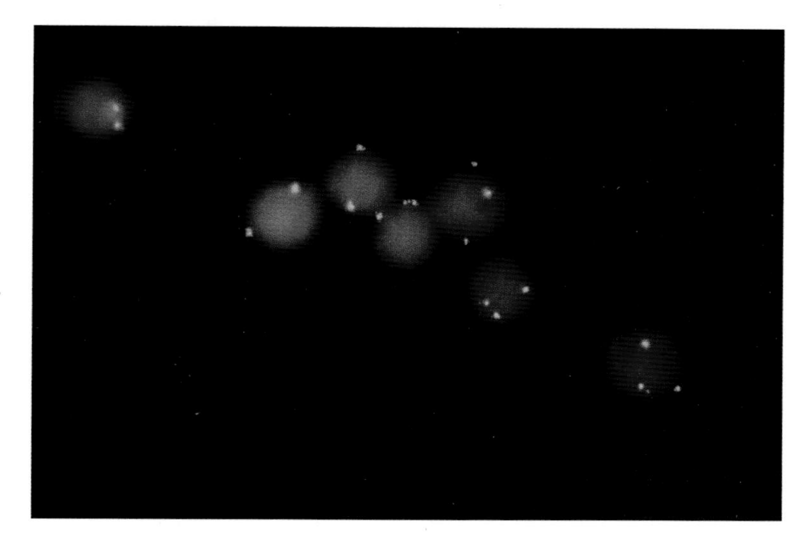

Figure 6.3 A bone marrow smear preparation shows hybridization of the α-satellite sequences specific for chromosome number 8. The chromosomes are counterstained with propidium iodide.

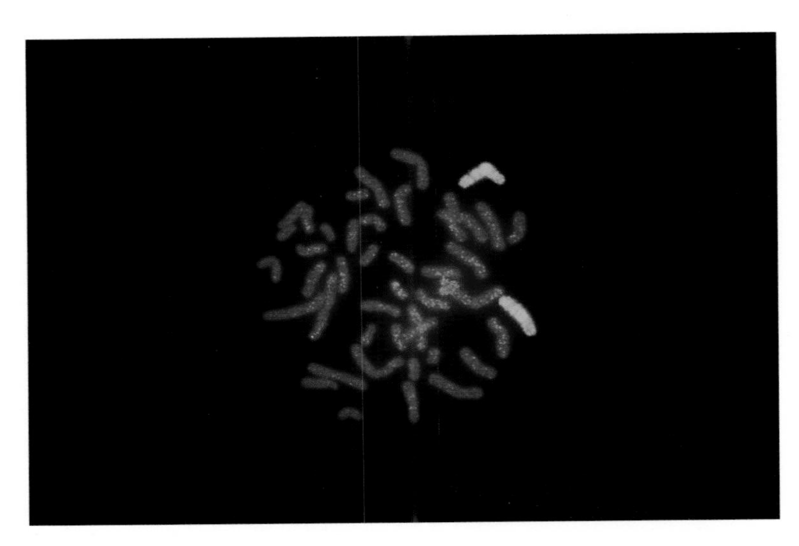

Figure 6.4 A human lymphocyte metaphase chromosome preparation shows hybridization of whole chromosome probes to chromosome 9. The cells are counterstained with propidium iodide.

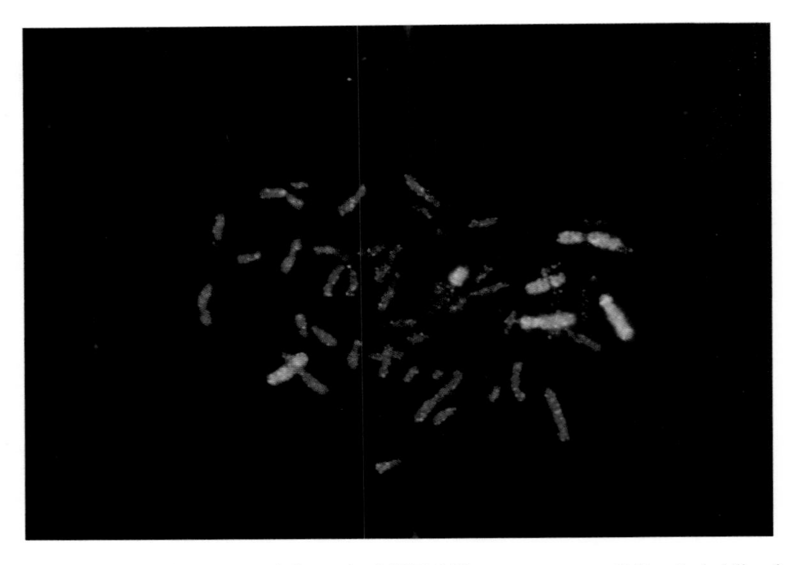

Figure 6.5 A metaphase spread from the MCF-10 breast cancer cell line hybridized with whole chromosome probes to chromosome 3 in red and 9 in green illustrating chromosomal rearrangements between these chromosomes. The spread is counterstained with DAPI.

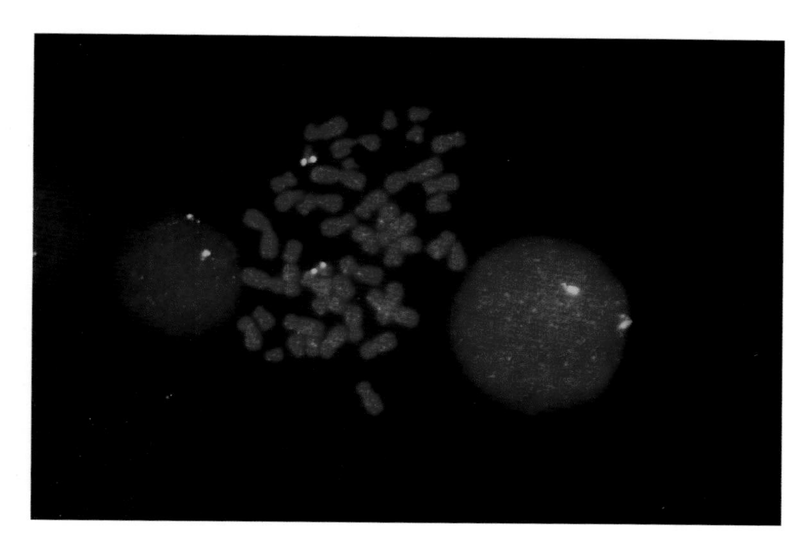

Figure 6.6 A metaphase spread and interphase cell that have been hybridized with a unique sequence probe to a region on the q arm of chromosome 21.

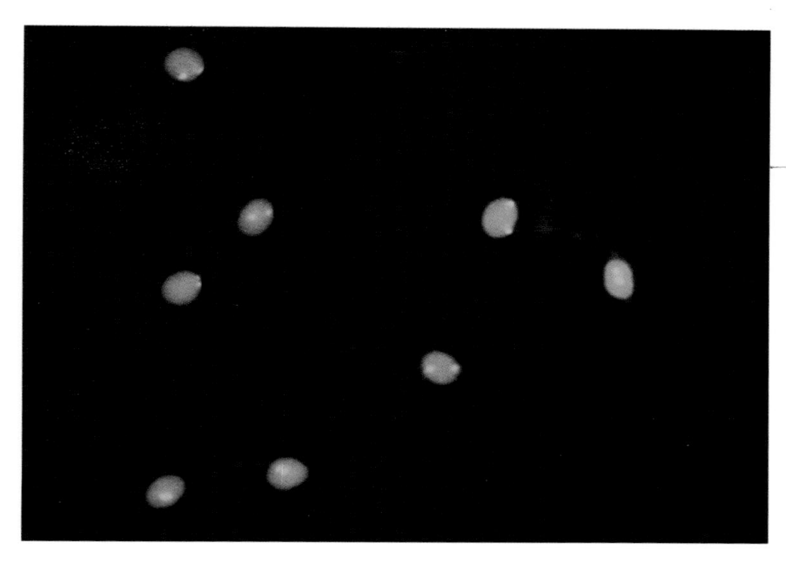

Figure 6.8 Cells obtained from a bladder wash that have been cytocentrifuged onto a slide. The cells have been hybridized with a fluorescein-labeled α-satellite probe to the chromosome 8 centromere. The cells are counterstained with propidium iodide.

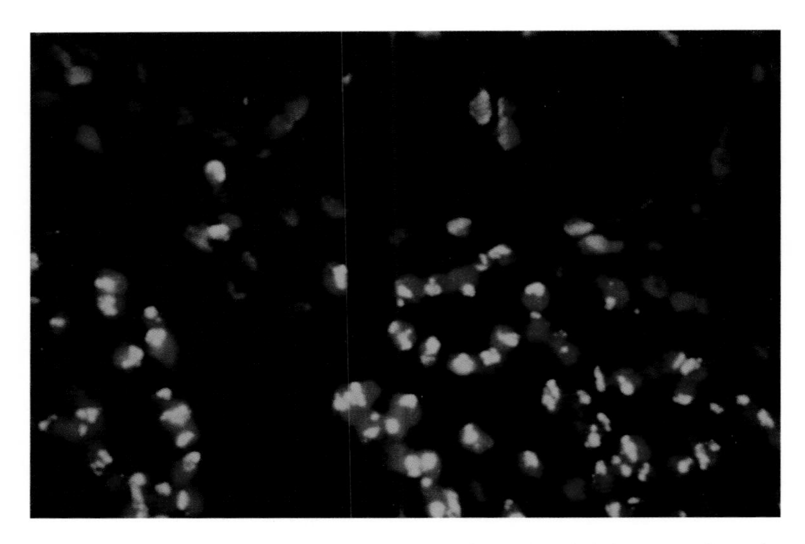

Figure 6.9 The sample is a formalin-fixed paraffin-embedded tissue section of a neuroblastoma hybridized with an *N-myc* probe labeled with fluorescein. Note the intensity and quantity of yellow staining in tumor cells are counterstained with propidium iodide.

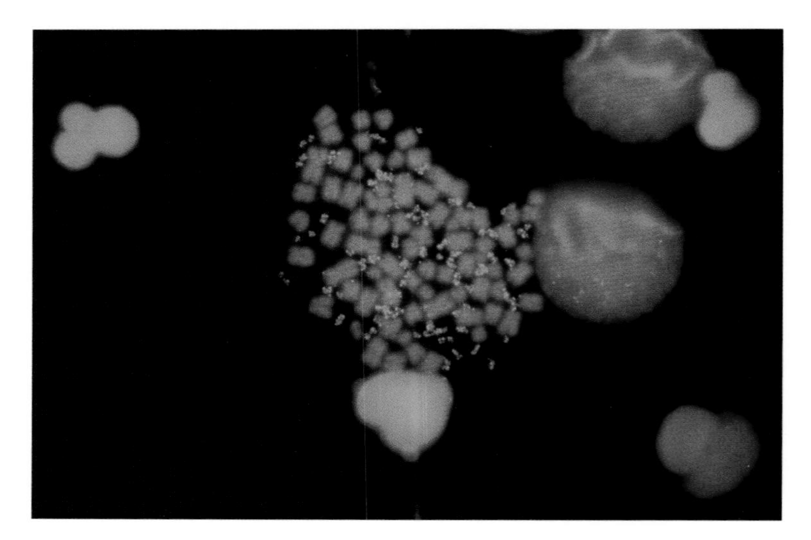

Figure 6.10 The bright yellow staining of extrachromosomal double minute amplifications with the *c-myc* probe in a breast tumor cell line. The cell is counterstained with propidium iodide.

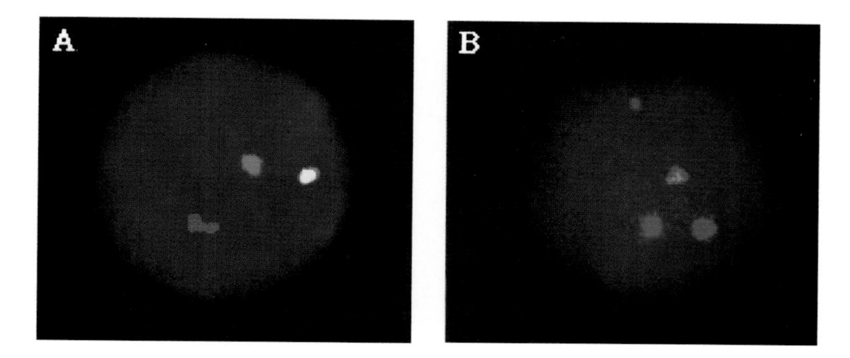

Figure 6.11 (A) Interphase fluorescent in situ hybridization showing an abnormal cell with three fluorescent signals representing one normal copy each of the *brc* and *abl* genes, and a fusion *bcr/abl* gene. **(B)** Interphase fluorescent in situ hybridization showing a normal cell with four fluorescent signals representing two normal copies each of the *brc* and *abl* genes.

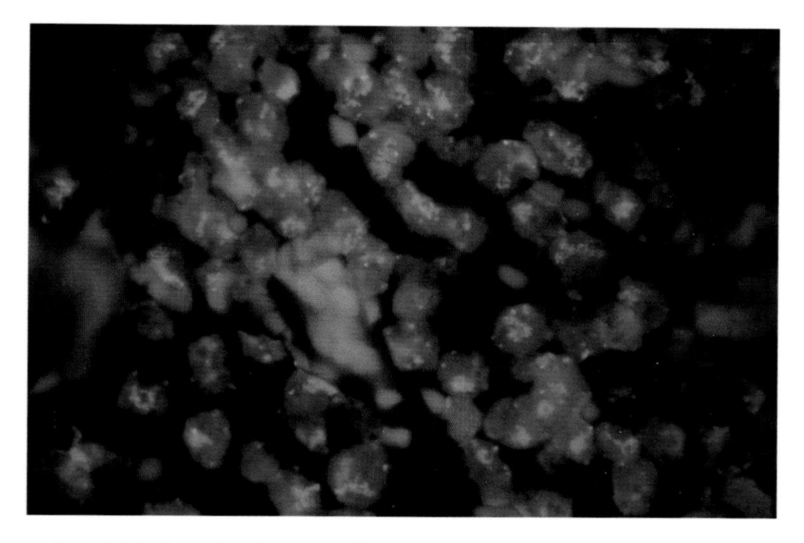

Figure 6.12 This formalin-fixed paraffin-embedded breast tumor is hybridized with a probe to the *HER-2/neu* gene, which is labeled with fluorescein and appears greenish in this photo. The amplifications of the gene are pictured in contrast to the two Texas Red labeled α-satellite probes to chromosome 17. The cells are counterstained with DAPI.

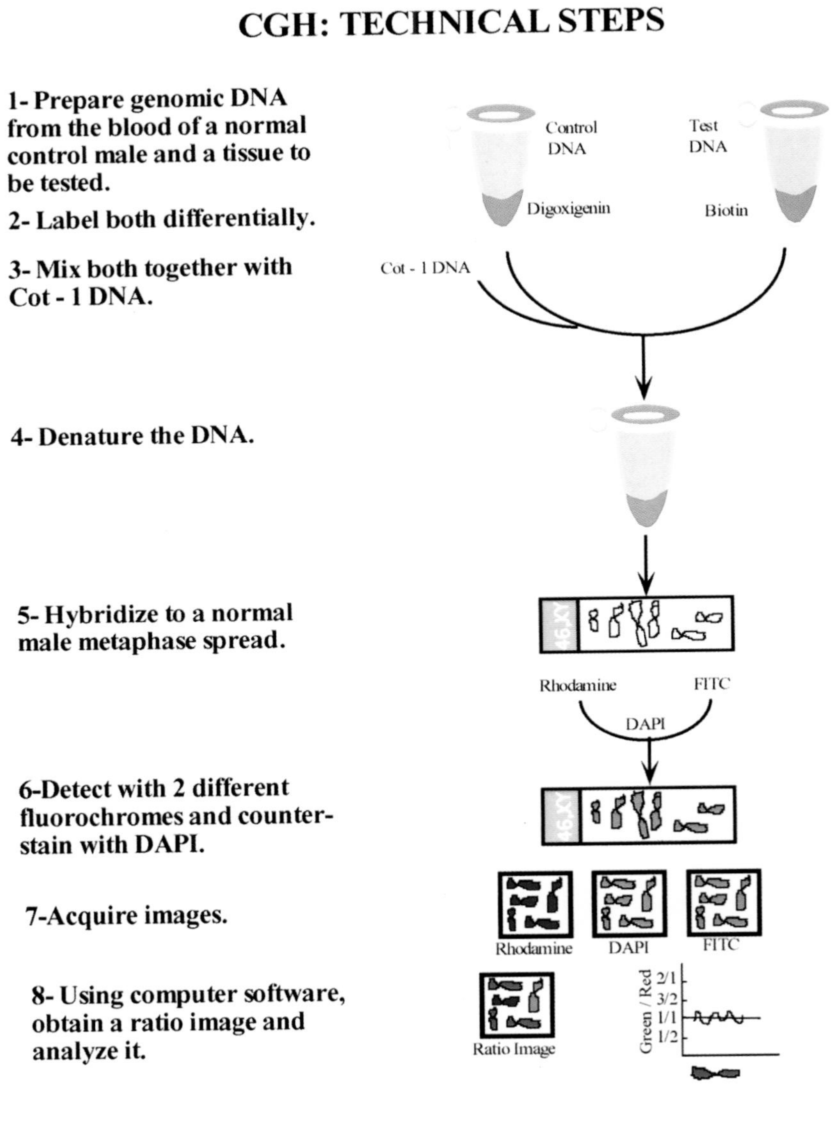

CGH: TECHNICAL STEPS

1- Prepare genomic DNA from the blood of a normal control male and a tissue to be tested.

2- Label both differentially.

3- Mix both together with Cot - 1 DNA.

4- Denature the DNA.

5- Hybridize to a normal male metaphase spread.

6-Detect with 2 different fluorochromes and counter-stain with DAPI.

7-Acquire images.

8- Using computer software, obtain a ratio image and analyze it.

Control DNA
Test DNA
Digoxigenin
Biotin
Cot - 1 DNA
46,XY
Rhodamine
FITC
DAPI
46,XY
Rhodamine
DAPI
FITC
Ratio Image
Green / Red
2/1
3/2
1/1
1/2

Figure 6.13 The steps of a comparative genomic hybridization (CGH) experiment.

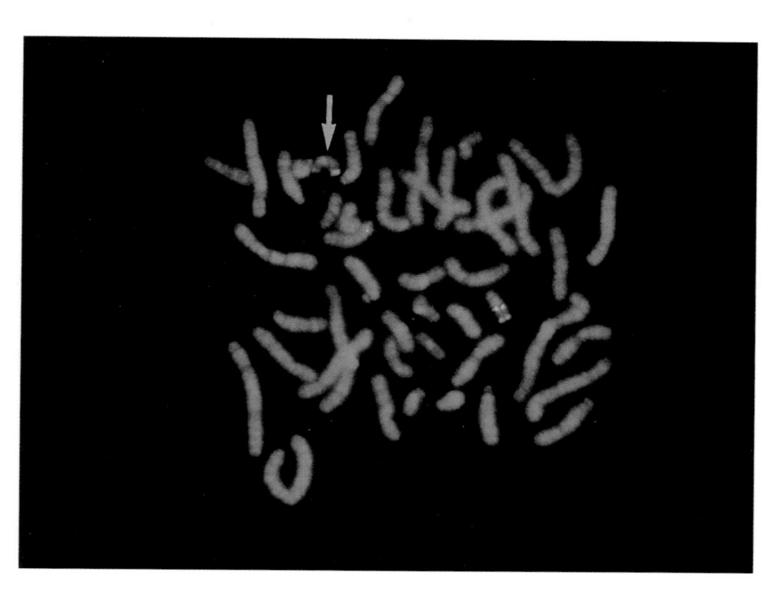

Figure 7.6 Detection of microdeletion involving chromosome 22 by FISH. A probe for a control region is stained green, whereas a probe for the deleted region is stained red. The deleted chromosome (arrow) stains only with a control probe.

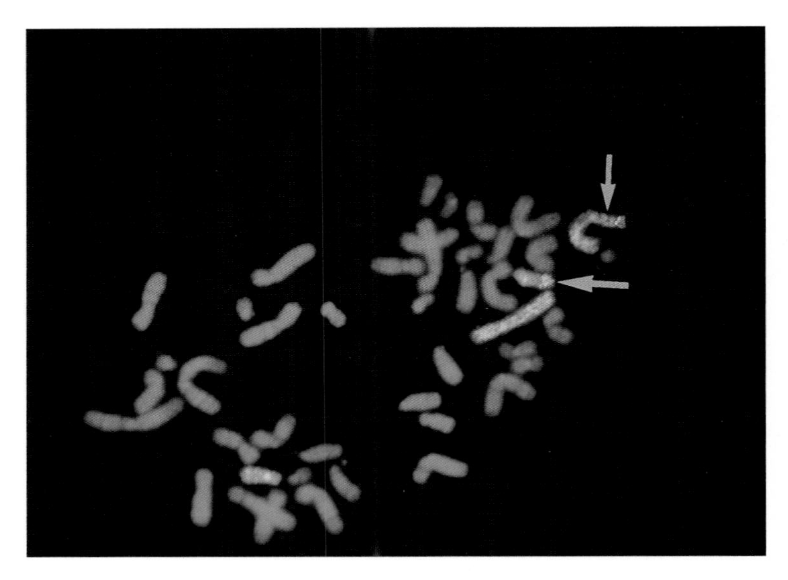

Figure 7.7 Chromosome painting of a balanced translocation involving chromosomes 2 and 17. The translocation chromosomes are indicated by arrows. (Courtesy of Dr. Stanislawa Weremowicz.)

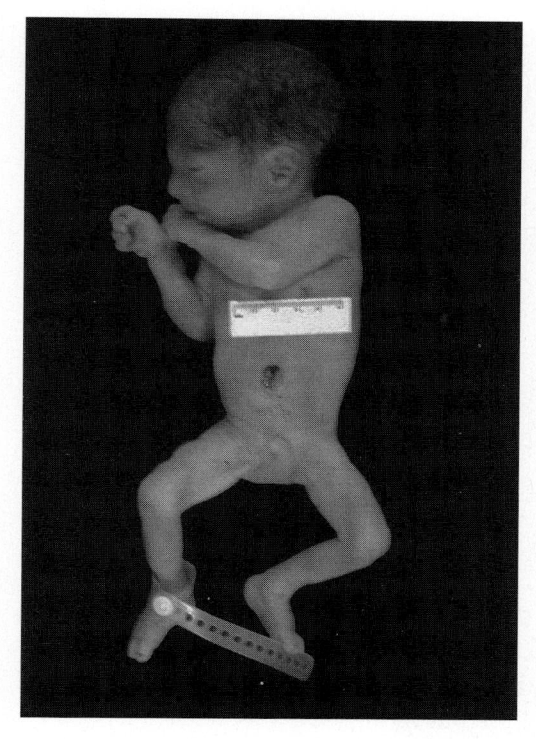

Figure 3 Postmortem photograph of a premature infant with trisomy 18, born at 26–27 weeks estimated gestational age, who lived for 1 h. Note characteristic features of growth retardation, prominent occiput, micrognathia, malformed ear, short palpebral fissures, and clenched, overlapping fingers.

dromes are described in Table 2. More recently delineated microdeletion syndromes will be described below.

C. Multiple Congenital Anomalies

Loss or gain of genetic material is usually not well tolerated, and results in phenotypic changes including congenital anomalies and developmental impairment. Although some of these result in characteristic syndromes, in most cases it is difficult to predict the exact chromosome region involved in an abnormality based on phenotype. Furthermore, chromosomal imbalance due to meiotic segregation of a balanced translocation can result in aneuploidy for two or more chromosomal segments, leading to a complex phenotype that includes effects from both chromosome regions.

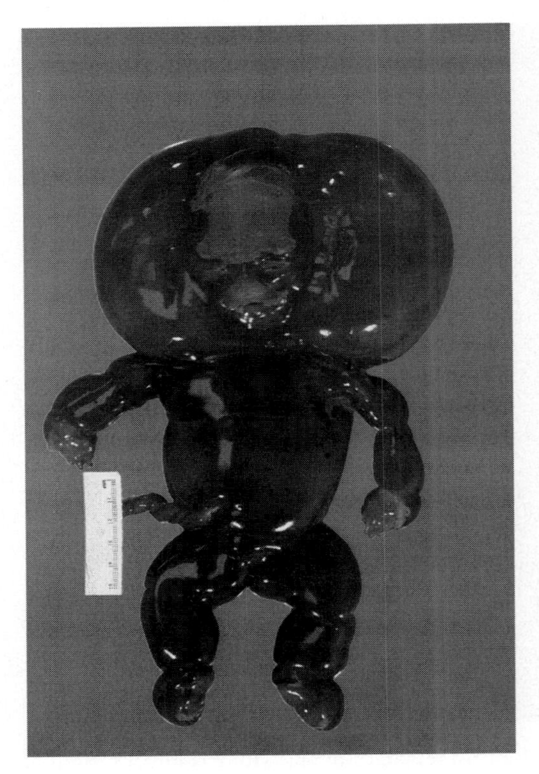

Figure 4 Spontaneously aborted 2-week severely macerated fetus with 45,X karyotype (Turner syndrome). Note characteristic features of massive bilateral cervical cystic hygromas and hydrops. The "webbed" neck of liveborn Turner syndrome girls is evidence of cystic hygromas, which resolve before birth. Approximately 99% of 45,X conceptuses abort spontaneously. The cytogenetic evaluation was obtained from a specimen of placenta; tissues from severely macerated fetuses virtually never produce dividing cells for cytogenetic analysis.

Clinical indications for chromosomal analysis can be correspondingly difficult to define. A good rule of thumb is to consider chromosomal analysis in any individual who has congenital anomalies in two or more systems in the absence of an alternative genetic diagnosis. Chromosome studies may also be indicated in individuals with pervasive developmental delay with or without features of autism. Some of these may have subtle chromosomal changes, or supernumerary marker chromosomes such as the so-called inverted duplicated chromosome 15 (17).

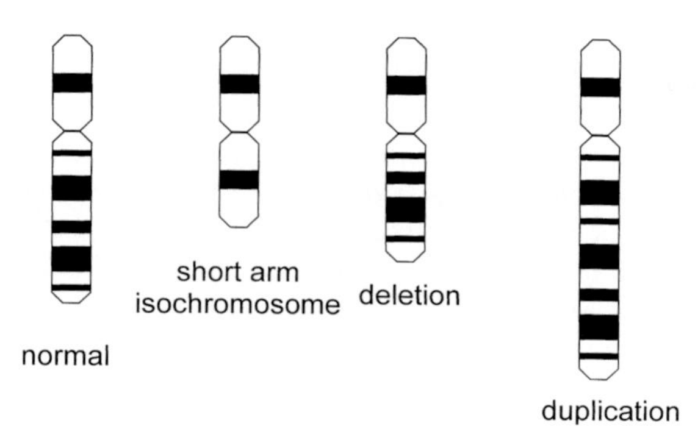

Figure 5 Diagram of normal chromosome, isochromosome for short arm, deletion, and duplication.

Table 2 Major Deletion or Duplication Syndrome (Excluding Microdeletions)

Syndrome	Major features	Chromosomal basis
Wolf-Hirschhorn	Facial anomalies, mental retardation	Del(4p)
Cri-du-chat	Cat-like cry, craniofacial anomalies, mental retardation	Del(5p)
Trisomy 9p	Craniofacial anomalies, mental retardation	Trisomy 9p
Deletion 9p	Trigoncephaly, mental retardation	Del(9p)
Tetrasomy 12p, Pallister-Killian syndrome	Hypotonia, coarse facies, developmental impairment	Mosaic tetrasomy 12p (usually not found in peripheral blood)
Deletion 13q	Holoprosencephaly, hypoplastic thumbs	Del(13q)
Deletion 18q	Midface hypoplasia, mental retardation	Del(18q)
Cat-eye syndrome	Iris coloboma; anal atresia	Duplication of 22q11 region

D. Recurrent Miscarriage or Family History

The most common outcome of chromosomal abnormality in an embryo or a fetus is miscarriage. Therefore, couples in which one partner carries a balanced chromosomal rearrangement such as a translocation or inversion will often experience recurrent first-trimester miscarriages (18,19). It is common practice to offer cytogenetic testing to both partners after two or more first-trimester miscarriages unless an alternative cause is apparent. In about 4% of such couples, one partner will be found to have a balanced rearrangement. Recognition of a balanced chromosome rearrangement in an individual provides a basis for genetic counseling and offering prenatal diagnosis. If there is a family history of chromosome rearrangement, the partner with this history can be offered chromosomal analysis to determine if he or she has inherited the rearrangement.

E. Microdeletion Syndromes

Through the early 1990s, cytogenetic analysis was limited to that which could be visualized by conventional staining and microscopic examination. Several syndromes had been identified as being associated with very small chromosome deletions, for example, WAGR (Wilms' tumor, aniridia, genitourinary malformations, mental retardation) (20) and Prader-Willi syndrome (21–23), but visible deletions were found in only a minority of patients. The advent of FISH has provided a new tool that has vastly changed the approach to molecular diagnosis of these disorders. FISH analysis is able to reveal submicroscopic deletions by using cloned DNA probes from the deletion critical region (see Color Plate 7.6). Most of these deletions are not cytogenetically visible by conventional microscopy. A list of major microdeletion syndromes is provided in Table 3. Recognition of these syndromes has vastly increased the power of cytogenetic analysis; the full scope of chromosome microdeletions remains unexplored.

F. Special Cytogenetic Studies

There exist a number of special modes of cytogenetic analysis that are used to identify specific genetic disorders. These should be available in the cytogenetics laboratory in most cases, or samples may be sent to referral laboratories. Because the clinical indications are specialized, it is important for the laboratory to work closely with the referring clinician to recognize the need to pursue special studies.

Table 3 Major Chromosome Microdeletion Syndromes

Syndrome	Major features	Chromosome region
Williams	Characteristic facies, supravalvar aortic stenosis, developmental impairment	7q11.23
Langer-Giedion	Exostoses, abnormal facies, developmental delay	8q24.1
WAGR	Wilms' tumor, aniridia, genitourinary dysplasia, mental retardation	11p13
Retinoblastoma/MR	Retinoblastoma, mental retardation	13q14
Prader-Willi	Hypotonia, developmental delay, obesity	15q12
Angelman	Seizures, abnormal movements, mental retardation	15q12
α-Thalassemia/MR	α-Thalassemia, mental retardation	16p13.3
Rubinstein-Taybi	Microcephaly, characteristic facies, mental retardation	16p13.3
Smith-Magenis	Mental retardation, characteristic facies	17p11.2
Miller-Dieker	Lissencephaly, characteristic facies	17p13.3
Charcot-Marie-Tooth Ia/hereditary susceptibility to pressure palsies	Peripheral neuropathy	CMT: dup(17p12) HSPP: del(17p12)
Alagille	Intrahepatic biliary atresia, peripheral pulmonic stenosis, characteristic facies	20p11.23
DiGeorge/velo-cardio-facial	Palatal anomalies, conotruncal cardiac anomalies, thymic hypoplasia, parathyroid hypoplasia	22q11.2
Steroid sulfatase deficiency/ Kallman syndrome	Ichthyosis; anosmia	Xp22.3

G. High-Resolution Cytogenetic Analysis

Standard chromosomal analysis is performed at the 400–550 band level in most laboratories. Higher-resolution analysis at greater than the 800 band level is possible and may provide advantages in specific clinical situations. Such high-resolution studies have the potential to reveal more subtle deletions or rearrangements, but are technically challenging since chromosomes tend to overlap one another when highly extended. High-resolution analysis is particularly suited for focused study of a particular chromosome region suspected of being abnormal because of clinical indications.

H. Special Banding Studies

There exist a wide variety of banding methods aside from GTG banding, the standard used in most laboratories. Additional banding methods may be used to complement GTG banding in characterizing abnormal chromosomes. A list of available techniques and some advantages to their use is provided in Table 4. Further discussion of banding and staining can be found elsewhere in this volume in Chapter 5 by Mark.

I. Studies for Mosaicism

Mosaicism for a chromosomal abnormality may be detected from peripheral blood lymphocyte analysis or prenatal cytogenetic analysis. When found in

Table 4 Alternative Means of Chromosome Staining and Clinical Applications

Method	Stain	Application
G-banding	Giemsa	Routine analysis
Q-banding	Quinacrine	Routine analysis, polymorphic variants, Y chromosome analysis
R-banding	Various	Accentuates ends of chromosomes; usually used as adjunct to G- or Q-banding
C-banding	Giemsa	Identifies heterochromatin
NOR staining	Silver	Stains active nucleolus organizer regions on acrocentric chromosomes
Distamycin-DAPI	Distamycin-DAPI	5-Methylcytosine enhances staining; used to identify chromosome 15

a peripheral blood study of a phenotypically affected child, the mosaicism must be interpreted with caution. Although the normal cells might be expected to ameliorate the phenotype, this is unpredictable. The proportion of abnormal cells in critical tissues such as heart or brain cannot be determined and may differ from that in lymphocytes. Also, in some instances there may be selection against the trisomic cells in lymphocytes, but not in other tissues. Some chromosomal abnormalities, for example, tetrasomy 12p, found in Pallister-Killian syndrome, are usually lost very early from peripheral blood lymphocytes, and can only be detected by skin fibroblast analysis postnatally (24,25).

J. FISH

FISH has seen broad application in clinical cytogenetics in recent years. Site-specific DNA probes are used routinely to detect microdeletions that are invisible with routine chromosomal staining. Cocktails of chromosome-specific DNA probes can be used to "paint" the entire chromosome. This is useful in metaphase preparations to identify the composition of rearranged chromosomes (26) (see Color Plate 7.7). In some cases, painting studies have revealed that apparently balanced translocations between two chromosomes may in fact involve three or more chromosomes. These complex translocations can cause genetic imbalance and may otherwise escape detection. Recently, DNA probes have been developed for the telomere regions of individual chromosomes (27). These can be used to reveal translocations involving chromosome ends that would escape detection by classical cytogenetics.

FISH can also be used for interphase chromosomal analysis. Chromosome-specific probes stain the nuclear domain occupied by a specific chromosome and can be used to identify aneuploidy in interphase cells. This has been used as a rapid screen for aneuploidy in cases of suspected mosaicism (28). It can also provide a rapid analysis of uncultured cells in cases of suspected trisomy to allow for a rapid diagnosis in time to make decisions about surgical treatments. Interphase FISH has also seen somewhat controversial application for rapid prenatal diagnosis (29–31). There is the potential for diagnostic errors, particularly false negative results, and FISH does not detect rearrangements in chromosomes other than those probed, or subtle rearrangements in general.

K. Chromosome Microdissection

The identification of marker chromosomes using chromosome painting requires the use of multiple painting probes, which can be time-consuming and expensive. Recently, technology has been developed that allows painting probes to be obtained directly from the marker chromosome. DNA is isolated from the marker from several cells (approximately five) by chromosome microdissection (32). This DNA is then used as a substrate for PCR using random primers. The PCR products are labeled with fluorescent tags and used in FISH studies on normal chromosomes. The pattern of normal chromosome staining in turn identifies the chromosomes that have contributed material to the marker chromosome.

L. Chromosome Breakage Studies

A number of genetically determined conditions predispose to spontaneous chromosome breakage (Table 5). These generally represent mutations in genes that control DNA replication or repair and the cell cycle. Affected individuals have various signs or symptoms, including in most instances a predisposition to malignancy. The advent of DNA tests for many chromosome breakage syndromes have obviated the need to do clinical studies of chromosome breakage. For example, individuals with Fanconi anemia exhibit a marked increase in chromosome breakage when cells are treated with the alkylating agent diepoxybutane (DEB) or with the cross-linking agent mitomycin C (33). Measurement of sensitivity of cells to breakage with these drugs remains a commonly used test for Fanconi anemia.

Table 5 Characteristics of Major DNA Repair Syndromes

Syndrome	Features	Genetic cause
Ataxia-telangiectasia	Ataxia, telangiectasia, immune deficiency, lymphoma	Mutations in ATM gene
Bloom syndrome	Short stature, photosensitive skin, risk of malignancy, increased frequency of sister chromatid exchange	Mutations in helicase gene
Fanconi anemia	Congenital anomalies, aplastic anemia, chromosome breakage increased with alkylating agents or mitomycin C	Genetically heterogeneous
Xeroderma pigmentosum	Photosensitive skin, freckling, skin cancer	Defective repair of UV-induced DNA damage
Cockayne syndrome	Photosensitive skin, white matter degeneration, cerebral calcification, dwarfism	Defective repair of UV-induced DNA damage

M. Sister Chromatid Exchange Analysis

Sister chromatid exchange can be visualized in metaphase cells that have been grown for two cell cycles in the presence of 5-bromodeoxyuridine (BrdU) (34). BrdU substitutes for thymidine. Each chromosome strand is half-substituted after one cell cycle. After two rounds of division, one chromatid of each chromosome is doubly substituted and one is singly substituted (Fig. 8). The strands can be differentially stained by the fluorescent dye Hoechst 33258, or by Giemsa after light exposure of the Hoechst-stained chromosomes. Cells from individuals with Bloom syndrome (see Table 5) exhibit a many-fold increase in the frequency of sister chromatid exchange, which is useful in laboratory diagnosis of this condition (35).

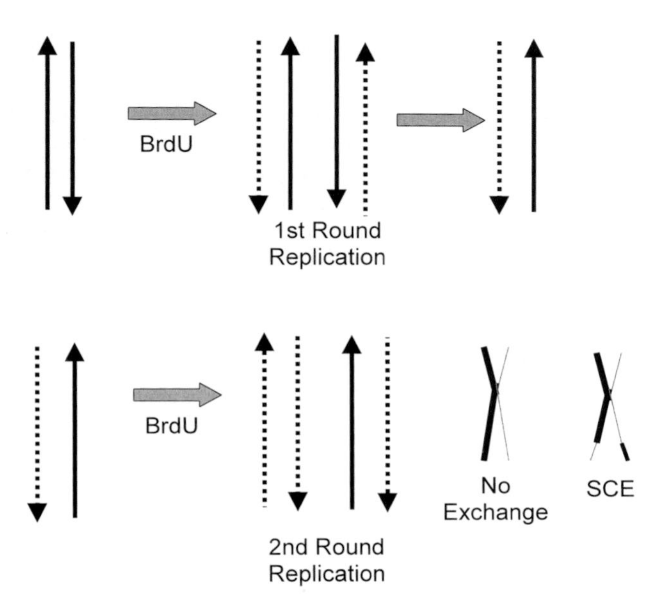

Figure 8 Diagram of sister chromatid differentiation using BrdU. Chromosomes are replicated in the presence of 5-bromodeoxyuridine (BrdU) for two rounds of replication. The BrdU incorporates into the DNA in the place of thymidine. After one round of replication, one of the two DNA strands is substituted with BrdU (dashed line). After the second round, one chromatid of a chromosome is substituted on one strand, whereas the other is substituted in both strands. When the chromosome is stained with the dye Hoechst 33258, fluorescence is brighter in the half-substituted strand than in the doubly substituted strand, allowing for sister chromatid differentiation. This makes it possible to visualize a sister chromatid exchange (right).

N. Imprinted Genes and Uniparental Disomy

Recent studies in developmental biology have revealed that some genes are not equally expressed from both the maternal and paternal copies. This is referred to as "genetic imprinting" (36). The phenomenon is clinically important because it means that loss of one copy of a gene may have different effects depending upon whether the copy lost is the one normally expressed or the one inactivated. This is graphically illustrated by the example of Prader-Willi and Angelman syndromes. These are distinct disorders, yet approximately two-thirds of affected individuals have an apparently identical microdeletion of chromosome 15q (23,37). There are imprinted genes in this region, so that a deletion of the paternal copy results in Prader-Willi syndrome and a deletion of the maternal copy results in Angelman syndrome. The remaining one-third of Prader-Willi patients and about 5% of Angelman patients are affected due to uniparental disomy, another mechanism that is dependent on imprinting. Here, both copies of chromosome 15 are inherited from the same parent (the mother for Prader-Willi syndrome, the father for Angelman syndrome). This most likely occurs when a trisomic zygote is "rescued" by loss of one chromosome due to mitotic nondisjunction in the embryo (Fig. 9). If the lost chromosome is from the parent who contributed only one copy of chromosome 15, uniparental disomy results; if the lost chromosome was from the parent whose genes remain active in the imprinted region, there is a consequential absence of expression of imprinted genes in the conceptus.

The number and location of all imprinted genes are not known, but it appears that not all chromosomes include imprinted genes (38). Uniparental

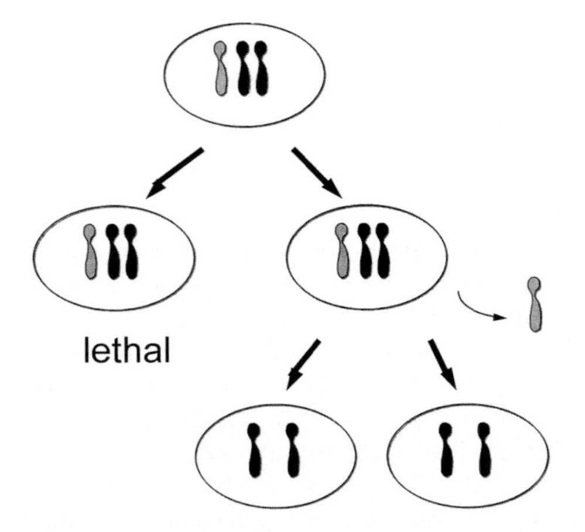

Figure 9 Mechanism of uniparental disomy formation by "trisomy rescue."

Table 6 Major Chromosome Regions Associated with Genomic Imprinting

Chromosome region	Clinical effects
15q12	Prader-Willi or Angelman syndrome
11p	Beckwith-Wiedemann syndrome
7q	Short stature (?Russell-Silver syndrome)
14	Short stature, developmental delay

disomy should be suspected when a syndrome known to depend on imprinting (Table 6) is seen in the absence of a deletion or gene mutation. It should also be explored in instances of mosaicism in prenatal specimens, especially if the mosaicism is seen in cultured chorionic villus cells but not amniotic fluid cells. Finally, it should be considered in individuals with familial balanced translocations involving chromosomes known to contain imprinted genes, for example, Robertsonian translocations involving chromosome 14. Such translocations are subject to nondisjunction (see below); rescue of such an embryo by mitotic nondisjunction can restore a balanced karyotype but leads to uniparental disomy.

Although uniparental disomy is a cytogenetic phenomenon, uniparental disomy usually cannot be detected by cytogenetic studies. Evaluation for uniparental disomy generally requires use of DNA polymorphisms, looking for biparental inheritance of distinct alleles. In some cases, imprinting results in methylation of DNA sequences that can be detected by Southern analysis (Fig. 10) (39).

III. COUNSELING ISSUES IN CLINICAL CYTOGENETICS

The identification of a chromosome abnormality raises important questions for counseling the family. This includes issues of prognosis, management, and genetic recurrence risk. Although prenatal diagnostic options are covered in Chapter 8 by Lamb and Miller, some of the important counseling issues in clinical cytogenetics will be covered in this section.

A. Chromosome Aneuploidy

Chromosome aneuploidy results from nondisjunction either in meiosis or mitosis (Fig. 11). For trisomy 21, nondisjunction occurs most commonly in the oocyte during the first meiotic division. The causes of such nondisjunc-

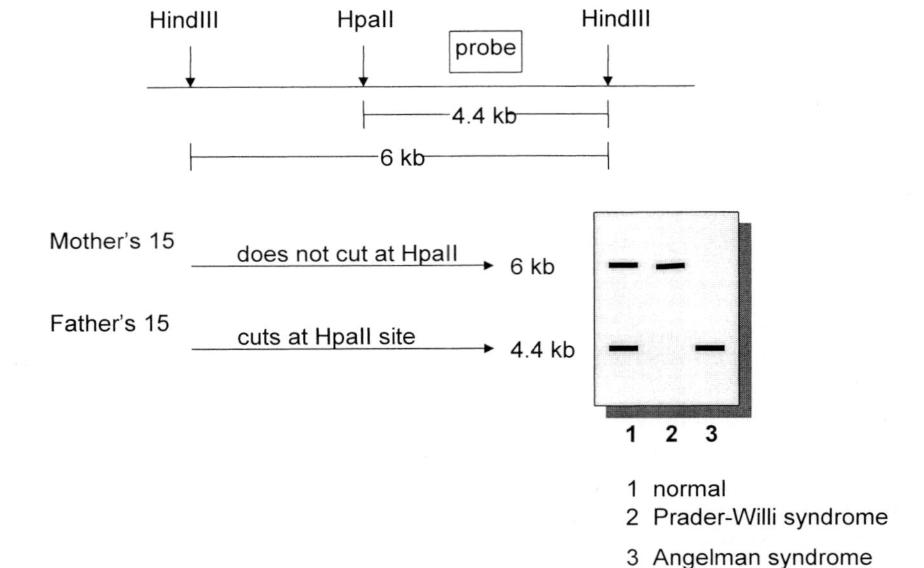

Figure 10 Detection of uniparental disomy by methylation analysis. *HindIII* cuts the DNA into a 6.0-kb fragment. *HpaII* cuts only if the DNA is not methylated. The probe hybridizes to a 6.0-kb fragment on the methylated, maternally derived chromosome 15 and a 4.4-kb fragment on the nonomethylated, paternally derived chromosome 15. Individuals with Angelman syndrome are missing a maternal fragment and those with Prader-Willi syndrome a paternal fragment, in either disorder due to either deletion or uniparental disomy.

tion are unknown; the only risk factor discovered after four decades of study has been maternal age (40). The risk of nondisjunction increases dramatically as a function of maternal age, with the curve beginning to rise after about 35 years of age (Fig. 12). Because of this, it is standard to offer prenatal testing by amniocentesis or chorionic villus sampling to pregnant women 35 years or older at the time of their due dates.

Nevertheless, most pregnancies occur among younger women, and hence most aneuploidy births occur in younger women. Most of these appear to be sporadic events of unknown cause. Rare families have been reported with multiple instances of aneuploid offspring, suggestive of a genetic factor for nondisjunction (41). Empirical data indicate a recurrence risk of approximately 1% for couples who have had an aneuploid offspring (42).

The maternal age effect applies to all autosomal trisomies and to XXY and XXX syndromes. XYY must result from paternal nondisjunction. The

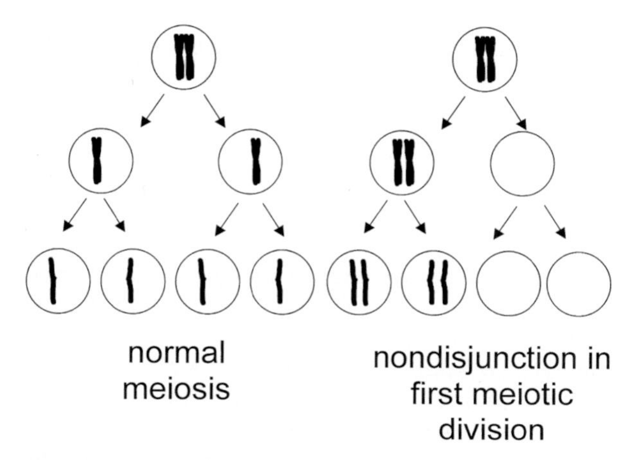

Figure 11 Nondisjunction in first meiotic division resulting in production of germ cells disomic or nullisomic for a chromosome. Nondisjunction occurs when two chromosomes segregate to the same pole.

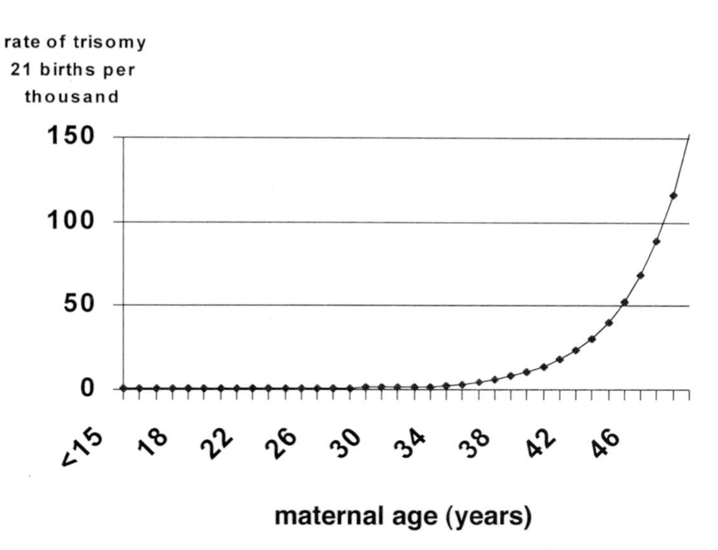

Figure 12 Frequency of trisomy 21 births as a function of maternal age.

45,X karyotype is not subject to the maternal age effect. This may be due to the fact that a high proportion of 45,X individuals are mosaic for a structurally abnormal X or Y, indicating that nondisjunction is not the primary genetic cause of the 45,X karyotype (43).

B. Chromosome Rearrangements

Balanced chromosome rearrangements may come to attention after an analysis for recurrent miscarriage, after the birth of a child with an unbalanced rearrangement, or on the basis of a family history of a balanced rearrangement. The major risk conferred to such a carrier is that germ cells may contain unbalanced segregants of the involved chromosomes. Balanced translocations pair to form a cross-shaped figure in meiosis (Fig. 13). Alternate segregation results in either the balanced translocation products or the normal chromosomes going to the germ cells. In adjacent type I segregation, homologous centromeres separate, whereas in adjacent type II segregation, nonhomologous centromeres separate. In either case germ cells get an unbalanced contribution from the two chromosomes. Moreover, any three of the involved chromosomes may segregate together to a germ cell resulting

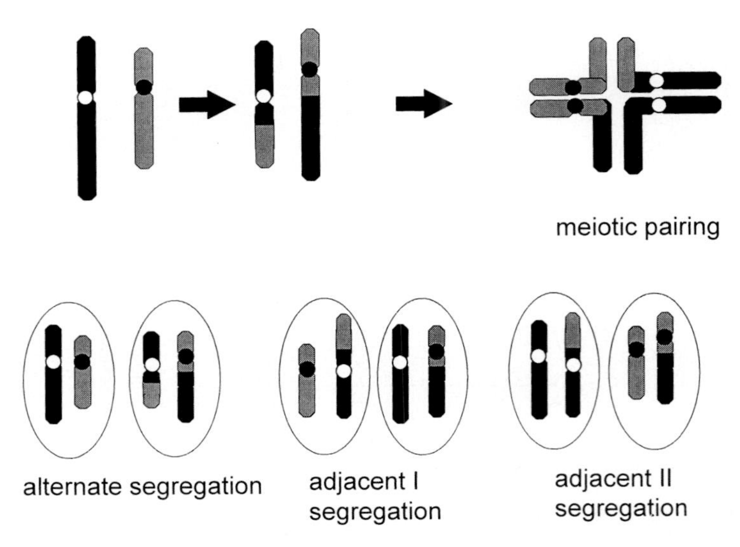

Figure 13 Modes of segregation of a balanced reciprocal translocation. Centromeres are depicted by circles (black or white). Although meiotic pairing actually occurs after chromosome replication, chromosomes are depicted as single stranded to simplify the diagram.

in an extra chromosome in the germ cell (3:1 segregation). The risks of these various outcomes depend on the break points of the rearrangement on the two chromosomes. Empirical data exist for some common translocations (44). Some of the unbalanced karyotypes may be incompatible with fetal development and result in miscarriage, but others may result in the birth of a child with multiple congenital anomalies.

Translocations between acrocentric chromosomes that fuse at their centromeres are referred to as Robertsonian translocations. A carrier for a Robertsonian translocation between chromosome 21 and another acrocentric can have a child with Down syndrome if both the translocation chromosome and the normal 21 chromosome segregate to the same cell at meiosis. Carriers of Robertsonian translocations can have offspring with trisomy for other acrocentric chromosomes, particularly chromosome 13. The risk appears to be lower than for trisomy 21, however. This may be due to the fact Roberstonian translocations involving the larger acrocentrics tend to disjoin from their normal counterparts more reliably, or to the fact that trisomy for acrocentrics other than 13 and 21 tend to be lethal in early development.

Chromosome inversions can also predispose to unbalanced chromosomes in germ cells, due to crossing over within the limits of the inversion loop formed when the inverted and noninverted homologs pair. When the inversion includes the centromere (pericentric inversion), the results are chromosomes with duplicated ends and some deleted material, which may lead to viable but abnormal fetuses (Fig. 14). If the inversion occurs outside

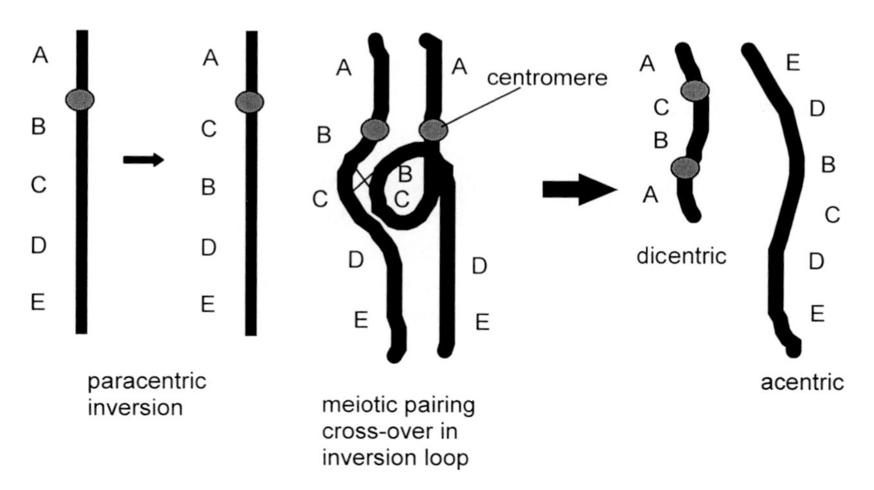

Figure 14 Crossing over within inversion loop in paracentric inversion results in dicentric or acentric fragments.

the centromere (paracentric inversion), the resultant chromosomes following crossing over are either acentric or dicentric, and hence tend to be unstable, leading to nonviable products of conception.

Balanced rearrangements may be seen in prenatal specimens or in phenotypically abnormal individuals. Finding the same rearrangement in a parent and a child tends to be reassuring, although it does not guarantee that the child will be unaffected. If both parents' chromosomes are normal, the rearrangement in the child is de novo. Empirical data indicate that de novo balanced translocations identified prenatally are associated with approximately a 6–7% risk of phenotypic effects (45). Such effects may be based on disruption of a gene by the rearrangement, position effect, or the possibility of uniparental disomy (see above).

C. Microdeletion Syndromes

Most microdeletion syndromes occur sporadically, with minimal recurrence risk to both parents. There are rare examples of subtle translocations that may exist in a parent and lead to microdeletion in an offspring. Also, some microdeletion syndromes, for example velo-cardio-facial syndrome (associated with deletion in chromosome 22) can be compatible with reproduction, and can be transmitted as autosomal dominant traits. Parents of a child with a microdeletion should be offered chromosomal and FISH analysis to rule out subtle translocation or microdeletion.

D. Marker Chromosomes

A marker chromosome is defined as a structurally rearranged chromosome, no part of which can be identified. The advent of FISH, with its ability to identify chromosome regions, will likely render this definition obsolete. Nevertheless, there are individuals with very small supernumerary chromosomes in whom it may be difficult to determine whether the extra chromosome is phenotypically significant. In some families such marker chromosomes may be passed from generation to generation without apparent ill effects. Karyotyping both parents is therefore indicated if such a marker is found in a prenatal or postnatal study. Finding the same marker in a parent is reassuring, whereas if the marker is de novo, counseling is more difficult. For small supernumerary chromosomes derived from chromosome 15, there are data showing that phenotypic effects are related to the size of the marker and the degree of trisomy for chromosome 15 (17).

E. Mosaicism

Mosaicism can arise from mitotic nondisjunction leading to ''rescue'' of a trisomic conceptus, or by mitotic nondisjunction in a disomic embryo leading to a trisomic cell line. The former indicates that nondisjunction occurred in a germ cell, and presumably represents a typical example of meiotic nondisjunction, associated with maternal age-related and empirical (1%) recurrence risk. The latter is less likely subject to these risks, but usually cannot be distinguished by routine cytogenetic methods. An individual who is mosaic him/herself faces a risk of transmission of the trisomy to a germ cell and should be counseled accordingly.

IV. CONCLUSION

In the decades since methods for cytogenetic analysis have become available, new waves of discovery of syndromes due to chromosomal abnormalities have followed each technical advance. There remains a significant proportion of individuals with multiple congenital anomalies or developmental impairment of unknown cause, however. The most recent technical advance has been a convergence of molecular genetic and cytogenetic study through FISH. Molecular cytogenetics provides a much higher-resolution study of chromosome structure and allows chromosomal rearrangements to be identified in nondividing cells. It is likely that new syndromes will be delineated due to very subtle changes of chromosome structure in the near future, further expanding the clinical role of cytogenetic analysis.

REFERENCES

1. American College of Medical Genetics. Standards and Guidelines: Clinical Genetics Laboratories. Bethesda, MD: American College of Medical Genetics, 1993.
2. Jacobs PA, Melville M, Ratcliffe S, Keay AJ, Syme J. A cytogenetic survey of 11,680 newborn infants. Ann Hum Genet 1974; 37:359–376.
3. Lubs HA, Ruddle FH. Chromosomal abnormalities in the human population: estimation of rates based on New Haven newborn study. Science 1970; 169: 495–497.
4. Carr DH. Chromosome studies in spontaneous abortions. Obstet Gynecol 1965; 26:308–326.
5. Boue J, Boue A, Lazar P. The epidemiology of human spontaneous abortions with chromosomal anomalies. In: Blandau RJ, ed. Aging Gametes International Symposium, Seattle, 1973. Basel: Karger, 1975: 330–348.

6. Shen JJ, Williams BJ, Zipursky A, et al. Cytogenetic and molecular studies of Down syndrome individuals with leukemia. Am J Hum Genet 1995; 56:915–925.
7. Lai F, Williams RS. A prospective study of Alzheimer disease in Down syndrome. Arch Neurol 1989; 46:849–853.
8. Mutton D, Alberman E, Hook EB. Cytogenetic and epidemiological findings in Down syndrome, England and Wales 1989 to 1993. National Down Syndrome Cytogenetic Register and the Association of Clinical Cytogeneticists. J Med Genet 1996; 33:387–394.
9. Galt J, Boyd E, Connor JM, Ferguson-Smith MA. Isolation of chromosome 21-specific DNA probes and their use in the analysis of nondisjunction in Down syndrome. Hum Genet 1989; 81:113–119.
10. Antonarakis SE, Adelsberger PA, Petersen MB, Binkert F, Schinzel AA. Analysis of DNA polymorphisms suggests that most de novo dup(21q) chromosomes in patients with down syndrome are isochromosomes and not translocations. Am J Hum Genet 1990; 47:968–972.
11. Embleton ND, Wyllie JP, Wright MJ, Burn J, Hunter S. Natural history of trisomy 18. Arch Dis Child Fetal Neonatal 1996; 75:F38–F41.
12. Wyllie JP, Wright MJ, Burn J, Hunter S. Natural history of trisomy 13. Arch Dis Child 1994; 71:343–345.
13. Saenger P. Current concepts—Turner's syndrome. N Engl J Med 1996; 335: 1749–1754.
14. Page DC. Y chromosome sequences in Turner's syndrome and risk of gonadoblastoma or virilisation. Lancet 1994; 343:240.
15. Linden MG, Bender BG, Harmon RJ, Mrazek DA, Robinson A. 47,XXX: what is the prognosis. Pediatrics 1988; 82:619–630.
16. Fryns JP, Kleczkowska A, Kubien E, Van den Berghe H. XYY syndrome and other Y chromosome polysomies. Mental status and psychosocial functioning. Genet Couns 1995; 6:197–206.
17. Cheng S-D, Spinner NB, Zackai EH, Knoll JHM. Cytogenetic and molecular characterization of inverted duplicated chromosomes 15 from 11 patients. Am J Hum Genet 1994; 55:753–759.
18. Katz VL, Kuller AL. Recurrent miscarriage. Am J Perinatol 1994; 11:386–397.
19. Flint S, Gibb DM. Recurrent second trimester miscarriage. Curr Opin Obstet Gynecol 1996; 8:449–453.
20. Fantes JA, Bickmore WA, Fletcher JM, Ballesta F, Hanson IM, van Heyningen V. Submicroscopic deletions at the WAGR locus, revealed by nonradioactive in situ hybridization. Am J Hum Genet 1992; 51:1286–1294.
21. Butler MG. High resolution chromosome analysis and fluorescent in situ hybridization in patients referred for Prader-Willi or Angelman syndrome. Am J Med Genet 1995; 56:420–422.
22. Wenger SL, Cummins JH. Fluorescent in situ hybridization for evaluation of Prader-Willi and Angelman syndromes. Am J Med Genet 1995; 57:639.
23. Knoll JHM, Nicholls RD, Magenis RE, Graham JM Jr, Lalande M, Latt SA. Angelman and Prader-Willi syndromes share a common chromosome 15 de-

letion but differ in parental origin of the deletion. Am J Med Genet 1989; 32: 285–290.
24. Butler MG, Dev VG. Pallister-Killian syndrome detected by fluorescent in situ hybridization. Am J Med Genet 1995; 57:498–500.
25. Speleman F, Leroy JG, Van Roy N, de Paepe A, Suijkerbuijk R, Brunner H. et al. Pallister-Killian syndrome: characterization of the isochromosome 12p by fluorescent *in situ* hybridization. Am J Med Genet 1991; 41:381–387.
26. Daniel A, Malafiej P, Preece K, Chia N, Nelson J, Smith M. Identification of marker chromosomes in thirteen patients using FISH probing. Am J Med Genet 1994; 53:8–18.
27. NIH and Institute of Molecular Medicine Collaboration. A complete set of human telomeric probes and their clinical application. Nature Genet 1996; 14: 86–89.
28. Lomax BL, Kalousek DK, Kuchinka BD, Barrett IJ, Harrison KJ, Safavi H. The utilization of interphase cytogenetic analysis for the detection of mosaicism. Hum Genet 1994; 93:243–247.
29. Schwartz S. Efficacy and applicability of interphase fluorescent in situ hybridization for prenatal diagnosis. Am J Hum Genet 1993; 52:851–853.
30. Lüdecke H-J, Senger G, Claussen U, Horsthemke B. Cloning defined regions of the human genome by microdissection of banded chromosomes and enzymatic amplification. Nature 1989; 338:348–350.
31. Mark HFL, Jenkins R, Miller WA. Current applications of molecular cytogenetic technologies. Ann Clin Lab Sci 1997; 47–56.
32. Senger G, Lüdecke H-J, Horsthemke B, Claussen U. Microdissection of banded human chromosomes. Hum Genet 1990; 84:507–511.
33. Auerbach AD. Fanconi anemia. Dermatol Clin 1995; 13:41–49.
34. Latt SA. Localization of sister chromatid exchanges in human chromosomes. Science 1974; 185:74–76.
35. Chaganti RS, Schonberg S, German J. A manyfold increase in sister chromatid exchanges in Bloom's syndrome lymphocytes. Proc Natl Acad Sci USA 1974; 71:4508–4512.
36. Sapienza C. Genome imprinting: an overview. Dev Genet 1995; 17:185–187.
37. Donlon TA. Similar molecular deletions on chromosome 15q11.2 are encountered in both the Prader-Willi and Angelman syndromes. Hum Genet 1988; 80: 322–328.
38. Ledbetter DH, Engel E. Uniparental disomy in humans: development of an imprinting map and its implications for prenatal diagnosis. Hum Mol Genet 1995; 4:1757–1764.
39. Dittrich B, Buiting K, Horsthemke B. PW71 methylation test for Prader-Willi and Angelman syndromes. Am J Med Genet 1996; 16:196–199.
40. Hook EB, Cross PK, Schreinemachers MS, Chromosomal abnormality rates at amniocentesis and in live-born infants. JAMA 1983; 249:2034–2038.
41. Krishna Murthy DS, Farag TI. Recurrent regular trisomy-21 in two Bedouin families. Parental mosaicism versus genetic predisposition. Ann Genet 1995; 38:217–224.

42. Lister TJ, Frota-Pessoa O. Recurrence risk for Down syndrome. Hum Genet 1980; 55:203–208.
43. Zinn AR, Page DC, Fisher EMC. Turner syndrome: the case of the missing sex chromosome. Trends Genet 1993; 9:90–93.
44. Daniel A, Hook EB, Wulf G. Risks of unbalanced progeny at amniocentesis to carriers of chromosome rearrangements: data from United States and Canadian laboratories. Am J Med Genet 1989; 33:14–53.
45. Warburton D. De novo balanced chromosome rearrangements and extra marker chromosomes identified at prenatal diagnosis: clinical significance and distribution of breakpoints. Am J Hum Genet 1991; 49:995–1013.

8
Prenatal Chromosome Diagnosis

Allen N. Lamb
Genzyme Genetics, Santa Fe, New Mexico

Wayne A. Miller
Perinatal Diagnosis Center, Lexington, Massachusetts

I. HISTORICAL INTRODUCTION

The first reported attempts to utilize fetal and/or placental tissue for prenatal genetic diagnosis occurred in 1955, when Serr et al. (1) and Fuchs and Riis (2) reported the ability to determine fetal sex through sex chromatin studies on amniotic fluid cells. Eleven years later, Steele and Breg (3) reported the successful culture of amniotic fluid cells and its use to examine fetal chromosome status. The next year Jacobson and Barter (4) applied the technique to diagnose a D/D translocation in a fetus. In 1968, both Valenti et al. (5) and Nadler (6) reported the diagnosis of a fetus with Down syndrome in utero. In the same year, Nadler (6) reported the in-utero diagnosis of a fetus affected with galactosemia. Chorionic villus sampling (CVS) was first reported in the late 1960s (7). However, technical difficulties with sampling relegated the technique to obscurity until ultrasound improvements allowed for direct monitoring of the procedure. Additional improvements, such as the development of a flexible catheter for transcervical sampling and a protocol for transabdominal sampling, markedly improved the success rate of CVS sampling and the safety of the procedure. Fetal blood sampling and biopsy of other fetal tissue began with fetoscopy (8) but have proven to be more successful and safer when performed under ultrasound guidance.

II. CLINICAL INDICATIONS FOR PRENATAL CHROMOSOME DIAGNOSIS

A. Advanced Maternal Age

Advanced maternal age is the most common indication for performing prenatal chromosome diagnosis. The risk of detecting a fetus with Down syndrome or other aneuploidy at the time of the procedure, or at time of birth, increases with the mother's age (9). The age when the majority of women in the United States are offered the option of prenatal chromosome diagnosis is 35 years at the time of delivery, although arguments can be made for either earlier or later maternal ages. Many health-care insurers now must approve this procedure, and some even require a positive result from a maternal serum triple screen test before approving amniocentesis for women 35 years and older. Testing all women 35 years of age and older would detect 20–30% of all Down syndrome pregnancies (10,11). Differences in the quoted rate of chromosome abnormalities at chorionic villus sampling (CVS) (10–12 weeks), at the time of amniocentesis (16–20 weeks), and at delivery are due to interval fetal loss for 45,X, trisomy 13, trisomy 18, and trisomy 21 conceptions (9,13). For example, the rate of chromosome abnormalities per 1000 pregnancies for a woman at age 35 is 8.8 at CVS, 7.7 at amniocentesis, and 5 at birth (9).

B. Abnormal Maternal Serum Screen

To detect Down syndrome fetuses in women under the age of 35 years, maternal serum screening with multiple analytes including AFP, human chorionic gonadotropin (hCG), beta subunit hCG, and unconjugated estriol (uE3) in some combination is now offered to all pregnant women. Without going into the details of the analysis protocol, low levels of AFP and uE3 and high levels of HCG or its subunits are associated with an increased risk for fetal Down syndrome, and low levels of all these analytes are associated with an increased risk for fetal trisomy 18 (10,11). Although the quoted rate of detection of Down syndrome fetuses for a screening program is about 60% with a 5% false positive rate, each of these percentiles varies greatly at different maternal ages. Applying the most common protocol yields a detection rate of 40% with a false positive rate of 3.5% at maternal age 20 and a detection rate of 58% with a false positive rate of 15% at maternal age 34. Other unbalanced and balanced chromosome rearrangements will be found fortuitously during cytogenetic analysis for cases referred due to an increased risk for a chromosome abnormality during maternal serum screening (11).

C. Abnormalities Detected by Ultrasound

Ultrasonography may lead to amniocentesis because many of the fetal abnormalities detected are associated with an increased risk for a chromosome abnormality (12). The reasons for routine use of ultrasonography include fetal dating; examination of the fetus prior to and during CVS, amniocentesis, and fetal blood sampling procedures; routine examination of the fetus at 15–20 weeks; examination after a positive maternal serum screen; a previous child with a neural tube defect or chromosome abnormality; and third-trimester examinations for medical management.

D. Previous Child with a Chromosome Abnormality

A couple with a previous child or fetus with trisomy 21 has an increased risk of approximately 1% of having another child with any trisomy/chromosome abnormality (9,13). Prenatal data do not exist to support this conclusion for other numerical abnormalities, but a cautious approach is advised and prenatal diagnosis should be recommended in these cases (9). There is controversy as to whether this approach should also be applied when trisomic abortuses are identified. To date, there is no strong evidence that women who experience first-trimester trisomic fetal losses are at an increased risk for trisomic live births. However, as for trisomies other than trisomy 21, a cautious approach that offers prenatal diagnosis could be considered (9).

E. Parent a Carrier of a Balanced Rearrangement

As there is generally no gain or loss of genetic material associated with a balanced chromosome rearrangement, the carrier parent is usually phenotypically normal. There is, however, an increased chance of producing an unbalanced gamete with too little or too much chromosomal material, resulting in a spontaneous abortion or a child with a chromosome abnormality. The risk of an unbalanced gamete resulting in a child with a chromosome abnormality varies with the type of rearrangement, the chromosomes involved, the size of the segments involved, the mode of ascertainment, and the sex of the carrier. When an amniotic fluid sample is received with the clinical indication that a parent carries a balanced rearrangement, the medical records, and if possible karyotypes, should be reviewed prior to the initiation of the testing. If a subtle, or potentially subtle, rearrangement is present, karyotypes must be reviewed. Otherwise, the carrier parent should have a repeat karyotype performed by the laboratory performing the prenatal chromosome analysis for comparison with the fetal karyotype. The use of FISH should be explored if the rearrangement is likely to be more difficult to visualize on the lower-resolution fetal chromosomes.

If the parental and fetal studies appear to have significantly different breakpoints, a repeat sample and karyotype from the parent who carries the translocation should also be considered. Breakpoint differences are most often not real, but are due to the higher resolution achieved on the parental blood chromosomes and differences between cytogeneticists.

1. Reciprocal Translocations

Reciprocal translocations are the result of two chromosome breaks and a subsequent exchange of material between two nonhomologous chromosomes. During meiosis, the chromosomes in a reciprocal translocation carrier form a quadrivalent instead of the usual bivalent. The mode of segregation of the chromosomes determines the gametes formed, as discussed in the chapter by Korf and Schneider in this volume [for a detailed review with illustrations, see Gardner and Sutherland (14), and for additional discussion see (15,16,19)].

Risk estimates for balanced and unbalanced offspring (summarized below) can be uncertain at best, even after a detailed analysis. Most translocation carriers have children with normal chromosomes or with a balanced translocation. However, infertility, pregnancy loss, and children with an unbalanced translocation are all observed for carriers of balanced translocations. In general, the risk of a live birth with an unbalanced translocation is in the range of 5–30% for adjacent (1 and 2) 2:2 segregation when the familial translocation is first discovered in the pedigree by an unbalanced live birth. It is closer to 5% when larger segments are involved in the translocation and closer to 30% when small segments are involved. For 3:1 segregation, a female carrier has a risk of 10–20% and a male carrier a risk of 1–3%. When the translocation carrier is ascertained by some other means, such as multiple abortions or by chance, the risk is lower for a live birth with an unbalanced rearrangement (i.e., 0–5% for 2:2 segregation and 0–1% for 3:1 segregation (9).

There are three types of gametes that form from 2:2 segregation: (a) alternate, which results in balanced gametes with either normal chromosomes or with the balanced translocation; (b) adjacent-1 (one centromere of each type in the quadrivalent in one gamete), which results in unbalanced gametes; and (c) adjacent-2 (two of the same centromere in one gamete), which also results in unbalanced gametes. Adjacent-1 is the most frequent mode of abnormal segregation seen in offspring of translocation carriers (14). The 3:1 mode of segregation results in only unbalanced gametes, with the subsequent embryo having either 47 or 45 chromosomes. When the two normal chromosomes and one of the derivative translocation chromosomes go to one gamete, the resulting embryo has tertiary trisomy, and when both translocation chromosomes segregate with a normal chromosome, it is referred to as interchange trisomy. Tertiary monosomy is rarely observed, and interchange monosomy and 4:0 segregation are never viable.

The above describes the possible segregation patterns of the chromosomes involved in the rearrangement and allows for a standardization of the discussion with the patient. Where does that leave the patient who wants to know the risk for a specific translocation? Is it possible to determine the risk for all possible outcomes? A detailed presentation can be found in Hook (10) and in Gardner and Sutherland (14). In general, the risk is inversely related to the size of the segments involved, measured as a percentage of the total haploid autosome length. For 3:1 segregation, the factors that lead to a higher risk include involvement of an acrocentric chromosome, small translocated segments, and a large discrepancy in size between the two chromosomes. The best estimate of risk would come from the pedigree of the translocation carrier. The next best approach would be to base the risk on the data from other families with similar translocations reported in the medical literature. For example, the t(11;22)(q23;q11.2) translocation has been reported in many different families worldwide and empirical estimates are available (17). If similar translocations cannot be found, rough estimates can be derived from the work of Daniel and co-workers (15,16). In general, the larger the unbalanced segment, the smaller the risk. If the unbalanced segment is small, i.e., less than 1% of the haploid autosome length, the risk is closer to 30%. If it is large, i.e., greater than 2% of the haploid autosome length, then the risk is closer to 5%.

2. Robertsonian Translocations

Robertsonian translocations involve the apparent centromeric fusion of two acrocentric chromosomes. The chromosome number is reduced from 46 to 45 by this fusion, resulting in a chromosome that consists of the long arms of both acrocentric chromosomes. The loss of the short arm region that contains ribosomal genes (found in the cytological stalks) and repetitive DNA sequences is, in general, inconsequential. Thus, a balanced carrier of a Robertsonian translocation has only 45 chromosomes. The best data regarding risk are derived from Boue and Gallano (summarized in Refs. 9, 13, and 14). These data describe the risk that a fetus with an unbalanced karyotype will be detected at amniocentesis (16–20 weeks). No unbalanced offspring were detected for the carriers of translocations involving only the D-group chromosomes. From experience, we know that the risk for trisomy 13, for example, is not zero for carriers of der(13;14), but that the risk is extremely low (9). Significant risks for live births with unbalanced chromosome constitutions are seen for carriers when G-group chromosomes are involved, specifically, chromosome 21. Furthermore, the risk for trisomy 21 is higher when a female carries a 14;21 translocation chromosome (15%) than when a male is the carrier (3%) (9,13). This difference in risk based on the sex of the carrier is supported by data presented by Daniel et al. (16). Insufficient data are available for the other Robertsonian translocations because they are less frequent than the der(14;21). Isochromosomes, such as 21/21 or 13/13, represent special cases and carriers cannot have a cytogenetically normal fetus.

Uniparental disomy (UPD) is a potential concern for offspring of carriers of Robertsonian translocations that involve chromosomes 14 and/or 15, or when de novo Robertsonian translocations involving these chromosomes are encountered during prenatal diagnosis. There is no apparent abnormal phenotype for maternal or paternal UPD associated with chromosome 13, 21, or 22. UPD is the inheritance of a chromosome pair from only one parent, with no contribution from the other parent. The problems associated with this are potentially due to imprinting. Imprinting refers to the fact that some genes are expressed only on the paternal homolog and other genes are expressed only on the maternal homolog. UPD can occur by nondisjunction involving the Robertsonian translocation [for example, a der(13;14)] and one of the two chromosomes involved in the Robertsonian translocation [for example, a chromosome 14]; the gamete would contain the der(13;14) and the free-standing 14. This would result in trisomy 14 upon fertilization with a normal gamete. UPD can occur when there is trisomy rescue and one free-standing chromosome 14 is lost. If the free-standing 14 remaining is from the same parent as the Robertsonian translocation, then this results in uniparental disomy (43,46).

The best examples of imprinted diseases in humans are Prader-Willi syndrome (PWS) and Angelman syndrome (AS) (44). These syndromes are distinct phenotypically and can be due to an interstitial deletion of the proximal long arm of chromosome 15 in 70% of cases of both syndromes. For PWS, the paternally derived chromosome 15 exhibits the deletion and for AS it is the maternally derived chromosome 15 that is deleted. About 25% of the cases of PWS are due to maternal UPD, but less than 5% of the cases of AS are due to paternal UPD. Paternal and maternal UPD for chromosome 14 have also been observed and both have been reported to be associated with an abnormal phenotype (43,46). UPD for chromosomes 14 and 15 have been observed in both inherited and de novo cases involving Robertsonian translocations with chromosome 14 and/or 15.

The frequency of UPD is currently unknown but appears to be low, making counseling regarding risks difficult. One ongoing study (45) has found UPD in 3 of 99 prenatal cases. UPD was observed only in the cases involving isochromosome, i.e., 13;13 and 14;14. Therefore, the risk appears to be very high when homologous chromosomes are involved and low when nonhomologous chromosomes are involved. Additional data are needed, but testing for UPD should be considered when Robertsonian translocations involving these chromosomes are encountered during prenatal testing (43,45,46).

3. Inversions

An inversion results when two breaks occur in a single chromosome and the segment between the two breaks flips or reverses direction. If the inversion includes the centromere, with one break in the long arm and one in the short arm, it is called a pericentric inversion. If it involves just a single arm of the chromosome, it is a paracentric inversion. During meiosis an inversion loop is formed to allow for the pairing of the homologs. For pericentric inversions, crossing-over within the inversion loop results in duplication/deficient gametes and may lead to unbalanced offspring. For paracentric inversions, crossing-over theoretically results in acentric and dicentric products that are not compatible with viable offspring (however, see below).

A common pericentric inversion seen in the southwestern United States is an inversion 8 observed predominantly in the Hispanic population of southern Colorado and northern New Mexico (18). Crossing-over in carriers of this inversion 8 results in the recombinant 8 phenotype. Risk figures for the recombinant 8 have been established due to extensive family pedigrees with this inversion (18).

What is the risk for an inversion carrier to produce offspring with a recombinant chromosome? As with translocations, the family pedigree provides the best estimate of risks for the carriers of inversions. Information from the literature about families with similar inversions can also provide risk estimates. In general, the longer the segment involved, the greater is the chance that crossing-over will occur and result in a potentially viable but unbalanced recombinant chromosome. Again, as with translocations, the presence of previous offspring with a recombinant chromosome is associated with a higher risk of recurrence, estimated to be in the range of 5–10% (14). If there are no viable offspring with a recombinant chromosome, the risk to carriers is 1–3% (14). The risks for live birth with unbalanced chromosome constitutions associated with paracentric inversions has generally been considered very low or even zero because dicentric and acentric fragments are formed when crossing-over occurs in these cases. A recent multicenter study confirmed that carriers of paracentric inversions have a low risk of having offspring with a recombinant chromosome (19). Therefore, prenatal chromosome diagnosis should be offered to all carriers of paracentric inversions as well as to carriers of pericentric inversions.

There are several exceptions to offering prenatal chromosome diagnosis to carriers of inversions (14). These exceptions include the common inversion of chromosome 9, inv(9)(p11.2q13), and the less common inversion of chromosome 1, both involving the inversion of the heterochromatin from the long arm into the short arm. These are both considered harmless variant chromosomes. The common inversion of the Y chromosome is also considered a harmless variant chromosome. While prenatal diagnosis is not recommended for carriers of the inversion Y chromosome, paternal chromosome studies are usually recommended when this chromosome is encountered during prenatal diagnosis. This is due to concern over a de novo versus an inherited event. If the inversion is de novo (paternity may be questioned), further investigation with SRY and centromeric and Yqh probes may be undertaken to ensure that this represents the standard inversion Y. Two other inversions are often encountered during prenatal diagnosis: inv(10)(p11.2q21.2) and inv(2)(p11.2q13). A recombinant chromosome has not been observed in the offspring of the carriers of either of these inversions, suggesting that prenatal diagnosis may not be necessary for these inversion carriers (14,20).

III. SPECIMEN TYPES

A. Amniotic Fluid

1. Traditional Amniocentesis at 15 Weeks or More

Mid-trimester sampling is the most frequent time for obtaining amniotic fluid. Details from the obstetrician's viewpoint can be found in recent reviews (21–23). The first few milliliters of the sample should be discarded, as this is the portion of the sample that is most likely to contain maternal cells (blood or cells from the abdominal wall or myometrium). For second-trimester amniocentesis, 20–30 ml of amniotic fluid should be collected. The volume of sample collected can affect turnaround time (TAT). When smaller volumes are obtained, TAT may be increased by 1–2 days. The specimen should be transferred to transport tubes that have been quality-control checked. We recommend the use of clear conical 15-ml culture centrifuge tubes. These are preferable to the dark brown (or other opaque) tubes that come in some preassembled amniocentesis trays because the sample volume and conditions can be assessed without transferring the sample to another tube. The tubes should be clearly labeled with the patient's name and other identifiers. Some offices show the labeled specimen tubes to the patient and have her initial the tubes and/or laboratory requisition as a quality-control step to avoid labeling errors. The sample should be transported to the laboratory at ambient temperature, with care taken to avoid temperature extremes. Amniocytes survive well when refrigerated (4°C), and may be stored at this temperature if delivery to the laboratory is delayed and temperature control is a problem. However, the sample must not be frozen, as this will disrupt the cells. Cells can sometimes withstand a "soft" freeze, so the laboratory should always try to establish a culture from a sample regardless of delays or temperature exposure. Grossly bloody amniotic fluid samples can take longer to grow, as the blood can interfere with cell attachment. If clots are present, the probability of culture failure increases greatly, as the clots can trap amniocytes. Multiple gestations need to be clearly indicated on all paperwork and on the sample tubes; the fetuses should always be designated the same by the obstetrician and by the laboratory.

2. Early Amniocentesis at 14 Weeks or Less

Early amniocentesis is similar to traditional amniocentesis, except that the volume of amniotic fluid removed is less (21). Usually 1 ml is removed for each week of gestation. Culture success rate is slightly reduced and TAT is 1–2 days longer (24). Early amniocentesis may not be indicated if a large volume of sample is needed for other tests.

B. Chorionic Villus Sampling

Chorionic villus sampling (CVS) is usually performed between 10 and 12 weeks gestation (21,23,25). Because of the concern about limb-reduction defects (21,26), most centers recommend that the procedure not be performed before 10 weeks except for certain clinical indications.

Early in development, the blastocyst differentiates into (a) the inner cell mass, which develops into the fetus, and into (b) the trophoblastic layer, which forms nonfetal structures such as amnion, chorion, and placenta. CVS collects the trophoblastic layer, which can be further broken down into the cytotrophoblastic layer and the mesenchymal core. The villi are cleaned to remove maternal decidua before processing for direct and short-term cultures. The cytotrophoblastic layer contains spontaneously dividing cells and is used for direct chromosome preparations. The cells of the mesenchymal core are disassociated from the villi by enzyme digestion and the mesenchymal cells are set up in culture. Direct chromosome preparations are usually of poor quality when compared to those obtained from short-term cultures. As results from the cultures can usually be obtained in 5–7 days, direct preparations are generally performed for specific medical management situations. As the cells from the cytotrophoblastic layer and the mesenchymal core are derived from different cell layers, direct and cultured preparations may yield different cytogenetic results. When this occurs, a follow-up amniocentesis is often needed to clarify the results.

C. Fetal Blood Sampling

Percutaneous umbilical blood sampling (PUBS), also referred to as cordocentesis, involves ultrasound-guided sampling of fetal blood from the umbilical cord (21,27). It is usually performed after 18 weeks of gestation. The fetal origin of the blood sample should be verified. The American College of Medical Genetics (ACMG) guidelines (28) requires that the cytogenetics laboratory must know if sample verification has been performed before reporting the final results on these specimens.

Fetal blood sampling for chromosome analysis has been used to help clarify chromosome mosaicism observed in amniocyte and CVS cultures and to provide rapid results when patients present late in the second trimester. It can also be performed in the third trimester when decisions concerning the mode of delivery need to be made. Cytogenetic results can be obtained within 48–72 h from short-term PHA-stimulated lymphocyte cultures. FISH can also be performed on the nondividing interphase cells.

IV. SAMPLE PROCESSING AND CELL CULTURING TECHNIQUES

Regulations and guidelines from the College of American Pathologists (CAP) and the ACMG exist for accessioning, culture setup, culturing, and harvesting (28,29). In general, each laboratory must develop protocols for reviewing each sample received, for determining which tests are to be performed, assigning specimen numbers, and processing samples without mixup. In addition, culture media must be quality-control checked to find the medium that best supports the growth of the specific specimen type. Turnaround time and failure rates should be actively monitored. Most laboratories use a supplemented media such as Amniomax (GIBCO/LTI, Gaithersburg, MD) and/or Chang (Irving Scientific, CA) for both amniocyte and CVS culture. In general, at least four cultures for amniocytes and CVS samples and at least two cultures for blood samples are established if the specimen volume is adequate. Both A-side cultures and both B-side cultures (representing the two original sample tubes for amniocytes) are split between at least two incubators.

There are two primary culture methods for both amniocytes and CVS samples: in situ and flask (30,31). The in situ method allows cells to attach and grow on coverslips as individual colonies, and the sample is usually harvested without subculturing. Analysis is performed on metaphases found in the in situ colonies. This method usually results in a faster TAT than the flask methods, as subculturing with subsequent slide making is avoided.

There are also other benefits in the analysis and interpretation of results based on the in situ versus the flask methods (see next section).

V. CHROMOSOME ANALYSIS

A. Banding Methods

The various banding and staining methods are reviewed in Chapter 5 by Mark. Most U.S. laboratories perform prenatal chromosome diagnosis on G-banded preparations, with a banding resolution of between 400 and 550 bands per haploid genome.

B. Guidelines for Chromosome Analysis

Regulations and guidelines from CAP and ACMG are available for the analysis of amniotic fluid, CVS, and fetal blood specimens (28,29).

C. Review of Cases

Most laboratories have several layers of review of karyotypes before the final results are signed out. This can include one or two technologists reading the cases, a potentially different individual karyotyping the cases, a supervisor review, and a director review. The final review by the director must be a band-by-band analysis of the karyotypes.

D. Common Autosomal Abnormalities

The common nonmosaic chromosome abnormalities involving the autosomes encountered during analysis of amniocytes include trisomy 21, trisomy 18, and trisomy 13. Rarely, are trisomy 8, trisomy 9, and trisomy 22, or other trisomies encountered. The three most common autosomal trisomies are also much less frequently observed in a mosaic state. In contrast, trisomies 8, 9, and 22 are more likely to be observed in mosaic states rather than as nonmosaic trisomies. All autosomes can be observed as true mosaics, as defined by the presence of an abnormal cell line in more than one culture. Hsu (32) presents a summary of the results of the chromosomes most frequently observed in true mosaicism, excluding trisomies 13, 18, 20, and 21. These latter trisomies are summarized in Hsu (13,33). Some of these, such as trisomy 8, present special problems when encountered prenatally because they have been described in liveborns with phenotypic abnormalities (34). Others, such as trisomy 20, have been rarely associated with abnormal phenotypes. Trisomy 20 is seen frequently enough in prenatal samples that there is a literature to guide one through these cases (13).

E. Common Sex Chromosome Abnormalities

Turner syndrome (45,X) is the most commonly encountered sex chromosome abnormality in cases tested because of abnormal ultrasound findings. 45,X is also seen more frequently than expected in women referred for testing because of abnormal serum screening results. XXY, XYY, and XXX fetuses are also identified, but at maternal age-specific rates, as their detection is not enhanced by abnormal ultrasound observations and/or positive maternal serum screening results.

The observation of a 45,X karyotype without abnormal ultrasound findings demands a cautious approach when reporting results. The referring physician should be consulted as to the ultrasound results and the sex of the fetus. If male genitalia are observed on ultrasound, there should be a reexamination of the fetal karyotype for any unusual-looking chromosome variants involving acrocentric chromosomes (sometimes these may not be apparent); these could contain a translocated Y chromosome. If necessary, FISH should be performed with SRY.

Sex discrepancy is sometimes observed between karyotype and ultrasound findings. Many cytogeneticists have had the experience of receiving a call from an obstetrician stating that the laboratory reported a normal female, but male genitalia were observed on an ultrasound examination. A review of laboratory documentation (including analysis sheets and slides for transcriptional errors, original specimen tubes, setup and harvesting records that include neighboring patients) should be performed to eliminate the possibility of laboratory error. Frequently a second ultrasound examination can reveal that the sex assignment on an earlier ultrasound was incorrect. If laboratory error is not apparent and an ultrasound is repeated and confirms the discrepancy between the genotype and phenotype, a repeat cytogenetic study should be considered. If this confirms the original result, other tests should be considered. A common finding is an XX male due to the translocation of SRY to the distal short arm of an X chromosome. This can be verified by FISH analysis with a probe containing the SRY gene. If the FISH finding is negative, other alternatives must be explored, such as genital ambiguity due to congenital adrenal hyperplasia, testicular feminization, or other sex-reversal situations.

A useful reference for the phenotype associated with sex chromosome abnormalities is by Robinson and co-workers (35).

F. Structural Abnormalities

When an apparently balanced structural chromosome abnormality is discovered unexpectedly during cytogenetic analysis, parental karyotyping is needed to determine if the rearrangement is familial or de novo in origin. If the rearrangement is familial, the carrier parent is an apparently phenotypically normal healthy individual, and the breakpoints appear to be identical to those seen in the fetus, the most likely outcome will be a normal baby. If it is de novo, then the possibility of false paternity needs to be raised, and the data presented by Warburton (36) for the risks for de novo rearrangements should be reviewed for the patient.

G. Chromosome Variants

Centromeric and acrocentric short arm variants may be encountered during chromosome analysis (13). If the general appearance of a centromeric variant is unusual or is present on a chromosome that usually does not exhibit such a variation, parental studies should be performed. Also, additional slides can be prepared and held for other studies, such as C-banding and/or FISH. The combination of C-bands and FISH should be considered if the additional material is C-band positive but does not consist of just alphoid DNA. If the pregnancy is late in the second trimester near the limit of pregnancy terminations, additional studies may be initiated before parental studies are completed. If time is not limited, the tests can be performed after parental studies, if still necessary. Sometimes a variant in fetal cells can appear different than what is seen in the parental blood cells. This is most likely due to differences in staining properties or chromosome condensation between cultured AF or CVS cells and blood cells.

Variants of the short arm region of the acrocentric chromosomes can also present a challenge. C-banding and a variety of FISH probes may need to be performed depending on the appearance of the region in question. If cytologic satellites are still present, the questionable region is most likely a variant. It is often useful to perform FISH studies to verify this interpretation. A useful probe is a ribosomal RNA gene probe that hybridizes to both active and inactive ribosomal genes in the stalk region on acrocentric chromosomes. This is preferred to Ag-NOR staining, as the silver staining detects only the active NOR regions. Therefore, with Ag-NOR, the origin of unstained regions is frequently still in question, depending on the pattern of silver staining. FISH studies with a rRNA gene probe also have an advantage in that a chromosome-specific centromeric probe can be included to ensure positive identification of the chromosome involved. Additional FISH probes, such as the beta-satellite and chromosome 15-specific classical satellite III probes, can also be very useful.

Chromosome 15 short arm variants encountered are usually one of two types. The presence or absence of stalks and satellites distal to the extra short arm material provides a useful guide to identify the extra material (AN Lamb and CH Lytle, unpublished data). If stalks and satellites are present, the extra material is most likely classical satellite III material. If stalks and satellites are absent, the extra material is most likely derived from the Yqh region (AN Lamb and CH Lytle, unpublished data). Both of these situations usually represent normal variant chromosomes, although material of Yqh origin can be of concern. In this case, it must be determined that active Y chromosome material is not present, especially in a 46,XX fetus.

As discussed above, some rare 45,X cases may have Y-chromosome material translocated to the short arm of an acrocentric chromosome. As these can appear to be typical variant chromosomes, the key to detecting these cases is the lack of the usual ultrasound findings associated with monosomy X and the discrepancy in fetal sex identified by ultrasound. Depending on the ultrasound results, FISH using SRY and/or the Y centromere and Yqh probes may be necessary.

C-band and FISH studies may not define the exact nature of all variants. Therefore, if a variant is of concern, parental karyotyping should be performed. The possibility of a balanced or unbalanced translocation should be considered if the extra material cannot be defined. At times, it may only be possible to conclude that the additional material appears identical to that present in a parent, and that its nature and origin cannot be further defined.

H. Supernumerary Marker Chromosomes

Supernumerary marker chromosomes can present as (a) mosaic or nonmosaic, (b) small, medium, or fairly large (approximately G-group size), and (c) bisatellited, unisatellited, or nonsatellited. Parental chromosome studies are indicated when a marker is identified in prenatal studies to check for the presence of a marker in a mosaic or nonmosaic state in either parent.

C-banding should be performed on the majority of markers to help define the extent of the heterochromatin and euchromatin present, as this may provide the only useful information available for counseling. Also, the most useful guide in cases of de novo markers is the data of Warburton (36).

FISH technology now allows for the identification of the chromosomal origin of most marker chromosomes. However, even if the chromosomal origin is determined by FISH, the information may not provide useful prognostic information for counseling (with the exception of chromosome 15, see below). First, the number of published cases characterized by FISH is limited (38). In addition, probes adjacent to the centromere are not available to see if markers derived from the same chromosome contain similar sequences, and finally, the level and distribution of mosaicism may make comparison difficult.

The most useful reference for marker chromosomes characterized by FISH is a recent review (38) emphasizing randomly ascertained cases. Unfortunately, no genotype/phenotype correlation was found for chromosomes other than 15 and 22. Therefore, most situations will have to rely on the pre-FISH data, with risks for serious congenital anomalies estimated to be between 11 and 15% for de novo markers (36).

FISH is useful for providing prognostic information for the bisatellited marker chromosomes derived from chromosome 15. These markers, also referred to as inv dup(15) or psu idic(15;15) markers, are of two types: (a) those with the PWS/AS region present that are associated with an abnormal phenotype (39), and (b) those without the PWS/AS region and are associated with a normal phenotype, or less frequently with a PWS or AS phenotype (39,40). The majority of cases in the second category are normal, with an undetermined number associated with PWS or AS due to UPD (40). Evidently, the small dic(15;15) is what remains from trisomy rescue and serves to signal that UPD may be present. A recent study of 17 prenatal cases with de novo small marker chromosomes identified two cases with maternal UPD 15, for a frequency of 12% (40). UPD testing would also be recommended for unisatellited or nonsatellited chromosome 15 markers.

When a marker chromosome is derived from an X chromosome, as often seen in 45,X/46,X,+r(X), further characterization of the marker is needed to determine if XIST is present. Small ring X chromosomes that lack XIST and still have significant euchromatic material present have been associated with an abnormal phenotype that includes mental retardation (42). If XIST is present, then the ring is usually inactivated and is associated with Turner syndrome. If the marker is derived from Y chromosome, then in addition to the difficulties of the prediction of the phenotypic sex (35), a risk of gonadoblastoma must also be considered.

I. Mosacisim and Pseudomosaicism

Distinguishing between pseudomosaicism and mosaicism is of concern when an abnormal cell is found during prenatal chromosome diagnosis. It is often difficult to determine how much additional work should be performed. Useful guidelines have been published by Hsu et al. (33) and should be consulted.

VI. MOLECULAR TECHNIQUES

FISH is a useful adjunctive technique for cytogenetic analysis and has been referred to throughout this chapter and can also be found as the main topic in Chapters 6 (Blancato and Haddad) and 17 (Mark). Also, see Ref. 41.

Specific microdeletion syndromes, such as DiGeorge/velocardiofacial syndrome at 22q11.2, Williams syndrome at 7q11.23, Prader-Willi and Angelman syndromes at 15q11.2-q12, to mention some of the most common, may be tested for during prenatal diagnosis (41,47).

FISH, using painting probes and single-locus probes, is also used to characterize and/or clarify rearrangements encountered during prenatal diagnosis.

Interphase FISH can provide rapid numerical information under specific circumstances on amniocytes, CV cells, fetal blood cells, and other cells of fetal origin (41). The limitations of this approach should be made clear to the physician and patient.

VII. GENETIC COUNSELING FOR PRENATAL DIAGNOSIS

Genetic counseling is a communication process that deals with the human problems associated with the occurrence, or risk for occurrence, of a genetic disorder. The counselor attempts to assist the individual or family to (a) comprehend the medical facts including the prognosis for the disorder and the available management, (b) appreciate the way heredity contributes to the disorder and therefore to the risk for occurrence/recurrence, (c) understand the options for dealing with the risk, and (d) choose the course of action that seems appropriate in view of the identified risk and family goals and to act in accordance with that decision. The counseling process is undertaken with strict adherence to the principles of nondirectiveness and patient confidentiality.

The process of genetic counseling involves five basic steps: intake, establishment of diagnosis, family history, pregnancy and medical history, and counseling. During the intake, basic socioeconomic and demographic information is obtained as well as determination of the patient's reasons for seeking counseling and his/her/their expectations of the process. In the next stage, establishment of a diagnosis, the counselor attempts to confirm that the stated reason for the consultation is indeed an accurate representation of the risks faced by the individual or family. This may involve reviewing medical records, conducting a physical examination, or obtaining additional laboratory evaluations. Frequently an individual or family will present with the belief that a certain disorder or disease is the correct diagnosis when, upon further investigation, it is apparent that the actual problem is different. This may lead to a very different prognosis, marked alterations in the quoted occurrence/recurrence risk, different approaches to evaluation and management, and different mechanisms for avoidance of the problem in future offspring or other family members. The family history can also assist in the determination of the correct diagnosis, but may also identify other potential risks. Medical and pregnancy history, likewise, can provide insight into the diagnosis or indications of other areas of concern. After all the information described above is obtained, the counseling process seeks to provide both information and support.

In the context of prenatal genetic testing, genetic counseling attempts to achieve the genetic counseling goals delineated above by: (a) providing information, (b) helping the patient/couple make a decision about whether or not to pursue testing, (c) addressing the issue of waiting for results, and (d) helping the couple deal with abnormal results if they occur.

Information regarding prenatal genetic testing is provided through a combination of written materials and personal communication. The topics covered include a discussion of the procedures, risks of the procedures, the laboratory techniques utilized, the accuracy of the testing, and the options available if an abnormality is identified.

The counselor's primary objective is to involve the patient/couple in the decision-making process. In order to do this, the couple's attitude toward the pregnancy must be explored. Topics covered include whether the pregnancy was planned or not, the difficulty encountered in conceiving, whether there are other living children, and/or whether there have been previous pregnancy losses. Perception of the risk for occurrence/recurrence of the disorder or disease in question must be established as well as the perception of the burden of the disorder. It is important to determine if the perceived burden is based on personal experience or on information provided by others. An attempt is made to determine whether the patient's/couple's reaction is based on an emotional response or measured consideration of the numerical risk/perceived burden and to reassure the patient/couple that both should be taken into consideration in their decision-making process.

Integral to this process is a presentation of the background risk for birth defects and genetic disorders. The population risk for birth defects is believed to be in the range of 2–4%. The patient's risk needs to be presented and evaluated in this context.

The patient/couple is also prepared for the possibility of an abnormal result and made aware of their options if a problem is identified. As part of this "results" part of the genetic counseling there is usually a discussion about the time required to complete the studies requested, how results are going to be transmitted, plans for a subsequent meeting if an abnormal result is obtained or if there are questions about the result.

Perhaps the most difficult aspect of genetic counseling is the nondirective position that needs to be adopted by the counselor. In order to deflect the frequently asked "What would you do?" or "What do you think I should do?," the counselor should develop the concept of the best choice for the patient or couple given the situation at hand. It is frequently necessary to stress that none of the choices may be obviously preferable, but that one option will usually evolve as the best course of action given all considerations.

When abnormal results are obtained, the options of termination of the pregnancy, continuing the pregnancy and placing the child for adoption, and continuing the pregnancy and keeping the baby should be explored. The counselor will find it necessary to provide additional information regarding the diagnosis and its prognosis in order to allow the patient/couple to make a decision on how to proceed. Frequently, additional resources, such as providers who specialize in the care of affected individuals, will be needed to provide additional information regarding the diagnosis. Once the decision about whether or not to continue the pregnancy is made, the counselor frequently will need to provide emotional support and will need to identify resources to assist the patient/couple in coping with the situation.

In counseling for screening rather than diagnostic tests, such as the multiple-marker biochemical test for Down syndrome, the counselor needs to stress the difference between a screening test and a diagnostic test. Part of the counseling in this situation should be to make the patient aware that the screening result may place her in a situation where she must make decisions about whether or not to undergo invasive diagnostic testing under severe time constraints. The potential for abnormal results from the diagnostic testing and the issues that such results would bring up, such as whether or not to continue with the pregnancy, what alterations need to be made in her obstetrical care, and whether the conduct and location of labor and delivery need to be changed, also need to be addressed. Frequently, patients tell us that they would not have agreed to the screening test if they had been aware of the "slippery slope" of additional testing and decision making, regarding their pregnancy that has resulted.

Similarly, patients opting to "screen" the well-being of their fetus through ultrasound need to be aware not only of the potential for abnormal findings but also the possibility of identifying normal variations that may imply an increased risk for certain birth defects. Again, patients need to be aware that these findings may place them in the uncomfortable position of having to make decisions in a very short period of time that they may have otherwise wished to avoid. With ultrasound this problem is heightened because of the parental bonding with the fetus that commonly occurs during the ultrasound study.

With any test, diagnostic or screening, the accuracy, sensitivity, specificity, and limitations must be well delineated. Provision of a fair appraisal of the risk associated with the procedure is mandatory. Alternatives in evaluation/diagnosis need to be explored. When these have been done within the context of risk identification and perception of burden described above, informed consent to either accept or decline a test can be obtained.

REFERENCES

1. Serr DM, Sachs L, Danon M. Diagnosis of sex before birth using dell from the amniotic fluid. Bull Res Council Isr 1955; 58:137.
2. Fuchs F, Riis P. Antenatal sex determination. Nature 1956; 177:330.
3. Steele MW, Breg WR. Chromosome analysis of human amniotic-fluid cells. Lancet 1966; i:383–385.
4. Jacobson CB, Barter RN. Intrauterine diagnosis and management of genetic defects. Am J Obstet Gynecol 1967; 99:795–807.
5. Valenti C, Schutta EJ, Kehaty T. Prenatal diagnosis of Down's syndrome. Lancet 1968; 2:220.
6. Nadler HL. Antenatal detection of hereditary disorders. Pediatrics 1968; 42: 912.
7. Mohr J. Foetal genetic diagnosis: development of techniques for early sampling of foetal cells. Acta Pathol Microbiol Scand 1968; 73:73.
8. Hobbins JC, Mahoney MJ. In utero diagnosis of hemoglobinopathies: technique for obtaining fetal blood. N Engl J Med 1974; 290:1065–1067.
9. Hook EB. Prevalence, risks and recurrence. In: Brock DJH, Rodeck CH, Ferguson-Smith MA, eds. Prenatal Diagnosis and Screening. New York: Churchill Livingstone, 1992:351–392.
10. Wald NJ, Kennard A. Prenatal screening for neural tube defects and Down syndrome. In: Rimoin DL, Connor JM, Pyeritz RE, eds. Emery and Rimoin's Principles and Practices of Medical Genetics, 3rd ed. New York: Churchill Livingstone, 1996:545–562.
11. Haddow JE, Palomaki GE. Prenatal screening for Down syndrome. In: Simpson JL, Elias S, eds. Essentials of Prenatal Diagnosis. New York: Churchill Livingstone, 1995.
12. Ville YG, Nicolaides KH, Campbell S. Prenatal diagnosis of fetal malformations by ultrasound. In: Milunsky A, ed. Genetic Disorders and the Fetus, 4th ed. Baltimore: Johns Hopkins University Press, 1998:750–811.
13. Hsu LYF. Prenatal diagnosis of chromosomal abnormalities through amniocentesis. In: Milunsky A, ed. Genetic Disorders and the Fetus, 4th ed. Baltimore: Johns Hopkins University Press, 1998:179–248.
14. Gardner RJM, Sutherland GR. Chromosome Abnormalities and Genetic Counseling, 2nd ed. New York: Oxford University Press, 1996.

15. Daniel A, Hook EB, Wulf G. Collaborative U.S.A. data on prenatal diagnosis for parental carriers of chromosome rearrangements: risks of unbalanced progeny. In: Daniel A, ed. The Cytogenetics of Mammalian Autosomal Rearrangements. New York: Alan R. Liss, 1988:73–162.
16. Daniel A, Hook EB, Wulf G. Risks of unbalanced progeny at amniocentesis to carriers of chromosome rearrangements: data from United States and Canadian Laboratories. Am J Med Genet 1989; 31:14–53.
17. Iselius L, Lindsten J, Aurias A, et al. The 11q;22q translocation: a collaborative study of 20 new cases and analysis of 110 families. Hum Genet 1983; 64:343–355.
18. Sujansky E, Smith ACM, Prescott KE, Freehauf CL, Clericuzio C, Robinson A. Natural history of the recombinant (8) syndrome. Am J Med Genet 1993; 47:512–525.
19. Pettanatti MJ, Rao PN, Phelan MC, et al. Paracentric inversions in humans: a review of 446 paracentric inversions with presentation of 120 new cases. Am J Med Genet 1995; 55:171–187.
20. Collinson MN, Fisher AM, Walker J, Currie J, Williams L, Roberts P. Inv(10)(p11.2q21.2), a variant chromosome. Hum Genet 1997; 101:175–180.
21. Elias S, Simpson JL. Prenatal diagnosis. In: Rimoin DL, Connor JM, Pyeritz RE, eds. Emery and Rimoin's Principles and Practices of Medical Genetics, 3rd ed. New York: Churchill Livingstone, 1996:563–580.
22. MacLachlan NA, Amniocentesis. In: Brock DJH, Rodeck CH, Ferguson-Smith MA, eds. Prenatal Diagnosis and Screening. New York: Churchill Livingstone, 1992:13–24.
23. Stranc LC, Evans JA, Hamerton JL. Chorionic villus sampling and amniocentesis for prenatal diagnosis. The Lancet 1997; 349:711–714.
24. Lockwood DH, Neu RL. Cytogenetic analysis of 1375 amniotic fluid specimens from pregnancies with gestational age less than 14 weeks. Prenat Diagn 1993; 13:801–805.
25. Silverman NS, Wapner RF. Chorionic villus sampling. In: Brock DJH, Rodeck CH, Ferguson-Smith MA, eds. Prenatal Diagnosis and Screening. New York: Churchill Livingstone, 1992:25–38.
26. Brambati B, Tului L. Prenatal genetic diagnosis through chorionic villus sampling. In: Milunsky A, ed. Genetic Disorders and the Fetus, 4th ed. Baltimore: Johns Hopkins University Press, 1998:150–178.
27. Nicolini U, Rodeck CH. Fetal blood and tissue sampling. In: Brock DJH, Rodeck CH, Ferguson-Smith MA, eds. Prenatal Diagnosis and Screening. New York: Churchill Livingstone, 1992:39–50.
28. American College of Medical Genetics. Standards and Guidelines: Clinical Genetics Laboratories. Bethesda, MD: ACMG, 1984.
29. College of American Pathologists, Commission on Laboratory Accreditation. Inspection Checklist, Cytogenetics. Northfield, IL: CAP, 1997.
30. Gosden C. Cell culture. In: Brock DJH, Rodeck CH, Ferguson-Smith MA, eds. Prenatal Diagnosis and Screening. New York: Churchill Livingstone, 1992:85–98.

31. Hoehn H. Fluid cell culture. In: Milunsky A, ed. Genetic Disorders and the Fetus, 4th ed. Baltimore: Johns Hopkins University Press, 1998:128–149.
32. Hsu LYF, Yu MT, Neu RL, et al. Rare trisomy mosaicism diagnosed in amniocytes, involving an autosome other than chromosomes 13, 18, 20, and 21: karyotype/phenotype correlations. Prenat Diagn 1997; 17:201–242.
33. Hsu LYF, Kaffe S, Jenkins EC, et al. Proposed guidelines for diagnosis of chromosome mosaicism in amniocytes based on data derived from chromosome mosaicism and pseudomosaicism. Prenat Diagn 1992; 12:555–573.
34. Sanford Hanna J, Neu RL, Barton JR. Difficulties in prenatal detection of mosaic trisomy 8. Prenat Diagn 1996; 15:1196–1197.
35. Robinson A, Linden MG, Bender B. Prenatal diagnosis of sex chromosome abnormalities. In: Milunsky A, ed. Genetic Disorders and the Fetus, 4th ed. Baltimore: Johns Hopkins University Press, 1998:249–285.
36. Warburton D. De novo balanced chromosome rearrangements and extra marker chromosomes identified at prenatal diagnosis: clinical significance and distribution of breakpoints. Am J Hum Genet 1991; 49:995–1013.
38. Crolla JA. FISH and molecular studies of autosomal supernumerary marker chromosomes excluding those derived from chromosome 15: II. Review of the literature. Am J Med Genet 1998; 75:367–381.
39. Webb T. Inv dup(15) supernumerary marker chromosomes. J Met Genet 1994; 31:585–594.
40. Christian SL, Mills P, Das S, Ledbetter DH. High risk of uniparental disomy associated with amniotic fluid containing de novo small supernumerary marker chromosomes. Am J Hum Genet Suppl 1998; 63:A11(#52).
41. Schwartz S. Molecular cytogenetics and prenatal diagnosis. In: Milunsky A, ed. Genetic Disorders and the Fetus, 4th ed. Baltimore: Johns Hopkins University Press, 1998:286–313.
42. Jani MM, Torchia BS, Pai GS, Migeon BR. Molecular characterization of tiny ring X chromosomes from females with functional X chromosome disomy and lack of cis X inactivation. Genomics 1995; 27:182–188.
43. Engel E. Uniparental disomies in unselected populations. Am J Hum Genet 1998; 63:962–966.
44. Nicholls RD, Saitoh S, Horsthemke B. Imprinting in Prader-Willi and Angelman syndromes. Trends Genet 1998; 14:194–200.
45. Shaffer L. Workshop on uniparental disomy. Am Soc Hum Genet, Denver, CO, Oct 1998.
46. Ledbetter DH, Engel E. Uniparental disomy in humans: development of an imprinting map and its implications for prenatal diagnosis. Hum Mol Genet 1995; 4, Review:1757–1764.
47. Spinner N, Emanuel B. Deletions and other structural abnormalities of the autosomes. In: Rimoin DL, Connor JM, Pyeritz RE, eds. Emery and Rimoin's Principles and Practices of Medical Genetics, 3rd ed. New York: Churchill Livingstone, 1996:999–1025.

9

Cytogenetics of Reproductive Wastage: From Conception to Birth

Dorothy Warburton
College of Physicians and Surgeons, Columbia University, New York, New York

Although cytogenetic abnormalities contribute significantly to human morbidity and mortality as causes of neonatal death, mental retardation, congenital abnormalities, reproductive problems, and cancer, their major concentration is found among those human conceptions which do not survive to term. Embryonic or fetal loss due to developmental abnormality reduces the incidence of cytogenetically detectable genetic abnormalities from as many as 30% of all conceptions to approximately 0.5% of livebirths. In this chapter we shall examine methods for studying human gametes and products of conception, the incidence and types of cytogenetic anomalies detected, and their pathologic correlates.

I. CHROMOSOME STUDIES OF HUMAN GAMETES

It has been estimated that half of all human conceptions are lost before pregnancy is usually recognized, i.e., the time of the first missed menstrual period, when the embryo is 3–4 weeks of age. Data using early pregnancy detection by serum HCG indicate that 20% of all pregnancies which are implanted may be lost before a positive urine test would occur. Estimates of losses before implantation must be based on very indirect inference, but tend to be of the same order of magnitude (1). Because of the difficulties of obtaining and analyzing such early material, there are no direct studies of the incidence of cytogenetic abnormalities in these early losses.

The best estimate of the frequency of cytogenetic abnormalities at conception would therefore come from knowledge of their frequency in gametes. Such studies have been technically challenging, although developments over the last 15 years have begun to provide reliable information. Sperm DNA exists in a highly condensed state, and chromosomes are visible only after fertilization, when they condense in the pronucleus. Oocytes are difficult to retrieve, and remain arrested in meiotic prophase until the time of ovulation, when they can be most easily examined in metaphase II. However, at this stage the chromosomes are highly contracted and difficult to count and identify.

A. Methods of Sperm Analysis

The first studies of chromosome abnormalities in human sperm were made possible because of the development of the ''humster'' technique, originally described by Rudak et al. (2) and used most extensively by two research groups led by Martin (3) and Brandriff (4). In this procedure, hamster oocytes obtained by superovulation are fertilized by human sperm in vitro, after digestion of the zona pellucida by trypsin (Fig. 1). After 24 h in culture the fertilized ova are spread, fixed onto slides, and banded by quinacrine or Giemsa banding. Both egg and sperm chromosomes can then be observed in metaphase. While difficult and labor intensive, this technique allows a complete karyotypic analysis of the sperm chromosome complement (Fig. 2).

Recently, the development of fluorescent in situ hybridization (FISH) probes for aneuploidy detection have enabled studies of specific aneuploidies in very large numbers of sperm. This process requires pretreatment of sperm to decondense the very contracted chromatin, followed by FISH with chromosome-specific probes (5). Multicolor FISH permits the examination of several chromosome pairs at one time, and also allows a distinction between diploid sperm and those with aneuploidy. In general this technique is not suitable for the analysis of chromosome rearrangements, although prespecified chromosome rearrangements can be studied with suitable probes (6).

B. Methods of Oocyte Analysis

Oocytes enter meiosis during fetal life, when synapsis, recombination, and chiasmata formation all occur in months 4 and 5 of gestation. Very few attempts have been made to analyze human oocytes at this stage, due to the difficulties of resolving individual chromosome pairs in the early stages, and

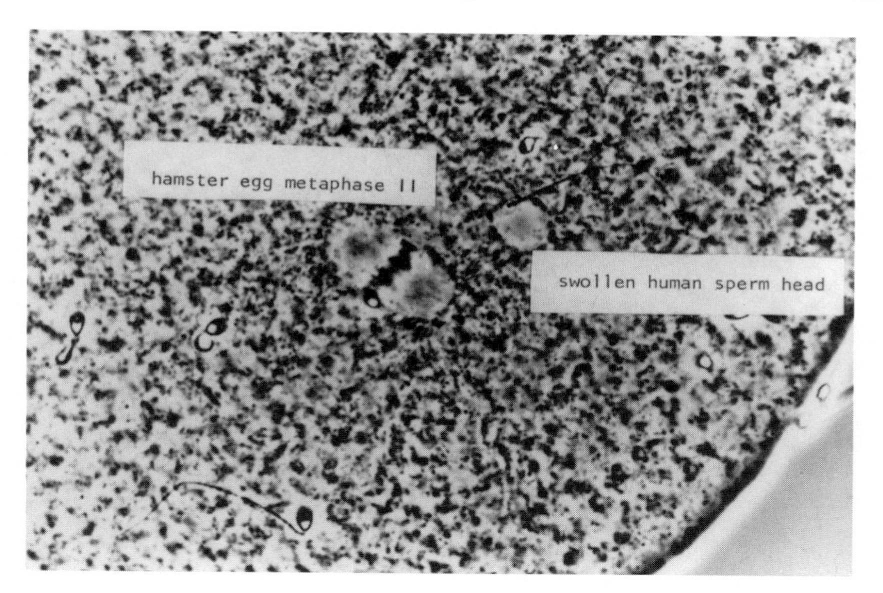

Figure 1 Hamster egg which has been penetrated by a human spermatazoan. The hamster pronucleus is undergoing meiosis II. The sperm head is beginning to enlarge, and the tail is visible inside the egg. (From M. T. Yu and D. Warburton, unpublished.)

the indistinct morphology of later stages. Recently, the use of chromosome-specific FISH paints has made it possible to follow individual chromosome pairs in meiotic prophase (7). This technique can provide information concerning chromosome pairing and recombination in human oocytes. For example, an examination of pairing in oocytes with trisomy 18 showed evidence of trivalent formations as well as bivalents plus a univalent (8).

The widespread use of in vitro fertilization (IVF) now provides an opportunity to study the chromosome complements of large numbers of mature human oocytes for the first time. Unfertilized oocytes retrieved for IVF can be observed in metaphase II at culture times varying from 24 to 72 h. The best chromosome preparations are made through the adaptation of a method described by Mikamo and Yamiguchi (9), where oocytes are dried onto slides through a process of gradual fixation. These chromosome preparations are still far from optimal. The chromosomes are highly contracted and banding is not possible, so chromosomes can only be identified by group. However, special staining or FISH techniques can be used to identify particular chromosome pairs (10).

Chromosome-specific FISH allows for the detection of aneuploidy in oocytes without the necessity for chromosome preparations. Observations on

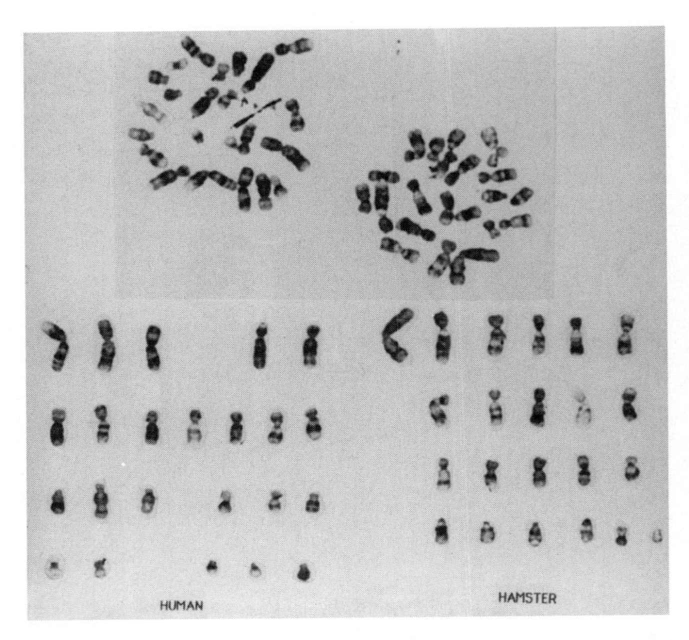

Figure 2 Metaphase chromosomes and karyotype from a "humster" preparation, 24 h after fertilization of the hamster egg by the human sperm. Both hamster and human chromosome sets remain distinct. (From M. T. Yu and D. Warburton, unpublished.)

both egg and polar body are sometimes possible, providing a confirmation of aneuploidy in both cases (11,12). The limitation of this technique is that an analysis is currently possible for only a few chromosomes at a time.

C. Chromosome Aberrations in Sperm

1. Incidence of Chromosomally Abnormal Sperm

The "humster" technique is labor intensive and difficult, and has therefore been applied on a large scale by only a small number of laboratories. Their efforts have provided chromosome preparations from approximately 20,000 sperm from normal males (13,14). Although there are some differences both among laboratories and among males, aneuploidy is found in about 1–2% of sperm, while structural abnormalities are found in about 8% of sperm (3). The frequency of overall sperm aneuploidy does not appear to depend upon paternal age. While the total number of aneuploid sperm analyzed is quite small, trisomy 1, 21, and the sex chromosomes appear to occur more frequently than trisomy for other chromosomes. A potential problem with this technique is that it can examine only sperm capable of fertilization (albeit of hamster oocytes).

FISH analysis of human sperm has been used by several groups to study aneuploidy for particular chromosome pairs. Major variation among studies was initially a problem due to technical factors in distinguishing diploidy from aneuploidy, but more recent results using multicolor FISH have been quite consistent. Diploidy is observed in about 0.2% of sperm. The frequency of any particular chromosomal disomy is approximately 0.1%, which is consistent with the overall aneuploidy rate of about 2% found in the chromosomal analyses. XY and YY disomy are about twice as common as disomy for other chromosomes (15,16). There are also reports of an effect of paternal age on sex chromosome aneuploidy (16,17), and of an increase of aneuploidy and other anomalies in sperm of males treated with chemotherapy (18). It is also interesting that significant male-to-male variation in the incidence of aneuploid sperm has been detected (19). This had previously been shown for structural aberrations using sperm chromosome preparations (4).

2. Sperm Analysis in Male Translocation Carriers

Carriers of Robertsonian or reciprocal translocations may produce chromosomally unbalanced gametes through the process of segregation during meiosis. Since many of the unbalanced chromosome constitutions will act as early embryonic lethals, the frequencies of various possible segregation products cannot be determined from observations at the time of prenatal diagnosis or livebirth. However, the ability to analyze individual sperm karyotypes via the humster assay allows direct observation of the segregation patterns of balanced translocations. This is of particular interest because male Robertsonian translocation carriers have a relatively low risk (1–2%) for chromosomally unbalanced offspring (20).

Sperm complements have been studied so far from six Robertsonian translocation carriers. In all cases the majority of sperm (72–97%) originated from alternate segregation. Normal and balanced sperm chromosome complements are equally frequent, as expected theoretically (14). The frequency of unbalanced sperm resulting from adjacent segregation was found to be compatible with the low frequency of observed unbalanced offspring, after accounting for the total lethality of some of the karyotypes and the reduced viability of the others.

For carriers of reciprocal translocations, about half of the sperm carried unbalanced complements. Products of adjacent-1 segregation were the most commonly observed, but products of adjacent-2 and 3:1 segregation were also seen frequently. Again, the higher frequency of unbalanced sperm than of unbalanced offspring is explained by reduced viability of unbalanced conceptions. The existence of large numbers of locus-specific FISH probes now makes it possible to design a clinically useful probe set to analyze the percentage of unbalanced sperm in any given male translocation carrier.

D. Chromosome Aberrations in Oocytes

Since the introduction of in vitro fertilization, the incidence of aneuploidy in metaphase II oocytes has been reported from many studies, with widely variant results (13). These inconsistencies are probably the result of the difficulty of interpreting the chromosome preparations, particularly in early studies where spreading was a problem. More recently, several groups (10,21,22) using an improved spreading method (9) have reported on several hundred metaphase II oocytes. About 10% of unfertilized oocytes are diploid; it is unknown whether these oocytes are less likely to be fertilized. The frequency of chromosomally aberrant haploid oocytes was 30% in the Angell study, 18% in the Kamiguchi study, and 17% in the Lim study. Since the frequency of unbalanced oocytes increases with maternal age, differences between studies may reflect differences in the age distribution of the subjects.

However, there is also disagreement concerning the predominant abnormality which is observed in these abnormal oocytes. Extra or missing chromosomes usually occur in oocytes due to nondisjunction occurring in either meiosis I or meiosis II. When an oocyte is observed at metaphase II, meiosis I nondisjunction occurring according to the classical model should be observed as an extra univalent (due to premature separation and random chromosome movement to the poles) or as an extra bivalent (due to "true" nondisjunction, in which chromosome pairs fail to separate at meiosis I). Neither of these types of oocytes was observed by Angell, who found only extra or missing half-univalents (chromatids) (Fig. 3). She interpreted this as the result of premature separation of the chromosome pair (bivalent) during meiosis I, with the chromatids of the unpaired univalents subsequently separating at meiosis I. The other two studies found some oocytes with extra whole univalents as well as those with chromatid aberrations, but interpreted some of the latter as artifacts. The differences in these studies are hard to explain, but may result from factors such as length of oocyte culture or the hormonal stimulation regimens for superovulation.

Angell (10) estimated that the abnormal oocytes she observed at metaphase II should result in approximately 7% disomic oocytes after completion of the second meiotic division. This is only slightly higher than estimates of the frequency of trisomy in all recognized conceptions based on

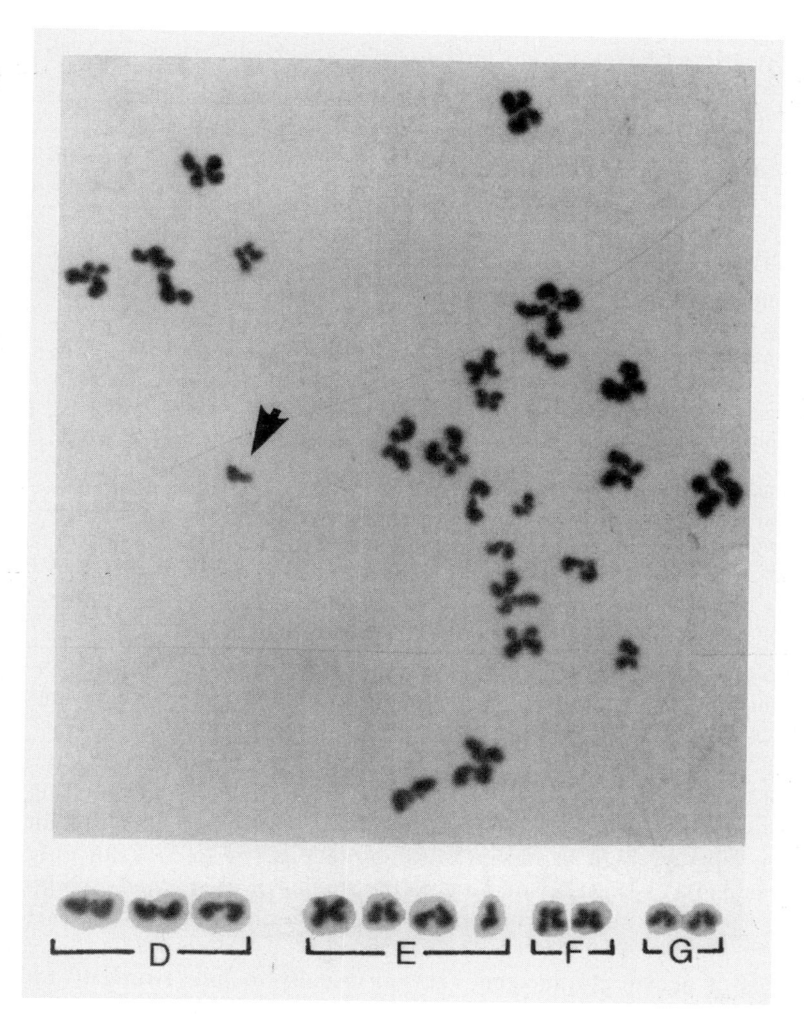

Figure 3 Metaphase II chromosomes from a human oocyte showing prematurely separated chromatids. (From Ref. 10, University of Chicago Press, © 1997 by The American Society of Human Genetics. All rights reserved.)

observations in spontaneous abortions (see below and Table 1), and similar to that estimated from only women over 35 years of age. Her data also showed that abnormalities involving chromosome 16 and the acrocentric chromosomes predominate among abnormal oocytes, which is compatible with the evidence from spontaneous abortions that these are the most com-

Table 1 Estimates of Frequency of Chromosomally Abnormal Conceptions from Gametes and Embryos

	Percent chromosome abnormality	Percent aneuploidy[a]	Percent trisomy (or estimate)
Sperm	10	2	1
Oocytes	30	15–20	7
Preimplantation embryos ("good")	30–40	15–20	8–10
Induced abortions:			
All	3–7	5	4
Maternal age > 35	10–15	9–14	9–14
Early spontaneous abortions			
All	8	5	4
Maternal age > 35	9–10	8	7

[a]Includes monosomy X.
Sources: Refs. 3, 10, 14, 21, 22, 27, 30, 33, 53.

mon aneuploidies. These oocyte analyses also are consistent with the expectation that the incidence of abnormal oocytes should increase with maternal age.

FISH analysis of oocytes and polar bodies has been performed on large numbers of oocytes, and provides evidence for both whole extra or missing univalents as well as chromatid abnormalities (11,23). However, it is difficult to reconcile the very high incidence of abnormalities (up to 33% for only chromosome pairs 13,18,21,X and Y) with the studies of oocyte and embryo karyotypes. This may reflect problems with the interpretation of the FISH data.

Studies of oocyte chromosomes do not usually include structural rearrangements because of the difficulty of performing detailed karyotype analyses on these specimens. Analysis of the parental origin of de novo chromosome rearrangements occurring in live births has consistently found a higher incidence of paternal origin, suggesting that the structural aberrations may occur considerably less often in oocytes than in sperm. On the other hand, studies of the parental origin of aneuploid conceptions found that 90% or more are due to an error occurring in the oocyte, consistent with the higher frequency of aneuploidy in oocytes.

Interpretation of the available data on oocytes must take into account that all studies to date have used material which may not be representative of the normal population. Oocytes retrieved for IVF are derived from women with reproductive problems, and who are older than the average for child-bearing. The oocytes which can be studied are also selected, since they are usually those remaining unfertilized. It is unknown whether such selection reflects genetic differences among oocytes.

II. CYTOGENETIC STUDIES OF EARLY HUMAN EMBRYOS

In vitro fertilization offers an opportunity to study not only oocytes, but also early preimplantation human embryos. Knowledge of the high frequency of chromosome aberrations in recognized human conceptions, and of the high rate of chromosomally abnormal gametes, leads to the expectation that the frequency of cytogenetic anomalies in early human embryos will be high. When used in conjunction with in vitro fertilization, preimplantation cytogenetic diagnosis has the potential to screen out aneuploid embryos before transfer.

A. Methods of Study of Preimplantation Embryos

The selection of embryos for cytogenetic analysis varies from study to study, but often involves using only those judged least likely to be successfully implanted, and commonly with abnormal morphology. However, several groups of apparently normal embryos have also been studied. For example, discarded male embryos can be studied when preimplantation diagnosis of sex was used to select for females because of the risk of X-linked disease (24). Standard cytogenetic preparations after colcemid treatment are difficult, albeit feasible, although the rate of successful karyotyping is only of the order of 30%. FISH analysis, using a limited number of chromosome-specific probes, has been developed for use in preimplantation diagnosis of aneuploidy. For this purpose only a single interphase cell from the cleavage embryo (25,26) is analyzed, in order to preserve the embryo for implantation. When used on discarded whole embryos, FISH is also useful to examine mosaicism, since almost all cells of the blastocyst are informative.

B. Incidence and Distribution of Chromosome Aberrations in Preimplantation Embryos

Most studies agree that the rate of chromosome abnormalities is greatest in embryos rated as poor-quality on the basis of their morphology or abnormal rates of development. Up to 85% of such abnormal embryos have major chromosome aberrations, such as haploidy or polyploidy, chromosome fragmentation, and mosaicism (27). Pooled results from a number of different studies showed that among 502 "good-quality" embryos which were successfully karyotyped, the frequency of chromosome abnormalities was 37% (27). Haploidy or polyploidy was found in 12%, aneuploidy in 15%, mosaicism in 9%, and structural abnormalities in 1%. As with the oocyte studies, a predominance of E-group aneuploidy and trisomy 16 was found in studies where this was examined (28). The numbers of hyperdiploid and hypodiploid embryos were quite similar, suggesting that the lack of hypodiploidy in recognized conceptions reflects selection at the implantation stage.

Larger numbers of preimplantation embryos have been studied using FISH, looking for aneuploidy with probes for a limited number of chromosomes. Chromosomes 1, 13, 16, 18, 21, X, and Y have all been examined in various combinations. Again, there is consistent evidence that arrested and morphologically abnormal embryos have the highest rates of chromosome abnormality, including a very large proportion with apparent mosaicism (24). Much of this mosaicism involves polyploid or haploid mosaicism in arrested or morphologically poor embryos and is associated with multinucleate cells. "Chaotic" embryos, with multiple different chromosome complements, are common among arrested or poorly developing embryos (as many as 80%), and are most frequent in those conceptions involving oocytes from older women (24,29). Even among "good-quality" embryos, as many as 40% were determined to have diploid/aneuploid mosaicism in some FISH studies (29).

The rates of aneuploidy and mosaicism observed in the FISH analyses are much higher than predicted from rates in either recognized spontaneous abortions or in gametes. Much of the mosaicism is hypothesized to occur because of mitotic irregularities in the developing embryos, and this may well be a function of the in vitro developmental environment. It is also likely that the FISH technique is producing some artifacts, since the aneuploidy and mosaicism rates using only probes for a few chromosomes exceed the total observed studies using full chromosome spreads. It is hard to generalize from these studies to normal in vivo development. However, the observed mosaicism has serious implications for the use of a single blastomere for preimplantation diagnosis of aneuploidy by FISH or for DNA diagnosis of single-gene disorders.

III. CYTOGENETIC STUDIES OF EARLY RECOGNIZED PREGNANCIES: INDUCED ABORTIONS

Early pregnancies terminated at 5–8 weeks by induced abortion for nonmedical reasons represent a relatively unselected group of early recognized human conceptions. Cytogenetic analysis of such pregnancies has been carried out in several areas of the world, and indicates that approximately 5–7% are chromosomally abnormal (30–32). This is also the estimate obtained by extrapolating from data on spontaneous abortions (see below and Table 1). This is much lower than the 30% abnormality rate observed in preimplantation embryos from IVF. The frequency of chromosome abnormality in induced abortions increases substantially with increasing maternal age, reaching 10–15% in women over 35 (31,33). This higher figure may be a more appropriate comparison to the IVF pregnancies, which are derived from a population of older women. Nevertheless, if the data on preimplantation embryos and oocytes are correct, there must be a substantial loss of chromosomally abnormal conceptions before the time of recognized pregnancy.

IV. CYTOGENETIC STUDIES OF SPONTANEOUS ABORTIONS

The high incidence of chromosome abnormalities among recognized pregnancies lost as spontaneous abortions has been known since the pioneering studies of Carr (34) and the Boués (35). A large number of cytogenetic surveys of spontaneous abortions have now been carried out, not only to study spontaneous losses per se, but also because such pregnancies are the most abundant source of chromosome abnormalities for epidemiologic (36) or genetic (37) analysis. The gestational age of pregnancies identified as spontaneous abortions ranges from about 4 weeks (2 weeks after the first missed menstrual period) to the lower limit of fetal viability outside the uterus. The definition of the latter has been variously defined among studies, ranging between 20 and 28 weeks.

A. Methods of Study of Spontaneous Abortion Material

Chromosome analysis of spontaneous abortions is usually carried out by short-term culture of fetal tissue. Embryonic or fetal parts may be used, if present, but often in early losses only embryonic membrane or placental villi are available for culture. The embryo may be rudimentary, fragmented, or macerated, and often is not found among the products of a dilation and curettage procedure, which is often the source of cultured tissue. Success rates in obtaining karyotypes from cultured fetal tissue are generally in the range from 60% to 90% [75% is the standard set by the American College of Medical Genetics (38)]. Failures are usually due to lack of viable tissue or contamination by bacteria or yeast, since the specimen is usually not collected under sterile conditions. While viable cells may persist for up to 5 days after expulsion, the freshness of the specimen and the length of retention in utero is also important. A common problem is the presence of only decidual tissue in a specimen that is submitted for culture, so that careful dissection of the specimen and rejection of those not containing villi or membranes is necessary. Failure to do so will be reflected in an excess of normal female karyotypes from products of conception.

A second method of chromosome analysis from spontaneous abortions is to carry out "direct" preparations utilizing uncultured villi, as is done for chorionic villus sampling (39). This has the advantages of being faster, not dependent upon the ability of the cells to grow in culture, and less liable to maternal cell contamination. However, the preparations are not usually of high quality and require access to very fresh and viable material.

B. Overall Incidence of Chromosome Abnormalities in Spontaneous Abortions

Numerous studies have demonstrated that approximately 50% of early spontaneous abortions (gestations up to 15 weeks) will have a chromosome abnormality, while for later abortions (15–24 weeks) the frequency is approximately 20%. Table 2 shows the frequency of various types of anomaly in three series of consecutive spontaneous abortions: one from New York City, in which fetal deaths up to 28 weeks were included as spontaneous abortions

Table 2 Distribution of Karyotypes in Spontaneous Abortions: Data from Three Studies

	New York City (Ref. 20)	Hawaii (Ref. 41)	Germany (Ref. 39)
Sample	4–28 wks	<500 g	5–25 wks
Karyotype	% of total (n = 3300)	% of total (n = 1000)	% of total (n = 750)
Normal	60.2	53.7	49.3
Trisomy (including mosaics and doubles)	22.6	22.7	31.5
Monosomy X	6.1	11.2	5.3
Triploidy	6.0	7.0	6.2
Tetraploidy	2.4	3.3	4.6
Unbalanced rearrangement	1.2	1.9	2.2
Balanced rearrangement	0.3	0.2	0.1
Other	0.6	0.2	0.8

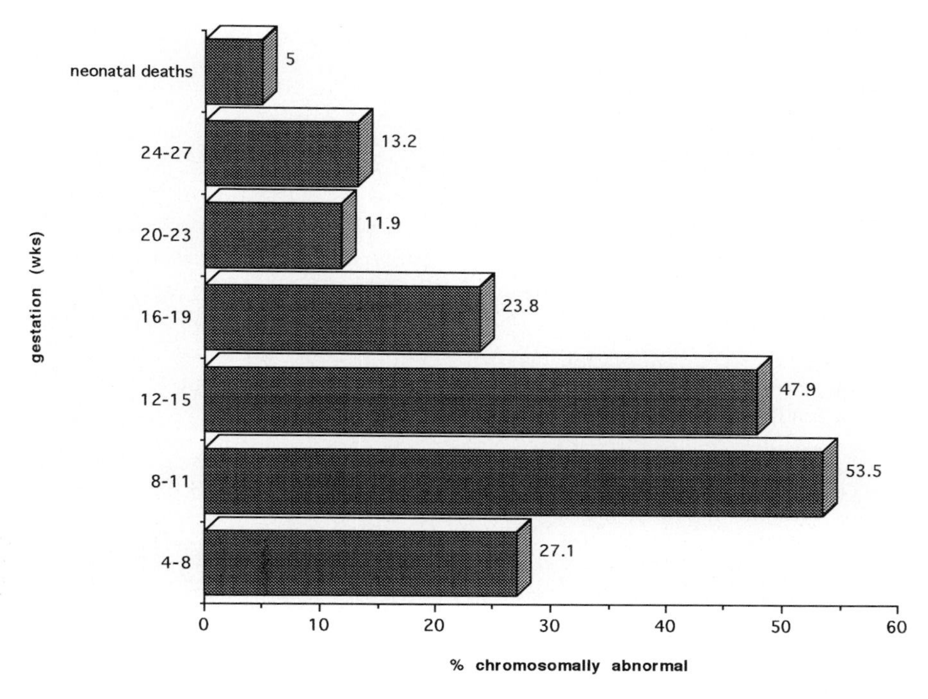

Figure 4 Percentage of chromosomal abnormalities in fetal deaths, by gestation. (Data from Ref. 71.)

(40), one from Hawaii in which specimens with fetuses over 500 g were excluded (41), and one from Germany (39) in which direct preparations were made from chorionic villi in specimens up to 25 weeks of gestation. The results of studies in all parts of the world have been remarkably consistent, suggesting that neither ethnic nor environmental variables are significant contributors to the distribution. Figure 4 shows the New York data subdivided by gestational age. In the following sections each type of chromosome abnormality found in spontaneous abortions is discussed individually, with respect to frequency, origin, and associated clinical or pathologic features.

C. Abnormalities of Number of Chromosome Sets (Ploidy)

1. Tetraploidy

Tetraploidy, the presence of four copies of each chromosome, is found in 2–3% of spontaneous abortions. At term it occurs only very rarely, most

Figure 5 Intact ''empty'' sac in spontaneous abortion with tetraploidy. (From Ref. 40.)

often as a diploid/tetraploid mosaicism. Tetraploidy can occur as a cultural artifact after long-term culture, and can also occur normally in some trophoblast cells in vivo. Its frequency among spontaneous abortions may thus be overestimated. It is also difficult to distinguish true tetraploidy/diploidy mosaicism from a cultural artifact.

Tetraploid specimens almost always have either a 92,XXXX or 92,XXYY sex chromosome constitution, apparently arising through a failed early mitotic division. Rare 92,XXXY tetraploid embryos have been found, due to trispermy (42). While the existence of a few apparently nonmosaic tetraploid full-term births suggests that this karyotype may sometimes be compatible with relatively normal development, 75% of spontaneous abortions with a tetraploid karyotype occur in specimens consisting of an apparently empty intact sac (Fig. 5), and almost all the rest contain only a very small, disorganized embryo (40).

2. Triploidy

Triploidy, the presence of three copies of each chromosome, is found in 6–7% of spontaneous abortions, implying a frequency of at least 1% of conceptions. The frequency is five times higher in IVF conceptions, which may reflect abnormal fertilization conditions favoring dispermy, or indicate a high percentage of loss before recognized conception.

Triploid spontaneous abortions have a very broad distribution of gestation lengths and specimen types. However, the majority consist of first-trimester losses with normal-appearing embryos of less than 30 mm in length, usually with development somewhat retarded for gestational age (40). A smaller number (7% of specimens in the New York series) present as fetuses over 30 mm in length. One well-documented presentation of a triploid conception is the so-called partial mole: a combination of cystic and normal villi (Fig. 6) in a placenta accompanied by a gestational sac. The frequency with which this occurs among triploid spontaneous abortions has been reported to be as high as 85% (43) and as low as 15% (44). In the unselected New York series, cystic villi were found in 40% of triploids, often accompanied by a well-developed embryo or fetus (40). Even though cystic villi are commonly found in triploid conceptions, they cannot be considered diagnostic of triploidy, since they are also commonly seen with other chromosome complements such as trisomy 16 and normal karyotypes.

Initial studies of parental origin using chromosomal polymorphisms indicated that about two-thirds of triploid conceptions contain two paternal chromosome sets (diandry), most often due to dispermy, while the other one-third have two maternal chromosome sets (digyny), most often due to fertilization of a diploid egg which has retained a polar body (43). These same studies found a high incidence of the partial mole phenotype, which was most strongly associated with diandry. Later studies found a lower percentage of cystic villi among triploids (44–46) and also a higher proportion of digynic triploids, using molecular DNA markers rather than chromosomal polymorphisms to define the parent of origin. The differences among studies in the frequency of diandry and of partial moles may reflect differences in the gestation distribution or types of specimen being examined, since abortions of late gestation presenting with a fetus are more likely to be digynic (46,47).

While dispermy should lead to XYY triploids, these are detected only very rarely. This chromosome complement is apparently incompatible with long-term embryonic development. The ratio of XXY/XXX triploids varies considerably among studies. In a New York series the ratio was 1.2:1, while in a Hawaiian series it was 3:1 (41). This may again result from differing gestational age distributions, since XXY triploids are more likely to be diandric.

Differences in fetal phenotype have also been defined between diandric and digynic triploids (46–48). Diandric triploid fetuses have large cystic placentas, rarely survive into the second trimester, and when they do, have

(a)

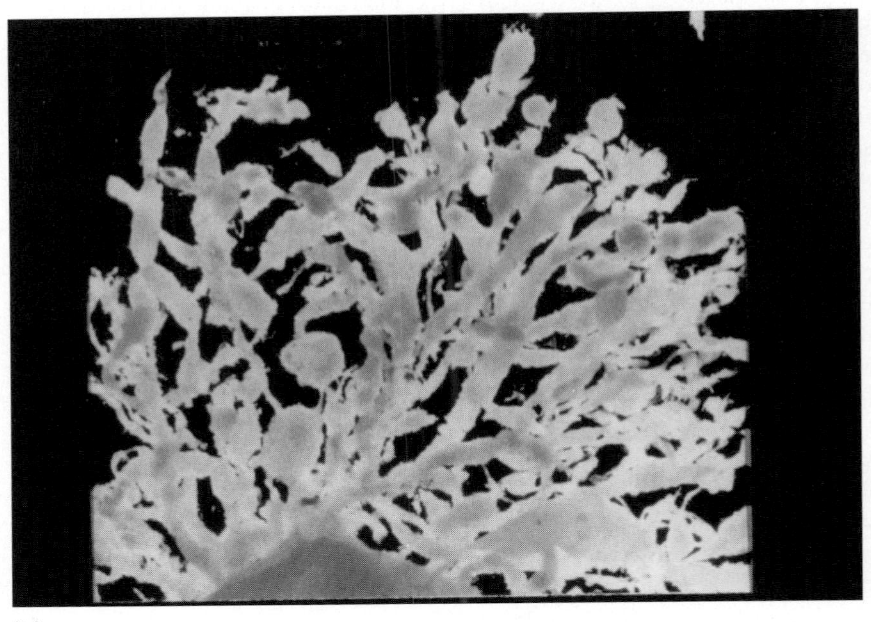
(b)

Figure 6 Placenta showing (a) cystic and (b) noncystic regions from spontaneous abortion with triploidy. (From Ref. 40.)

relatively normal fetal proportions—the so-called type I phenotype. Digynic triploid fetuses have relatively small placentas and are more likely to survive into the second trimester. They are growth-retarded, with relative macrocephaly—the type II phenotype (49). The phenotypic differences between diandric and digynic triploids have been interpreted to reflect genomic imprinting, consistent with data in the mouse which indicate a major role for the paternal genome in the development of the placenta, and for the maternal genome in the development of the embryo.

3. Haploidy

Haploidy is inconsistent with long-term embryonic development, but can be found in embryos which develop as far as the blastocyst stage after in vitro fertilization (28).

4. Uniparental Diploidy (Diandry and Digyny)

Conceptions which are normally diploid but contain two copies of one parental genome are expected to be highly abnormal because of the phenomenon of imprinting. Diandric diploidy leads to the development of a hydatidiform mole, with a potential for malignant trophoblastic disease. It is presumed to arise when an X-bearing sperm fertilizes an oocyte without a functional chromosome set. Most commonly the paternal haploid genome then undergoes chromosome replication without cell division, but dispermy may also lead to a diploid zygote (50). Most hydatidiform moles have an XX chromosome complement, although XY moles can result from dispermy.

Digynic diploidy occurs in benign ovarian teratomas, which are capable of disorganized tissue and organ differentiation. These usually have an XX karyotype, and arise from a single oocyte after failure of either meiosis I or meiosis II (51).

As in triploidy, the nature of uniparental diploid development indicates the importance of the paternal genome for placental development, and of the maternal genome for fetal development.

D. Trisomy

Aneuploidy, the presence of extra or missing chromosomes, is the most common abnormality found in spontaneous abortions. It accounts for about half of all chromosomally abnormal abortions, and therefore for about 25% of early embryonic and fetal deaths (52).

The genesis of aneuploidy is usually meiotic nondisjunction, an event expected to lead to trisomic and monosomic conceptions with equal frequency. However, even among early spontaneous abortions, autosomal monosomy is almost never seen. Only monosomy 21 has been found, occurring in about 1 in 1000 karyotyped spontaneous abortions, and occasionally compatible with survival to term in the mosaic state. The lack of any other monosomies in recognized human conceptions, together with the documentation of nullisomy in gametes and monosomy in early embryos, indicates that autosomal monosomies probably do not survive implantation, as has been shown experimentally in the mouse.

Trisomy has been documented in spontaneous abortions for every human chromosome, although the proportion of trisomy varies a great deal from chromosome to chromosome (53). Figure 7 shows the distribution of trisomies in two large series of spontaneous abortions. Some of the differences in trisomy frequency probably reflect the relatively poor development of conceptions with some trisomies, such as 1, 11, 17, and 19, or their failure to support growth in culture. Other differences, such as the very high frequency of trisomy 16 (one-third of all recognized trisomic conceptions) and the relatively high frequency of trisomy for acrocentric chromosomes, almost certainly indicate differences in their frequency of origin. This is supported by the studies on oocytes and early embryos, which have also found

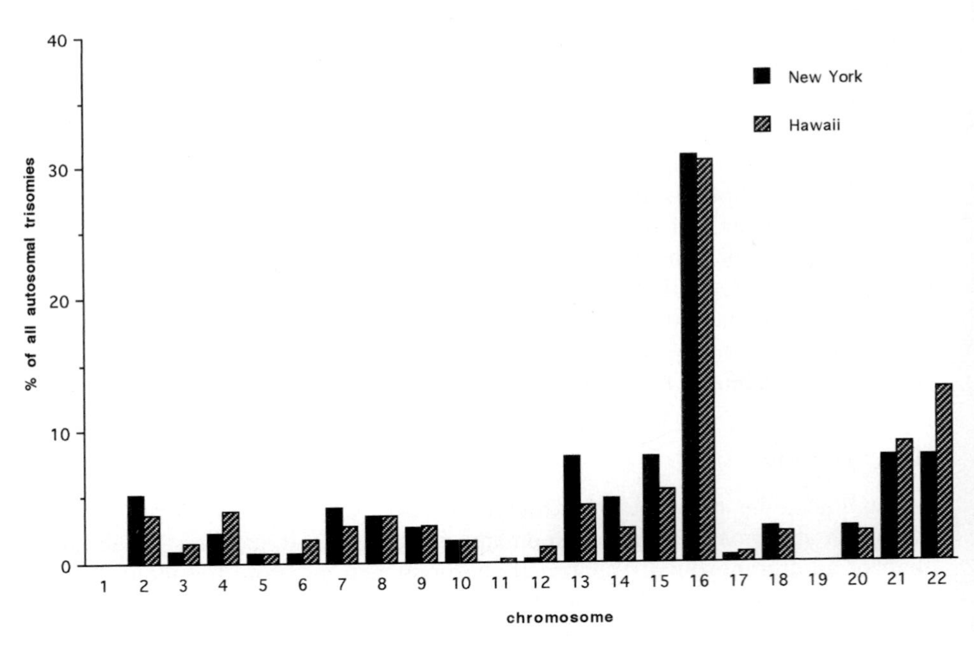

Figure 7 Frequency of trisomy for each human chromosome in two large surveys of spontaneous abortion in New York City and Hawaii. (From Ref. 53.)

a relatively high proportion of these trisomies (10,28,54). The reason for these differences in frequency is unknown. It should be noted that the trisomies 13, 18, and 21, which are seen in term births, do not occur more frequently in spontaneous abortions than trisomies for similar chromosomes which are always embryonic lethals: for example, trisomies 21 and 22 are equally frequent at conception, but only trisomy 21 can survive gestation.

All trisomies, with the exception of those for the largest chromosomes, are associated with increased maternal age, the biggest effect being seen for the smallest chromosomes (53,55) (Table 3). Because of this strong association with maternal age, the chance that any given spontaneous abortion is trisomic increases significantly with maternal age: e.g., in the New York study, trisomy occurred in 15% of miscarriages in women of 20–24, and in 55% of miscarriages in women over 40 (Table 4).

Studies on the mechanism of origin of autosomal trisomy have demonstrated that about 95% are due to maternal meiotic nondisjunction (52). For trisomy 21, about two-thirds are classified as maternal meiosis I errors, and one-third as maternal meiosis II errors (56). The preponderance of maternal error agrees with the much higher frequency of aneuploidy seen in oocytes as compared to sperm. These meiotic errors are associated with changes in the recombination patterns of the nondisjoined chromosome pair. The effect of maternal age seems to render chromosome pairs with either too few or too many chiasmata susceptible to nondisjunction (56). However, the mode of action of such a mechanism is still unknown. Trisomy 16, the

Table 3 Mean Maternal Age for Spontaneous Abortions with Known Karyotype

	New York series (Ref. 40)	Hawaii series (Ref. 55)
Chromosomally normal	27.2	27.0
Trisomy:		
2–5	26.8	27.7
6–12	30.3	28.7
13–15	31.9	31.7
16	29.4	30.0
17–20	33.4	32.8
21–22	33.4	30.0
Triploidy	26.4	26.5
Monosomy X	25.0	26.9
Tetraploidy	26.0	28.4
Live births from same population	26.2	26.4

Table 4 Frequency of Trisomy in Spontaneous Abortions by Maternal Age Group

	Maternal age (yrs)					
	≤20 (n = 293)	20–24 (n = 591)	25–29 (n = 664)	30–34 (n = 597)	35–39 (n = 317)	40+ (n = 107)
Percent chromosomally abnormal	36.9	36.2	36.0	43.7	47.0	58.9
Percent trisomic	12.3	14.9	17.3	21.3	37.2	55.1

Data from J. Kline and D. Warburton, unpublished.

most common trisomy in spontaneous abortions, is almost always of maternal meiosis I origin, and is also associated with a decreased rate of pericentric recombination (37).

Trisomies can have a wide range of pathologic presentations, ranging from late abortions with a well-developed fetus to an empty gestational sac. Cystic villi are also quite common, especially in trisomy 16. Not surprisingly, those trisomies which sometimes survive gestation occur more frequently in late abortions, while the lethal trisomies rarely survive into the second trimester and are usually associated with only rudimentary embryonic development (Fig. 8). It can be estimated that about 75% of known trisomy 21 conceptions, 95% of trisomy 18 conceptions, and 97% of trisomy 13 conceptions end as embryonic or fetal deaths. After a prenatal diagnosis of trisomy 21 by amniocentesis, almost 40% of pregnancies will end in spontaneous fetal death (57).

Approximately 10% of cultured trisomic abortions are mosaic with a normal cell line (40). The fact that approximately equal numbers of male and female mosaics are found indicates that this is not due simply to maternal cell contamination. Trisomy mosaicism is also seen in 1–2% of cho-

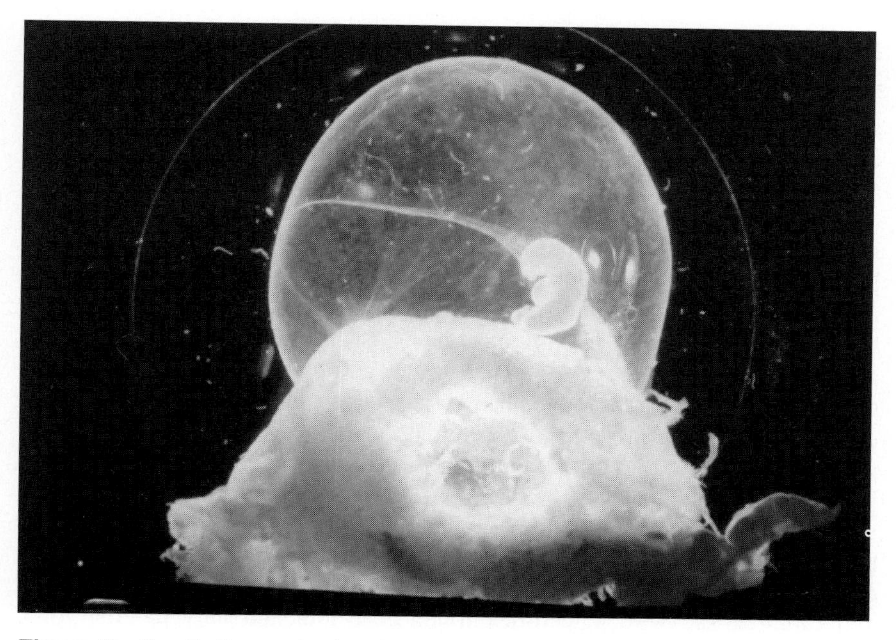

Figure 8 Small disorganized embryo in pregnancy with trisomy 22. (From Ref. 40.)

rionic villus samples, and may take the form of confined placental mosaicism, where abnormal cells are not found in the fetus. This has led to the term "zygote rescue" to describe the phenomenon where an originally trisomic conception develops a normal cell line through mitotic nondisjunction. Since the placenta is commonly the source of cultured tissue in abortion specimens, confined placental trisomy will contribute to the estimate of trisomy among products of conception. While this may overestimate the actual proportion of trisomic embryos, it is not an important source of bias since it is most likely that (a) trisomy in the placenta is the cause of the embryonic loss and (b) the original zygote was trisomic and arose by meiotic nondisjunction.

A question of some practical importance is whether the existence of a trisomic spontaneous abortion predicts an increased likelihood of a trisomic live birth in the same woman. The existing data do not support the idea that there are couples who have an increased risk for trisomy in general (58). However, the data are not sufficient to rule out an increased risk for any particular trisomy. Prenatal diagnosis should therefore be offered to women with documented trisomy 13, 18, or 21 pregnancies, regardless of whether they are spontaneous abortions or term births.

E. Monosomy X

This karyotype is found in about 7% of early spontaneous abortions. Although surviving fetuses with monosomy X have only the nonlethal features of Turner syndrome, the relative frequencies of monosomy X in spontaneous abortions and term pregnancies indicate that 99.7% of all monosomy X conceptions die in utero (52). A high proportion of live births with monosomy X are demonstrably mosaic, with a second cell line containing all or part of an X or a Y. Molecular methods have identified even more cryptic mosaicism, though not in the majority of cases (59–61). It has therefore been hypothesized that true monosomy X is an embryonic lethal, and all cases of Turner syndrome are actually mosaic, or were during early embryonic development. Embryonic survival is then assumed to depend upon two copies of a gene residing in the pseudoautosomal region of the X and Y. A number of potential "Turner syndrome" candidate genes have been identified, but none has yet been verified.

Embryonic development in monosomy X appears to continue relatively normally until about 6 weeks, after which development stops and embryonic or fetal death occurs. A characteristic pathologic presentation is a placenta and ruptured sac with a well-developed cord attached only to embryonic fragments (62) (Fig. 9). In some studies the same presentation has been found in unruptured sacs, suggesting that the fetus has been resorbed into

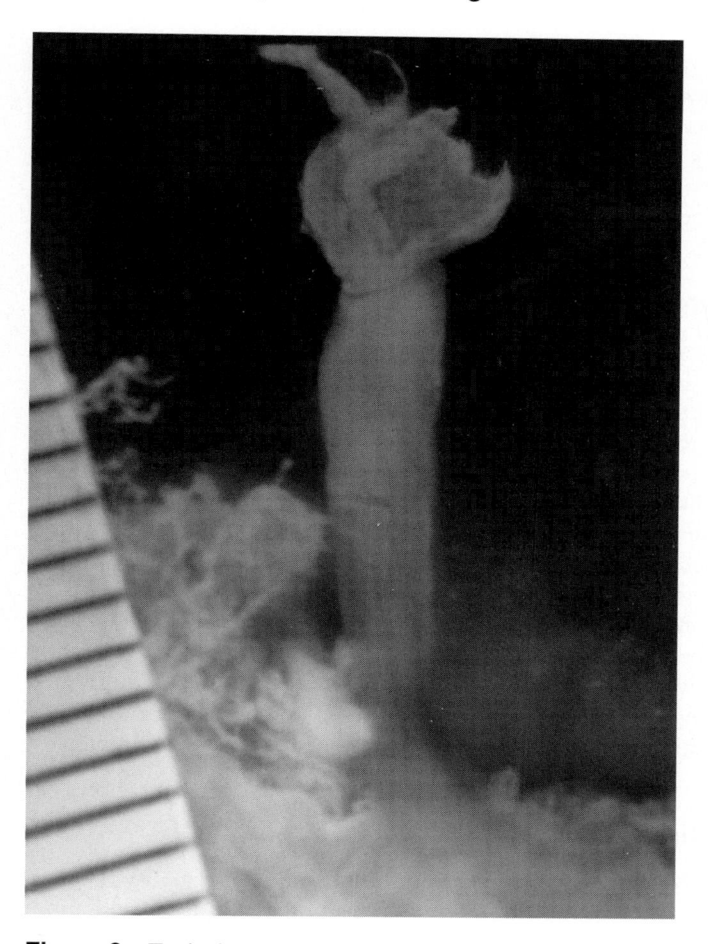

Figure 9 Typical specimen from early spontaneous abortion with monosomy X, showing well-developed placenta and cord, with only fragments of embryonic tissue at the end of the cord. (From Ref. 40.)

the amniotic fluid after death (35). Villi are often hypoplastic. Fetuses surviving into the second trimester have a recognizable phenotype compatible with features of Turner syndrome: fetal hydrops, cystic hygromata, and often congenital heart defects.

Monosomy X is associated with a relatively young maternal age (39,40), but not with advanced maternal age. It is thought to arise usually by loss of a sex chromosome in mitosis of the early embryo, rather than by meiotic nondisjunction. In about 80% of cases the single X chromosome is derived from the mother (63), in both live-born cases and spontaneous abortions. Attempts to define differences in phenotype between live-born cases with a paternal or maternal X chromosome have been unsuccessful, except for a recent study describing behavioral differences (64).

Though few data exist, it does not appear that there is an increased recurrence risk for pregnancies with monosomy X.

F. Chromosome Rearrangements

Chromosome rearrangements occur in only a small proportion of spontaneous abortions. Unbalanced rearrangements, which can be assumed to be the cause of the abortion, are seen in 1–2% of specimens (40,41). Apparently balanced rearrangements are not found in elevated frequency in spontaneous abortions as compared to live births (65), and are thus unlikely to be related to the cause of the abortion.

Although unbalanced rearrangements are rare causes of spontaneous abortion, their recognition is significant. Therefore, it is standard clinical practice to carry out chromosome analysis on a couple who have experienced two or three reproductive losses. About 4% of such couples will be found to carry a balanced rearrangement (66), which can lead to a conception with an unbalanced karyotype. Since some unbalanced segregants may be able to survive gestation but have multiple congenital anomalies and mental retardation, prenatal diagnosis would be recommended for such couples.

G. Multiple Chromosome Anomalies

Given the high frequency of each type of chromosome abnormality in spontaneous abortion, one would expect to find occasional conceptions with more than one anomaly. Thus triploids with an additional or missing chromosome can occur, as can combinations of monosomy X and other anomalies. Double and triple trisomies also occur in about 1% of spontaneous abortions.

H. Pathologic Correlations with Karyotype of Spontaneous Abortions

Because of the expense and sometimes impossibility of karyotyping spontaneous abortion specimens, it would be very useful to be able to make predictions about the karyotype merely by examining specimens from the products of conception. As noted above, there are some characteristics of the specimen which tend to occur nonrandomly among particular karyotypes, e.g., cystic villi with triploidy, intact empty sac with tetraploidy, sac with cord and no embryo in monosomy X. However, the ability of the specimen type to predict the karyotype is very poor (45,67). For example, in the New York City study, 75% of specimens with tetraploidy had an intact empty sac, but only 7% of specimens with an intact empty sac were tetraploid. Similarly, 40% of triploid specimens had cystic villi, but only 25% of specimens with cystic villi were triploid. The single exception is that a chromosomally normal abortion can be predicted from the presence of a well-developed, normal-appearing fetus over 9 weeks of gestation, or signs of such presence such as anucleate fetal erythrocytes (45,68).

I. Recurrent Spontaneous Abortion

Patients are usually classified as having recurrent spontaneous abortion after three or more losses, but sometimes two losses is used as the criterion. Spontaneous abortion is a common event, occurring in at least 15% of recognized pregnancies for multiple different causes. Therefore, many patients with "recurrent abortion" do not actually have a recurrent cause for their fetal losses, but have simply encountered bad luck. When multiple spontaneous abortions from the same woman are karyotyped, it is not unusual for them to have a different chromosome status, e.g., triploidy, trisomy, and a normal karyotype. Since triploidy and trisomy arise through dissimilar mechanisms, they are unlikely to be related. There is no good evidence that the same type of chromosomal abnormality tends to recur in subsequent pregnancies (58).

Evidence from several different sources indicates that women with the highest risk for another abortion are those with chromosomally normal spontaneous abortions occurring in the second trimester (58,69,70). Such women are also at increased risk for premature births, suggesting a process interfering with the normal physiology of gestation. Women with chromosomally abnormal repeated abortions do not appear to have an elevated risk of abortion above that associated with their age. Figure 10 shows that, in younger women, the percentage of chromosomally normal spontaneous abortions increases with the number of previous abortions, reaching about 80% after two or more. However, in older women the maternal age association with trisomic conceptions overwhelms this effect, so that the percentage of chromosomally normal abortions is only 52% after two or more previous abortions.

It is difficult to come to a firm conclusion as to whether chromosome analysis of specimens from spontaneous abortions is clinically warranted. In practice it is common to request karyotyping of spontaneous abortion specimens in women who are having a second or third loss, and to offer parental karyotyping to the couples involved when fetal karyotyping is not possible.

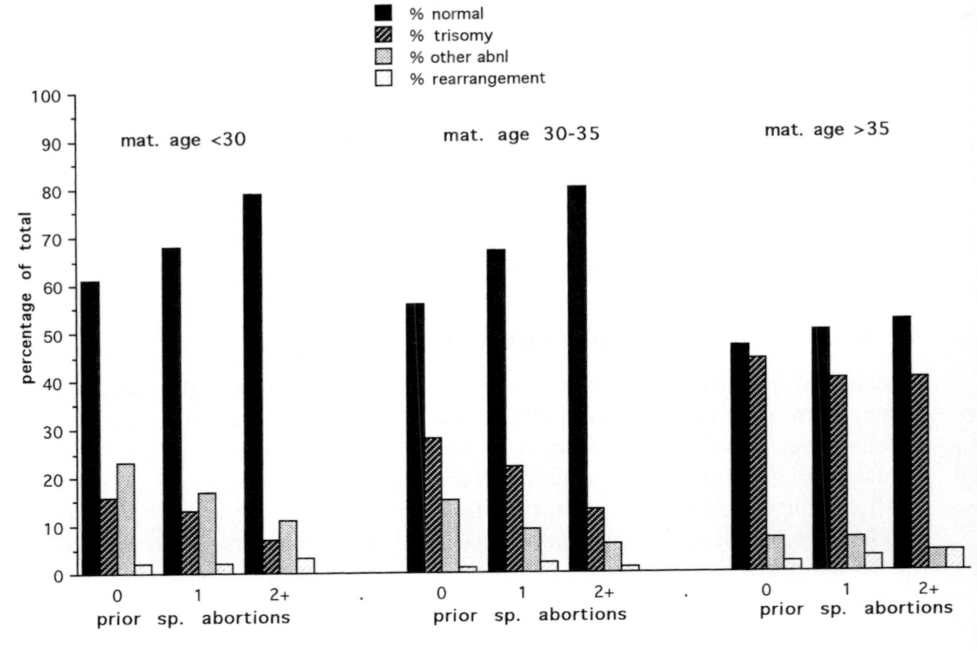

Figure 10 Karyotype of spontaneous abortions, by maternal age and reproductive history (From J. Kline and D. Warburton, unpublished data.)

One argument in favor of karyotyping is similar to that which can be made concerning autopsy in general. The patient suffering a spontaneous abortion, particularly if this has occurred before, seeks a cause for what has happened. Karyotyping of the specimen will provide this answer in about 50% of cases, and even more in older women where maternal age-related trisomy is the most common reason for abortion. Through adequate genetic counseling, information about an abnormal karyotype may allay guilt and anxiety about events during the pregnancy which were construed as causal, and provide reassurance about future pregnancies. Of course, patients whose spontaneous abortions have normal karyotypes cannot receive the same reassurances, since the cause will remain unknown, and recurrence risks will be higher. Of more practical value is the fact that patients whose losses have abnormal karyotypes need not undergo further testing to search for hormonal, anatomic, or immunologic reasons for the spontaneous abortion, nor need they undertake any special procedures for either conception or maintenance of future pregnancies.

V. CYTOGENETIC STUDIES OF STILLBIRTHS AND NEONATAL DEATHS

Fetal death occurring after the period of fetal viability will be classified as either a neonatal death or a stillbirth. The incidence of chromosome abnormalities is still relatively high in such deaths, although the range of karyotypic abnormalities is much smaller than in earlier embryonic and fetal deaths. In the New York City study which karyotyped losses for for up to 28 weeks of gestation, 12% of fetal deaths between 20 and 28 weeks had a chromosome abnormality, usually with a viable trisomy, triploidy, or monosomy X (71). Several studies have documented a rate of 6–7% for chromosome abnormalities in neonatal deaths and stillbirths over 28 weeks of gestation (72). Cytogenetic studies of such cases are thus indicated.

VI. CONCLUSION

Table 1 summarizes the rates of chromosome abnormalities from conception to birth, based on estimates in gametes, embryos, induced and spontaneous fetal deaths, and stillbirths and neonatal deaths. The results are quite compatible across studies: for example, the estimate of 2% sperm disomy and 7% oocyte disomy agrees with an 8–10% rate of trisomy in early preimplantation embryos. The estimates based on recognized conceptions are very much lower than the rates in preimplantation embryos. This may reflect a substantial reduction due to loss before implantation or in the first 3 weeks of conception. However, the rates from IVF embryos may not represent the in-vivo situation, and also are derived from a group of women much older than the average.

If we begin life at conception, chromosome anomalies are the single most important cause of human mortality. Approximately half of human conceptions do not survive gestation, and estimates from gametes and early human embryos imply that at least 60% of these losses, at least in older women, result from cytogenetically detectable abnormalities. This very high incidence of cytogenetic abnormalities appears to be peculiar to humans, at least among those mammals which have been examined. It has been speculated that such a high rate of embryonic loss may have been an advantage in primitive societies as a natural method of limiting family size, in a species where no limitations are placed on conception by mating patterns, and where the successful rearing of one offspring requires a very large commitment of time and effort.

REFERENCES

1. Kline J, Stein Z, Susser M. Conception to Birth. Epidemiology of Prenatal Development. New York: Oxford University Press, 1989:53–56.
2. Rudak E, Jacobs PA, Yanagimachi R. The chromosome constitution of human spermatazoa: a method of direct analysis. Nature 1978; 274:911–913.
3. Martin RH, Ko E, Rademaker A. Distribution of aneuploidy in human gametes: comparison between sperm and oocytes. Am J Med Genet 1991; 39:321–331.
4. Brandriff B, Gordon L, Ashworth L, Watchmaker G, Moore D, Wyrobek AJ, Carrano AV. Chromosomes of human sperm: variability among normal individuals. Hum Genet 1985; 70:18–24.
5. Williams BJ, Bellenger CA, Malter HE, Bishop F, Tucher M, Zwingmen TA, Hassold TJ. Nondisjunction in human sperm: results of fluorescence in situ hybridization using two and three probes. Hum Mol Genet 1993; 2:1929–1936.
6. Hummelen PV, Lowe XR, Wyrobek AJ. Simultaneous detection of structural and numerical chromosome abnormalities in sperm of healthy men by multicolor fluorescent in situ hybridization. Hum Genet 1996; 98:608–615.
7. Cheng EY, Gartler SM. A fluorescent in situ hybridization analysis of X chromosome pairing in early human female meiosis. Hum Genet 1994; 94:389–394.
8. Cheng EY, Chen YJ, Gartler SM. Chromosome painting analysis of early oogenesis in human trisomy 18. Cytogenet Cell Genet 1995; 70:205–210.
9. Mikamo K, Kamiguchi Y. A new assessment system for chromosomal mutagenicity using oocytes and early zygotes in the Chinese hamster. In: Ishihara T, Sasaki M, eds. Radiation-Induced Chromosomal Damage in Man. New York: Alan R. Liss, 1983:411–432.
10. Angell R. First-meiotic division nondisjunction in human oocytes. Am J Hum Genet 1997; 61:23–39.
11. Dailey T, Dale B, Cohen J, Munné S. Association between nondisjunction and maternal age in meiosis II human oocytes. Am J Hum Genet 1996; 59:176–184.
12. Verlinsky Y, Cieslak J, Freidine M, Ivakhnenko V, Wolf G, Kovalinskaya L, White M, Lifchez A, Kaplan B, Moise J, Valle J, Ginsberg N, Strom C, Kuliev A. Polar body diagnosis of common aneuploidies by FISH. J Assist Reprod Genet 1996; 13:157–162.
13. Jacobs P. The chromosome complement of human gametes. Oxf Rev Reprod Biol 1992; 14:48–72.
14. Guttenbach M, Engel W, Schmid M. Analysis of structural and numerical abnormalities in sperm of normal men and carriers of constitutional chromosome aberrations. A review. Hum Genet 1997; 100:1–21.
15. Spriggs EL, Rademaker AW, Martin RH. Aneuploidy in human sperm: results of two- and three-color fluorescence in situ hybridization using centromeric probes for chromosomes 1, 12, 15, 18, X and Y. Cytogenet Cell Genet 1995; 71:47–53.
16. Robbins WA, Baulch JE, Moore D, Weier H-U, Blakey D, Wyrobek AJ. Three-probe fluorescence in situ hybridization to assess chromosome X, Y and 8

aneuploidy in sperm of 14 men from two healthy groups: evidence for a paternal age effect on sperm aneuploidy. Reprod Fertil Dev 1995; 7:799–809.
17. Griffin DK, Abruzzo MA, Millie EA. Non-disjunction in human sperm: evidence for an effect of increasing paternal age. Hum Mol Genet 1995; 4:2227–2232.
18. Robbins WA, Meistrich ML, Moore D, Hagemeister FB, Weier WU, Cassel MJ, Wilson G, Eskenazi B, Wyrobek JA. Chemotherapy induces transient sex chromosomal and autosomal aneuploidy in human sperm. Nature Genet 1997; 16:74–78.
19. Spriggs EL, Rademaker AW, Martin RH. Aneuploidy in human sperm: the use of multicolor FISH to test various theories of nondisjunction. Am J Hum Genet 1996; 58:356–362.
20. Daniel A, Hook EB, Wolf G. Risks of unbalanced progeny at amniocentesis to carriers of chromosome rearrangements: data from United States and Canadian laboratories. Am J Med Genet 1989; 31:14–53.
21. Kamiguchi Y, Rosenbusch B, Sterizk K, Mikamo K. Chromosome analysis of unfertilized human oocytes prepared by a gradual fixation-air drying method. Hum Genet 1993; 90:533–541.
22. Lim A, Ho A, Tsakok M. Chromosomes of oocytes failing in vitro fertilization. Hum Reprod 1995; 10:2570–2575.
23. Verlinsky Y, Cieslak J, Ivakhnenko V, Lifchez A, Strom C, Kuliev A. Birth of healthy children after preimplantation diagnosis of common aneuploidies by polar body fluorescent in situ analysis. Preimplantation Genet Group Fertil Steril 1996; 66:126–129.
24. Delhanty JDA, Harper JC, Ao A, Handyside AH, Winston RML. Multicolor FISH detects frequent chromosomal mosaicism and chaotic division in normal preimplantation embryos from fertile patients. Hum Genet 1997; 99:755–760.
25. Munné S, Weier HU. Simultaneous enumeration of chromosomes 13, 18, 21, X and Y in interphase cells for preimplantation genetic diagnosis of aneuploidy. Cytogenet Cell Genet 1996; 75:263–270.
26. Harper CJ, Dawson K, Delhanty JD, Winston RM. The use of fluorescent in-situ hybridization (FISH) for the analysis of in-vitro fertilization embryos: a diagnostic tool for the infertile couple. Hum Reprod 1995; 10:3255–3258.
27. Pellestor F. The cytogenetic analysis of human zygotes and preimplantation embryos. Hum Reprod Update 1995; 1:581–585.
28. Jamieson ME, Coutis JRT, Connor JM. The chromosome complement of human preimplantation embryos fertilized in vitro. Hum Reprod 1994; 9:709–715.
29. Munné S, Alikani M, Tomkin G, Grifo J, Cohen J. Embryo morphology, developmental rates and maternal age are correlated with chromosome abnormalities. Fertil Steril 1995; 64:382–391.
30. Kajii T, Ohama K, Mikamo K. Anatomic and chromosomal anomalies in 944 induced abortuses. Hum Genet 1978; 43:247–258.
31. Zhou XT, Hong HS, Wong SG, Shen QE, Fu XW, Cui YQ. Chromosome abnormalities in early pregnancy analyzed by direct chromosome preparation of chorionic villi. Hum Genet 1989; 83:277–279.

32. Yamamoto M, Watanabe G. Epidemiology of gross chromosomal anomalies in the early embryonic stage of pregnancy. In: Klingberg, MA, Weatherall JA, eds. Epidemiologic Methods for Detection of Teratogens. Basel: S. Karger, 1979:101–106.
33. Hoshi N, Hanatani K, Kishida T, Sagawa T, Fujimoto S. Chromosomal analysis in 894 induced abortuses from women of advanced age in relation to gestational weeks and fetal sex ratio. J Obstet Gynaecol Res 1997; 23:1–7.
34. Carr DH. Chromosome anomalies as a cause of spontaneous abortion. Am J Obstet Gynecol 1967; 97:283–293.
35. Boué J, Philippe E, Giroud A, Boué A. Phenotypic expression of lethal chromosomal anomalies in human abortuses. Teratology 1976; 14:3–20.
36. Kline J, Stein ZA. Environmental causes of aneuploidy: why so elusive? In: Dellarco VL, Voytek PE, Hollaender A, eds. Aneuploidy: Etiology and Mechanisms. New York: Plenum Press, 1985:149–164.
37. Hassold T, Merrill M, Adkins K, Freeman S, Sherman S. Recombination and maternal-age dependent non-disjunction: molecular studies of trisomy 16. Am J Hum Genet 1995; 57:867–874.
38. Committee LP. Standards and Guidelines for Clinical Genetics Laboratories. 2nd ed. Bethesda: The American College of Medical Genetics, 1999.
39. Eiben B, Bartels I, Barh-Porsh S, Borgmann S, Gatz G, Gellert G, Coebel R, Hammans W, Hentemann M, Osmers R, Rauskolb R, Hansmann I. Cytogenetic analysis of 750 spontaneous abortions with the direct preparation method of chorionic villi and its implications for studying genetic causes of fetal wastage. Am J Hum Genet 1990; 47:656–663.
40. Warburton D, Byrne J, Canki N. Chromosomal Anomalies and Prenatal Development: An Atlas. New York: Oxford University Press, 1991.
41. Hassold T, Chen N, Funkhouser J, Jooss T, Manuel B, Matsuura J, Matsuyama A, Wilson C, Yamane JA, Jacobs PA. A cytogenetic study of 1000 spontaneous abortions. Ann Hum Genet 1980; 44:151–178.
42. Surti U, Szulman A, Wagner K, Leppert M, O'Brien SJ. Tetraploid partial hydatidiform moles: two cases with a triple paternal contribution and an XXXY karyotype. Hum Genet 1986; 72:15–21.
43. Jacobs PA, Szulman AE, Funkhouser J, Matsuura JS, Wilson CC. Human triploidy: relationship between parental origin of the additional haploid complement and development of partial hydatidiform mole. Ann Hum Genet 1982; 46:223–231.
44. McFadden D, Pantzar T. Placental pathology of triploidy. Hum Pathol 1996; 27:1018–1020.
45. Genest DR, Roberts D, Boyd T, Bieber RR. Fetoplacental histology as a predictor of karyotype: a controlled study of spontaneous first trimester abortions. Hum Pathol 1995; 26:201–209.
46. Dietzsch E, Ramsay M, Christianson AL, Henderson BD, Ravel TJD. Maternal origin of extra haploid set of chromosomes in third trimester triploid fetuses. Am J Med Genet 1995; 58:360–364.
47. Miny P, Koppers B, Dworniczak B, Bogdanova N, Holzgreve W, Tercanli S, Basaran S, Rehder H, Exeler R, Horst J. Parental origin of the extra haploid

chromosome set in triploidies diagnosed prenatally. Am J Med Genet 1995; 57:102–106.

48. McFadden DE, Kwong LC, Yam IY, Langlois S. Parental origin of triploidy in human fetuses: evidence for genomic imprinting. Hum Genet 1993; 92:465–469.
49. Lockwood C, Scioscia A, Stiller R, Hobbins J. Sonographic features of the triploid fetus. Am J Obstet Gynecol 1987; 157:285–287.
50. Lawler SD, Fisher RA, Dent J. A prospective genetic study of complete and partial hydatidiform moles. Am J Obstet Gynecol 1991; 164:1270–1277.
51. Surti U, Hoffner L, Chakravarti A, Ferrell RE. Genetics and biology of human ovarian teratomas. I. Cytogenetic analysis and mechanism of origin. Am J Hum Genet 1990; 47:635–643.
52. Hassold T, Abruzzo M, Adkins K, Griffin D, Merrill M, Millie D, Saker D, Shen J, Zaragoza M. Human aneuploidy: incidence, origin and etiology. Environ Mol Mutagen 1996; 28:167–175.
53. Warburton D, Kinney A. Chromosomal differences in susceptibility to meiotic aneuploidy. Environ Mol Mutagen 1996; 28:237–247.
54. Munné S, Sultan K, Weier H-U, Grifo J, Cohen C, Rosenwales Z. Assessment of numeric abnormalities of X, Y, 18 and 16 chromosomes in preimplantation human embryos before transfer. Am J Obstet Gynec 1995; 172:1191–1201.
55. Hassold T, Warburton D, Kline J, Stein Z. The relationship of maternal age and trisomy among trisomic spontaneous abortions. Am J Hum Genet 1984; 36:1349–1356.
56. Lamb NE, Freeman SB, Austin-Savage A, Pettay D, Taft L, Hersey J, Gu Y, Shen J, Saker D, May KM, Avramopoulos D, Petersen MB, Hallberg A, Mikkelsen M, Hassold TJ, Sherman SL. Susceptible chiasmata configurations of chromosome 21 predispose to non-disjunction in both maternal meiosis I and meiosis II. Nature Genet 1996; 14:400–405.
57. Hook EB, Mutton DE, Ide R, Alberman E, Bobrow M. The natural history of Down syndrome conceptuses diagnosed prenatally that are not electively terminated. Am J Hum Genet 1995; 57:875–881.
58. Warburton D, Kline J, Stein Z, Hutzler M, Chin A, Hassold T. Does the karyotype of a spontaneous abortion predict the karyotype of a subsequent abortion: Evidence from 273 women with two karyotyped spontaneous abortions. Am J Hum Genet 1987; 41:465–483.
59. Binder G, Koch A, Wajs E, Ranke MB. Nested polymerase chain reaction study of 53 cases with Turner's syndrome: is cytogenetically undetected Y mosaicism common? J Clin Endocrinol Metab 1995; 80:3532–3536.
60. Yorifuji T, Muroi J, Kawai M, Sasaki H, Momoi T, Furusho K. PCR-based detection of mosaicism in Turner syndrome patients. Hum Genet 1997; 99:62–65.
61. Fernandez R, Mendez J, Pasaro E. Turner syndrome: a study of chromosomal mosaicism. Hum Genet 1996; 98:29–35.
62. Canki N, Warburton D, Byrne J. Morphological characteristics of monosomy X in spontaneous abortion. Ann Genet 1988; 31:4–13.

63. Hassold T, Pettay D, Robinson A, Uchida I. Molecular studies of parental origin and mosaicism in 45,X conceptuses. Hum Genet 1992; 89:647–652.
64. Skuse DH, James RS, Bishop DV, Coppin B, Dalton P, Aamodt-Leeper G, Bacarese-Hamilton M, Creswell C, McGurk R, Jacobs PA. Evidence from Turner's syndrome of an imprinted X-linked locus affecting cognitive function. Nature 1997; 387:705–708.
65. Warburton D. De novo structural rearrangements: implications for prenatal diagnosis. In: Willey AM, Kelly S, Porter IH, eds. Clinical Genetics: Problems in Diagnosis and Counselling. New York: Academic Press, 1982:63–75.
66. Simpson JL, Meyers CM, Martin AO, Elias S, Ober C. Translocations are infrequent among couples having repeated spontaneous abortions but no other abnormal pregnancies. Fertil Steril 1989; 5:811–814.
67. Martinazzi S, Zampieri A, Todeschin P, Vegetti PL, Ceriani L, Brianza MR, Crivelli F, Martinazzi M. Correlation of the histological and cytogenetic pictures in a placental tissue from early abortion. Does immunohistochemistry have a role? Pathologica 1996; 88:275–288.
68. Byrne J, Warburton D, Kline J, Blanc W, Stein Z. Morphology of early fetal deaths and their chromosomal characteristics. Teratology 1985; 32:297–315.
69. Strobino BR, Fox HF, Kline J. Characteristics of women with recurrent spontaneous abortions and women with favourable reproductive histories. Am J Publ Hlth 1986; 76:986–991.
70. Boué J, Boué A, Lazar P. Retrospective and prospective epidemiological studies of 1500 karyotyped spontaneous human abortions. Teratology 1975; 12: 11–26.
71. Warburton D, Kline J, Stein Z, Strobino B. Cytogenetic abnormalities in spontaneous abortions of recognized conceptions. In: Porter IH, Willey A, eds. Perinatal Genetics: Diagnosis and Treatment. New York: Academic Press, 1986:133–148.
72. Angell RR, Sandison A, Bain AD. Chromosome variation in perinatal mortality: a survey of 500 cases. J Med Genet 1984; 21:39–44.

APPENDIX: GLOSSARY OF TERMS

adjacent-1 segregation

2:2 Meiotic segregation of the tetravalent formation occurring in reciprocal balanced translocation carriers, such that homologous centromeres segregate to opposite poles. The resulting gametes will each have one rearranged chromosome, and will be unbalanced.

adjacent-2 segregation

2:2 Meiotic segregation of the tetravalent formation occurring in reciprocal, balanced translocation carriers, such that homologous centromeres segregate to the same poles. The resulting gametes will each have one rearranged chromosome, and will usually be even more unbalanced than after adjacent-1 segregation.

alternate segregation

2:2 Meiotic segregation of the tetravalent formation occurring in reciprocal balanced translocation carriers such that both normal chromosomes go to one pole, and both rearranged chromosomes go to the other pole. The resulting gametes will both be balanced, one perfectly normal, and one with a balanced translocation.

aneuploidy

An abnormal number of individual chromosomes, as opposed to an abnormal number of chromosome sets.

bivalent

In meiosis, the structure consisting of paired homologous chromosomes, usually held together by chiasmata.

chromatid

After chromosome replication, each sister chromosome, as seen in mitosis or meiosis, is called a chromatid.

cryptic mosaicism

Mosaicism which is not revealed by standard cytogenetic analysis.

cystic villi

Placental villi in which the blood vessels and connective tissue of the core are hypoplastic or absent, and the villous cavity is swollen and fluid-filled.

diandry

The presence of a diploid chromosome complement in which both copies of each chromosome are derived from the paternal genome.

digyny

The presence of a diploid chromosome complement in which both copies of each chromosome are derived from the maternal genome.

dispermy

The simultaneous fertilization of an oocyte by two sperm.

haploidy

The presence of a single copy of each chromosome: a single chromosome set.

"humster" technique

A technique for examining sperm chromosome complements, in which human sperm are used to fertilize superovulated hamster oocytes after removal of the zona pellucida. Chromosomes are examined in the sperm pronucleus just prior to the first mitotic division of the zygote.

meiosis I error

Misdivision during the first (reduction) division of meiosis, resulting in the failure of each member of the bivalent to migrate to opposite poles.

meiosis II error

Misdivision which occurs during the second division, resulting in the failure of the two duplicated chromatids to migrate to opposite poles.

monosomy
The presence of a single copy of an individual chromosome.

partial mole
A conception in which some but not all portions of the placenta contain cystic villi, and where there is also evidence of fetal membranes and/or a developing embryo.

polyploidy
The presence of more than two copies of every chromosome.

preimplantation diagnosis
Genetic diagnosis carried out on an in vitro fertilized embryo before it is implanted into the uterus.

recognized conception
A pregnancy which has persisted long enough to cause a missed menstrual cycle, and in which a positive urine pregnancy test is achieved.

Robertsonian translocation
The fusion of two acrocentric chromosomes in the centromere or short-arm regions, resulting in a single functional chromosome unit.

tetraploidy
The presence of four copies of every chromosome.

triploidy
The presence of three copies of every chromosome.

trisomy
The presence of three copies of an individual chromosome.

univalent
One-half of the bivalent pair present in meiosis: each univalent consists of two identical chromatids.

zygote rescue
The formation, from a zygote with a chromosomal imbalance incapable of embryonic survival, of an embryo with a chromosome constitution capable of completing development. For example, a trisomic zygote may develop a normal cell line through mitotic nondisjunction.

10
Cytogenetics of Male Infertility

Hon Fong L. Mark
Brown University School of Medicine and Rhode Island Department of Health, Providence, and KRAM Corporation, Barrington, Rhode Island
Mark Sigman
Brown University School of Medicine, Providence, Rhode Island

I. INTRODUCTION

Each year approximately 15% of couples trying to conceive for the first time will be unable to do so. A sole male factor is responsible in 20% of cases, while in an additional 30% the male factor is contributory. The evaluation of the infertile male proceeds as with the evaluation of any other medical problem. A thorough history and physical examination is performed, followed by appropriate laboratory tests. Ideally, this evaluation identifies the etiology of the male's infertility and leads to appropriate and specific therapy. Unfortunately, in many cases, the etiology of the male's infertility remains unknown.

Recent advancements in molecular biological techniques have resulted in new inroads being made into the etiology of male infertility. The genes responsible for spermatogenesis and mutations in these genes causing infertility have now been identified. As further progress in this area is made, the pathologist and cytogeneticist will be called on more frequently to aid in the diagnosis and the management of the infertile couple.

II. OVERVIEW OF SPERMATOGENESIS

A. Endocrinology

The adequate production of sperm by the testis requires an intact functioning hypothalamic pituitary gonadal axis. Hypothalamic neurosecretory cells produce gonadotropin-releasing hormone (GnRH). This peptide hormone is secreted in a pulsatile pattern every 70–90 min into the portal circulation by which it reaches the pituitary gland. This results in the stimulation of cells in the anterior pituitary (the gonadotrophs), causing secretion of follicle-stimulating hormone (FSH) and luteinizing hormone (LH). The gonadotropins travel in the peripheral circulation to the testis, where they bind to specific cell surface receptors. Leydig cells, present in the interstitial tissue of the testis, have cell surface receptors which bind LH, resulting in an increase in intracellular cyclic adenosine monophosphate (cAMP). This causes an increase in testosterone production by the Leydig cell. The initiation of spermatogenesis requires testosterone, while the maturation of spermatids into mature sperm (spermiogenesis) will not take place without FSH. Sertoli cells, which are found within the seminiferous tubules, have cell surface receptors which bind FSH. This interaction results in the stimulation of spermatogenesis to completion. Testosterone diffuses passively into the circulation, where it results in a negative feedback of LH secretion at the hypothalamic pituitary level. Sertoli cells produce inhibin which is composed of one alpha subunit and one beta subunit. Pituitary gonadotrophs contain specific receptors for inhibin. The binding of this hormone results in an inhibition of FSH secretion. With a normal functioning hypothalamic pituitary gonadal axis and normal spermatogenesis, gonadotropin and steroid hormone levels are precisely regulated. Hypothalamic or pituitary dysfunction results in decreased levels of FSH, LH, and testosterone. This results in a lack of stimulation of the testicles, decreased sperm production, and small-sized testes. On the other hand, if Sertoli cell and Leydig cell function are inadequate, there is a lack of negative feedback to the pituitary, resulting in elevated levels of FSH and LH as is found in primary testicular failure.

B. Spermatogenesis

The testis consists primarily of seminiferous tubules that form loops which connect into a single duct, the tubulus rectus. This tubule connects with the rete testis, which connects to several efferent ductules entering the caput of the epididymis and the epididymal duct. The seminiferous tubules contain germ cells and Sertoli cells. The Sertoli cells have complicated cytoplasmic extensions that surround the germ cells and connect with other Sertoli cells by tight junctional complexes. It is this connection which forms the blood–testis barrier separating the seminiferous tubule into basal and adluminal compartments. Sertoli cells are complex secretory cells which produce differing compounds into the basal and adluminal compartments.

Like somatic cells, germ cells undergo mitosis. Unlike other body cells, they also undergo meiosis. The stem cells of the germ cell line are the spermatogonia. Four spermatogonial cell types have been identified in the human, termed A_{long}, A_{dark}, A_{pale}, and B. Type B spermatogonia are derived from type A. However, it is not clear which type A spermatogonia is the initial or stem cell of the germ line. Mitosis of this stem cell replenishes the spermatogonial population. At some point, some type A spermatogonia differentiate into type B spermatogonia. Mitosis of type B spermatogonia produces preleptotene spermatocytes. The basal compartment of the seminiferous tubule contains both types of spermatogonia as well as preleptotene spermatocytes. As the preleptotene diploid (2*N*) spermatocytes differentiate and replicate their DNA in preparation for meiosis, Sertoli cell-type junctional complexes form behind them, transporting them from the basal to the adluminal compartment. The leptotene primary spermatocytes then pass through the zygotene and pachytene stages, during which they contain a tetraploid amount (4*N*) of DNA (two copies of maternal and paternal chromosomes). It is during this latter stage (sometimes called the tetrad stage) when crossing over of chromosomes occurs. The pachytene spermatocytes proceed through the diplotene phase, followed by the first meiotic division, forming secondary spermatocytes containing a diploid amount of DNA (two copies of either paternal or maternal chromosomes). This is quickly followed by the second meiotic division, which produces haploid round spermatids with one copy of either paternal or maternal chromosome for each specific chromosome. The dividing germ cells maintain cytoplasmic connections as they proceed through meiosis and differentiation. Spermiogenesis consists of the maturation of round spermatids into elongated mature spermatozoa. During this process, the nucleus of the spermatid elongates, the cytoplasm is eliminated from the germ cell, and the acrosome and flagellum form. A diagram summarizing spermatogenesis in the human male is depicted in Fig. 1.

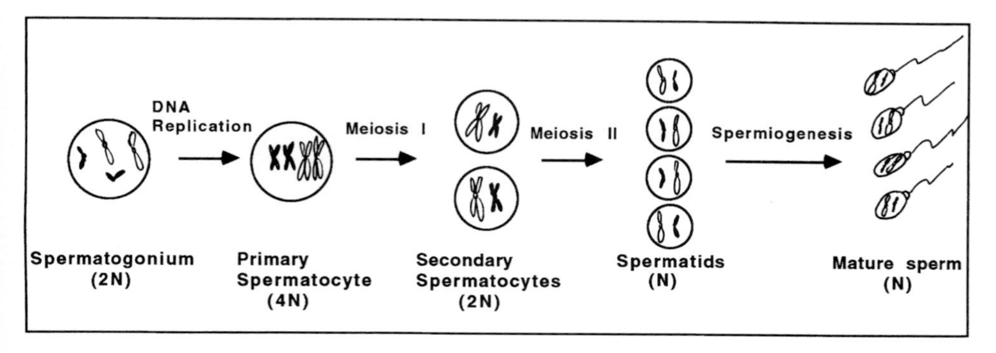

Figure 1 Spermatogenesis.

A careful histologic examination of human seminiferous tubules reveals defined patterns of maturing germ cells (Fig. 2). These have been grouped into six stages of spermatogenesis. While a transverse section through a seminiferous tubule reveals that all portions of the wall are in the same stage of spermatogenesis in the rat, an examination of the human tubule reveals sections of differing stages of spermatogenesis. The wave of spermatogenesis in the human tubule is arranged in a helical pattern along the length of the seminiferous tubule. The complete spermatogenic cycle lasts approximately 74 days. Sperm transport through the epididymis takes

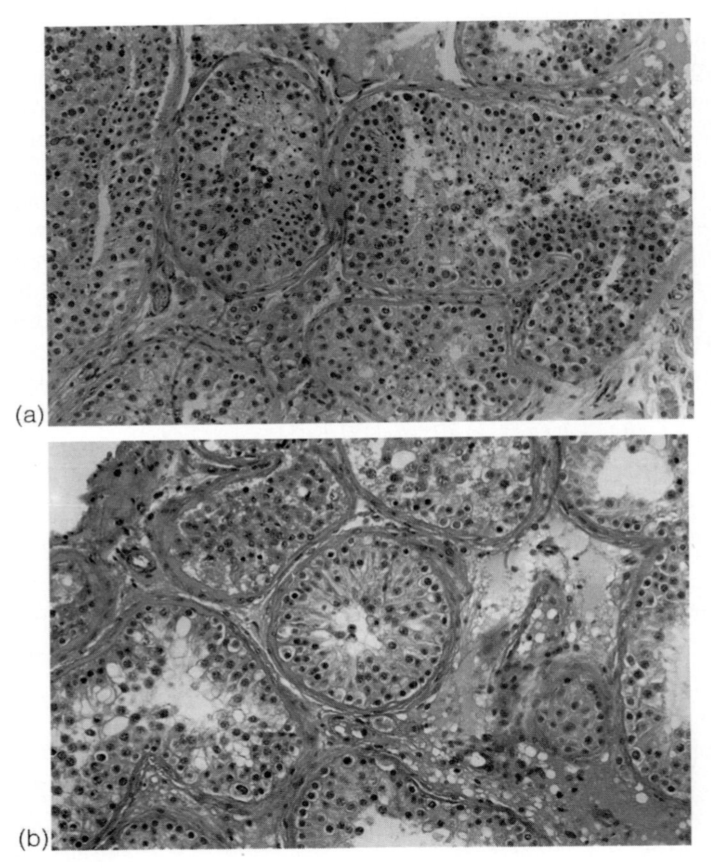

Figure 2 Hematoxylin and eosin sections of human testis: (a) normal spermatogenesis; (b) complete early maturation arrest demonstrating the presence of spermatogonia and spermatocytes but no spermatids; (c) hypospermatogenesis with all stages of germ cells present but in reduced numbers; (d) Sertoli-only syndrome.

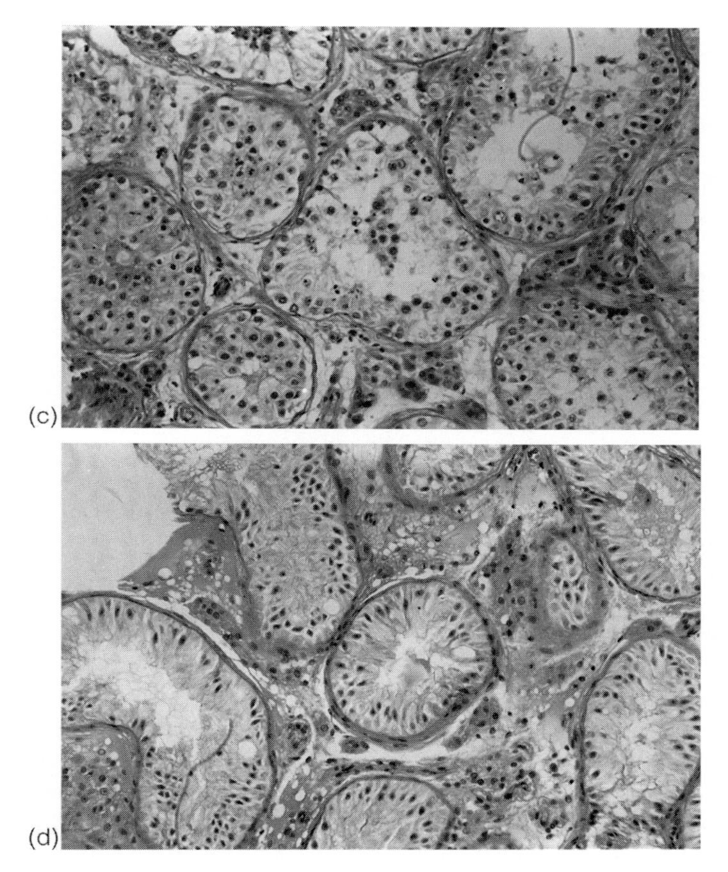

Figure 2 Continued

2–6 days, with sperm storage occurring primarily in the distal epididymis and convoluted vas deferens.

III. OVERVIEW OF MALE INFERTILITY

The workup of the infertile male begins with a thorough history and physical examination followed by appropriate laboratory tests. Following this, the patient is placed into an etiologic category based on the physician's understanding of the cause of the male's infertility. This may lead to specific therapy such as treatment of a genital tract infection, correction of the varicocele, or nonspecific treatment such as empiric medical therapy or in vitro

Table 1 Minimal Levels of Adequacy for a Semen Analysis

Volume	1.5–5.5 ml
Sperm concentration	20×10^6 to 250×10^6 sperm/ml
Motility	>60%
Forward progression	>2.0 (scale 1–4)
Normal morphologic forms	>14%
pH	7.2–7.8
Minimal sperm agglutination	Less than 1 million WBC/ml

fertilization. Important components of the history include the duration of infertility as well as any prior pregnancies and treatments. In addition, the couple's sexual history must be obtained, as well as a review of the male's childhood and pubertal developments. Current medical conditions should be explored since some of them, such as diabetes, may affect fertility. The patient's surgical history as well as recent systemic infections and exposure to gonadotoxins are also important. Finally, a review of systems and of family history, looking for evidence of sexual differentiation disorders or cystic fibrosis, should be explored. The physical exam focuses on a thorough examination of the genitalia. Laboratory testing is an essential component of the evaluation of the infertile male. While the semen analysis forms the cornerstone of this evaluation, it is not a perfect test of fertility. Indeed, there is not any one test of male fertility. The interpretation of the semen analysis depends on how it is collected and analyzed and how its normal values are defined. A commonly used set of criteria for minimum parameters of adequacy is presented in Table 1. There are many other laboratory tests for the evaluation of the infertile male which should be used on a case-by-case basis (Table 2).

Table 2 Ancillary Tests of Male Fertility

Antisperm antibody
Sperm viability
Semen white blood cell staining
Hypoosmotic sperm swelling
Hemizona assay
Sperm penetration assay
Semen culture
Acrosome reaction evaluation
Sperm cervical mucus interaction assay

Table 3 Etiology of Male Infertility

Category	Percent
Varicocele	38.17
Idiopathic	24.78
Obstruction	13.15
Normal	9.86
Cryptorchidism	3.58
Antisperm antibodies	2.52
Ejaculatory dysfunction	1.29
Drug	1.00
Endocrinopathy	1.00
WBC	1.00
Sexual dysfunction	0.59
Testicular failure	0.47
Genetic	0.23
Ultrastructural	0.23
Sertoli-only	0.23
Cancer	0.18
Heat	0.06
Radiation	0.06
Systemic illness	0.06
Testis cancer	0.06
Other	1.48

Following the completion of the history, physical, and laboratory studies, the patient is placed into an etiologic category based on the physician's understanding of the cause of the male's infertility. A list of etiologies in a large infertility clinic population is presented in Table 3. Of interest is the observation that in this cohort of patients, genetics accounts for only 0.23% of the cases.

IV. GENETICS OF SPERMATOGENESIS

The normal embryologic development of the male phenotype requires adequate gonadal function. Gonadal differentiation occurs during the fifth through seventh weeks of gestation. In the presence of proper genetic factors, the undifferentiated gonad develops into a testis. In the absence of these factors, ovarian development occurs. A functional Y chromosome (Fig. 3) is necessary for proper male development. The human Y chromosome is divided into the short arm (Yp) and the long arm (Yq) as shown in Fig. 4. All

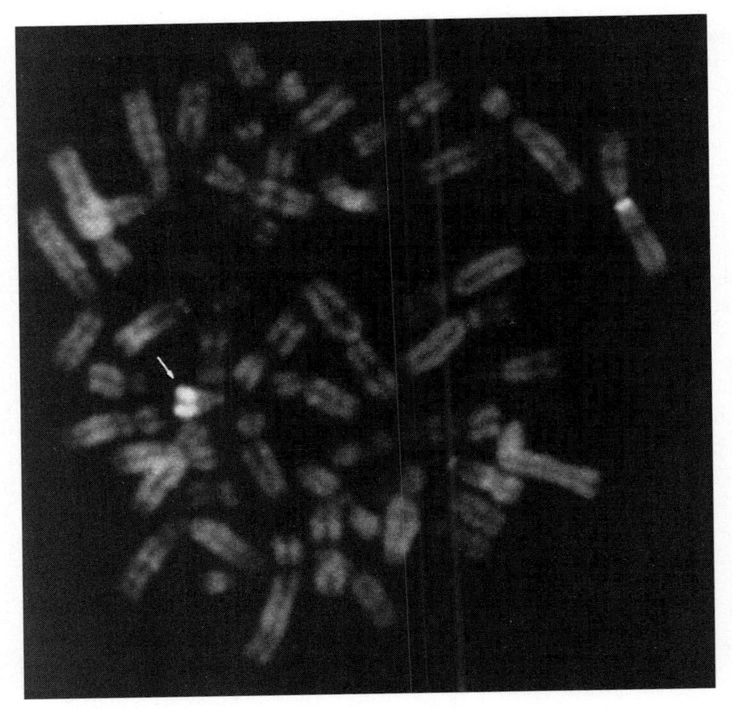

Figure 3 The human Y chromosome (arrow) illustrated by QFQ banding.

of Yp and the proximal portion of Yq are euchromatic, whereas distal Yq is heterochromatic. Two genetic factors have been identified as necessary for proper testicular development and spermatogenesis. Differentiation of the gonad into a testis requires a testis-determining factor which has been identified as SRY (1). This gene is found on distal Yp. The presence of this gene is necessary, but not sufficient, for testicular development.

Genetic studies of azoospermic men have revealed deletions of Yq. The genetic factor responsible for spermatogenesis within the testis has been termed the azoospermia factor (AZF). Two genes thought to be the spermatogenesis factor gene have been mapped to interval 6 within Yq: deleted in azoospermia (DAZ) and Y-chromosome RNA-recognition motive (YRRM, also known as RNA-binding motif or RBM). Two sequences of cDNAs for YRRM have been identified: YRRM1 and YRRM2. Only YRRM1 is thought to be present in males and involved in spermatogenesis. The nucleotide sequence YRRM1 seems to encode for an RNA-binding protein (2).

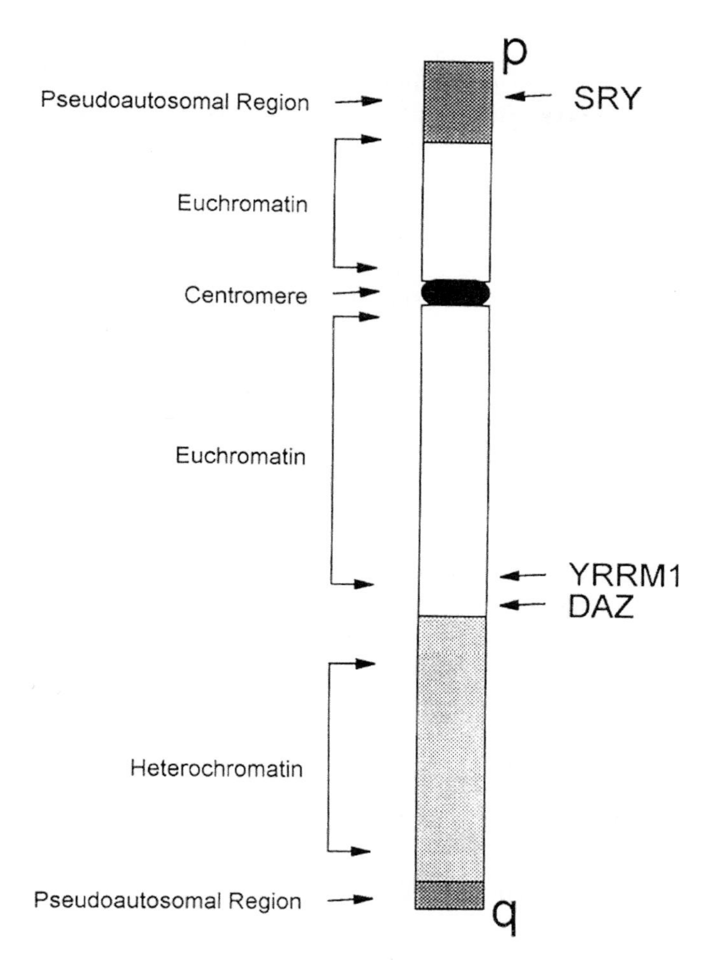

Figure 4 Diagrammatic representation of the human Y chromosome.

DAZ also appears to encode an RNA-binding protein (3). Recent studies have elucidated the possible evolution of the DAZ gene. A homolog of DAZ, termed DAZH (also called DAZLA), has been found on human chromosome 3. Of interest, DAZH is found in most species, whereas DAZ has been identified only in New World monkeys and humans. This suggests that DAZH is the original gene which was transposed from an autosome to the Y chromosome during primate evolution. Thus, in the mouse, DAZH is present while DAZ is absent. The transposed gene appears to have undergone amplification and pruning. There are at least two or three DAZ copies on the Y chromosome which demonstrate a 99.9% sequence identity. The products of DAZH and DAZ are quite similar with the exception that in DAZ there is a 24-amino-acid sequence that is randomly repeated. The total DAZ unit contains 26 exons and pseudoexons, whereas DAZH only contains 11 exons (4). Northern blotting and in situ hybridization with DAZ probes have determined that DAZ gene expression occurs only in adult testes and at lower levels in adult ovaries. There is no expression in other human tissues. The expression appears to be highest in spermatogonia, with expression also occurring in leptotene/zygotene (premeiotic) primary spermatocytes. No expression in pachytene or later spermatocytes or other testis cells, such as Leydig or Sertoli cells, has been identified (5). While deletions of DAZ have been identified in infertile males, no point mutations have been found to date.

Microdeletions of the long arm of the Y chromosome are found in approximately 13% of azoospermic men (3,6). While some men are found to have deletions of the DAZ gene, others are found to have an intact DAZ gene but a deletion of YRRM. These data clearly suggest that there must be more than one gene involved in spermatogenesis. Of interest, testicular biopsies of patients with DAZ deletions have revealed a range of patterns from a complete absence of germ cells (Sertoli-only syndrome) through early spermatogenic arrest and occasionally late maturation arrest with the presence of spermatids. Recently, some men with severe oligospermia have been found to have DAZ deletions (6). Thus, spermatogenesis can occur in the absence of DAZ, although apparently at very low rates. Studies of nonselected males presenting for infertility evaluations have demonstrated Y chromosome microdeletions in only 7% of patients (7).

V. GENETIC EVALUATION OF THE INFERTILE MALE

Genetic abnormalities associated with infertility may be evaluated by different approaches. Gross structural or numerical chromosomal abnormalities are the most commonly evaluated defects through the use of karyotype analysis. Recently, microdeletions have been evaluated through a polymerase chain reaction (PCR) of sequence tagged sites on the Y chromosome. These studies are done on DNA obtained from peripheral white blood cells. Finally, studies of germ cells are beginning to be performed.

A. Numerical and Structural Chromosomal Abnormalities

Studies on populations of infertile patients have demonstrated an approximately 5% incidence of chromosomal abnormalities (8–10). The majority of these involve the sex chromosomes. This is significantly higher than the approximately 0.4% incidence of chromosomal abnormalities in newborn infants (11). As the sperm count becomes lower, the incidence of abnormalities increases. Oligospermic men have approximately a 4.6% incidence of chromosomal abnormalities, while azoospermic men have an approximately 14% incidence (12–14). Of interest, sex chromosomal abnormalities predominate in azoospermic men, while autosomal abnormalities predominate in oligospermic men.

B. Sex Chromosomal Abnormalities

1. Klinefelter Syndrome

The most common sex chromosomal abnormality in infertile men is Klinefelter syndrome. This syndrome occurs 1 in 600 live male births and is associated with small firm testes, gynecomastia, and azoospermia. The chromosomal complement in the most common pattern is 47,XXY (Fig. 5),

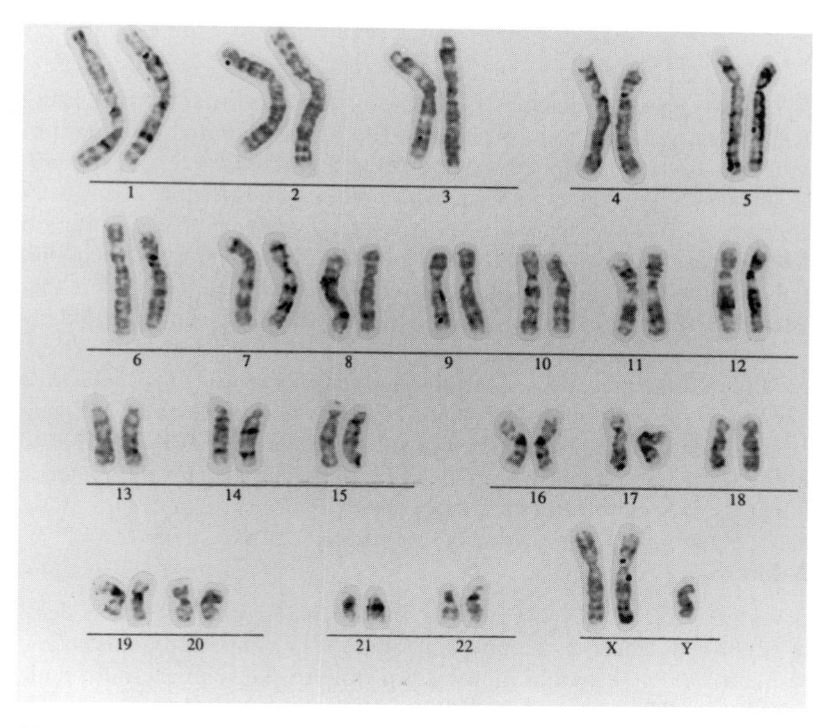

Figure 5 The XXY karyotype.

which occurs in 90% of cases, while a mosaic pattern 46,XY/47,XXY occurs in 10%. The extra X chromosome can derive from either the maternal or paternal line and is due to nondisjunction during meiosis. Low-level sperm production may be found in mosaic patients, while most 47,XXY patients are azoospermic. Histologic examination of the testes demonstrates seminiferous tubule hyalinization and Leydig cell hyperplasia. Patients with Klinefelter syndrome also demonstrate increased height, obesity, lower-extremity varicosities, diabetes, as well as an increased incidence of achondroplasia, nonseminomatous extragonadal germ cell tumors, breast carcinoma, and leukemia. While it has often been thought that spermatozoa from mosaic Klinefelter patients were likely derived from the normal 46,XY cell line, recent studies have suggested that there is an increased incidence for hyperhaploid 24,XY spermatozoa in these patients (15,16). Thus, it appears that 47,XXY cells are able to go through meiosis and form spermatozoa. Recently, spermatozoa have been extracted from the testes of pure XXY patients and used for in vitro fertilization (IVF) combined with intracytoplasmic sperm injection (ICSI) (17). This raises the possibility that genetic abnormalities may be passed on through these techniques.

2. XYY Syndrome

The XYY karyotype is found in 0.1–0.4% of newborn infants. This karyotype had at one point in time been linked to aggressive and criminal behavior. Patients with the XYY constitution (Fig. 6) are characteristically tall. Semen analyses range from severe oligospermia to azoospermia. It has been generally thought that the spermatozoa from these patients arise from germ cells that have lost the extra chromosome and therefore are normally haploid. Recent fluorescent in situ hybridization (FISH) studies have shown that while many of the spermatozoa carry a normal haploid genotype, others carry extra chromosomes, demonstrating that some XYY germ cells are capable of undergoing meiosis and producing sperm (18,19). Testicular histology ranges from maturation arrest to complete germ cell aplasia. Occasional cases demonstrate seminiferous tubule sclerosis. While LH and testosterone levels are typically normal, FSH levels may be elevated in those patients with severe defects in spermatogenesis.

3. XX Male

Patients with the XX male syndrome (sex reversal syndrome) typically demonstrate small firm testes, gynecomastia, small to normal-sized penises, and azoospermia (20,21). Testicular biopsies typically reveal seminiferous tubule sclerosis. Gonadotropins are often elevated with decreased testosterone levels. In contrast to Klinefelter syndrome, these patients often are shorter than

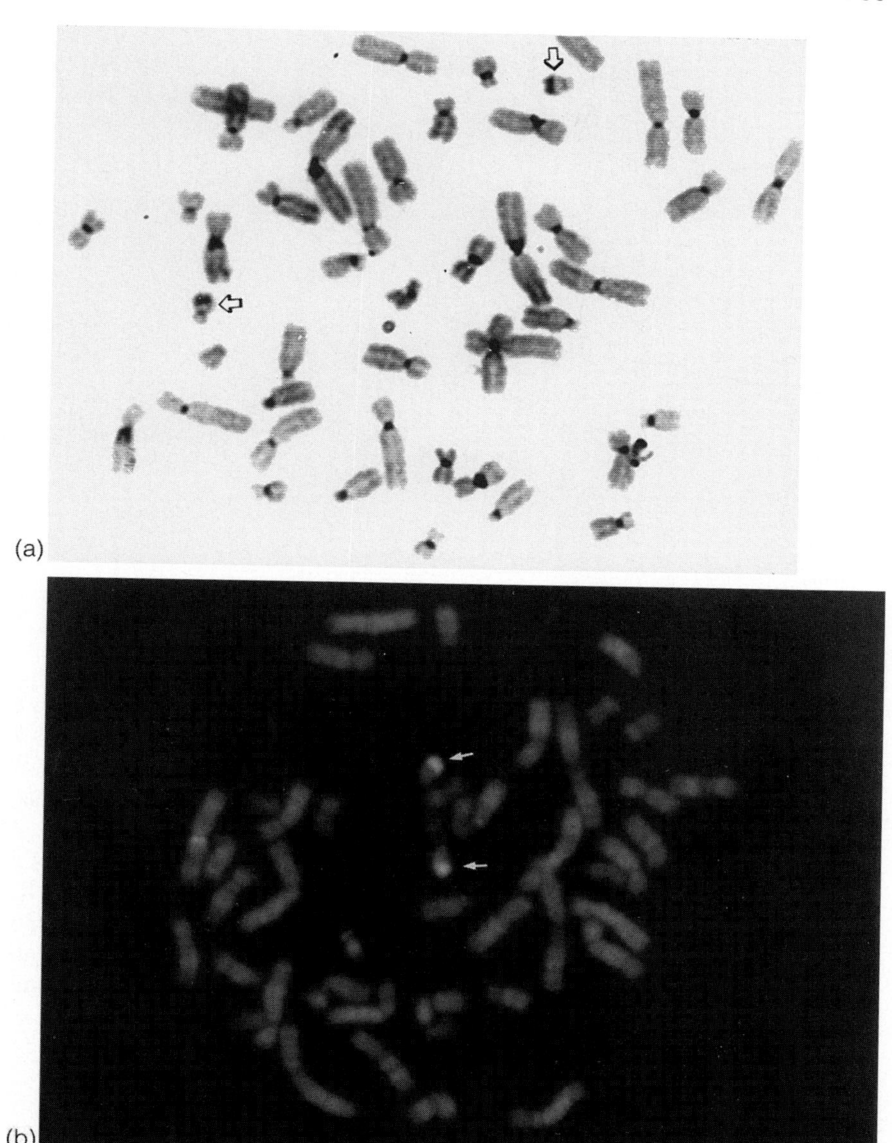

Figure 6 Cells of an XYY individual stained by different techniques. (a) Metaphase cell, CBG technique; (b) metaphase cell, QFQ technique; (c) interphase cell, fluorescent Y body technique.

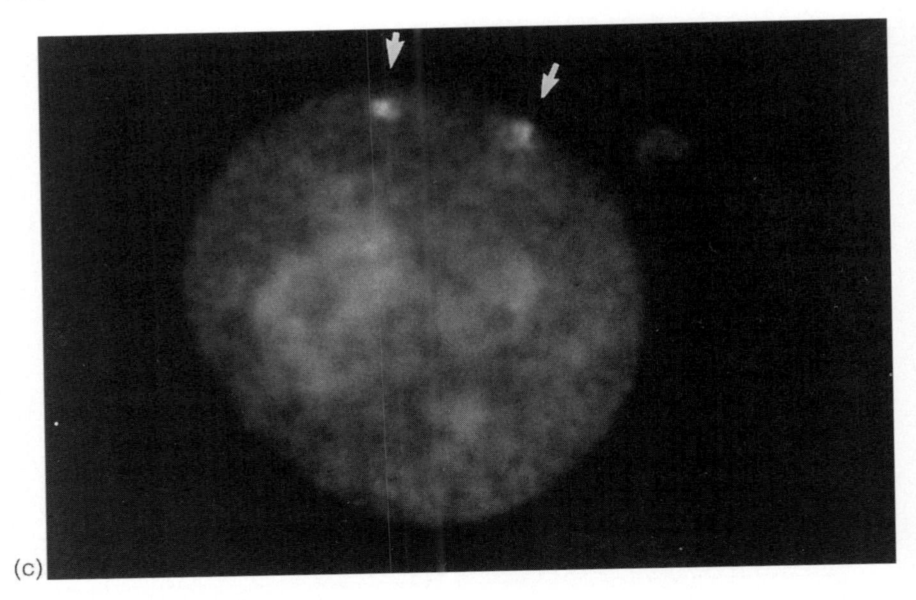

(c)

Figure 6 Continued

average, have no increased incidence of mental deficiency, and demonstrate an increased incidence of hypospadias. Cytogenetic analysis demonstrates a 46,XX karyotype. It appears that this syndrome may be caused by one of several genetic abnormalities. Typically, molecular biologic mapping has demonstrated a translocation of a portion of the short arm (Yp) of the Y chromosome to the X chromosome. This material often contains the SRY (sex-determining region of the Y or testis-determining factor) gene (22–24). The presence of SRY in the absence of AZF would result in testes that do not produce sperm. Other patients may demonstrate SRY translocated to an autosome. In some patients, no Y material has been identified, indicating that this is a genetically heterogeneous class of patients (20).

4. Mixed Gonadal Dysgenesis

Patients with mixed gonadal dysgenesis demonstrate phenotypes ranging from female through ambiguous genitalia to male patterns (25). Approximately two-thirds of patients demonstrate a 45,X/46,XY genotype, while one-third are 46,XY. These patients demonstrate a unilateral testis with a contralateral streak gonad. The testis most commonly is abdominal, as is the streak gonad, but in some cases both gonads may be scrotal (26). Testicular histology demonstrates Leydig cells with a Sertoli-only pattern. The streak gonad has an increased incidence of tumors, most commonly gonadoblastomas. The utility of FISH for studying this class of patients is discussed elsewhere in this volume.

5. Other Y-Chromosome Structural Abnormalities

Pericentric inversions, deletions, dicentric Y chromosomes, and ring Y chromosomes have been associated with azoospermia (27,28). In some of these cases, a structural abnormality appears to interfere with the pairing of the X and Y chromosomes in metaphase of meiosis I, resulting in a breakdown of spermatogenesis. As mentioned previously, microdeletions involving AZF in deletion interval 6 are commonly found in infertile men.

6. Sex Chromosomal Reciprocal Translocations

A variety of reciprocal translocations between an autosome and a sex chromosome have been found in studies of infertile men. Both translocations involving X chromosome and autosomes as well as Y chromosome and autosomes have been reported (8,13,29). These abnormalities may result in an activation of X chromosome genes translocated to autosomes which may impair spermatogenesis (30). Similarly, translocations involving a Y chromosome may inactivate Y chromosomal genes or interfere with meiotic pairing.

C. Abnormalities in Autosomal Chromosomes

1. Reciprocal Translocations

An increased incidence of reciprocal translocations are found in infertile men as compared to newborns. Azoospermic men demonstrate an incidence of these translocations that is approximately nine times higher than in control populations. A somewhat lesser incidence is found in oligospermic patients (13,14). Translocations involving chromosomes 1 and 9, 1 and 16, 1 and 18, 3 and 4, as well as others have been reported (31). While the mechanism by which reciprocal translocations of autosomes interfere with spermatogenesis is not completely clear, there is some evidence that the translocated chromosomes may form centromeric associations with the XY bivalent at the pachytene stage in meiosis I. This may interfere with X chromosomal inactivation in the primary spermatocyte (32). A lack of X chromosomal inactivation may interfere with spermatogenesis.

2. Robertsonian Translocations

Robertsonian translocations involve the fusion of two chromosomes with distally located centromeres (acrocentric chromosomes) at the centromeric regions. Of interest, these translocations have been more commonly found in oligospermic patients than in azoospermic patients (12,14,33). One common Robertsonian translocation involves chromosomes 13 and 14. The mechanism of impairment in spermatogenesis may be similar to that with reciprocal autosomal translocations in which the translocated chromosome associates with the XY bivalent (34).

3. Chromosomal Segment Inversions

Both paracentric and pericentric inversions of a variety of chromosomes have been associated with male infertility (13,14,33). Paracentric inversions of chromosome 1 usually result in male infertility. Studies suggest that the cause may be due to a defect in synapses of the involved chromosomes during metaphase of meiosis I, eventually resulting in impaired spermatogenesis (35). As discussed elsewhere in this book, meiotic crossing over within the limits of paracentric inversions leads to acentric fragments and dicentric bridges. Crossing over within the limits of pericentric inversions, on the other hand, leads to duplications and deficiencies in chromosomal materials. Genetically unbalanced gametes may not be viable and/or functional.

4. Other Autosomal Chromosomal Abnormalities

A variety of chromosomal abnormalities have been reported in studies of infertile males, including supernumerary marker chromosomes and ring chromosomes (8,27). These abnormalities are generally quite rare, and their relationship to infertility is unclear. With the advent of FISH and other FISH-related technologies, as discussed in the chapters by Blancato and Haddad, and Mark, the frequency of this category is expected to decrease. An increased number of markers and ring chromosomes will be delineated using one or more of these technologies.

D. Genetic Abnormalities Localized to Germ Cells

Infertile patients with normal peripheral blood karyotypes have been found to have meiotic abnormalities (8,12,36). Reasons can vary. One explanation may be a sampling error. For example, a genetic abnormality may have been present in a percentage of cells throughout the body (mosaicism), but was by chance not detected in the routine 20-cell chromosome analysis of the blood. Another explanation may be that true confined gonadal mosaicism exist, i.e., the genetic abnormality is found only in the gonadal and not in any somatic (body) tissues. Disturbances during various stages of development may be responsible. A defect in the mechanism of pairing, for example, may lead to mitotic and meiotic nondisjunction in the gonadal tissues and give rise to numerical chromosomal abnormalities. Abnormalities such as the presence of univalents, asymmetrical or fragmented bivalents, low chiasma counts, as well as polyploidy have been identified in this group of patients (37). The exact incidence of these abnormalities in the infertile population is not clear, due to differences in the techniques used in each study. Although in theory meiotic abnormalities of germ cells may be studied by a number of techniques, this kind of study is by no means easy to perform in practice. Testicular biopsy is an invasive procedure and is not routinely utilized for infertile patients.

Studies of individual spermatozoal chromosomal complements have recently been performed. Sperm nuclei have been difficult to study because of the condensed nature of the genetic material. Allowing spermatozoa to fertilize zona-free hamster oocytes results in a decondensation of the genetic material, making it available for genetic study (38). This approach is limited since one can only study patients with sperm of sufficient quality to fertilize hamster oocytes, which precludes its use in severe male factor cases. In addition, those sperm causing fertilization of hamster oocytes may not be representative of the total spermatozoal pool. Recently, FISH (see Color Plate 10.7) has been used to study individual spermatozoa utilizing chromosome-specific chromosome-enumeration DNA probes (See, for example, Refs. 39,40). These studies allow for a determination of numerical chromosomal abnormalities for the probed chromosomes. No information is obtained on the other chromosomes, and structural defects in the studied chromosomes may be missed. An additional limitation is that to perform FISH, the sperm are processed, making it difficult to correlate FISH results with individual sperm morphologic or motion parameters. Recently, structural abnormalities have been detected by using two probes to the same chromosome (41). Finally, FISH technology of sperm pronuclei in zona-free hamster eggs has been utilized to examine the sperm with fertilizing capacity. Unlike the earlier examination of chromosome complements with the zona-free hamster eggs, this technique does not require an evaluation of metaphase chromosomes and thus is more easily applied to a large number of spermatozoa. Recently, primed in situ labeling (PRINS) has been developed for use with spermatozoa (42). In this technique, chromosome-specific primers are used to amplify the probed chromosome segment with labeled nucleotides. While initially able to probe only one chromosome at a time, a triple PRINS technique has now been developed that allows for the identification of three chromosomes simultaneously within a 3-h period (43). This technique appears to be more rapid than FISH. We are likely to see further improvements in both of these technologies over time.

Studies of spermatozoa from fertile men have demonstrated chromosomal abnormalities in approximately 10% of spermatozoa. There are conflicting data as to whether the incidence of chromosomal abnormalities increases with paternal age (44,45). Recent studies have found no relationship between paternal age and X:Y ratio of sperm or diploidy. However, an increased frequency of YY sperm and sperm disomic for chromosome number 1 have been found. In addition, an increase in structural abnormalities has been found to correlate with an increase in paternal age (46,47). An increased frequency of sex chromosomal aneuploidy has been found as compared with autosomal aneuploidy in sperm from fertile donors (48). Studies of infertile populations have demonstrated a significant increase in the frequency of aneuploidies as compared to normal fertile donors. Increased frequencies of diploidy, autosomal disomy, numerical sex chromosomal abnormalities, and autosomal nullisomy have been demonstrated (49,50). Of significance, peripheral blood karyotypes in all donors and patients in these studies were completely normal.

VI. TREATMENT

Current data suggest that patients with genetic causes of infertility demonstrate severe oligospermia or azoospermia. There are no treatments to improve sperm production in this class of patients. However, there are treatments that utilize patients' germ cells to produce conceptions. In-vitro fertilization (IVF) combined with intracytoplasmic sperm injection (ICSI) has revolutionized the management of severely oligospermic men. In this technique, women are induced to superovulate with gonadotropins. This results in the maturation of multiple ova. The ova are retrieved through ultrasound-guided needle aspiration. Individual spermatozoa are then selected and micro-injected into ova. This bypasses the need for the sperm to bind to and penetrate the zona pellucida and the oolemma. The fertilized eggs are allowed to divide in vitro for approximately 48 h, following which they are placed transcervically into the women's uterus. Pregnancy rates are approximately 25% per cycle. Miscarriages occur in approximately 30% of cases, which is similar to the rates following pregnancy through natural intercourse. It appears that pregnancy rates are independent of sperm quality. In-vitro fertilization with ICSI may also be applied to azoospermic patients. Azoospermia may be due to one of two causes. Obstructive azoospermia occurs in patients with normal spermatogenesis but an obstructed genital ductal system. In these patients, sperm may be retrived from the epididymis or vas deferens proximal to the obstruction and used for IVF with ICSI. Nonobstructive azoospermia is due to absent spermatogenesis with a patent ductal system. Previously, there had been no treatment for this class of patients beyond donor insemination or adoption. Recently, individual spermatozoa have been found following microdissection or homogenization of testicular tissue from nonobstructive azoospermic patients. Spermatozoa are identified in approximately 60% of these patients. Of significance, spermatozoa have been identified even when a permanent histologic examination of testicular tissue revealed no evidence of spermatozoa. Spermatozoa have even been identified in 15–20% of patients with Sertoli-only syndrome. Finally, pregnancies have been produced in some cases by the injection of round spermatids into ova. This procedure raises the possibility of using immature germ cells in patients with complete maturation arrest.

The discovery of genetic causes of some cases of male infertility as well as the increased incidence of chromosomal abnormalities in spermatozoa of infertile men has raised the concern that the use of sperm from these patients may result in the transmission of genetic abnormalities to children produced through these techniques. In cases of sex chromosome aneuploidy (such as Klinefelter syndrome or XXY males), the risk of transmission is currently unknown. Current data suggest that the majority of spermatozoa from these patients may be normal; however, some of the karyotypically abnormal spermatogonia are able to produce sperm and thus yield an increased risk of genetic transmission. Certainly, autosomal structural abnormalities are associated with an increased risk of congenital abnormalities as well as miscarriages. Recent data on AZF suggest that the utilization of sperm retrieved from patients with AZF deletion (such as DAZ) will likely lead to transmission of infertility to male offspring. Finally, the significance of chromosomal abnormalities found in spermatozoa from infertile men is unclear. Early studies have reported a 1.2% incidence of karyotype abnormalities found through prenatal diagnosis of pregnancies derived from ICSI (51). This is no different than the expected incidence in the general population. In cases in which either partner carries a known genetic disease that can be analyzed, embryos produced through IVF may undergo blastomere biopsy with follow-up genetic analysis. In this procedure, individual blastomeres are removed from developing embryos prior to embryo transfer back into the uterus. PCR is performed to determine whether or not that embryo carries the genetic trait. Only embryos free of the genetic trait are then transferred back into the uterus. This technique has been used successfully to perform sex selection to avoid the transmission of sex-linked genetic diseases. As more and more data on the genetic etiology of male infertility emerges, it has become imperative that physicians treating these couples have a firm grasp of the genetic implications of these techniques. Genetic counseling has become an integral part of the management of these couples. While there is currently no treatment to improve sperm production, rapid advances in gene therapy raise the possibility that curative therapy may be available in the not too distant future.

ACKNOWLEDGMENTS

We thank Sue Batalon for word processing and Dr. Roger Mark for reading the drafts of the manuscript. The support of the staffs of our laboratories are acknowledged.

REFERENCES

1. Hawkins JR. Cloning and mutational analysis of SRY. Horm Res 1992; 38: 222–225.
2. Najmabadi H, Chai N, Kapali A, et al. Genomic structure of a Y-specific ribonucleic acid binding motif-containing gene: a putative candidate for a subset of male infertility. J Clin Endocrinol Metab 1996; 81:2159–2164.
3. Reijho R, Lee T, Salo P, Alagappan R, Brown LG, Rosenberg M, Rozen S, Jaffe T, Straus D, Hovatta O, Chapelle A, Silber S, Page D. Diverse spermatogenic defects in humans caused by Y chromosome deletions encompassing a novel RNA binding protein gene. Nat Genet 1995; 10:383–393.
4. Saxena R, Brown LG, Hawkins T, Alagappan RK, Skaletsky H, Reeve MP, Reijo R, Rozen S, Dinulos MB, Disteche CM, Page DC. The DAZ gene cluster on the human Y chromosome arose from an autosomal gene that was transposed, repeatedly amplified and pruned. Nat Genet 1996; 14:292–299.
5. Menke DB, Mutter GL, Age DC. Expression of DAZ, an azoospermia factor candidate in human spermatogonia. Am J Hum Genet 1997; 60:241–244.
6. Nakahori Y, Kuroki Y, Komaki R, Kondoh N, Namiki M, Iwamoto T, Toda T, Kobayashi K. The Y chromosome region essential for spermatogenesis. Horm Res 1996; 46(suppl 1):20–23.
7. Pryor JL, Kent-First M, Muallem A, Vanbergen AH, Nolten WE, Meisner L, Roberts KP. Mircodeletions in the Y chromosome of infertile men. N Engl J Med 1997; 336:534–539.
8. Chandley AC. The chromosomal basis of human infertility. Br Med Bull 1979; 35:181–186.
9. Zuffardi O, Tiepolo L. Frequencies and types of chromosome abnormalities associated with human male infertility. In: Crosignani PG, Rubsin BL, eds. Genetic Control of Gamete Production and Function. Serono Clinical Collogquia on Reproduction III. London, UK: Academic Press and Grune & Stratton, 1982:261–273.
10. Yoshida A, Tamayama T, Nagao K, et al. A cytogenetic survey of 1007 infertile males. Contracept Fertil Sex 1995; 103a.

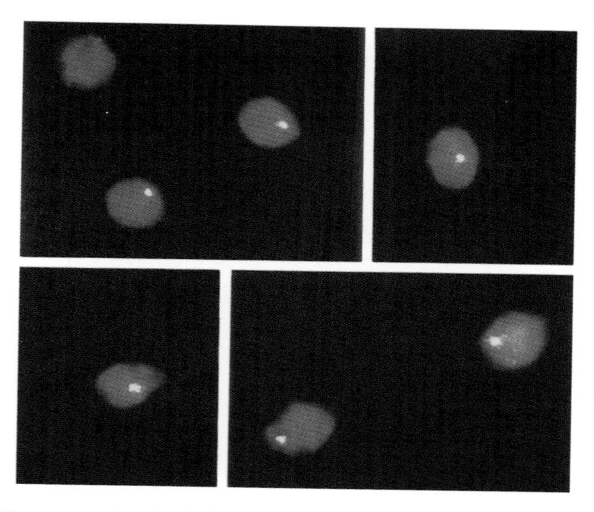

Figure 10.7 Fluorescent in situ hybridization of sperm using a chromosome 8-specific probe: four representative fields showing positive hybridization signals in normal haploid sperm. (Reprinted with permission from Ref. 39.)

11. Walzer S, Gerald P. A chromosome survey of 13,751 male newborns. In Hook EB, Porter H, eds. Population Cytogenetics: Studies in Humans. New York: Academic Press, 1977:45–61.
12. Hendry WF, Polani PE, Pugh ACB, et al. 200 infertile males: correlations of chromosome, histological, endocrine and clinical studies. Br J Urol 1976; 899–908.
13. Retief AE, Van Zyl JA, Menkveld R, et al. Chromosome studies in 496 infertile males with a sperm count below 10 million/ml. Hum Genet 1984; 66:162–164.
14. Bourrouillou G, Dastugue N, Colombies P. Chromosome studies in 952 infertile males with a sperm count below 10 million/ml. Hum Genet 1985; 71:366–367.
15. Chevret E, Rousseaux S, Monteil M, Usson Y, Cozzi J, Pelletier R, Sele B. Increased incidence of hyperhaploid 24,XY spermatozoa detected by three-colour FISH in a 46,XY/47,XXY male. Hum Genet 1996; 97(2):171–175.
16. Cozzi J, Chevret E, Rousseaux S, Pelletier R, Benitz V, Jalbert H, Sele B. Achievement of meiosis in XXY germ cells: study of 543 sperm karyotypes from an XY/XXY mosaic patient. Hum Genet 1994; 93(1):32–34.
17. Tournaye H, Staessen C, Liebaers I, Van Assche E, Devroey P, Bonduelle M, Van Steirteghem A. Testicular sperm recovery in nine 47,XXY Klinefelter patients. Hum Reprod 1996; 11(8):1644–1649.
18. Blanco J, Rubio C, Simon C, Egozcue J, Vidal F. Increased incidence of disomic sperm nuclei in a 47,XYY male assessed by fluorescent in situ hybridization (FISH). Hum Genet 1997; 99(3):413–416.
19. Chevret E, Rousseaux S, Monteil M, Usson Y, Cozzi J, Pelletier R, Sele B. Meiotic behavior of sex chromosomes investigated by three-colour FISH on 35,142 sperm nuclei from two 47,XYY males. Hum Genet 1997; 99(3):407–412.
20. Lopez M, Torres L, Mendez JP, Cervantes A, Perez-Palacios G, Erickson RP, Alfaro G, Kofman-Alfaro S. Clinical traits and molecular findings in 46,XX males. Clin Genet 1995; 48(1):29–34.
21. Birnbacher R, Frisch H. Hormonal, genetic and clinical findings in an XX male. Horm Res 1995; 43(5):213–215.
22. Nomura Y, Yagi H, Onigata K, Yutani S, Nagashima T, Ogawa R, Nagashima K, Morikawa A. A sex reversal infant with XX karyotype and complete male external genitalia. Acta Paediatr Jpn 1995; 37(6):706–709.
23. Lopez M, Torres L, Mendez JP, Cervantes A, Perez-Palacios G, Erickson RP, Alfaro G, Kofman-Alfaro S. Clinical traits and molecular findings in 46,XX males. Clin Genet 1995; 48(1):29–34.
24. Birnbacher R, Frisch H. Hormonal, genetic and clinical findings in an XX male. Horm Res 1995; 43(5):213–215.
25. Borer JG, Nitti VW, Glassberg KI. Mixed gonadal dysgenesis and dysgenetic male pseudohermaphroditism. J Urol 1995; 153(4):1267–1273.
26. Gasso-Matoses M, Pico-Alfonso A, Fernandez-Garcia J, Lobato-Encinas, Mira-Llinares A. 45,X/46,XY gonadal dysgenesis in an infertile adult male. Urol Int 1992; 48(2):239–241.

27. Mc Iree ME, Price WH, Court Brown WM, et al. Chromosome studies on testicular cells from 50 subfertile men. Lancet 1966; 2(7454):69–71.
28. Chandley AC, Edmond P. Meiotic studies on a subfertile patient with a ring Y chromosome. Cytogenetics 1971; 10(4):295–304.
29. Madan K. Balanced structural changes involving the human X: effect on sexual phenotype. Hum Genet 1983; 63(3):216–221.
30. Lifschytz E, Lindsley DL. The role of X chromosome inactivation during spermatogenesis. Proc Natl Acad Sci USA 1972; 69:182–186.
31. Matsuda T, Horii Y, Hayashi K, Yoshida O. Quantitative analysis of seminiferous epithelium in subfertile carriers of chromosomal translocations. Int J Fertil 1991; 36(6):344–351.
32. Forejt J. Nonrandom association between a specific autosome and the X chromosome in meiosis of the male mouse: possible consequences of homologous centromere separation. Cytogenet Cell Genet 1974; 13:369–383.
33. Micic M, Micic S, Diklic V. Chromosomal constitution of infertile men. Clin Genet 1984; 25:33–36.
34. Johannisson R, Schwinger E, Wolff HH, et al. The effect of 13;14 Robertsonian translocations on germ-cell differentiation in infertile males. Cytogenet Cell Genet 1993; 63:151–155.
35. Chandley AC, McBeath S, Speed RM, et al. Pericentric inversion in human chromosome 1 and the risk for male sterility. J Med Genet 1987; 24:325–334.
36. Egozcue J, Templado C, Vidal F, et al. Meiotic studies in a series of 1100 infertile and sterile males. Hum Genet 1983; 65:185–188.
37. Koulischer L, Schoysman R, Gillerot Y. Chromosomes meiotiques et infertilite masculine: evaluation des resultats. J Genet Hum 1982; 30:81–89.
38. Rudak E, Jacobs E, Yanagimachi R. Direct analysis of the chromosome constitution of human spermatozoa. Nature 1978; 274:911–913.
39. Mark HF, Sheu M, Lopes L, Campagnone J, Takezawa K, Santoro K, Sigman M. Towards the development of a fluorescent in situ hybridization based genotoxicologic assay for aneuploidy detection in sperm. Cytobios 1995; 82(330): 171–180.
40. Pieters MH, Geraedts JP, Meyer H, et al. Human gametes and zygotes studied by nonradioactive in situ hybridization. Cytogenet Cell Genet 1990; 53:15–19.
41. Van Hummelen P, Lowe XR, Wyrobek AJ. Simultaneous detection of structural and numerical chromosome abnormalities in sperm of healthy men by multicolor fluorescence in situ hybridization. Hum Genet 1996; 98(5):608–615.
42. Pellestor F, Girardet A, Lefort G, et al. PRINS as a method for rapid chromosomal labeling on human spermatozoa. Mol Reprod Dev 1995; 40:333–337.
43. Coignet L, Girardet A, Andreo B, Charlieu JP, Pellestor F. Double and triple in situ chromosomal labeling of human spermatozoa by PRINS. Cytogenet Cell Genet 1996; 73(4):300–303.
44. Rosenbusch B, Strehler E, Sterzik K. Cytogenetics of human spermatozoa: correlations with sperm morphology and age of fertile men. Fertil Steril 1992; 58(5):1071–1072.

45. Martin RH, Rademaker AW, Hildebrand K, et al. Variation in the frequency and type of sperm chromosomal abnormalities among normal men. Hum Genet 1987; 77:108–114.
46. Martin RH, Spriggs E, Ko E, Rademaker AW. The relationship between paternal age, sex ratios, and aneuploidy frequencies in human sperm, as assessed by multicolor FISH. Am J Hum Genet 1995; 57(6):1395–1399.
47. Martin RH, Rademaker AW. The effect of age on the frequency of sperm chromosomal abnormalities in normal men. Am J Hum Genet 1987; 41(3): 484–492.
48. Martin RH, Spriggs E, Rademaker AW. Multicolor fluorescence in situ hybridization analysis of aneuploidy and diploidy frequencies in 225,846 sperm from 10 normal men. Biol Reprod 1996; 54(2):394–398.
49. Moosani N, Pattinson HA, Carter MD, Cox DM, Rademaker AW, Martin RH. Chromosomal analysis of sperm from men with idiopathic infertility using sperm karyotyping and fluorescence in situ hybridization. Fertil Steril 1995; 64(4):811–817.
50. Pang MG, Zachowski JL, Hoegerman SF, et al. Detection by fluorescence in situ hybridization of chromosome 7, 11, 12, 18 X and Y abnormalities in sperm from oligoasthenoteratazoospermic patients of an in vitro fertilization program. J Assist Reprod Genet 1995; 12(suppl):OC 105.
51. Bonduelle M, Wilikens A, Buysse A, Van Assche E, Wisanto A, Devroey P, Van Steireghem AC, Liebaers I. Prospective follow-up study of 877 children born after intracytoplasmic sperm injection (ICSI), with ejaculated epididymal and testicular spermatozoa and after replacement of cryopreserved embryos obtained after ICSI. Hum Reprod 1997; 11(suppl 4): 131–155.

APPENDIX: GLOSSARY OF TERMS

acrosome
A vesicle that contains enzymes in the head of spermatozoa.

aneuploidy
The state of having an abnormal unbalanced set of chromosomes.

aplasia
The condition in which an organ or tissue fails to develop.

autosomal
Related to the body; refers to a non-sex chromosome.

AZF
Azoospermic factor. The genetic factor implicated to be responsible for spermatogenesis in the testis.

azoospermia
A condition in which the male has a complete lack of spermatozoa in his semen.

bivalent
Two paired homologous chromosomes in meiosis.

blastomere biopsy
A procedure that involves the removal of a sample of tissue from the dividing zygote for diagnostic analysis.

cAMP
Cyclic adenosine monophosphate. A secondary messenger in the signal transduction cascade that signals the activation of various enzymatic mechanisms in respective target cells.

chiasma
The cytologically identifiable point of crossover between members of a homologous pair of chromosomes during meiosis.

DAZ
A gene mapped to interval 6 of region 1, band 1, on the long arm of the Y chromosome, which has been found to be deleted in some cases of azoospermia.

DAZLA
The homolog of the DAZ gene, also known as DAZH, found on chromosome 3.

deletion
The loss of a segment of a chromosome.

diakinesis
The stage of prophase I in meiosis in which the chiasmata terminalize, the nucleolus dissipates, and the nuclear membrane disappears.

dicentric Y chromosome
A Y chromosome with two centromeres.

diploid
Having two sets of chromosomes.

diplotene
The stage of prophase I in meiosis in which the bivalent chromosomes, which are still attached at their chiasmata, begin to separate from each other in preparation for the division of the cytoplasm into two cells.

distal
At the terminal end, away from the point of origin.

epididymis
A 6-ft-long tubule through which sperm must travel before they develop a motile capability.

euchromatin
Chromatin that is universally perceived as gene-bearing, in contrast to heterochromatin.

FSH

Follicle-stimulating hormone. A glycoproteinous, gonadotropic hormone secreted by the pituitary gland which activates target tissues, such as the testes, through a secondary messenger called cyclic adenosine monophosphate (cAMP).

germ cells

Cells of the reproductive system that carry the genetic information to be bequeathed to offspring.

GnRH

Gonadotropin-releasing hormone. A 10-amino-acid peptide hormone that is secreted by the hypothalamus to stimulate the anterior pituitary gland to secrete two other gonadotropic hormones, follicle-stimulating hormone (FSH) and luteinizing hormone (LH).

gonadal mosaicism

An individual who possesses genetically distinct populations of cells in the gonadal and the body cells.

gynecomastia

The condition in which the male mammary glands overdevelop.

haploid

Having one set of chromosomes.

heterochromatin

Chromatin that is tightly coiled and devoid of genes, usually stained positive with C-banding, which can be constitutive heterochromatin or facultative heterochromatin.

hyperhaploid

Having one or more chromosomes than the haploid set.

hypospadias

An abnormality in the development of the male urethra that leads to its opening on the underside of the penis.

ICSI

Intracytoplasmic sperm injection. A type of in vitro fertilization procedure in which the sperm is injected into the cytoplasm of the egg.

inhibin

A hormone released by the Sertoli cells in response to adequate sperm production that causes a decrease in the release of FSH by the anterior pituitary.

in vitro fertilization

Artificial fertilization of the sperm and egg in a test tube, as opposed to fertilization as a result of sexual intercourse.

leptotene

The stage of prophase I in meiosis in which the chromatin condenses into visible chromosomes.

Leydig cells
The cells located in the interstitium of the testis that secrete testosterone.

LH
Luteinizing hormone. A glycoproteinous gonadotropic hormone secreted by the pituitary gland that stimulates the Leydig cells to secrete testosterone.

meiosis
The process that occurs in the gonads whereby the number of chromosomes is reduced to half its diploid number.

mitosis
The process whereby daughter cells are produced which are genetically identical to the parent cells.

oligospermic
The state of producing fewer than the normal number of sperm.

pachytene
The stage of prophase I in meiosis in which the chromosomes continue to condense and in which crossing over occurs.

paracentric inversion
An inversion is a structural chromosomal rearrangement in which a chromosomal segment is broken off and rejoined to the chromosome, but in reversed sequence. In a paracentric inversion, the two breaks occur on one chromosome arm on the same side of the centromere. The shape of the chromosome is not altered as a result of the inversion.

pericentric inversion
An inversion is a structural chromosomal rearrangement in which a chromosomal segment is broken off and rejoined to the chromosome, but in reversed sequence. In a pericentric inversion, the two breaks occur one on each side of the centromere. The shape of the chromosome may be altered as a result of this inversion.

polyploidy
The state of having many sets of chromosomes.

primary spermatocytes
The male germ cells that arise directly from the development of spermatogonia and through further division give rise to secondary spermatocytes.

PRINS
Primed in situ. A DNA labeling technique that is based on the annealing of specific oligonucleotide primers to individual chromosomes and subsequent primer extension by DNA polymerase.

proximal
At the point of origin or attachment, as opposed to distal.

RBM
RNA binding motif. Another name for YRRM, the Y-chromosome RNA-recognition motive.

reciprocal translocation
A chromosomal structural rearrangement in which a segment of one chromosome is exchanged with a segment from another chromosome.

ring Y chromosome
A circular-like Y chromosome.

Robertsonian translocation
A structural chromosomal abnormality that involves the centric fusion of the long arms of two acrocentric chromosomes, first described by W. R. B. Robertson in 1916.

secondary spermatocytes
The male germ cell precursors that arise directly from the nuclear division of primary spermatocytes through meiosis I.

seminiferous tubules
The coiled tubular structures that make up the majority of the testis.

Sertoli cells
Large cells that nurture the development of spermatozoa.

somatic cells
Cells of the body.

sperm
Mature spermatids.

spermatids
Male germ cells containing a haploid set of 23 single chromosomes that are formed as a result of the second meiotic division.

spermatogenesis
The male meiotic process leading to the formation of sperm.

spermiogenesis
The differentiation process in male development whereby spermatids become mature sperm.

SRY
Sex-determining region of the Y. The gene that codes for the testis-determining factor (TDF).

supernumerary chromosome
A marker chromosome usually referring to an extra, small chromosome, the origin of which cannot be identified by current cytogenetic technology.

superovulate
To release more than one ovum at one time during each female monthly cycle.

synapsis
The close pairing of homologous pairs of chromosomes during meiosis in preparation for crossing over.

testosterone
The androgen responsible for the development of male secondary sexual characteristics.

tetraploid
Having four sets of chromosomes.

tubulus rectus
The straight portion of the seminiferous tubule that connects to the rete testis.

univalent
One-half of a bivalent present in meiosis, often refers to a single unpaired chromosome.

varicocele
A dilitation of the testicular veins. This condition may cause infertility.

YRRM
Y-chromosome RNA-recognition motive. The gene mapped to interval 6 of region 1, band 1, on the long arm of the Y chromosome.

zona pellucida
A uniform, noncellular layer that surrounds an oocyte.

zygotene
The stage in prophase I of meiosis in which the nucleolus remains visible and in which the homologous chromosomes synapse.

11
Recurring Cytogenetic Abnormalities in the Myeloid Hematopoietic Disorders

John Anastasi and Diane Roulston
The University of Chicago Medical Center, Chicago, Illinois

I. INTRODUCTION

Over the past 25 years, cytogenetic studies of the myeloid disorders have played a major role in the advancement of our knowledge of genetic and chromosomal abnormalities and their relationship to the development of neoplastic disease. Due to the easy accessibility of specimens from blood or bone marrow, and the relatively uncomplicated karyotypic findings in this group of diseases compared to solid tumors, the myeloid disorders have provided a hotbed for new discoveries. A large number of specific, recurring structural or numerical abnormalities have been identified and correlated with sometimes specific and unique pathologic findings and important clinical features associated with the myriad of myeloid diseases. More important, however, recognition of the abnormalities has fostered the discovery of a long list of newly recognized genes, which have been identified through the molecular analysis of breakpoints involved in structural chromosomal changes. Expectations are high that understanding alterations in these involved genes will elucidate mechanisms of malignant transformation, and the means to correct or counteract the genetic defects underlying the diseases. Unfortunately, there still remains much to do before the present state of knowledge of chromosomes and genes is translated to new therapeutics.

Although the impact of scientific work in this area has yet to be fully realized, cytogenetic evaluation of myeloid malignancies has important clinical significance at the present time. Cytogenetic evaluation of myeloid disorders can be useful as a diagnostic tool, as a means of gaining prognostic information, and as a means of identifying markers for the malignant clone. The diagnostic utility of cytogenetic evaluation comes first in the association of specific chromosomal abnormalities with specific clinicopathologic disorders, such as the t(15;17) in acute promyelocytic leukemia (APL). Second, it comes with the ability to identify a proliferation as clonal, in situations in which a neoplastic process is being considered in the differential diagnosis along with a reactive one. For example, the findings of a clone with a gain of chromosome 9 can be useful diagnostically in a patient with polycythemia, when polycythemia vera (P vera) is being considered in the differential with reactive erythroid hyperplasia. The prognostic utility of cytogenetic analysis has increasing importance in issues related to selection of therapy. This includes the choice of a specific treatment protocol or the decision for, and the timing of, bone marrow or hematopoietic stem cell transplantation. The identification of cytogenetic abnormalities at diagnosis also has importance for the evaluation of the response to therapy and the identification of an early reemergence of disease. For example, an initial cytogenetic evaluation at diagnosis can identify abnormalities which can be later used for molecular or molecular cytogenetic monitoring during and after treatment.

In this chapter, we attempt to discuss the cytogenetic abnormalities associated with myeloid malignancies from two perspectives. In the first, we list and present the cytogenetic abnormalities commonly seen in the entire group of disorders, which includes the acute myeloid leukemias (AML), the myelodysplastic syndromes (MDS), and the chronic myeloproliferative disorders (CMPD) (see Table 1) (1–4). We took this approach because there is considerable overlap of the cytogenetic anomalies within these three groups, making it difficult to discuss a certain abnormality under a single disease entity. In this section, we present the abnormalities from a point of view where the abnormality is the primary focus, and the association with the disease is secondary. We divide the abnormalities into four groups: (a) the Philadelphia (Ph) chromosome, t(9;22); (b) the common recurring translocations and inversions; (c) the common recurring gains, losses, and deletions; and (d) the less common abnormalities. We discuss the abnormalities, first, with a general comment of their overall significance, then with a brief mention of what is known at the molecular level. We then review the chromosome change itself and follow this with some information regarding the associated pathology and clinical significance. Because we present only a fraction of the knowledge about the recurring abnormalities, we refer the reader to numerous references and reviews with more detailed summaries.

In the second section, we attempt to discuss clinicopathologic correlations and diagnostic issues from a broader perspective, in which cytoge-

Table 1 The Myeloid Disorders

Acute myeloid leukemia (AML) (revised FAB schema, including other proposed entities)	
M0	Minimally differentiated AML
M1	AML without maturation
M2	AML with maturation
M3	Acute promyelocytic leukemia (APL)
M4	Acute myelomonocytic leukemia (AMML)
M4eo	AMML with abnormal eosinophils
M5A	Acute monoblastic leukemia
M5B	Acute monocytic leukemia
M6	Erythroleukemia [A: pure, or B]
M7	Megakaryoblastic leukemia
Other proposed entities	
AUL	Acute undifferentiated leukemia
AML with trilineage dysplasia	
Acute basophilic leukemia	
Myeloid sarcoma (granulocytic sarcoma, chloroma)	
Acute panmyelosis (acute myelofibrosis, acute myelodysplasia)	
Myelodysplastic syndromes (MDS) (FAB schema, including other proposed entities)	
RA	Refractory anemia
RARS	Refractory anemia with ringed sideroblasts
RCMD	Refractory cytopenia with multilineage dysplasia
RAEB	Refractory anemia with excess of blasts
RAEBT	Refractory anemia with excess of blasts in transformation
CMML[a]	Chronic myelomonocytic leukemia (also noted as CMMoL)
Other proposed entities	
MDS with fibrosis	
"Atypical CML" (Ph-negative, *bcr/abl*-negative)	
Therapy-related disorders	
t-MDS/t-AML	Therapy-related MDS or therapy-related AML
Chronic myeloproliferative disorders (CMPD)	
CML	Chronic myelogenous (or myeloid) leukemia (also referred to as CGL, chronic granulocytic leukemia)
ET	Essential thrombocythemia
P vera	Polycythemia vera
CIMF	Chronic idiopathic myelofibrosis with myeloid metaplasia (or MMM, myelosclerosis with myeloid metaplasia, or agnogenic myeloid metaplasia)
CNL	Chronic neutrophilic leukemia

[a]Can be considered a CMPD with myelodysplastic features. In juvenile patients: JMML, formerly "juvenile CML" or "juvenile CMML."
Sources: Refs. 1–4.

netic findings supplement the more routine clinical and pathologic evaluations for the specific disease types. Among the issues we present in this section are differential diagnostic issues for which cytogenetic information is useful, correlations with currently used morphologic classification systems, and correlations with clinical and prognostic features. In discussing cytogenetic abnormalities and the myeloid disorders using these two approaches, we wish to emphasize that the utility of cytogenetic analysis comes from the back-and-forth discussion among diagnostician, cytogeneticist, and clinician, who each brings his or her own perspective to the diagnosis, understanding, and treatment of these disorders.

II. THE RECURRING ABNORMALITIES

A. The Philadelphia Chromosome, t(9;22)(q34;q11.2)

1. General

The Philadelphia (Ph) chromosome has a unique place in the history of medical discovery. This chromosomal abnormality was the first to be recognized as a recurring abnormality in a human neoplasm, it was the first chromosomal abnormality which was recognized as a translocation of genetic material from one chromosome to another, and it was the first anomaly to be elucidated at a molecular level as giving rise to a chimeric gene capable of producing a chimeric protein. The initial association of a chromosome abnormality with chronic myelogenous leukemia (CML) came in 1960, when Nowell and Hungerford described an abnormal G-group chromosome with an apparent deletion of part of its long arm in metaphase preparations from patients with CML (5). This abnormal chromosome later became known as the Philadelphia chromosome (Ph), in reference to the city in which it was described. Although there was some debate as to the origin of the Ph chromosome (6,7), it was eventually recognized as chromosome 22, after the introduction of chromosomal banding techniques in the early 1970s permitted identification of individual chromosomes (8). Banding also permitted Rowley to show, in 1973, that the abnormality was not a deletion, but a balanced reciprocal translocation involving the distal segment of the long arm of chromosome 22 and the distal portion of the long arm of chromosome 9 (9). The reciprocal nature of the translocation was further clarified in the 1980s when molecular studies established that the proto-oncogene *abl*, normally located on chromosome 9 (10), was transferred to a specific region on chromosome 22, named the breakpoint cluster region (bcr) within a gene called *bcr* (11,12), and that a reciprocal exchange occurred between chromosomes 22 and 9. The resulting hybrid gene on chromosome 22 (*bcr/abl*) was later shown to be capable of being transcribed to mRNA and then translated into a chimeric protein, which eventually was shown to have abnormal tyrosine kinase activity and oncogenic potential (13–15).

The Ph chromosome or its molecular equivalent, the *bcr/abl* fusion gene, is seen in virtually all cases of CML. It is also seen in a subset of acute lymphoblastic leukemia (Ph+ ALL, see Chapter 12 by Raimondi) (16). Although the Ph chromosome has been reported in other entities, such as essential thrombocytopenia (17–19), in most instances the diagnosis in these other disorders must be considered questionable. The identification of the t(9;22) or its underlying molecular alteration is critical in the differential diagnosis of CML, and is a tool for evaluating response to therapy and residual disease. Hopefully, knowledge of the associated molecular alterations and the subsequent biochemical changes will lead eventually to improved modalities for treatment.

2. Molecular Biology

Although some intriguing data suggest that there may be an initial clonal expansion in hematopoietic elements in which the Ph chromosome develops in a subclone (20,21), the juxtaposition of the *abl* gene on chromosome 9 next to the *bcr* gene on chromosome 22 is generally regarded as the underlying molecular initiating event in CML. How the Ph chromosome translocation actually comes about is not known, but it is thought to occur in a pluripotent hematopoietic stem cell as a breakage and rejoining of genetic material. The break in the *abl* gene occurs between exons 1a and 1b, so almost the entire gene is transferred to chromosome 22 (22). The breakpoint in the *bcr* gene in CML (major *bcr*, or M*bcr*) is usually between exons b2 and b3 or between b3 and b4. Due to the fusion, the proximal *bcr* sequences on chromosome 22 are positioned next to the distal *abl* sequences, and an aberrant *bcr/abl* mRNA is transcribed. Regardless of which of the two breaks occurs in the *bcr*, the transcribed mRNA is translated to a p210 *bcr/abl* protein. This protein has abnormal tyrosine kinase activity which somehow results in the myeloid expansion and the features of the disease. The exact mechanism through which this occurs, however, is not known. There is some evidence that the abnormal p210 tyrosine kinase activity is related to a defective adhesion capacity of the affected progenitor cell, and that this results in kinetic abnormalities and the expansion in hematopoiesis (23); however, this is still somewhat speculative. Other possible mechanisms involve the activation of *RAS*, which is involved in hematopoietic growth factor pathways; induction of *MYC* expression, which is involved in transformation; and activation of other pathways involved in cell proliferation (24). In Ph+ ALL, the *bcr/abl* translocation can give rise to the same p210 protein, but more commonly there is a different breakpoint in *bcr*, within the first intron (minor *bcr*, or m*bcr*), which gives rise to a slightly different chimeric protein with smaller size (p190 *bcr/abl* protein) (25). Whether the differences in Ph+ ALL and CML lie in the different molecular events, or whether they are due to similar mutations occurring in different progenitor cells with different lineage potential, has yet to be determined (26–28).

3. Chromosomes

Currently it is generally accepted that CML is defined by the *bcr/abl* fusion underlying the t(9;22). However, it is important to recognize that the molecular fusion may not be evident at the level of the chromosomes in all cases. Thus, although the t(9;22) is considered a hallmark of CML, a small percentage of cases (~2%) do not have the typical Ph chromosome, but the fusion gene is present due to a molecular-level rearrangement. It should be recognized, however, that the term "Ph-negative CML" can be the source of considerable confusion. In the past, the term encompassed not only those cases which were Ph-negative and *bcr/abl*-positive, but also cases which mimicked typical CML pathologically but which were actually *bcr/abl*-negative, as would be later discovered. Such disorders include some cases of chronic myelomonocytic leukemia (CMML), the poorly understood disorder of "atypical CML" and other myeloproliferative disorders (29). The importance in separating these latter diseases from Ph-negative but *bcr/abl*+ CML lies in different treatment approaches and differences in survival patterns. The term "Ph-negative CML" should probably be reserved for those rare cases of typical CML in which the Ph chromosome is absent, but where *bcr/abl* rearrangements can be confirmed through molecular techniques (30). Recent fluorescence in situ hybridization (FISH) studies of cases with apparently normal chromosomes 9 and 22 have shown the *bcr/abl* rearrangement on the otherwise normal-appearing chromosome 22 (31). However, similar studies have shown the *bcr/abl* rearrangement on a normal-appearing chromosome 9, in some rare cases (32). The other disorders which are Ph-negative but also *bcr/abl*-negative should probably not be referred to as "CML," although for some of these disorders the confusing terms of "atypical CML" and "juvenile CML" (Section II.C.1) seem to persist.

In most cases of Ph+ CML, the chromosome abnormality is the t(9;22)(q34;q11.2) (see Fig. 1). Variant translocations make up 4–8% of cases (33). The variant translocations are now all recognized as "complex" in that they involve chromosomes 9, 22, and a third, or, sometimes, a fourth chromosome. In a three-way variant translocation, three chromosome breaks occur simultaneously: the material from the 9q34→qter region, containing *abl*, translocates to the *bcr* gene on 22q, and the 22q11.2→qter material translocates to a third chromosome. The material from the third chromosome

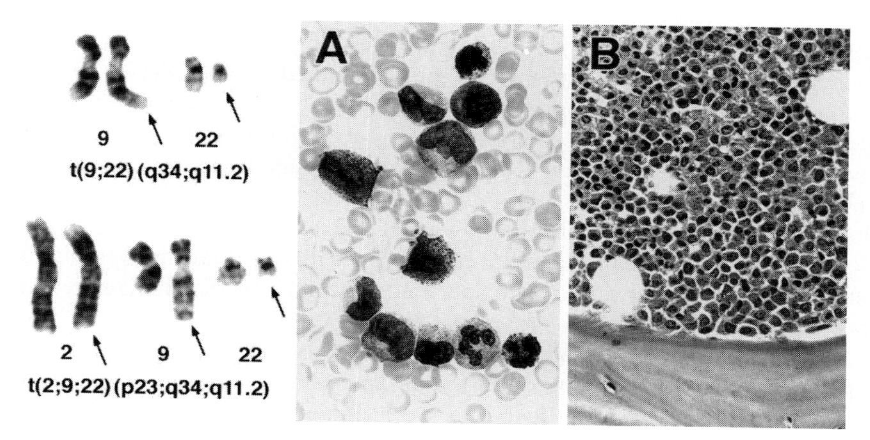

Figure 1 Partial karyotypes showing the t(9;22) typical of CML (above), and a three-way translocation which also produces a Ph chromosome (below) (see text for description). A. Peripheral blood smear showing granulocytosis, with left shift, "myelocyte bulge," and basophilia. B. Bone marrow section showing a hypercellular bone marrow with marked granulocytic proliferation with maturation.

translocates to the 9q34 breakpoint. Thus, the translocation creates a typical Ph chromosome, with production of a *bcr/abl* fusion gene as observed in the typical t(9;22), but the der(9) contains material from a different chromosome (see Fig. 1). Another type of variant translocation, initially referred to as a "simple" variant, is somewhat deceptive in that the typical Ph chromosome is present but chromosome 9 appears normal, and another chromosome may have the 22q11.2→qter material. However, by in situ hybridization, Hagemeijer et al. (34) demonstrated that the normal-appearing chromosome 9 homologs are in fact involved, and have undergone a translocation with another chromosome due to a complex rearrangement. These rearrangements can be three-way translocations or two-step rearrangements where the der(9)t(9;22) undergoes a second rearrangement with another chromosome. A "masked" Ph chromosome refers to the rare condition in which the Ph chromosome is not apparent due to a second step which adds material to the der(22)t(9;22) in the complex exchange of genetic material. Cases with a "masked" Ph chromosome would need to be identified using molecular or molecular cytogenetic techniques (30).

Chromosomal abnormalities other than the t(9;22) are not uncommon in CML. Such secondary changes can be seen at diagnosis and during the chronic phase, but are more common in the accelerated and blast phases of the disease. Whether they are useful for prognosis is debatable. When a t(9;

22) is documented at diagnosis, during the chronic phase, a subsequent secondary change can herald the onset of the accelerated or blast phase (35,36). Secondary changes can be grouped into common or uncommon types, or, according to Mitelman, into groups which proceed through a major or minor route of karyotypic evolution (37,38). The secondary changes can also be sorted into those associated with either a myeloid or lymphoid blast phase. However, this latter grouping is somewhat inexact, probably because the blast phase in CML can be multilineal (e.g., with myeloid and lymphoid markers), and because the lineage specificity of many of the changes is not absolute.

The most frequently seen secondary changes are a gain of an additional copy of the Ph chromosome (+Ph), +8, an isochromosome for the long arm of chromosome 17 [i(17q)], and +19. Taken together, these changes occur in 71% of cases with secondary abnormalities. They are frequently seen together, with a gain of the Ph chromosome with +8 being the most common combination. The less common abnormalities include −7, −17, +17, +21, −Y, and the t(3;21)(q26.2;q22) which is seen during blast phase. These occur in about 15% of all cases with secondary change. Rarely, other abnormalities can be seen with the Ph chromosome, including almost any of the recurring abnormalities associated with AML or ALL. Thus, in CML, one might see a subclone with a t(9;22) together with the inv(16); this is associated with a myelomonocytic blast phase of CML (39). Similarly, in lymphoid blast phase, the t(9;22) can occur together with hyperdiploidy (>50 chromosomes), with gains of specific chromosomes typical of hyperdiploid ALL.

As mentioned above, it is somewhat debatable whether a secondary abnormality noted at diagnosis is prognostic of a worsening disease. It does appear that a change in the karyotype is a sign of acceleration or impending blast crisis (40). The impression that the secondary changes are related directly to the development of blast transformation in all cases may also be oversimplified. In one study employing FISH analysis on Wright-stained bone marrow aspirate slides, the secondary change was not always found in the blastic cells (41). On the other hand, when a recurring translocation occurs as a secondary change in CML, it would seem likely that its molecular consequence would be the same as when it occurs in a de novo acute leukemia.

4. Pathology

The pathologic features of CML are well known, and have been reviewed in numerous texts and review articles (42,43). The peripheral blood shows a granulocytosis comprised of granulocytic elements of all stages of maturation. There is usually a "bulge" in myelocytes (%myelocytes > %metamyelocytes), absolute basophilia, and frequently thrombocytosis (Fig. 1A). The marrow is usually hypercellular with granulocytic hyperplasia (Fig. 1B), erythroid hypoplasia, basophilia, and megakaryocytic hyperplasia in some cases. Some investigators have attempted to subclassify different types of chronic-phase CML, but the correlations with *bcr* breakpoints, or other clinical parameters such as survival, are controversial (44–46). The accelerated phase of CML is poorly defined, but is generally considered to occur when myeloblasts are greater than 20%, but less than 30%, of the nucleated marrow elements. However, a discussion is ongoing concerning the possibility of lowering the blast count for the accelerated phase designation (47). Other, less clear criteria are the presence of marked basophilia, increased fibrosis, and changes in the clinical status of the patient. The blast phase of CML is defined as >30% myeloid blasts, or >20% lymphoid blasts, but lower blast counts are also being considered. In the myeloid blast phase, the blastic component can be difficult to define, as the blast phase is frequently multilineal. The lymphoid blast phase is usually of the precursor B-cell type although a T-lymphoid blast phase can occur.

5. Clinical

Because of the well-established clinical course of CML and established treatment protocols, the correct diagnosis is of utmost importance (3). CMML, other MPDs (including some unusual types), a leukemoid reaction, or MDS/AML must be excluded from the differential. Cytogenetic, molecular, or molecular cytogenetic studies are essential in all patients. Such studies can be useful not only in confirming the diagnosis of CML, but also in evaluating therapy, and in monitoring residual disease and early relapse (48,49). However, use of these techniques must be undertaken with caution. In fact, correlation of molecular findings with clinical and pathologic parameters is always important. For example, persistence of *bcr/abl* transcripts after bone marrow transplantation as identified by PCR may not always herald relapsed disease. Apparently, residual, terminally differentiated (nonclonogenic) cells, such as histiocytes, can persist in the marrow for at least 1 year and can result in a positive PCR reaction for *bcr/abl* transcripts (50).

B. The Common Recurring Translocations and Inversions

Three recurring structural abnormalities, t(8;21), t(15;17), and inv(16)/t(16;16), are the most common balanced rearrangements seen in myeloid malignancies and are present exclusively in specific morphologic subtypes of

AML. Together, they account for over 30% of the cytogenetic abnormalities in AML. Translocations of 11q23 are also among the most common anomalies in myeloid malignancies, but are present in a wide variety of disorders, including MDS, AML, ALL, and leukemias with lineage infidelity, such as biphenotypic, mixed lineage, or bilineal processes.

1. t(8;21)(q22;q22) and Other 21q22 Translocations

t(8;21)(q22;q22) (see reviews 51,52)

General. The balanced t(8;21)(q22;q22), first described by Rowley et al. in 1973 (53), is now recognized as the most common structural abnormality in AML (54). It occurs in about 10–20% of all cases of AML, and in approximately 10–30% of cases subclassified as M2 according to the FAB (1). The abnormality is important because it involves a molecular pathway critical in hematopoiesis, and because it is associated with a leukemic process that has distinctive morphologic and clinical features. The significant characteristics include the molecular targeting of a critical hematopoietic transcription factor, the core binding factor (CBF), which is also rearranged with other translocations such as the t(3;21) and inv(16)/t(16;16) (51,52); an array of morphologic features which can be used to recognize the entity even when the chromosomal change is not seen (55); and the usual clinical features of a high incidence of extramedullary involvement at presentation or relapse (56), and a good response to chemotherapy with high remission rate and long survival (57). A fascinating additional feature of the t(8;21)-associated leukemia is the persistence of PCR-detected transcripts for the associated fusion gene years after complete clinical remission is attained (58).

Molecular. Initial cytogenetic studies of leukemias with a t(8;21) abnormality, including those with complex karyotypes, indicated that the derivative 8 was conserved in all cases, and likely a factor in leukemogenesis (59). It is now known that the t(8;21) results in the disruption of the *AML1* gene on chromosome 21, and the *ETO* gene on chromosome 8, forming a new chimeric gene, *AML1/ETO*, on the der(8) (60,61). The *AML1* gene has been found to encode a protein which is identical to the alpha2 subunit of a heterodimeric transcription factor called CBF (CBFalpha2), and is thus now referred to as *CBFA2*. The gene encoding the beta subunit of CBF at 16q22 is disrupted by the inv(16), the second most common cytogenetic abnormality in AML (Section II.B.2). Although the function of the *ETO* gene product is not known, the *AML1/ETO* fusion creates a protein that interferes in *CBFA2*-dependent transcription (62). Genes induced by *CBF* include genes encoding cytokines (GM-CSF, IL3), cell surface differentiation markers (TCR, CD2epsilon), cytokine receptors (CSF1 receptor), and enzymes (myloperoxidase, neutrophil elastase, granzyme B serine protease). The exact means through which disruption of *CBFA2* results in leukemogenesis is not known precisely, but the importance of CBF in hematopoiesis appears central (63).

Chromosomes. The t(8;21) (see Fig. 2) can be seen alone as a simple or occasionally three-way translocation, which conserves the der(8) with the *AML1/ETO* fusion gene. The most common secondary change is loss of a sex chromosome: 39% have loss of the Y, and 16% have loss of the X chromosome (64). While loss of a sex chromosome was initially thought to convey a poorer prognosis, this has not been entirely substantiated.

Pathology. Most cases with t(8;21) can be classified as AML with maturation (M2), although only 25–30% of M2 subtypes have this translocation. Some cases have been classified as M4. The t(8;21)-associated leukemias have several distinctive morphologic features. These include M2 morphology, myeloperoxidase-positive blasts with long, slender Auer rods, bone marrow eosinophilia, and cytoplasmic atypia in the granular elements represented by cytoplasmic globules, vacuoles, and a waxy-orange or salmon-colored granulation with centripetal localization (Fig. 2A, B, C) (65). Recently it has been shown that these morphologic features can be used to detect cases without cytogenetic evidence of the t(8;21), but with expression of *AML1/ETO* by RT-PCR (55), in a manner analogous to Ph-negative, *bcr/*

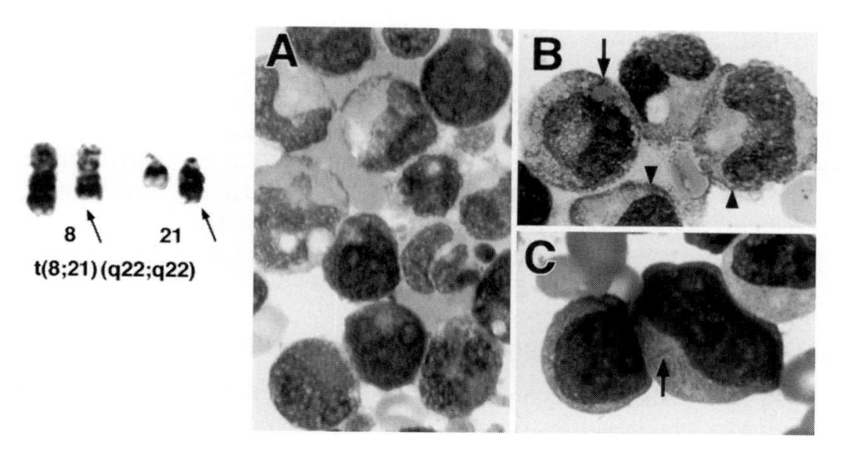

Figure 2 Partial karyotype showing the t(8;21) associated with AML-M2. A. Bone marrow aspirate with the typical morphologic features of AML with maturation. B. Cells with peripheral basophilia (arrow heads), vacuolization, and granulation abnormalities including cytoplasmic globules (arrow). C. A blast with thin Auer rod (arrow).

abl+ CML (Section II.A). Such cases were initially estimated to account for 10% of all cases, but may occur less frequently. It is important to recognize that the t(8;21) can be seen in cases with less than the 30% blasts required for a diagnosis of AML. Such cases have prompted discussion as to whether they should be considered evolving AML or MDS (Section III.A.1) (66–69).

Clinical. Clinically, the t(8;21) is generally associated with a favorable prognosis. Although this association is well established in adults, it may not be true for children (70). As noted above, there is an increased incidence of extramedullary disease (granulocytic or myeloid sarcoma) with this anomaly. The extramedullary presentation can be seen both at onset of the disease and at relapse (56). An interesting finding in patients with the t(8;21) is long-term remission with persistence of *AML1/ETO* mRNA transcripts identified by RT-PCR. This can be seen in patients for up to 10 years after onset of clinical remission (58). The significance of the persistent positive PCR result is not clear, but does not appear to be inconsistent with prolonged and continued remission. The cell reservoir of the persistent *AML1/ETO* transcripts has not been identified, although some have presented evidence that it is a primitive hematopoietic progenitor (71).

Other 21q22 Translocations. Other recurring translocations of 21q22 that are associated with myeloid malignancies are relatively rare; of these, the t(3;21)(q26;q22) is best characterized (72,73). The t(3;21) has been observed in CML as a secondary change associated with the blast phase, in t-MDS/t-AML, and in one case of de novo AML. When present in t-MDS/t-AML, the translocation is usually seen after therapy targeting DNA topoisomerase II (Section III.C). The translocation involves *CBFA2* (*AML1*) from 21q and *EVI1*, as well as two additional, unrelated genes on 3q26 (51). Additional, rare 21q22 abnormalities which involve *CBFA2*(*AML1*) have been reported (51,74) and, as for the t(3;21), appear to be predominant among patients who have had therapy with topoisomerase II-targeting agents (75,76). Another translocation, the t(16;21)(p13;q22), involves *ERG*, which is distal to *AML1* on 21q, and *FUS* on 16p.

2. inv(16)(p13q22), t(16;16)(p13;q22) (see review 77)

General. Arthur and Bloomfield were the first to describe leukemias associated with abnormalities of 16q22 (78). They reported five cases in acute leukemias classified as M2 or M4 with the notable feature of increased eosinophils. Later, it was recognized that two abnormalities of 16q22, that is, the inv(16)(p13q22) and its variant, the t(16;16)(p13;q22), were actually quite common in AML, and were associated with a particular subtype of acute myelomonocytic leukemia (M4) with abnormal but not necessarily increased eosinophils (79). The two rearrangements involve the same breakpoints, and both give rise to the same fusion gene and fusion protein (see below). Taken together, they occur in about 2000 patients per year and account for 8% of new cases of AML (80). Interestingly, the inv(16)/t(16; 16) leukemias have a number of similarities with those associated with t(8; 21). At the molecular level, both types involve the core binding factor (CBF), a heterodimeric transcription factor important in cell differentiation and hematopoiesis; both have a distinctive morphologic phenotype, and both have a good overall prognosis.

Molecular. The important breakpoint in the inversion and translocation occurs at 16q22 in the coding region for the beta subunit of the core binding factor (*CBFB*). This region is fused to the myosin heavy-chain gene (*MYH11*) on 16p13 (81). The fusion gene *CBFB/MYH11* encodes a protein with true chimeric properties, as it can associate with CBFA2 and form multimers, as do the normal products of *CBFB* and *MYH11*, respectively. Apparently, each of these properties of the chimeric protein is needed for leukemic transformation. Models for the molecular mechanism of action of the abnormal protein include those in which the CBFB/MYH11 multimeric protein superactivates CBF, traps CBFA2 (preventing its access to DNA binding sites), or interferes with the normal function of CBF through a stearic hindrance (80,81). The exact mechanism through which the fusion product causes the leukemia is not known.

Chromosomes. The inv(16)/t(16;16) (Fig. 3) can occur alone, but frequently is found with additional karyotypic abnormalities including trisomy 22, trisomy 8, or trisomy 21, in decreasing order of frequency (82). It is interesting that a gain of chromosome 22 is not observed as a secondary change in any other AML subtype. The inv(16) can also occur as a secondary change in the myelomonocytic blast transformation of CML, although this is rare. The blast phase has morphologic features similar to those of the de novo acute process (39).

Pathology. The inv(16)/t(16;16) occurs in a distinctive subset of acute myelomonocytic leukemia in which eosinophils are variable in number but consistently abnormal (83). The abnormality is best appreciated in eosinophils at immature stages of maturation. The immature forms are large and have prominent, atypical basophilic granules which are frequently clustered, of variable size, and intermixed with typical secondary eosinophilic granules (see Fig. 3). Normal immature eosinophils, which also have basophilic granules, can easily be mistaken for the atypical forms, particularly if the cytologic preparations and staining are less than optimal. One feature that can sometimes help in distinguishing the abnormal forms is their cy-

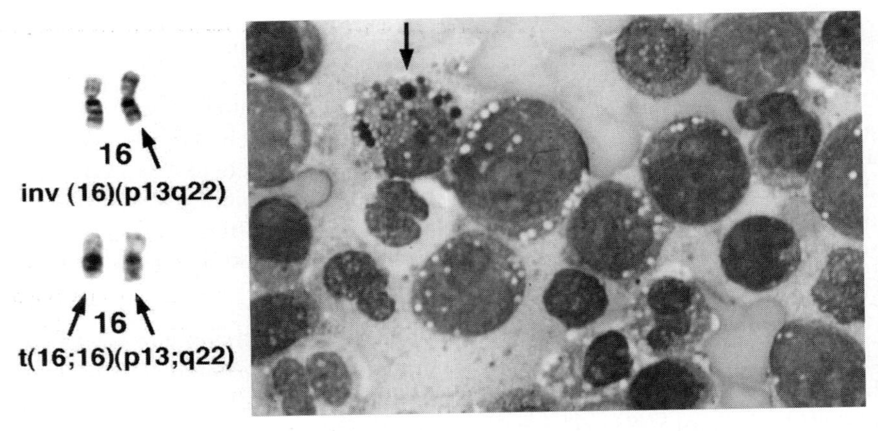

Figure 3 Partial karyotype of the inv(16) (above), and t(16;16) (below) associated with AML-M4eo. A. Bone marrow aspirate showing morphologic features of a typical case of AML-M4eo with myelomonocytic blasts and immature eosinophils with abnormal basophilic granules (arrow). Note the irregular size and abnormal clustering of the basophilic granules.

tochemical profile. The abnormal eosinophils are chloroacetate esterase-positive, whereas normal eosinophils are not (84). In addition to the abnormal eosinophils, the usual other features of M4 are present, with some dysplasia, predominantly in the granulocytic and monocytic lineages. The relationship between the eosinophils and the more actively proliferating myelomonocytic element is not known, although the eosinophils have been demonstrated to be part of the abnormal clone (85). This may be why reappearance of the abnormal eosinophils after remission, or persistence of the abnormal forms, are predictors of relapsing disease.

Clinical. Patients with these cytogenetic abnormalities more frequently have CNS involvement at presentation or at relapse (86). High leukocyte counts at presentation are not uncommon, but despite this, complete remission rates are among the highest for all types of AML (87,88).

3. t(15;17)(q22;q11.2-12) and Variants t(11;17)(q23;q11.2-12), t(5;17)(q34;q11.2-12) (see reviews 89–91)

General. The t(15;17) is seen exclusively in acute promyelocytic leukemia (APL, M3). Although the translocation was initially believed to be present in all cases of APL (92), it is now recognized that there are also rare variant translocations seen in less than 2% of cases. The latter include t(11;

17)(q23;q11.2-12) and t(5;17)(q34;q11.2-12) (93–95). Precise identification of APL associated with the usual t(15;17) is important because this leukemia has been shown to be sensitive to differentiation therapy with all-trans retinoic acid (ATRA) (96). Other subtypes of AML and the rare cases of "APL-like" disorders associated with the variant translocations apparently do not respond to this treatment (97).

Molecular. The mechanism of action of the various translocations has been the subject of intense investigation (91,98). The translocations are believed to be responsible for a block in both maturation of the neoplastic promyelocytes as well as in their proliferation/growth advantage. At the molecular level, the breakpoint on chromosome 17 lies within the retinoic acid receptor alpha (*RARA*) gene. This gene is a member of a superfamily of nuclear hormone receptors which have a number of properties. The gene product has ligand-binding properties (e.g., for binding to retinoic acid), DNA-binding properties (e.g., for binding to promoter regions of target genes), and dimerization properties (e.g., to permit dimerization with other classes of the nuclear hormone receptors, referred to as RXR). Breakpoints in the partner chromosomes 15, 11, and 5 involve the *PML* (promyelocytic leukemia), *PLZF* (promyelocytic zinc finger), and *NPM* (nucleophosmin) genes, respectively. Expression of the *PML* and *PLZF* genes is believed to be related to the control of cell growth and tumor suppression, at least as evaluated in cell lines; the function of the *NPM* product is not entirely known, but it may be involved in transcription. Translocations of chromosome 17 with the various partners result in fusion genes and fusion products denoted as *X/RARA* (i.e., *PML/RARA*, *PLZF/RARA*, and *NPM/RARA*). The fusion products apparently heterodimerize with the normal allelic products of the *PML*, *PLZF*, and *NPM* genes and block their normal action. This results in loss of the control of cell proliferation in a dominant negative effect. Alteration of the *RARA* component of the *X/RARA* products is believed to result in the arrest of the cells at the promyelocyte stage of differentiation. The function of the products of the reciprocal *RARA/X* genes is not known.

Chromosomes. The t(15;17) was initially described by Rowley et al. in the mid-1970s (99,100). In the initial description, the breakpoint on chromosome 17 was denoted as band q21, but is now recognized as involving q11.2-12. The t(15;17) (see Fig. 4) is usually seen alone, but in about 10% of cases it is present with trisomy 8. The variant translocations are quite rare; in fact, only a few cases of t(5;17) have been reported. Even more rare are cases of APL with no recognizable t(15;17), but with molecularly detected *PML/RARA* (101). These would be analogous to Ph-negative, *bcr/abl*+ CML.

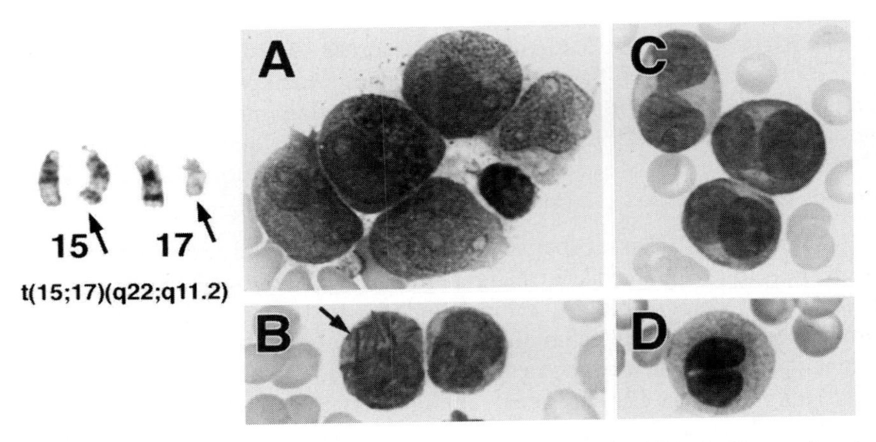

Figure 4 A partial karyotype of the t(15;17) observed in APL. A. Promyelocytes from a typical case of APL with light to heavy granulation in the cytoplasm. B. A cell with multiple Auer rods (arrow). C. Cells from a hypogranular variant (M3var). Note bilobed, reniform, or dumbbell-shaped nuclei. D. Abnormal peripheral blood granulocyte from a patient with APL treated with ATRA.

Pathology. Acute promyelocytic leukemia is fairly distinctive (102,103), but occasionally it may be difficult to diagnose unequivocally. APL can be separated morphologically into a common type and a variant form. These do not correspond to those with typical and variant translocations. The common morphologic type is characterized by hypergranular promyelocytes with cytologic characteristics which vary somewhat among cases and even within a single case. At one extreme, the promyelocytes have large, densely packed azurophilic granules which can cover the nucleus. At the other end of the spectrum, granules can be fine and dustlike (Fig. 4A). A characteristic cytologic feature in APL is the presence of cells with Auer rods, Auer bodies, and sometimes with clusters of Auer rods, referred to as ''faggot'' (bundle of sticks) cells (Fig. 4B). Rare cases may show basophilic differentiation (104). The morphologic variant form of APL is important to recognize, as it can resemble acute monoblastic leukemia. In this so-called hypogranular variant (M3var) (105), the cells resemble monocytes with folded nuclei. A helpful cytologic clue that the cells are abnormal promyelocytes, rather than monoblasts, is the dumbbell or reniform shape of the nuclei of some of the cells (Fig. 4C), as well as the presence of Auer rods. Cytochemical reactions and flow immunophenotyping may also be helpful in clarifying the nature of the variant form as well as helping confirm the diagnosis of the typical form. Cytochemically, promyelocytes show strong myeloperoxidase reactivity. This is usually much more intense than what would be expected in monoblastic leukemias. Esterase staining for the nonspecific esterases is also usually negative, but can be positive in about 20% of cases (106). Immunophenotypically, cases of APL usually show CD13, CD15, and CD33. Characteristically, most cases (75%) are HLA-DR negative. Some cases can show expression of the pan-T-cell marker CD2 (T11). If aspirated material is not available, the histologic features of cells with moderately abundant cytoplasm lead to a differential including mast cell disease and hairy cell leukemia. These latter entities can be easily sorted out with immunohistochemical studies.

It is important to be aware of the cytologic changes induced with ATRA therapy, as these may be confusing to interpret (107). ATRA induces cell differentiation, causing the APL cells to mature to segmented neutrophils. However, the mature segmented forms derived from the neoplastic clone are atypical, frequently with abnormal nuclear lobation and abnormal cytoplasmic granulation (Fig. 4D). These cells can be seen in the peripheral blood or marrow for up to a few months after ATRA treatment. Their presence does not necessarily indicate residual acute disease.

Clinical. Patients with the typical type of APL usually present with low peripheral white blood cell counts. This is not the case with M3var, as the white count is usually elevated. Either morphologic type can present with low platelets and possibly with disseminated intravascular coagulation (DIC) (91). This is one reason why an early, rapid, and accurate diagnosis is necessary. Another reason is that patients with APL are now treated routinely with ATRA (108). Although this therapy was found to be useful before the disease was understood to involve the retinoic acid receptor gene, it has since been shown that high doses of ATRA can reverse the effect of the *PML/RARA* gene product. ATRA can induce patients into a complete remission, but additional conventional chemotherapy is still needed to keep patients in remission permanently. An interesting finding in patients treated with ATRA therapy is the development of a marked hyperleukocytosis, the cause of which is not known (109). Also, patients can develop the so-called ATRA syndrome (110), which can result in a potentially fatal pulmonary capillary leakage. This latter clinical syndrome is associated with fever, respiratory distress, pulmonary infiltrates, and pleural effusions.

Patients with t(15;17) have a favorable prognosis, with complete remission rates of about 70% and overall 5-year survival of 30–40% (88). As mentioned above, cases of APL with the rare variant translocations do not respond to ATRA therapy. Cases of these are too few to provide remission and survival information.

4. t(11q23) and Trisomy 11

t(11q23)

General. Recurring translocations involving chromosome 11, band q23, are of great interest for several reasons. First, over 25 different recurring abnormalities (mostly translocations) involve 11q23 and, thus, along with band 14q32, it is one of the bands most frequently involved in rearrangements in human tumor cells (111). Second, abnormalities of 11q23 are somewhat unusual, in that they are seen in both myeloid and lymphoid disorders, as well as in leukemias with lineage infidelity, such as mixed lineage (biphenotypic) leukemias, bilineal leukemias, and leukemias which undergo lineage switch (112,113). Third, abnormalities of 11q23 have unusual clinical features, as they are seen as the most common abnormality in infant leukemias (114,115), and in adults they are commonly associated with previous therapy, specifically, with agents which target topoisomerase II (116). In the myeloid disorders, most of the abnormalities of 11q23 are associated with leukemias with monocytic lineage. Berger and co-workers were first to report a higher-than-expected frequency of abnormalities of the long arm of chromosome 11 (11q) in patients with acute monocytic leukemia, and, in an expanded series, investigators emphasized an especially strong association between abnormalities of 11q and the poorly differentiated form, acute monoblastic leukemia (M5A) (54,117,118).

Molecular. Several groups of investigators identified the gene located at the breakpoint at 11q23 involved in these recurring translocations. The gene, *MLL* (mixed lineage leukemia, or myeloid-lymphoid leukemia; also *HRX* or *ALL-1*) is large, with multiple transcripts ranging in size from 11–12.5 kb. Motifs characteristic of transcription factors are encoded (AT hooks and zinc finger domains), and the gene has significant homology to the *Drosophila* trithorax gene, which encodes a known transcription factor that is important in development. Many of the partner genes that contribute to production of a fusion gene have been cloned and characterized. An apparent lack of significant similarities among the many partner genes, with only a few exceptions, has been noted (119,120). The manner in which the fusion proteins mediate malignant transformation remains poorly understood.

The breakpoints in *MLL* cluster in an 8.3-kb region. Thus, the 11q23 translocations can be identified by Southern blot analysis using a cDNA clone that contains the exons contained in this region (121). These probes detect the rearrangements in all patients with the more common recurring 11q23 translocations [i.e., the t(4;11), t(9;11), t(6;11), t(11;19)(q23;p13.3), and t(11;19)(q23;p13.1)] as well as those *MLL* rearrangements that are less common or rare.

Chromosomes. The breakpoint in the more common 11q23 translocation partners in AML include 1q21, 2p21, 6q27, 9p22, 10p13, 17q25,

19p13.3, and 19p13.1, with the t(9;11) being the most frequent (see Fig. 5). In ALL they include 1p32, 4q21, and 19p13.3 (Chapter 12 by Raimondi). The der(11) has been determined to be the "critical recombinant chromosome," i.e., that which produces the fusion gene most significant in leukemogenesis, because it remains the recipient of material from the recurring translocation partner in three-way translocations (122). The identification of the partner chromosome involved in some translocations may present a challenge for standard cytogenetic analysis, due to the relatively small amount of material on 11q that is translocated. The t(10;11) can also be difficult to detect because it is frequently produced by a more complex rearrangement, such as an insertion or a three-way translocation (123–125). In such subtle cases, FISH analysis using a probe for *MLL* that spans the 11q23 breakpoint can be useful to confirm the translocation partner. However, for cases with the t(10;11), a probe for the fusion gene partner *AF10* might also be required, due to the frequent involvement of additional chromosomes.

Pathology. The translocations of 11q23 occur in ALL and AML. One common translocation in infants, the t(4;11)(q21;q23), usually has a lymphoblastic phenotype, although the leukemia cells may express some myeloid surface markers. In some cases, variable numbers of monocytoid blast cells have been identified (126). Other translocations, such as the t(9;11)(p22;q23) and t(11;19)(q23;p13.1), are common in leukemias of monocytic lineage (127). Abnormalities of 11q23 are seen in about 35% of M5 patients (Fig. 5), and slightly less than one-half of patients with M5A

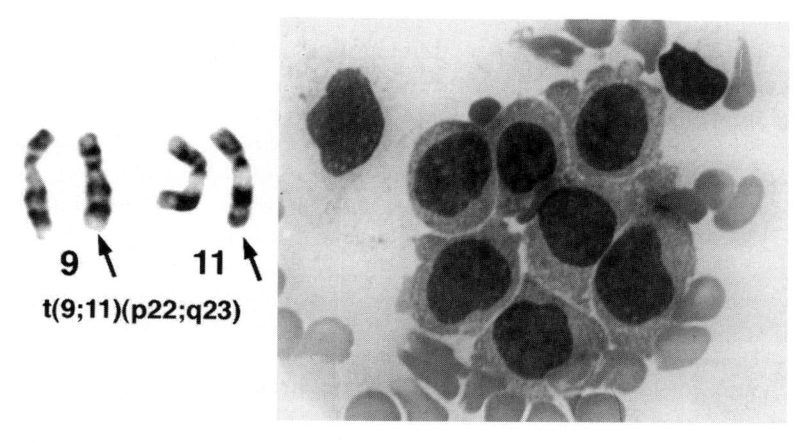

Figure 5 A partial karyotype showing the recurring abnormality t(9;11) with *MLL* involvement. Right panel, Cells from a patient with monoblastic leukemia.

(118,128). The occurrence of 11q23 abnormalities in both AML and ALL suggests that a gene at 11q23 may be involved in determining the differentiation of primitive hematopoietic stem cells into lymphoblasts or monoblasts, or that it may be a gene that is active in both cell lineages. Cases with mixed lineage (biphenotypic processes), two lineages (bilineal processes), or those undergoing a lineage switch can present diagnostic problems. Such leukemias commonly have translocations of 11q23 (112,113).

Clinical. Translocations involving 11q23 have a very unusual age distribution; they comprise about two-thirds of chromosomal abnormalities in leukemia cells of children <1 year of age (128). In one study of infants with ALL, rearrangements were detected in 70% of cases, including some that were normal or inadequate by standard cytogenetic analysis (115). Patients with AML and ALL who have 11q23 abnormalities have aggressive clinical features and often present with hyperleukocytosis and early involvement of the central nervous system (121). These abnormalities with rearrangements involving *MLL* identify patients with a poor outcome. A notable exception, however, is the t(9;11), particularly in children, which is associated with a 3-year event-free survival of 56%. This compares favorably to children with other 11q23 translocations (11%) and to all other children with AML (70).

t(11q23) and Therapy-related MDS/AML. Of increasing interest have been characteristic chromosomal abnormalities found in patients who develop a therapy-related MDS or AML (t-MDS/t-AML) after chemotherapy and/or radiotherapy for an earlier disorder (116,129,130) (see Section III.C). An estimated 10–20% of cases of AML may be secondary or therapy-related (129,131). Several clinical and biological subtypes of t-MDS/t-AML have been recognized. The type associated with abnormalities of 11q23 was first observed among patients receiving extremely high cumulative doses of etoposide for lung cancer (132), but also has been seen subsequently in patients receiving other drugs known to inhibit topoisomerase II (e.g., teniposide, doxorubicin) (129,131). These secondary leukemias are characterized cytogenetically primarily by balanced translocations involving 11q23, such as the t(9;11) (132), but abnormalities of 21q22, such as the t(3;21)(q26;q22) are seen (131,133) (Section II.B.1). In a University of Chicago series of 270 consecutive patients with t-MDS/t-AML, the t(11q23) has been found in 3.7% and the t(21q22) has been found in 3.3% of patients (Le Beau et al., unpublished results). Patients with the 11q23-related t-MDS/t-AML have a relatively short latency period (often only 1 or 2 years from the time of therapy), and present with overt leukemia, often with the similar morphology seen in de novo AML associated with 11q23, (i.e., cases with M4 or M5 morphology). Also, as in de novo AML, the *MLL* gene is involved in the 11q23 translocations (134). Patients with 11q23-related t-MDS/t-AML have a favorable response to treatment relative to those with trilineage dysplasia and with abnormalities of chromosomes 5 and 7 (Sections II.C.1 and III.C).

Trisomy 11

General. Although trisomies are common in hematologic and nonhematologic malignancies, the molecular mechanism leading to the neoplastic transformation in cases with trisomy has been largely unknown. Trisomy 11 in acute myeloid leukemia has recently been found to have a partial duplication of the *MLL* (*ALL1*) gene located at 11q23 (135–137). This finding constitutes the first identification of a specific gene abnormality associated with a recurrent trisomy in human disease; however, whether other trisomies will be associated with specific gene rearrangements, or a more general gene dosage effect, is unknown.

Molecular. Recurrent translocations of 11q23 in AML involve more than 25 partner chromosomal regions and result in fusion genes and fusion proteins that are likely responsible for the malignant transformation (see above). However, in addition to translocations, rearrangements of *MLL* can be detected in cases lacking cytogenetic evidence of 11q23 abnormalities (135–137). *MLL* gene rearrangements were discovered among cases with +11 and in some cases reported to have normal karyotypes by standard cytogenetic analysis. Further studies have shown that rearrangements of *MLL* were present in 90% of cases of acute leukemias with +11 as a sole abnormality, whereas they were seen in 11% of cases of AML in which trisomy 11 was present in a complex karyotype. The *MLL* rearrangement in these cases is due to a partial duplication of the gene, in the area spanning exons 2 through 6 or 2 through 8. The duplication of this region gives rise to an mRNA capable of encoding an abnormal, partially duplicated protein (136,138).

Chromosomes. Trisomy 11 is more commonly seen alone and occurs as a sole abnormality in 1% of all AML cases (139).

Pathology. As a sole abnormality, trisomy 11 appears to be restricted to myeloid malignancies and is relatively uncommon in both MDS and AML. There has been no association of +11 with any particular type of MDS or AML, as it has been noted in RA, RARS, RAEB, CMML, and in AML of the M1, M2, and M4 subtypes (139,140). In the acute leukemias, a +11 is associated with dysplasia. In the cases of +11 with duplication of the *MLL* gene, all were classified as M1 or M2. Immunophenotypically, one report found an association of +11 with a stem cell phenotype, with blasts expressing HLA-DR, CD34 and CD15, CD33, and/or CD13 (141). However, this phenotype is not specific to cases with this chromosomal change.

Clinical. Leukemias with +11 have an unfavorable prognosis (142).

C. Common Gains, Losses, and Deletions

1. −5/del(5q) and −7/del(7q) Including "5q− Syndrome" and "Monosomy 7 Syndrome"

−5/del(5q) and −7/del(7q) (see reviews 143,144)

General. Loss of the entire chromosome 5 or deletions of its long arm, and loss of chromosome 7 or deletions of its long arm, commonly occur together, and are addressed concurrently in this section. The abnormalities are seen in a wide variety of myeloid malignancies including MDS, AML, and CMPD. However, they are most common in the therapy-related disorders, t-MDS/t-AML, and particularly in those following treatment with alkylating agents. In t-MDS/t-AML, abnormalities of 5 and/or 7 are the most common chromosomal change, and are present in about 75% of cases following alkylating agent therapy (145). Morphologically, −5/del(5q) and −7/del(7q) are associated with trilineage dysplasia, and clinically, the abnormalities are associated with poor response to therapy, and poor survival. The so-called "5q− syndrome," where the del(5q) occurs as a sole abnormality, is a distinct entity and is discussed separately (see below). The "monosomy 7 syndrome," a debated entity associated with CMML in pediatric patients, is also discussed separately (see below).

Molecular/cytogenetic. Loss of genetic material is thought to be of primary importance in the pathogenesis of the −5/del(5q)- and −7/del(7q)-associated diseases. However, neither the loss of genetic material on 5q nor the loss on 7q is well understood at the molecular level, and the search continues for a better understanding of the molecular pathogenesis. Recurring translocations of the regions of interest are not common, and the few which do occur have not been useful for the localization of a critical gene involved in the deletions.

The extent of the loss of material on 5q varies considerably among cases, but the deletions are interstitial and the breakpoints most frequently observed are del(5)(q13q31). Numerous studies have attempted to delineate the smallest commonly deleted segment, and it appears that band 5q31 is the major critical region, as it is deleted in all cases of AML examined and in all MDS cases with the exception of a few cases of the 5q− syndrome. This region contains numerous genes important to hematopoiesis, including *GM-CSF*(*CSF2*), interleukin genes (*IL3*, *IL4*, *IL5*, and *IL9*), *CSF1R*, *CD14*, *EGR1*, and *IRF1* (146). A number of molecular studies have been undertaken in attempts to implicate a specific gene. Although one study focused attention on *IRF1*, a gene with antioncogenic activity (147), this was later shown to be located outside of the commonly deleted segment (148). More recently, investigators have narrowed the critical region to a 1- to 1.5-Mb region which includes *EGR1* (149), but mutations of the known genes within this region have not been detected. Thus, a putative tumor suppressor gene on 5q remains to be identified.

It should be noted that there may be more than one region on 5q that is involved in leukemogenesis. A more distal region at 5q33 and a more proximal region at 5q13.1 have been identified as other critical loci (150,151). However, the 5q33 locus may be more closely associated with the "5q− syndrome" than with the more common cases of MDS, AML, and t-MDS/t-AML.

Analogous to the situation for chromosome 5, loss of material on 7q varies considerably but is interstitial. Two distinct regions that are likely to contain relevant genes have been identified (152,153). One commonly deleted segment is located at 7q22. This occurs more frequently than the second area, which maps from 7q31 to 7q36, with a commonly deleted region at 7q32-33. A recent study suggested that there may be more than one region of importance within 7q22. A proximal region correlated with del(7q) was seen in two patients with CML, and a distal region was seen in 17 patients with MDS/AML. The underlying molecular defect, another putative tumor-suppressor gene, has yet to be identified.

The abnormalities of 5 and 7 can occur alone but, as noted above, they more often are seen together (Fig. 6), and in complex karyotypes (154,155). The other chromosomal abnormalities commonly seen with −5/

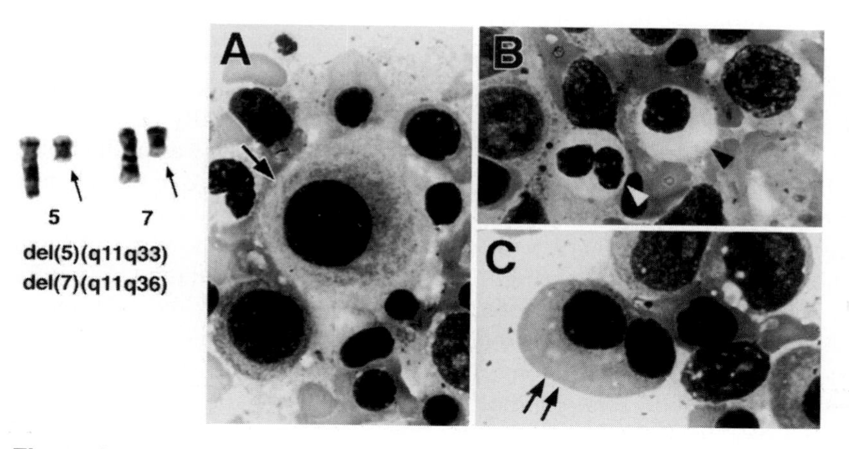

Figure 6 Partial karyotype from a single metaphase cell with both a del(5q) and del(7q) from a patient with t-MDS. These were part of a complex karyotype. An example of t-MDS exhibiting marked trilineage dysplasia. A. micromegakaryocyte (arrow). B. Dysplastic granulocytes (arrowheads) with pale cytoplasm and nuclear hyposegmentation. C. Erythroid cell (double arrow) exhibiting dysplasia with nuclear to cytoplasmic dysynchrony.

del(5q) and −7/del(7q) include those of chromosomes 17, 8, 20, and 12. As a secondary change, loss of chromosome 7 frequently occurs with the inv(3)/t(3;3). Monosomy 7 can be seen as a secondary change in the blast phase of CML, and can also occur with the t(9;22) in ALL, where it signifies a very poor prognosis.

Pathology. Although the spectrum of disease associated with −5/del(5q) and −7/del(7q) is wide, in most cases trilineage dysplasia is a common, but not specific, morphologic feature. In t-MDS/t-AML, trilineage dysplasia is usually prominent, and is characterized by micromegakaryocytes, pale-staining granulocytes with hypercondensed chromatin and nuclear segmentation abnormalities, and erythroid precursors with abnormal, sometimes budded nuclei, and nuclear-to-cytoplasmic dysynchrony (Fig. 6A, B, C) (154). Cases of t-MDS also not infrequently have some fibrosis and a lower cellularity than does primary MDS. Cases of de novo AML with −5/del(5q) and/or −7/del(7q) usually have background trilineage dysplasia. This may be an indication that they evolved from a silent or rapidly evolving MDS. Cases of de novo M6, which frequently exhibit −5/del(5q) and/or −7/del(7q) (Section III.A.2), commonly have severe trilineage dysplasia (156).

Clinical. Whether present in MDS, AML, or t-MDS/t-AML, −5/del(5q) and −7/(del7q) are associated with a poor response to therapy and poor survival. In MDS, the abnormalities help classify cases into the "poor outcome" group, whereas in AML they help classify cases as "unfavorable." In AML, −7 is also associated with fever and infection.

"5q− Syndrome." The "5q− syndrome" seems to evoke considerable confusion in its distinction from other types of MDS associated with del(5q). The "5q− syndrome" is a distinct entity, originally described by Van den Berghe et al. in 1974 (157), and delineated in more detail by Sokal et al. a year later (158). Patients with the 5q− syndrome are predominantly female and have macrocytic anemia, normal or increased platelets, only mild leukopenia and a bone marrow characterized by increased numbers of small monolobed megakaryocytes (Fig. 7), and fewer than 20% blasts (157–160). The deletion of the long arm of chromosome 5 is usually the sole abnormality, and patients have a relatively good prognosis. Patients with del(5q) who do not have this syndrome have a more typical MDS/t-MDS, with multiple cytogenetic abnormalities and an overall poor prognosis. As mentioned above, the deleted region in the "5q− syndrome" may be more distal than the commonly deleted region in the more typical cases with del(5q), although this still needs clarification.

"Monosomy 7 Syndrome." Another disorder, which also generates some confusion, is the so-called "monosomy 7 syndrome" (161–164). This is a myelodysplastic syndrome which occurs in pediatric patients and resem-

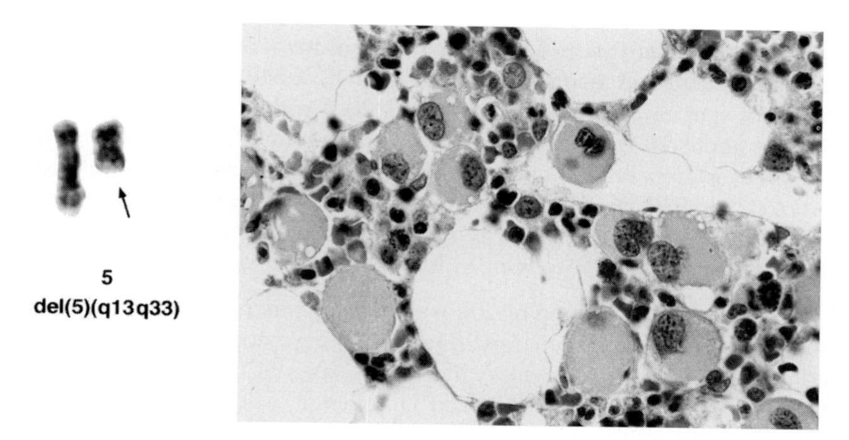

Figure 7 The left panel shows a partial karyotype of a del(5q) from a woman diagnosed with RA and the "5q− syndrome." The right panel shows section of bone marrow showing increased numbers of small (round) hypolobated megakaryocytes.

bles chronic myelomonocytic leukemia (CMML). The debated issue concerning the "monosomy 7 syndrome" is whether the syndrome is distinct from other cases of CMML in young patients (now referred to as juvenile CMML or JMML), which do not have monosomy 7 (164). Both disorders occur with an increased frequency in patients with neurofibromatosis type 1 (NF1). Both occur in young males, who frequently present with hepatosplenomegaly, skin rashes, and who are found to have high hemoglobin F (hgb F) levels. However, in patients without the monosomy 7, there are higher hgb F levels, more pronounced lymphadenopathy, and more severe skin rashes. In patients who do have monosomy 7, there is an increased frequency of leukopenia and bacterial infections. Some investigators believe the two entities are distinct but related, sharing some of the steps commonly seen in malignant transformation. Apparently, some cases of JMML present without monosomy 7, but can acquire it later in the course of the disease (163).

It should be noted that in the past there was considerable confusion with terminology, as some cases of JMML were called "juvenile CML." This required them to be distinguished from the typical or "adult type" of CML (Ph+, *bcr/abl*+) which is seen rarely in children.

2. del(20q) (see review 165)

General. Deletions of the long arm of chromosome 20 are seen in approximately 5% of patients with myeloid malignancies (166). Although originally described in P vera (167), the deletion now has been observed in a wide range of myeloid disorders, including MDS, CMPD, and AML (168). Although not specific, the association of del(20q) with P vera is still valid, however, as about 10% of cases of this disorder show the abnormality. In fact, these cases constitute about 35% of all reported cases with the abnormality (169). The del(20q) is believed to arise in a primitive hematopoietic cell, capable of both myeloid and lymphoid lineage potential (170). However, the abnormality is rarely seen in lymphoid malignancies.

Molecular. The deletion of 20q was initially reported to be a terminal deletion. However, in a study employing molecular cytogenetic techniques, Roulston et al. found it to be interstitial in all cases (171). The commonly deleted segment is 20q11.2 to q12, and is believed, as for other deletions, to harbor a tumor-suppressor gene. The region of deletion was more recently refined by microsatellite polymorphism to an 8-Mb region within 20q12 (172). Genes within this deleted segment include topoisomerase 1 (*TOP1*), phospholipase c gamma (*PLC1*), hepatocyte nuclear factor 4 (*HNF4*), and adenosine deaminase (*ADA*), as well as many as yet unidentified gene sequences.

Chromosomes. Because of the small size of chromosome 20 and the lack of a distinctive banding pattern on its long arm, an interstitial deletion of 20q is difficult to recognize through cytogenetic analysis alone. As mentioned above, molecular techniques, such as FISH analysis, were helpful in identifying the abnormality as interstitial. Deletions of 20q can occur alone, as a primary abnormality, but frequently they are seen with other chromosomal changes typical of myeloid disorders. The secondary abnormalities commonly identified include del(5q) and +8. Less common secondary changes include del(13q), +21 and −7 (168).

Pathology. Although seen in a wide range of myeloid malignancies, del(20q) has an interestingly high frequency in P vera, chronic idiopathic myelofibrosis (CIMF), and refractory anemia (RA). This spectrum of diseases has led to the suggestion that the deletion has a disproportionate effect on erythroid and megakaryocytic stem cells. In one study, dysmegakaryocytopoiesis and dyserythropoiesis were found in almost all cases with del(20q), whereas dysgranulopoiesis was never seen as a striking feature (168). Although the del(20q) may have a more recognizable effect on erythroid and megakaryocytic elements, it is believed to arise in a pluripotent stem cell with both myeloid and lymphoid lineage potential. The abnormality has been found in Epstein-Barr virus immortalized B-lymphoblastoid cell lines developed from lymphocytes of patients with del(20q) and MDS (170). Interestingly, however, the deletion was not seen in the peripheral blood lymphocytes from these patients. This has led to the speculation that although a multipotent stem cell is effected, it may not be able to produce normal mature lymphoid cells, or it may produce lymphoid cells with defective or short lifespans.

Clinical. Initially, the presence of del(20q) in the variety of diseases in which it occurs did not appear to have any specific clinical correlations. In one study, no clinical differences were noted in patients with del(20q) when they were compared to other patients with the same disease, without del(20q) (173). In another report, however, when the del(20q) was seen in MDS, it was associated with a high rate of transformation to AML and poor survival (174). More recently, it has been recognized that when del(20q) is present alone, it is associated with a good prognosis (Section III.B.2). Apparently, the poor prognosis associated with del(20q) may be more related to associated abnormalities of chromosomes 5 and/or 7 that were seen as secondary changes.

3. Trisomy 8

General. Trisomy 8 is the most frequently observed trisomy in hematopoietic malignancies (175,176). It occurs predominantly, but not exclusively, in myeloid disorders. A gain of chromosome 8 can occur as a sole abnormality or as a secondary change, and is seen in MDS, AML, and CMPD. A +8 occurs as the sole abnormality in about 5–10% of AML cases and in 2.8% of MDS cases. In CML, +8 is the most frequent secondary chromosomal change (37,38).

Molecular. The molecular basis through which +8 is related to myeloid malignancies is unknown. The general model that gains and losses of whole chromosomes might be explained by a gain or amplification of a mutated oncogene has been supported by the recent revelations concerning trisomy 11 and tandem duplication of the *MLL* (*ALL*) gene (Section II.B.4). However, the underlying molecular mechanism might also involve a dosage effect of a wild-type allele, such as *MYC* on chromosome 8. An underlying molecular abnormality has not been recognized for +8.

Chromosomes. When a +8 occurs as a sole abnormality, it frequently is seen as a mosaic with normal cells. Whether this is an indication that the trisomy is a late event in the neoplastic process is difficult to determine, due to problems in evaluating clonality in karyotypically normal myeloid cells. Unusual findings concerning trisomy 8 include cases with fluctuating trisomy, transitory trisomy, sporadic trisomy, and trisomy 8 in a clone unrelated to the primary clone (177); the latter is not uncommonly seen in MDS.

Pathology. As noted, a +8 can be seen in MDS, AML, and CMPD. A large set of data from one lab has shown that, in AML, trisomy 8 has its highest frequency as the sole abnormality in M5 (10.4%), followed by M7 (7.3%), M2 (6.2%), M4 (4.7%), and M1 (4.7%) (177). In MDS it is seen more frequently in RARS and in CMML. As a secondary abnormality, the +8 is most common in CML, as it occurs in 21.7% of cases. Interphase FISH studies of CML and MDS have demonstrated that +8 occurs in all types of nonlymphoid cells, but not in lymphocytes (178–180). In AML, +8 is observed as a secondary change in 31% of cases with a der(1;7), 17% of cases with a t(9;11), 11.8% of cases with an inv(16), 11.4% of cases with t(6;9), and in 10% of M3 cases with a t(15;17) (88). The occurrence of +8 together with +9 seems to be specific for P vera (181).

Clinical. Correlation of clinical outcome and a gain of chromosome 8 in acute leukemia has been hampered by studies in which cases with trisomy 8 as a sole abnormality are examined together with those in which it appears as a secondary change. When patients with +8 as a sole change were evaluated, a complete remission rate of 51% was reported (182), and overall survival was 9–11 months. In MDS, +8 is an indicator of poor prognosis (183,184) (Section III.B.2).

D. Less Common Recurring Abnormalities

1. inv(3)/t(3;3)(q21;q26)

General. inv(3)/t(3;3) were first described in AML associated with increased platelets (185,186), but have since been found in a wide variety of subtypes of AML, and have occasionally been observed in MDS (187). In AML, the inv(3)/t(3;3) occurs in only 1% or 2% of all cases. A particular morphologic correlation of note is the association with increased and dysplastic micromegakaryocytes, but it appears that not all patients with inv(3)/t(3;3) have increased platelets.

Molecular. Molecular studies of inv(3)/t(3;3) have shown involvement of the human ecotropic virus integration site-1 (*EVI1*) gene, which maps to 3q26 (188). Although *EVI1* is not expressed in normal hematopoietic cells, it is overexpressed in inv(3)/t(3;3)-associated malignancies (189,190). Interestingly, the breakpoints within *EVI1* differ in the inversion and translocation and occur at the 3′ and 5′ ends of the gene, respectively. *EVI1* encodes a transcription factor. A candidate oncogene at the 3q21 breakpoint has not yet been found.

Chromosomes. Abnormalities of the long arm of chromosome 3 at q21 or q26 include the more common inv(3) and t(3;3) (Fig. 8), as well as less common translocations involving a number of partner chromosomes [i.e., t(3;12)(q26;p13), t(1;3)(p36;q21)]. The inv(3) and t(3;3), with identical

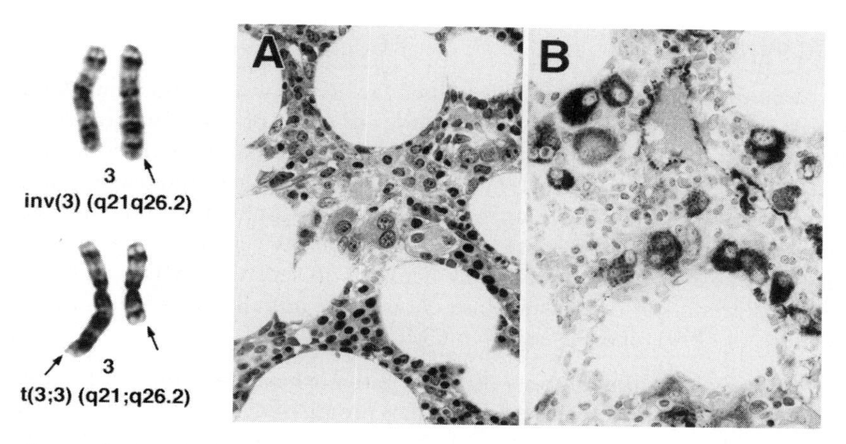

Figure 8 Partial karyotype showing an inv(3) (above), and the t(3;3) (below). A. Hematoxylin and eosin (H&E)-stained bone marrow section from a patient with MDS associated with inv(3) and an increased number of small megakaryocytes. B. Same bone marrow stained with anti-Factor VIII, to accentuate the numerous micromegakaryocytes.

chromosomal breakpoints, are clearly related at the molecular level, but this may not be the case with the other abnormalities. Secondary cytogenetic changes occur in over half of all cases with the inv(3)/t(3;3), and the most frequent is −7 (88).

Pathology. The morphologic features of cases showing inv(3)/t(3;3) are quite variable and include MDS (RAEB/RAEBT and CMML) and AML with M1, M2, M4, M6, or M7 morphology (186). A specific feature of many cases is the presence of abnormal megakaryocytes. The megakaryocytes are usually quite prominent, and markedly dysplastic with tiny micromegakaryocytic forms (Fig. 8). Multilineage dysplasia has also been an association found in many cases, but this may be related to the commonly associated −7.

Clinical. As noted above, the inv(3)/t(3;3)-associated disorders were initially reported to be associated with increased platelets, although in more recent studies this has not been seen in all patients (186). Discrepancies in reporting may be due to the fact that some patients develop the thrombocytosis not at diagnosis but only later in the disease. Although the gene for thrombopoietin maps to 3q27 and was a logical possible cause for elevated platelets, a recent study demonstrated that the thrombopoietin gene is not involved and is not responsible for the thrombocytosis (191,192). The prognosis of acute leukemias with inv(3)/t(3;3) is generally poor (193).

2. t(6;9)(p23;q34)

General. The t(6;9), first described by Rowley and Potter in 1976, was noted in two patients from a large series of cases of acute myelogenous leukemia (194). It is now recognized as a rare abnormality, accounting for less than 1% of all cases of AML (195–198). The t(6;9) is of interest because the associated leukemias have an accompanying basophilia. Although most frequently noted in AML, some cases of t(6;9) have a preceding myelodysplastic phase, so that patients may actually present with MDS, or AML with a background of dysplasia. Rare cases were initially reported as acute myelofibrosis (199), and Ph-negative CML (200).

Molecular. Although only 50 or so patients have been reported in the literature, work at the molecular level has progressed quite rapidly. As in most myeloid leukemias associated with reciprocal translocations, the rearrangement produces a fusion gene and a putative fusion protein. The t(6;9) fuses the *CAN* gene located at 9q34, with the 3′ end of the *DEK* gene at 6p23 (201). A fusion mRNA is produced which encodes for a protein, although the function of the fusion protein has yet to be elucidated.

Chromosomes. In many cases, the t(6;9) is found as the sole abnormality, implicating it directly in leukemogenesis. Not uncommonly, however, secondary changes are seen. Most frequently the secondary changes are gains of chromosomes 8 and 13 (88). However, both +8 and +13 are common secondary changes in myeloid leukemia in general.

Pathology. The association of t(6;9) with leukemias with basophilia was first recognized by Pearson et al. (196), and the observation has since been confirmed by a number of reports. Bodger et al., using the MAC (morphology and chromosomes) technique, have demonstrated the t(6;9) to be in cells immunophenotypically identified as basophils (202). Most cases of t(6;9) leukemia have been categorized as M1, M2, or M4 according to the FAB schema, although there is some discussion concerning the establishment of a new category of acute basophilic leukemia. Cases usually have dysplasia in at least the granulocytic and erythroid series, with ringed sideroblasts seen as part of the latter. Immunophenotypically, according to a recent report of 8 cases, there is expression of CD13, CD33, and HLA-DR, but no CD34 reactivity (195). Some earlier reports had noted that there is TdT expression as well, although this has not been a feature of all cases. Because of the basophilia, differential diagnosis considerations must include CML in blast crisis and acute promyelocytic leukemia with basophilic differentiation, although the latter is rare.

Clinical. The small number of patients reported with this abnormality precludes definitive statements concerning clinical features. However, most of the cases have been in younger patients, and most have had a poor outcome (88).

3. t(1;22)(p13;q13)

General. The t(1;22)(p13;q13) appears to be restricted to cases of acute megakaryoblastic leukemia occurring in the very young (203). The abnormality has not been reported in other leukemias, other hematopoietic disorders, or in other malignancies. In infants, the abnormality is present in as high as 67–100% of cases of M7, although, in general, it occurs in about a third of patients with this type of leukemia.

Molecular. To date, there has been no molecular characterization of the translocation. In fact, there has been no agreement as to which derivative chromosome is critical. The observation that some cases of megakaryoblastic leukemia have abnormalities of 22q13, without the t(1;22), indicates that chromosome 22 may harbor a critical gene (204). On the other hand, many cases show a gain of the der(1) involved in the translocation [+der(1)t(1;22)], suggesting, by analogy to a +Ph, that the der(1) may be important in the molecular transformation (203). This analogy may not entirely hold, however. The +der(1) also creates partial trisomy for 1q, and this has been observed in nearly all types of neoplasia and may be the important aspect of the +der(1).

Chromosomes. Although in many cases the t(1;22) is the sole abnormality (Fig. 9), in others there is a complex karyotype, usually hyperdiploid, in which there may be a duplication of the derivative chromosome 1, as discussed above. In cases with a hyperdiploid karyotype, the most frequently observed trisomies are those of chromosomes 21, 19, 6, and 7. In the cases with trisomy 21, the trisomy was not found to be constitutional (205,206).

Pathology. The t(1;22) is restricted to acute megakaryoblastic leukemia and can be a useful marker for this disorder in cases that are otherwise difficult to diagnose (207). Not infrequently, the blast cells may appear primitive and may resemble metastatic carcinoma, neuroblastoma, or even sarcoma (Fig. 9). Cytochemical and immunophenotypic evaluations may be hampered, as the bone marrow biopsies are not infrequently fibrotic. When cells can be obtained, the typical case of M7, including those with t(1;22), is negative for myeloperoxidase and is negative, or minimally positive, for nonspecific esterase reactions. Blasts can usually be identified by positive staining for one of the immunomarkers of megakaryocytic differentiation (i.e., CD41, CD42, and CD61). Ultrastructural detection of platelet peroxidase can also be useful, but the test is difficult to perform.

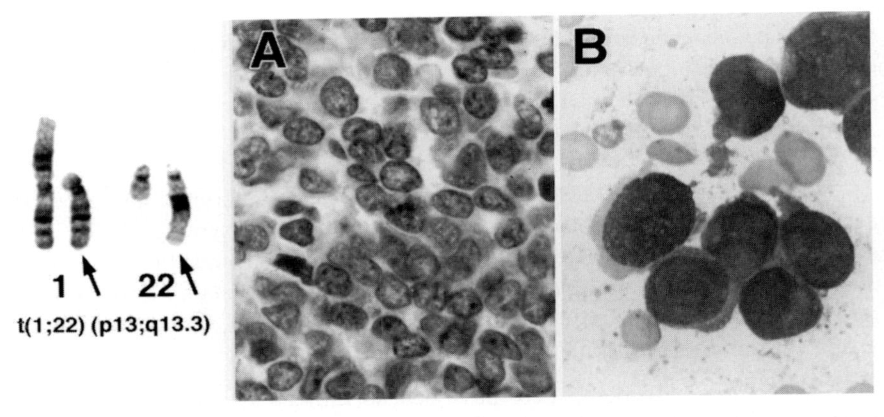

Figure 9 Partial karyotype showing the t(1;22) associated with AML-M7. A. Section of bone marrow showing an undifferentiated malignancy. B. Bone marrow aspirate showing primitive blast cells. Note that some of the cells have pseudopods. The blasts in this case were negative for the myeloperoxidase and nonspecific esterase reactions. Flow immunophenotyping showed them to be CD61+.

Clinical. Most cases of t(1;22) acute megakaryoblastic leukemia are in infants below the age of 12 months. According to one report, the median age of patients was only 6 months. Rare cases have been diagnosed at birth. Infants with t(1;22) leukemias present with sometimes severe hepatosplenomegaly and cytopenias, although the platelet count can be normal or elevated. Outcome is variable and difficult to comment upon, due to the few cases reported.

4. t(8;16)(p11;p13) and Other 8p11 Translocations

t(8;16)

General. The reciprocal t(8;16)(p11;p13) is a relatively rare abnormality in AML and in t-AML, occurring in less than 1–2% of cases (208). However, it is of interest because of recent advances at the molecular level, and because of an unusual morphologic phenotype characterized by a myelomonocytic process associated with erythro- or hemophagocytosis. The other translocations of 8p11 are less frequently seen and are associated with a different clinicopathologic syndrome (see below).

Molecular. Recent studies of the breakpoints on chromosomes 8 and 16 indicate that the genes involved are *MOZ* and *CBP*, respectively, which, when fused, form a MOZ/CBP protein (209). The latter has been shown to play a dominant role in leukemogenesis. The *MOZ* gene encodes a protein which contains multiple zinc fingers and has histone acetlytransferase activity, this is believed to facilitate gene transcription by relaxing the DNA helix. The CBP is a transcriptional co-activator that facilitates the transcriptional activation of many target genes. Moreover, CBP also has histone acetyltransferase activity. The leukemic transformation is believed to occur through aberrant chromatin acetylation.

Chromosomes. Most patients with 8p11 abnormalities have the t(8;16)(p11;p13) (Fig. 10). Other abnormalities involving this region and that are associated with similar pathologic features include t(8;22)(p11;q13) and t(8;19)(p11;q13).

Pathology. The t(8;16) leukemias are characterized as de novo leukemias with M4, M5, or M4/M5 morphology according to the FAB schema, with differences apparently due to varying numbers of monocytes/blasts in the individual cases. In some patients, the myeloperoxidase reaction is positive and can be detected with light microscopy, although in others it can only be recognized ultrastructurally. Nonspecific esterase activity is at least weakly positive in most cases; combined granulocytic and monocytic esterase studies show dual staining. Erythrophagocytosis or hemophagocytosis is reported as a distinctive feature in most cases (Fig. 10A, B, C).

Clinical. Many patients (40%) presenting with t(8;16)-related leukemias are young (less than 17 years); hepatosplenomegaly is not uncommon (88).

Other Translocations Involving 8p11. There are other 8p11 abnormalities that together constitute a disorder referred to as the "8p11 myelo-

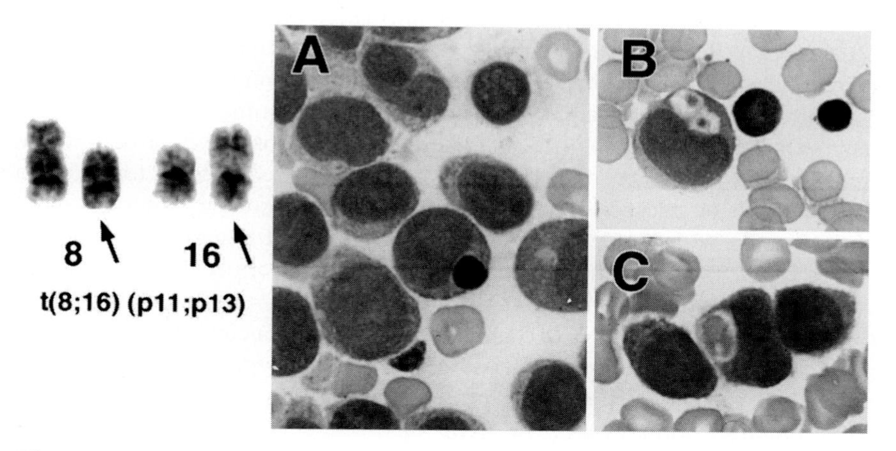

Figure 10 Partial karyotype showing the t(8;16) seen in AML M4 or M5. A, B, and C. Myelomonocytic leukemia with blastic elements showing phagocytosis of nucleated cells, erythrocytes, and platelets.

proliferative syndrome'' (210,211). These occur less frequently than the t(8;16) and include t(8;13)(p11;q11), t(8;9)(p11;q32), and t(6;8)(q27;p11). The "8p11 myeloproliferative syndrome'' is characterized by weight loss, malaise, generalized lymphadenopathy, and a chronic myeloproliferation associated with eosinophilia in the blood and bone marrow. The myeloproliferative process can transform to an acute myelomonocytic leukemia, and some patients can develop a T-cell non-Hodgkin's lymphoma. The relationship between the myeloproliferative process and the T-cell lymphoma is not clear, but some believe the two are related and are part of the same malignancy.

5. der(1;7)(q10;q10)

General. The der(1;7) is an unbalanced translocation which results in the gain of 1q and a loss of 7q. The abnormality is associated with myelodysplastic syndromes which are commonly secondary to radiation or chemotherapy (212–214). Because the breakpoints are so close to the centromere on each chromosome, there has been some confusion regarding the mechanism of the rearrangement.

Molecular. The underlying molecular basis of the simultaneous gain of 1q and loss of 7q is not known, although partial trisomy for 1q is common in neoplasia and loss of 7q is common in myeloid disorders, especially MDS and AML. Initial studies demonstrated amplification of *EGFR* (epidermal growth factor receptor), the cellular homolog of the v-erb B oncogene, at 7p12-13 (215). However, this was later found not to be the case (216).

Chromosomes. Originally, the abnormality was denoted as 46,XX, or 46,XY,−7, +der(1)t(1;7) (p11;p11), due to the assumption that the centromere of the rearranged chromosome was derived entirely from chromosome 1. More recently, using alpha-satellite probes for chromosomes 1 and 7, it has been shown that the entire arm from each chromosome is present in the translocation (217–220). Thus, the karyotype is now considered to be der(1;7)(q10;q10), in accordance with ISCN (1995). Secondary chromosomal changes seen with the der(1;7) include trisomy 8 and trisomy 21.

Pathology. Initially, the abnormality was described in cases designated as myelofibrosis with myeloid metaplasia (212). Subsequently, it was reported in a wider range of hematopoietic malignancies. At present, the der(1;7) is believed to be most strongly associated with a myelodysplastic syndrome which is frequently found in patients with a history of previous radiation, alkylating agents, or toxic environmental exposure (213,221). Some develop acute leukemias, more frequently of the M4 type.

Clinical. Similar to other cases of t-MDS/t-AML, those with the der(1;7) are associated with poor survival (221).

6. del(12p) and i(12p)

del(12p)

General. Translocations and deletions of the short arm of chromosome 12 are seen in a variety of hematopoietic malignancies including myeloid and lymphoid disorders. In ALL, the t(12;21) or its associated molecular alteration, *ETV6/CBFA2* (formerly *TEL/AML1*), is the most common abnormality in childhood ALL (Chapter 12). In the myeloid disorders, 12p involvement includes deletions and a variety of translocations of a growing number of partner chromosomes. The associated myeloid diseases include MDS, AML, t-MDS/t-AML, and CMPD (222,223). In AML, the abnormalities of 12p account for approximately 0.5–2% of all cases, although this may be an underestimation.

Molecular. It is not clear, but it seems unlikely that the various 12p deletions and translocations involve the same region in a single gene or the same gene in all cases. The *ETV6* gene (formerly *TEL*) and *CDKN1B* (formerly *KIP1*) are present in the smallest region of overlapping deletions, suggesting the possibility that each may act as a tumor-suppressor gene (224,225). *ETV6* is a member of the *ETS* gene family of transcription factors (ETs variant gene 6), and has been shown to be involved in producing a chimeric gene and chimeric protein in some translocations. In the t(5; 12)(q33;p13), the first translocation found to produce a *TEL* fusion gene, the chimeric protein involves the helix-loop-helix domain of *ETV6* and the transmembrane and tyrosine kinase domain of *PDGRFB* on chromosome 5 (226). In the t(3;12)(q26;p13), there is a fusion of *ETV6* and *EVI1* on 3q26 (227), which is involved in the inv(3)/t(3;3) and the t(3;21) as discussed in Section II.B.1. In the t(12;22)(p13;q11), there is a fusion of *ETV6* and *MN1*, a candidate gene for meningioma (228). Data suggest that the *ETV6* may have a DNA-binding domain and a protein dimerization domain, which each may separately be involved in myeloid leukemogenesis. The *CDKN1B* gene product is believed to play a negative role in the growth-regulatory pathways in some hematopoietic cell lineages (229). This obviously makes it an attractive candidate as a tumor suppressor. Other genes on 12p have yet to be identified.

Chromosomes. In a survey of cases with del(12p) using FISH, the smallest area of deletion is bordered by *ETV6* on the telomeric side and *CDKN1B* on the centromeric side (224). Since the deletion may be small and even submicroscopic, molecular studies are showing that it may be missed by routine cytogenetic analysis (225). Cryptic deletions may also be present in translocations of 12p and, conversely, cryptic translocations may be present in some cases with apparent deletions of 12p. For the myeloid disorders, the translocations involve a growing number of partner chromosomes, including chromosomes 2, 3, 4, 5, 7, 13, and 22 (230,231). The list obviously becomes longer when translocations associated with the lymphoid malignancies are included. Translocations in the myeloid disorders can sometimes be complex and involve a third chromosome [e.g., t(9;15;12)] (232). The abnormalities of 12p can occur alone, but are commonly seen in complex karyotypes with other cytogenetic findings associated with the myeloid malignancies. The latter is particularly true regarding del(12p).

Pathology. There does not seem to be any morphologic or cytologic feature which is common to the 12p aberrations, although some cases may show prominent basophila (222). The first description of a translocation involving 12p was the t(12;22)(p13;q11) in a disorder that was felt to be CML (233). The resulting 22q− chromosome was initially thought to be an alternative Philadelphia chromosome. Presently, 12p abnormalities are noted to occur in "atypical CML" and in other disorders that can mimic CML. However, they occur in CML only as an infrequent secondary change with the classic t(9;22) or *bcr/abl* fusion. Abnormalities of 12p have been reported in MDS, AML, t-MDS/t-AML, and in other CMPDs. The spectrum of disorders within these entities is quite broad. In MDS, 12p abnormalities have been found in both low- and high-grade disease. In AML, they have been reported in cases with M0, M1, M2, M4, M5, and M6 subtypes. In therapy-related disorders, they have been seen in both t-MDS and t-AML after radiation, alkylating agents, or environmental exposure to toxic agents (234). In the CMPDs, they have been reported in P vera, CMML, and "atypical CML." In CMML, there is a recurring abnormality that involves chromosome t(5;12)(q33;q13). It is interesting that in this disease the underlying molecular alteration involves *ETV6* and the gene for the platelet-derived growth factor (*PDGRFB*).

Clinical. The wide spectrum of myeloid disorders associated with 12p abnormalities precludes any comment regarding common clinical features or survival, although poor survival has been noted in some case studies.

i(12p)

General. An isochromosome of 12p is the most common abnormality in germ-cell tumors (235,236). However, the abnormality is of interest because of the association of germ-cell tumors with leukemia (237–239). Although leukemias in patients with germ-cell tumors are not common, they are important to recognize because the leukemia can present without any evidence of the actual underlying malignancy. Diagnosticians should be aware that patients with leukemias found to have i(12p) must be evaluated for the presence of an underlying germ-cell tumor and that patients with germ-cell tumors need to be observed closely for the possible development of leukemia.

Molecular. Isochromosomes are not uncommon in cancer. For example, in CML the i(17q) is one of the most common secondary chromosomal changes. The molecular basis for malignant transformation of disorders with isochromosomes is not known. However, it is obvious that the isochromosome leads to both a loss and gain of genetic material. Thus, a simultaneous loss of a tumor-suppressor gene and a gain of oncogenic sequences may be the cause of malignant transformation. Although all of the mechanisms through which an isochromosome may develop are not entirely characterized, Mukherjie et al. have illustrated that the i(12p) in germ-cell tumors is often formed by a nonreciprocal centromeric interchange between sister chromatids of homologous chromosomes, and not through a misdivision of the centromere (240).

Chromosomes. In the leukemias identified with germ-cell tumors and i(12p), the abnormality is commonly seen within a complex karyotype. Multiple copies of the i(12p) are frequent, and, in germ-cell tumors, multiple copies are believed to be associated with a poor response to therapy.

Pathology. The leukemic disorders associated with germ-cell tumors can be of any type, including MDS, AML, and ALL. In AML, the unusual morphologic types are seen more frequently; these include megakaryoblastic leukemia, erythroleukemia, and acute undifferentiated leukemia. Recent evidence suggests that the leukemias arise from the hematopoietic elements within the germ-cell tumor itself (241).

Clinical. Acute leukemias developing in germ-cell malignancies occur more commonly in young men who have mediastinal tumors that contain areas of yolk sac neoplasms. The leukemias can occur simultaneously, subsequent to, or even before the germ-cell tumor is evident. If subsequent to the primary malignancy, they occur quickly, usually by 2 months. Because patients with germ-cell tumors may already have received therapy, the possibility of a t-AML/t-MDS must be considered if leukemia develops. However, the usual 2-month period in which the leukemias develop is too short a latency period for a therapy-related leukemia, and the presence of the i(12p) as well as the absence of the frequently seen therapy-related abnormalities of chromosomes 5 and 7 speak against a therapy-related process.

7. Trisomy 21

General. A gain of chromosome 21 is one of the most frequently observed trisomies in myeloid malignancies (242). A +21 is not at all specific for a myeloid phenotype, as it is also quite common in acute lymphoblastic leukemia, usually as a secondary abnormality or as part of the hyperdiploid karyotype associated with good prognosis (243). For myeloid disorders, the trisomy is of particular interest because of the high incidence of acute myeloid leukemias in patients with a constitutional trisomy 21 (Down syndrome) (244). Although it has been speculated that myeloid leukemia associated with the constitutional and acquired trisomy 21 might share some underlying commonality, it is now recognized that they are quite different, at least from the clinical and pathologic standpoints (245). However, it has yet to be determined whether the underlying molecular mechanism of leukemogenesis is similar in the two.

Down Syndrome. In Down syndrome, there is a 10- to 20-fold increased risk for the development of acute leukemia, and a particularly high relative frequency for myeloid malignancies (244). The myeloid leukemias tend to occur early in life, frequently before the age of 1–2 years, and are quite sensitive to treatment. There is an excellent overall complete remission rate. The myeloid leukemias in Down syndrome children are most frequently megakaryoblastic (246–249). They can show exclusive megakaryoblastic proliferations, but frequently show erythroid involvement as well. The mast cell lineage has also been implicated, but only from in vitro culturing experiments, which used IL-3 and the Steel Factor (250). The leukemias may be preceded by a transient myeloproliferative disorder (TMD) characterized by a high white blood cell count, thrombocytopenia, and a high blast count (usually >30%). The process is morphologically indistinguishable from AML, but usually occurs in the neonatal period and undergoes spontaneous remission in 2–14 weeks. Approximately 25% of patients with TMD will develop AML within 3 years.

The Down syndrome leukemias are associated with trisomy 8 or monosomy 7 in about half of the cases. The underlying mechanism of leukemic transformation is not known, but it has been speculated that it is related to the genetic alterations associated with the constitutional trisomy 21. Genes in the region of chromosome 21 critical for the development of Down syndrome include those known to play a role in the leukemogenesis of other leukemias. The *CBFA2* gene associated with the t(8;21) or the t(3;21) (see Section II.B.1) is one such gene. The ETS-related gene *ERG* involved in the t(16;21) is another (251), and *ETS2* is a third identified oncogene on chromosome 21.

It is important to note that 1–2% of patients with Down syndrome show mosaicism, with many of their cells having a normal diploid karyotype rather than trisomy 21. The finding of a +21 clone in an infant or child with a leukemia process should raise the question of whether the +21 represents mosaicism for a constitutional +21 line.

Myeloid Leukemia with Acquired Trisomy 21. The myeloid malignancies associated with a gain of chromosome 21 have a distribution unlike those in Down syndrome, in that they are rarely, if ever, megakaryoblastic.

In a review of 37 cases culled from 1187 consecutive patients with myeloid disorders, Cortes et al. found a +21 in nearly all types of MDS (RAEB, RAEBT, RA, and CMML) and in all types of acute myeloid leukemia except M7 (252). A gain of chromosome 21 almost always occurred with other abnormalities, the most frequent of which were loss of chromosome 5 or 7. Gain of chromosome 21 can also be a secondary change with the abnormalities t(15;17), inv(16), t(8;21), and +8. The clinical features and complete response rates were similar to those of patients with these other abnormalities but without trisomy 21. Thus, the trisomy probably does not strongly influence the clinical outcome when it is present as an acquired abnormality.

8. Other Abnormalities: −Y, t(3;5), +9, +13, i(17q)

−Y. Loss of the Y chromosome is a common occurrence in elderly males and can be seen in the hematopoietic malignancies in such patients. However, it has no apparent clinical relevance (253). Loss of the Y chromosome is a frequent secondary change in t(8;21) or *AML1/ETO*-associated leukemias (see Section II.B.1).

t(3;5). The t(3;5)(q25.1;q34) is rare, but is of interest because of the progress that has been made in identifying the genes involved in the translocation. The abnormality is seen in MDS and in AML of most types except M3. A possible association with Sweet's syndrome has been noted (254). The breakpoint on chromosome 3 differs from those in inv(3)/t(3;3). The genes involved have been identified and found to produce a fusion gene *NPM/MLF1* (255).

+9. A gain of chromosome 9 is a recurring abnormality in P vera, where it commonly occurs with a gain of chromosome 8 (256).

+13. A gain of chromosome 13 has been reported to occur in undifferentiated leukemias (AUL) (Section III.A.1), but can also be seen in myeloid and lymphoid disorders. The myeloid diseases include cases of AML of any subtype, but M0 may be most common (257–259). This abnormality is associated with a poor outcome.

i(17q). An isochromosome of 17q, i(17q) is a common secondary change in CML. However, it can also be seen as a sole abnormality in rare cases of MDS (260,261). The MDS is unusual, as thrombocytosis and granulocytosis are both present. Such features suggest that the cases should be considered as borderline between MDS and CMPD. Frequently the MDS associated with i(17q) progresses to AML. It is notable that *Tp53* is located on the short arm of chromosome 17 and that it is altered in about 30% of cases of CML in myeloid blast crisis with i(17q). It is not clear whether *Tp53* is altered in the rare cases of MDS associated with i(17q).

III. CYTOGENETICS IN THE DIFFERENTIAL DIAGNOSIS, AND MORPHOLOGIC, CLINICAL, AND PROGNOSTIC CORRELATIONS

A. AML

The frequency with which clonal karyotypic abnormalities are detected in AML appears to have increased steadily over the past 15 years. This has been due to improved techniques for culturing neoplastic cells, refinements in banding, and the increased awareness of common subtle abnormalities, such as the inv(16), which otherwise could easily be overlooked. At the Fourth International Workshop on Chromosomes in Leukemia in 1982 (54), the frequency of clonal abnormalities was close to 50%, whereas currently it is about 75% in most laboratories. It will be interesting to see if a significant number of cases with a normal karyotype will be found to have abnormalities with the use of new technologies such as FISH, or spectral karyotyping (262).

The cytogenetic abnormalities in AML can be divided into the broad groups of those with numerical and structural abnormalities. The latter group includes the common recurring abnormalities for which there is increasing information regarding underlying molecular alterations as discussed in Section II.B. The three most common translocations collectively account for over 30% of all cases. The numerical abnormalities have defied understanding at the molecular level, with the exception of trisomy 11 which is associated with the partial duplication of *MLL* (see Section II.B.4). Numerical abnormalities, together with cases showing normal karyotypes, account for approximately 45% of all cases (263).

1. Cytogenetics in the Differential Diagnosis of AML

The diagnosis of acute myeloid leukemia usually rests on the morphologic, cytochemical, and immunophenotypic features of the individual case. Cytogenetic analysis is useful in confirming cases of APL (M3) and in recognizing important prognostic subtypes as discussed below (Section III.A.2). There are rare cases, however, where cytogenetic findings might be essential for the diagnosis, as the only major indication that the malignancy is of myeloid origin. For example, the authors have seen a primitive sarcomatous-like lesion in the bone of a young child which lacked most of the morphologic or the usual cytochemical and immunologic markers of AML, but which was subsequently shown to have the specific t(1;22)(p13;q13) associated with megakaryoblastic leukemia (M7) (Section II.D.3). Conversely, the finding of a cytogenetic abnormality specific for a nonhematopoietic malignancy, such as neuroblastoma, might be critical in a difficult case in which acute leukemia cannot entirely be ruled out (264).

Whether cytogenetic findings might be helpful diagnostically in malignancies in which the differential diagnosis of ALL versus AML cannot be resolved is a question that has not been fully answered. In a small study of nine cases of primitive acute leukemia of uncertain lineage, only one case was found to have a cytogenetic abnormality suggestive of a certain lineage (265). Interestingly, one case had a del(5q) and, in retrospect, was also noted to have associated myelodysplasia. The majority of the cases in the series had abnormalities not specific for either myeloid or lymphoid disorders. These included three cases with +13, an abnormality previously shown to be associated with undifferentiated leukemia (258,266), but which may also be seen in AML and ALL. Except for the common translocations, which would be unlikely in a diagnostically challenging case, few cytogenetic abnormalities can serve as a definitive indicator of lymphoid or myeloid lineage. Even monosomy 7 and trisomy 8, which are very common in the myeloid disorders, occasionally can be seen in lymphoid neoplasia.

Although difficulty in resolving lineage is not frequently aided by cytogenetic findings, confusion over lineage in acute leukemias can sometimes be explained with the discovery of an 11q23 abnormality. Translocations of the 11q23-associated *MLL* gene are frequently observed in leukemias showing mixed lineage, multiple lineage, or lineage switch (112,113). Other cases with ambiguous lineage might have the t(9;22)(q34;q11.2) (267).

The distinction of AML from MDS is also one that is usually made by morphology. Cytogenetic findings do not typically play a role in distinguishing between the two entities. In fact, because there is considerable overlap between the cytogenetic findings in AML and MDS, in a questionable case, cytogenetic evaluation would usually be of little help (e.g., −5/del(5q) and −7/del(7q) occur in both AML and MDS). However, one question that has received some attention involves cases with blasts counts <30%, in keeping with MDS, but that have translocations common for AML. This scenario can be seen with a number of abnormalities, including the t(8;21), and inv(16)/t(16;16). The specific issue as to whether cases with a t(8;21), but with <30% blasts, should be classified as MDS or AML has been addressed in a number of communications. Some authors believe that such a case should be considered as an "evolving acute leukemia" or as an "early presentation" of an acute leukemia, and should be treated with chemotherapeutic regimens for AML; others believe that t(8;21) may indeed have an indolent myelodysplastic phase (66–69).

Although there is some overlap between the abnormalities seen in AML and CMPD, in most instances the patient's history or clinical evidence for a chronic disease will be obvious and will suffice to aid in distinguishing a de novo AML from a blastic transformation of a CMPD. The difficult diagnosis of one poorly understood entity, acute panmyelosis (acute myelofibrosis) (268), which, nosologically, is at the border of AML, CMPD, and MDS, would unlikely be aided by cytogenetic findings; there do not appear to be any specific recurring abnormalities to support or exclude a diagnosis. Another issue concerning cytogenetics in the differential diagnosis of AML versus CMPD arises when the Philadelphia chromosome is observed at diagnosis, in cases which appear to be AML. It has been difficult to determine whether such cases represent true de novo leukemias (Ph+ AML) or cases of CML in the blast phase, with an undiagnosed or very short chronic phase. In some cases the features of chronic-phase CML in the background of the acute leukemia, or subsequent reversion to chronic-phase CML after treatment of the acute disease, may help resolve the issue (26,267).

Knowledge of the initial karyotypic findings identified before treatment in a case of AML can be useful in helping to evaluate posttherapy specimens. Abnormalities from the initial karyotype can be useful for identifying residual disease, or for detecting early relapse after therapy. If appropriate probes are available, the targeted abnormality can be searched for with molecular or molecular-cytogenetic techniques (49,269). With the use of new growth factor therapies, such as megakaryocyte growth and differentiation factor (MGDF), some posttherapeutic bone marrows are becoming increasingly difficult to evaluate morphologically because of increased regenerative blastic cells. The recognition of residual clonality can be very useful in the interpretation of a morphologically equivocal case. A few caveats must be taken into account, however. There is a well-known persistence of fusion mRNA transcripts (detectable by PCR) in patients in long-term remission from *AML1/ETO* leukemias (58), and there is a known persistence of clonal hematopoiesis during remission (from acute disease) in some patients with AML associated with underlying myelodysplasia (270). Care must be taken in interpreting the presence of residual clonality, and in some instances it may be useful to know whether the residual clonal cells are blasts (50).

2. Cytogenetics and Morphology in AML (see Table 2)

A discussion of the correlation of cytogenetic findings with morphology in AML is made somewhat difficult due to the ongoing reexamination of the FAB classification scheme for AML and discussions concerning proposals for revising it, or by ongoing efforts to establish an entirely new classification system (271–276). Although useful for many years as a framework for the morphologic subclassification of AML, the FAB system is being

Table 2 Morphology and Cytogenetic Correlations in AML

Morphology	Recurring cytogenetic abnormality[a]		
FAB subtype (% of all AML)		[% of subtype]	% of all AML
M0 (<1%)			
M1 (~20%)			
M2 (~25%)	**t(8;21)**	**[~30%]**	**~8%**
M3 (~10%)	**t(15;17)**	**[98%]**	**~10%**
M4eo (~10%)	**inv(16)/t(16;16)**	**[100%]**	**~10%**
M4 (~25%)		[30%] (M4, M5)	11%
M5 (~5%)	11q23		
M6 (~3%)	−5/del(5q), −7/del(7q)	[~70%]	
M7 (<1%)	**t(1;22)**	**[~30%]**	**<1%**
Other associations			
Basophilia	t(6;9)		1%
Hemophagocytosis	t(8;16)		<1%
Abnormal megakaryocytes	inv(3)/t(3;3)		2%
All cases	+8		13%
	−7		9%
	−5/del(5q)		10%
	del(20q)		5%
	del(12p)/t(12p)		2%

[a]Boldface indicates a specific association.

criticized for, among other things, its somewhat poor correlation with associated cytogenetic findings.

Basically, only three of the FAB morphologic subtypes have a strong correlation with a specific cytogenetic feature; these are M3, M4eo, and a subset of M2. The M3 subtype of acute promyelocytic leukemia has the t(15;17) in over 98% of cases, with the remainder having variant translocations involving 17q, and very rare cases with no recognizable t(15;17) but with molecular *PML/RARA* fusion (Section II.B.3). Currently, there are no consistent, specific morphologic features which correlate with the variant translocations in M3. The subtype of M4 leukemia associated with abnormal eosinophils (M4eo) correlates with the presence of an inv(16)/t(16;16) in essentially all cases (Section II.B.2). Although some cases with inv(16)/t(16; 16) have been reported to have M2 morphology or M4 morphology without abnormal eosinophils, this might be questioned. In M4eo, monocytes sometimes do not exhibit the expected classic nonspecific esterase reaction. Recognition of the abnormal eosinophils can sometimes be difficult due to their low numbers. In fact, one common misconception is that M4eo denotes an acute myelomonocytic leukemia with increased eosinophils. In fact, the distinction is in reference to *abnormal* eosinophils, which can sometimes be infrequent. A subset of AML with maturation, accounting for about 25–30% of cases of M2, correlates with the t(8;21). This has characteristic morphologic features which can be correlated with the *AML1/ETO* molecular rearrangement, even in the absence of the cytogenetic translocation (Section II.B.1).

A significant number of cases with M4 and M5 morphology show a translocation of 11q23 (see Section II.B.4), but there are no particular features to predict which cases will have this abnormality. Furthermore, the 11q23 partner chromosome that is involved varies considerably, and 11q23 rearrangements are seen in other types of myeloid leukemia as well. Erythroid leukemias (M6) typically show abnormalities of chromosome 5 and/or 7. In one study, such anomalies were usually part of a complex karyotype and were present in over 70% of cases (156). Megakaryoblastic leukemia (M7) has been associated with the t(1;22), which is quite specific for this disease in very young patients (Section II.D.3). However, M7 leukemia is heterogeneous and can have a gain of chromosome 21, t(3;3)/inv(3), or loss of chromosome 7 (204). Other morphologic features showing correlation with cytogenetic findings include an association of basophila with the t(6;9) (Section II.D.2), hemophagocytosis with the t(8;16) (Section II.D.4), and increased small megakaryocytes and thrombocytosis with the inv(3)/t(3;3) (see Section II.D.1). The extent of the correlation between the morphologic features and the cytogenetic finding has yet to be determined.

3. Clinical and Prognostic Correlations

Despite the sometimes inexact correlation of cytogenetic findings with morphology, the karyotype at diagnosis has prognostic significance that can be independent of other clinical features. The karyotype may also have clinical significance in alerting the clinician to the possibility of a specific presentation or complication. For example, the inv(16)/t(16;16) is associated with increased CNS involvement, not only at diagnosis, but also at relapse (Section II.B.2). The t(8;21) is associated with a higher incidence of granulocytic sarcoma at diagnosis and also at relapse (Section II.B.1). Both the t(15;17) and the t(8;16) is associated with disseminated intravascular coagulation, while a t(3;5)(q25.1;q34) is associated with Sweet's syndrome (263). The finding of an i(12p) in an acute leukemia, although rare, should alert the clinician or diagnostician to the likelihood of an underlying germ-cell tumor (Section II.C.3).

The independent prognostic significance of karyotypic findings at diagnosis has been well known for some time. Data presented at the Sixth International Workshop on Chromosomes in Leukemia (277) supported and refined observations made at earlier workshops and illustrated that cytogenetic findings in AML can be separated into three groups: favorable risk, standard risk and unfavorable risk. The favorable risk group includes cases with the common translocations, t(8;21), inv(16)/t(16;16), and t(15;17) [in addition to the t(9;11), but not necessarily the other 11q23 abnormalities]. The standard risk group includes cases with trisomy 8 and cases with a normal karyotype. The unfavorable group includes cases with −5/del(5q) and/or −7/del(7q) as well as those with the inv(3)/t(3;3). In some protocols, the identification of patients with favorable cytogenetic features will indicate selection of intensive consolidation chemotherapy with the intent for cure, whereas those with less favorable cytogenetic findings would be treated with more investigational agents or considered for transplantation in first remission.

B. Primary MDS

Clonal karyotypic abnormalities are less common in the myelodysplastic disorders than in AML. At least 40% and possibly up to 70% of MDS cases will have a detectable cytogenetic abnormality (278–283). The abnormalities show significant overlap with many of those seen in AML. This is consistent with the fact that in approximately 20% of all cases, MDS evolves to AML. However, abnormalities specific to AML, not seen in MDS, are notable. The recurring abnormalities in MDS include deletions and gains/losses of entire chromosomes for the most part, with far fewer translocations than are typically seen in AML (see Table 3). Sometimes unrelated clones are detected in MDS, and karyotypic evolution can occur during the course of the disease.

1. Cytogenetics in the Differential Diagnosis of MDS

The diagnosis of MDS is usually based on cytomorphologic features of the peripheral blood, bone marrow aspirate, and bone core biopsy, taken together with knowledge of the clinical history and other relevant laboratory findings. However, the diagnostically useful features of myelodysplasia and increased blasts can be subtle and may overlap findings in some reactive bone marrow conditions, vitamin deficiencies, and bone marrow toxicities, such as in cases involving heavy-metal exposure. Features of MDS can also overlap those seen in congenital dyserythropoietic syndromes (284). In such difficult cases, the finding of a clonal karyotypic abnormality at diagnosis is a very helpful adjunct to routine pathology for distinguishing between a reactive nonclonal

Table 3 Cytogenetic Abnormalities in MDS

Numerical abnormalities/ deletions		Balanced translocations		Other	
del(5q)	(27%)	der(1;7)(q10;q10)	(2%)	i(17q)	(5%)
+8	(19%)	t(1;3)(p36;q21)	(<1%)	inv(3)(q21q26)	1%)
−7	(15%)	t(3;3)(q21;q26)	(<1%)		
del(11q)	(7%)	t(6;9)(p23;q34)	1(<1%)		
+7	(<5%)				
del(12p)	(5%)				
del(20q)	(5%)				
del(7q)	(4%)				
del(13q)	(2%)				

disorder from a clonal MDS. The absence of a cytogenetic anomaly, however, is not useful in the exclusion of MDS, as many cases of MDS lack a clonal abnormality.

Hypocellular myelodysplastic disorders are not uncommon, and can present a problem in their differentiation from aplastic anemia (285–288). The marked dyserythropoiesis which can be seen in aplastic anemia can further complicate the distinction. The finding of a common karyotypic abnormality usually seen in MDS [e.g., −7 or del(7q)] is a useful indicator supporting the diagnosis of MDS. However, the finding of any clonal abnormality may not be sufficient. Recently, a few clonal abnormalities such as trisomy 6 and t(1;20) have been reported to show a clinicopathologic picture more like aplastic anemia than MDS (287,288).

The distinction between MDS and CMPD is usually clear-cut, but cases of MDS with fibrosis can be difficult to evaluate (289,290). Although some cytogenetic abnormalities have been noted to be associated with MDS with fibrosis, these are not specific and can be seen in the other myeloid disorders (291). Issues regarding the utility of cytogenetics in separating MDS from AML were discussed in Section III.A.2.

2. Morphologic and Clinical Correlations

In MDS, there are no common, or unique, morphologic-cytogenetic associations comparable to those of APL and the t(15;17) in AML. However, there are some morphologic associations. These include sideroblastosis with deletions of 11q (292,293), ringed sideroblasts with Xq13 abnormalities (294), and prominent dyserythropoiesis with deletions of 20q (173). Furthermore, there is a correlation between the frequency of chromosomal abnormalities and the FAB subtypes of MDS. Only 30–50% of cases of the low-grade RA or RARS show clonal abnormalities, whereas closer to 60–70% of cases of the high-grade RAEB and RAEB-T show an abnormality. In CMML the frequency is 30–40% (295). Correlation of karyotypic abnormalities with prognosis has been demonstrated repeatedly: patients with karyotypic abnormalities (particularly, complex abnormalities) have a shorter survival and a higher incidence of progression to AML than those without. Recently, three groups of patients with MDS could be subdivided into good, intermediate, and poor outcomes based on the variables of cytogenetic abnormalities, percentage of bone marrow blasts, and number of cytopenias (282). The cytogenetic findings in the "good outcome" group included normal karyotypes, −Y alone, del(5q) alone, or del(20q) alone. Those in the "poor outcome" group included complex karyotypes with three or more abnormalities, which frequently included those of chromosomes 5 and/or 7. The "intermediate group" included those cases with "other abnormalities."

Two specific clinicopathologic entities within the spectrum of MDS are the so-called 5q− syndrome and the monosomy 7 syndrome. These were both discussed in Section II.C.1 above.

C. t-MDS/t-AML (see review 129)

MDS and AML developing in patients who have been previously treated with chemotherapeutic agents or ionizing radiation are, by definition, considered therapy-related leukemic processes, t-MDS/t-AML (130,154). These secondary disorders account for approximately 10–20% of all cases of leukemia and are likely increasing in frequency due to a number of causes, including an increased use of a widening spectrum of therapeutic drugs capable of damaging DNA, the higher intensities at which these drugs are used, and the longer survival of patients receiving the treatments. Two broad categories of t-MDS/t-AML are recognized (Table 4): those following alkylating agents and/or irradiation, and those following treatment with drugs known to inhibit topoisomerase II, such as epipodophyllotoxins and anthracyclines (131).

The t-MDS/t-AML associated with alkylating agents and/or irradiation commonly occur about 4–6 years following the therapy, and present with ineffective hematopoiesis, trilineage dysplasia, and a myelodysplastic syndrome which will likely culminate with an acute leukemia. The t-MDS are more commonly hypocellular and are associated with fibrosis, rather than primary MDS. The t-AML are more frequently of the M6 or M7 FAB subtype than are the de novo cases (154). Abnormalities of chromosome 5 and or 7 (see Section II.C.1) are preferentially involved, and these are frequently

Table 4 Cytogenetic Subsets in t-MDS/t-AML

	−5/del(5q) and −7/del(7q)	11q23 abnormalities
Treatment	Alkylating agents	Topoisomerase II inhibitors
Age	Older patients	Younger patients
Onset after therapy	~5 years	1–2 years
Associated dysplasia/MDS	Yes	No
FAB subtype	Variable	M4 or M5
Response to therapy	Poor	Favorable

present in a complex karyotype. The der(1;7) is also associated with a previous therapy with alkylating agents (Section II.D.5).

The secondary leukemic disorders following drugs inhibiting topoisomerase II have a different clinical pattern (131,132). They occur generally between 1 and 3 years following therapy, develop with an abrupt onset (i.e., not associated with a preleukemic phase) and are frequently of the monocytic or myelomonocytic type. These leukemias have a high association with abnormalities of 11q23 and *MLL* rearrangements (Section II.B.4), although less frequently they can show 21q22 abnormalities and *CBFA2* rearrangements such as t(8;21) and t(3;21) (Section II.B.1).

Although t-MDS/t-AML have the above characteristics, a diagnosis is generally based on the usual morphologic, cytochemical, and immunophenotypic criterias for AML or MDS, and on the appropriate documented history of previous therapy. It should be noted that although t-MDS/t-AML are more common after therapy for solid tumors, and other malignancies such as Hodgkin's disease, non-Hodgkin's lymphoma, and multiple myeloma, they can also occur after therapy for autoimmune disease or organ transplantation.

D. Chronic Myeloproliferative Disorders (see review 296)

A discussion of the cytogenetic findings in CML was presented in the section dealing with the Philadelphia chromosome and will not be repeated here (see Section I.A). In contrast to CML, in which the t(9;22) or the resultant *bcr/abl* rearrangement is associated with essentially all cases, the other myeloproliferative disorders have no defining karyotypic or genetic abnormalities. In fact, most of the chromosome anomalies seen in the other CMPDs are nonspecific, and are observed in the other types of myeloid disorders as well (Table 5). The most common abnormalities reported are listed in Table

Table 5 Common Chromosomal Abnormalities in CMPD (excluding CML) (sole finding at diagnosis)

Numerical abnormalities	Structural abnormalities	Commonly deleted segment
+8	del(20q)	q12
−7	del(13q)	q14
+21	del(5q)	q31
+9	del(12p)	p12
+13	del(7q)	q22
	i(17q)	

4 and include trisomy 8, monosomy 7, trisomy 21, del(20q), del(13q), and trisomy 9. Karyotypic evolution can occur in a minority of cases (12%) and may correlate with disease progression.

1. Polycythemia Vera

Based on a compilation of data from three large studies the rate of clonal karyotypic abnormalities in P vera is 41% (297–299). This may be an overestimation, however, due to inclusion of both early and late cases in the analysis. Abnormalities are more frequent as disease progresses and, according to a study by Diez-Martin, they increase from only 19% in uncomplicated P vera, to 40–70% in P vera with myelofibrosis or postpolycythemia myeloid metaplasia. Essentially all cases of polycythemia associated with acute leukemia have clonal karyotypic findings. The presence of a cytogenetic abnormality at diagnosis is believed to be associated with a worse prognosis, although this view is not accepted by all. At diagnosis the more common abnormalities seen are del(20q), +8, and +9. The +8 and +9 are commonly seen together. Abnormalities which have been suggested to be specific for P vera are del(1)(p11), del(3)(p11p14), t(1;6)(q11;p21), and t(1;9)(q10;p10), but they are relatively rare (296). It is important to note that some patients develop a chromosomally abnormal clone secondary to therapy. These are the expected abnormalities of chromosome 5 and/or 7 and der(1;7).

2. Chronic Idiopathic Myelofibrosis

Cytogenetic studies in CIMF have been difficult because the associated marrow fibrosis often prevents adequate cellular aspiration. In fact, in one study, over one-quarter of the cases had an inadequate number of metaphase cells for a complete cytogenetic analysis (300). Among the cases suitable for cytogenetic evaluation, the reported rate of clonal karyotypic abnormalities at diagnosis varied from 35% to 68% (296,300,301). The common abnormalities seen include −7, +8, +9, +21, and deletions of 1q, 20q, and 13q. The chromosome 13 abnormality involves a deletion of q12–q14, and is similar to that seen in other CMPDs with associated fibrosis. The deleted area includes the locus for the tumor-suppressor gene in retinoblastoma (*RB1*), but whether the loss of this gene plays a role in the etiology of CIMF has not yet been determined.

3. Essential Thrombocythemia

Among the CMPDs, essential thrombocythemia (ET) is least frequently associated with cytogenetic abnormalities. In an older series, only 5% of cases showed a clonal karyotypic change (302). Occasional reports of ET associated with the Philadelphia chromosome have been published, but most of these may represent CML with thrombocytosis. The possibility that some cases of ET have unusual *BCR* rearrangements is a recent observation which needs further investigation (303).

4. Cytogenetic Findings and the Differential Diagnosis of CMPD

The pathologic diagnosis of CML during the chronic phase is usually straightforward, as the typical findings in the peripheral blood and bone marrow make other entities unlikely. Nevertheless, the diagnosis of CML should always be confirmed with cytogenetic analysis, molecular studies, or molecular cytogenetic (FISH) evaluation. If a patient presents in the accelerated phase of CML, dysplasia, increased blasts, and fibrosis are frequently seen, making the distinction of CML from MDS (or from the other CMPDs) more difficult. In this situation, cytogenetic or related molecular studies are even more critical. In the rare instances when CML presents in myeloid (or even lymphoid) blast phase, the distinction between acute leukemia and CML is still possible, as some cases can show a background of CML. The issue of whether there are truly cases of Ph+ AMLs has been mentioned above (Section II.A.1).

Because the CMPDs in their early phases often are difficult to distinguish from reactive bone marrows (e.g., those with erythroid or megakaryocytic hyperplasia), well-defined clinical criteria have been established for diagnostic purposes. In particularly difficult cases, the finding of a clonal cytogenetic abnormality, such as +8 and +9, will help confirm the diagnosis. Distinctions between the non-CML CMPDs are frequently subtle, and to date there are no specific cytogenetic findings which can help subclassify a difficult case, although +8 together with + 9 would support a diagnosis of P vera, as would the rare specific structural abnormalities listed above (see Section III.D.1).

IV. CONCLUSION

In this chapter we have attempted to summarize the major recurring cytogenetic abnormalities in the myeloid hematopoietic disorders. We first presented the chromosomal abnormalities and discussed what is known about their molecular, pathologic, and clinical associations. Second, we attempted to discuss, with a broader perspective, the currently understood diagnostic utility that chromosomal anomalies have in clinical practice. As is obvious from the comparative lengths of the sections, there is considerably more basic information about the abnormalities than there is current clinical relevance. However, only with a more complete understanding of the genetic, chromosomal, cytomorphologic, and physiologic aspects of the myeloid disorders will there be progress with improved modalities of therapy for patients who suffer from these frequently fatal diseases. Cloning and characterization of tumor-suppressor genes from the critical regions of deletions, and understanding the role of fusion proteins in leukemic transformation, will be the next immediate areas to develop in order to reach this goal. Once there are specific therapies for specific disorders (e.g., like ATRA for APL), precise molecular, cytogenetic, and pathologic diagnosis will be of critical importance.

ACKNOWLEDGMENTS

The authors thank Drs. Michelle M. Le Beau and James W. Vardiman for their helpful comments on the manuscript, and Miss Penny Lindberg for her help with typing the references.

REFERENCES

1. Bennett JM, Catovsky D, Daniel M-T, Flandrin G, Galton DAG, Gralnick HR, Sultan C. Proposed revised criteria for the classification of acute myeloid leukemia. A report of the French-American-British Cooperative Group. Ann Intern Med 1985; 103:620–625.
2. Bennett JM, Catovsky D, Daniel M-T, Flandrin G, Galton DAG, Gralnick HR, Sultan C. Proposals for the classification of the myelodysplastic syndromes. Br J Haematol 1982; 51:189–199.

3. Bennett JM, Catovsky D, Daniel MT, Flandrin G, Galton DAG, Gralnick H, Sultan C, Cox C. The chronic myeloid leukaemias: guidelines for distinguishing chronic granulocytic, atypical chronic myeloid and chronic myelomonocytic leukaemia: proposals by the French-American-British Cooperative Leukaemia Group. Br J Haematol 1994; 87:746–754.
4. World Health Organization proposals for the classification of the hematopoietic neoplasms. Presented at the United States and Canadian Academy of Pathology meeting, Society for Hematopathology, Orlando, FL, March 2, 1997. Am J Surg Pathol 1997; 21:114–121.
5. Nowell PC, Hungerford DA. A minute chromosome in human chronic granulocytic leukemia. Science 1960; 132:1497–1501.
6. Court Brown WM, Tough IM. Cytogenetic studies in chronic myeloid leukemia. Adv Cancer Res 1963; 7:351–381.
7. Sandberg AA. Cytogenetics of the leukemias and lymphomas. In: Luderer AA, Weetall HH, eds. The Human Oncogeneic Viruses. Clifton, NJ: Humana Press, 1986: 1–41.
8. Prieto F, Egozcue J, Forteza G, Marco F. Identification of the Philadelphia (Ph^1) chromosome. Blood 1970; 35:23–38.
9. Rowley JD. A new consistent chromosome abnormality in chronic myelogenous leukaemia detected by quinacrine fluorescence and Giemsa staining. Nature 1973; 243:290–293.
10. Heisterkamp N, Stephenson JR, Groffen J, Hansen PF, de Klein A, Bartram CR, Grosveld G. Localization of the c-abl oncogene adjacent to a translocation breakpoint in chronic myelocytic leukaemia. Nature 1983; 306:239–242.
11. Heisterkamp N, Stam K, Groffen J, De Klein A, Grosveld G. Structural organization of the bcr gene and its role in the Ph^1 translocations. Nature 1985; 315:758–761.
12. Groffen J, Stephenson JR, Heisterkamp N, De Klein A, Bartram CR, Grosveld G. Philadelphia chromosomal breakpoints are clustered within a limited region, bcr, on chromosome 22. Cell 1984; 36:93–99.
13. Gale RP, Canaani E. An 8 kilobase abl RNA transcript in chronic myelogenous leukemia. Proc Natl Acad Sci USA 1984; 81:5648–5652.
14. Stam K, Heisterkamp N, Grosveld G, de Klein A, Verma RS, Coileman M, Dosik H, Groffen J. Evidence for a new chimeric bcr-abl mRNA in patients with chronic myelogenous leukemia and the Philadelphia chromosome. N Engl J Med 1985; 313:1429–1433.
15. Konopka JB, Watanabe SM, Singer JW, Collins SJ, Witte ON. Cell lines and clinical isolates derived from Ph positive chronic myelogenous leukemia patients express abl proteins with a common structural alteration. Proc Natl Acad Sci USA 1985; 82:1810–1814.
16. Bloomfield CD, Goldman AI, Alimena G, Berger R, Borgstrom GH, Brandt L, Catovsky D, de la Chapelle A, Dewald GW, Garson OM, Garwicz S, Golomb HM, Hossfeld DK, Lawler SD, Mitelman F, Nilsson P, Pierre RV, Philip P, Prigogina E, Rowley JD, Sakurai M, Sandberg AA, Secker Walker LM, Tricot G, Van Den Berghe H, Van Orshoven A, Vuopio P, Whang-Peng

J. Chromosomal abnormalities identify high-risk and low-risk patients with acute lymphoblastic leukemia. Blood 1986; 67:415–420.

17. Michiels JJ, Prins E, Hagermeijer A, Brederoo P, van der Meulen J, van Vliet HH, Abels J. Philadelphia chromosome positive thrombocythemia and megakaryoblastic leukemia. Am J Clin Pathol 1987; 88:645–652.
18. Paietta E, Rosen N, Roberts M, Papenhausen P, Wiernik PH. Philadelphia chromosome positive essential thrombocythemia evolving into lymphoid blast crisis. Cancer Genet Cytogenet 1987; 25:227–231.
19. Stoll DB, Peterson P, Esten R, Laszlo J, Pisciotta AV, Ellis JT, White P, Vaidya K, Bozdech M, Murphy S. Clinical presentation and natural history of patients with essential thrombocythemia and the Philadelphia chromosome. Am J Hematol 1988; 27:77–83.
20. Fialkow PJ, Martin PJ, Najfeld V, Penfold GF, Jacobson RJ, Hansen JA. Evidence of a multistep pathogenesis of chronic myelogenous leukemia. Blood 1981; 58:158–163.
21. Raskind WH, Ferraris AM, Najfeld V, Jacobson RJ, Moohr JW, Fialkow PJ. Further evidence for the existence of a clonal Ph-negative stage in some cases of Ph-positive chronic myelocytic leukemia. Leukemia 1993; 7:1163–1167.
22. Ferrajoli A, Fizzotti M, Liberati AM, Grignani F. Chronic myelogenous leukemia: an update on the biological findings and therapeutic approaches. Crit Rev Oncol Hematol 1996; 22:151–174.
23. Gordon MY, Downing CR, Riley GP, Goldman JM, Greaves MF. Altered adhesive interactions with marrow stroma of haematopoietic progenitor cells in chronic myeloid leukaemia. Nature 1987; 328:342–344.
24. Gordon MY, Goldman JM. Cellular and molecular mechanisms in chronic myeloid leukaemia: biology and treatment. Br J Haematol 1996; 95:10–20.
25. Clark SS, Mc Laughlin J, Crist WM, Champlin TR, Witte ON. Unique forms of the abl tyrosine kinase distinguish Ph positive CML from Ph positive ALL. Science 1987; 235:85–88.
26. Anastasi J, Feng J, Dickstein JI, Le Beau MM, Rubin CM, Larson RA, Rowley JD, Vardiman JW. Lineage involvement by *BCR-ABL* in Ph+ lymphoblastic leukemias: chronic myelogenous leukemia presenting in lymphoid blast phase vs Ph+ acute lymphoblastic leukemia. Leukemia 1996; 10:795–802.
27. Secker-Walker LM, Craig JM. Prognostic implications of breakpoint and lineage heterogeneity in Philadelphia-positive acute lymphoblastic leukemia: a review. Leukemia 1993; 7:147–151.
28. Haferlach T, Winkemann M, Ramm-Petersen L, Meeder M, Schoch R, Weber-Matthiesen K, Schlegelberger B, Schoch C, Ludwig WD, Hiddemann W, Loffler H. New insights into the biology of Philadelphia-chromosome-positive acute lymphoblastic leukaemia using a combination of May-Grunwald-Giemsa staining and fluorescence in situ hybridization techniques at the single cell level. Br J Haematol 1997; 99:452–459.
29. Pugh WC, Pearson M, Vardiman JW, Rowley JD. Philadelphia chromosome-negative chronic myelogenous leukaemia: a morphological reassessment. Br J Haematol 1984; 60:457–467.

30. Ganesan TS, Rassool F, Guo A-P, Th'ng KH, Dowding C, Hibbin JA, Young BD, White H, Kumaran TO, Galton DA, Goldman JM. Rearrangement of the *bcr* gene in Philadelphia-chromosome-negative chronic myeloid leukemia. Blood 1986; 68:957–960.
31. Lazaridou A, Chase A, Melo J, Garicochea B, Diamond J, Goldman J. Lack of reciprocal translocation in BCR-ABL positive Ph-negative chronic myeloid leukaemia. Leukemia 1994; 8:454–457.
32. Hagemeijer A, Buijs A, Smit E, Jansen B, Creemers G-J, Van der Plas D, Grosveld G. Translocation of BCR to chromosome 9: A new cytogenetic variant detected by FISH in two Ph-negative, BCR-positive patients with chronic myeloid leukemia. Genes Chromosom Cancer 1993; 8:237–245.
33. Rowley JD, Testa JR. Chromosome abnormalities in malignant hematologic diseases. Adv Cancer Res 1982; 36:103–148.
34. Hagemeijer A, Bartram CR, Smit EME, van Agthoven AJ, Bootsma D. Is the chromosomal region 9q34 always involved in variants of the Ph^1 translocation? Cancer Genet Cytogenet 1984; 13:1–16.
35. Sokal JE, Gomez GA, Baccarani M, Tura S, Clarkson BD, Cervantes F, Rozman C, Carbonell F, Anger B, Heimpel H. Prognostic significance of additional cytogenetic abnormalities at diagnosis of Philadelphia chromosome-positive chronic granulocytic leukemia. Blood 1988; 72:294–298.
36. Swolin B, Weinfield A, Westin J, Waldenstrom J, Magnusson B. Karyotypic evolution in Ph-positive myeloid leukemia in relation to management and disease progression. Cancer Genet Cytogenet 1985; 18:65–79.
37. Bernstein R. Cytogenetics of chronic myelogenous leukemia. Semin Hematol 88; 25:20–34.
38. Mitelman F. The cytogenetic scenario of chronic myeloid leukemia. Leuk Lymph 1993; 11;suppl 1:11–13.
39. Enright H, Weisdorf D, Peterson L, Rydell RE, Kaplan ME, Arthur DC. Inversion of chromosome 16 and dysplastic eosinophils in accelerated phase of chronic myeloid leukemia. Leukemia 1992; 6:381–384.
40. Whang-Peng J, Canellos GP, Carbone PP, Tjio JH. Clinical implications of cytogenetic variants in chronic myelocytic leukemia. Blood 1968; 32:755–759.
41. Anastasi J, Feng J, Le Beau MM, Larson RA, Rowley JD, Vardiman JW. The relationship between secondary chromosomal abnormalities and blast transformation in chronic myelogenous leukemia. Leukemia 1995; 9:628–633.
42. Vardiman JW. Chronic myelogenous leukemia and the myeloproliferative disorders. In: Knowles DM, ed. Neoplastic Hematopathology. Baltimore: Williams & Wilkins, 1992:1405–1438.
43. Dickstein JI, Vardiman JW. Hematopathologic findings in the myeloproliferative disorders. Semin Oncol 1995; 22:355–373.
44. Georgii A, Vykoupil KF, Buhr T, Choritz H, Kaloutsi V, Werner M. Chronic myeloproliferative disorders in bone marrow biopsies. Pathol Res Pract 1990; 186:3–27.

45. Lorand-Metze I, Vassalo J, Souza C. Histological and cytological heterogeneity of bone marrow in Philadelphia-positive chronic myelogenous leukemia at diagnosis. Br J Haematol 1987; 67:45–49.
46. Knox WF, Bhavnani M, Davson J, Geary CG. Histological classification of chronic granulocytic leukemia. Clin Lab Haematol 1984; 6:171–175.
47. Cervantes F, Lopez-Guillermo A, Terol M-J, Rozman C, Montserat E. An assessment of the clinicohematological criteria for the accelerated phase of chronic myeloid leukemia. Eur J Haematol 1996; 57:286–291.
48. Stock W, Westbrook CA, Peterson B, Arthur DC, Szatrowki TP, Silver RT, Sher DA, Wu D, Le Beau MM, Schiffer CA, Bloomfield CD. Value of molecular monitoring during the treatment of chronic myeloid leukemia: a Cancer and Leukemia Group B Study. J Clin Oncol 1997; 15:26–36.
49. Hochhaus A, Reiter A, Skladny H, Reichert A, Saussele S, Hehlmann R. Molecular monitoring of residual disease in chronic myelogenous leukemia patients after therapy. Rec Res Cancer Res 1998; 144:36–44.
50. Anastasi JA, Musvee T, Roulston D, Domer PH, Larson RA, Vardiman JW. Pseudo-Gaucher histiocytes identified up to 1 year after transplantation for CML are *BCR/ABL*-positive. Leukemia 1998; 12:233–237.
51. Nucifora G, Rowley JD. AML1 and the 8;21 and 3;21 translocation in acute and chronic myeloid leukemia. Blood 1995; 86:1–14.
52. Hiebert SW, Downing JR, Lenny N, Meuers S. Transcriptional regulaton by the t(8;21) fusion protein, AML-1/ETO. Curr Top Micro Immunol 1996; 211: 253–258.
53. Rowley JD. Identification of a translocation with quinacrine fluorescence in a patient with acute leukemia. Ann Genet 1973; 16:109–112.
54. Anonymous. Fourth International Workshop on Chromosomes in Leukemia. Cancer Genet Cytogenet 1984; 11:249–360.
55. Nucifora G, Dickstein JI, Torbenson V, Roulston D, Rowley JD, Vardiman JW. Correlation between cell morphology and expression of the AML1/ETO chimeric transcripts in patients without the t(8;21). Leukemia 1994; 8:1533–1538.
56. Byrd JC, Edenfield WJ, Shields DJ, Dawson NA. Extramedullary myeloid cell tumors in acute nonlymphocytic leukemia: a clinical review. J Clin Oncol 1995; 13:1800–1816.
57. Schiffer CA, Lee EJ, Tomiyasu T, Wiernik PH, Testa JR. Prognostic impact of cytogenetic abnormalities in patients with de novo acute nonlymphocytic leukemia. Blood 1989; 73:263–270.
58. Nucifora G, Larson RA, Rowley JD. Persistence of the 8;21 translocation in patients with AML-M2 in long term remission. Blood 1993; 82:712–715.
59. Rowley JD. Identification of the constant chromosome regions involved in human hematologic malignant disease. Science 1982; 216:749–751.
60. Erickson P, Gao J, Chang KS, Look T, Whisenant E, Raimondi S, Lasher R, Trujillo J, Rowley J, Drabkin H. Identification of breakpoints in t(8;21) acute myelogenous leukemia and isolation of a fusion transcript, AML1/ETO with similarity to a *Drosophila* segmentation gene, *runt*. Blood 1992; 80:1825–1831.

61. Nucifora G, Birn DJ, Erickson P, Gao J, Le Beau MM, Drabkin HA, Rowley JD. Detection of DNA rearrangements in the AML1 and ETO loci and of an AML1/ETO fusion mRNA in patients with t(8;21) myeloid leukemia. Blood 1993; 81:883–888.
62. Meyers S, Lenny N, Hiebert SW. The t(8;21) fusion protein interferes with AML1B-dependent transcriptional activity. Mol Cell Biol 1993;; 15:1974–1982.
63. Orkin SH. Transcriptional factors and hematopoietic development. J Biol Chem 1995; 270:4955–4958.
64. Johansson B, Mertens F, Mitelman F. Secondary chromosomal abnormalities in acute leukemias. Leukemia 1994; 8:953–962.
65. Bitter MA, Le Beau MM, Rowley JD, Larson RA, Golomb HG, Vardiman JW. Associations between morphology, karyotype, and clinical features in myeloid leukemias. Hum Pathol 1987; 18:211–225.
66. Berger R, Hillion J, Janvier D, Chen Z, Bussel A. t(8;21) prior to acute leukemia. Cancer Genet Cytogenet 1993; 70:125–126.
67. Hamblin TJ. Pseudomyelodysplastic syndrome with t(8;21). Leukemia Res 1995; 18:767–768.
68. Taj AS, Ross FM, Vickers M, Choudhury DN, Harvey JF, Barber JC, Barton C, Smith AG. t(8;21) myelodysplasia, an early presentation of M2 AML. Br J Haematol 1995; 89:890–892.
69. Xue Y, Yu F, Zhou Z, Guo Y, Xie X, Lin B. Translocation (8;21) in oligoblastic leukemia: is this a true myelodysplastic syndrome? Leukemia Res 1995; 18:761–765.
70. Martinez-Climent JA, Lane NJ, Rubin CM, Morgan E, Johnstone HS, Mick R, Murphy SB, Vardiman JW, Larson RA, Le Beau MM, Rowley JD. Clinical and prognostic significance of chromosomal abnormalities in childhood acute myeloid leukemia. Leukemia 1995; 9:95–101.
71. Miyamoto T, Nagfuji K, Akashi K, Harada M, Kyo T, Akashi T, Takenaka K, Mizuno S, Gondo H, Okamura T, Dohy H, Niho Y. Persistence of multipotent progenitors expressing AML1/ETO transcripts in long-term remission patients with t(8;21) acute myelogenous leukemia. Blood 1996; 87:4789–4796.
72. Rubin CM, Larson RA, Bitter MA, Carrino JJ, Le Beau MM, Diaz MO, Rowley JD. Association of a chromosomal 3;21 translocation with the blast phase of chromic myelogenous leukemia. Blood 1987; 70:1338–1342.
73. Rubin CM, Larson RA, Anastasi J, Winter JN, Thangavelu M, Vardiman JW, Rowley JE, Le Beau MM. t(3;21)(q26;q22): a recurring chromosomal abnormality in therapy-related myelodysplastic syndromes and acute myeloid leukemia. Blood 1990; 76:2590–2596.
74. Shimada M, Ohtsuka E, Shimizu T, Matsumoto T, Matsushita K, Tanimoto F, Kajii T. A recurrent translocation, t(16;21)(q24;q22), associated with acute myelogenous leukemia: identification by fluorescence in situ hybridization. Cancer Genet Cytogenet 1997; 96:102–105.
75. Roulston D, Espinosa R III, Nucifora G, Larson RA, Le Beau MM, Rowley

JD. *CBFA2(AML1)* translocations with novel partner chromosomes in myeloid leukemias: associations with prior therapy. Blood 1998; 92:2879–2885.

76. Gamou T, Kitamura E, Hosoda F, Shimizu K, Shinohara K, Hayashi Y, Nagase T, Yokoyama Y, Ohki M. The partner gene *AML1* in t(12;21) myeloid malignancies is a novel member of the *MTG8(ETO)* family. Blood 1998; 91:4028–4037.
77. Liu PP, Hajra A, Wijmenga C, Collins FS. Molecular pathogenesis of the chromosome 16 in the M4Eo subtype of acute myeloid leukemia. Blood 1995; 85:2289–2302.
78. Arthur DC, Bloomfield CD. Partial deletion of the long arm of chromosome 16 and bone marrow eosinophilia in acute non-lymphocytic leukemia: A new association. Blood 1983; 61:994–998.
79. Le Beau MM, Larson RA, Bitter MA, Vardiman JW, Golomb HM, Rowley JD. Association of an inversion of chromosome 16 with abnormal marrow eosinophils in acute myelomonocytic leukemia. N Engl J Med 1984; 309: 630–636.
80. Hajra A, Liu PP, Collins FS. Transforming properties of the leukemia inv(16) fusion gene CBFb-MYH11. Curr Top Micro Immunol 1996; 211:289–298.
81. Liu P, Tarle SA, Hajra A, Claxton DF, Marlton P, Freedman M, Siciliano MJ, Collins FS. Fusion between transcription factor CBF beta/PEBP2 beta and a myosin heavy chain in acute myeloid leukemia. Science 1993; 261:1041–1044.
82. Marlton P, Keating M, Kantarjian H, Pierce S, O'Brien S, Freireich EJ, Estey E. Cytogenetic and clinical correlates in AML patients with abnormalities of chromosome 16. Leukemia 1995; 9:965–971.
83. Bitter MA, Le Beau MM, Larson RA, Rosner MC, Golomb HM, Rowley JD, Vardiman JW. A morphologic and cytochemical study of acute myelomonocytic leukemia with abnormal marrow eosinophils associated with inv(16)(p13q22). Am J Clin Pathol 1984; 81:733–741.
84. Bitter MA, Le Beau MM. Comparison of eosinophils in acute non-lymphocytic leukemia associated with the t(8;21) and inv(16). Br J Haematol 1985; 59:185–188.
85. Haferlach T, Winkermann M, Loffler H, Schoch R, Gassmann W, Fonatsch C, Schoch C, Poetsch M, Weber-Matthiesen K, Schlegelberger B. The abnormal eosinophils are part of the leukemia cell population in acute myelomonocytic leukemia with abnormal eosinophils (AML M4eo) and carry the pericentric inversion 16: A combination of May-Grunwald-Giemsa staining and fluorescence in situ hybridization. Blood 1996; 87:2459–2463.
86. Holmes R, Keating MJ, Cork A, Broach Y, Trujillo J, Dalton WT Jr, McCredie KB, Freireich EJ. A unique pattern of central nervous system relapse in acute myelomonocytic leukemia associated with inv(16)(p13q22). Blood 1985; 65: 1071–1078.
87. Larson RA, Williams SF, Le Beau MM, Bitter MA, Vardiman JW, Rowley JD. Acute myelomonocytic leukemia with abnormal eosinophils and inv(16) or t(16;16) has a favorable prognosis. Blood 1986; 68:1242–1249.

88. Mrozek K, Heinonen K, de la Chapelle A, Bloomfield CD. Clinical significance of cytogenetics in acute myeloid leukemia. Semin Oncol 1997; 24:17–31.
89. Avvisati G, Wouter ten Cate J, Mandelli F. Acute promyelocytic leukemia. Br J Haematol 1992; 81:315–320.
90. Warrell RP Jr, de The H, Wang Z-Y, Degos L. Acute promyelocytic leukemia. N Engl J Med 1993; 329:177–189.
91. Grignani F, Fagioli M, Alcalay M, Longo L, Pandolfi PP, Donti E, Biondi A, Lo Coco F, Grignani F, Pelicci PG. Acute promyelocytic leukemia: from genetics to treatment. Blood 1994; 83:10–25.
92. Larson RA, Kondo K, Vardiman JW, Butler AE, Golomb HM, Rowley JD. Evidence for a 15;17 translocation in every patient with acute promyelocytic leukemia. Am J Med 1984; 76:827–841.
93. Najfeld V, Scalise A, Troy K. A new variant translocation 11;17 in a patient with acute promyelocytic leukemia together with t(7;12). Cancer Genet Cytogenet 1989; 43:103–108.
94. Licht JD, Chomienne C, Goy A, Scott AA, Head DR, Michaux JL, Wu Y, DeBlasio A, Willer WH Jr, Zelenetz AD, Willman CL, Chen Z, Chen S-J, Zelent A, McIntyre E, Veil A, Cortes J, Kantarjian H, Waxman S. Clinical and molecular characterization of a rare syndrome of acute promyelocytic leukemia associated with translocation (11;17). Blood 1995; 85:1083–1094.
95. Corey S, Locker J, Oliveri DR, Shekhter-Levin S, Redner RL, Penchansky L, Gollin SM. A non-classical translocation involving 17q12 (retinoic receptor-alpha) in acute promyelocytic leukemia with atypical features. Leukemia 1994; 8:1350–1353.
96. Huang ME, Ye YC, Chen SR, Chai JR, Lu JX, Zhoa L, Gu LJ, Wang ZY. Use of all-trans retinoic acid in the treatment of acute promyelocytic leukemia. Blood 1988; 72:567–572.
97. Guidez F, Huang W, Tong J-H, Dubois C, Balitrand N, Waxman S, Michaux JL, Martiat P, Degos L, Chen Z, Chomienne C. Poor response to all-trans retinoic acid therapy in a t(11;17) *PLZ/RARA* patient. Leukemia 1994; 8:312–317.
98. Pandolfi PP. *PML*, *PLZF*, and *NPM* genes in the molecular pathogenesis of acute promyelocytic leukemia. Haematologica 1996; 81:472–482.
99. Golomb HM, Rowley J, Vardiman J, Baron J, Locker G, Krasnow S. Partial deletion of long arm of chromosome 17; a specific abnormality in acute promyelocytic leukemia? Arch Intern Med 1976; 136:825–828.
100. Rowley JD, Golomb HM, Dougherty C. 15/17 translocation: a consistent chromosomal change in acute promyelocytic leukemia. Lancet 1977; 1:549–550.
101. Grimwade D, Gorman P, Duprez E, Howe K, Langabeer S, Oliver F, Walker H, Culligan D, Waters J, Pomfret M, Goldstone A, Burnett A, Freemont P, Sheer D, Solomon E. Characterization of cryptic rearrangements and variant translocations in acute promyelocytic leukemia. Blood 1997; 90:4876–4885.

102. Bennett JM, Catovsky D, Daniel MT, Flandrin G, Galton DA, Gralnick HR, Sultan C. Proposals for the classification of acute leukemias: French-American-British Cooperative Group. Br J Haematol 1976; 33:451–458.
103. Castoldi GL, Liso V, Specchia G, Tomasi P. Acute promyelocytic leukemia: morphologic aspects. Leukemia 1994; 8:S27–S32.
104. Tallman MS, Hakimian D, Rubin C, Reisel H, Variakojis D. Basophilic differentiation in acute promyelocytic leukemia. Leukemia 1993; 7:521–526.
105. Golomb HG, Rowley JD, Vardiman JW, Testa JR, Butler A. "Microgranular" acute promyelocytic leukemia: a distinct clinical, ultrastructural and cytogenetic entity. Blood 1980; 55:253–259.
106. Davey FR, Davis RB, MacCallum JM. Morphologic and cytochemical characteristics of acute promyelocytic leukemia. Am J Hematol 1989; 30:221–227.
107. Castaigne S, Chomienne C, Daniel MT, Ballerini P, Berger R, Fenaux P, Degos L. All-trans retinoic acid as a differentiation therapy for acute promyelocytic leukemia. I. Clinical results. Blood 1990; 76:1704–1709.
108. Tallman MS, Andersen JW, Schiffer CA, Appelbaum FR, Feusner JH, Ogdon A, Shepherd L, Willman C, Bloomfield CD, Rowe JM, Wiernik PH. All-trans-retinoic acid in acute promyelocytic leukemia. N Engl J Med 1997; 377: 1021–1028.
109. Castaigne S, Chomienne C, Feneaux P, Daniel MT, Degos L. Hyperleukocytosis during all-trans retinoic acid for acute promyelocytic leukemia. Blood 1990; 76(suppl):260a.
110. Frankel SR, Eadley A, Lauwers G, Weiss M, Warrell RP Jr. The "retinoic acid syndrome" in acute promyelocytic leukemia. Ann Intern Med 1992; 117: 292–296.
111. Mitelman F, Kaneko Y, Berger R. Report of the committee on chromosome changes in neoplasia. In: Cuttichia AJ, Pearson PL, eds. Human Gene Mapping 1993. Baltimore: Johns Hopkins Univ Press, 1994; 773–812.
112. Hayashi Y, Sugita K, Nakazawa S, Abe T, Kojima S, Inaba T, Hanada R, Yamamoto K. Karyotypic patterns in acute mixed lineage leukemia. Leukemia 1990; 4:121–126.
113. Drexler HG, Theil E, Ludwig WD. Review of the incidence and clinical relevance of myeloid-antigen positive acute lymphoblastic leukemia. Leukemia 1991; 5:637–645.
114. Rowley JD. Consistent chromosome abnormalities in human leukemia and lymphoma. Cancer Invest 1983; 1:267–280.
115. Chen C-S, Sorenson PHB, Domer PH, Reaman GH, Korsmeyer SJ, Heerema NA, Hammond GD, Kersey JH. Molecular rearrangements on chromosome 11q23 predominate in infant acute lymphoblastic leukemia and are associated with specific biologic variables and a poor outcome. Blood 1993; 81:2386–2393.
116. Smith MA, McCaffrey RP, Karp JE. The secondary leukemias: challenges and research directions. J Natl Can Inst 1996; 88:407–418.
117. Berger R, Bernheim A, Weh HJ, Daniel M-T, Flandrin G. Cytogenetic studies on acute monocytic leukemia. Leukemia Res 1980; 4:119–127.

118. Berger R, Berheim A, Siqaux F, Daniel MT, Valensi F, Flandrin G. Acute monocytic leukemia chromosome studies. Leukemia Res 1982; 6:17–26.
119. Rowley JD. Rearrangements involving chromosome band 11q23 in acute leukemia. Cancer Biol 1993; 4:377–385.
120. Bernard OA, Berger R. Molecular basis of 11q23 rearrangements in hematopoietic malignant proliferations. Genes Chromosom Cancer 1995; 13:75–85.
121. Thirman MJ, Gill HJ, Burnett RC, Mbangkollo D, McCabe NR, Kobayashi H, Ziemin van der Poel S, Kaneko Y, Morgan R, Sandberg AA, Chaganti RSK, Larson RA, Le Beau MM, Diaz MO, Rowley JD. Rearrangement of the *MLL* gene in acute lymphoblastic and acute myeloid leukemias with 11q23 chromosomal translocations. N Engl J Med 1993; 329:909–914.
122. Rowley JD. The der(11) chromosome contains the critical breakpoint junction in the 4;11, 9;11, and 11;19 translocations in acute leukemia. Genes Chromosom Cancer 1992; 5:264–266.
123. Le Beau MM, Bitter MA, Kaneko Y, Ueshima Y, Rowley JD. Insertion (10; 11)(p11;q23q24) in two cases of acute monocytic leukemia. Leukemia Res 1985; 9:605–611.
124. Chaplin T, Ayton P, Bernard OA, Saha V, Della Valle V, Hillion J, Gregorini A, Lillington D, Berger R, Young BD. A novel class of zinc finger/leucine zipper genes identified from the molecular cloning of the t(10;11) in acute leukemia. Blood 1995; 85:1435–1441.
125. Beverloo HB, Le Coniat M, Wijsman J, Lillington DM, Bernard O, de Klein A, van Wering E, Welborn J, Young BD, Hagemeijer A, Berger R. Breakpoint heterogeneity in t(10;11) AML-M4/M5 resulting in fusion of *AF10* and *MLL* is resolved by fluorescent in situ hybridization analysis. Cancer Res 1995; 55: 4220–4224.
126. Bitter MA, Le Beau MM, Rowley JD, Larson RA, Golomb HM, Vardiman JW. Associations between morphology, karyotype, and clinical features in myeloid leukemias. Hum Pathol 1987; 18:211–225.
127. Hagemeijer A, Hahlen K, Sizoo W, Abels J. Translocation (9;11)(p21;q23) in three cases of acute monoblastic leukemia. Cancer Genet Cytogenet 1982; 5: 95–105.
128. Kaneko Y, Maseki N, Takasak N, Sakurai M, Hayashi Y, Nakazawa S, Mori T, Sakurai M, Takeda T, Shikano T, Hiyoshi Y. Clinical and hematologic characteristics in acute leukemia with 11q23 translocations. Blood 1986; 67: 484–491.
129. Thirman MJ, Larson RA. Therapy-related myeloid leukemia. Hematol Oncol Clin N Am 1996; 10:293–320.
130. Rowley JD, Golomb HM, Vardiman JW. Nonrandom chromosome abnormalities in acute leukemia and dysmyelopoietic syndrome in patients with previously treated malignant disease. Blood 1981; 58:759–767.
131. Pedersen-Bjergaard J, Rowley JD. The balanced and the unbalanced chromosome aberrations of acute myeloid leukemia may develop in different ways and may contribute differently to malignant transformation. Blood 1994; 83: 2780–2786.

132. Ratain MJ, Rowley JD. Therapy-related acute myeloid leukemia secondary to inhibitors of topoisomerase II. From the bedside to the target genes. Ann Oncol 1992; 3:107–111.
133. Pedersen-Bjergaard J, Philip P. Balanced translocations involving chromosome bands 11q23 and 21q22 are highly characteristic of myelodysplasia and leukemia following therapy with cytostatic agents targeting DNA-topoisomerase II. Blood 1991; 78:1147–1148.
134. Gill Super HJ, McCabe NR, Thirman MJ, Larson RA, Le Beau MM, Pedersen-Bjergaad J, Philip P, Diaz MO, Rowley JD. Rearrangements of the *MLL* gene in therapy-related acute myeloid leukemia in patients previously treated with agents targeting DNA-topoisomerase II. Blood 1993; 82:3705–3711.
135. Caligiuri MA, Schichman SA, Strout MP, Mrozek K, Baer MR, Frankel SR, Barcos M, Herzig GP, Croce CM, Bloomfield CD. Molecular rearrangement of the ALL-1 gene in acute myeloid leukemia without evidence of 11q23 chromosomal translocations. Cancer Res 1994; 54:370–373.
136. Schichman SA, Caligiuri MA, Strout MP, Carter SL, Gu Y, Canaani E, Bloomfield CD, Croce CM. ALL-1 tandem duplication in acute myeloid leukemia with a normal karyotype involves homologous recombination between Alu elements. Cancer Res 1994; 54:4277–4280.
137. Schichman SA, Caligiuri MA, Gu Y, Strout MP, Canaani E, Bloomfield CD, Croce CM. ALL-1 patrial duplication in acute leukemia. Proc Natl Acad Sci USA 1994; 91:6236–6239.
138. Caligiuri MA, Strout MP, Oberkircher AR, Yu F, de la Chapelle A, Bloomfield CD. The partial tandem duplication of ALL1 in acute myeloid leukemia with normal cytogenetics or trisomy 11 is restricted to one chromosome. Proc Natl Acad Sci USA 1997; 94:3899–3902.
139. Bilhou-Nabera C, Leseve JF, Marit G, Lafage M, Dastugue N, Goullin B, Arnoulet C, Stoppa AM, Huguet F, Attal M, Lacombe F, Broustet A, Reiffers J, Bernard P. Trisomy 11 in myeloid leukemia: ten cases. Leukemia 1994; 8: 2240–2241.
140. Ohyashiki K, Ohyashiki JH, Iwabuchi A, Toyama K, Fukui M, Yoshitake H, Okajima S. Trisomy 11 in non-lymphocytic neoplasia. Br J Haematol 1988; 69:420
141. Slovak ML, Traweek ST, Willman CL, Head DR, Kopecky KJ, Magenis RE, Appelbaum FR, Forman SJ. Trisomy 11: an association with stem/progenitor cell immunophenotype. Br J Haematol 1995; 90:266–273.
142. Heinonen K, Mrozek K, Lawrence D. Trisomy 11 as the sole karyotypic abnormality identifies a group of older patients with acute myeloid leukemia (AML) with FABM2 or M! and unfavorable clinical outcome. Results of CALGB 8461 (abstr). Proc Am Assoc Cancer Res 1996; 37:186–187.
143. Van den Berghe H, Michaux L. 5q−, twenty-five years later: a synopsis. Cancer Genet Cytogenet 1997; 94:1–7.
144. Johnson E, Cotter FE. Monosomy 7 and 7q−-associated with myeloid malignancy. Blood Rev 1997; 11:46–55.
145. Neuman WL, Rubin CM, Rios RB, Larson RA, Le Beau MM, Rowley JD, Vardiman JW, Schwartz JL, Farber RA. Chromosomal loss and deletion are

the most common mechanisms for loss of heterozygosity from chromosome 5 and 7 in malignant myeloid disorders. Blood 1992; 79:1501–1510.

146. Le Beau MM, Espinoza R III, Neuman WL, Stock W, Roulston D, Larson RA, Keinanen M, Westbrook C. Cytogenetic and molecular delineation of the smallest commonly deleted region of chromosome 5 in malignant myeloid disease. Proc Natl Acad Sci USA 1993; 90:5484–5488.
147. Willman CL, Sever CE, Pallavicini MG, Harada H, Tanaka N, Slovak ML, Yamamoto H, Harada K, Meeker T, List AF. Taniguchi T. Deletion of the IRF-1, mapping to chromosome 5q31.1 in human leukemia and preleukemic myelodysplasia. Science 1993; 259:968–971.
148. Boultwood J, Fidler C, Lewis S, MacCarthy KA, Sheridan H, Kelly S, Oscier D, Buckle VJ, Wainscoat JKS. Allelic loss of *IRF1* in myelodysplasia and acute myeloid leukemia. Retention of IRF1 on the 5q− chromosome in some patients with the 5q− syndrome. Blood 1993; 82:2611–2616.
149. Zhao N, Stoffel A, Wang PW, Eisenbart JD, Espinosa R, Larson RA, Le Beau MM. Molecular delineation of the smallest commonly deleted region of chromosome 5 in malignant myeloid diseases to 1-1.5 Mb and preparation of a PAC-based physical map. Proc Natl Acad Sci USA 1997; 94:6948–6953.
150. Nagarajan L, Zavadil J, Claxton D, Lu X, Fairman J, Warrington JA, Wasmuth JJ, Chinault AC, Sever CE, Slovak ML, Willman CL, Deisseroth AB. Consistent loss of the D5S89 locus mapping telomeric to the interleukin gene cluster and centromeric to ERG-1 in patients with 5q− chromosome. Blood 1994; 83:199–208.
151. Fairman J, Wang RY, Liang H, Zhao L, Saltman D, Liang JC, Nagarajan L. Translocations and deletions of 5q13.1 in myelodysplasia and acute myelogenous leukemia: evidence for a novel critical locus. Blood 1996; 88:2259–2266.
152. Le Beau MM, Espinosa R III, Davis EM, Eisenbart JD, Larson RA, Green ED. Cytogenetic and molecular delineation of a region of chromosome 7 commonly deleted in malignant myeloid diseases. Blood 1996; 88:1930–1935.
153. Fischer K, Frohling S, Scherer SW, McAllister-Brown J, Scholl C, Stilgenbauer S, Tsui LC, Lichter P, Dohner H. Molecular cytogenetic delineation of deletions and translocations involving chromosome bank 7q22 in myeloid leukemias. Blood 1997; 89:2036–2041.
154. Le Beau MM, Albain KS, Larson RA, Vardiman JW, Davis EM, Blough RR, Golomb HM, Rowley JD. Clinical and cytogenetic correlations in 63 patients with therapy-related myelodysplastic syndromes and acute nonlymphocytic leukemia: further evidence for characteristic abnormalities of chromosome no. 5 and 7. J Clin Oncol 1986; 4:325–435.
155. Velloso ERP, Michaux L, Ferrant A, Hernandez JM, Meeus P, Dierlamm J, Criel A, Louwagie A, Verhoff G, Boogaerts M, Michaux JL, Bosly A, Mecucci C, Van den Berghe H. Deletions of the long arm of chromosome 7 in myeloid disorders: loss of band 7q22 implies worst prognosis. Br J Haematol 1996; 92:574–581.

156. Olopade OI, Thangavelu M, Larson RA, Mick R, Kowal-Vern A, Schumacher HR, Le Beau MM, Vardiman JW, Rowley JD. Clinical, morphologic, and cytogenetic characteristics of 26 patients with acute erythroleukemia. Blood 1992; 80:2873–2882.
157. Van den Berghe H, Cassiman JJ, David G, Fryns S, Michaux JL, Sokal G. Distinct hematological disorder with the deletion of the long arm of no. 5 chromosome. Nature 1974; 251:437–438.
158. Sokal G, Michaux JL, Van den Berghe H, Cordier A, Rodhain J, Ferrant A, Moriau M, De Bruyere M, Sonnet J. A new hematological syndrome with a distinct karyotype: the 5q− chromosome. Blood 1975; 46:519–533.
159. Teerenhovi L. Specificity of haematological indicators for '5q− syndrome' in patients with myelodysplastic syndromes. Eur J Haematol 1987; 39:326–330.
160. Boultwood J, Lewis S, Wainscoat JS. The 5q− syndrome. Blood 1994; 84: 3253–3260.
161. Sieff CA, Chessels JM, Harvey BA, Pickthall VJ, Lawler SD. Monosomy 7 in childhood: a myeloproliferative disorder. Br J Haematol 1981; 49:235–249.
162. Gadner H, Haas OA. Experience in pediatric myelodysplastic syndromes. Hematol Oncol Clin N Am 1992; 6:655–672.
163. Butcher M, Frenck R, Emperor J, Paderanga D, Maybee D, Olson K, Shannon K. Molecular evidence that childhood monosomy 7 syndrome is distinct from juvenile chronic myelogenous leukemia and other childhood myeloproliferative disorders. Genes Chromosom Cancer 1995; 12:50–57.
164. Arico M, Biondi A, Pui C-H. Juvenile myelomonocytic leukemia. Blood 1997; 90:479–488.
165. Asimakopoulos FA, Green AR. Deletions of chromosome 20q and the pathogenesis of myeloproliferative disorders. Br J Haematol 1996; 95:219–226.
166. Testa JR, Kinnealey A, Rowley JD, Golde DW, Potter D. Deletion of the long arm of chromosome 20 [del(20)(q11)] in myeloid disorders. Blood 1978; 52: 868–877.
167. Reeves BR, Lobb DS, Lawler SD. Identity of the abnormal F-group chromosome associated with polycythemia vera. Humangenetik 1972; 14:159–161.
168. Kurtin PJ, Dewald GW, Shields DJ, Hanson CA. Hematologic disorders associated with deletions of chromosome 20q: a clinicopathologic study of 107 patients. Am J Clin Pathol 1996; 106:680–688.
169. Aatola M, Armstrong E, Teerenhovi L, Borgstrom GH. Clinical significance of the del(20q) chromosome in hematologic disorders. Cancer Genet Cytogenet 1992; 62:75–80.
170. White NJ, Nacheva E, Asimakopoulos FA, Bloxham D, Paul B, Green AR. Deletion of chromosome 20q in myelodysplasia can occur in a multipotent precursor of both myeloid cells and B cells. Blood 1994; 83:2809–2816.
171. Roulston D, Espinosa R III, Stoffel M, Bell GI, Le Beau MM. Molecular genetics of myeloid leukemia: identification of the commonly deleted segment of chromosome 20. Blood 1993; 82:3424–3429.

172. Wang PW, Iannantuoni K, Davis EM, Espinosa R III, Stoffel M, Le Beau MM. Refinement of the commonly deleted segment in myeloid leukemias with a del(20q). Genes Chromosom Cancer 1998; 21:75–81.
173. Davis MP, Dewald GW, Pierre RV, Hoagland HC. Hematologic manifestations associated with deletions of the long arm of chromosome 20. Cancer Genet Cytogenet 1984; 12:63–71.
174. Campbell LJ, Carson OM. The prognostic significance of deletion of the long arm of chromosome 20 in myeloid disorders. Leukemia 1994; 8:67–71.
175. Rowley JD. Nonrandom chromosomal abnormalities in hematologic disorders of man. Proc Natl Acad Sci USA 1975; 72:152–156.
176. Heim S, Mitelman F. Numerical chromosome aberrations in human neoplasia. Cancer Genet Gytogenet 1986; 22:99–108.
177. Secker-Walker LM, Fitchett M. Constitutional and acquired trisomy 8. Leukemia Res 1995; 19:737–740.
178. Anastasi J, Feng J, Le Beau MM, Larson RA, Rowley JD, Vardiman JW. Cytogenetic clonality in myelodysplastic syndromes studied with fluorescence in situ hybridization: lineage, response to growth factors, and clone expansion. Blood 1993; 81:1580–1585.
179. Nguyen PL, Arthur DC, Litz CE, Brunning RD. Fluorescence in situ hybridization (FISH) detection in myeloid cells in chronic myeloid leukemia (CML): a study of archival blood and bone marrow smears. Leukemia 1994; 8:1654–1662.
180. Knuutila S. Lineage specificity in hematological neoplasms. Br J Haematol 1997; 96:2–11.
181. Diez-Martin JL, Graham DL, Petitt RM, Dewald G. Chromosome studies in 104 patients with polycythemia vera. Mayo Clin Proc 1991; 66:287–299.
182. Byrd JC, Lawrence D, Arthur DC. Acute myeloid leukemia (AML) patients with pretreatment isolated trisomy 8 are rarely cured with chemotherapy: results of CALGB 8461. Proc Am Assoc Cancer Res 1996; 37:188.
183. Geddes AA, Bowen DT, Jacobs A. Clonal karyotype abnormalities and clinical progression in the myelodysplastic syndrome. Br J Haematol 1990; 76: 194–202.
184. Parlier V, van Mele G, Beris P, Schmidt PM, Tobler A, Haller E, Bellomo MJ. Hematologic, clinical and cytogenetic analysis in 109 patients with primary myelodysplastic syndrome: prognostic significance of morphology and chromosome findings. Cancer Genet Cytogenet 1994; 78:219–231.
185. Bitter MA, Neilly ME, Le Beau MM, Pearson MG, Rowley JD. Rearrangements of chromosome 3 involving bands 3q21 and 3q26 are associated with normal or elevated platelet counts in acute nonlymphocytic leukemia. Blood 1985; 66:1362–1370.
186. Jenkins RB, Tefferi A, Solberg LA, Dewald GW. Acute leukemia with abnormal thrombopoiesis and inversions of chromosome 3. Cancer Genet Cytogenet 1989;39:167–179.
187. Secker-Walker LM, Mehta A, Bain B. Abnormalities of 3q21 and 3q26 in myeloid malignancy: a United Kingdom Cancer Cytogenetic Group study. Br J Haematol 1995; 91:490–501.

188. Morishita K, Parganas E, Bartholomew C, Sacchi N, Valentine MB, Raimondi SC, Le Beau MM, Ihle JN. The human Evi-1 gene is located on chromosome 3q24-q26 but is not rearranged in three cases of acute nonlymphocytic leukemia containing t(3;5)(q25;q32) translocations. Oncogene Res 1990; 5:221–231.
189. Fichelson S, Dryfus E, Berger R, Melle J, Bastard C, Miclea JM, Gisselbrecht S. Evi-1 expression in leukemia patients with rearrangements of the 3q25-3q28 chromosome region. Leukemia 1992; 6:93–99.
190. Morishita K, Parganas E, Willman CL, Whittaker MH, Drabkin H, Oval J, Taetle R, Valentine MB, Ihle N. Activation of EVI1 gene expression in human acute myelogenous leukemias by translocations spanning 300–400 kilobases on chromosome band 3q26. Proc Natl Acad Sci USA 1992; 89:3937–3941.
191. Suzukawa K, Satoh H, Taniwaki M, Yokota J, Morishita K. The human thrombopoietic gene is located on chromosome 3q26.3-3q27, but is not transcriptionally activated in leukemic cells with 3q21 and 3q26 abnormalities (3q21-3q26 syndrome). Leukemia 1995; 9:1328–1331.
192. Bouscary D, Fontenay-Roupie M, Chretien S, Hardy AC, Viguie F, Picard F, Melle J, Dreyfus F. Thrombopoietin is not responsible for the thrombocytosis observed in patients with acute myeloid leukemias and the 3q21-3q26 syndrome. Br J Haematol 1995; 91:425–427.
193. Fontsch C, Gudat H, Lengfelder E, Wandt H, Silling-Engelhardt G, Ludwig WD, Thiel E, Freund M, Bodenbstein H, Schwieder G, Gruneisen A, Aul C, Schnittger S, Rieder H, Haase D, Hild E. Correlation of cytogenetic findings with clinical features in 18 patients with inv(3)(q21q26) or t(3;3)(q21;q26). Leukemia 1994; 8:1318–1326.
194. Rowley JD, Potter D. Chromosomal banding patterns in acute nonlymphocytic leukemia. Blood 1976; 47:705–721.
195. Alsabeh R, Brynes RK, Slovak M, Arber DA. Acute leukemia with t(6;9)(p23; q34): association with myelodysplasia, basophilia, and initial CD34-negative immunophenotype. Am J Clin Pathol 1997; 107:430–437.
196. Pearson MG, Vardiman JW, Le Beau MM, Rowley JD, Schwartz S, Kerman SL, Cohen MM, Fleischman EW, Prigogina EL. Increased numbers of marrow basophils may be associated with a t(6;9) i ANLL. Am J Hematol 1985; 18: 393–403.
197. Soekarman D, Von Lindern M, Van der Plas DC, Selleri L, Bartam CR, Martiat P, Culligan D, Padua RA, Hasper-Voogt KP, Haegemeijer A, Grasveld G. Dek-Can rearrangement in translocation t(6;9)(p23;q34). Leukemia 1992; 6:489–494.
198. Sandberg AA, Morgan R, McCallister JA, Kaiser-McCaw B, Hect F. Acute myeloblastic leukemia (AML) with t(6;9)(p23;q34): a specific subgroup of AML? Cancer Genet Cytogenet 1983; 10:139–142.
199. Cuneo A, Kerim S, Vandenberghe E, Van Orshoven A, Rodhain J, Bosly A, Zachee P, Louwagie A, Michaux JL, Dal Cin P, Van Den Berghe H. Translocation t(6;9) occurring in acute myelofibrosis, myelocysplastic syndrome, and acute nonlymphocytic leukemia suggests multipotent stem cell involvement. Cancer Genet Cytogenetic 1989; 42:209–219.

200. Soekarman D, Von Lindern M, Daenen S, de Jong B, Fonatsch C, Heinze B, Bartram C, Hagemeijer A, Grosveld G. The translocation (6;9)(p23;q34) shows consistent rearrangement of two genes and defines a myeloproliferative disorder with specific features. Blood 1992; 79:2990–2997.
201. Von Lindern M, Fornerod M, Soekarman N, van Baal S, Jaegle M, Heigemijer A, Bootsma D, Grosveld G. Translocation t(6;9) in acute non-lymphocytic leukemia results in the formation of a DEK-CAN fusion gene. Baillieres Clin Naematol 1992; 5:857–879.
202. Bodger MP, Morris CM, Kennedy MA, Bowen JA, Hilton JM, Fitzgerald PH. Basophils of (Bsp-1+) derive from the leukemia clone in human myeloid leukemias involving the chromosome breakpoint 9q34. Blood 1989; 73:777–781.
203. Lion T, Hass OA. Acute megakaryoblastic leukemia with the t(1;22)(p13; q13). Leukemia Lymphoma 1993; 11:15–20.
204. Lu G, Altman AJ, Benn PA. Review of the cytogenic changes in acute megakaryoblastic leukemia: one disease or several. Cancer Genet Cytogenet 1993; 67:81–89.
205. Carroll A, Civin C, Scheider N, Dahl G, Pappo A, Bowman P, Emami A, Gross S, Alvarado C, Phillips C, Krischer J, Crist W, Head D, Gresik M, Ravindranath Y, Weinstein H. The t(1;22)(p13;q13) is nonrandom and restricted to infants with acute megakaryoblastic leukemia: a pediatric oncology group study. Blood 1991; 78:748–752.
206. Sait SN, Brecher ML, Green DM, Sandberg AA. Translocation t(1;22) in congenital acute megakaryoblastic leukemia. Cancer Genet Cytogenet 1988; 34:277–288.
207. Chan WC, Carroll A, Alvarado CS, Phillips S, Gonzalez-Crussi F, Kurczynski E, Pappo A, Emami A, Bowman P, Head DR. Acute megakaryoblastic leukemia in infants with t(1;22)(p13;q13) abnormality. Am J Clin Pathol 1992; 98:214–221.
208. Stark B, Resnitzky P, Jeison M, Luria D, Blau O, Avigad S, Shaft D, Kodman Y, Gobuzov R, Ash S, Stein J, Yaniv I, Barak Y, Zaizov R. A distinct subtype of M4/M5 acute myeloblastic leukemia (AML) associated with t(8;16)(p11; p13) in a patient with the variant t(8;19)(p11;q13)—case report and review of the literature. Leukemia Res 1995;19:367–379.
209. Borrow J, Stanton VP Jr, Andresen JM, Becher R, Behm FG, Chaganti RSK, Civin CI, Disteche C, Dube I, Frischauf AM, Horsman D, Mitelman F, Volinia S, Watmore AE, Housman DE. The translocation t(8;16)(p11;p13) of acute myeloid leukemia fuses a putative acetyltransferase to the CREB-binding protein. Nat Genet 1996; 14:33–41.
210. Macdonald D, Aguiar RCT, Mason PJ, Goldman JM, Cross NCP. A new myeloproliferative disorder associated with chromosomal translocations involving 8p11: a review. Leukemia 1995; 9:1628–1630.
211. Aguiar RC, Chase A, Coulthard S, Macdonald DHC, Carapeti M, Reiter A, Sohal J, Lennard A, Goldman JM, Cross NCP. Abnormalities of chromosome band 8p11 in leukemia: two clinical syndromes can be distinguished on the basis of MOZ involvement. Blood 1997; 90:3130–3135.

212. Geraedts JPM, den Ottolander GJ, Ploem JE, Muntinghe OG. An identical translocation between chromosome 1 and 7 in three patients with myelofibrosis and myeloid metaplasia. Br J Haematol 1980; 44:569–575.
213. Scheres JMJC, Hustinx TWJ, Geraedts JPM, Leeksma HW, Meltzer PS. Translocation 1;7 in hematologic disorders: a brief review of 22 cases. Cancer Genet Cytogenet 1985; 18:207–213.
214. Yunis JJ, Rydell RE, Oken MM, Arnesen MA, Mayer MG, Lobell M. Refined chromosome analysis as an indicator in de novo myelodysplastic syndromes. Blood 1986; 67:1721–1730.
215. Woloschak GE, Dewald GW, Bahn RS, Kyle RA, Gripp PR, Ash RC. Amplification of RNA and DNA specific for erb B in unbalanced 1;7 chromosomal translocation associated with myelodysplastic syndrome. J Cell Biochem 1986; 32:23–34.
216. Abrahamson GM, Rack K, Buckle VJ, Oscier DG, Kelly S, Wainscoat JS. Translocation 1;7 in hematologic disorders—a report of three further cases: absence of amplification of the gene for the epidermal growth factor receptor. Cancer Genet Cytogenet 1991; 53:91–95.
217. Nederlof PM, van der Flier S, Raap AK, Tanke HJ, van der Ploeg M, Kornips F, Geraedts JPM. Detection of chromosome aberrations in interphase tumor nuclei by nonradioactive in situ hybridization. Cancer Genet Cytogenetic 1989; 42:87–98.
218. Alitalo T, Willard HF, de la Chapelle A. Determination of the breakpoints of 1;7 translocation in myelodysplastic syndrome by in situ hybridization using chromosome-specific alpha satellite DNA from human chromosome 1 and 7. Cancer Genet Cytogenet 1989; 50:49–53.
219. Speleman F, Mangelshots K, Vercruyssen M, Dal Lin P, Aventin A, Offner F, Laureys G, Van den Berghe H, Leroy J. Analysis of whole-arm translocations in malignant blood cells by non-isotopic in situ hybridization. Cytogenet Cell Genet 1991; 56:14–17.
220. Hoo J, Szego K, Jones B. Confirmation of centromeric fusion in 7p/1q translocation associated with myelodysplastic syndrome. Cancer Genet Cytogenet 1992; 64:186–188.
221. Horiike S, Taniwaki M, Misawa S, Nishigaki H, Okuda T, Yokota S, Kashima K, Inazawa J, Abe T. The unbalanced 1;7 translocation in de novo myelodysplastic syndrome and its clinical implication. Cancer 1990; 65:1350–1354.
222. Berger R, Bernheim A, Le Coniat M, Vecchione D, Pacot A, Daniel M-T, Flandrin G. Abnormalities of the short arm of chromosome 12 in acute nonlymphocytic leukemia and dysmyelopoietic syndrome. Cancer Genet Cytogenet 1986; 19:281–289.
223. Weh HJ, Hossfeld DK. 12p− chromosome in patients with acute myelocytic leukemia or myelodysplastic syndromes following exposure to mutagenic agents (letter). Cancer Genet Cytogenet 1986; 19:355–356.
224. Sato Y, Suto Y, Pietenpol J, Golub TR, Gilliland DG, Davis EM, Le Beau MM, Roberts JM, Vogelstein B, Rowley JD, Bohlander SK. TEL and KIP1 define the smallest region of deletions on 12p13 in hematopoietic malignancies. Blood 1995; 86:1525–1533.

225. Andreasson P, Johansson B, Arheden K, Billstrom R, Mitelman F, Hoglund M. Deletions of CDKN1B and ETV6 in acute myeloid leukemia and myelodysplastic syndromes without cytogenetic evidence of 12p abnormalities. Genes Chromosomes Cancer 1997; 19:77–83.
226. Golub TR, Barker GF, Lovett M, Gilliland DG. Fusion of PDGF receptor b to a novel ets-like gene, tel, in chronic myelomonocytic leukemia with t(5; 12) chromosomal translocation. Cell 1994; 77:307–316.
227. Peeters P, Wlodarska I, Baens M, Criel A, Selleslag D, Hagemeijer A, Van den Berghe H, Marynen P. Fusion of ETV6 to MDS1/EV11 as a result of t(3:12)(q26;p13) in myeloproliferative disorders. Cancer Res 1997; 57:564–569.
228. Buijs A, Sherr S, van Baal S, van Bezouw S, van der Plas D, van Kessel AG, Riegman P, Deprez RL, Zwarthoff E, Hagemeijer A, Grosveld G. Translocation (12;22)(p13;q11) in myeloproliferative disorders results in fusion of the ETS-like TEL gene on 12p13 to the MN1 gene on 22q11. Oncogene 1995; 10:1511–1519.
229. Hoglund M, Johansson B, Pedersen-Bjergaard J, Marynen P, Mitelman F. Molecular characterization of 12p abnormalities in hematologic malignancies: deletion of KIP1, rearrangement of TEL, and amplification of CCND2, Blood 1996; 87:324–330.
230. Berger R, Le Coniat M, Lacronique V, Daniel MT, Lessard M, Berthou C, Marynen P, Bernard O. Chromosome abnormalities of the short arm of chromosome 12 in hematopoietic malignancies: a report including three novel translocations involving the TEL/ETV6 gene. Leukemia 1997; 11:1400–1403.
231. Tosi S, Guidici G, Mosna F, Harbott J, Specchia G, Grosveld G, Privitera E, Kearney L, Biondi A, Cazzaniga G. Identification of new partner chromosomes involved in fusions with the ETV6 (TEL) gene in hematologic malignancies. Genes Chromosom Cancer 1998; 21:223–229.
232. Peeters P, Raynaud SD, Cools J, Wlodarska I, Grosgeorge J, Philip P, Monpoux F, Van Rompaey L, Baens M, Van den Berghe H, Marynen P. Fusion of TEL, the ETS-variant gene 6 (ETV6), to the receptor-associated kinase. JAK2 as a result of t(9;12) in a lymphoid and t(9;15;12) in a myeloid leukemia. Blood 1997; 90:2535–2540.
233. Engel E, McGee BJ, Myers BJ, Flexner JM, Krantz SB. Chromosome banding patterns of 49 cases of chronic myelocytic leukemia (letter). N Engl J Med 1977; 296:1295.
234. Wilmoth D, Feder M, Finan J, Nowell P. Preleukemia and leukemia with 12p and 19q+ chromosome alterations following Alkeran therapy. Cancer Genet Cytogenet 1985; 15:95–98.
235. Atkin NB, Baker MC. i(12p): specific chromosomal marker in seminoma and malignant teratoma of the testis? Cancer Genet Cytogenet 1983; 10:199–204.
236. Bosl GJ, Dmitrovsky E, Reuter VE, Samaniego F, Rodriquez E, Geller NL, Chaganti RS. Isochromosome of chromosome 12: clinically useful marker for male germ cell tumors. J Natl Cancer Inst 1989; 81:1874–1878.

237. Chaganti RSK, Ladanyi M, Samaniego F, Offit K, Reuter VE, Jhanwar SC, Bosl GJ. Leukemia differentiation of a mediastinal germ cell tumor. Genes Chromosomes Cancer 1989; 1:83–87.
238. Ladanyi M, Samaniego F, Reuter VE, Motzer RJ, Jhanwar SC, Bosl GJ, Chaganti RS. Cytogenetic and immunohistochemical evidence for the germ cell origin of a subset of acute leukemias associated with mediastinal germ cell tumors. J Natl Cancer Inst 1990; 32:221–227.
239. Nichols CR, Roth BJ, Heerema N, Griep J, Tricot G. Hematopoietic neoplasia associated with primary mediastinal germ cell tumors. N Engl J Med 1990; 322:1425–1429.
240. Mukherjee AB, Murty VV, Rodriguez E, Reuter VE, Bosl GJ, Chaganti RS. Detection and analysis of origin of i(12p), a diagnostic marker of human male germ cell tumors by fluorescence in situ hybridization. Genes Chromosom Cancer 1991; 3:300–307.
241. Orazi A, Neiman RS, Ulbright TM, Heerema NA, John K, Nichols CR. Hematopoietic precursor cells within the yolk sac tumor component are the source of secondary hematopoietic malignancies in patients with mediastinal germ cell tumors. Cancer 1993; 71:3873–3881.
242. Mitelman F, Heim S, Mandahl N. Trisomy 21 in neoplastic cells. Am J Med Genet Suppl 1990; 7:262–266.
243. Watson MS, Carroll AJ, Shuster JJ, Steuber CP, Borowitz MJ, Behm FG, Pullen DJ, Land VJ. Trisomy 21 in childhood acute lymphoblastic leukemia: a Pediatric Oncology Group study (8602). Blood 1993; 82:3098–3102.
244. Fong CT, Brodeur GM. Down's syndrome and leukemia: epidemiology, genetics, cytogenetics and mechanism of leukemogenesis. Cancer Genet Cytogenet 1987; 28:55–76.
245. Dewald GW, Diez-Martin JL, Steffen SL, Jenkins RB, Stupca PJ, Burgert EO Jr. Hematologic disorders in 13 patients with acquired trisomy 21 and 13 individual with Down syndrome. Am J Med Genet Suppl 1990; 7:247–250.
246. Litz CE, Davies S, Brunning RD, Kueck B, Parkin JL, Gajl Peczalska KJ, Arthur DC. Acute leukemia and the transient myeloproliferative disorder associated with Down syndrome: morphology, immunophenotype and cytogenetic manifestations. Leukemia 1995; 9:1432–1439.
247. Zipursky A, Thorner P, De Harven E, Christensen H, Doyle J. Myelodysplasia and acute megakaryoblastic leukemia in Down's syndrome. Leukemia Res 1994; 18:163–171.
248. Creutzig U, Ritter J, Vornoor J, Ludwig WD, Niemeyer C, Reinisch I, Stollmann-Gibbels B, Zimmermann M. Myelodysplasia and acute myelogenous leukemia in Down's syndrome. A report of 40 children of the AML-BFM study group. Leukemia 1996; 10:677–686.
249. Shen JJ, Williams BJ, Zipursky A, Doyle J, Sherman SL, Jacobs PA, Shugar AL, Soukup SW, Hassold TJ. Cytogenetic and molecular study of Down syndrome individuals with leukemia. Am J Hum Genet 1995; 56:915–925.
250. Suda T, Suda J, Miura Y, Hayashi Y, Eguchi M, Tadokoro K, Saito M. Clonal analysis of basophil differentiation in bone marrow cultures from a Down's syndrome patient with megaloblastic leukemia. Blood 1985;66:1278–1283.

251. Papas TS, Watson DK, Fujiwara S, Seth AK, Fisher RJ, Bhat NK, Mavrothalassitis G, Koizumi S, Sacchi N, Jorcyk CL, Schweinfest CW, Kottaridis D, Ascoine R. ETS family of genes in leukemia and Down syndrome. Am J Med Genet Suppl 1990; 7:251–261.
252. Cortes JE, Kantarjian H, O'Brien S, Keatting M, Pierce S, Frereich EJ, Estey E. Clinical and prognostic significance of trisomy 21 in adult patients with acute myelogenous leukemia and myelodysplastic syndromes. Leukemia 1995; 9:115–117.
253. United Kingdom Cytogenetic Group (UKCG). Loss of the Y chromosome from normal and neoplastic bone marrows. Genes Chrom Cancer 1992; 5: 83–88.
254. McCarthy CM, Cobcroft RG, Harris MG, Scott DC. Sweet's syndrome, acute leukemia and t(3;5). Cancer Genet Cytogenet 1987; 28:87–91.
255. Yoneda-Kato N, Look AT, Kirstein MN, Valentine MB, Raimondi SC, Cohen KJ, Carroll AJ, Morris SW. The t(3;5)(q25.1;q34) of myelodysplastic syndrome and acute myeloid leukemia produces a novel fusion gene, NPM-MLF1. Oncogene 1996; 12:265–275.
256. Cournoyer D, Noel P, Schmidt MA, Dewald GW. Trisomy 9 in hematologic disorders: possible associations with primary thrombocytosis. Cancer Genet Cytogenet 1987; 27:73–78.
257. Sreekantaiah C, Baer MR, Morgan S, Isaacs JD, Miller KB, Sandberg AA. Trisomy/tetrasomy 13 in seven cases of acute leukemia. Leukemia 1990; 4: 781–785.
258. Dohner H, Arthur DC, Ball ED, Sobol RE, Davey FR, Lawrence D, Gordon L, Patil SR, Surana RB, Testa JR, Vermer RS, Schiffer CA, Wurster-Hill PH, Bloomfield CD. Trisomy 13: a new recurring abnormality in acute leukemia. Blood 1990; 76:1614–1621.
259. Cuneo A, Michaux JL, Ferrant A, Van Hove L, Bosly A, Stul M, Dal Cin P, Van den Berghe E, Cassiman JJ, Negrini M, Piva N, Castoldi G, Van Den Bergh H. Correlation of cytogenetic and clinicobiologic features in adult acute myeloid leukemia expressing lymphoid markers. Blood 1992; 79:720–727.
260. Sole F, Tarabadella M, Granada I, Florensa L, Vallespi T, Ribera JM, Irriguible D, Milla F, Woessner S. Isochromosome 17q as a sole anomaly: a distinct myelodysplastic syndrome entity? Leukemia Res 1993; 17:717–720.
261. Becher R, Carbonell F, Bartram CR. Isochromosome 17q in Ph1-negative leukemia: a clinical, cytogenetic and molecular study. Blood 1990; 75:1679–1683.
262. Veldman T, Vignon C, Schrock E, Rowley JD, Reid T. Hidden chromosome abnormalities in hematological malignancies detected by multicolor spectral karyotyping. Nat Genet 1997; 15:406–410.
263. Heim S, Mitelman F. Cancer Cytogenetics. New York: Wiley 1995.
264. Boyd JE, Parmley RT, Landevin AM, Saldivar VA. Neuroblastoma presenting as acute monoblastic leukemia. J Ped Hematol Oncol 1996; 118:206–221.
265. Cuneo A, Ferrant A, Michaux JL, Bosly A, Chatelain B, Stul M, Dal Cin P, Dierlamm J, Cassiman J-J, Hossfels DK, Castoldi G, Van den Bergha H.

Cytogenetic and clinicobiological features of acute leukemia with stem cell phenotype: study of 9 cases. Cancer Genet Cytogenet 1996; 92:31–36.
266. Mertens F, Sallerfors B, Heim S, Johansson B, Kristoffersson U, Malm C, Mitelman F. Trisomy 13 as a primary chromosome aberration in acute leukemia. Cancer Genet Cytogenet 1991; 56:36–44.
267. Cuneo A, Ferrant A, Michaux JL, Demuynck H, Boogaerts M, Louwagie A, Doyen C, Stul M, Cassiman JJ, Dal Cin P, Castoldi G, Van den Berghe H. Philadelphia chromosome-positive acute myeloid leukemia: cytoimmunologic and cytogenetic features. Haematologica 1996; 81:423–427.
268. Bain BJ, Catofsky D, O'Brien M, Prentice HG, Lawlor Ee, Kumaran TD, McCann SR, Matutes E, Galton DA. Megakaryoblastic leukemia presenting as acute myelofibrosis—a study of four cases with the platelet-peroxidase reaction. Blood 1981; 58:206–213.
269. Zhao L, Chang KS, Estey EH, Hayes K, Deisseroth AB, Liang JC. Detection of residual leukemic cells in patients with acute promyelocytic leukemia by the fluorescence in situ hybridization method: potential for predicting relapse. Blood 1995; 85:495–499.
270. Gale RE, Wheadon H, Goldstone AH, Burnett AK, Linch DC. Frequency of clonal remission in acute myeloid leukemia. Lancet 1993; 341:138–142.
271. Paietta E. Proposals for the immunologic classification of acute leukemia. Leukemia 1995; 9:2147–2148.
272. Claxton DF, Albitar M. Acute leukemia: phenotyping and genotyping. Leukemia 1995; 9:2150–2152.
273. Macintyre E, Flandrin G. Biological classification of acute leukemia: federalization or centralization. Leukemia 1995; 9:2152–2154.
274. Pui CH, Campana D, Crist WM. Toward a useful classification of acute leukemia. Leukemia 1995; 9:2154–2157.
275. Bene MC, Castoldi G, Knapp W, Ludwig WD, Matutes E, Orfao A, van't Veer MB. Proposals for the immunologic classification of acute leukemia. Leukemia 1995; 9:1783–1786.
276. Head DR. Revised classification of acute myeloid leukemia. Leukemia 1996; 10:1826–1831.
277. Anonymous. Sixth International Workshop on Chromosomes in Leukemia. London, England, May 11–18, 1987. Selected papers. Cancer Genet Cytogenet 1989; 40:141–230.
278. Jacobs RH, Cornbleet M, Vardiman JW, Larson RA, Le Beau MM, Rowley JD. Prognostic implications of morphology and karyotype in primary myelodysplastic syndromes. Blood 1986; 67:1765–1772.
279. Anonymous: Recommendations for a morphologic, immunologic, and cytogenetic (MIC) working classification of the primary myelodysplastic syndromes and therapy-related myelodysplastic disorders. Cancer Genet Cytogenet 1988; 32:1–10.
280. Yunis JJ, Lobell M, Arneson MA, Oken MM, Mayer MG, Rydell RE, Brunning RD. Refined chromosome study helps define prognostic subgroups in most patients with primary myelodysplastic syndrome and acute myelogenous leukemia. Br J Haematol 1988; 68:189–194.

281. Morel P, Hebber M, Lai J-L, Duhamel A, Preudhomme C, Wattel E, Bauters F, Fenaux P. Cytogenetic analysis has strong independent prognostic value in de novo myelodysplastic syndromes and can be incorporated in a new scoring system: a report of 408 cases. Leukemia 1993; 7:1315–1323.
282. Toyama K, Ohyashiki K, Yoshida Y, Abe T, Asano S, Hirai H, Hirashima K, Hotta T, Kuramoto A, Kuriya S, Miyazaki T, Kakishita E, Mizoquchi H, Okada M, Shirakawa S, Takaku F, Tomonzga M, Uchino H, Yasunaga K, Nomura T. Clinical implications of chromosomal abnormalities in 401 patients with MDS: a multicentric study in Japan. Leukemia 1993; 7:499–508.
283. Greenberg P, Cox C, Le Beau MM, Fenaux P, Morel P, Sanz G, Sanz M, Vallespi T, Hamblin T, Oscier D, Ohyashiki K, Toyama K, Aul C, Mufti G, Gennett J. International scoring system for evaluating prognosis in myelodysplastic syndromes. Blood 1997; 89:2079–2088.
284. Brunning RD. Myelodysplastic syndromes. In: Knowles DM, ed. Neoplastic Hematopathology. Baltimore: Williams & Wilkins, 1992:1367–1404.
285. Applebaum FR, Barrall J, Storb R, Ramberg R, Doney K, Sale GE, Thomas ED: Clonal cytogenetic abnormalities in patients with otherwise typical aplastic anemia. Exp Hematol 1987; 15:1134–1139.
286. Tichelli A, Gratwohl A, Nissen C, Speck B. Late clonal complications in severe aplastic anemia. Leukemia Lymphoma 1994; 12:167–175.
287. Moormeier JA, Rubin CM, Le Beau MM, Vardiman JW, Larson RA, Winter JN. Trisomy 6: a recurrent cytogenetic abnormality associated with marrow hypoplasia. Blood 1991; 77:1397–1398.
288. Varma N, Varma S, Movafagh A, Garewal G. Unusual clonal cytogenetic abnormalities in aplastic anemia. Am J Hematol 1995; 49:256–257.
289. Rosati S, Anastasi J, Vardiman JW. Recurring diagnostic problems in the pathology of the myelodysplastic syndromes. Semin Hematol 1996; 33:111–126.
290. Maschek H, Georgii A, Kaloutsi V, Werner M, Bandecar K, Kressel MG, Choritz H, Freund M, Hufnagl D. Myelofibrosis in primary myelodysplastic syndromes: a retrospective study of 352 patients. Eur J Haematol 1992; 48: 208–214.
291. Ohyashiki K, Sasao I, Ohyashiki JH, Murakami T, Iwabuchi A, Tauchi T, Saito M, Nakazawa S, Serizawa H, Ebihara Y, Toyama K. Clinical and cytogenetic characteristics of myelodysplastic syndromes developing myelofibrosis. Cancer 1991; 68:178–183.
292. Mecucci C, Van Orshoven A, Vermaelen K, Michaux JL, Tricot G, Louwagie A, Delannoy A, Van den Berghe H. 11q− chromosome is associated with abnormal iron stores in myelodysplastic syndromes. Cancer Genet Cytogenet 1987; 27:39–44.
293. Vila L, Charrin C, Archimbaud E, Treille-Ritouet D, Fraisse J, Felman P, Fiere D, Germain D. Correlation between cytogenetics and morphology in myelodysplastic syndromes. Blut 1990; 60:223–227.
294. Dewald GW, Brecher M, Travis LB, Stupca PJ. Twenty-six patients with hematologic disorders and X chromosome abnormalities. Frequent

idic(X)(q13) chromosomes and Xq13 anomalies associated with pathologic ringed sideroblasts. Cancer Genet Cytogenet 1989; 42:173–185.
295. Mufti GJ. Chromosomal deletions in the myelodysplastic syndrome. Leukemia Res 1992; 16:35–41.
296. Dewald GW, Wright PI. Chromosome abnormalities in the myeloproliferative disorders. Semin Oncol 1995; 22:341–354.
297. Diez-Martin JL, Graham DL, Petit RM, Dewald GW. Chromosomal studies in 104 patients with polycythemia vera. Mayo Clin Proc 1991; 66:287–299.
298. Rege-Cambrin G, Mecucci C, Tricot G, Michaux JL, Louwagie A, Van Hove W, Francart H, Van den Berghe H. A chromosomal profile of polycythemia vera. Cancer Genet Cytogenet 1987; 25:233–245.
299. Swolin B, Weinfeld A, Westin J. A prospective long-term cytogenetic study in polycythemia vera in relation to treatment and clinical course. Blood 1988; 72:386–395.
300. Demroy JL, Dupriez B, Fenaux P, Lai JL, Beuscart R, Jovet JD, Deminatti M, Bauteis F. Cytogenetic studies and their prognostic significance in agnogenic myeloid metaplasia: a report on 47 cases. Blood 1988; 72:855–859.
301. Sandberg AA. The Chromosomes in Human Cancer and Leukemia. 2nd ed. New York: Elsevier, 1990.
302. Third International Workshop on Chromosomes in Leukemia, 1980. chromosomal abnormalities in acute lymphoblastic leukemia: structural and numerical changes in 234 cases. Cancer Genet Cytogent 1981; 4:101–110.
303. Mittre H, Leymarie P, Macro M, Leporrier M. A new case of chronic myeloid leukemia with c3/a2 bcr/abl junction. Is it really a distinct disease? (letter). Blood 1997; 89:4239–4240.

12
Cytogenetics of Lymphoid Neoplasias

Susana C. Raimondi
St. Jude Children's Research Hospital, Memphis, Tennessee

I. INTRODUCTION

During the past two decades, the close association between specific chromosomal abnormalities and various hematologic neoplasms has been established. The recent molecular cloning of most recurrent chromosome translocation breakpoints and the identification of the involved genes are helping to elucidate the mechanisms of leukemogenesis and aberrant regulation of leukemic cell growth. Understanding the role of the affected genes may provide insights into the altered biology of these cells and lead to better treatment (1–5). Perhaps more important, molecular diagnostic tests that are based on the sequence of chromosomal breakpoint regions are constantly being developed. In time, these new assays will augment the armamentarium that allows accurate identification of clonal rearrangements of genes and subclassification of diseases. Tables 1 and 2 present the most frequent karyotypic-morphologic and genetic changes associated with lymphoid neoplasias.

II. ACUTE LYMPHOBLASTIC LEUKEMIA

Unlike those in adults, the vast majority of leukemia cases in children are acute. Acute lymphoblastic leukemia (ALL) is the most common subtype in children, accounting for 75% of cases. Most of the chromosomal aberrations

Table 1 Recurrent Chromosome Changes in Acute Lymphoblastic Leukemia

Abnormality	Approximate incidence (%)		Specific phenotype	Chromosome bands/genes involved	
	In children	In adults			
t(1;19)(q23;p13.3)	5–6	3	Pre-B, 90% of cases	1q23/*PBX1*	19p13.3/*E2A*
t(9;22)(q34;q11.2)	3–5	30	B-lineage, 90% of cases	9q34/*ABL*	22q11.2/*bcr*
t(4;11)(q21;q23)	2	3–5	Early pre-B, 80% of cases	4q21/*AF4*	11q23/*MLL* (*ALL1*/*HRX*)
t(1;11)(p32;q23)	<1	<1	Early pre-B	1p32/*AF1P*	11q23/*MLL* (*ALL1*/*HRX*)
t(9;11)(p22;q23)	1	<1	Early pre-B	9p11/*AF9*	11q23/*MLL* (*ALL1*/*HRX*)
t(11;19)(q23;p13.3)	1	<1	Early pre-B, T-cell	11q23/*MLL* (*ALL1*/*HRX1*)	19p13.3/*ENL*
t(5;14)(q31;q32)	<1	<1	Eosinophilia	5q31/*IL3*	14q32/*IGH*
t(17;19)(q22;p13.3)	<1	Unknown	Early pre-B	17q22/*HLF*	19p13.3/*E2A*
t(12;21)(p13;q22)[a]	<1	<1	Early pre-B, Pre-B	12p13/*TEL* (*ETV6*)[a]	21q22/*AML1* (*CBFA2*)
t(9;12)(q34;p13)	<1	Unknown	Early pre-B, Pre-B	9q34/*ABL*	12p13/*TEL* (*ETV6*)
del(6q)	4–13	5	Nonspecific lineage	Unknown	
t/del(9p)	7–12	10–15	Nonspecific lineage	9p22/*p16* (*MTS1*) and *p15* (*MTS2*)	
t/del(11q)	3–5	6–7	Nonspecific lineage	11q23/*MLL* (*ALL1*/*HRX*)	
t/del(12p)	10–12	5	Nonspecific lineage	12p13/*TEL* (*ETV6*)[a]	
t(8;14)(q24.1;q32)	3	5	B-cell, 90% of cases	8q24.1/*MYC*	14q32/*IGH*

t(2;8)(p12;q24.1)	<1	<1	B-cell, 4–5% of cases	2p12/*IGK*	8q24.1/*MYC*
t(8;22)(q24.1;q11.2)	<1	<1	B-cell, 6–10% of cases	8q24.1/*MYC*	22q11.2/*IGL*
t(11;14)(p13;q11.2)	1	<1	T-cell, 7% of cases	11p13/*RBTN2* (*TTG2*)	14q11.2/*TCRAD*
t(11;14)(p15;q11.2)	<1	<1	T-cell, 1% of cases	11p15/*RBTN1* (*TTG1*)	14q11.2/*TCRAD*
t(10;14)(q24;q11.2)	1	3	T-cell, 5–14% of cases	10q24/*HOX11*	14q11.2/*TCRAD*
t(8;14)(q24.1;q11.2)	<1	<1	T-cell, 2% of cases	8q24.1/*MYC*	14q11.2/*TCRAD*
t(1;14)(p32-p34; q11.2)	<1	<1	T-cell, 3% of cases	1p32-p34/*TAL1* (*TCL5/SCL*)[b]	14q11.2/*TCRAD*
t(1;14)(p34;q11.2)	<1	<1	T-cell	1p34/*LCK*	14q11.2/*TCRAD*
inv(14)(q11.2q32) and t(14;14)(q11.2;q32)	<1	1	T-cell	14q11.2/*TCRAD*	14q32/*IGH*
t(1;3)(p32-p34;p21)	<1	<1	T-cell	1p32-p34/*TAL1* (*TCL5/SCL*)[b]	3p21/*TCTA*
t(1;7)(p32;q35)	<1	<1	T-cell	1p32-p34/*TAL1* (*TCL5/SCL*)[b]	7q35/*TCRB*
t(1;7)(p34;q35)	<1	<1	T-cell	1p34/*LCK*	7q35/*TCRB*
t(7;9)(q35;q34)	<1	<1	T-cell	7q35/*TCRB*	9q34/*TAN1*
t(7;9)(q35;q32)	<1	<1	T-cell	7q35/*TCRB*	9q32/*TAL2*
t(7;10)(q35;q24)	<1	<1	T-cell	7q35/*TCRB*	10q24/*HOX11*
t(7;11)(q35;q13)	<1	<1	T-cell	7q35/*TCRB*	11p13/*RBTN2* (*TTG2*)
inv(7)(p15q35)	<1	1	T-cell	7p15/*TCRG*	7q35/*TCRB*
t(7;19)(q35;p13)	<1	<1	T-cell	7q35/*TCRB*	19p13/*LYL1*

[a]*TEL* (*ETV6*) gene rearrangements by Southern blotting/RT-PCR positive for cryptic t(12;21) in children (16–25%) and adults (1%) of B-lineage immunophenotype.

[b]*TAL1* submicroscopic deletion in 9–26% of children and in 3–8% of adult patients with T-cell ALL.

Table 2 Recurrent Chromosomal Changes in Non-Hodgkin Lymphoma (NHL)[a]

Chromosomal abnormality	Incidence in children (%)	Incidence in Adults (%)	Main disease associations	Chromosomal bands/genes involved	
t(2;5)(p23;q35)	6–15	1–8	ALCL (Ki-1-positive)	2p23/*ALK*	5q35/*NPM*
t(1;2)(q25;p23)	Unknown	Unknown	ALCL	1q25/unknown	2p23/*ALK*
t(14;18)(q32;q21.3)	<1	35–40	FCL	14q32/*IGH*	18q21.3/*BCL2*
t(18;22)(q21.3;q11.2)	<1	1–3	FCL, 70–90% of cases; DLCL, 20% of cases	18q21.3/*BCL2*	22q11.2/*IGL*
t(11;14)(q13;q32)	<1	4–8	MCL, 50% of cases	11q13/*BCL1*	14q32/*IGH*
t(3;14)(q27;q32)	<1	7–12[b]	DLCL, 30–40% of cases[b]; FCL, 6–11% of cases[b]	3q27/*BCL6*	14q32/*IGH*
t(3;22)(q27;q11.2)	<1	<1	DLCL	3q27/*BCL6*	22q11.2/*IGL*
t(2;3)(p12;q27)	Unknown	<1	DLCL	2p12/*IGK*	3q27/*BCL6*
t(3;4)(q27;p11)	Unknown	<1	DLCL	3q27/*BCL6*	4p11/*TTF*
t(3;6)(q27;p21)	Unknown	<1	DLCL	3q27/*BCL6*	6p21/*H4* histone
t(3;11)(q27;q23.1)	Unknown	<1	DLCL	3q27/*BCL6*	11q23.1/*BOB1*/*OBF1*
t(2;18)(p12;q21.3)	Unknown	<1	FCL	2p12/*IGK*	18q21.3/*FVT1*
t(8;12)(q24.1;q22)	Unknown	<1	CLL	8q24.1/*MYC*	12q22/*BTG1*
t(8;13)(p11;q11)	Unknown	<1	NHL	8p11/unknown	13q11/unknown
t(8;17)(q24.1;q22)	Unknown	1.5	FCL	8q24.1/*MYC*	17q22/*BCL3*
t(9;14)(p13;q32)	Unknown	2	Plasmacytoid-NHL	9p13/*PAX5*	14q32/*IGH*
t(10;14)(q24;q32)	Unknown	<1	B-cell NHL	10q24/*LYT10*	14q32/*IGH*
t(11;14)(q23;q32)	Unknown	<1	NHL	11q23/*RCK*	14q32/*IGH*
t(11;14)(q23;q32)	Unknown	<1	NHL	11q23/*MLL* (*ALL1*/*HRX*)	14q32/*IGH*
t(11;14)(q23;q32)	Unknown	<1	LCL	11q23/*LPC*	14q32/*IGH*
t(11;18)(q21;q21.1)	Unknown	<1	MALT	11q21/unknown	18q21.1/unknown
t(14;19)(q32;q13.1)	Unknown	<1	CLL	14q32/*IGH*	19q13.1/*BCL3β*

ALCL, anaplastic large cell lymphoma; FCL, follicular cell lymphoma; DLCL, diffuse large cell lymphoma; MCL, mantle cell lymphoma; CLL, chronic lymphocytic leukemia; LCL, large cell lymphoma; MALT, mucosa-associated lymphoid tissue.

[a]Some of these rearrangements are variable and/or nonrestricted.

[b]The frequency and disease subtype include all cases with a 3q27/*BCL6* lesion.

seen in children with ALL have also been observed in adult patients. However, the incidence of several specific chromosomal aberrations varies between children and adults, and these differences, when known, will be noted.

Chromosomal analysis of leukemic blast cells is essential for the classification and treatment stratification of childhood ALL. Most of the recurrent chromosomal abnormalities affect outcome in children; this effect is less marked in adults. The characteristics of pediatric versus adult ALL have been attributed to differences in disease biology and treatment tolerances (6). The most frequent numeric and structural chromosomal findings and their clinical associations in patients with ALL will be described.

A. Numeric Chromosomal Changes

Based on the modal number of chromosomes, ALL is classified into five subtypes: hyperdiploid with more than 50 chromosomes (this category encompasses near-tetraploid, near-triploid, and hyperdiploid 51+ cases); hyperdiploid with 47–50 chromosomes; pseudodiploid (46 chromosomes, with structural or numeric abnormalities); diploid (46 normal chromosomes); and hypodiploid (fewer than 46 chromosomes) (Fig. 1). Ploidy is recognized as a distinct cytogenetic feature in ALL that is predictive of clinical outcome,

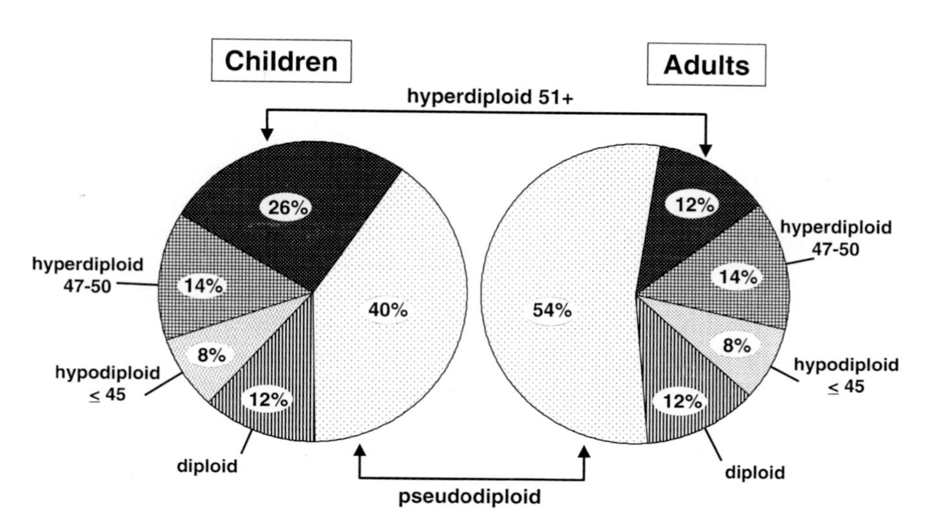

Figure 1 Incidence of ploidy subtypes of acute lymphoblastic leukemia. Note that only the frequencies of hyperdiploid 51+ and pseudodiploid cases vary between children and adults.

and this feature has been incorporated into stratification schemes for risk-specific therapy (7–15).

1. Near-Tetraploidy (Range in Modal Number of Chromosomes, 82–94)

Fewer than 1% of reported cases of ALL in children are near-tetraploid (16–18). These cases probably originate from endoreduplication, because in only a few instances were two copies of the abnormal chromosomes found in the near-tetraploid line. Near-tetraploidy is more likely to be associated with L2 morphology (30% of cases), T-cell immunophenotype (47%), and older age (median, 8.6 years) than are other ploidy groups of childhood ALL (19). In an adult series, near-tetraploidy was identified in 2% of cases, a high percentage (67%) of which expressed T-cell lineage markers, and one case was Philadelphia chromosome (Ph)-positive (20).

2. Near-Triploidy (Range in Modal Numbers, 69–81)

Near-triploidy in children with ALL is extremely rare (0.3%), whereas 3% of adult cases showed this pattern (20). In children, the presenting clinical features of this subgroup did not appear to differ from those of the general ALL population, except for a slight increase in pre-B immunophenotype (19).

3. Hyperdiploidy 51+ (Range in Modal Numbers, 51–68)

Childhood ALL. It is well known that patients with hyperdiploid 51+ ALL, a cohort comprising 25–30% of children with lymphoid leukemia, have more favorable presenting features and higher cure rates than do other major prognostic groups (9,10,12,21–26). Favorable presenting features commonly associated with hyperdiploid 51+ ALL include an early pre-B immunophenotype, lack of T cells, lower leukocyte counts, and age between 2 and 10 years. One presenting feature that is associated with poor outcome in this patient subgroup is the presence of an isochromosome 17 (27). Patients with hyperdiploidy can rapidly be identified by flow cytometric analysis of the leukemic cell DNA content. However, this test does not yield information about structural or specific numeric chromosomal abnormalities and therefore must be used in conjunction with conventional cytogenetic techniques (12,26,28). Investigators of the Pediatric Oncology Group have reported that the subset of cases among the hyperdiploid 53+ group whose blasts have trisomy of both chromosomes 4 and 10 has a superior prognosis (29,30).

Recently we studied the fully banded chromosomes of 182 children with hyperdiploid (51–67) ALL to better delineate the heterogeneity of this disease subtype (31). Whereas 46% of the cases had numeric changes exclusively, the remainder had both numeric and structural abnormalities. Addition of chromosome 21 was most frequent (97% of cases), followed by addition of chromosome 6 (86%), X (81%), 14 (80%), 4 (76%), 17 (68%), 18 (68%), 10 (56%), 8 (34%), or 5 (26%). The most common structural alterations were duplication of the 1q arm and isochromosome 17q, present in 25 (14%) and 9 (5%) cases, respectively. Chromosomal translocations, including t(1;19)(q23;p13.3) and t(9;22)(q34;q11.2), were detected in only 20% of the hyperdiploid cases, as compared with 50% of ALL cases in general. In this study, the presence or absence of structural abnormalities did not influence event-free survival (EFS), as assessed in the 168 patients who were enrolled in three successive protocols for children with newly diagnosed ALL. By contrast, patients with 51–55 chromosomes per leukemic cell (n = 105) appeared to fare worse than did those in the 56–67 subgroup (n = 63) (5-year EFS, 72% ± 5% [SE] versus 86% ± 5%; P = 0.04 by the stratified log-rank test) (31). Therefore, ALL defined by 51–55 chromosomes appears to be a clinicobiologic entity that is quite distinct from cases with higher modal numbers. Although our recent findings have not yet been confirmed by larger series, Whitehead et al. reported higher accumulations in vitro of methotrexate and its polyglutamates in cases with 56–65 chromosomes as compared to those in cases with 51–55 chromosomes (32). These results suggest that cases with 56–65 chromosomes are more susceptible to methotrexate-induced cytotoxicity.

In some instances, conventional cytogenetic techniques are not well suited to evaluating the chromosomes of hyperdiploid metaphases, because of the limited number of spread metaphase plates and poor chromosome morphology, especially in cases with a high modal chromosome number. To overcome these technical difficulties, the United Kingdom Cancer Cytogenetics Group recently proposed a strategy to detect hyperdiploidy in ALL by fluorescence in situ hybridization (FISH); these investigators used chromosome-specific centromeric probes to detect hyperdiploidy in interphase cells and to assign cases to a ploidy subgroup (33). Investigators of the Pediatric Oncology Group performed FISH techniques for rapid and accurate detection of trisomies of chromosomes 4 and 10 in hyperdiploid (>50) cases, in an attempt to identify patients who could be treated effectively with relatively nontoxic antimetabolite therapy (30). Furthermore, identification of numeric chromosomal abnormalities by interphase-FISH has aided detection of minimal residual disease and description of the size of the population of abnormal cells for cases in which the cytogenetic nature of the leukemic clone is known (30,34–40).

Adult ALL. Earlier studies of adult patients with ALL indicated that the incidence of hyperdiploidy 51+ was markedly lower (approximately

12%) than in children (25–30%) (8,41). This finding was confirmed in a recent study of 443 adult patients with ALL, in which the modal number was 51+ for 7% of the cases (20). In this study, as in childhood ALL, the median modal number was 55, with extra chromosomes and presenting clinical features comparable to those associated with low-risk ALL in children. In contrast to children with this subtype of ALL, a high proportion (37%) of adults with hyperdiploid ALL in this study were Ph chromosome-positive (20).

4. Hyperdiploidy 47-50 (Range in Modal Numbers, 47–50)

Childhood ALL. This group accounts for 10–15% of cases of childhood ALL and was initially recognized because of studies indicating that the presence of 47–50 chromosomes conferred an intermediate prognosis (8–11,13). Gains of almost every chromosome have been observed in such cases, with the most frequent addition of chromosomes 8, 10, and 21. Our analysis of 86 cases with 47–50 chromosomes revealed +21 as the most common addition ($n = 34$), followed by +X ($n = 18$), +8 ($n = 8$), and +10 ($n = 7$) (42). Trisomy 21 as the sole numeric abnormality of ALL in non-Down syndrome patients has been associated with a good prognosis (43,44). Trisomy 8 as the sole abnormality is infrequent in childhood ALL (0.3–1.2% of cases) and appears to be associated with a T-cell immunophenotype; its prognostic significance remains to be determined (45,46). In our series, the chromosomal regions most often affected by structural abnormalities (76% of the 86 cases) were 1q ($n = 13$), 6q ($n = 12$), 12p ($n = 18$), and 19p ($n = 9$). Patients with hyperdiploid 47–50 ALL in our Total Therapy Study XI had a favorable outcome (4-year EFS, 77% ± 11% SE) (14).

Adult ALL. In early reports of adults with ALL, the frequency of hyperdiploid 47–50 cases was 11–13%. However, in a recent study of adult patients, the incidence of hyperdiploid ALL (15%) was similar to that in children (14.4%) (8,41,42). Most cases had a modal number of 47 (71%), with chromosomes of 8, 21, 5, and 10 being most often involved. In this subgroup of adult patients, 40% had T-cell ALL, four of whom had +8 as the sole chromosomal abnormality (20).

5. Pseudodiploidy

Patients whose leukemic cell chromosomes contain numeric and/or structural rearrangements in the context of a diploid chromosome number (i.e., 46 chromosomes) comprise the largest and most heterogeneous ploidy group of ALL.

Childhood ALL. In children with ALL, the approximate incidence of pseudodiploidy is 40% (8,13). Pseudodiploid leukemic blast cells tend to be associated with bulky disease, high white blood cell counts, and elevated levels of serum lactic dehydrogenase (10,11,14). The outcome of therapy in this group of children was extremely poor until the development of modern multiagent chemotherapy. Currently, children with pseudodiploid ALL who are treated on appropriately intensified regimens can be expected to fare as well as those in most other risk groups (14,47–49). The exceptions are children with t(9;22) or t(4;11), which continue to mark highly drug-resistant disease; these cases will be discussed later.

Adult ALL. Pseudodiploidy is the largest ploidy group of adults with ALL, accounting for approximately 55–60% of patients. In one study, the Ph chromosome was identified in 40% of adults with pseudodiploid ALL (20).

6. Diploidy

The incidence of cases that lack apparent cytogenetic abnormalities in ALL varies. Whether so-called cytogenetically normal cases represent metaphase cells of suboptimal quality for analysis of subtle chromosomal rearrangements, mitotically inactive clones, or clones with submicroscopic genetic changes is uncertain. In many instances, current molecular cytogenetic techniques have identified chromosomal aberrations undetected by conventional cytogenetics, including a cryptic t(12;21)(p13;q22) (see 12p section) (30,50–53).

Childhood ALL. It is not unusual for 10–15% of cases in large childhood series to lack chromosomal abnormalities as determined by standard cytogenetic evaluation. The incidence of normal karyotypes in T-cell cases may be as high as 30% (54,55). The prognosis associated with diploid childhood ALL has been intermediate in most clinical trials (10,14).

Adult ALL. In one series by the French Hematologic Cytogenetics Group, about 15% of adult cases had a normal karyotype. These patients had a higher incidence of T-cell ALL (35%) compared to that for the study population overall (21%) (20).

7. Hypodiploidy

Based on the modal number of chromosomes, three subdivisions of the hypodiploid category of ALL are recognized: 41–45, 30–40, and <30 (near-haploid) chromosomes (56,57). Because of the limited number of cases reported for each of these categories, further study is needed to determine precisely the unique prognostic implications of these chromosomal loss syndromes in ALL.

Childhood ALL. Hypodiploidy (<46 chromosomes) is typically found in 7–8% of cases of childhood ALL; investigators at the Third International Workshop on Chromosomes in Leukemia studied 330 cases of ALL, of which 20 (6%) were hypodiploid (two near-haploid) (8). Most hypodiploid cases (80%) have a modal number of 45 and arise from the loss of a whole chromosome, unbalanced translocations, or the formation of dicentric chromosomes.

Among the 409 patients with newly diagnosed ALL who were studied at St. Jude Children's Research Hospital, 31 (7.6%) had a hypodiploid karyotype (58). The modal number was 45 in 26 cases, 28 in 2 cases, and 26, 36, and 43 in 1 case each. Hypodiploidy resulted from the loss of whole chromosomes in 20 patients, from unbalanced translocations in 7, and from the formation of dicentric chromosomes in 4. Chromosome 20 was lost most frequently (9 cases), and in 3 cases, its absence was the only abnormality; a similar finding was reported by Betts et al. (59). Recent reports have identified a dic(9;20)(p11;q11) in children and adults with monosomy 20 (60–62).

Despite relatively favorable presenting features (median leukocyte count, 12.7×10^9 L; median age, 5 years), patients with hypodiploidy did not fare well (4-year EFS, 46% ± 19% SE) on a protocol that was effective for other ploidy groups (14). This finding suggests that alternative therapy should be considered for children with this subtype of ALL. One exception to this recommendation is the hypodiploid subgroup with dic(9;12)(p11; p13), which is associated with an excellent prognosis (63,64).

Although very rare (<1%), near-haploid cases of childhood ALL have been studied extensively (57,58,65–67). The main clone contains at least one copy of each chromosome as well as a second copy of the sex chromosomes and chromosome 21 in most cases (90%). Other chromosomes found in two copies include 18 (65% of cases), 10 (45%), and 14 (45%) (57,58,67). In general, the morphologic features of cases with a modal number <30 are poorly defined, and few or no structural abnormalities can be discerned. Many near-haploid cases contain a second abnormal line, which typically has a hyperdiploid karyotype that represents a doubling of the chromosomes in the near-haploid line, suggesting a common precursor cell with a near-haploid karyotype (58,66). Near-haploid leukemia in children has been associated with poor prognosis (median survival, 10 months), even though complete remission was achieved in most cases (58,67). In rare instances, a presumed hyperdiploid karyotype is actually a case of near-haploidy that has gone undetected by conventional cytogenetics; DNA index analysis in such cases can aid proper risk assignment of these patients (68).

Adult ALL. As in children, hypodiploidy was found in about 7–8% of adult cases (20,41). The most common modal number was 45, and near-haploid cases were infrequently identified in these studies.

B. Structural Chromosomal Changes

Describing ALL by the types of structural abnormalities in the chromosomes of the leukemic clone has led to advances in our understanding of leukemia pathobiology and in formulating risk-specific therapy. Molecular analysis of genes adjacent to the breakpoints of these structural anomalies and studies on the functions of their protein products have helped to clarify the complex interactions that promote leukemogenesis and perpetuate the leukemic cell phenotype (Table 1). The more frequent structural abnormalities found in leukemic cells and their likely contribution to the disease process are the focus of this section.

1. t(1;19)(q23;p13.3)

Childhood ALL. With an overall incidence of 5–6% (69–74), the t(1;19) is found in 20–25% of pre-B (cIg-positive) cases and in <1% of cases with an early pre-B (cIg-negative) or T-cell immunophenotype (71,74–78). The t(1;19)(q23;p13.3) is the most common translocation in childhood ALL that is detected by conventional cytogenetics. This rearrangement occurs as a balanced translocation in 25% of the cases positive for t(1;19)(q23;p13.3) and as an unbalanced der(19)t(1;19)(q23;p13.3) in the remaining 75% (Fig. 2). Some leukemic cells contain both forms, suggesting that der(1) can be lost through clonal evolution without loss of the transformed phenotype. A recent retrospective collaborative group study of patients of all ages reported a significantly better outcome for those who had the unbalanced der(19)t(1;19) than for patients with the balanced t(1;19) form of ALL (79). However, we were unable to confirm these findings (74).

Although most t(1;19)-positive ALL cases in children are pseudodiploid and express cIg, 5–10% are cIg-negative and associated with hyperdiploidy (74,80). The presence of t(1;19) in pre-B cases correlates with recognized adverse prognostic features including higher leukocyte counts and DNA indexes ≥1.16 (76). It was previously thought that the t(1;19) accounted for most treatment failures in pre-B ALL (72,81). In an early study, we evaluated 112 cases of pre-B leukemia with banded karyotypes in which t(1;19) was associated, as expected, with an inferior outcome in patients treated on minimally or moderately intensified protocols (76). However, in a more recent study testing reinforced early treatment and rotational com-

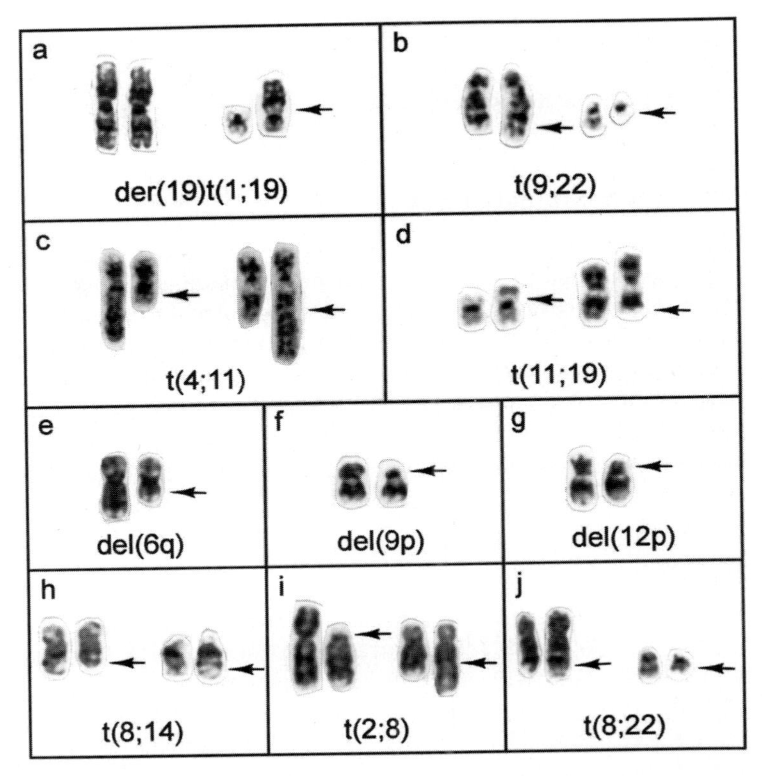

Figure 2 These partial karyotypes from G-banded metaphase plates show various structural chromosomal abnormalities in acute lymphoblastic leukemias: (a) der(19)t(1;19)(q23;p13.3); (b) t(9;22)(q34;q11.2); (c) t(4;11)(q21;q23); (d) t(11; 19)(q23;p13.3); (e) del(6q); (f) del(9p); (g) del(12p); (h) t(8;14)(q24.1;q32); (i) t(2; 8)(p12;q24.1); (j) t(8;22)(q24.1;q11.2).

bination chemotherapy, the 4-year EFS (79% ± 21% SE) for the subgroup of patients with t(1;19) rearrangements was comparable to that of the entire cohort (73% ± 4% SE) (14). This improvement illustrates that effective treatment may nullify the negative impact of chromosomal rearrangements in some cases of childhood pre-B ALL (76).

Adult ALL. In one study, approximately 3% of adults with ALL were positive for the t(1;19) (20). Most of the patients with t(1;19) in this series were younger (median age, 22 years), and their disease did not respond to therapy.

Molecular Consequences. The abnormalities t(1;19) and der(19)t(1;19) lead to fusion of the *E2A* and *PBX1* genes. The *E2A* gene (mapped to 19p13.3) encodes the Ig enhancer-binding factors E12 and E47 (82), and *PBX1*, a homeobox gene of unknown function, maps to 1q23 (83,84). In vitro, the fusion gene *E2A-PBX1* seems to be a potent oncogene (85). Molecular detection of t(1;19) is now feasible by reverse transcriptase (RNA-based) polymerase chain reaction (RT-PCR) analysis (80,86). Additional molecular evaluation of t(1;19) cases has identified heterogeneous breakpoints due to alternative splicing and variant fusion transcripts (87–89). RT-PCR technology has made it possible to distinguish cIg-positive cases with typical *E2A-PBX1* transcripts from cIg-negative cases in which the t(1;19) does not yield these characteristic transcripts (80). However, a t(1;22)(q21 or q23;p13) that resembles t(1;19) morphologically has recently been identified (90,91). Therefore, FISH analysis of cases with equivocal chromosome 19 markers is recommended to resolve diagnostic uncertainty, because such distinctions may be important in devising risk-specific therapy for patients with B-lineage ALL.

2. t(9;22)(q34;q11.2)

Childhood ALL. The Philadelphia chromosome, t(9;22), occurs in 3–5% of children with ALL (Fig. 2) (8,92–95). All reported studies confirm that the Ph chromosome is associated with an extremely poor prognosis. Clinically, children with Ph-positive blasts are older and have higher leukocyte counts, a larger percentage of circulating blasts, a higher frequency of FAB-L2 morphology, and an increased incidence of pseudodiploid karyotypes than those with Ph-negative disease. Most Ph-positive blast cells have a B-lineage immunophenotype, although isolated cases with a T-cell or mixed phenotype have been reported (93,95). In a large series of 57 children with Ph-positive ALL, partial or complete monosomy 7 was found in approximately 25% of patients, and the rate of induction failure (31%) was much higher among Ph-positive than among Ph-negative cases (96). The consistent lack of success in treating this form of ALL has prompted most investigators to consider bone marrow transplantation in first remission as a therapeutic option (97).

Adult ALL. The Ph chromosome is the most frequent rearrangement found in adults with ALL, affecting 17–30% of patients, most of whom have a dismal outcome (11,41,98,99). The relatively high incidence of the Ph translocation in adults may account in part for the generally less successful treatment of these ALL patients, as compared to the outcome of therapy for children with this disease. As reported by French investigators, additional abnormalities included partial or total monosomy 7 (17%), extra

Ph (17%), trisomy 8 (10%), and 9p rearrangements (12%) (20). The percentage of patients with Ph-positive ALL increased with age; the highest frequency was observed in the 40- to 50-year-old subgroup, in which half of the patients were Ph-positive. No Ph-positive T-cell lineage cases were observed in that study, although the proportion of patients showing expression of myeloid antigens was higher in adults with Ph-positive ALL (24%) than in Ph-negative ALL cases (9%). This proportion was even higher in cases of Ph-positive ALL with a monosomy 7 (43%) (20).

Molecular Consequences. Cytogenetically, the Ph chromosome in ALL is identical to that seen in chronic myelogenous leukemia (CML), in which the marker is retained throughout the course of the disease. By contrast, in patients with Ph-positive ALL, the Ph chromosome is no longer cytogenetically detectable after complete remission is achieved. In both Ph-positive ALL and CML, the translocation fuses *bcr* on chromosome 22q11.2 to *abl* on chromosome 9q34, but at the molecular level the breakpoints differ (100,101). In more than 90% of patients with CML and in 40–50% of adults with ALL, the translocation results in a 210-kD fusion protein. However, in 50% of adults and 80% of children with ALL, the Ph chromosome generally yields a 190-kD fusion protein (98,99,102–104). Both proteins exhibit enhanced tyrosine kinase activity and have been associated with leukemogenesis in a transgenic mouse model (105–107).

Case 1: Results from cytogenetics assays can alter treatment course

A 9-year-old girl was referred to our institution because of a history of fever and pain in her left arm and shoulder for the previous 2 days, as well as pain in her legs for the past 3 months. She presented with pancytopenia and a leukocyte count of 4×10^9/L; a bone marrow sample obtained at this time was 50% replaced with large blast cells. Immunophenotyping revealed that she had B-lineage ALL, and her DNA index was elevated (1.79). In light of these results, her young age, and low white cell count, she began therapy for low-risk ALL. However, conventional cytogenetics (Fig. 3) and FISH (see Color Plate 12.4) of cultured bone marrow cells revealed the Philadelphia chromosome [t(9;22)]. RT-PCR assay confirmed the presence of *bcr-abl* fusion transcripts. Because the t(9;22) is an adverse prognostic feature, her treatment was intensified. She completed therapy for high-risk ALL without complication and remains free of disease. In the absence of cytogenetics, she would have remained on low-risk therapy, thereby leaving her at risk for relapse and treatment-resistant disease.

Cloning of the t(9;22) breakpoint has provided DNA probes necessary for its molecular diagnosis by RT-PCR (108). Recently, German investigators

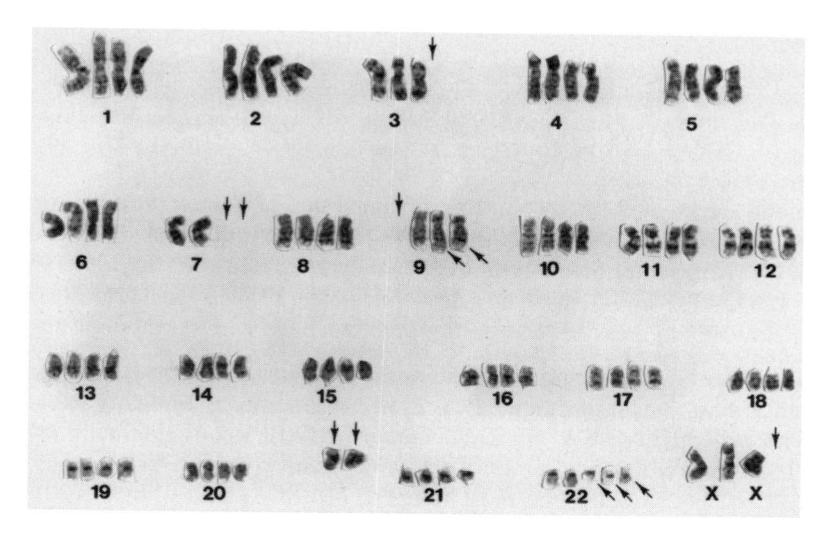

Figure 3 Case 1: conventional cytogenetics.

evaluated *bcr-abl* transcripts in 170 children with bone marrow relapse and found that 20 (12%) were positive for this marker; this rate is much higher than that reported for newly diagnosed cases of ALL (109). This finding reinforces the importance of early detection of Ph-positive cases as well as the need to establish novel chemotherapeutic approaches to treat this disease. To aid identification of Ph-positive cases, probes for FISH analysis are available commercially that can be used to detect and quantify the *bcr-abl* rearrangement in metaphase plates and interphase cells (110–115).

3. 11q23 Abnormalities

Childhood ALL. The q23 region of chromosome 11 is a relatively frequent site of structural rearrangement in children with ALL, and abnormalities of this region are detected in 4.5–5.7% of blast cells (116,117). Compared to their counterparts, children with 11q23 abnormalities are usually younger, have higher leukocyte counts, lack hyperdiploidy, have organomegaly and central nervous system involvement, are characterized by an early pre-B-cell immunophenotype and myeloid-related antigen expression, and have a very poor prognosis. In the majority of cases, translocations involving the 11q23 region result from exchanges with chromosome 4 [e.g., t(4;11)(q21;q23)], with an overall frequency of 2% (Fig. 2). Most of these children are younger than 1 year, frequently newborns, at diagnosis (118,119). Chromosomes 1, 9, 10, and 19 are also recurrently affected, leading to t(1;11)(p32;q23), t(9;11)(p22;q23), t(10;11)(p12;q23), and t(11;19)(q23;p13.3) (Fig. 2) (120). These and other random and nonrandom translocations that involve the 11q23 band [e.g., t(6;11)(q27;q23) and t(11;19)(q23;p13.1)] can also be found in cases of acute myelomonocytic and monocytic leukemia (121).

Among cases of ALL in children younger than 1 year, the leukemic cells show preferential involvement of the 11q23 region (up to 60% of cases), with t(4;11) being the most commonly observed, followed by t(11;19) (15,122–125). ALL in infants has a low incidence (3% of pediatric ALL cases), is clinically aggressive, and is strongly associated with poor prognosis.

Adult ALL. Overall, the frequency of t(4;11) in adults (5%) appears to be higher than that in children (2%). This abnormality is typically associated with several high-risk clinical features (11,20). Rearrangements involving 11q23 were observed in 33 of 443 (7%) adult patients in one series, with t(4;11)(q21;q23) being the most frequent (16 of 33 cases, 48%) (20). Other known recurrent translocations were found in three cases each and included t(9;11)(p21;q23), t(11;19)(q23;p13.3), and t(3;11)(q22;q23). In that study, all patients with rearrangements involving 11q23 other than t(4;11) had a dismal outcome, even though they did not present the high-risk factors associated with t(4;11).

Molecular Consequences. Several genes that encode proteins active in cell signaling have been localized to the 11q23 region (120,126). Molecular investigations of well-known 11q23 translocations have identified the gene that spans this region: *MLL* (for *m*yeloid-*l*ymphoid *l*eukemia or *m*ixed *l*ineage *l*eukemia; also named *ALL1*, *HRX*, and *HTRX*) (127–130). Cloning of the t(4;11) breakpoint showed that the *MLL* gene is juxtaposed to *AF4* (*a*cute leukemia *f*used to chromosome *4*) (129). RT-PCR detects this translocation in virtually all t(4;11)-positive cases (131,132). The status of *MLL* can be evaluated at the molecular level by using commercially available probes in Southern or FISH analyses. Molecular cytogenetic studies of cases with 11q23 translocations typically have shown involvement of the *MLL* gene, as has evaluation of additional cases in which such rearrangements had not been appreciated by conventional cytogenetic analysis (52,133–140).

Rearrangement of the *MLL* gene by translocation in children with ALL confers a poor prognosis, and stratifying therapy based on the presence or absence of this abnormality in leukemic cells has been proposed (123,141,142). By contrast, the few children with ALL who have deletions or inversions that affect the 11q23 band typically lack *MLL* gene rearrangement, have favorable clinical features, and generally have a good prognosis (143). Therefore, molecular techniques are needed to assess further the status of the *MLL* in cases with 11q23 abnormalities to stratify patients better for treatment (144).

In infants with ALL, the frequency of molecular rearrangement of *MLL* (75%) is even higher than that of cytogenetically identified 11q23 aberrations (50–60%). In some adults with AML, partial tandem duplication of the *MLL* gene has been observed in the absence of cytogenetic evidence of changes at 11q23 (145). This new type of gene defect is known as "self-fusion" and has been found in adult AML patients who have a normal karyotype or trisomy 11 (146,147).

Case 2: FISH is a useful adjunct to conventional cytogenetics

A 16-month-old boy presented to our institution with a 3-week history of lethargy and intermittent fever. During his diagnostic work-up, a bone marrow sample was submitted for cytogenetic analysis. Conventional cytogenetics (Fig. 5) revealed a pseudodiploid line that was characterized by a t(11;19)(q23;p13.3) rearrangement. Patients with ALL whose disease is characterized by a translocation affecting 11q23 have an adverse prognosis if the *MLL* gene is rearranged. Therefore, FISH analysis using a gene-specific probe (see Color Plate 12.6) revealed three *MLL*

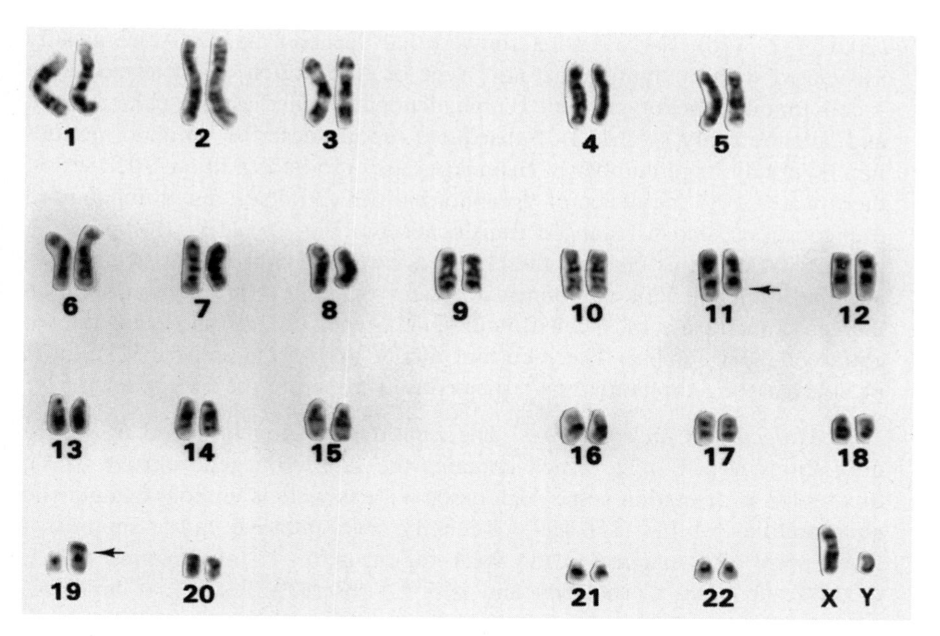

Figure 5 Case 2: conventional cytogenetics.

signals in the interphase nuclei, thereby documenting rearrangement. As a result, this patient promptly received treatment for high-risk ALL.

4. 6q Abnormalities

Reported in 4–13% of cases, deletion of the long arm of chromosome 6 is a relatively common finding in childhood ALL, a finding similar to the 5% of adults with ALL who have rearrangements or deletions of 6q (Fig. 2) (15,20). In a review of 412 consecutive cases of children with ALL with fully banded karyotypes, we identified 45 (11%) that had a 6q abnormality, of which 85% were deletions (148). The presenting features and EFS of children with a 6q abnormality were not demonstrably different from those of patients lacking this feature. No conclusive molecular evidence for loss of heterozygosity has been reported for 6q deletions (149,150). Recently, the human cyclin C (*CCNC*) gene has been localized to 6q21 (151). This gene was deleted in most of the ALL cases with cytogenetic deletion of 6q, but mutation of the remaining allele was not observed (151).

5. 9p Abnormalities

Abnormalities of the short arm of chromosome 9 have been observed in 7–12% of cases of childhood ALL and in 10–15% of adult ALL cases (20,41,152–156). Early reports showed that 9p abnormalities were usually associated with a ''lymphomatous'' type of ALL, often characterized by a T-cell immunophenotype with lymphadenopathy, mediastinal enlargement, and splenomegaly (152,153). Subsequent reports describe 9p abnormalities in a relatively large number of B-lineage cases (154,155). In our ALL series, there was a 10% incidence of 9p abnormalities (26 deletions, 9 unbalanced translocations, and 5 balanced translocations) (Fig. 2) (156). Compared to patients lacking this change, the children with 9p abnormalities were older and had higher leukocyte counts, a greater frequency of splenomegaly, an increased incidence of T-cell immunophenotype, and an increased rate of extramedullary relapse. The recurrent dic(7;9)(p13;p11) and dic(9;12)(p11; p13) are among the numerous translocations affecting the 9p region (157).

Molecular Consequences. The important region involved in 9p abnormalities is p21–p22, which contains the interferon gene cluster (*IFN*); this region is deleted in some ALL cases with as well as without cytogenetic abnormalities of 9p (158,159). Recently, two putative tumor-suppressor genes were also mapped to the 9p21 region, $p16^{INK4A}$ (also named *MTS1*, *CDKN2*, or *CDK4*) (160,161) and $p15^{INK4B}$ (or *MTS2*) (162). A high frequency of homozygous deletions and/or mutations of these genes has been reported in a variety of tumors, including pediatric ALL cases (163, 164), but currently their roles as tumor-suppressor genes remain controversial (165).

6. 12p Abnormalities

Chromosome 12p abnormalities are identified by conventional cytogenetics in 8–11% of cases of childhood ALL and in 5% of adults with ALL (20,166,167). In children, most 12p abnormalities involve translocation, including dic(9;12)(p11;p11) (63,168–170), dic(7;12)(p11;p11-12) (157,169), t(12;13)(p13;q14) (157,169,171), t(2;12)(q14;p13) (166,169), and t(12;17)(p13;q21) (172). Deletions of 12p typically comprise one-quarter of these chromosomal aberrations (Fig. 2). In our recent cytogenetic evaluation of 815 children with newly diagnosed ALL, a total of 104 structural chromosomal abnormalities of the 12p11–p13 region were observed in 94 (11.5%) cases (173). The 12p abnormalities included translocations ($n = 69$), deletions ($n = 29$), inversions ($n = 5$), and isochromosome 12q ($n = 1$). Overall, the 69 translocations that affected the 12p region were associated with 20 different chromosomes as reciprocal partners and with 43 distinct breakpoints. Compared to the ALL population in general, cases with 12p abnormalities had a similar distribution of immunophenotypes but a markedly lower frequency of hyperdiploidy >50 (7%). The clinical outcome of patients with 12p abnormalities did not differ significantly from that of patients who lacked this change (5-year EFS, 70% ± 5% versus 64% ± 2%, $p = 0.64$). Our findings confirmed that the most frequently observed recurrent 12p abnormality, dic(9;12)(p11;p12), confers an excellent prognosis (63,64,170).

Molecular Consequences. The search for the genes involved in 12p abnormalities led to the identification of *TEL* (renamed *ETV6*), which encodes an ETS-like putative transcription factor (174). This gene was subsequently found to be involved in a number of translocations in acute and chronic leukemias, and many of the resulting fusion transcripts [e.g., *TEL-PDGFRB*, t(5;12)(q33;p13) (174); *TEL-MDS1/EVI1*, t(3;12)(q26;p13) (175,176); *TEL-ABL*, t(9;12)(q34;p13) (177,178); *MN1-TEL*, t(12;22)(p13;q11) (179); *TEL-unknown*, t(10;12)(q24;p13) (180); and *TEL-AML1* (*ETV6-CBFA2*), t(12;21)(p13;q22) (181–183)] have been characterized.

Initially detected by FISH, the cryptic t(12;21)(p13;q22) is rarely identified by conventional cytogenetics (53,184). Recently, the resulting *TEL-AML1* (*ETV6-CBFA2*) fusion has been identified as the most frequent recurring chromosomal abnormality in childhood ALL, affecting 16–25% of B-lineage cases, and is associated with an excellent prognosis (182,183,185,186). In contrast, the *TEL-AML1* fusion gene is a rare event in adult ALL, found in only 1% of B-lineage cases (187,188). Of the cases in our recent study that had cytogenetically apparent aberrations of 12p, *TEL* rearrangement and *TEL-AML1* chimeric transcripts were detected in 56% (36 of 64) and 66% (25 of 38) pediatric cases, respectively (173). All but one *TEL-AML1*-positive cases had gene rearrangement, as determined by Southern and FISH analyses.

A limited number of additional cases with cytogenetically abnormal 12p regions but lacking t(12;21) have been reported (189). This result suggests that other genes important to leukemogenesis may be located in this region. Among the candidates, *CDKN1B* (*c*yclin-*d*ependent *k*inase *i*nhibitor *1B*, also named $p27^{KIP1}$) has been localized by FISH to 12p13. Although several hematologic disorders are associated with deletion of one *CDKN1B* allele, extensive sequence analysis has failed to reveal any mutation in the retained allele (190–194). Therefore, assuming that hemizygosity for *CDKN1B* is insufficient to cause growth alterations, as suggested by *CDKN1B* knock-out studies in mice (194), this gene can be eliminated as a potential target of leukemogenic rearrangement.

7. t(8;14)(q24.1;q32)

The t(8;14)(q24.1;q32) abnormality, followed by the less frequent variants t(2;8)(p12;q24.1) and t(8;22)(q24.1;q11.2) (Fig. 2), was the first immunophenotype-specific translocation to be identified in surface immunoglobulin-positive (sIg-positive) B-cell neoplasias, mainly Burkitt lymphoma (195,196). Although the overall frequency of 8q24.1 aberrations is approximately 3–5%, with slightly higher incidence in adults than in children, these translocations are found in 85–90% of cases of sIg-positive, B-cell ALL (197), which is generally considered to be a disseminated stage of Burkitt lymphoma (198). The t(8;14) has also been observed in other lymphomas, mainly those of the diffuse large cell and small noncleaved (non-Burkitt) cell types. Most t(8;14)-positive cases have additional abnormalities, frequently partial duplication in the long arm of chromosome 1 (199). In children, the use of newer regimens has improved the outcome of patients with these chromosomal abnormalities to approximately 60–70% (200). Similarly, adult patients with t(8;14) or its variants can achieve a long-term remission with appropriate therapy.

Molecular Consequences. Molecular studies have shown that the event central to all three of these translocations is the juxtaposition of the *MYC* (8q24.1) proto-oncogene locus with an immunoglobulin gene, the μ heavy-chain (*IGH*) (14q32), κ light-chain (*IGK*) (2p12), or λ light-chain (*IGL*) (22q11.2) locus (201,202). These rearrangements are thought to be mediated by errant recombinase enzyme activity during early lymphoid differentiation (203). The juxtaposition of Ig-gene regulatory sequences and

MYC, a transcription factor that is involved in the control of cell proliferation, leads to the dysregulated transcription of the fusion construct (204,205).

III. T-CELL LEUKEMIA AND LYMPHOBLASTIC LYMPHOMA

In children, cases of T-cell ALL and lymphoblastic lymphoma appear to be different clinical presentations of the same disease. T-cell childhood leukemia, accounting for about 15% of cases of ALL, has a relatively poor prognosis that can be attributed in part to hyperleukocytosis, frequent involvement of the CNS, mediastinal enlargement, and the lack of hyperdiploidy (206). Regardless of the high leukocyte count, fewer patients with T-cell immunophenotype disease have cytogenetically detectable abnormal clones compared to findings for those with other immunophenotypes of ALL (20,55,207,208). Approximately 30–40% of the abnormal karyotypes in cases of T-cell leukemia have nonrandom breakpoints within the 14q11.2, 7q35, or 7p15 regions, which contain the T-cell receptor (*TCR*) genes: $\alpha\delta$-chain locus (*TCRAD*), β-chain (*TCRB*), and γ-chain locus (*TCRG*), respectively (209). Like the translocations associated with cases of B-cell ALL and Burkitt lymphoma, these abnormalities appear to result from mistakes in the recombination process that leads to generation of functional antigen receptors. The basic transforming mechanism in each of these chromosomal rearrangements is the dysregulated expression of the reciprocal partner gene. The prominent recurring chromosomal abnormalities in T-cell ALL, identified equally in children and adults, are discussed in the following section (Table 1).

A. t(11;14)(p13;q11.2)

This translocation (present in 7% of cases) is among the most frequent nonrandom abnormalities that are detected in childhood T-cell ALL by conventional cytogenetics (55,210). At the molecular level, the breakpoint on chromosome 14 is within the $\alpha\delta$ locus of the *TCR* gene, and on chromosome 11 at p13, there is a breakpoint cluster region (T-ALL^{bcr}) (211–213). Another gene, *rhombotin-2* (*RBTN2*; also called the *T*-cell *t*ranslocation *g*ene *2*, *TTG2*) was identified to lie distal to T-ALL^{bcr} and is overexpressed in cases of T-cell ALL that carry the t(11;14)(p13;q11.2) (214,215).

B. t(11;14)(p15;q11.2)

Found in 1% of childhood T-cell neoplasias (55,210), the t(11;14) also disrupts the *TCRAD* gene on chromosome 14; the break on chromosome 11 occurs in a stage-specific differentiation gene (*TTG1*, or *RBTN1*) (216,217). The *rhombotin* gene products are members of a family of proteins that contain a cysteine-rich protein–protein interaction domain (214,217). Because *rhombotin* genes are not normally expressed in T cells, their translocation into *TCR* gene loci results in their aberrant expression. Therefore, members of the *rhombotin* gene family may be a class of transcription factors that have particular relevance in T-cell neoplasia (215,218).

C. t(10;14)(q24;q11.2)

This translocation was reported in approximately 5% of children with T-cell ALL or lymphomas and in 14% of adult patients (20,219). The t(10;14) affects the *TCRAD* locus on chromosome 14 and a breakpoint cluster region on chromosome 10 (220,221). The translocation deregulates the homeobox gene *HOX11* through illegitimate physiologic recombination with *TCRAD* (221–223). *HOX11* is normally expressed in nonhematopoietic cells only during embryogenesis. Because of the tight clustering of the chromosomal breakpoints, the t(10;14) can be detected by PCR analysis (224).

D. t(8;14)(q24.1;q11.2)

This translocation is commonly observed in childhood T-cell leukemias (2%) but is not restricted to this lineage (55,225). At the molecular level, the *TCRAD* gene is juxtaposed to the *MYC* oncogene, resulting in transcriptional deregulation of *MYC* (226).

E. 1p32-p34 Abnormalities

The t(1;14)(p32-p34;q11.2) has been observed in 3% of children with T-cell ALL (227). At the molecular level, *TAL1* (also named *SCL/TCL5*), located in the p32–p34 region of chromosome 1, is juxtaposed with the *TCRAD* chain locus on chromosome 14 (228,229). The *TAL1* gene, a member of the basic helix-loop-helix family of transcription factors, is not expressed in normal T lymphocytes (230,231). The *TAL1* gene is also dysregulated by juxtaposition with a *TCR* locus in other rare translocations, for example, the t(1;7)(p32;q35) (232), and by the t(1;3)(p34;p21), which fuses *TAL1* with a novel gene on chromosome 3, the *T-c*ell leukemia *t*ranslocation-*a*ssociated (*TCTA*) gene (233,234).

The most common disruption of the *TAL1* gene was identified as a site-specific interstitial deletion of chromosome 1 that juxtaposes the coding region of the *TAL1* gene with the promoter region of *SIL* (*S*CL *i*nterrupting *l*ocus) (235–237). The deletion, which occurs in approximately 9–26% of children with T-cell ALL, and in 3–8% of adult patients, is not visible by karyotypic analysis but is easily detected by Southern blotting or PCR analysis (51,236,238–240). Therefore, alteration of *TAL1*, either by translocation or other rearrangement, represents the most common genetic lesion associated with T-lineage leukemia. In these cases, the altered *TAL1* alleles are transcriptionally active, and inappropriate expression of TAL1 protein may contribute to leukemogenesis (229,241).

F. inv(14)(q11.2q32)

Inversion(14) is a rare finding in children with T-cell ALL (55). In adults, this abnormality has been seen in other T-cell malignancies such as leukemia-lymphoma syndrome, chronic lymphocytic leukemia, and prolymphocytic leukemia and in association with ataxia telangiectasia (242,243).

G. 7q35 (*TCRB*) or 7p15 (*TCRG*) Abnormalities

Although affected by chromosomal aberrations less often than is the αδ region of chromosome 14, the 7qter region, which contains the *TCRB* gene, participates in reciprocal exchanges with several chromosomes (244,245). Two infrequent translocations, the t(7;9)(q35;q32) and t(7;19)(q35;p13), result in dysregulated expression of *TAL2* or *LYL1*, respectively (246,247). A t(1;7)(p34;q35) results in the juxtaposition of *LCK* and *TCRB* (248). In addition, t(7;11)(q35;p13) and a t(7;10)(q35;q24), variants of the more frequently observed t(11;14)(p13;q11.2) and t(10;14)(q24;q11.2), respectively, also activate DNA-binding transcription factors (*RBTN2* and *HOX11*) (249,250). Recent cytogenetic studies have identified isochromosome 7q in cases of hepatosplenic γδ T-cell lymphoma (251–253). Chromosomal rearrangements rarely affect the *TCRG* locus.

IV. CHRONIC LYMPHOCYTIC LEUKEMIA

A. B-Cell Chronic Lymphocytic Leukemia

Chronic lymphocytic leukemia (CLL) is rare in children (254,255), but it is the most common leukemia in adults, comprising approximately 30% of all adult leukemia cases. In 90% of patients, CLL seems to result from a monoclonal neoplastic proliferation of small B-cell lymphocytes, which contain clonal chromosomal aberrations in approximately 40–50% of cases by conventional chromosome analysis (256–259). An extra copy of chromosome 12 (+12) is the most frequent finding in CLL and is found in about one-third of cases with abnormal metaphases on cytogenetic analysis (256,257).

In addition, aberrations of 11q, 13q, and 14q are observed with some frequency (258,259). FISH analysis with a centromeric probe for this chromosome is an excellent method for evaluating nondividing leukemic cells. Several studies have shown that FISH identified more cases of trisomy 12 in CLL patients than did conventional cytogenetics (260–262). Other chromosomal aberrations detected in B-cell CLL include t(11;14)(q13;q32) and t(14;19)(q32;q13.1), involving *IGH* and either *BCL1* or *BCL3* (263). Deletions of 13q that encompass the *RB1* gene and other sequences in 13q14 have also been detected in these leukemias (263).

B. T-Cell Chronic Lymphocytic Leukemia

Cases with a T-cell immunophenotype account for fewer than 5% of all cases of CLL. The most frequently observed changes are recombinations involving 14q11.2 (*TCRAD*) and 14q32 (*IGH*) and those that have already been described for T-cell ALL (264).

V. NON-HODGKIN LYMPHOMA

Non-Hodgkin lymphoma (NHL) comprises a heterogeneous group of neoplasms that arise from B or T lymphocytes (265,266), and nearly all patients with NHL have clonal karyotypic and molecular abnormalities in their blast cells (267–270). The karyotypes of NHL cases are usually complex and consist of a broad range of numeric and structural abnormalities. The discovery of recurrent chromosomal changes, development of Southern, FISH, and PCR techniques, and advances in immunologic methods for identifying B and T clonality have yielded highly sensitive molecular markers for characterizing cases of NHL. Although no genetic abnormality is specific for any one subtype, each has a nonrandom association with particular classifications of NHL, suggesting a correlation between the altered gene(s) and the histologic, phenotypic, and clinical features of the lymphoma (271).

Only limited cytogenetic data are available from children with NHL, and the subtypes seen in pediatric patients differ markedly from those seen in adults, likely due to the constitution of the immunologic system and the type of cells susceptible to transformation (272). In children, small noncleaved cell (Burkitt) lymphoma, T-cell lymphoblastic lymphoma, and diffuse large cell lymphoma are prevalent, collectively accounting for more than 90% of cases. In adults, Burkitt and lymphoblastic lymphomas are rare, and follicular center cell lymphoma (infrequent in children) predominates (273,274). Although the prevalence of NHL subtypes varies with age, the underlying molecular defects are relatively uniform within each histologic subtype.

With modern risk-based therapy, children with NHL fare about as well as do those with ALL. Because the treatment of NHL is dependent in part on the histologic subtype and disease stage, a staging work-up is essential prior to treatment. Therefore, cytogenetic findings need to be carefully correlated with pathologic, immunologic, and molecular features to confirm the proper diagnosis and staging of cases of NHL. In adults with NHL, the association between presenting features and outcome is controversial because of the complexity of lymphoma classification and the lack of standardized diagnostic criteria (275).

Numerous novel genes have been identified through cloning of translocation junctions, and most of these molecular studies have been performed in cell lines generated from patients with various subtypes of NHL. The biologic significance of these abnormalities is often not clearly understood. An early survey of cytogenetic deletion maps of NHL indicated that the chromosome segments or bands most frequently lost due to structural aberrations were 6q21–q27, 11q13–q25, and 14q24–q32 (276). The comparative genomic hybridization (CGH) technique is providing additional evidence of genomic imbalances in NHL (277–281). This review outlines the major subgroups of chromosomal abnormalities associated with NHL and their currently known molecular biologic consequences.

A. Numeric Chromosomal Changes

The low proliferation index and the technical difficulties in culturing solid tissue samples obtained from diverse lymphoma sites hamper the cytogenetic study of these neoplastic tissues. The use of FISH on interphase nuclei for detection of suspected changes in copy number is an excellent means of complementing conventional cytogenetics.

1. +3

Although unassociated with any particular subtype of lymphoma, a gain of chromosome 3 is a common finding in NHL, with frequencies of 5–24% (268,270,282,283). In one study, +3 was found in 30% of cases of NHL (9 of 30 patients) (284) and in 44% (12 of 27 cases) of B-immunoblastic NHL in another report (285). A high proportion of cases of small lymphocytic NHL and some lymphomas of mucosa-associated lymphoid tissue (MALT) demonstrate +3 abnormalities (286–289).

2. +7

Trisomy 7 is one of the most frequent numeric abnormalities observed in NHL and is often a secondary change. Trisomy 7 has often been detected in cases of adult T-cell leukemia/lymphomas from Japan (290,291).

3. +12

Found in 4% of adult NHL cases, +12 occurs in about one-third of small B-cell lymphocytic lymphomas and in some cases of CLL (292).

4. +18

About 10–20% of all cases of NHL have a gain of chromosome 18 (268,282,288,293). Slavutsky et al. noted that 25% (3 of 12 cases) of diffuse small cleaved cell lymphomas had a gain of chromosome 18 (294). Nashelsky et al. found +18 in 44% (12 of 27 cases) of B-immunoblastic NHL (285).

B. Structural Chromosomal Changes

1. 14q32 Abnormalities

This region is the most frequently observed chromosomal breakpoint in NHL, and numerous recurrent translocations have been identified. Several of these abnormalities, particularly those characteristic of B-cell leukemias and Burkitt lymphomas of FAB L3 morphology, have already been described in the section on ALL. Other recurrent structural abnormalities associated with NHL are discussed below. In most cases, the chromosomal partner can be established by conventional cytogenetics. When banding techniques alone cannot resolve the translocation partner, FISH with chromosome painting is currently used to determine the origin of the exchange DNA material (295,296).

2. t(2;5)(p23;q35)

This rearrangement is typically identified in cases of anaplastic large cell lymphoma (ALCL), mainly of the T-cell type, that express the CD30 (Ki-1) antigen (297–301). Between 12% and 50% of cases of ALCL show a t(2;5) (Fig. 7); the higher percentage is observed in pediatric series (302–305). The incidence of ALCL ranges from 6% to 15% in pediatric series and from 1% to 8% in adults (306), and these patients often respond well to appropriate treatment (299,304). The t(2;5) has also been found in cases of peripheral T-cell lymphoma and even in B-cell lymphomas that lack features characteristic of ALCL (304,307–309).

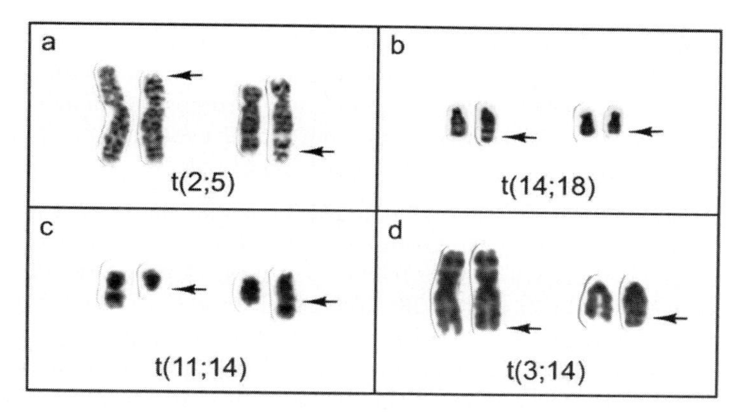

Figure 7 These partial karyotypes from G-banded metaphase plates show various structural chromosomal abnormalities in non-Hodgkin lymphoma: (a) t(2;5)(p23;q35); (b) t(14;18)(q32;q21.3); (c) t(11;14)(q13;q32); (d) t(3;14)(q27;q32).

Molecular Consequences. The t(2;5) fuses the ubiquitously expressed *n*ucleolar *p*hosphoprotein *n*ucleophosmin (*NPM*) gene (at 5q35) to the novel tyrosine kinase gene *ALK* (*a*naplastic *l*ymphoma *k*inase, at 2p23) (310). The variant t(1;2)(q25;p23) presumably joins the *ALK* gene to an unidentified gene with an active promoter on chromosome 1 (305,311). The RT-PCR assay is a rapid, accurate, and sensitive means of detecting *NPM-ALK* fusion transcripts (307,308). Recently, immunocytochemical and FISH studies have been used successfully to detect the t(2;5) (305,311–313).

3. t(14;18)(q32;q21.3)

Observed in 35–40% of adults with NHL, the t(14;18) (Fig. 7) is the most common translocation in adult B-cell lymphoma, but it is rarely seen in children. A t(14;18) is observed in 80–90% of cases of the cleaved, follicular subtype that is associated with an indolent clinical course. In cases of the diffuse variant and other histologies, this translocation is found less frequently (267–269,314–316). Overall, as many as 30% of cases of aggressive NHL have the t(14;18), and additional cases with the variant translocations t(2;18)(p12;q21.3) and t(18;22)(q21.3;q11.2) are identified infrequently.

In the majority of cases with t(14;18), secondary chromosomal abnormalities are also present. Gains of chromosome 7 or del(6q) have been reported to correlate with a more malignant phenotype (273). The t(14;18)(q32;q21.3) and t(8;14)(q24.1;q32) can occasionally be found together in high-grade lymphomas during the leukemic phase of the disease. This finding correlates with the histologic transformation of low-grade lymphoma to a clinically more aggressive high-grade lymphoma that is associated with a particularly poor prognosis (317–319). Therefore, the chromosomal abnormalities present in addition to t(14;18) result from clonal evolution of the tumor and are usually reflected by clinical progression of disease.

Molecular Consequences. The region 18q21.3 contains the *BCL2* (*B-cell lymphoma/leukemia 2*) oncogene. Overexpression of the *BCL2* product occurs when the gene is juxtaposed to the *IGH* enhancer on 14q32 or, rarely, to *IGK* (2p12) or *IGL* (22q11.2) (320–328). The BCL2 protein is an inhibitor of apoptosis (programmed cell death) that in turn leads to prolonged survival of B lymphocytes and accumulation of clonal cells in tissues such as the lymph nodes and spleen (329–331). McDonnell and Korsemeyer found that a large percentage of transgenic mice carrying t(14;18) rearrangements developed follicular hyperplasia that evolved into immunoblastic lymphoma (332). In half of these cases, rearrangement of *MYC* was present. This finding suggests that in certain aggressive cases of NHL, tumor progression may be associated with aberrant expression of specific genes that regulate cell death (*BCL2*) as well as cell growth (*MYC*) (332). Furthermore, this transgenic mouse model demonstrated that t(14;18) alone is insufficient for malignant transformation.

The t(14;18) can be detected in interphase nuclei by FISH (333), and RT-PCR offers a more sensitive means for identifying this translocation (334). The clinical relevance of these findings should be interpreted with caution, as t(14;18)-positive cells have been found in patients who have no evidence of recurrent lymphoma and in patients who have benign follicular hyperplasia (335,336).

4. t(11;14)(q13;q32)

The t(11;14) (Fig. 7) is seen in 4–8% of adult NHL cases but rarely in pediatric patients (337). Variant translocations involving the 11q13 breakpoint have also been reported (338,339). This rearrangement is strongly associated with centrocytic/mantle cell lymphoma, in which it is seen in as many as 70% of cases (340,341). In mantle cell lymphomas, large atypical cells containing hyperdiploid complex karyotypes have been associated with an aggressive clinical course (342). This translocation has been identified rarely in adult patients with B-cell CLL, multiple myeloma, and other types of NHL (343,344). Moreover, trisomy 12 appears to be a frequent secondary abnormality in t(11;14)-positive cases (345).

Molecular Consequences. The t(11;14) results in the juxtaposition of the *BCL1* locus at 11q13 with the *IGH* chain gene at 14q32, thereby leading to deregulation of the *PRAD1* (*CCND1*) gene within the *BCL1* locus (346,347). *PRAD1* encodes cyclin D_1, a cell cycle protein, which is undetectable in normal lymphoid cells but overexpressed in nearly all cases of mantle cell lymphomas (348,349). This protein helps cells to progress from G1 to S phase (347,350,351).

The molecular detection of this translocation by PCR requires multiple probes because the breakpoints in the *BCL1* locus are not tightly clustered (337,352–355). The translocation can often be demonstrated by Southern analysis of *BCL1* rearrangement, staining with cyclin D_1 antibodies, and interphase FISH (337,351,354,356). By using recently developed reagents for FISH, virtually all cases of mantle cell lymphomas have a detectable 11q13 rearrangement (357,358).

5. 3q27 Abnormalities

Since the late 1980s, 3q27 has been recognized as a recurrent breakpoint in NHL. By cytogenetics, 3q27 is affected in 7–12% of cases of B-lineage diffuse large cell lymphomas (359–362). The t(3;14)(q27;q32) (Fig. 7) and the variants t(3;22)(q27;q11.2) and t(2;3)(p12;q27) are among the most common recurrent translocations in malignant lymphomas. Although early reports primarily described cases with t(3;22), the more subtle t(3;14) is found most frequently in recent series, likely due to improved banding techniques and increased awareness among cytogeneticists (359,360,362). Therefore, translocations involving 3q27 affect several chromosomal partners, which include but are not limited to 14q32 (*IGH*), 2p12 (*IGK*), and 22q11.2 (*IGL*) (363–365). In addition, these 3q27 rearrangements are frequently observed with other aberrations, including t(8;14), t(14;18), and a 1q abnormality.

Molecular Consequences. The 3q27 region may be the site of a proto-oncogene analogous to the *MYC* gene in Burkitt lymphoma. The immunoglobulin locus at 14q32 provided a probe to clone the most frequent t(3; 14)(q27;q32). This same genomic region was cloned from several cases of diffuse cell lymphomas, all of which carried 3q27 alterations regardless of the partner chromosome involved (366–369). The gene identified was named *BCL6* (*BCL5*/*LAZ3*) and is expressed at very low levels in all tissues except B cells and skeletal muscle (366–369). *BCL6* encodes a zinc-finger protein (Kruppel family) that can regulate transcription (370). Reported in >30% of diffuse large cell lymphomas and in 6–11% of follicular cases, rearrangements of *BCL6* presumably lead to deregulated protein expression (361,368,371). Only half of the cases in which *BCL6* is rearranged have an aberration that can be detected by cytogenetic methods. Recent studies have shown that most cases of diffuse large cell lymphoma and many of those of follicular lymphoma carry structural alterations of the regulatory region of

BCL6 independent of rearrangement, suggesting that deregulation of *BCL6* may be important in lymphomagenesis (372,373).

Among the 3q27-associated translocations are those affecting regulatory sequences, other than immunoglobulin gene loci, on various partner chromosomes. A few of these translocations that have been cloned show juxtaposition of *BCL6* with the small G-protein gene *TTF* (for *t*ranslocation *t*hree *f*our) in the t(3;4)(q27;p11) (374), with a novel *H4* histone gene in the t(3;6)(q27;p21), or with the *BOB1/OBF1* gene (375), which encodes a B-cell-specific coactivator of octamer-binding transcription, in the t(3;11)(q27; q23.1) (376).

6. Additional Structural Abnormalities

The potential prognostic implications of several infrequent structural chromosomal abnormalities merits their mention. For example, the t(8;13)(p11; q11) has been identified in several patients who presented with lymphoblastic lymphoma, peripheral blood eosinophilia, and myeloid hyperplasia of the bone marrow. Most of these patients have subsequently developed myeloid leukemia (377,378). Aberrations of 7q (379) and 17p (380), although rare, have been associated with adverse outcome in cases of aggressive NHL. Associated with the adverse prognostic feature of *MLL* rearrangement in ALL, abnormalities of 11q23 have occasionally been observed in NHL (381). Three different rearranged genes have been identified in various NHL cases with a t(11;14)(q23;q32): the *RCK* gene in the RCK-8 cell line, the *MLL/HRX/ALL1* gene in a few NHL samples, and the *LPC* (*p*roprotein *c*onvertase family) gene in a case of mediastinal large cell lymphoma (134,381,382).

Two infrequent rearrangements, del(7)(q32) (383) and t(9;14)(p13;q32) (384), are highly correlated with small cell lymphocytic lymphomas of the plasmacytoid subtype. The t(9;14), identified in approximately 2% of cases of NHL, leads to deregulation of *PAX5*, a transcription factor, because of translocation of the *IGH* locus to lie adjacent to two alternative *PAX5* promoters (385,386).

In contrast to the low incidence of the aberrations mentioned previously in this section, structural abnormalities of chromosome 1 are found in many malignancies, including 25–50% of pediatric NHL cases. These abnormalities are thought to be secondary in nature and a component of clonal evolution (270). In a recent survey of cases of B-cell NHL, structural rearrangements of 1q21-q23 were described in 5–25% of cases, and those of 1p32-p36 were identified in 2–15% of cases (288). Deletion of 6q occurs frequently in cases of malignant lymphoma, and in most instances, it is but one of several aberrations. A del(6q) was reported in 17% of the 260 cases of NHL that were reevaluated during the Fifth International Workshop on Chromosomes in Leukemia-Lymphoma, but there was no correlation with any particular histologic subtype (270). Offit et al. reported that deletion of 6q is observed in cases of small cell lymphocytic NHL and defines a distinct clinicopathologic subset of NHL (387,388). Common regions of deletions are clustered, but an associated tumor-suppressor gene has not yet been identified (389). Recurrent abnormalities of 10q are also fairly frequent in NHL—approximately 11% of 159 consecutive NHL patients with abnormal cytogenetic findings had abnormalities of 10q23-25 (390). Finally, structural and numeric aberrations affecting the X chromosome are rather common in NHL (391).

The *d*iffuse *B*-cell *l*ymphoma oncogene (*DBL*, also called *MCF2*) has been mapped to Xq27 (392,393). The putative oncogene *MTCP1*, at Xq28, was cloned from t(X;14)(q28;q11)-positive cases of mature T-cell leukemia and from patients who also had ataxia telangiectasia (394).

VI. CYTOGENETIC EVOLUTION PATTERNS IN NHL

Frequently, lymphomas progress histologically and clinically. This progression is associated with the acquisition of subclones containing additional karyotypic changes that consist mainly of partial and total gains or losses of chromosomal material. The most frequent secondary aberrations in NHL are +X, −Y, dup(1q), del(6q), +7, and +12 (395). Occasionally, cases of NHL with t(14;18) acquire the Burkitt-associated t(8;14) and progress to high malignancy (319).

VII. SUMMARY

Cytogenetic and molecular classification of cases of leukemia and lymphoma is likely to improve the uniformity of diagnosis and the ability to predict therapeutic response and long-term prognosis. Recently developed cytogenetic and molecular biologic analyses provide the means to elucidate the underlying pathogenetic mechanisms and to disclose the differing biology of disease subtypes. Moreover, elucidation of the biochemical pathways involved in transformation will undoubtedly influence future treatment strategies.

ACKNOWLEDGMENT

I thank Dr. Amy L. B. Frazier for editing the manuscript, Dr. S. Mathew for the FISH figures, the cytogenetic technologists for their commitment to identifying the chromosomal aberrations, Dr. W. G. Sanger, University of Nebraska Medical Center, for the partial karyotypes presented in panels (b), (c), and (d) of Fig. 7, and the physicians for providing clinical correlations.

REFERENCES

1. Rabbitts TH. Translocations, master genes and differences between the origins of acute and chronic leukemias. Cell 1991; 67:641–644.
2. Sawyers CL, Denny CT, Witte ON. Leukemia and the disruption of normal hematopoiesis. Cell 1991; 64:337–350.
3. Cleary ML. Oncogenic conversion of transcription factors by chromosomal translocations. Cell 1991; 66:619–622.
4. Cline MJ. The molecular basis of leukemia. N Engl J Med 1994; 330:328–336.
5. Shivdasani R, Orkin S. The transcriptional control of hematopoiesis. Blood 1996; 87:4025–4039.
6. Copelan EA, McGuire EA. The biology and treatment of acute lymphoblastic leukemia in adults. Blood 1995; 85:1151–1168.
7. Secker-Walker LM, Lawler SD, Hardisty RM. Prognostic implications of chromosomal findings in acute lymphoblastic leukemia at diagnosis. Br Med J 1978; 2:1529–1530.
8. Third International Workshop on Chromosomes in Leukaemia (1980): clinical significance of chromosomal abnormalities in acute lymphoblastic leukaemia. Cancer Genet Cytogenet 1981; 4:111–137.
9. Williams DL, Tsiatis A, Brodeur GM, Look AT, Melvin SL, Bowman WP, Kalwinsky DK, Rivera G, Dahl GV. Prognostic importance of chromosome number in 136 untreated children with acute lymphoblastic leukemia. Blood 1982; 60:864–871.
10. Secker-Walker LM, Chessells JM, Stewart EL, Swansburry GJ, Richards S, Lawler SD. Chromosomes and other prognostic factors in acute lymphoblastic leukaemia: a long-term follow-up. Br J Haematol 1989; 72:336–342.
11. Bloomfield CD, Secker-Walker LM, Goldman AI, Van Den Berghe H, de la Chapelle A, Ruutu T, Alimena G, Garson OM, Colomb HM, Rowley JD, Kaneko Y, Whang-Peng J, Prigogina E, Philip P, Sandberg AA, Lawler SD, Mitelman F. Six-year follow-up of the significance of karyotype in acute lymphoblastic leukemia. From the Sixth International Workshop on Chromosomes in Leukemia 1987. Cancer Genet Cytogenet 1989; 40:171–185.
12. Look AT, Roberson PK, Williams DL, Rivera G, Bowman WP, Pui C-H, Ochs J, Abromowitch M, Kalwinsky D, Dahl GV, George S, Murphy SB. Prognostic importance of blast cell DNA content in childhood acute lymphoblastic leukemia. Blood 1985; 65:1079–1086.
13. Williams DL, Raimondi SC, Pui C-H, Rivera GK. Evolving chromosome patterns and new cytogenetic concepts in childhood acute lymphoblastic leukemia. In Gale RP, Hoelzer D, eds. Acute Lymphoblastic Leukemia: UCLA

Symposia on Molecular and Cellular Biology, New Series, Vol. 108. New York: Wiley-Liss, 1990:91–100.

14. Rivera GK, Raimondi SC, Hancock ML, Behm FG, Pui C-H, Abromowitch M, Mirro J Jr, Ochs J, Look AT, Williams DL, Murphy SB, Dahl GV, Kalwinsky DK, Evans WE, Kun LE, Simone JV, Crist WM. Improved outcome in childhood acute lymphoblastic leukaemia with reinforced early treatment and rotational combination chemotherapy. Lancet 1991; 337:61–66.
15. Raimondi SC. Current status of cytogenetic research in childhood acute lymphoblastic leukemia. Blood 1993; 81:2237–2251.
16. Heerema NA, Palmer CG, Baehner RL. Karyotypic and clinical findings in a consecutive series of children with acute lymphocytic leukemia. Cancer Genet Cytogenet 1985; 17:165–179.
17. Abe R, Raza A, Preisler HD, Tebbi CK, Sandberg AA. Chromosomes and causation of human cancer and leukemia. LIV. Near-tetraploidy in acute leukemia. Cancer Genet Cytogenet 1985; 14:45–59.
18. Heim S, Alimena G, Billstrom R, Diverio D, Kristoffersson U, Mandahl N, Nanni M, Mitelman F. Tetraploid karyotype (92,XXYY) in two patients with acute lymphoblastic leukemia. Cancer Genet Cytogenet 1987; 29:129–133.
19. Pui C-H, Carroll AJ, Head D, Raimondi SC, Shuster JJ, Crist WM, Link MP, Behm FG, Land VJ, Nash M, Pullen DJ, Look AT. Near-triploid and near-tetraploid acute lymphoblastic leukemia of childhood. Blood 1990; 76:590–596.
20. The Groupe Francais de Cytogenetique Hematologique. Cytogenetic abnormalities in adult acute lymphoblastic leukemia: correlations with hematologic findings and outcome. A collaborative study of the groupe francais de cytogenetique hematologique. Blood 1996; 87:3135–3142.
21. Kaneko Y, Rowley JD, Variakojis D, Chilcote RR, Check I, Sakurai M. Correlation of karyotype with clinical features in acute lymphoblastic leukaemia. Cancer Res 1982; 42:2918–2929.
22. Kowalczyk JR, Grossi M, Sandberg AA. Cytogenetic findings in childhood acute lymphoblastic leukemia. Cancer Genet Cytogenet 1985; 15:47–64.
23. Heinonen K, Rautonen J, Siimes MA, Knuutila S. Cytogenetic study of 105 children with acute lymphoblastic leukemia. Eur J Haematol 1988; 41:237–242.
24. Prigogina EL, Puchkova GP, Mayakova SA. Nonrandom chromosomal abnormalities in acute lymphoblastic leukemia of childhood. Cancer Genet Cytogenet 1988; 32:183–203.
25. Pui C-H, Raimondi SC, Dodge RK, Rivera GK, Fuchs LAH, Abromowitch M, Look AT, Furman WL, Crist WM, Williams DL. Prognostic importance of structural chromosomal abnormalities in children with hyperdiploid (>50 chromosomes) acute lymphoblastic leukemia. Blood 1989; 73:1963–1967.
26. Trueworthy R, Shuster J, Look T, Crist W, Borowitz M, Carroll A, Frankel L, Harris M, Wagner H, Haggard M, Mosijczuk A, Pullen J, Steuber P, Land V. Ploidy of lymphoblasts is the strongest predictor of treatment outcome in B-progenitor cell acute lymphoblastic leukemia of childhood: a Pediatric Oncology Group study. J Clin Oncol 1992; 10:606–613.

27. Pui C-H, Raimondi SC, Williams DL. Isochromosome 17q in childhood acute lymphoblastic leukemia: an adverse cytogenetic feature in association with hyperdiploidy? Leukemia 1988; 2:222–225.
28. Smets LA, Slater R, van Wering ER, van der Does-van den Berg A, Hart AAM, Veerman AJP, Kamps WA. DNA index and % S-phase cells determined in acute lymphoblastic leukemia of children: a report from studies ALL V, ALL VI, and ALL VII (1979–1991) of the Dutch childhood leukemia study group and the Netherlands workgroup on cancer genetics and cytogenetics. Med Pediat Oncol 1995; 25:437–444.
29. Harris MB, Shuster JJ, Carroll A, Look AT, Borowitz MJ, Crist WM, Nitschke R, Pullen J, Steuber CP, Land VJ. Trisomy of leukemic cell chromosomes 4 and 10 identifies children with B-progenitor cell acute lymphoblastic leukemia with a very low risk of treatment failure: a Pediatric Oncology Group study. Blood 1992; 79:3316–3324.
30. Martin PL, Look TA, Schnell S, Harris MB, Pullen J, Shuster JJ, Carroll AJ, Pettenati MJ, Rao PN. Comparison of fluorescence in situ hybridization, cytogenetic analysis, and DNA index analysis to detect chromosomes 4 and 10 aneuploidy in pediatric acute lymphoblastic leukemia: a Pediatric Oncology Group study. J Pediat Hematol Oncol 1996; 18:113–121.
31. Raimondi SC, Pui C-H, Hancock ML, Behm FG, Filatov L, Rivera GK. Heterogeneity of hyperdiploid (51-67) childhood acute lymphoblastic leukemia. Leukemia 1996; 10:213–224.
32. Whitehead VM, Vuchich MJ, Carroll AJ, Lauer SJ, Mahoney D, Shuster JJ, Payment C, Koch PA, Akabutu JJ, Bowen T, Kamen BA, Ravindranath Y, Emami A, Beardsley GP, Pullen DJ, Camitta B. Accumulation of methotrexate polyglutamates (MTX PGS) in lymphoblasts, ploidy and trisomy of both chromosomes 4 and 10 in childhood B-progenitor cell acute lymphoblastic leukemia (ALL): a Pediatric Oncology Group study. Blood 1994; 84(suppl 1): 515a.
33. Moorman AV, Clark R, Farrell DM, Hawkins JM, Martineau M, Secker-Walker LM. Probes for hidden hyperdiploidy in acute lymphoblastic leukaemia. Genes Chromosom Cancer 1996; 16:40–45.
34. Anastasi J, Thangavelu M, Vardiman JW, Hooberman AL, Bian ML, Larson RA, Le Beau MM. Interphase cytogenetic analysis detects minimal residual disease in a case of acute lymphoblastic leukemia and resolves the question of origin of relapse after allogeneic bone marrow transplantation. Blood 1991; 77:1087–1091.
35. Heerema NA, Argyropoulos G, Weetman R, Tricot G, Secker-Walker LM. Interphase in situ hybridization reveals minimal residual disease in early remission and return of the diagnostic clone in karyotypically normal relapse of acute lymphoblastic leukemia. Leukemia 1993; 7:537–543.
36. Romana SP, Cherif D, Le Coniat M, Derre J, Flexor M, Berger R. In situ hybridization to interphase nuclei in acute leukemia. Genes Chromosom Cancer 1993; 8:98–103.
37. Tosi S, Ritterbach J, Maglia O, Harbott J, Riehm H, Masera G, Biondi A, Lampert F. Double target in situ hybridization applied to the study of nu-

merical aberrations in childhood acute lymphoblastic leukemia. Cancer Genet Cytogenet 1994; 73:103–108.
38. Berger A, Strehl S, Hekele A, Ambros PF, Haas OA, Gadner H. Interphase cytogenetic study of childhood acute lymphoblastic leukemia. Med Pediat Oncol 1994; 23:413–421.
39. White DM, Crolla JA, Ross FM. Detection of minimal residual disease in childhood acute lymphoblastic leukaemia using fluorescence in-situ hybridization. Br J Haematol 1995; 91:1019–1024.
40. Heinonen K, Mahlamaki E. Detection of numerical chromosome abnormalities by FISH in childhood acute lymphoblastic leukemia. Cancer Genet Cytogenet 1996; 87:123–126.
41. Rieder H, Ludwig W-D, Gassmann W, Thiel E, Loffler H, Hoelzer D, Fonatsch C. Chromosomal abnormalities in adult acute lymphoblastic leukemia: results of the German ALL/AUL Study Group. Recent Results Cancer Res 1993; 131:133–148.
42. Raimondi SC, Roberson PK, Pui C-H, Behm FG, Rivera GK. Hyperdiploid (47-50) acute lymphoblastic leukemia in children. Blood 1992; 79:3245–3252.
43. Raimondi SC, Pui C-H, Head D, Behm F, Privitera E, Roberson PK, Rivera GK, Williams DL. Trisomy 21 as the sole acquired chromosomal abnormality in children with acute lymphoblastic leukemia. Leukemia 1992; 6:171–175.
44. Watson MS, Carroll AJ, Shuster JJ, Steuber CP, Borrowitz MJ, Behm FG, Pullen DJ, Land VJ. Trisomy 21 in childhood acute lymphoblastic leukemia: a Pediatric Oncology Group study (8602). Blood 1993; 10:3098–3102.
45. Garipidou V, Yamada T, Grant-Prentice H, Secker-Walker LM. Trisomy 8 in acute lymphoblastic leukemia (ALL): a case report and update of the literature. Leukemia 1990; 4:717–719.
46. Pettenati MJ, Rao N, Wofford M, Shuster JJ, Pullen DJ, Ling MP, Steuber CP, Carroll AJ. Presenting characteristics of trisomy 8 as the primary cytogenetic abnormality associated with childhood acute lymphoblastic leukemia. A Pediatric Oncology Group (POG) study (8600/8493). Cancer Genet Cytogenet 1994; 75:6–10.
47. Fletcher JA, Kimball VM, Lynch E, Donnelly M, Pavelka K, Gelber RD, Tantravahi R, Sallan SE. Prognostic implications of cytogenetic studies in an intensively treated group of children with acute lymphoblastic leukemia. Blood 1989;74:2130–2135.
48. Lampert F, Harbott J, Ritterbach J, Schellong G, Ritter J, Creutzig U, Riehm H, Reiter A. Karyotypes in acute childhood leukemias may lose prognostic significance with more intensive and specific chemotherapy. Cancer Genet Cytogenet 1991; 54:277–279.
49. Rubin CM, Le Beau MM, Mick R, Bitter MA, Nachman J, Rudinsky R, Appel HJ, Morgan E, Suarez CR, Schumacher HR, Subramanian U, Rowley JD. Impact of chromosomal translocations on prognosis in childhood acute lymphoblastic leukemia. J Clin Oncol 1991; 9:2183–2192.
50. Devaraj PE, Foroni L, Kitra-Roussos V, Secker-Walker LM. Detection of BCR-ABL and E2A-PBX1 fusion genes by RT-PCR in acute lymphoblastic

leukaemia with failed or normal cytogenetics. Br J Haematol 1995; 89:349–355.

51. Bash RO, Crist WM, Shuster JJ, Link MP, Amylon M, Pullen J, Carroll AJ, Buchanan GR, Smith RG, Baer R. Clinical features and outcome in T-cell acute lymphoblastic leukemia in childhood with respect to alterations at the TAL1 locus; a Pediatric Oncology Group study. Blood 1993; 81:2110–2117.
52. Griesinger F, Elfers H, Ludwig W-D, Falk M, Rieder H, Harbott J, Lampert F, Heinze B, Hoelzer D, Thiel E, Riehm H, Wormann B, Fonatsch C, Hiddemann W. Detection of HRX-FEL fusion transcripts in pre-pre-B-ALL with and without cytogenetic demonstration of t(4;11). Leukemia 1994; 8:542–548.
53. Romana SP, Poirel H, Leconiat M, Flexor M-A, Mauchauffè M, Jonveaux P, Macintyre EA, Berger R, Bernard OA. High frequency of t(12;21) in childhood B-lineage acute lymphoblastic leukemia. Blood 1995; 86:4263–4269.
54. Williams DL, Raimondi SC, Rivera G, George S, Berard CW, Murphy SB. Presence of clonal chromosome abnormalities in virtually all cases of acute lymphoblastic leukemia. N Engl J Med 1985; 313:640–641.
55. Raimondi SC, Behm FG, Roberson PK, Pui C-H, Rivera GK, Murphy SB, Williams DL. Cytogenetics of childhood T-cell leukemia. Blood 1988; 72: 1560–1566.
56. Callen DF, Raphael K, Michael PM, Garson OM. Acute lymphoblastic leukemia with a hypodiploid karyotype with less than 40 chromosomes: the basis for division into two subgroups. Leukemia 1989; 3:749–752.
57. Pui C-H, Carroll AJ, Raimondi SC, Land VJ, Crist WM, Shuster JJ, Williams DL, Pullen DJ, Borowitz MJ, Behm FG, Look AT. Clinical presentation, karyotypic characterization, and treatment outcome of childhood acute lymphoblastic leukemia with a near-haploid or hypodiploid <45 line. Blood 1990; 75:1170–1177.
58. Pui C-H, Williams DL, Raimondi SC, Rivera GK, Look AT, Dodge RK, George SL, Behm FG, Crist WM, Murphy SB. Hypodiploidy is associated with a poor prognosis in childhood acute lymphoblastic leukemia. Blood 1987;70:247–253.
59. Betts DR, Kingston JE, Dorey EL, Young BD, Webb D, Katz FE, Gibbons B. Monosomy 20: A nonrandom finding in childhood acute lymphoblastic leukemia. Genes Chromosom Cancer 1990; 2:182–185.
60. Rieder H, Schnittger S, Bodenstein H, Schwonzen M, Wormann B, Berkovic D, Ludwig W-D, Hoelzer D, Fonatsch C. dic(9;20): a new recurrent chromosome abnormality in adult acute lymphoblastic leukemia. Genes Chromosom Cancer 1995; 13:54–61.
61. Slater R, Smit E, Kroes W, Jotterand Bellomo M, Muhlematter D, Harbott J, Behrendt H, Hahlen K, Veerman AJP, Hagemeijer A. A non-random chromosome abnormality found in precursor-B lineage acute lymphoblastic leukaemia: dic(9;20)(p1?3;q11). Leukemia 1995; 9:1613–1619.
62. Heerema NA, Maben KD, Bernstein J, Breitfeld PP, Neiman RS, Vance GH. Dicentric (9;20)(p11;q11) identified by fluorescence in situ hybridization in

four pediatric acute lymphoblastic leukemia patients. Cancer Genet Cytogenet 1996; 92:111–115.

63. Huret J-L, Heerema NA, Brizard A, Provisor AJ, Benz-Lemoine E, Fuilhot F, Savage JRK, Tanzer J. Two additional cases of t dic(9;12) in acute lymphocytic leukemia (ALL): prognosis in ALL with dic(9;12). Leukemia 1990; 4:423–425.
64. Mahmoud H, Carroll AJ, Behm F, Raimondi SC, Schuster J, Borowitz M, Land V, Pullen DJ, Vietti TJ, Crist W. The non-random dic(9;12) translocation in acute lymphoblastic leukemia is associated with B-progenitor phenotype and an excellent prognosis. Leukemia 1992; 6:703–707.
65. Brodeur GM, Williams DL, Look AT, Bowman WP, Kalwinsky DK. Near-haploid acute lymphoblastic leukemia: a unique subgroup with a poor prognosis? Blood 1981; 58:14–19.
66. Tallents S, Forster DC, Garson OM, Michael PM, Briggs P, Brodie GN, Pilkington G, Januszewicz E. Hybrid biphenotypic acute leukemia with extreme hypodiploidy. Pathology 1987; 19:197–200.
67. Gibbons B, MacCallum P, Watts E, Rohatiner AZS, Webb D, Katz FE, Secker-Walker LM, Temperley IJ, Harrison CJ, Campbell RHA, Nash R, Broadbent V, Chessells JM. Near haploid acute lymphoblastic leukemia: seven new cases and a review of the literature. Leukemia 1991; 5:738–743.
68. Onodera N, McCabe NR, Nachman JB, Johnson FL, Le Beau MM, Rowley JD, Rubin CM. Hyperdiploidy arising from near-haploidy in childhood acute lymphoblastic leukemia. Genes Chromosom Cancer 1992; 4:331–336.
69. Williams DL, Look AT, Melvin SL, Roberson PK, Dahl G, Flake T, Stass S. New chromosomal translocations correlate with specific immunophenotypes of childhood acute lymphoblastic leukemia. Cell 1984; 36:101–109.
70. Carroll AJ, Crist WM, Parmley RT, Roper M, Cooper MD, Finley WH. Pre-B cell leukemia associated with chromosome translocation 1;19. Blood 1984; 63:721–724.
71. Michael PM, Levin MD, Garson OM. Translocation 1;19—a new cytogenetic abnormality in acute lymphocytic leukemia. Cancer Genet Cytogenet 1984; 12:333–341.
72. Shikano T, Kaneko Y, Takazawa M, Ueno N, Ohkawa M, Fujimoto T. Balanced and unbalanced 1;19 translocation-associated acute lymphoblastic leukemias. Cancer 1986; 58:2239–2243.
73. Lai JL, Fenaux P, Estienne MH, Huart JJ, Savary JB, Lepelley P, Jouet JP, Nelken B, Bauters F, Deminatii M. Translocation t(1;19)(q23;p13) in acute lymphoblastic leukemia: a report on six new cases and an unusual t(17;19)(q11;q13), with special reference to prognostic factors. Cancer Genet Cytogenet 1989; 37:9–17.
74. Pui C-H, Raimondi SC, Hancock ML, Rivera GK, Ribeiro RC, Mahmoud HH, Sandlund JT, Crist WM, Behm FG. Immunologic, cytogenetic, and clinical characterization of childhood acute lymphoblastic leukemia with the t(1;19)(q23;p13) or its derivative. J Clin Oncol 1994; 12:2601–2606.
75. Crist W, Boyett J, Jackson J, Vietti T, Borowitz M, Chauvenet A, Winick N, Ragab A, Mahoney D, Head D, Lyer R, Wagner H, Pullen J. Prognostic

importance of the pre-B-cell immunophenotype and other presenting features in B-lineage childhood acute lymphoblastic leukemia: a Pediatric Oncology Group study. Blood 1989; 74:1252–1259.
76. Raimondi SC, Behm FG, Roberson PK, Williams DL, Pui C-H, Crist WM, Look AT, Rivera GK. Cytogenetics of pre-B acute lymphoblastic leukemia with emphasis on prognostic implications of the t(1;19). J Clin Oncol 1990; 8:1380–1388.
77. Yamada T, Craig JM, Hawkins JM, Janossy G, Secker-Walter LM. Molecular investigation of 19p13 in standard and variant translocations: The E12 probe recognizes the 19p13 breakpoint in cases with t(1;19) and acute leukemia other than pre-B immunophenotype. Leukemia 1991; 5:36–40.
78. Troussard X, Rimokh R, Valensi F, Leboeuf C, Fenneteau O, Guitard A-M, Manel A-M, Schillinger F, Leglise C, Brizard A, Gardais J, Lessard M, Flandrin G, Macintyre E. Heterogeneity of t(1;19)(q23;p13) acute leukaemias. Br J Haematol 1995; 89:516–526.
79. Secker-Walker LM, Berger R, Fenaux P, Lai JL, Nelken B, Garson M, Michael PM, Hagemeijer A, Harrison CJ, Kaneko Y, Rubin CM. Prognostic significance of the balanced t(1;19) and unbalanced der(19)t(1;19) translocations in acute lymphoblastic leukemia. Leukemia 1992; 6:363–369.
80. Privitera E, Kamps MP, Hayashi Y, Inaba T, Shapiro LH, Raimondi SC, Behm F, Hendershot L, Carroll AJ, Baltimore D, Look AT. Different molecular consequences of the 1;19 chromosomal translocation in childhood B-cell precursor acute lymphoblastic leukemia. Blood 1992; 79:1781–1788.
81. Crist WM, Carroll AJ, Shuster JJ, Behm FG, Whitehead M, Vietti TJ, Look AT, Hahoney D, Ragab A, Pullen DJ, Land VJ. Poor prognosis of children with pre-B acute lymphoblastic leukemia is associated with the t(1;19)(q23;p13). A Pediatric Oncology Group study. Blood 1990; 76:117–122.
82. Mellentin JD, Murre C, Donlon TA, McCaw PS, Smith SD, Carroll AJ, McDonald ME, Baltimore D, Cleary ML. The gene for enhancer binding proteins E12/E47 lies at the t(1;19) breakpoint in acute leukemias. Science 1989; 246:379–382.
83. Nourse J, Mellentin JD, Galili N, Wilkinson J, Stanbridge E, Smith SD, Clearly ML. Chromosomal translocation t(1;19) results in synthesis of a homeobox fusion mRNA that codes for a potential chimeric transcription factor. Cell 1990; 60:535–545.
84. Kamps MP, Murre C, Sun X-H, Baltimore D. A new homeobox gene contributes the DNA binding domain of the t(1;19) translocation protein in pre-B ALL. Cell 1990; 60:547–555.
85. Hunger SP. Chromosomal translocations involving the E2A gene in acute lymphoblastic leukemia: clinical features and molecular pathogenesis. Blood 1996; 87:1211–1224.
86. Hunger SP, Galili N, Carroll AJ, Crist WM, Link MP, Cleary ML. The t(1;19)(q23;p13) results in consistent fusion of E2A and PBX1 coding sequences in acute lymphoblastic leukemias. Blood 1991; 77:687–693.
87. Izraeli S, Kovar H, Gadner H, Lion T. Unexpected heterogeneity in E2A/PBX1 fusion messenger RNA detected by the polymerase chain reaction in

pediatric patients with acute lymphoblastic leukemia. Blood 1992; 80:1413–1417.
88. Numata S-I, Kato K, Horibe K. New E2A/PBX1 fusion transcript in a patient with t(1;19)(q23;p13) acute lymphoblastic leukemia. Leukemia 1993; 7: 1441–1444.
89. Privitera E, Luciano A, Ronchetti D, Arico M, Santostasi T, Basso G, Biondi A. Molecular variants of the 1;19 chromosomal translocation in pediatric acute lymphoblastic leukemia (ALL). Leukemia 1994; 8:554–559.
90. Rieder H, Kolbus H, Koop U, Ludwig WD, Gibbels BS, Fonatsch CC. Translocation t(1;22) mimicking t(1;19) in a child with acute lymphoblastic leukemia as revealed by chromosome painting. Leukemia 1993; 7:1663–1666.
91. Filatov LV, Behm FG, Pui C-H, Head DR, Downing JR, Raimondi SC. Childhood acute lymphoblastic leukemia with equivocal chromosome markers of the t(1;19) translocation. Genes Chromosom Cancer 1995; 13:99–103.
92. Ribeiro RC, Abromowitch M, Raimondi SC, Murphy SB, Behm F, Williams DL. Clinical and biologic hallmarks of the Philadelphia chromosome in childhood acute lymphoblastic leukemia. Blood 1987; 70:948–953.
93. Crist W, Carroll A, Shuster J, Jackson J, Head D, Borowitz M, Behm F, Link M, Steuber P, Ragab A, Hirt A, Brock B, Land V, Pullen J. Philadelphia chromosome positive childhood acute lymphoblastic leukemia: clinical and cytogenetic characteristics and treatment outcome. A Pediatric Oncology Group study. Blood 1990; 76:489–494.
94. Fletcher JA, Lynch EA, Kimball VM, Donnelly M, Tantravahi R, Sallan SE. Translocation (9;22) is associated with extremely poor prognosis in intensively treated children with acute lymphoblastic leukemia. Blood 1991; 77: 435–439.
95. Schlieben S, Borkhardt A, Reinisch I, Ritterback J, Janssen JWG, Ratei R, Schrappe M, Repp R, Zimmermann M, Kabisch H, Janka-Schaub G, Bartram CR, Ludwig W-D, Riehm H, Lampert F, Harbott J. Incidence and clinical outcome of children with BCR/ABL-positive acute lymphoblastic leukemia (ALL). A prospective RT-PCR study based on 673 patients enrolled in the German pediatric multicenter therapy trials ALL-BFM-90 and CoALL-05-92. Leukemia 1996; 10:957–963.
96. Russo C, Carroll A, Kohler S, Borowitz M, Amylon M, Homans A, Kedar A, Shuster J, Land V, Crist W, Pullen J, Link M. Philadelphia chromosome and monosomy 7 in childhood acute lymphoblastic leukemia: a Pediatric Oncology Group study. Blood 1991; 77:1050–1056.
97. Roberts WM, Rivera G, Raimondi SC, Santana VM, Sandlund JT, Crist WM, Pui C-H. Intensive chemotherapy for Philadelphia-chromosome positive acute lymphoblastic leukemia. Lancet 1994; 343:331–332.
98. Secker-Walker LM, Craig JM, Hawkins JM, Hoffbrand AV. Philadelphia positive acute lymphoblastic leukemia in adults: age distribution, BCR breakpoint and prognostic significance. Leukemia 1991; 5:196–199.
99. Westbrook CA, Hoobermann AL, Spino C, Dodge RK, Larson RA, Davey F, Wurster-Hill DH, Sobol RE, Schiffer C, Bloomfield CD. Clinical significance

of the BCR-ABL fusion gene in adult acute lymphoblastic leukemia: a Cancer and Leukemia Group B study. Blood 1992; 80:2983–2990.

100. Hermans A, Heisterkamp N, von Lindern M, van Baal S, Meijer D, van der Plas D, Wiedemann LM, Groffen J, Bootsma D, Grosveld G. Unique fusion of BCR and c-ABL genes in Philadelphia positive acute lymphoblastic leukemia. Cell 1987; 51:33–40.
101. Clark SS, McLaughlin J, Crist WM, Champlin R, Witte ON. Unique forms of the abl tyrosine kinase distinguish Ph1-positive CML from Ph1-positive ALL. Science 1987; 235:85–88.
102. Clark SS, Crist WM, Witte ON. Molecular pathogenesis of Ph-positive leukemias. Annu Rev Med 1989; 40:113–122.
103. Kantarjian HM, Talpaz M, Dhingra K, Estey E, Keating MJ, Ku S, Trujillo J, Hug Y, Stass S, Kurzrock R. Significance of the P210 versus P190 molecular abnormalities in adults with Philadelphia chromosome-positive acute leukemia. Blood 1991; 79:2411–2418.
104. Specchia G, Mininni D, Guerrasio A, Palumbo G, Pastore D, Liso V. Ph-positive acute lymphoblastic leukemia in adults: molecular and clinical studies. Leukemia Lymphoma. 1995; 18:37–42.
105. Lugo TG, Pendergast A-M, Muller AJ, Witte ON. Tyrosine kinase activity and transformation potency of bcr-abl oncogene products. Science 1990; 247: 1079–1082.
106. Daley GQ, Van Etten RA, Baltimore D. Induction of chronic myelogenous leukemia in mice by the p210 bcr-abl gene of the Philadelphia chromosome. Science 1990; 247:824–830.
107. Voncken JW, Kaartinen V, Pattengale PK, Germeraad WTV, Groffen J, Heisterkamp N. BCR/ABL p210 and p190 cause distinct leukemia in transgenic mice. Blood 1995; 86:4603–4611.
108. Hooberman AL, Westbrook CA. Molecular diagnosis of the Philadelphia chromosome in acute lymphoblastic leukemia. Leukemia Lymphoma 1989; 1:3–10.
109. Beyermann B, Agthe AG, Adams H-P, Seeger K, Linderkamp C, Goetze G, Ludwig W-D, Henze G for the BFM Relapse Study Group. Clinical features and outcome of children with first marrow relapse of acute lymphoblastic leukemia expressing BCR-ABL fusion transcripts. Blood 1996; 87:1532–1538.
110. Arnoldus EPJ, Wiegant J, Noordermeer IA, Wessels JW, Beverstock GC, Grosveld GC, van der Ploeg M, Raap AK. Detection of the Philadelphia chromosome in interphase nuclei. Cytogenet Cell Genet 1990; 54:108–111.
111. Tkachuk DC, Westbrook CA, Andreeff M, Donlon TA, Cleary ML, Suryanarayan K, Homge M, Redner A, Gray F, Pinkel D. Detection of *bcr-abl* fusion in chronic myelogenous leukemia by in situ hybridization. Science 1990; 250:559–562.
112. Anastasi J, Feng J, Dickstein JI, Le Beau MM, Rubin CM, Larson RA, Rowley JD, Vardiman JW. Lineage involvement by BCR/ABL in Ph+ lymphoblastic leukemias: chronic myelogenous leukemia presenting in lymphoid

blast phase vs. Ph+ acute lymphoblastic leukemia. Leukemia 1996; 10:795–802.

113. Bentz M, Cabot G, Moos M, Speicher MR, Ganser A, Lichter P, Dohner H. Detection of chimeric BCR-ABL genes on bone marrow samples and blood smears in chronic myeloid and acute lymphoblastic leukemia by in situ hybridization. Blood 1994; 83:1922–1928.
114. Latham JAE, Bown N, Cain G, Middleton P, Kernaham J, Windebank K, Reid MM. Molecular heterogeneity of variant Philadelphia translocations in childhood acute lymphoblastic leukaemia. Leukemia 1994; 8:292–294.
115. Seong DC, Kantarjian HM, Ro JY, Talpaz M, Xu J, Robinson JR, Deisseroth AB, Champlin RE, Siciliano MJ. Hypermetaphase fluorescence in situ hybridization for quantitative monitoring of Philadelphia chromosome-positive cells in patients with chronic myelogenous leukemia during treatment. Blood 1995; 86:2343–2349.
116. Kaneko Y, Maseki N, Takasaki N, Sakurai M, Hayashi Y, Nakazawa S, Mori T, Sakurai M, Takeda T, Shikano T, Hiyoshi Y. Clinical and hematologic characteristics in acute leukemia with 11q23 translocations. Blood 1986; 67:484–491.
117. Raimondi SC, Peiper SC, Kitchingman GR, Behm FG, Williams DL, Hancock ML, Mirro J. Childhood acute lymphoblastic leukemia with chromosomal breakpoints at 11q23. Blood 1989; 73:1627–1634.
118. Lampert F, Harbott J, Ludwig W-D, Bartram C-R, Ritter J, Gerein U, Neidhardt M, Mertens R, Graf N, Riehm H. Acute leukemia with chromosome translocation (4;11): 7 new patients and analysis of 71 cases. Blut 1987; 54:325–335.
119. Pui C-H, Frankel LS, Carroll AJ, Raimondi SC, Shuster JJ, Head DR, Crist WM, Land VJ, Pullen DJ, Steuber CP, Behm FG, Borowitz MJ. Clinical characteristics and treatment outcome of childhood acute lymphoblastic leukemia with the t(4;11)(q21;q23): a collaborative study of 40 cases. Blood 1991; 77:440–447.
120. Human Gene Mapping-11: Eleventh International Workshop on Human Gene Mapping. Cytogenet Cell Genet 1991; 58:Nos 1–4.
121. Raimondi SC, Kalwinsky DK, Hayashi Y, Behm FG, Mirro J, Williams DL. Cytogenetics of childhood acute nonlymphocytic leukemia. Cancer Genet Cytogenet 1989; 40:13–27.
122. Pui C-H, Raimondi SC, Murphy SB, Ribeiro RC, Kalwinsky DK, Dahl GV, Crist WM, Williams DL. An analysis of leukemic cell chromosomal features in infants. Blood 1987; 9:1289–1293.
123. Chen C-S, Sorensen PHB, Domer PH, Reaman GH, Korsmeyer SJ, Heerema NA, Hammond GD, Kersey JH. Molecular rearrangements on chromosome 11q23 predominate in infant acute lymphoblastic leukemia and are associated with specific biologic variables and poor outcome. Blood 1993; 81:2386–2393.
124. Cimino G, Lo Coco F, Biondi A, Elia L, Luciano A, Croce CM, Masera G, Mandelli F, Canaani E. ALL-1 gene at chromosome 11q23 is consistently altered in acute leukemia of early infancy. Blood 1993; 82:544–546.

125. Rubnitz JE, Link MP, Shuster JJ, Carroll AJ, Hakami N, Frankel LS, Pullen DJ, Cleary ML. Frequency and prognostic significance of HRX rearrangements in infant acute lymphoblastic leukemia: a Pediatric Oncology Group study. Blood 1994; 84:570–573.
126. Chromosome Workshop: report of the Fifth International Workshop on Human Chromosome 11 Mapping. Cytogenet Cell Genet 1996; 74:1–56.
127. Zieman-van der Poel S, McCabe NR, Gill HJ, Espinosa R, Patel S, Harden A, Rubinelli P, Smith SD, Le Beau MM, Rowley JD, Diaz MO. Identification of a gene, MLL, that spans the breakpoint in 11q23 translocations associated with human leukemias. Proc Natl Acad Sci USA 1991; 88:10735–10739.
128. Cimino G, Moir DT, Canaani O, Williams K, Crist WM, Katzav S, Cannizzaro L, Lange B, Nowell PC, Croce CM, Canaani E. Cloning of ALL-1, the locus involved in leukemias with the t(4;11)(q21;q23), t(9;11)(p22;q23), and t(11;19)(q23;p13) chromosome translocations. Cancer Res 1991; 51:6712–6714.
129. Gu Y, Nakamura T, Alder H, Prasad R, Canaani O, Cimino G, Croce CM, Canaani E. The t(4;11) chromosome translocation of human acute leukemias fuses the ALL-1 gene, related to *Drosophila trithorax*, to the AF-4 gene. Cell 1992; 71:701–708.
130. Tkachuk DC, Kohler S, Cleary ML. Involvement of a homolog of *Drosophila trithorax* by 11q23 chromosomal translocations in acute leukemias. Cell 1992; 71:691–700.
131. Domer Ph, Fakharzadeh SS, Chen C-S, Jockel J, Johansen L, Silverman GA, Kersey JH, Korsmeyer SJ. Acute mixed-lineage leukemia t(4;11)(q21;q23) generates an MLL-AF4 fusion product. Proc Natl Acad Sci USA 1993; 90:7884–7888.
132. Downing JR, Pullen DR, Raimondi SC, Carroll AJ, Curcio-Brint AM, Motroni TA, Hulshof MG, Pullen DJ, Domer PH. The der(11)-encoded MLL-AF-4 fusion transcript is consistently detected in t(4;11)(q21;q23)-containing acute lymphoblastic leukemia. Blood 1994; 83:330–335.
133. Rowley JD, Diaz MO, Espinosa R, Patel YD, van Melle E, Ziemin S, Taillon-Miller P, Lichter P, Evans GA, Kersey JH, Ward DC, Domer PH, Le Beau MM. Mapping chromosome band 11q23 in human acute leukemia with biotinylated probes: identification of 11q23 translocation breakpoints with a yeast artificial chromosome. Proc Natl Acad Sci USA 1990; 87:9358–9362.
134. Thirman MJ, Gill HJ, Burnett RC, Mbangkollo D, McCabe NR, Kobayashi H, Ziemin-Van der Poel S, Kaneko Y, Morgan R, Sandberg AA, Chaganti RSK, Larson RA, Le Beau MM, Diaz MO, Rowley JD. Rearrangement of the MLL gene in acute lymphoblastic and acute myeloid leukemias with 11q23 chromosomal translocations. N Engl J Med 1993; 329:909–914.
135. Kobayashi H, Espinosa R III, Thirman MJ, Fernald AA, Shannon K, Diaz MO, Le Beau MM, Rowley JD. Do terminal deletions of 11q23 exist? Identificaton of undetected translocations with FISH. Genes Chromosom Cancer 1993; 7:204–208.
136. Kobayashi H, Espinosa R III, Fernald AA, Begy C, Diaz MO, Le Beau MM, Rowley JD. Analysis of deletions of the long arm of chromosome 11 in

hematological malignancies with fluorescence in situ hybridization. Genes Chromosom Cancer 1993; 8:246–252.
137. Caligiuri MA, Schichman SA, Strout MP, Mrozek K, Baer MR, Frankel SR, Barcos M, Herzig GP, Croce CM, Bloomfield CD. Molecular rearrangement of the ALL-1 gene in acute myeloid leukemia without cytogenetic evidence of 11q23 chromosomal translocations. Cancer Res 1994; 54:370–373.
138. Schichman SA, Caligiuri MA, Gu Y, Strout MP, Canaani E, Bloomfield CD, Croce CM. ALL-1 partial duplication in acute leukemia. Proc Natl Acad Sci USA 1994; 91:6236–6239.
139. Martinez-Climent JA, Thirman MJ, Espinosa R III, Le Beau MM, Rowley JD. Detection of 11q23/MLL rearrangements in infant leukemias with fluorescence in situ hybridization and molecular analysis. Leukemia 1995; 9: 1299–1304.
140. Taki T, Ida K, Bessho F, Hanada R, Kikuchi A, Yamamoto K, Sako M, Tsuchida M, Seto M, Ueda R, Hayashi Y. Frequency and clinical significance of the MLL gene rearrangements in infant acute leukemia. Leukemia 1996; 10: 1303–1307.
141. Pui C-H, Behm FG, Downing JR, Hancock ML, Shurtleff SA, Ribeiro RC, Head DR, Mahmoud HH, Sandlund JT, Furman WL, Roberts WM, Crist WM, Raimondi SC. 11q23/MLL rearrangement confers a poor prognosis in infants with acute lymphoblastic leukemia. J Clin Oncol 1994; 12:909–915.
142. Behm FG, Raimondi SC, Frestedt JL, Liu Q, Crist WM, Downing JR, Rivera GK, Kersey JH, Pui C-H. Rearrangement of MLL confers a poor prognosis in childhood acute lymphoblastic leukemia, regardless of presenting age. Blood 1996; 7:2870–2877.
143. Raimondi SC, Frestedt JL, Pui C-H, Downing JR, Head DR, Kersey JH, Behm FG. Acute lymphoblastic leukemias with deletion of 11q23 or a novel inversion (11)(p13q23) lack MLL gene rearrangements and have favorable clinical features. Blood 1995; 86:1881–1886.
144. Smith M, Arthur D, Camitta B, Carroll D, Crist W, Gaynor P, Golbar R, Heerema N, Korn EL, Link M, Murphy S, Pui C-H, Pullen J, Reaman G, Sallan SE, Sather H, Shuster J, Simon R, Trigg M, Tuburgen D, Uckun F, Ungerloider R. Uniform approach to risk classification and treatment assignment for children with acute lymphoblastic leukemia. J Clin Oncol 1996; 14: 18–24.
145. Schichman SA, Caligiuri MA, Gu Y, Strout MP, Canaani E, Bloomfield CD, Croce CM. ALL-1 partial duplication in acute leukemia. Proc Natl Acad Sci USA 1994; 91:6236–6239.
146. Yu M, Honoki K, Andersen J, Paietta E, Nam SK, Yunis JJ. MLL tandem duplication and multiple splicing in adult acute myeloid leukemia with normal karyotype. Leukemia 1996; 10:774–780.
147. Caligiuri MA, Strout MP, Schichman SA, Mrozek K, Arthur DC, Herzig GP, Baer MR, Schiffer CA, Heinonen K, Knuutila S, Nousiainen T, Ruutu T, Block AMW, Schulman P, Pedersen-Bjergaard J, Croce MC, Bloomfield CD. Partial tandem duplication of ALL1 as a recurrent molecular de-

fect in acute myeloid leukemia with trisomy 11. Cancer Res 1996; 56:1418–1425.

148. Hayashi Y, Raimondi SC, Look AT, Behm FG, Kitchingman GR, Pui C-H, Rivera GK, Williams DL. Abnormalities of the long arm of chromosome 6 in childhood acute lymphoblastic leukemia. Blood 1990; 76:1626–1630.
149. Barletta C, Pelicci P-G, Kenyon LC, Smith SD, Dalla-Favera R. Relationship between the c-myb locus and the 6q− chromosomal aberration in leukemias and lymphomas. Science 1987; 235:1064–1067.
150. Ohyashiki K, Ohyashiki JH, Kinniburgh AJ, Toyama K, Ito H, Minowada J, Sandberg AA. myb oncogene in human hematopoietic neoplasia with 6q− anomaly. Cancer Genet Cytogenet 1988; 33:83–92.
151. Li H, Lahti JM, Valentine M, Saito M, Reed SI, Look AT, Kidd VJ. Molecular cloning and chromosomal localization of the human cyclin C (CCNC) and cyclin E (CCNE) genes: deletion of the CCNC gene in human tumors. Genomics 1996; 32:253–259.
152. Kowalczyk J, Sandberg AA. A possible subgroup of ALL with 9p−. Cancer Genet Cytogenet 1983; 9:383–385.
153. Chilcote RR, Brown E, Rowley JD. Lymphoblastic leukemia with lymphomatous features associated with abnormalities of the short arm of chromosome 9. N Engl J Med 1985; 313:286–291.
154. Carroll AJ, Castleberry RP, Crist WM. Lack of association between abnormalities of the chromosome 9 short arm and either "lymphomatous" features of T-cell phenotype in childhood acute lymphocytic leukemia. Blood 1987; 69:735–738.
155. Pollak C, Hagemeijer A. Abnormalities of the short arm of chromosome 9 with partial loss of material in hematological disorders. Leukemia 1987; 1: 541–548.
156. Murphy SB, Raimondi SC, Rivera GK, Crone M, Dodge RK, Behm FG, Pui C-H, Williams DL. Nonrandom abnormalities of chromosome 9p in childhood acute lymphoblastic leukemia: association with high-risk clinical features. Blood 1989; 74:409–415.
157. Raimondi SC, Privitera E, Williams DL, Look AT, Behm F, Rivera GK, Crist WM, Pui C-H. New recurring chromosomal translocations in childhood acute lymphoblastic leukemia. Blood 1991; 77:2016–2022.
158. Diaz MO, Rubin CM, Harden A, Ziemin S, Larson RA, LeBeau MM, Rowley JD. Deletions of interferon genes in acute lymphoblastic leukemia. N Engl J Med 1990; 322:77–82.
159. Middleton PG, Prince RA, Williamson IK, Taylor PRA, Reid MM, Jackson GH, Katz F, Chessells JM, Proctor SJ. Alpha interferon gene deletions in adults, children and infants with acute lymphoblastic leukemia. Leukemia 1991; 5:680–682.
160. Kamb A, Gruis NA, Weaver-Feldhaus J, Liu Q, Harshman K, Tavtigian SV, Stockert E, Day RS III, Johnson BE, Skolnick MH. A cell cycle regulator potentially involved in genesis of many tumor types. Science 1994; 264:436–440.

161. Nobori T, Miura K, Wu DJ, Lois A, Takabayashi K, Carson DA. Deletions of the cyclin-dependent kinase 4 inhibitor gene in multiple human cancers. Nature 1994; 368:753–756.
162. Hannon GJ, Beach D. $p15^{INK4B}$ is a potential effector of TGF-β-induced cell cycle arrest. Nature 1994; 371:257–261.
163. Okuda T, Shurtleff SA, Valentine MB, Raimondi SC, Head DR, Behm F, Curcio-Brint AM, Liu Q, Pui C-H, Sherr CJ, Beach D, Look AT, Downing JR. Frequent deletion of $p16^{INK4a}$/MTS1 and $p15^{INK4b}$/MTS5 in pediatric acute lymphoblastic leukemia. Blood 1995; 85:2321–2330.
164. Kees UR, Ranford PR, Hatzis M. Deletions of the p16 gene in pediatric leukemia and corresponding cell lines. Oncogene 1996; 12:2235–2239.
165. Quesnel B, Preudhomme C, Fenaux P. $p16^{ink4a}$ gene and hematological malignancies. Leukemia Lymphoma 1996; 22:11–24.
166. Raimondi SC, Williams DL, Callihan T, Peiper S, Rivera GK, Murphy SB. Nonrandom involvement of the 12p12 breakpoint in chromosome abnormalities of childhood acute lymphoblastic leukemia. Blood 1986; 68:69–75.
167. van der Plas DC, Dekker I, Hagemeijer A, Hooijkaas H, Hahlen K. 12p chromosomal aberrations in precursor B childhood acute lymphoblastic leukemia predict an increased risk of relapse in the central nervous system and are associated with typical blast cell morphology. Leukemia 1994; 8:2041–2046.
168. Carroll AJ, Raimondi SC, Williams DL, Behm FG, Borowitz M, Castleberry RP, Harris MB, Patterson RB, Pullen DJ, Crist WM. t dic(9;12). A nonrandom chromosome abnormality in childhood B-cell precursor acute lymphoblastic leukemia: a Pediatric Oncology Group study. Blood 1987; 70:1962–1965.
169. United Kingdom Cancer Cytogenetics Group (UKCCG). Translocations involving 9p and/or 12p in acute lymphoblastic leukemia. Genes Chromosom Cancer 1992; 5:255–259.
170. Behrendt H, Charrin C, Gibbons B, Harrison CJ, Hawkins JM, Heerema NA, Horschler-Botel B, Huret JL, Lai JL, Lampert F, Nelken B, Perot C, Ritterbach J, Schlegelberger B, Secker-Walker L, Slater R, Slovak ML, Tanzer J, Van Den Akker J. Dicentric (9;12) in acute lymphocytic leukemia and other hematological malignancies: report from a dic(9;12) study group. Leukemia 1995; 9:102–106.
171. Keene P, Mendelow B, Pinto MR, Bezwoda W, MacDougall L, Falkson G, Ruff P, Bernstein R. Abnormalities of chromosome 12p13 and malignant proliferation of eosinophils: a nonrandom association. Br J Haematol 1987; 67: 25–31.
172. Krance RA, Raimondi SC, Dubowy R, Estrada J, Borowitz M, Behm F, Land VJ, Pullen J, Carroll AJ. t(12;17)(p13;q21) in early pre-B acute lymphoid leukemia. Leukemia 1992; 6:251–255.
173. Raimondi SC, Shurtleff SA, Downing JR, Rubnitz J, Mathew S, Hancock M, Pui C-H, Rivera GK, Grosveld GC, Behm FG. 12p abnormalities and the *TEL* gene (*ETV6*) in childhood acute lymphoblastic leukemia. Blood 1997; 90: 4559–4560.

174. Golub TR, Barker GF, Lovett M, Gilliland DG. Fusion of PDGF receptor β to a novel ets-like gene, tel, in chronic myelomonocytic leukemia with t(5; 12) chromosomal translocation. Cell 1994; 77:307–316.
175. Raynaud SD, Baens M, Grosgeorge J, Rodgers K, Reid CDL, Dainton M, Dyer M, Fuzibet JG, Gratecos N, Taillan B, Ayraud N, Marynen P. Fluorescence in situ hybridization analysis of t(3;12)(q26;p13): a recurring chromosomal abnormality involving the TEL gene (ETV6) in myelodysplastic syndromes. Blood 1996; 88:682–689.
176. Peeters P, Wlodarska I, Baens M, Criel A, Selleslag D, Hagemeijer A, Van den Berghe H, Marynen P. Fusion of ETV6 to MDS1/EVI1 as a result of t(3; 12)(q26;p13) in myeloproliferative disorders. Cancer Res 1997; 57:564–569.
177. Janssen JWG, Ridge SA, Papadopoulos P, Cotter F, Ludwig W-D, Fonatsch C, Rieder H, Ostertag W, Bartram CR, Wiedemann LM. The fusion of TEL and ABL in human acute lymphoblastic leukaemia is a rare event. Br J Haematol 1995;90:222–224.
178. Papadopoulos P, Ridge SA, Boucher CA, Stocking C, Wiedemann LM. The novel activation of ABL by fusion to an ets-related gene, TEL. Cancer Res 1995; 55:34–38.
179. Buijs A, Sherr S, van Baal S, van Bezouw S, van der Plas D, van Kessel AG, Riegman P, Deprez RL, Zwarthoff E, Hagemeijer A, Grosveld G. Translocation (12;22)(p13;q11) in myeloproliferative disorders results in fusion of the ETS-like TEL gene on 12p13 to the MN1 gene on 22q11. Oncogene 1995; 10:1511–1519.
180. Wlodarska I, Mecucci C, Marynen P, Guo C, Franckx D, La Starza R, Avetin A, Bosly A, Martelli MF, Cassiman JJ, Van den Berghe H. TEL gene is involved in myelodysplastic syndromes with either the typical t(5;12)(q33; p13) translocation or its variant t(10;12)(q24;p13). Blood 1995; 85:2848–2852.
181. Golub TR, Barker GF, Bohlander SK, Hiebert SW, Ward DC, Bray-Ward P, Morgan E, Raimondi SC, Rowley JD, Gilliland DG. Fusion of the TEL gene on 12p13 to the AML1 gene on 21q22 in acute lymphoblastic leukemia. Proc Natl Acad Sci USA 1995; 92:4917–4921.
182. Romana SP, Le Coniat M, Berger R. t(12;21): a new recurrent translocation in acute lymphoblastic leukemia. Genes Chromosom Cancer 1994; 9:186–191.
183. Shurtleff SA, Buijs A, Behm FG, Rubnitz JE, Raimondi SC, Hancock ML, Chan GC-F, Pui C-H, Grosveld G, Downing JR. TEL/AML1 fusion resulting from a cryptic t(12;21) is the most common genetic lesion in pediatric ALL and defines a subgroup of patients with an excellent prognosis. Leukemia 1995; 9:1985–1989.
184. Filatov LV, Saito M, Behm FG, Rivera GK, Raimondi SC. A subtle deletion of 12p by routine cytogenetics is found to be a translocation to 21q by fluorescence in situ hybridization: t(12;21)(p13;q22). Cancer Genet Cytogenet 1996; 89:136–140.
185. McLean TW, Ringold S, Neuberg D, Stegmaier K, Tantravahi R, Ritz J, Koeffler HP, Takeuchi S, Janssen JWG, Seriu T, Bartram CR, Sallan SE, Gilliland

DG, Golub TR. TEL/AML-1 dimerizes and is associated with a favorable outcome in childhood acute lymphoblastic leukemia. Blood 1996; 88:4252–4258.

186. Rubnitz JE, Downing JR, Pui C-H, Shurtleff SA, Raimondi SC, Evans WE, Head DR, Crist WM, Rivera GK, Hancock ML, Boyett JM, Buijs A, Grosveld G, Behm FG. TEL gene rearrangement in acute lymphoblastic leukemia: a new genetic marker with prognostic significance. J Clin Oncol 1997; 15: 1150–1157.
187. Raynaud S, Mauvieux L, Cayuela JM, Bastard C, Bilhou-Nabera C, Debuire B, Bories D, Boucheix C, Charrin C, Fiere D, Gabert J. TEL/AML1 fusion gene is a rare event in adult acute lymphoblastic leukemia. Leukemia 1996; 10:1529–1530.
188. Shih L-Y, Chou T-B, Liang D-C, Tzeng Y-S, Rubnitz JE, Downing JR, Pui C-H. Lack of TEL-AML1 fusion transcript resulting from a cryptic t(12;21) in adult B lineage acute lymphoblastic leukemia in Taiwan. Leukemia 1996; 10:1456–1458.
189. Kobayashi H, Satake N, Maseki N, Sakashita A, Kaneko Y. The der(21)t(12; 21) chromosome is always formed in a 12;21 translocation associated with childhood acute lymphoblastic leukemia. Br J Haematol 1996; 94:105–111.
190. Pietenpol JA, Bohlander SK, Sato Y, Papadopoulos N, Liu B, Friedman C, Trask BJ, Roberts JM, Kinzler KW, Rowley JD, Vogelstein B. Assignment of the human $p27^{Kip1}$ gene to 12p13 and its analysis in leukemias. Cancer Res 1995; 55:1206–1210.
191. Cavè H, Gerard B, Martin E, Guidal C, Devaux I, Weissenbach, J, Elion J, Vilmer E, Grandchamp B. Loss of heterozygosity in the chromosomal region 12p12-13 is very common in childhood acute lymphoblastic leukemia and permits the precise localization of a tumor-suppressor gene distinct from $p27^{KIP1}$. Blood 1995; 86:3869–3875.
192. Ponce-Castaneda MV, Lee M-H, Latres E, Polyak K, Lacombe L, Montgomery K, Mathew S, Krauter K, Sheinfield J, Massague J, Cordon-Cardo C. $p27^{Kip1}$: chromosomal mapping to 12p12-12p13.1 and absence of mutations in human tumors. Cancer Res 1995; 55:1211–1214.
193. Takeuchi S, Bartram CR, Miller CW, Reiter A, Seriu T, Zimmermann M, Schrappe M, Mori N, Slater J, Miyoshi I, Koeffler HP. Acute lymphoblastic leukemia of childhood: identification of two distinct regions of delection on the short arm of chromosome 12 in the region of TEL and KIP1. Blood 1996; 87:3368–3374.
194. Fero ML, Rivkin M, Tasch M, Porter P, Carow CE, Firpo E, Polyak K, Tsai L-H, Broudy V, Pelmutter RM, Kaushansky K, Roberts JM. A syndrome of multiorgan hyperplasia with features of gigantism, tumorigenesis, and female sterility in $p27^{kip1}$-deficient mice. Cell 1996; 85:733–744.
195. Manolov G, Manolova Y. Marker band in one chromosome 14 from Burkitt lymphomas. Nature 1972; 237:33–34.
196. Berger R, Bernheim A, Brouet JC, Daniel MT, Flandrin G. t(8;14) translocation in Burkitt's type of lymphoblastic leukaemia (L3). Br J Haematol 1979; 43:87–90.

197. Lai JL, Fenaux P, Zandecki M, Nelken B, Huart JJ, Deminatti M. Cytogenetic studies in 30 patients with Burkitt's lymphoma of L3 acute lymphoblastic leukemia with special reference to additional chromosome abnormalities. Ann Genet 1989; 32:26–32.
198. Magrath IT, Ziegler JL. Bone marrow involvement in Burkitt's lymphoma and its relationships to acute B-cell leukemia. Leukemia Res 1979; 4:33–59.
199. Solomons HD, Mendelow B, Levinstein A, Johnson A-E, Bernstein R. Burkitt cell leukemia with abnormality of chromosome No. 1. Am J Hematol 1983; 14:297–300.
200. Bowman WP, Shuster JJ, Cook B, Griffin T, Behm F, Pullen J, Link M, Head D, Carroll A, Berard C, Murphy S. Improved survival for children with B-cell acute lymphoblastic leukemia and stage IV small noncleaved-cell lymphoma: a Pediatric Oncology Group study. J Clin Oncol 1996; 14:1252–1261.
201. Dalla-Favera R, Bregni M, Erikson J, Patterson D, Gallo RC, Croce CM. Human c-myc oncogene is located on the region of chromosome 8 that is translocated in Burkitt lymphoma cells. Proc Natl Acad Sci USA 1982; 79: 7824–7827.
202. Croce CM, Nowell PC. Molecular basis of human B cell neoplasia. Blood 1985; 65:1–7.
203. Rabbitts TH, Forster A, Hamlyn P, Baer R. Effect of somatic mutation within translocated c-myc genes in Burkitt's lymphoma. Nature 1984; 309:592–597.
204. Blackwood EM, Eisenman RN. Max: a helix-loop-helix zipper protein that forms a sequence-specific DNA-binding complex with Myc. Science 1991; 251:1211–1217.
205. Ayers DE, Kretzner L, Eisenman RN. MAD: a heterodimeric partner for MAX that antagonizes MYC transcriptional activity. Cell 1993; 72:211–222.
206. Pui C-H, Behm FG, Singh B, Schell MJ, Williams DL, Rivera GK, Kalwinsky DK, Sandlund JT, Crist WM, Raimondi SC. Heterogeneity of presenting features and their relation to treatment outcome in 120 children with T-cell acute lymphoblastic leukemia. Blood 1990; 75:174–179.
207. Berger R, Le Coniat M, Vecchione D, Derre J, Chen SJ. Cytogenetic studies of 44 T-cell acute lymphoblastic leukemias. Cancer Genet Cytogenet 1990; 44:69–75.
208. Kaneko Y, Frizzera G, Shikano T, Kobayashi H, Maseki N, Sakurai M. Chromosomal and immunophenotypic patterns in T cell acute lymphoblastic leukemia (T ALL) and lymphoblastic lymphoma (LBL). Leukemia 1989; 3:886–892.
209. Croce CM, Isobe M, Palumbo A, Puck J, Ming J, Tweardy D, Erikson J, David M, Rovera G. Gene for alpha-chain of human T-cell receptor: location of chromosome 14 region involved in T-cell neoplasms. Science 1985; 227: 1044–1047.
210. Ribeiro RC, Raimondi SC, Behm FG, Cherrie J, Crist WM, Pui C-H. Clinical and biologic features of childhood T-cell leukemia with the t(11;14). Blood 1991; 78:466–470.
211. Erikson J, Williams DL, Finan J, Nowell PC, Croce CM. Locus of the α-

chain of the T-cell receptor is split by chromosome translocation in T-cell leukemias. Science 1985; 229:784–786.

212. Boehm T, Baer R, Lavenir I, Forster A, Waters JJ, Nacheva E, Rabbitts TH. The mechanism of chromosomal translocation t(11;14) involving the T-cell receptor δ-locus on human chromosome 14q11 and a transcribed region of chromosome 11p15. EMBO J 1988; 7:385–394.
213. Yoffe G, Schneider N, Van Dyk L, Yang CY-C, Siciliano M, Buchanan G, Capra JD, Baer R. The chromosome translocation (11;14)(p13;q11) associated with T-cell acute lymphocytic leukemia: an 11p13 breakpoint cluster region. Blood 1989; 74:374–379.
214. Boehm T, Foroni L, Kaneko Y, Perutz MF, Rabbitts TH. The rhombotin family of cysteine-rich LIM-domain oncogenes: distinct members are involved in T-cell translocations to human chromosomes 11p15 and 11p13. Proc Natl Acad Sci USA 1991; 88:4367–4371.
215. Royer-Pokora B, Loos U, Ludwig W-D. TTG-2, a new gene encoding a cysteine-rich protein with the LIM motif, is overexpressed in acute T-cell leukaemia with the t(11;14)(p13;q11). Oncogene 1991; 6:1887–1893.
216. Le Beau MM, McKeithan TW, Shima EA, Goldman-Leikin RE, Chan SJ, Bell GI, Rowley JD, Diaz MO. T-cell receptor α-chain gene is split in a human T-cell leukemia cell line with a t(11;14)(p15;q11). Proc Natl Acad Sci USA 1986; 83:9744–9748.
217. McGuire EA, Hockett RD, Pollock KM, Bartholdi MF, O'Brien SJ, Korsmeyer SJ. The t(11;14)(p15;q11) in a T-cell acute lymphoblastic leukemia cell line activates multiple transcripts, including Ttg-1, a gene encoding a potential zinc finger protein. Mol Cell Biol 1989; 9:2124–2132.
218. Neale GAM, Rehg JE, Goorha RM. Ectopic expression of rhombotin-2 causes selective expansion of CD4- CD8- lymphocytes in the thymus and T-cell tumors in transgenic mice. Blood 1995; 86:3060–3071.
219. Dube ID, Raimondi SC, Pi D, Kalousek DK. A new translocation, t(10; 14)(q24;q11), in T-cell neoplasia. Blood 1986; 67:1181–1184.
220. Zutter M, Hockett RD, Roberts CWM, McGuire EA, Bloomstone J, Morton CC, Deaven LL, Crist WM, Carroll AJ, Korsmeyer S. The t(10;14)(q24;q11) of T-cell acute lymphoblastic leukemia juxtaposes the δ T-cell receptor with TCL-3, a conserved and activated locus at 10q24. Proc Natl Acad Sci USA 1990; 87:3161–3165.
221. Lu M, Dube ID, Raimondi SC, Carroll A, Zhao Y, Minden M, Sutherland P. Molecular characterization of the t(10;14) translocation breakpoints in T-cell acute lymphoblastic leukemia: further evidence for illegitimate physiological recombination: Genes Chromosom Cancer 1990; 2:217–222.
222. Hatano M, Roberts CWM, Minden M, Crist WM, Korsmeyer SJ. Deregulation of a homeobox gene, HOX11, by the t(10;14) in T-cell leukemia. Science 1991; 253:79–82.
223. Dube ID, Kamel-Reid S, Yuan CC, Lu M, Wu X, Corpus G, Raimondi SC, Crist WM, Carroll AJ, Minowada J, Baker JB. A novel human homeobox gene lies at the chromosome 10 breakpoint in lymphoid neoplasias with chromosomal translocation t(10;14). Blood 1991; 78:2996–3003.

224. Kagan J, Finger LR, Besa E, Croce CM. Detection of minimal residual disease in leukemic patients with the t(10;14)(q24;q11) chromosomal translocation. Cancer Res 1990; 50:5240–5244.
225. Lange BJ, Raimondi SC, Heerema N, Nowell PC, Minowada J, Steinherz PE, Arenson EB, O'Connor R, Santoli D. Pediatric leukemia/lymphoma with t(8; 14)(q24;q11). Leukemia 1992; 6:613–618.
226. Erikson J, Finger L, Sun L, Ar-Rushdi A, Nishikura K, Minowada J, Finan J, Emanuel BS, Nowell PC, Croce CM. Deregulation of c-myc by translocation of the α-locus of T-cell receptor in T-cell leukemias. Science 1986; 232:884–886.
227. Carroll AJ, Crist WM, Link MP, Amylon MD, Pullen JP, Ragab AH, Buchanan GR, Wimmer RS, Vietti TJ. The t(1;14)(p34;q11) is non-random and restricted to T-cell acute lymphoblastic leukemia. Blood 1990; 76:1220–1224.
228. Finger LR, Kagan J, Christopher G, Kurtzberg J, Hershfield MS, Nowell PC, Croce CM. Involvement of the Tcl5 gene on human chromosome 1 in T-cell leukemia and melanoma. Proc Natl Acad Sci USA 1989; 86:5039–5043.
229. Begley CG, Aplan PD, Davey MP, Nakahara K, Tchorz K, Kurtzberg J, Hershfield MS, Haynes BF, Cohen DI, Waldmann TA, Kirsch IR. Chromosomal translocation in a human leukemic stem-cell line disrupts the T-cell antigen receptor delta-chain diversity region and results in a previously unreported fusion transcript. Proc Natl Acad Sci USA 1989; 86:2031–2035.
230. Chen Q, Cheng J-T, Tsai L-H, Schneider N, Buchanan G, Carroll W, Crist W, Ozanne A, Siciliano MJ, Baer R. The tal-1 gene undergoes chromosome translocation in T cell leukemia and potentially encodes a helix-loop-helix protein. EMBO J 1990; 9:415–424.
231. Begley CG, Aplan PD, Denning SM, Haynes BF, Waldmann TA, Kirsch IR. The gene SCL is expressed during early hematopoiesis and encodes a differentiation-related DNA-binding motif. Proc Natl Acad Sci USA 1989; 86: 10128–10132.
232. Fitzgerald TJ, Neale GAM, Raimondi SC, Goorah RM. c-tal, a helix-loop-helix protein, is juxtaposed to the T-cell receptor-β chain gene by a reciprocal chromosomal translocation: t(1;7)(p32;q35). Blood 1991; 78:2686–2695.
233. Aplan PD, Lombardi DP, Reaman GH, Sather HN, Hammond GD, Kirsch IR. Involvement of the putative hematopoietic transcription factor SCL in T-cell acute lymphoblastic leukemia. Blood 1992; 79:1327–1333.
234. Aplan PD, Johnson BE, Russell E, Chervinsky DS, Kirsch IR. Cloning and characterization of TCTA, a gene located at the site of a t(1;3) translocation. Cancer Res 1995; 55:1917–1921.
235. Aplan PD, Lombardi DP, Ginsberg AM, Cossman J, Bertness VL, Kirsch IR. Disruption of the human SCL locus by "illegitimate" V(D)J recombinase activity. Science 1990; 250:1426–1429.
236. Brown L, Cheng J-T, Chen Q, Siciliano MJ, Crist W, Buchanan G, Baer R. Site-specific recombination of the tal-1 gene is a common occurrence in human T cell leukemia. EMBO J 1990; 9:3343–3351.

237. Bernard O, Lecointe N, Jonveaux P, Souyri M, Mauchauffe M, Berger R, Larsen CJ, Mathieu-Mahul D. Two site-specific deletions and t(1;14) translocation restricted to human T-cell acute leukemias disrupt the 5′ part of the tal-1 gene. Oncogene 1991; 6:1477–1488.
238. Kikuchi A, Hayashi Y, Kobayashi S, Hanada R, Moriwaki K, Yamamoto K, Fukimoto J-I, Kaneko Y, Yamamori S. Clinical significance of TAL1 gene alteration in childhood T-cell acute lymphoblastic leukemia and lymphoma. Leukemia 1993; 7:933–938.
239. Janssen JWG, Ludwig WD, Sterry W, Bartram CR. SIL-TAL deletion in T-cell acute lymphoblastic leukemia. Leukemia 1993; 7:1204–1210.
240. Kwong YL, Chan D, Liang R. SIL/TAL1 recombination in adult T-acute lymphoblastic leukemia and T-lymphoblastic lymphoma. Cancer Genet Cytogenet 1995; 85:159–160.
241. Elwood NJ, Cook WD, Metcalf D, Begley CG. SCL, the gene implicated in human T-cell leukemia, is oncogenic in a murine T-lymphocyte cell line. Oncogene 1993; 8:3093–3101.
242. Trent JM, Kaneko Y, Mitelman F. Report of the committee on structural chromosome changes in neoplasia. Cytogenet Cell Genet 1989; 51:533–562.
243. Fujita K, Fukuhara S, Nasu K, Yamabe H, Tomono N, Inamoto Y, Shimazaki C, Ohno H, Doi S, Ueshema Y, Uchino H. Recurrent chromosome abnormalities in adult T-cell lymphomas of peripheral T-cell origin. Int J Cancer 1986; 37:517–524.
244. Raimondi SC, Pui C-H, Behm FG, Williams DL. 7q32-q36 translocation in childhood T cell leukemia: cytogenetic evidence for involvement of the T cell receptor β-chain gene. Blood 1987; 69:131–134.
245. Smith SD, Morgan R, Gemmell R, Amylon MD, Link MP, Linker C, Hecht BK, Warnke R, Glader BE, Hecht F. Clinical and biologic characterization of T-cell neoplasias with rerarrangements of chromosome 7 band q34. Blood 1988; 71:395–402.
246. Xia Y, Brown L, Yang CY-C, Tsan JT, Siciliano MJ, Espinosa R, Le Beau MM, Baer RJ. TAL2, a helix-loop-helix gene activated by the (7;9)(q34;q32) translocation in human T-cell leukemia. Proc Natl Acad Sci USA 1991; 88: 11416–11420.
247. Mellentin JD, Smith SD, Cleary ML. Lyl-1, a novel gene altered by chromosomal translocation in T cell leukemia, codes for a protein with a helix-loop-helix DNA binding motif. Cell 1989; 58:77–83.
248. Burnett RC, Thirman MJ, Rowley JD, Diaz MO. Molecular analysis of the T-cell acute lymphoblastic leukemia-associated t(1;7)(p34;q34) that fuses LCK and TCRB. Blood 1994; 84:1232–1236.
249. Sanchez-Garcia I, Kaneko Y, Gonzalez-Sarmiento R, Campbell K, White L, Boehm T, Rabbitts TH. A study of chromosome 11p13 translocations involving TCR β and TCR δ in human T cell leukaemia. Oncogene 1991; 6:577–582.
250. Kennedy MA, Gonzalez-Sarmiento R, Kees UR, Lampert F, Dear N, Boehm T, Rabbitts TH. HOX11, a homeobox-containing T-cell oncogene on human chromosome 10q24. Proc Natl Acad Sci USA 1991; 88:8900–8904.

251. Wang CC, Tien HF, Lin MT, Su IJ, Wang CH, Chuang SM, Shen MC, Liu CH. Consistent presence of isochromosome 7q in hepatosplenic T gamma/delta lymphoma: a new cytogenetic-clinicopathologic entity. Genes Chromosom Cancer 1995; 12:161–164.
252. Cooke CB, Krenacs L, Stetler-Stevenson M, Greiner TC, Raffeld M, Kingma DW, Abruzzo L, Frantz C, Kaviani M, Jaffe ES. Hepatosplenic T-cell lymphoma: a distinct clinicopathologic entity of cytotoxic γδ T-cell origin. Blood 1996; 88:4265–4274.
253. Jonveaux P, Daniel MT, Martel V, Maarek O, Berger R. Isochromosome 7q and trisomy 8 are consistent primary, non-random chromosomal abnormalities associated with hepatosplenic T γ/δ lymphoma. Leukemia 1996; 10:1453–1455.
254. Sonnier JA, Buchanan GR, Howard-Peebles PN, Rutledge J, Smith RG. Chromosomal translocation involving the immunoglobulin kappa-chain and heavy-chain loci in a child with chronic lymphocytic leukemia. N Engl J Med 1983; 309:590–594.
255. Yoffe G, Howard-Peeble PN, Smith RG, Tucker PW, Buchanan GR. Childhood chronic lymphocytic leukemia with t(2;14) translocation. J Pediatr 1990; 116:114–117.
256. Juliusson G, Oscier DG, Fitchett M, Ross FM, Stockdill G, Mackie MJ, Parker AC, Castoldi GL, Cuneo A, Knuutila S, Elonen E, Gahrton G. Prognostic subgroups in B-cell chronic lymphocytic leukemia defined by specific chromosomal abnormalities. N Engl J Med 1990; 323:720–724.
257. Dierlamm J, Michaux L, Criel A, Wlodarska I, Van Der Berghe, Hossfeld DK. Genetic abnormalities in chronic lymphocytic leukemia and their clinical prognostic implications. Cancer Genet Cytogenet 1997; 94:27–35.
258. Dohner H, Stilgenbauer S, James MR, Benner A, Weilguni T, Bentz M, Fischer K, Hunstein W, Lichter P. 11q deletions identify a new subset of B-cell chronic lymphocytic leukemia characterized by extensive nodal involvement and inferior prognosis. Blood 1997; 89:2516–2522.
259. Crossen P. Genes and chromosomes in chronic B-cell leukemia. Cancer Genet Cytogenet 1997; 94:44–51.
260. Perez-Losada A, Wessman M, Tiainen M, Hopman AHN, Willard HF, Sole F, Caballin MR, Woessner S, Knuutila A. Trisomy 12 in chronic lymphocytic leukemia: an interphase cytogenetic study. Blood 1991; 78:775–779.
261. Cuneo A, Wlodarska I, Sayed Aly M, Piva N, Carli MG, Fagioli F, Tallarico A, Pazzi I, Ferrari L, Cassiman JJ, Van den Berghe H, Castoldi GL. Nonradioactive in situ hybridization for the detection and monitoring of trisomy 12 in B-cell chronic lymphocytic leukaemia. Br J Haematol 1992; 81:192–196.
262. Garcia-Marco JA, Price CM, Catovsky D. Interphase cytogenetics in chronic lymphocytic leukemia. Cancer Genet Cytogenet 1997; 94:52–58.
263. Cuneo A, Bigoni R, Negrini M, Bullrich F, Veronese ML, Roberti MG, Bardi A, Rigolin M, Cavazzini P, Croce CM, Castoldi G. Cytogenetic and interphase cytogenetic characterization of atypical chronic lymphocytic leukemia carrying BCL1 translocation. Cancer Res 1997; 57:1144–1150.

264. Hoyer JD, Ross CW, Li C-Y, Witzig Te, Gascoyne RD, Dewald GW, Hanson CA. True T-cell chronic lymphocytic leukemia: a morphologic and immunophenotypic study of 25 cases. Blood 1995; 86:1163–1169.
265. Non-Hodgkin's lymphoma pathologic classification project: National Cancer Institute sponsored study of classifications of non-Hodgkin's lymphomas: summary and description of a Working Formulation for clinical usage. Cancer 1982; 49:2112–2135.
266. Harris NL, Jaffe ES, Stein H, Banks PM, Chan JKC, Cleary ML, Delsol G, De Wolf-Peeters C, Falini B, Gatter KC, Grogan TM, Isaacson PG, Knowles DM, Mason DY, Muller-Hermelink HK, Pileri SA, Piris MA, Ralfkiaer E, Warnke RA. A revised European-American classification of lymphoid neoplasms: a proposal from the International Lymphoma Study Group. Blood 1994; 84:1361–1392.
267. Yunis JJ, Oken MM, Kaplan ME, Ensrud KM, Howe RR, Theologides A. Distinctive chromosomal abnormalities in histologic subtypes of non-Hodgkin's lymphoma. N Engl J Med 1982; 307:1231–1236.
268. Bloomfield CD, Arthur DC, Frizzera G, Levine EG, Peterson BA, Gajl-Peczalska KJ. Non-random chromosome abnormalities in lymphoma. Cancer Res 1983; 43:2975–2984.
269. Yunis JJ, Oken MM, Theologides A, Howe RB, Kaplan ME. Recurrent chromosomal defects are found in most patients with non-Hodgkin's lymphoma. Cancer Genet Cytogenet 1984; 13:17–28.
270. The Fifth International Workshop on Chromosomes in Leukemia-Lymphoma. Correlation of chromosome abnormalities with histologic and immunologic characteristics in non-Hodgkin's lymphoma and adult T-cell leukemia-lymphoma. Blood 1987; 70:1554–1564.
271. Dyer MJS. The new genetics of non-Hodgkin lymphoma. Eur J Cancer 1990; 26:1099–1102.
272. Sandlund JT, Downing JR, Crist WM. Non-Hodgkins's lymphoma in childhood. N Engl J Med 1996; 334:1238–1248.
273. Offit K, Wong G, Filippa DA, Tao Y, Chaganti RSK. Cytogenetic analysis of 434 consecutively ascertained specimens of non-Hodgkin's lymphoma: clinical correlations. Blood 1991; 77:1508–1515.
274. Pileri SA, Ceccarelli C, Sabatinni E, Santini D, Leone O, Damiani S, Leoncini L, Falini B. Molecular findings and classification of malignant lymphomas. Acta Haematol 1996; 95:181–187.
275. Shipp MA. Prognostic factors in aggressive non-Hodgkin's lymphoma: who has "high-risk" disease? Blood 1994; 83:1165–1173.
276. Johansson B, Mertens F, Mitelman F. Cytogenetic deletion maps of hematologic neoplasms: circumstantial evidence for tumor suppressor loci. Genes Chromosom Cancer 1993; 8:205–218.
277. Bentz M, Huck K, du Manoir S, Joos S, Werner CA, Fischer K, Dohner H, Lichter P. Comparative genomic hybridization in chronic B-cell leukemias reveals a high incidence of chromosomal gains and losses. Blood 1995; 85: 3610–3618.

278. Bentz M, Werner CA, Dohner H, Joos S, Barth TFE, Siebert R, Schroder M, Stilgenbauer S, Fischer K, Moller P, Lichter P. High incidence of chromosomal imbalances and gene amplifications in the classical follicular variant of follicle center lymphoma. Blood 1996; 88:1437–1444.
279. Joos S, Otano-Joos MI, Ziegler S, Bruderlein S, Du Manoir S, Bentz M, Moller P, Lichter P. Primary mediastinal (thymic) B-cell lymphoma is characterized by gains of chromosomal material including 9p and the REL gene. Blood 1996; 87:1571–1578.
280. Houldsworth D, Mathew S, Rao PH, Dyomina K, Louie DC, Parsa N, Offit K, Chaganti RSK. Rel proto-oncogene is frequently amplified in extranodal diffuse large cell lymphoma. Blood 1996; 87:25–29.
281. Monni O, Joensuu H, Franssila K, Knuutila S. DNA copy number changes in diffuse large B-cell lymphoma—comparative genomic hybridization study. Blood 1996; 87:5269–5278.
282. Koduru PRK, Filippa DA, Richardson ME, Jhanwar SC, Chaganti SR, Koziner B, Clarkson BD, Lieberman PH, Chaganti RSK. Cytogenetic and histologic correlation in malignant lymphoma. Blood 1987; 69:97–102.
283. Dierlamm J, Michaux L, Wlodarska I, Pittaluga S, Zeller W, Stul M, Criel A, Thomas J, Boogaerts M, Delaere P, Cassiman J-J, De Wolf-Peeters C, Mecucci C, Van den Berghe H. Trisomy 3 in marginal zone B-cell lymphoma: a study based on cytogenetic analysis and fluorescence in situ hybridization. Br J Haematol 1996; 93:242–249.
284. Chenevix-Trench G. The molecular genetics of human non-Hodgkin's lymphoma. Cancer Genet Cytogenet 1987; 27:191–213.
285. Nashelsky MB, Hess MM, Weisenburger DD, Pierson JL, Bast MA, Armitage JO, Sanger WG. Cytogenetic abnormalities in B-immunoblastic lymphoma. Leukemia Lymphoma 1994; 14:415–420.
286. Offit K, Jhanwar SC, Ladanyi M, Filippa DA, Chaganti RSK. Cytogenetic analysis of 434 consecutively ascertained specimens of non-Hodgkin's lymphoma: correlations between recurrent aberrations, histology, and exposure to cytotoxic treatment. Genes Chromosom Cancer 1991; 3:189–201.
287. Wotherspoon AC, Pan L, Diss TC, Isaacson PG. Cytogenetic study of B-cell lymphoma of mucosa-associated lymphoid tissue. Cancer Genet Cytogenet 1992; 58:35–38.
288. Dierlamm J, Pittaluga S, Wlodarska I, Stul M, Thomas J, Boogaerts M, Michaux L, Driessen A, Mecucci C, Cassiman JJ, De Wolf-Peeters C, Van den Berghe H. Marginal zone B-cell lymphomas of different sites share similar cytogenetic and morphologic features. Blood 1996; 87:299–307.
289. Wotherspoon AC, Finn TM, Isaacson PG. Trisomy 3 in low grade B-cell lymphomas of mucosa-associated lymphoid tissue. Blood 1995; 85:2000–2004.
290. Ueshima Y, Fukuhara S, Hattori T, Uchiyama T, Takatsuki K, Uchino H. Chromosome studies in adult T-cell leukemia in Japan: significance of trisomy 7. Blood 1981; 58:420–425.
291. Kamada N, Sakurai M, Miyamoto K, Sanada I, Sadamori N, Fukuhara S, Abe S, Shiraishi Y, Abe T, Kaneko Y, Shimoyama M. Chromosome abnormalities

in adult T-cell leukemia/lymphoma: a karyotype review committee report. Cancer Res 1992; 52:1481–1493.

292. Offit K, Chaganti RSK. Chromosomal aberrations in non-Hodgkin's lymphoma: biologic and clinical correlations. Hematol Oncol Clin N Am 1991; 5:853–869.
293. Chenevix-Trench G, Brown JA, Tyler GB, Behm FG. Chromosome analysis of 30 cases of non-Hodgkin's lymphoma. Med Oncol Tumor Pharmacother 1988; 5:17–32.
294. Slavutsky I, Labal de Vinuesa M, Estevez ME, Sen L, Dupont J, Larripa I. Chromosome studies in human hematologic diseases: non-Hodgkin's lymphoma. Haematologica 1987; 72:25–37.
295. Bajalica S, Serensen A-G, Pedersen NT, Hein S, Brondum-Nielsen K. Chromosome painting as a supplement to cytogenetic banding analysis in non-Hodgkin's lymphoma. Genes Chromosom Cancer 1993; 7:231–239.
296. Hammond DW, Hancock BW, Goyns MR. Analysis of 14q+ derivative chromosomes in non-Hodgkin's lymphomas by fluorescence in situ hybridization. Leukemia Lymphoma 1995; 20:111–117.
297. Fischer P, Nacheva E, Mason DY, Sherrington PD, Hoyle C, Hayhoe FG, Karpas A. A Ki-1 (CD30)-positive human cell line (Karpas 299) established from a high-grade non-Hodgkin's lymphoma, showing a 2;5 translocation and rearrangement of the T-cell receptor β-chain gene. Blood 1988; 72:234–240.
298. Rimokh R, Magaud JP, Berger F, Samarut J, Coiffier B, Germain D, Mason DY. A translocation involving a specific breakpoint on chromosome 5 is characteristic of anaplastic large-cell lymphoma (Ki-1 lymphoma). Br J Haematol 1989; 71:31–36.
299. Le Beau MM, Bitter MA, Larson RA, Doane LA, Ellis ED, Franklin WA, Rubin CM, Kadim ME, Vardiman JW. The t(2;5)(p23;q35): a recurring chromosomal abnormality in Ki-1 positive anaplastic large-cell lymphoma. Leukemia 1989; 3:866–870.
300. Mason DY, Bastard C, Rimokh R, Dastugue N, Huret JL, Kristoffersson U, Magaud JP, Nezelof C, Tilly H, Vannier JP, Hemet J, Warnke R. CD30-positive large cell lymphomas (''Ki-1 lymphoma'') are associated with a chromosomal translocation involving 5q35. Br J Haematol 1990; 74:161–168.
301. Greer JP, Kinney MC, Collins RD, Salhany KE, Wolff SN, Hainsworth JD, Flexner JM, Stein RS. Clinical features of 31 patients with Ki-1 anaplastic large-cell lymphoma. J Clin Oncol 1991; 9:539–547.
302. Bullrich F, Morris SW, Hummel M, Pileri S, Stein H, Croce CM. Nucleophosmin (NPM) gene rearrangement in Ki-1-positive lymphomas. Cancer Res 1994; 54:2873–2877.
303. Lopategui JR, Sun LH, Chan JKC, Gaffey MJ, Frierson HF, Glackin C, Weiss LM. Low frequency association of the t(2;5)(p23;q35) chromosomal translocation with CD30+ lymphomas from American and Asian patients. Am J Pathol 1995; 146:323–328.
304. Sandlund JT, Pui CH, Roberts M, Santana VM, Morris SW, Berard CW, Hutchison RE, Ribeiro RC, Mahmoud H, Crist WM, Heim M, Raimondi SC.

Clinicopathologic features and treatment outcome of children with large-cell lymphoma and the t(2;5)(p23;q35). Blood 1994; 84:2467–2471.

305. Lamant L, Meggetto F, Saati TA, Brugieres L, de Paillerets BB, Dastugue N, Bernheim A, Rubie H, Terrier-Lacombe MJ, Robert A, Brousset P, Rigal F, Schlaifer D, Shiota M, Mori S, Delsol G. High incidence of the t(2;5)(p23; q35) translocation in anaplastic large cell lymphoma and its lack of detection in Hodgkin's disease. Comparison of cytogenetic analysis, reverse transcriptase-polymerase chain reaction, and P-80 immunostaining. Blood 1996; 87: 284–291.
306. Massimino M, Gasparini M, Giardini R. Ki-1 (CD30) anaplastic large-cell lymphoma in children. Ann Oncol 1995; 6:915–920.
307. Wood GS, Hardman DL, Boni R, Dummer R, Kim Y-H, Smoller BR, Takeshita M, Kikuchi M, Burg G. Lack of the t(2;5) or other mutations resulting in expression of anaplastic lymphoma kinase catalytic domain in CD30+ primary cutaneous lymphoproliferative disorders in Hodgkin's disease. Blood 1996; 88:1765–1770.
308. Downing JR, Shurtleff SA, Zielenska M, Curcio-Brint AM, Behm FG, Head DR, Sandlund JT, Weisenburger DD, Kossakowska AE, Thorner P, Lorenzana A, Ladanyi M, Morris SW. Molecular detection of the (2;5) translocation of non-Hodgkin's lymphoma by reverse transcriptase-polymerase chain reaction. Blood 1995; 85:3416–3422.
309. Wellmann A, Otsuki T, Vogelbruch M, Clark HM, Jaffe ES, Raffeld M. Analysis of the t(2;5)(p23;q35) translocation by reverse transcription-polymerase chain reaction in CD30+ anaplastic large-cell lymphomas, in other non-Hodgkin's lymphomas of T-cell phenotype and in Hodgkin's disease. Blood 1995; 86:2321–2328.
310. Morris SW, Kirstein MK, Valentine MB, Dittmer KG, Shapiro DN, Saltman DL, Look AT. Fusion of a kinase gene, ALK, to a nuclear protein gene, NPM, in non-Hodgkin's lymphomas. Science 1994; 263:1281–1284.
311. Pulford K, Lamant L, Morris SW, Butler LH, Wood KM, Stroud D, Delsol G, Mason DY. Detection of anaplastic lymphoma kinase (ALK) and nucleolar protein nucleophosmin (NPM)-ALK proteins in normal and neoplastic cells with the monoclonal antibody ALK1. Blood 1997; 89:1394–1404.
312. Shiota M, Fujimoto J, Taneka M, Satoh H, Ichinohasama R, Abe M, Nakano M, Yamamoto T, Mori S. Diagnosis of t(2;5)(p23;q35)-associated Ki-1 lymphoma with immunohistochemistry. Blood 1994; 84:3684–3652.
313. Mathew P, Sanger WG, Weisenburger DD, Valentine M, Valentine V, Pickering D, Higgins C, Hess M, Cui X, Srivastava DK, Morris SW. Detection of the t(2;5)(p23;q35) and NPM-ALK fusion in non-Hodgkin's lymphoma by two-color fluorescence in situ hybridization. Blood 1997; 89:1678–1685.
314. Fukuhara S, Rowley JD, Variakojis D, Golomb HM. Chromosome abnormalities in poorly differentiated lymphocytic lymphoma. Cancer Res 1979; 39:3119–3128.
315. Levine EG, Arthur DC, Frizzera G, Peterson BA, Hurd DD, Bloomfield CD. There are differences in cytogenetic abnormalities among histologic subtypes of the non-Hodgkin's lymphomas. Blood 1985; 66:1414–1422.

316. Yunis JJ, Frizzera G, Oken MM, McKenna J, Theologides A, Arnesen M. Multiple recurrent genomic defects in follicular lymphoma. A possible model for cancer. N Engl J Med 1987; 316:79–84.
317. Gauwerky CE, Haluska FG, Tsujimoto Y, Nowell PC, Croce CM. Evolution of B-cell malignancy: pre-B-cell leukemia resulting from MYC activation in a B-cell neoplasm with rearranged BCL2 gene. Proc Natl Acad Sci USA 1988; 85:8548–8552.
318. Thangavelu M, Olopade O, Beckman E, Vardiman JW, Larson RA, McKeithan TW, Le Beau MM, Rowley JD. Clinical, morphologic and cytogenetic characteristics of patients with lymphoid malignancies characterized by both t(14;18)(q32;q21) and t(8;14)(q24;q32) or t(8;22)(q24;q11). Genes Chromosom Cancer 1990; 2:147–158.
319. Karsan A, Gascoyne RD, Coupland RW, Shepherd JD, Phillips GL, Horsman DE. Combination of t(14;18) and a Burkitt's type translocation in B-cell malignancies. Leuk Lymphoma 1993; 10:433–441.
320. Tsujimoto Y, Finger LR, Yunis J, Nowell PC, Croce ML. Cloning of the chromosome breakpoint of neoplastic B cells with the t(14;18) chromosome translocation. Science 1984; 226:1097–1099.
321. Cleary ML, Smith SD, Sklar J. Cloning and structural analysis of cDNAs for bcl-2 and a hybrid bcl-2/immunoglobulin transcript resulting from the t(14; 18) translocation. Cell 1986; 47:19–28.
322. Weiss LM, Warnke RA, Sklar J, Cleary ML. Molecular analysis of the t(14; 18) chromosomal translocation in malignant lymphomas. N Engl J Med 1987; 317:1185–1189.
323. Osada H, Seto M, Ueda R, Emi N, Takagi N, Obata Y, Suchi T, Takahashi T. Bcl-2 gene rearrangement analysis in Japanese B cell lymphoma; novel bcl-2 recombination with immuloglobulin kappa chain gene. Jpn Cancer Res 1989; 80:711–715.
324. Aventin A, Mecucci C, Guanyabens C, Brunet S, Soler J, Bordes R, Van den Berghe H. Variant translocation in a Burkitt conversion of follicular lymphoma. Br J Haematol 1990; 74:367–370.
325. Bertheas MF, Rimokh R, Berger F, Gaucherand M, Machado P, Vasselon C, Calmardoriol P, Jaubert J, Guyotat D, Magaud JP. Molecular study of a variant translocation t(2;18)(p11;q21) in a follicular lymphoma. Br J Haematol 1991; 78:132–134.
326. Hillion J, Mecucci C, Aventin A, Leroux D, Wlodarska I, Van Den Berghe H, Larsen C-J. A variant translocation t(2;18) in follicular lymphoma involves the 5′ end of BCL2 and Igκ light chain. Oncogene 1991; 6:169–172.
327. Adachi M, Tefferi A, Greipp PR, Kipps TJ, Tsujimoto Y. Preferential linkage of BCL2 to immunoglobulin light chain gene in chronic lymphocytic leukemia. J Exp Med 1990; 171:559–564.
328. Seite P, Leroux D, Hillion J, Berger R, Mathieu-Mahul D, Larsen CJ. Molecular analysis of a variant t(18;22) in a lymphocytic lymphoma. Genes Chromosom Cancer 1993; 6:39–44.
329. Tsujimoto Y, Gorham J, Cossman J, Jaffe E, Croce CM. The t(14;18) chro-

mosome translocations involved in B-cell neoplasms result from mistakes in VDJ joining. Science 1985; 229:1390–1393.

330. Korsmeyer SJ. Bcl-2 initiates a new category of oncogenes: regulators of cell death. Blood 1992; 80:879–886.
331. Leoncini L, Del Vecchio MT, Megha T, Barbieri P, Pileri S, Sabattini E, Tosi P, Kraft R, Cottier H. Correlations between apoptotic and proliferative indices in malignant non-Hodgkin's lymphomas. Am J Pathol 1993; 142:755–763.
332. McDonnell TJ, Korsmeyer SJ. Progression from lymphoid hyperplasia to high-grade malignant lymphoma in mice transgenic for the t(14;18). Nature 1991; 349:254–256.
333. Taniwaki M, Sliverman GA, Nishida K, Horiike S, Misawa S, Shimazaki C, Miura I, Nagai M, Abe M, Fukuhara S, Kashima K. Translocations and amplifications of the BCL2 gene are detected in interphase nuclei of non-Hodgkin's lymphomas by in situ hybridization with yeast artificial chromosomes. Blood 1996; 86:1481–1486.
334. Gribben JG, Freedman AS, Woo SD, Blake K, Shu RS, Freeman G, Longtine JA, Pinkus GS, Nadler LM. All advanced stage non-Hodgkin's lymphomas with a polymerase chain reaction amplifiable breakpoint of bcl-2 have residual cells containing the bcl-2 rearrangement at evaluation and after treatment. Blood 1991; 78:3275–3280.
335. Price CGA, Meerabux J, Murtagh S, Cotter FE, Rohatiner AZS, Young BD, Lister TA. The significance of circulating cells carrying t(14;18) in long remission from follicular lymphoma. J Clin Oncol 1991; 9:1527–1532.
336. Limpens J, de Jong D, van Krieken JHJM, Price CGA, Young BD, van Ommen G-JB, Kluin PM. Bcl-2/JH rearrangements in benign lymphoid tissues with follicular hyperplasia. Oncogene 1991; 6:2271–2276.
337. Weisenburger DD, Armitage JO. Mantle cell lymphoma—an entity comes of age. Blood 1996; 87:4483–4494.
338. Vandenberghe E, de Wolf-Peeters C, Wlodarska I, Stul M, Louwagie A, Verhoef G, Thomas J, Criel A, Cassiman JJ, Mecucci C, van den Berghe H. Chromosome 11q rearrangements in B non-Hodgkin's lymphoma. Br J Haematol 1992; 81:212–217.
339. Komatsu H, Iida S, Yamamoto K, Mikuni C, Nitta M, Takahashi T, Ueda R, Seto M. A variant chromosome translocation at 11q13 identifying PRAD1/cyclin D1 as the bcl-1 gene. Blood 1994; 84:1226–1231.
340. Ott MM, Ott G, Kuse R, Porowski P, Gunzer U, Feller AC, Muller-Hermelink HK. The anaplastic variant of centrocytic lymphoma is marked by frequent rearrangements of the bcl-1 gene and high proliferation indices. Histopathology 1994; 24:329–334.
341. Argatoff LH, Connors JM, Klasa RJ, Horsman DE, Gascoyne RD. Mantle cell lymphoma: a clinicopathologic study of 80 cases. Blood 1997;89:2067–2078.
342. Daniel MT, Tigaud I, Flexor MA, Nogueira ME, Berger R, Jonveaux P. Leukaemic non-Hodgkin's lymphomas with hyperdiploid cells and t(11;14)(q13;q32): a subtype of mantle cell lymphoma? Br J Haematol 1995; 90:77–84.

343. Brito-Babapulle V, Ellis J, Matutes E, Oscier D, Khokhar T, MacLennan K, Catovsky D. Chromosome translocation t(11;14)(q13;q32) in chronic lymphoid disorders. Genes Chromosom Cancer 1992; 5:158–165.
344. Bosch F, Jares P, Campo E, Lopez-Guillermo A, Piris MA, Villamor N, Tassies D, Jaffe ES, Montserrat E, Rozman C, Cardesa A. PRAD1/cyclin D1 gene overexpression in chronic lymphoproliferative disorders: a highly specific marker of mantle cell lymphoma. Blood 1994; 84:2726–2732.
345. Neilson JR, Cai M, Bienz N, Leyland MJ. Leukaemic mantle cell lymphoma with t(11;14) and trisomy 12 showing clinical features of stage AO B-cell chronic lymphocytic leukemia. J Clin Pathol: Mol Pathol 1995; 48:M165–M166.
346. Tsujimoto Y, Finger L, Yunis J, Nowell PC, Croce CM. Cloning of the chromosome breakpoint of neoplastic B-cells with the t(14;18) chromosome translocation. Science 1984; 226:1097–1099.
347. Rosenberg CL, Wong E, Petty EM, Bale AE, Tsujimoto Y, Harris NL, Arnold A. PRAD1, a candidate BCL1 oncogene: mapping and expression in centrocytic lymphoma. Proc Natl Acad Sci USA 1991; 88:9638–9642.
348. Motokura T, Bloom T, Kim HG, Juppner H, Ruderman JV, Kronenberg HM, Arnold A. A novel cyclin encoded by a bcl-1 linked candidate oncogene. Nature 1991; 350:512–515.
349. Oka K, Ohno T, Kita K, Yamaguchi M, Takakura N, Nishii K, Miwa H, Shirakawa S. PRAD1 gene overexpression in mantle cell lymphoma but not in other low-grade B-cell lymphomas, including extranodal lymphoma. Br J Haematol 1994; 86:786–791.
350. Withers DA, Harvey RC, Faust JB, Melnyk O, Carey K, Meeker TC. Characterization of a candidate bcl-1 gene. Mol Cell Biol 1991; 11:4846–4853.
351. Zukerberg LR, Yang WI, Arnold A, Harris NL. Cyclin D1 expression in non-Hodgkin's lymphoma: detection by immunohistochemistry. Am J Clin Pathol 1995; 103:756–760.
352. de Boer CJ, Loyson S, Kluin PM, Kluin-Nelemans HC, Schuuring E, van Krieken JH. Multiple breakpoints within the BCL-1 locus in B-cell lymphoma: rearrangements of the cyclin D1 gene. Cancer Res 1993; 53:4148–4152.
353. Williams ME, Swerdlow SH, Meeker TC. Chromosome t(11;14)(q13;q32) breakpoints in centrocytic lymphoma are highly localized at the bcl-1 major translocation cluster. Leukemia 1993; 7:1437–1440.
354. Rimokh R, Berger F, Delsol G, Digonnet I, Rouault JD, Tigaud JD, Gadoux M, Coiffier B, Bryon PA, Magaud JP. Detection of the chromosomal translocation in t(11;14) by polymerase chain reaction in mantle cell lymphomas. Blood 1994; 83:1871–1875.
355. Luthra R, Hai S, Pugh WC. Polymerase chain reaction detection of the t(11;14) translocation involving the bcl-1 major translocation cluster in mantle cell lymphoma. Diag Mol Pathol 1995; 4:4–7.
356. Zucca E, Soldati G, Schlegelberger B, Booth MJ, Weber-Matthiesen K, Cav-

alli F, Cotter FE. Detection of chromosome 11 alterations in blood and bone marrow by interphase cytogenetics in mantle cell lymphoma. Br J Haematol 1995; 89:665–668.

357. Coignet LJA, Schuuring E, Kibbelaar RE, Raap TK, Kleiverda KK, Bertheas M-F, Wiegant J, Beverstock G, Kluin PM. Detection of 11q13 rearrangements in hematologic neoplasms by double-color fluorescence in situ hybridization. Blood 1996; 87:1512–1519.

358. Vaandrager J-W, Schuuring E, Zwikstra E, de Boer CJ, Kleiverda KK, van Krieken JH, Kluin-Nelemans HC, van Ommen G-J, Raap AK, Kluin PM. Direct visualization of dispersed 11q13 chromosomal translocations in mantle cell lymphoma by multicolor DNA fiber fluorescence in situ hybridization. Blood 1996; 88:1177–1182.

359. Offit K, Jhanwar SC, Ebrahim SAD, Filippa D, Clarkson BD, Chaganti RSK. t(3;22)(q27;q11): a novel translocation associated with diffuse non-Hodgkin's lymphoma. Blood 1989; 74:1876–1879.

360. Bastard C, Tilly H, Lenormand B, Bigorgne C, Boulet D, Kunlin A, Monoconduit M, Piguet H. Translocation involving band 3q27 and Ig gene regions in non-Hodgkin's lymphoma. Blood 1992; 79:2527–2531.

361. Lo Coco F, Ye BH, Lista F, Corradini P, Offit K, Knowles DM, Chaganti RSK, Dalla-Favera R. Rearrangements of bcl-6 gene in diffuse large cell non-Hodgkin's lymphoma. Blood 1994; 83:1757–1759.

362. Horsman DE, McNeil BK, Anderson M, Shenkier T, Gascoyne RD. Frequent association of t(13;14) or variant with other lymphoma-specific translocations. Br J Haematol 1995; 89:569–575.

363. Ye BH, Lista F, Lo Coco F, Knowles DM, Offit K, Chaganti RSK, Dalla-Favera R. Alterations of a zinc finger-encoding gene, Bcl-6, in diffuse large-cell lymphoma. Science 1993; 262:747–750.

364. Bastard C, Deweindt C, Kerckaert JP, Lenormand B, Rossi A, Pezella F, Fruchart C, Duval C, Manconduit M, Tilly H. LAZ3 rearrangements in non-Hodgkin's lymphoma correlation with histology, immunophenotype, karyotype, and clinical outcome in 217 patients. Blood 1994; 83:2423–2427.

365. Wlodarska I, Mecucci C, Stul M, Michaux L, Pittaluga S, Hernandez JM, Cassiman J-J, De Wolf-Peeters C, Van den Berghe H. Fluorescence in situ hybridization identifies new chromosomal changes involving 3q27 in non-Hodgkin's lymphomas with BCL6/LAZ3 rearrangement. Genes Chromosom Cancer 1995; 14:1–7.

366. Baron BW, Nucifora G, McCabe N, Espinosa R, Le Beau MM, McKeithan TW. Identification of the gene associated with the recurring chromosomal translocations t(3;14)(q27;q32) and t(3;22)(q27;q11) in B-cell lymphomas. Proc Natl Acad Sci USA 1993; 90:5262–5266.

367. Deweindt C, Kerckaer JP, Tilly H, Quief S, Nguyen VC, Bastard C. Cloning of a breakpoint cluster region at band 3q27 involved in human non-Hodgkin's lymphoma. Genes Chromosom Cancer 1993; 8:149–154.

368. Ye BH, Rao PH, Chaganti RSK, Dalla-Favera R. Cloning of bcl-6, the locus

involved in chromosome translocations affecting band 3q27 in B-cell lymphoma. Cancer Res 1993; 53:2732–2735.
369. Miki T, Kawamata N, Arai A, Ohashi K, Nakamura Y, Kato A, Hirosawa S, Aoki N. Molecular cloning of the breakpoint for 3q27 translocation in B-cell lymphomas and leukemias. Blood 1994; 83:217–222.
370. Chang C-C, Ye BH, Chaganti RSK, Dalla-Favera R. BCL-6, a POZ/zinc-finger protein, is a sequence-specific transcriptional repressor. Proc Natl Acad Sci USA 1996; 93:6947–6952.
371. Otsuki T, Yano T, Clark HM, Bastard C, Kerckaert JP, Jaffe ES, Raffeld M. Analysis of LAZ3 (BCL6) status in B-cell non-Hodgkin's lymphoma: results of rearrangements and gene expression studies, and a mutational analysis of coding region sequences. Blood 1995; 85:2877–2884.
372. Offit K, Lo Coco F, Louie DC, Parsa NZ, Leung D, Portlock C, Ye BH, Lista F, Filippa DA, Rosenbaum A, Ladanyi M, Jhanwar S, Dalla-Favera R, Chaganti RSK. Rearrangement of the bcl-6 gene as a prognostic marker in diffuse large cell lymphoma. N Engl J Med 1994; 331:74–80.
373. Migliazza A, Martinotti S, Chen W, Fusco C, Ye BH, Knowles DM, Offit K, Chaganti RSK, Dalla-Favera R. Frequent somatic hypermutation of the 5′ noncoding region of the BCL6 gene in B-cell lymphoma. Proc Natl Acad Sci USA 1995; 92:12520–12524.
374. Dallery E, Galiegue-Zouitina S, Collyn-d'Hooghe M, Quief S, Denis C, Hildebrand M-P, Lantoine D, Deweindt C, Tilly H, Bastard C, Kerckaert J-P. TTF, a gene encoding a novel small G protein, fuses to the lymphoma-associated LAZ3 gene by t(3;4) chromosomal translocation. Oncogene 1995; 10:2171–2178.
375. Akasaka T, Miura I, Takahashi N, Akasaka H, Yonetani N, Ohno H, Fukuhara S, Okuma M. A recurring translocation, t(3;6)(q27;p21), in non-Hodgkin's lymphoma results in replacement of the 5′ regulatory region of BCL6 with a novel H4 histone gene. Cancer Res 1997; 57:7–12.
376. Galiegue-Zouitina S, Quief S, Hildebrand M-P, Denis C, Lecocq G, Collyn-d'Hooghe M, Bastard C, Yuille M, Dyer MJS, Kerckaert JP. The B cell transcriptional coactivator *BOB1/OBF1* gene fuses to the *LAZ3/BCL6* gene by t(3;11)(q27;q23.1) chromosomal translocation in a B cell leukemia line (Karpas 231). Leukemia 1996;10:579–587.
377. Abruzzo LV, Jaffe ES, Cotelingam JD, Whang-Peng J, Del Duca V, Medeiros LJ. T-cell lymphoblastic lymphoma with eosinophilia associated with subsequent myeloid malignancy. Am J Surg Pathol 1992; 16:236–245.
378. Inhorn RC, Aster JC, Roach SA, Slapak CA, Soiffer R, Tantravahi R, Stone RM. A syndrome of lymphoblastic lymphoma, eosinophilia, and myeloid hyperplasia/malignancy associated with t(8;13)(p11;q11): description of a distinctive clinicopathologic entity. Blood 1995; 85:1881–1887.
379. Cabanillas F, Pathak S, Grant G, Hagemeister FB, McLaughlin P, Swan F, Rodriguez MA, Trujillo J, Cork A, Butler JJ, Katz R, Bourne S, Freireich EJ. Refractoriness to chemotherapy and poor survival related to abnormalities of chromosomes 17 and 7 in lymphoma. Am J Med 1989; 87:167–172.

380. Rodriguez MA, Ford RJ, Goodacre A, Selvanayagam P, Cabanillas F, Deisseroth AB. Chromosome 17p and p53 changes in lymphoma. Br J Haematol 1991; 79:575–582.
381. Meerabux J, Yaspo M-L, Roebroek AJ, Van de Ven WJM, Lister TA, Young BD. A new member of the proprotein convertase gene family (*LPC*) is located at a chromosome translocation breakpoint in lymphomas. Cancer Res 1996; 56:448–451.
382. Akao Y, Seto M, Yamamoto K, Iida S, Nakazawa S, Inazawa J, Abe T, Takahashi T, Ueda R. The *RCK* gene associated with t(11;14) translocation is distinct from the *MLL/ALL-1* gene with (4;11) and t(11;19) translocations. Cancer Res 1992; 52:6083–6087.
383. Offit K, Louie DC, Parsa NZ, Noy A, Chaganti RSK. Del(7)(q32) is associated with a subset of small lymphocytic lymphoma with plasmacytoid features. Blood 1995; 86:2365–2370.
384. Offit K, Parsa NZ, Filippa D, Jhanwar SC, Chaganti RSK. t(9;14)(p13;q32) denotes a subset of low-grade non-Hodgkin's lymphoma with plasmacytoid differentiation. Blood 1992; 80:2594–2599.
385. Iida S, Rao PH, Nallasivam P, Hibshoosh H, Butler M, Louie DC, Dyomin V, Ohno H, Chaganti RS, Dalla-Favera R. The t(9;14)(p13;q32) chromosomal translocation associated with lymphoplasmacytoid lymphoma involves the *PAX-5* gene. Blood 1996; 88:4110–4117.
386. Busslinger M, Klix N, Pfeffer P, Graninger PG, Kozmik Z. Deregulation of *PAX-5* by translocation of the Eμ enhancer of the *IgH* locus adjacent to two alternative *PAX-5* promoters in a large-cell lymphoma. Proc Natl Acad Sci USA 1996; 93:6129–6134.
387. Offit K, Parsa NZ, Gaidano G, Filippa DA, Louie D, Pan D, Jhanwar SC, Dalla-Favera R, Chaganti RSK. 6q deletions define clinico-pathologic subsets of non-Hodgkin's lymphoma. Blood 1993; 82:2157–2162.
388. Offit K, Louie DC, Parsa NZ, Filippa D, Gangi M, Siebert R, Chaganti RSK. Clinical and morphologic features of B-cell small lymphocytic lymphoma with del(6)(q21q23). Blood 1994; 83:2611–2618.
389. Menasce LP, Orphanos V, Santibanez-Koref M, Boyle JM, Harrison CJ. Common region of deletion on the long arm of chromosome 6 in non-Hodgkin's lymphoma and acute lymphoblastic leukaemia. Genes Chromosom Cancer 1994; 10:286–288.
390. Speaks SL, Sanger WG, Masih AS, Harrington DS, Hess M, Armitage JO. Recurrent abnormalities of chromosome bands 10q23-25 in non-Hodgkin's lymphoma. Genes Chromosom Cancer 1992; 5:239–243.
391. Goyns MH, Hammond DW, Harrison CJ, Menasce LP, Ross FM, Hancock BW. Structural abnormalities of the X chromosome in non-Hodgkin's lymphoma. Leukemia 1993; 7:848–852.
392. Eva A, Aaronson SA. Isolation of a new human oncogene from a diffuse B-cell lymphoma. Nature 1985; 316:273–275.
393. Eva A, Vecchio G, Rao CD, Tronick SR, Aaronson SA. The predicted *DBL* oncogene product defines a distinct class of transforming proteins. Proc Natl Acad Sci USA 1988; 85:2061–2065.

394. Madani A, Choukroun V, Soulier J, Cacheux V, Claisse J-F, Valensi F, Daliphard S, Cazin B, Levy V, Leblond V, Daniel M-T, Sigaux F, Stern M-H. Expression of $p13^{MTCP1}$ is restricted to mature T-cell proliferations with t(X; 14) translocations. Blood 1996; 87:1923–1927.
395. Johansson B, Mertens F, Mitelman F. Cytogenetic evolution patterns in non-Hodgkin's lymphoma. Blood 1995; 86:3905–3914.

13

Chromosome Abnormalities in Solid Tumors: An Overview

Avery A. Sandberg
St. Joseph's Hospital and Medical Center, Phoenix, Arizona
Zhong Chen
University of Utah School of Medicine, Salt Lake City, Utah

I. INTRODUCTION

During the past two decades, evidence has accumulated indicating that tumor formation is often associated with the occurrence of nonrandom chromosomal abnormalities (1,2). Hematopoietic neoplasms (see chapters by Raimondi, and also Anastasi and Roulston in this volume) account for less than 10% of human cancers, but karyotypic data regarding leukemias and lymphomas predominate, with solid tumors representing only about 15% of the total available data (3).

Nonetheless, our modest knowledge of tumor cytogenetics is already proving useful in the understanding of cancer pathogenesis, as well as in the diagnosis and prognosis of these malignancies. Generally, carcinomas, some lymphomas, and multiple myeloma are associated with a number of karyotypic changes compatible with a stepwise process of orchestrated genetic changes necessary for the cell to progress from proliferation to malignant transformation to invasiveness and metastases. However, in most sarcomas and leukemias, relatively simple karyotypes can be observed, commonly including translocations and less commonly inversions or insertions, which can be diagnostic of these malignancies. Chromosomal deletions are common occurrences in epithelial adenocarcinomas, such as those of the large bowel, lung, breast, and prostate. As new and additional chromosome changes develop in any condition, the disease tends to progress, and an evaluation of therapeutic approaches is required (1,2,4,5).

The pathologist and the tumor pathology play a key role in tumor cytogenetics (see also the chapter by Khorsand in this volume). The pathologist is often responsible for the handling, selection, and disposition of the tissue samples for cytogenetic analysis. These are crucial steps, as the tissue submitted for cytogenetic analysis must be viable, representative of the tumor, and not fixed. Furthermore, the correct diagnosis and precise histologic definition of the tumor is even more crucial, since they serve as a basis for correlating cytogenetic events with specific tumor types. Without the cooperation and collaboration of the pathologist, it is not possible to more fully and usefully understand and apply the cytogenetic data characterizing various tumors.

II. ADVANCES IN THE CYTOGENETIC METHODOLOGY FOR SOLID TUMORS

As stated in the introduction, the body of information regarding tumor cytogenetics is comparatively small as compared with that of the leukemias and lymphomas, due to a number of reasons: (a) many tumors, particularly the benign ones, have a low proliferative index, necessitating cell culture which may be unsuccessful or may lead to overgrowth by normal stromal cells such as fibroblasts; (b) extensive necrosis or contamination by a coexistent infection may result in a low yield of sufficient metaphases for analysis; (c) the morphology of the chromosomes in hematologic malignancies is often more optimal for detailed analysis than in solid tumors; and (d) a host of chromosomal aberrations are frequently seen in solid tumors, particularly adenocarcinomas, making it more difficult to discern the primary abnormality (6,7). At present, it is usually impossible to obtain successful chromosome preparations from every tumor studied. The yield depends on the source of the tumor (e.g., solid tumor versus effusion), the site of the tumor, and a number of other parameters. In our laboratory, the yield with solid tumors is at present about 60%, which is at least twice as high as that 10 years ago. In some tumors (e.g., testes, kidney) the rate is more than 90%, whereas in others it is still relatively low (e.g., about 30–40% in prostate and bladder). Adequate banding is still a difficult chore in some tumors, because of the condensed and fuzzy nature of the chromosomes. Despite these obstacles encountered with the karyotypic evaluation of solid tumors, refinements in cytogenetic techniques have led to an increasing number of reported studies.

A. Culturing Technologies

Detailed methodologies for the examination of chromosomes in various solid tumors can be found in the literature (1,8,9). The description here represents a summary of these advances.

Tumor tissue may be transported in a small amount (i.e., 5–10 ml) of sterile saline solution or phosphate-buffered saline (PBS) in a sterile container suitable for transfer. For those specimens whose transport requires a greater length of time (i.e., 5–48 h), medium containing serum is preferable. RPMI-1640 medium supplemented with fetal calf serum (FCS) and antibiotics has often been used (10). L-15 medium can also serve as a transport medium (11). Specimen samples may be kept at room temperature or refrigerated.

Four general techniques of processing cells from solid tumors for chromosome analysis are in present use: the direct method, suspension culture, mixed colony cultures, and in situ culture and harvest techniques (1). Specimens are prepared for culture by disaggregation techniques, which include mechanical and/or enzymatic processes. Enzymatic disaggregation procedures with collagenase II and DNase I are preferred over the mechanical method of mincing the tumor with scissors or scalpel or digestion by trypsin. The former methods yield a larger number of viable cells and the quality of banding is superior. In addition, excellent results can frequently be obtained with long-term (i.e., 16 h to 6 days) incubation with these enzymes (1,8,9).

The direct method involves incubating the disaggregated tumor samples in culture medium containing colchicine for a few hours at 37°C in order to collect metaphases from the in vivo dividing cells. The drawbacks of this method include an often insufficient quantity of metaphases and inferior banding quality.

Suspension cultures involve incubation of the disaggregated tumor specimen in a serum-supplemented culture medium for 24–72 h before harvest. The short-term suspension culture method has been applied to a number of different neoplasms, but its success depends on the quantity of viable cells and their in vivo spontaneous mitotic activity.

One of the most efficient methods of culturing is the mixed colony culture. With this method, a monolayer of cells is established on a polystyrene substrate and fed with culture medium supplemented with FCS and antibiotics (10). Chemical supplements, such as insulin and glutathione, can be added to the culture medium to help improve tumor growth, but these supplements are not essential (1,8). The cells are usually removed from the culture flask with an enzymatic treatment. Optimal success with this technique requires frequent and careful observation of the culture process to establish the timing and length of colchicine exposure. Overgrowth of cultures by normal stromal fibroblasts that adhere more readily to plastic and are stimulated by high concentrations of serum is a problem frequently encountered with this method (1,8).

The in situ method involves the establishment of a monolayer of cells in culture, but the cells are normally grown and harvested in situ on a microscope slide or coverslip (1,8). The advantage of the in situ method includes conservation of technologist time and of cells that otherwise may have been lost during the centrifugation and piping steps involved with standard harvesting techniques.

Each of the above methods has its advantages and disadvantages. Selection of which technique to use depends on the intrinsic qualities of the tumor sample (i.e., sample size, degree of necrosis, and in vitro growth rate).

Cytogenetic banding studies can be performed on solid tumors grown in nude mice (1,12), an approach that has to be taken with some tumors that fail to grow in vitro. Although it is possible that karyotypes obtained on such material do not necessarily represent the dominant one in vivo, the results undoubtedly reflect those of at least some of the cells of the original tumor. The presence of mouse cells can usually be established through their distinctive karyotype. When tumor cells are present in the marrow, the specimen should be processed as in leukemia.

B. Special Conditions and Mitogenic Agents

Since most human solid tumors require culturing of the affected cells for optimal cytogenetic analysis, methods leading to growth and division of these tumors have received much attention. For example, with the use of defined or partially defined media most human lung cancers can be successfully cultured (Table 1).

Some tumors are more difficult to culture than others, with prostatic cancers being among the most difficult. An improved technique for short-term culturing of human prostate adenocarcinoma and its cytogenetic analysis has been described (14). The method is based on: (a) prolonged mild collagenase treatment; (b) careful washing and repeated centrifugation and sedimentation of the disaggregated material to isolate viable prostatic epithelial cells; (c) short-term culture on collagen R-coated (Serva, Heidelberg, Germany) chamber slides with PFMR-4 medium (SBL, Stockholm, Sweden)

Table 1 Growth Factor Requirements for Culture of the Major Types of Lung Cancer

All require insulin, transferrin, and a steroid hormone
Non-small cell lung cancer: epidermal growth factor (EGF)
Small cell lung cancer: selenium, estradiol
Adenocarcinoma: selenium, ethanolamine, albumin, T3
Squamous cell: serum, cholera toxins, low Ca^{2+}, feeder layers

Source: From Ref. 13.

supplemented with mitogenic factors; and (d) daily inspection of the cultured cells to determine the optimal time for harvesting.

Another type of tumor difficult to culture and analyze cytogenetically is transitional cell carcinoma of the bladder. A method that allows routine cytogenetic analysis of small cytoscopic biopsies from urothelial tumors has been described (15). This method is based on prolonged mild collagenase treatment, a 12–16 h culture, and harvesting procedures adapted to give maximal metaphase recovery.

The use of phorbol-12,13-dibutyrate as a mitogen appears to be useful in the cytogenetic analysis of colon tumors (16). This compound, alone or in combination with other agents (e.g., calcium ionophore A23187), may be applicable to many other tumors that are difficult to karyotype because of the dearth of mitoses.

Cell culture synchronization is a method frequently used in combination with standard techniques. The general opinion holds that synchronization of cells in culture should increase the number of metaphases suitable for cytogenetic analysis. The method most widely used to date is that of methotrexate (MTX) block and thymidine release (17).

Though some mitogens capable of stimulating specific cells are available, including a small number of malignant cells (e.g., B-cell mitogens for the leukemic cells in chronic lymphocytic leukemia (CLL) and giant cell tumor cell line (GCT)-conditioned media for myeloid leukemia cells), a critical need exists for mitogens specific for various cancers, lymphomas, and sarcomas. It is possible that some of the growth factors, both natural and synthetic, may prove useful in that regard, although to date they have not been tried extensively in various conditions. It is also possible that growth inhibitors of some cells (e.g., fibroblasts) may also be useful in suppressing the growth of such cells so that the less mitotically active tumor cells have a chance to grow in vitro; a cogent example is cancer of the prostate, in which the normal stromal cells tend to overgrow the cancer cells. It is apparent that much remains to be explored in establishing factors with inhibitor or suppressor activity in the cytogenetic examination of tumor cells.

III. CHROMOSOME CHANGES IN BENIGN TUMORS

Although for a number of technical reasons the establishment of meaningful cytogenetic findings in solid tumors lagged behind that in the leukemias, this situation has been radically altered in recent years with the description of specific and recurrent changes in an array of benign and malignant tumors, particularly those of the bones and soft tissues.

The presence of karyotypic changes is not necessarily indicative of the existence of malignancy within a tumor. Generally, the chromosome number of benign tumors is at or near the diploid range, whereas that of many malignant tumors is near triploid or near tetraploid (1). An interesting fact of chromosome changes in solid tumors has been the demonstration that benign tumors (e.g., lipomas, leiomyomas (Table 2)) are associated with recurrent anomalies, particularly reciprocal translocations, specific for these tumors. These findings have provided another approach to the differential diagnosis of these benign tumors versus their malignant homologs. A case in point is the primary change in myxoid liposarcomas (LPS): the translocation (12;16)(q13;p11). A similar but not identical translocation has been found in benign lipomas. High-resolution karyotypes have revealed that the breakpoint in lipomas is probably at 12q14 (1). Here we have one of the first clinical fruits of solid tumor cytogenetics: under the pathologist's microscope it may be difficult and sometimes impossible to differentiate certain lipomas from some liposarcomas, whereas karyotyping can readily differentiate the two tumors.

Compared with malignant tumors, the chromosome changes in benign tumors are often relatively simple and less complex. Although reciprocal translocations have frequently been found in benign tumors, these abnormalities are rare in solid cancers and especially so in carcinomas (1,9,18). The structural changes that are present in most malignant tumors usually consist of unbalanced translocations, deletions, and isochromosomes that result in the loss or duplication of segments of chromosomes. Thus, there may be fundamental differences in the pathways along which benign tumors develop as compared with malignant tumors. An example of a difference may be the need for malignant tumors to undergo a greater number of successive genetic changes.

The increase in the body of information on the chromosomal constitution of benign tumors has resulted in an increase in the use of karyotypic analysis to delineate subgroups in tumors of similar histology. Examples include the cytogenetic subtypes of lipoma involving translocations or insertions on 12q, deletions only, and chromosome loss or rearrangement on 13q. Uterine leiomyomas also demonstrate chromosomal alterations including deletions on 7q and 13q, anomalies of chromosome 1, translocations (e.g., 12;14), and other changes involving chromosome 12 (9). Cytogenetic analysis also implies that tumors with similar cytogenetic changes may share a similar etiology and biologic behavior. A case in point is the 12q14-15 abnormality in uterine leiomyoma. The 12q14-15 region is also the location of consistent rearrangements in other benign solid tumors, including lipomas and pleomorphic adenomas of the salivary gland (1,9). Rearrangements of

Table 2 Chromosomal Changes Diagnostic of the Tumor Involved

Tumor	Chromosomal abnormality
Translocations	
Myxoid liposarcoma	t(12;16)(q13;p11)
Rhabdomyosarcoma (alveolar)	t(2;13)(q37;q14) or t(1;13)(p13;q14)
Leiomyoma	t(12;14)(q14-15;q23-24)
Lipoma	t(12;V)(q14;V)[a]
Ewing sarcoma/peripheral neuroectodermal tumors (PNETs)	t(11;22)(q24;q12), t(21;22)(q22;q12), t(7;22)(p22;q12)
Synovial sarcoma	t(X;18)(p11;q11)
Pleomorphic adenoma	t(3;8)(p21;q12) or t(9;12)(p13;q13)
Giant cell fibroblastoma	t(17;22)(q21;q13)
Clear cell sarcoma (malignant melanoma of soft parts)	t(12;22)(q13;q12 or 13)
Desmoplastic small round-cell tumor	t(11;22)(p13;q11.2 or 12)
Extraskeletal myxoid chondrosarcoma	t(9;22)(q22-q31;q12)
Deletions or other changes	
Germ cell tumors	i(12)(p10)
Meningioma	−22/del(22q)
Retinoblastoma	del(13)(q13q31)
Wilms tumor	del(11p)
Kidney tumors	del(3p), der(3p)

[a]V, variable chromosomes.

12q13-15 have been also reported in pulmonary chondroid hamartoma (19,20), endometrial polyps (9,21), epithelial breast tumors (22), hemangiopericytoma (23), and an aggressive angiomyxoma (24). These tumors are all mesenchymally derived and benign. Therefore, a single gene involved in mesenchyme differentiation and growth could be responsible for the development of these multiple tumor types (25). Studies at the ultrastructural level might show differences in tumor histology not visible at the light-microscopic level. Thus, chromosome changes in benign proliferations aid in the identification of genes whose function is related to growth rather than to malignant transformation of cells. These changes also reflect alternative pathways in neoplastic transformation of cells.

Evidence to support this conclusion is the demonstration that a gene of the CCAAT/enhancer binding protein (^{c}/EBP) family, CHOP, which maps to the 12q13 and is assumed to be involved in adipocyte differentiation, is rearranged in the t(12;16) translocation in malignant myxoid liposarcoma (LPS). In contrast, no tumor without a t(12;16) showed aberrant CHOP restriction digest patterns on Southern blot analysis, including some benign tumors, such as lipomas with various cytogenetic aberrations of 12q13-15, uterine lipomyomas with t(12;14)(q14-15;q23-24), and hemangiopericytomas and chondromas, both of which also had 12q13 changes (9,26). These findings demonstrate that the 12q breakpoint of the t(12;16) in myxoid LPS differs from that in the other benign tumors investigated, in which no cytogenetically visible differences in breakpoint position could be seen. The breakpoints in myxoid LPS cluster to the 5′ region of the gene. Thus, the CHOP rearrangement is specific for myxoid LPS. On the other hand, a link between a member of the HMG gene family and benign tumor development has been reported (27,28).

In summary, the specificity with which breaks cluster to specific regions on certain chromosomes suggests that the karyotype can be used to differentiate tumors with an apparently similar histology, to determine diversity, and to create subgroups in human malignancies. With the analysis of more benign neoplasms, further chromosomal categorization may be possible.

The significance of these specific changes in benign tumors is not yet clear. These changes may reflect differences in tumor etiology, site, histology, or biology. Careful follow-up of patients with these changes should reveal differences, if any, in clinical behavior with respect to recurrence rate or malignant transformation, although the latter rarely occurs.

Another interesting phenomenon in tumor cytogenetics is the observation of normal tissues with aneuploidy, such as trisomy 10 in non-neoplastic kidney tissue (29,30) and trisomy 7 in non-neoplastic cells from the brain (31), kidney (29,32), liver (33), and lung (34), as well as in atherosclerotic plaques (35), Dupuytren's contracture (36), gliosis (37), Peyronie's disease (38), placenta (39), and rheumatoid and pigmented villonodular synovitis (40,41). A recent study showed that patients with renal cell carcinomas contained a high percentage of CD4 positive cells with trisomy 7, indicating that the trisomic cells were tumor-infiltrating T-helper cells (30). These findings indicate that cells with trisomy 7 may reflect a naturally occurring mosaicism, or represent a common cell type present at different locations and, therefore, have no pathogenetic importance in the neoplastic process.

IV. CHROMOSOME CHANGES IN MALIGNANT TUMORS

As mentioned previously, most malignant mesenchymal tumors demonstrate relatively simple karyotypic abnormalities, such as translocations, inversions, or insertions, which can be diagnostic in pathologically confusing tumors (Table 2). In contrast, cytogenetic findings in common cancers of epithelial origin (e.g., breast, lung, colon, prostate, and bladder) are often numerous, complex, and have lower diagnostic specificity than karyotypic changes seen in sarcomas. Chromosomal deletions are common occurrences in these tumors (1,9). These diverse changes have been related to a stepwise process of genetic changes thought to characterize these cancers. According to this concept, the initial genetic change leads to increased cell growth followed by a sequence of changes that subsequently result in malignant transformation and ultimately in metastatic spread (see Figures 13.1, 13.2, and 13.3).

The events associated with this concept are compatible with the multiplicity of chromosomal changes seen in adenocarcinomas, although some of these events may be irrelevant to the basic process and cytogenetically invisible. It appears that this orchestrated multistage cascade of genetic events necessary for the genesis of adenocarcinomas is unique to each tumor type.

Although much remains to be proved to authenticate the scheme of genetic changes for every kind of adenocarcinoma, and although the nature of the changes themselves must be established on a firmer basis, an understanding of the cascade of genetic changes may suggest a means for interrupting this chain of events leading to cancer, as well as for diagnosing the early stages of tumorigenesis.

Certain karyotypic features that are to some extent independent of the modal chromosome number should be noted. First is the presence of mark-

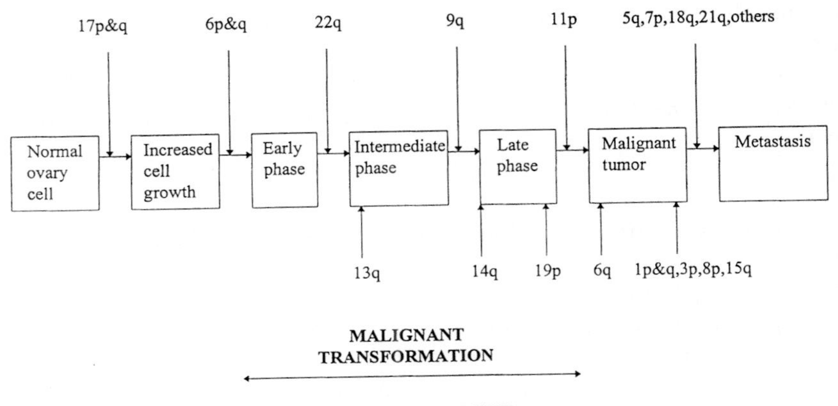

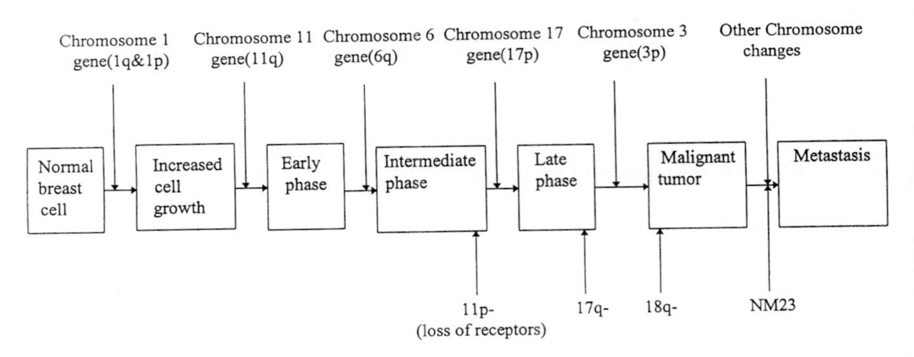

Figure 1 Suggested stepwise cytogenetic pathway in the development of ovarian and breast cancers. The cytogenetic events consist primarily of chromosome deletions with loss of tumor-suppressor genes. The ordered accumulation of these changes gives rise to tumors that eventually become malignant. With each progressive genetic change, the affected cell becomes less responsive to normal control mechanisms, ultimately resulting in a tumor capable of invasion and metastasis. The chromosomes that are shown to be involved are based on our present knowledge, and these schemes or parts of them should be modified or abandoned as more cogent information is obtained on genetic events in these cancers. Furthermore, the schemes shown may apply to only a subtype of these cancers. In addition to the deletions shown in this figure, loss of heterozygosity in these cancers may be evident only upon molecular analysis.

ers. Markers are very commonly and perhaps universally present in tumors with hyperdiploid chromosome numbers (1). A small number of chromosomes tend to be nonrandomly involved in structural changes in carcinomas. Examples include the involvement of chromosomes 1 and 11 in most carcinomas, chromosome 10 in carcinoma of the prostate, and chromosome 18 in colorectal carcinoma (1,18). It is also noteworthy that the changes affecting a particular chromosome in a given type of tumor show considerable variations. In particular, these alterations consist of translocations with varying partners and breakpoints, deletions, duplications, and isochromosomes. A critical region of the chromosome may, however, be consistently lost or duplicated as a result of the changes.

The question of whether diploid cancers exist has become essentially semantic. Genetic changes, whether at the microscopic level and hence of a cytogenetic nature or at the molecular level, characterize all cancers examined to date. Thus, a cancer may appear diploid cytogenetically but be genetically abnormal through a variety of mechanisms.

In summary, in presenting the cytogenetic data on malignant solid tumors, we would like to emphasize the application of these data in several areas:

1. Defining cytogenetic subsets within putative histologically homogeneous tumor types [e.g., isochromosome (5p) or abnormalities of 5q, trisomy 7, monosomy 9, and deletion of chromosome 11 (11p-) in bladder cancer] (1,18)
2. Creating etiologic connections between histologically diverse tumors [e.g., t(11;22) in Ewing sarcoma and a number of neuroectodermal tumors] (1,9)
3. Highlighting specific changes in histological subtypes [e.g., t(12;16) in myxoid liposarcoma, t(2;13) in alveolar rhabdomyosarcoma, and t(X;18) in synovial sarcoma)] (1,9)
4. Suggesting a primary site when a specific cytogenetic change is found in a metastasis or in bone marrow
5. Applying the knowledge of the cytogenetic changes in solid tumor classification, diagnosis, and causation
6. Establishing the molecular events underlying these tumors based on the cytogenetic findings [e.g., FUS-CHOP in t(12;16) in myxoid liposarcoma, FLI1-EWS in t(11;22) in Ewing sarcoma, and loss of APC/TP53/DCC in del(5q)/(17p)/(18q) in colon cancer] (1,4)

Thus, the combination of cytogenetic and molecular genetics may ultimately serve as a guide for the diagnosis of these tumors.

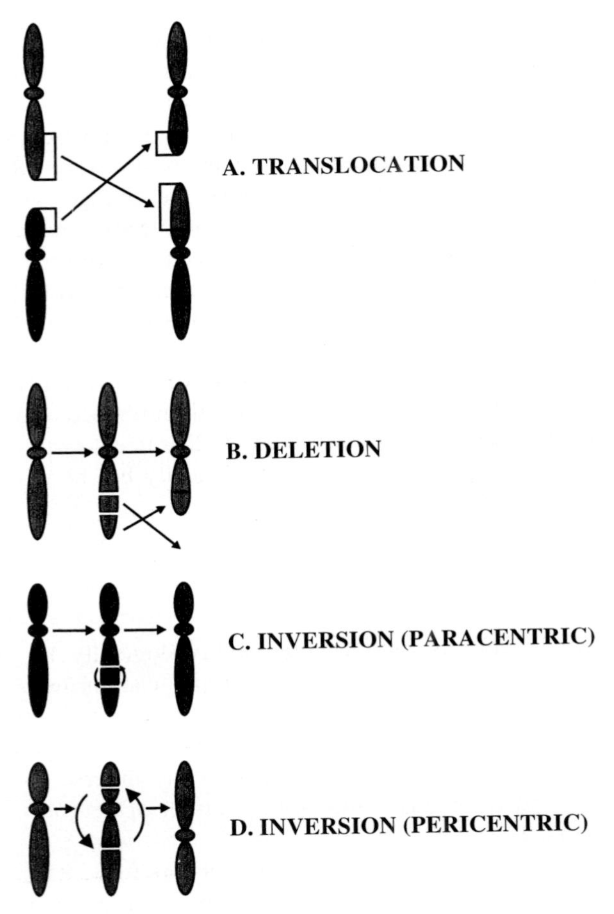

Figure 2 Schematic representations of chromosome changes seen often in human neoplasia (1,2). (A) Translocation. Shown in the figure is an exchange of genetic material between the long arm of the chromosome on the left and the short arm of the chromosome on the right. As a consequence of this exchange of genetic material between two (or more) chromosomes, and the juxtaposition of regions, aberrant (fusion) genes, known as oncogenes, may result and lead to abnormal cellular growth. A cogent example is the Philadelphia (Ph) chromosome, which results from a translocation (9;22)(q34;q11) and involves the oncogene abl on chromosome 9 and the bcr gene on chromosome 22, forming the fusion abl/bcr gene. Translocations are particularly common in leukemias and sarcomas. (B) Deletion. Loss of chromosome material, often associated with loss of heterozygosity (LOH), is common in adenocarcinomas. The amount of material deleted may vary from a whole chromosome to a whole or part of a chromosomal arm to submicroscopic loss that is not discernible microscopically. Shown in the figure is a loss of material in the

V. CLINICAL APPLICATION OF CHROMOSOME CHANGES IN SOLID TUMORS

As described previously, most human cancers have abnormal numeric and/or morphologic chromosomal changes by karyotypic analysis. These chromosomal changes are of an extremely protean nature, and it is rare for two cancers, particularly carcinomas, to have an exactly identical karyotypic makeup. The presence of any karyotypic change, particularly morphologic (including markers), indicates that a cell is cancerous (Tables 2 and 3). Therefore, the utilization of karyotypic examination, particularly in the diagnosis of unclear cases of cancer, in conjunction with other cytologic and clinical criteria, may become an important tool in the diagnosis of neoplasias in human subjects. Molecular cytogenetics also provides a rational approach to evaluate the prognosis and subsequent therapy for patients with solid tumors. Several examples deserve mention here.

A. Neuroblastoma

A common childhood tumor of the peripheral nervous system, neuroblastoma is associated with deletions of the short arm of chromosome 1 and amplification of the MYCN gene located on the short arm of chromosome 2. Amplification of the MYCN gene is often associated with a poor outcome for the patient (42), whereas hyperdiploidy or triploidy without a del(1p) or MYCN amplification is found in patients younger than 1 year and with a good prognosis. Near-diploidy or near-tetraploidy without a del(1p) or MYCN amplification is usually correlated with a slowly progressive course of the disease, whereas the cases in which these changes are present [i.e., del(1p) and MYCN amplification] are often associated with a rapidly fatal outcome.

B. Breast Cancer

Breast cancer affects one in eight American women. A number of markers have been correlated with clinical outcome. The most common site of am-

---

chromosome long arm significant enough to result in a shortened chromosome. (C, D) Inversions. Shown in C is a paracentric inversion of a segment of the long arm of the chromosome. In D is shown a pericentric inversion involving the centromeric region. The inversion depicted in D often leads to a grossly appearing chromosome. In both instances, the inversions lead to the juxtaposition of two new regions and possible eventual genesis of abnormal fusion genes, as in the case of translocations.

A

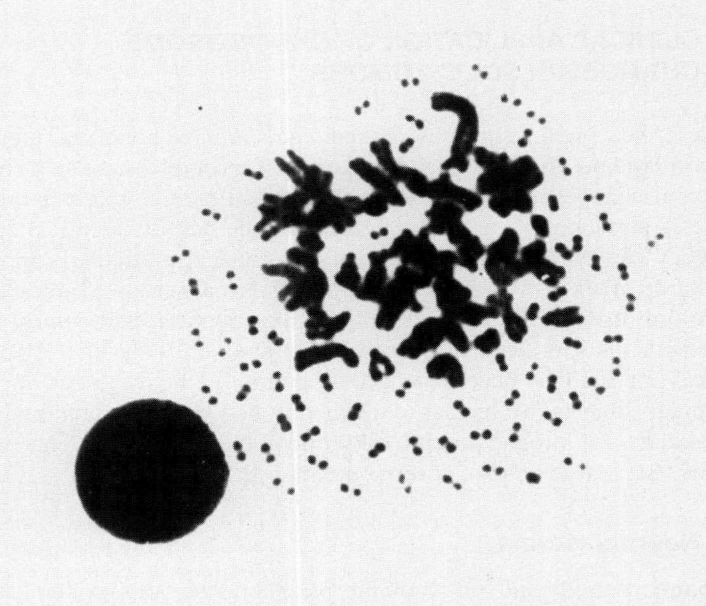

B

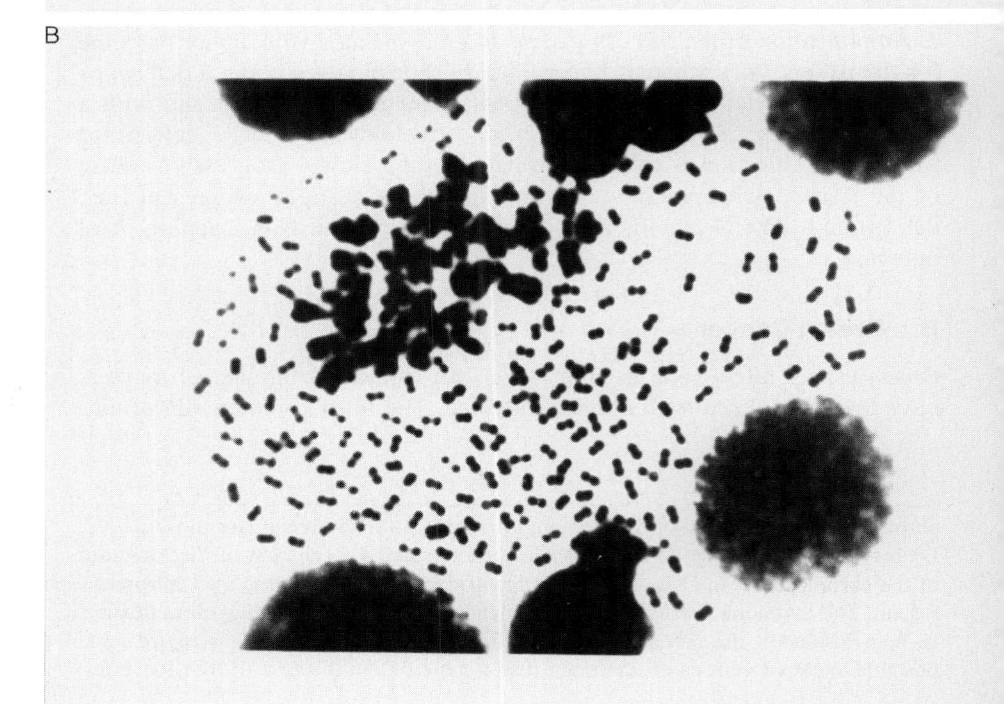

plification is the HER-2/neu(erbB-2) gene. This amplification is associated with an increased rate of relapse and poor survival in 25–30% of cases, particularly node-negative patients (43). Overexpression of HER-2/neu has also been demonstrated in about 30% of patients with ovarian cancer, in which, as in breast cancer, overexpression correlates with poor prognosis (43). Moreover, overexpression of the NM23 gene is correlated with low metastatic potential. In contrast, overexpression of the epidermal growth factor receptor protein and cathepsin D gene correlates with a poor prognosis in node-negative breast cancer cases. Studies have shown that amplification of the MYC oncogene is one of the most common genetic alterations (about 30%) in breast cancer. Amplification of the MYC gene is clearly associated with a poor prognosis (44) and with postmenopausal disease (44). Expression of MYC may alter cellular resistance to cisplatin and other DNA strand-scission-inducing drugs and may suppress differentiation in association with decreased collagen gene transcription (44). However, MYC amplification is clearly not associated with HER-2/neu amplification in primary breast cancer (44).

The cytogenetics of breast cancer is further discussed in the chapter by Wolman in this volume.

C. Colorectal Cancer

Cytogenetic analysis of colorectal carcinomas showed frequent deletion of 5q, 12p, and 18q; LOH analysis showed preferential loss in these same chromosome arms. Small colonic adenomas are most commonly associated with changes in 5q (i.e., APC gene). About 50% of larger adenomas of intermediate stage have point mutations in the K-RAS gene, whereas late adenomas have alterations in 17p (TP53 gene) and 18q (DCC gene) in about 25% and 50% of cases, respectively (45). Thus, the practical benefits of recognition of critical early genetic events in colon cancer could lead to the establishment of new screening tests and more adequate early management.

←

Figure 3 Double minute chromosomes (dmin) are seen in some leukemias and cancers. In the two examples shown here from a case of neuroblastoma, the number of dmin is large and unusual. Dmin are associated with gene amplification, which may be responsible for resistance to therapy (1). Homogeneously staining regions (hsr) are the equivalent of dmin (1).

Table 3 Some of the More Recurrent but Not Necessarily Diagnostic Chromosomal Changes Observed in Tumors

Tumor	Chromosomal changes
Alveolar soft part sarcoma	t(17q25)
Aggressive angiomyxoma	t(12q14-15)
Bladder cancer	i(5)(p10)/del(5q), +7, del(8p), −9/del(9q), del(11p)
Breast cancer	t/del(1)(p35), del(3p), t/del(16q), del(17q), del(6)(q21-22), +7, +18, +20
Colon cancer	Structural changes of chromosomes 1, 17, and 18, +7, +12, del(5q), del(10q)
Dermatofibrosarcoma protuberans	Rings (sequences 17q and 22q)
Desmoid tumors	+8, +20
Fibrosarcoma (congenital)	Combination of trisomies (8,11,17,20)
Glioma	Double minute chromosomes, +7, −10
Hemangiopericytoma	Structural changes of 12q13-15, 3p12-21, 19q13
Kidney adenoma	+7, +17, −Y
Kidney carcinoma	t(9;14)(q13;q22), t(X;1)(p11.2;q21)
Leiomyoma	+12, del(7)(q21.2q31.2), del/inv(1q), −22
Leiomyosarcoma	Monosomy 1p12-1pter with hypodiploidy
Lipoma	del(13)(q12q22), der(6)(p21-23), der(11)(q13)
Lipoblastoma	t(8q11-13)
Liposarcoma (well differentiated and pleomorphic)	Telomere associations, rings, markers
Malignant fibrous histiocytoma	Telomere associations, rings, dicentric chromosomes, double minute chromosomes, homogeneoulsy staining regions
Malignant melanoma	t/del(1)(p12-22), t(1;19)(q12;p13), t/del(6q)/ i(6)(p10), +7, del(9p)
Medulloblastoma	i(17)(q10)
Mesothelioma	del(1p), del(3p), del(6q), del(9p), t(12p13), −22, −18
Neuroblastoma	del(1)(p31-32) Double minute chromosomes Homogeneously staining regions
Ovary thecoma	+12
Ovarian carcinoma	Structural change of chromosome 1, 17q−, complex rearrangements
Parosteal osteosarcoma	+r(12)
Osteosarcoma	del(6q)
Prostate cancer	del(7)(q22), del(10)(q24), +7, double minute chromosomes, −Y
Retinoblastoma	Structural changes of chromosome 1, i(6)(p10)
Salivary gland mucoepidermoid carcinoma	del(6q); t(11;19)(q14-21;p12-13)
Salivary gland adenocarcinoma	del(6q)

Table 3 *Continued*

Tumor	Chromosomal changes
Small cell lung cancer	del(3)(p14p23), del(17p), del(5q), del(13q)
Testicular tumors	Structural and numerical changes of chromosome 1
Thyroid adenoma	+5, +7, +12, structural changes of 19q
Thyroid papillary carcinoma	del(10q), inv(10)(q11q21)
Uterine cancer	Structural and numerical changes
Uveal melanoma	−3, trisomy 8q, +i(8)(q10)
Wilms tumor	Structural changes of chromosomes 1 and 16q, +12

D. Lung Cancer

Delineation of the molecular changes associated with the carcinogenic processes of initiation, promotion, and progression provides information which may eventually benefit the early detection, prevention, and treatment of lung cancer. Genetic alterations in genes such as K-RAS and TP53 might be early events in lung carcinogenesis (46), and thus be potentially useful markers for screening high-risk populations such as smokers and radon-exposed individuals. An increased understanding of the changes associated with later stages of lung cancer development will be helpful in predicting prognosis and developing rational treatment protocols. The development of drugs and pharmacologic interventions has been demonstrated experimentally with selective inhibitors of RAS farnesyltransferases (46).

E. Renal Cell Cancer (RCC)

Molecular cytogenetic analysis may provide powerful parameters for the precise diagnosis and prognosis of small renal cell tumors. Fluorescent in situ hybridization (FISH) analysis of tumor cells using appropriate genetic markers is a good approach for identifying genetic types of renal cancer. Detection of trisomies 16, 3q, 8, 12, or 20 indicates malignant potential in papillary RCC, whereas DNA losses at 14q, 6q, 8p, and 9 in nonpapillary RCC are strongly associated with an aggressive growth and metastatic potential. Generally, the prognosis is much better for a nonpapillary RCC localized to the kidney without complex chromosome changes (47).

Thus, potential applications of chromosomal findings to human cancer therapy have probably not been fully explored. With newly developed techniques, the study of larger series of patients, and meaningful correlations of the various clinical, cytologic, and karyotypic parameters, may lead to a more realistic and frequent utilization of chromosomal data in the therapy of human neoplasia.

VI. MOLECULAR CYTOGENETICS

An important recent technical advance in cytogenetics has been the introduction of FISH techniques. FISH analysis has become an essential procedure that complements classic cytogenetic studies in various forms of cancers (1,2,4,5,9). Until recently, molecular cytogenetic analyses with centromeric probes have been used primarily to establish numerical chromosomal changes. Such changes (e.g., +7 and −9) in bladder cancer can also be established in the nuclei of nondividing cells using FISH technology. Thus, FISH may reveal changes when conventional cytogenetic results are normal or the material is insufficient or unsuitable for conventional cytogenetic analysis.

FISH can also be performed on fresh tumor tissue, on exfoliative tumor cells, and, most importantly, on embedded and archival specimens. One practical application of FISH analysis has been developed in the authors' laboratory to detect the presence of cancer cells in the urine of patients with bladder cancer and in the pleural and ascitic fluids of cancer patients. This important information can be used in the diagnosis and follow-up of these conditions without having to resort to invasive procedures (48,49). The utilization of chromosome cosmid and painting probes has proven to be of value not only in detecting and confirming established translocations but also in identifying translocations involving chromosomes other than or in addition to those seen with banding techniques. Moreover, FISH plays an increasingly important role in a variety of research areas, including tumor biology, gene amplification, and gene mapping. The applications of FISH are further explored in a chapter by Mark in this volume.

Another new molecular cytogenetic technique, termed comparative genomic hybridization (CGH), has recently been developed for detecting chromosomal imbalances in tumor genomes (50). CGH is based on two-color FISH. Equal amounts of differentially labeled tumor DNA and normal DNA are mixed together and hybridized, under conditions of Cot 1 DNA suppression, to normal metaphase spreads. In a single experiment, CGH identifies DNA gains and losses and maps these variations to metaphase chromosomes. DNA extracted from either fresh or frozen tissue, cell lines, as well as from formalin-fixed, paraffin-embedded samples is suitable for CGH (51). CGH becomes particularly advantageous when structural analysis of chromosomal changes in cancers are severely limited by the banding quality. Of note is that CGH is an effective screening method for describing and establishing a phenotype/genotype correlation in solid tumor progression. Several examples of chromosomal aberrations that define specific stages in tumor progression have already been established in brain, colon, prostate, cervix, and breast carcinogenesis (50).

Cancer cytogenetics is often hampered by low mitotic indices, poor-quality metaphase spreads, and the presence of complex marker chromosomes. A newly developed technique, multicolor spectral karyotyping (SKY), may have the ability to overcome these obstacles. This technique combines Fourier spectroscopy, charge-coupled device (CCD) imaging, and optical microscopy to measure chromosome-specific spectra after FISH with differently labeled painting probes (52,53). This technique was reported to be in excellent agreement with results from previously performed FISH experiments and banding analysis. Currently, work is underway to generate a multicolor banding pattern (bar code) of the human chromosome complement by using chromosome arm- and band-specific painting probes in order to identify intrachromosomal anomalies. Therefore, it appears that SKY may be a very promising approach to the rapid and automatic karyotyping of neoplastic cells in human beings. A thorough discussion of CGH and SKY is presented in the chapter by Green et al. in this volume. With these rapidly evolving molecular cytogenetic technologies, it is not unrealistic to expect that the nature of many, if not all, of the chromosomal abnormalities encountered in various solid tumors will be clearly elucidated in the not too distant future.

REFERENCES

1. Sandberg AA. The Chromosomes in Human Cancer and Leukemia. 2d ed. New York: Elsevier Science Publishing Co., 1990.
2. Sandberg AA, Chen Z. Cytogenetics of cancer. Cancer cytogenetics: nomenclature and clinical applications. In: Kurzrock R, Talpaz M, eds. Molecular Biology in Cancer Medicine. London: Martin Dunitz, 1995:55–84.
3. Mitelman F. Catalog of Chromosome Aberrations in Cancer. 3d ed. New York: Alan R. Liss, 1988.
4. Sandberg AA, Chen Z. Cancer cytogenetics and molecular genetics: clinical implications (review). Int J Oncol 1995; 7:1241–1251.
5. Sandberg AA, Chen Z. Cytogenetics of acute leukemia. In: Wiernik PH, Canellos GP, Dutcher JP, Kyle RA, eds. Neoplastic Diseases of the Blood. 3d ed. New York: Churchill Livingstone, 1996:249–269.
6. Teyssier JR. The chromosomal analysis of human solid tumors. A triple challenge. Cancer Genet Cytogenet 1989; 37:103–125.
7. Sandberg AA, Turc-Carel C. The cytogenetics of solid tumors. Relation to diagnosis, classification and pathology. Cancer 1987; 59:387–395.

8. Sandberg AA, Bridge JA. Techniques in cancer cytogenetics: an overview and update. Cancer Invest 1992; 10:163–172.
9. Sandberg AA, Bridge JA. The Cytogenetics of Bone and Soft Tissue Tumors. Austin: RG Landes, 1994.
10. Gibas LM, Gibas Z, Sandberg AA. Technical aspects of cytogenetic analysis of human solid tumors. Karyogram 1984; 10:25–27.
11. Leibovitz A. Development of tumor cell lines. Cancer Genet Cytogenet 1986; 19:11–19.
12. Povlsen CO, Visefeldt J, Rygaard J, Jensen G. Growth patterns and chromosome constitutions of human malignant tumours after long-term serial transplantation in nude mice. Acta Pathol Microbiol Immunol Scand [A] 1975; 83: 709–716.
13. Gazdar AF, Oie HK. I. Advances in cell culture. Cell culture methods for human lung cancer. Cancer Genet Cytogenet 1986; 19:5–10.
14. Limon J, Lundgren R, Elfving P, Heim S, Kristoffersson U, Mandahl N, Mitelman F. An improved technique for short-term culturing of human prostatic adenocarcinoma tissue for cytogenetic analysis. Cancer Genet Cytogenet 1990; 46:191–200.
15. Fraser C, Sullivan LD, Kalousek DK. A routine method for cytogenetic analysis of small urinary bladder tumor biopsies. Cancer Genet Cytogenet 1987; 29: 103–108.
16. Eiseman E, Luck JB, Mills AS, Brown JA, Westin EH. Use of phorbol-12,13 dibutyrate as a mitogen in the cytogenetic analysis of tumors with low mitotic indices. Cancer Genet Cytogenet 1988; 34:165–175.
17. Hagemeijer A, Smit EME, Bootsman D. Improved identification of chromosomes of leukemic cells in methotrexate-treated cultures. Cytogenet Cell Genet 1979; 23:208–212.
18. Atkin NB. Solid tumor cytogenetics: progress since 1979. Cancer Genet Cytogenet 1989; 40;3–12.
19. Dal Cin P, Kools P, De Jonge I, Moerman P, Van de Ven W, Van den Berghe H. Rearrangements of 12q14-15 in pulmonary chondroid hamartoma. Genes Chromosom Cancer 1993; 8:131–133.
20. Fletcher JA, Longtine J, Wallace K, Mentzer SJ, Sugarbaker DJ. Cytogenetic and histologic findings in 17 pulmonary chondroid hamartomas: evidence for a pathogenetic relationship with lipomas and leiomyomas. Genes Chromosom Cancer 1995; 12:220–223.
21. Vanni R, Dal Cin P, Marras S, Moerman P, Andria M, Valdes E, Deprest J, Van den Berghe H. Endometrial polyp: another benign tumor characterized by 12q13-15 changes. Cancer Genet Cytogenet 1993; 68:32–33.
22. Rohen C, Bonk U, Staats B, Bartnitzke S, Bullerdiek J. Two human breast tumors with translocations involving 12q13-15 as the sole cytogenetic abnormality. Cancer Genet Cytogenet 1993; 69:68–71.
23. Mandahl N, Örndahl C, Heim S, Willén H, Rydholm A, Bauer HC, Mitelman F. Aberrations of chromosome segment 12q13-15 characterize a subgroup of hemangiopericytomas. Cancer 1993; 71:3009–3013.

24. Kazmierczak B, Wanschura S, Meyer-Bolte K, Caselitz J, Meister P, Bartnitzke S, Van de Ven W, Bullerdiek J. Cytogenetic and molecular analysis of an aggressive angiomyxoma. Am J Pathol 1995; 147:580–585.
25. Schoenberg Fejzo M, Yoon S-J, Montgomery KT, Rein MS, Weremowicz S, Krauter KS, Dorman TE, Fletcher JA, Mao J, Moir DT, Kuchlerlapati RS, Morton CC. Identification of a YAC spanning the translocation breakpoints in uterine leiomyomata, pulmonary chondroid hamartoma, and lipoma. Physical mapping of the 12q14-q15 breakpoint region in uterine leiomyomata. Genomics 1995; 26:265–271.
26. Åman P, Ron D, Mandahl N, Fioretos T, Heim S, Arheden K, Willén H, Rydholm A, Mitelman F. Rearrangement of the transcription factor gene CHOP in myxoid liposarcomas with t(12;16)(q13;p11). Genes Chromosom Cancer 1992; 5:278–285.
27. Schoenberg Fejzo M, Ashar HR, Krauter KS, Powell WL, Rein MS, Weremowicz S, Yoon S-J, Kucherlapati RS, Chada K, Morton CC. Translocation breakpoints upstream of the HMGIC gene in uterine leiomyomata suggest dysregulation of this gene by a mechanism different from that in lipomas. Genes Chromosom Cancer 1996; 17:1–6.
28. Schoenmakers EFPM, Wanschura S, Mols R, Bullerdiek J, Van den Berghe H, Van de Ven WJM. Recurrent rearrangements in the high modality group protein gene, HMGI-C, in benign mesenchymal tumours. Nature Genet 1995; 10: 436–444.
29. Elfving P, Cigudosa JC, Lundgren R, Limon J, Mandahl N, Kristoffersson U, Heim S, Mitelman F. Trisomy 7, trisomy 10, and loss of the Y chromosome in short-term cultures of normal kidney tissue. Cytogenet Cell Genet 1990; 53: 123–125.
30. Dal Cin P, Aly MS, Delabie J, Ceuppens JL, Van Gool S, Van Damme B, Baert L, Van Poppel H, Van den Berghe H. Trisomy 7 and trisomy 10 characterize subpopulations of tumor-infiltrating lymphocytes in kidney tumors and in the surrounding kidney tissue. Proc Natl Acad Sci USA 1992; 89:9744–9748.
31. Heim S, Mandahl N, Jin Y, Strömblad S, Lindström E, Salford LG, Mitelman F. Trisomy 7 and sex chromosome loss in human brain tissue. Cytogenet Cell Genet 1989; 52:136–138.
32. Elfving P, Åman P, Mandahl N, Lundgren R, Mitelman F. Trisomy 7 in nonneoplastic epithelial kidney cells. Cytogenet Cell Genet 1995; 69:90–96.
33. Bardi G, Johansson B, Pandis N, Heim S, Mandahl N, Hägerstrand I, Holmin T, Andrén-Sandberg Å, Mitelman F. Trisomy 7 in nonneoplastic focal steatosis of the liver. Cancer Genet Cytogenet 1992; 63:22–24.
34. Lee JS, Pathak S, Hopwood V, Thomasovic B, Mullins TD, Baker FL, Spitzer G, Neidhart J. Involvement of chromosome 7 in primary lung tumor and nonmalignant normal lung tissue. Cancer Res 1987; 47:6349–6352.
35. Vanni R, Cossu L, Licheri S. Atherosclerotic plaque as a benign tumor? Cancer Genet Cytogenet 1990; 47:273–274.
36. Wurster-Hill DH, Brown F, Park JP, Gibson SH. Cytogenetic studies in Dupuytren contracture. Am J Hum Genet 1988; 43:285–292.

37. Moertel CA, Dahl RJ, Stalboerger PG, Kimmel DW, Scheithauer BW, Jenkins RB. Gliosis specimens contain clonal cytogenetic abnormalities. Cancer Genet Cytogenet 1993; 67;21–27.
38. Somers KD, Winters BA, Dawson DM, Leffell MS, Wright GL Jr., Devine CJ Jr., Gilbert DA, Horton CE. Chromosome abnormalities in Peyronie's disease. J Urol 1987; 137:672–675.
39. Delozier-Blanchet CD, Engel E, Extermann P, Pastori B. Trisomy 7 in chorionic villi: follow-up studies of pregnancy, normal child, and placental clonal anomalies. Prenat Diagn 1988; 8:281–286.
40. Fletcher JA, Henkle C, Atkins L, Rosenberg AE, Morton CC. Trisomy 5 and trisomy 7 are nonrandom aberrations in pigmented villonodular synovitis: confirmation of trisomy 7 in uncultured cells. Genes Chromosom Cancer 1992; 4: 264–266.
41. Mertens F, Örndahl C, Mandahl N, Heim S, Bauer HFC, Rydholm A, Tufvesson A, Willén H, Mitelman F. Chromosome aberrations in tenosynovial giant cell tumors and nontumorous synovial tissue. Genes Chromosom Cancer 1993; 6:212–217.
42. Brodeur GM, Castleberry RP. Neuroblastoma. In: Pizzo PA, Paplack DG, eds. Principles and Practice of Pediatric Oncology. 2d ed. Philadelphia: Lippincott, 1993:749–763.
43. Slamon DJ, Godolphin W, Jones LA, Holt JA, Wong SG, Keith DE, Levin WJ, Stuart SG, Udove J, Ullrich A, Press MF. Studies of the HER-2/neu protooncogene in human breast and ovarian cancer. Science 1989; 244:707–712.
44. Dickson RB. The molecular basis of breast cancer. In: Kurzrock R, Talpaz M, eds. Molecular Biology: Cancer Medicine. London: Martin Dunitz, 1995: 241–272.
45. Rowley JD, Aster JC, Sklar J. The clinical applications of new DNA diagnostic technology on the management of cancer patients. JAMA 1993; 17:2331–2337.
46. Sabichi AL, Bober MA, Birrer MJ. Lung cancer. In: Kurzrock R, Talpaz M, eds. Molecular Biology: Cancer Medicine. London: Martin Dunitz, 1995: 223–240.
47. Kovacs G. The value of molecular genetic analysis in the diagnosis and prognosis of renal cell tumors. World J Urol 1994; 12:64–68.
48. Meloni AM, Peier AM, Haddad FS, Powell IJ, Block AMW, Huben RP, Todd I, Potter W, Sandberg AA. A new approach in the diagnosis and follow-up of bladder cancer: FISH analysis of urine, bladder washings, and tumors. Cancer Genet Cytogenet 1993; 71:105–118.
49. Chen Z, Wang DD, Peier A, Stone JF, Sandberg AA. FISH in the evaluation of pleural and ascitic fluids. Cancer Genet Cytogenet 1995; 84:116–119.
50. Veldman T, Heselmeyer K, Schröck E, Ried T. Comparative genomic hybridization: a new approach for the study of copy number changes in tumor genomes. Appl Cytogenet 1996; 22(4):117–122.
51. Speicher MR, du Manoir S, Schröck E, Holtgreve Grez H, Schoell B, Lengauer C, Cremer T, Ried T. Molecular cytogenetic analysis of formalin fixed, paraffin

embedded solid tumors by comparative genomic hybridization after universal DNA amplification. Mol Genet 1993; 2:1907–1914.
52. Liyanage M, Coleman A, du Manoir S, Veldman T, McCormack S, Dickson RB, Barlow C, Wynshaw-Boris A, Janz S, Wienberg J, Ferguson-Smith MA, Schröck E, Ried T. Multicolour spectral karyotyping of mouse chromosomes. Nature Genet 1996; 14:312–315.
53. Schröck E, du Manoir S, Veldman T, Schoell B, Wienberg J, Ferguson-Smith MA, Ning Y, Ledbetter DH, Bar-Am I, Soenksen D, Garini Y, Ried T. Multicolor spectral karyotyping of human chromosomes. Science 1996; 273:494–497.

14
Chromosome Instability and Fragile Sites

Herman E. Wyandt and Roger V. Lebo
Boston University School of Medicine, Boston, Massachusetts

Vijay S. Tonk
Texas Tech University Health Science Center, Lubbock, Texas

I. INTRODUCTION

The rapid progress in molecular genetics over the past two decades is changing our view of human genome organization, of which chromosomes are the most visible manifestation. We are just learning the importance of heterogeneity in the distribution and function of nucleotide sequences in eukaryotic genomes that have evolved mechanisms to maintain genomic stability over hundreds of millions of years of natural selection. Radiation, chemicals, intrinsic errors in metabolism or in the cell cycle, and foreign DNA such as phages, retroviruses, and adenoviruses act on eukaryotic chromosomes, causing polyploidy, aneuploidy, breaks, incomplete replication, and rearrangements, all of which restructure the genome. Major errors in chromosome distribution or imbalance are eliminated from the time of oogenesis or spermatogenesis, through conception, embryogenesis, and fetal development. As discussed in detail in the chapter by Warburton in this volume, 15% of all conceptions end in miscarriage (1). As high as 70% of first-trimester miscarriages have chromosomal aneuploidy or polyploidy (2). By the time of birth, less than 0.6% of newborns have a microscopically detectable chromosome abnormality (3). Even with this attrition, environmental and genetic factors continue to modify DNA and chromosomes to contribute to somatic chromosome instability. Some structural rearrangements, such as dicentric chromosomes, rings, and acentric fragments, are mechanically less able to segregate correctly during mitosis and tend to be lost. Entire chromosomes or chromosome sets are also lost or gained by errors in distribution during meiosis or mitosis. In most cases these gains or losses result either in cell death or, more rarely, in mosaicism. Still more rarely, loss or gain can result in genome instability, and occasional cells lose cell cycle control, leading to cancer.

Errors at the molecular level, which include missing, inserted, altered, or cross-linked bases, and single-strand breaks can also disrupt genomic stability. Both prokaryotes and eukaryotes have evolved complex mechanisms to repair or prevent the majority of these molecular errors. Logically, it follows that inherited genetic defects in DNA repair mechanisms will result in more generalized genomic or chromosome instability as well as increased cancer risk. Progress in understanding DNA repair defects in chromosome instability syndromes and more recently in familial and sporadic forms of ovarian, breast, and colon cancers has been especially significant.

A. Definition of Chromosome Instability

Chromosome instability might best be defined as a rate of mutation, chromosome breakage, rearrangement, or imbalance that is higher than normal, either ubiquitously (in all tissues) or in specific cell types under environmental conditions that are similar for both normal and unstable cells. The most familiar forms of chromosome instability are in cancers and leukemias, which have long been recognized to be clonal in origin. The process of tumorigenesis is usually an evolving chromosome instability which begins as a chromosome abnormality or mutation in a single cell. Breast, ovarian and colon cancers, squamous cell carcinomas, neuroblastomas, and other cancers, in their malignant stages, typically have multiple chromosome abnormalities (Fig. 1) with proliferation and waning of subclones with different chromosome constitutions. However, such instability needs to be distinguished from the process of malignancy. The ability of malignant cells to disseminate and proliferate in a particular cellular environment appears to involve both genetic and nongenetic factors (4). Not all malignancies are associated with complex chromosome abnormalities, nor are all tumors with complex chromosome abnormalities necessarily malignant (5,6). Nevertheless, very specific chromosomal changes are often associated with the initiation of specific cancers or leukemias, which then develop additional multiple chromosome abnormalities that are nonspecific. One of the more significant challenges of cancer cytogenetics over the past several decades

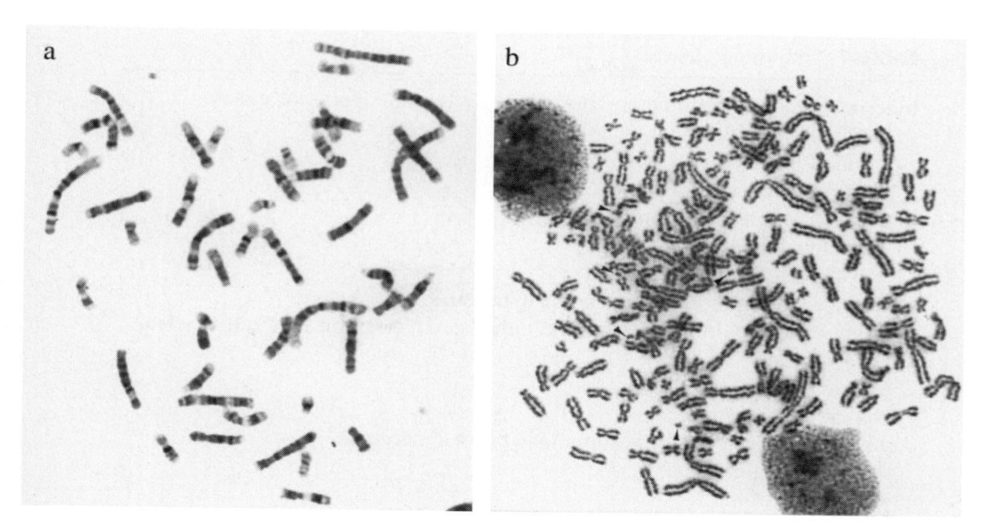

Figure 1 (a) Metaphase with 46 chromosomes from normal human subject (male). (b) Example of metaphase with polyploidy and multiple chromosome abnormalities from cultured breast cancer cells with extra and missing chromosomes, rearrangements, and double minutes (arrows).

has been to determine whether specific chromosomal changes are the cause of disease or its effect. Almost all such changes that can be shown to be causally related are due to specific oncogene loss or alteration (7,8).

Not only does somatic chromosome instability lead to cancer, but more recently it has become clear that there are ''hotspots'' for chromosome mutations and rearrangements. The fact that certain rearrangements recur reveals something about the structural heterogeneity of the genome. In some cases these ''hotspots'' or regions of fragility represent the same sites involved in rearrangements or genetic changes in cancer. Other sites may not be cancer sites, but are sites of more subtle genome instability, which manifest as a genetic disease in offspring after passing through maternal or paternal meiosis (9).

II. CHROMOSOME INSTABILITY IN CANCER

The variety of intrinsic and environmental factors that can cause chromosome breaks, rearrangements, and DNA mutations has already been mentioned. Nevertheless, the majority of such occurrences do not result in chromosome instability or in cancer. Usually, in order for neoplasia to occur, a second mutation must occur at the same locus on the homologous chro-

Table 1 Inherited Cancers

Inherited cancer syndromes involving single tumor-suppressor genes
Familial retinoblastoma
Familial adenomatous polyposis
Multiple endocrine neoplasia
Aniridia Wilms syndrome
Neurofibromatosis
Von Hippel-Lindau syndrome
Atypical multiple mole melanoma syndrome
Familial cancers with clustering: individual predisposition may not be clear
Breast cancer
Ovarian cancer
Nonpolyposis colon cancer
Autosomal recessive chromosome instability syndromes
Ataxia telangiectasia
Bloom syndrome
Cockayne syndrome
Fanconi anemia
Nijmegen syndrome
Roberts syndrome
Werner syndrome
Xerodema pigmentosum

mosome. If this confers a proliferative advantage to the cell, a clone giving rise to a sporadic form of cancer may develop. Tumors that arise in this way can arise more readily when the first hit has been inherited. The number of cancers demonstrated to have a heritable component is growing rapidly (10) and can be categorized into three groups (Table 1): (a) mutations in tumor-suppressor genes related to specific types of cancer or syndromes; (b) inherited (familial) predisposition to cancer, which is usually more difficult to demonstrate genetically and often involves more than one type of cancer within a family; and (c) autosomal recessive syndromes with defective DNA repair, with a high risk of developing a variety of cancers. In fact, groups (b) and (c) may have more similarities than differences.

A. Tumor-Suppressor Genes

In 1971 the two-hit hypothesis was proposed by Knudson (11) to explain familial and sporadic cases of retinoblastoma (Rb), in which one allele at the Rb gene locus on chromosome 13 is defective either in a germline or in a somatic cell. The second allele is subsequently mutated, lost, or replaced by the already mutant allele (loss of heterozygosity) (Fig. 2a), resulting in the tumorigenesis. The Rb gene is referred to as a "tumor-suppressor gene" because one normally functioning gene prevents tumorigenesis. In retinoblastoma families, individuals with constitutional deletion or mutation of the Rb gene have a 90% chance for a "second hit" that will result in tumors, usually bilaterally (13,14). Sporadic cases of Rb are more frequent than familial cases, constituting 88–90% of all cases. These also have mutations of the Rb gene on 13q (Fig. 2b). In contrast to familial cases, sporadic cases usually develop only unilateral retinoblastoma.

Tumor-suppressor genes have been subsequently demonstrated for a number of other cancers, including Wilms tumor (15), multiple mole melanoma (16), acoustic neuroma (neurofibromatosis, type I) (17), and others (Table 2) (15–30). As for retinoblastoma, both sporadic and familial forms occur, with sporadic forms being more frequent. Most forms of cancer, such as breast and colon cancer, have implicated more than one tumor-suppressor gene and/or oncogene. Familial patterns are consequently more difficult to determine for phenotypes caused by mutations at more than one locus.

B. Tumor Progression and Metastasis

Retinoblastoma and the other tumors mentioned typically show multiple chromosome abnormalities. Cytogenetically visible abnormalities of chromosome 13 are present in about 20% of retinoblastoma tumors. Loss of heterozygosity (LOH) at the Rb locus can occur by a variety of mechanisms, including chromosome loss or deletion (Fig. 2b). A variety of other chromosome abnormalities are more likely to be encountered in the actual tumor, the most frequent being loss or gain of all or part of chromosome 6, trisomy for 1q, loss of chromosome 16, and the formation of double minutes (DMs) or homogeneously staining regions (HSRs) (14). DMs and single HSRs result in the increase in copy number of small chromosomal regions. Mutation of the Rb gene has also been associated with other forms of cancer-like osteosarcoma and small-cell lung carcinoma, while family members who carry the Rb gene mutation frequently develop a second primary tumor. However, LOH in the Rb locus alone is not sufficient for these other cancers to develop outside the retina. Additional LOH or chromosome change is required. For example, small-cell lung carcinoma requires mutation of the p53 oncogene on chromosome 17 and deletion of the short arm of chromosome 3 (13,14).

Recent studies of metastases also indicate that very specific chromosome changes are typically required for cells to successfully invade and proliferate outside the primary tumor site. While metastasis usually evolves from a single focus within the primary tumor (31–33), tumors representing

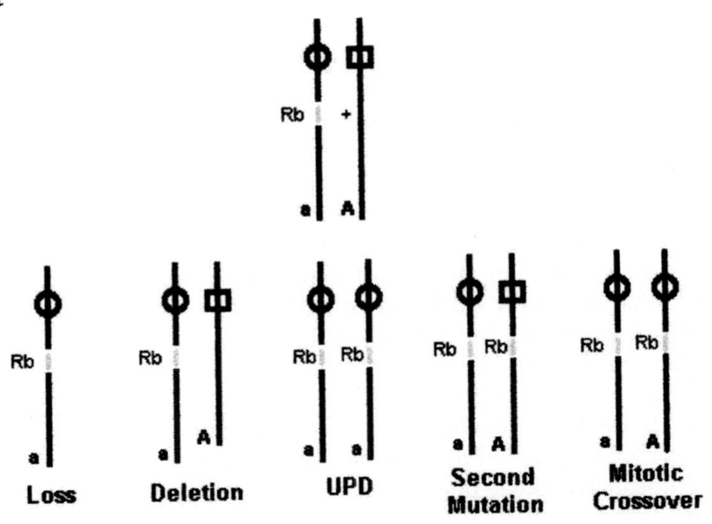

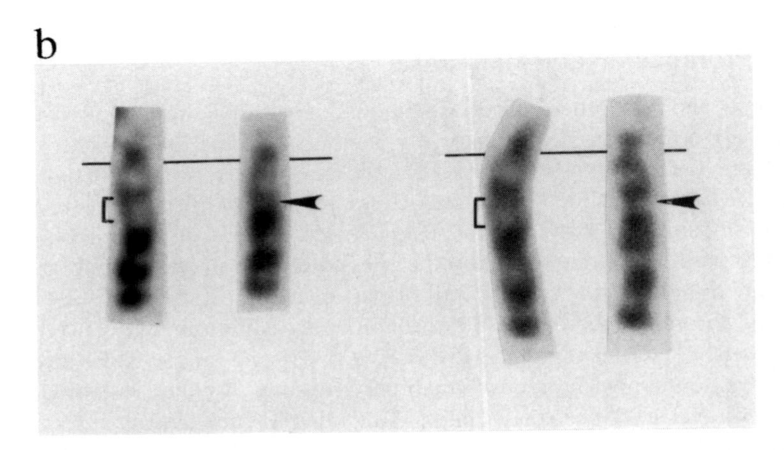

Figure 2 (a) Schematic diagram (adapted from Ref. 12) of mechanisms for loss of heterozygosity (LOH) of the retinoblastoma tumor-suppressor gene locus on chromosome band 13q14. Starting with a heterozygote which has a mutation (Rb) at one locus and a normal wild-type (+) allele, LOH can occur (i) by loss of an entire chromosome 13; (ii) by deletion of the + allele from the normal chromosome 13; (iii) by uniparental disomy (UPD), i.e., loss of the normal chromosome 13 and replacement with two copies of a chromosome 13 with the Rb mutation (UPD can occur in a variety of ways, usually by mitotic nondisjunction giving rise to trisomy, followed by loss of the normal chromosome 13); (iv) occurrence of a second somatic mutation which converts the wild-type (+) allele to a Rb allele; or (v) crossing over between homologs, resulting in the proximal part of chromosome 13 becoming homozygous for the Rb allele. (b) Two pairs of chromosome 13s showing a deletion of 13q14 in the right-hand chromosome 13 of each pair. Deleted region is bracketed in the left-hand member of each pair.

Table 2 Tumor-Suppressor Genes Involved in Human Neoplasms

Gene	Chromosome location[a]	Neoplasms associated with somatic mutations	Neoplasms associated with inherited germline mutations
VHL	3p25 (18)	Hemangioblastoma, renal carcinoma	von Hippel-Lindau disease; retinal and cerebellar hemangioblastomas; renal cell carcinomas; angiomas and cysts of visceral organs
APC/MCC	5q21 (19)	Carcinomas of colon, stomach, and pancreas	FAP; colon cancer
p16	9p21 (16)	Cutaneous malignant melanoma	Atypical multiple-mole melanoma syndrome
Ptch	9q22.3 (20)	Basal cell carcinoma	Basal cell nevus syndrome
TSC1	9q34 (21)	Angiomyolipoma, angiofibroma	Tuberous sclerosis
PTEN	10q23 (22)	Glioblastomas, advanced prostate cancer, breast cancer	Not reported
Ext2	11p12 (23)	Exostoses	Familial exostoses
WT1	11p13 (15)	Wilms tumor, hepatoblastoma	WAGR syndrome
BRCA2	13q12.3 (24)	Breast cancer	Familial breast cancer
Rb	13q14 (25)	Retinoblastoma, osteosarcoma, breast, prostate, bladder, and lung cancers	Retinoblastoma; osteosarcoma
TSC2	16p13.3 (26)	Angiomyolipoma, angiofibroma	Tuberous sclerosis
p53	17p13.1 (27)	Most human cancers	Li Fraumeni syndrome; breast and adrenal cortex cancers; sarcomas, leukemias and brain tumors
NF1	17q11.2 (17)	Schwannomas, meningiomas, ependymomas	Neurofibromatosis type I; neural tumors
BRCA1	17q21 (28)	Breast and ovarian cancer	Familial breast and ovarian cancer
DCC	18q21 (29)	Carcinomas of colon and stomach	
NF2	22q12 (30)	Schwannoma and meningioma	Neurofibromatosis type II; central acoustic schwanomas; meningiomas

[a]Reference number in parentheses.

different metastases within the same person show considerable heterogeneity, often have a narrower range in DNA content by flow cytometry (5), and share more specific chromosome abnormalities than the parent tumor (34,35). In general, the more complex the karyotype or the greater the chromosomal imbalance in a solid tumor, the more aggressive the malignancy and the more rapid the metastasis (Table 3) (32,34–47). Polyploidization also frequently precedes aggressive malignancy and rapid metastasis to other sites (34,36,46). In contrast, diploid tumors tend to have simpler chromosome abnormalities, remain more or less diploid, are less aggressive in the various stages of malignancy, and metastasize more slowly to other sites (35,46).

C. Colon Cancer: A Paradigm for Chromosome Instability and Tumor Progression

Colon or colorectal cancer is a classic example of multiple genetic and/or chromosome changes that culminate in malignancy. The majority of colorectal cancer is thought to develop from adenomatous polyps arising in the colon glandular epithelium. A number of histologic features are predictive of the likelihood of a particular adenoma containing cancerous foci or progressing to cancer (48,49). It is estimated that 20% of the North American Caucasian population carries genes which predispose to adenomas and that 8–17% of persons over 40 have adenomatous polyps, 10% of whom will develop colorectal cancer or adenomatous polyposis coli (APC) (50). Approximately 1 in 10,000 individuals in the United States has familial adenomatous polyposis (FAP) (48,49,51). This autosomal dominant disorder is characterized by high penetrance and development of thousands of colorectal polyps during childhood, some of which inevitably become malignant. In FAP, carcinoma may arise at anytime after late childhood, with usual progression to adenocarcinoma in the third or fourth decade (52). FAP accompanied by dental anomalies and jaw cysts (sebaceous cysts, fibromas, lipomas, or osteomas), originally thought to be a separate condition called Gardner syndrome, is now considered to be a variant of FAP (50,52). Other variant forms of FAP include Turcot syndrome with brain tumors (50), and attenuated APC with a smaller number of polyps (10 to >100). Forms with non-neoplastic adenomatous polyps include Peutz-Jeugers syndrome, juvenile polyposis, and Cowden syndrome (50). HNPCC (heritable nonpolyposis colon cancer) develops from adenomas without polyposis and may involve cancers of the stomach, endometrium, or uroendothelium (51).

The discovery of the first important gene in colon cancer was aided by the study of a single patient with a contiguous gene syndrome that included familial multiple polyposis coli, mental retardation, and deletion in

Table 3 Recent Studies of Chromosome Aberrations That Specifically Appear to Be Associated with Tumor Progression or Metastasis

Type of tumor[a]	Chromosome change	Tumor behavior
Breast (34,35)	Triploidy: LOH of 3p, 13q, and 17p	Skeletal metastasis: rapid tumor progression
	Diploidy: t(X;14), −19; del(3)	Slow tumor progression
Colorectal (32,36)	Tetraploidy: del(17p)	Parenchymal metastasis
	LOH of p53 and DCC	Brain
	LOH of 13q and 14q	Liver
Small cell lung carcinoma (37,38)	Deletions: 2q, 9p, 18q, and 22q	Brain metastasis
Meningioma (39)	Abnormalities of 1, 3, and 6	Rapid tumor progression
	Loss of 22	Benign tumor behavior
Malignant melanoma (40–42)	Deletions: 9p, 10q, 18q	Tumor progression
	Deletion or loss of 1 and 6	Lung metastasis (suppressed by adding of 1 or 6 to cell line)
Prostate cancer (43–45)	Deletion or loss of 8p23-q21, 10q, 11, and 17	Lung metastases
Squamous cell carcinoma (46)	Loss of 9	Nonmetastasizing
	Tetrasomy or polysomy with +7, +17, or −18 and complex chromosome abnormalities	Metastasizing with over expression of p53
Osteosarcoma (47)	Complex abnormalities and deletions or loss: del(6)(q21), del(9)(p21), −10, −13, del(17)(p21), and −20	High-grade
	Diploid: t(5;10), add(Xp)	Low-grade

[a]Reference numbers in parentheses.

the long arm of chromosome 5, bands 5q15-5q22 (19). This case allowed the APC gene to be mapped to 5q21-22 (53,54) and led to its identification and characterization (55,56). The APC gene is large, consisting of 15 exons that encode a 2843-amino acid protein. About two-thirds of families with FAP or Gardner syndrome have mutations that inactivate the APC gene.

Although mutation of APC by itself does not result in carcinoma, it does affect cell proliferation and microadenoma formation. LOH of APC occurs in a high proportion of early adenomas. Numerous studies have shown that loss of genetic material occurs on different chromosomes, including a second hit on APC on chromosome 5, loss of markers on chromosomes 17 or 18, and mutations of the K-ras gene on chromosome 12. These losses may occur in any order, but each additional loss results in an increasing degree of malignancy and tumor progression through several stages (Fig. 3). In addition to the loss of heterozygosity at APC, at least two other tumor-suppressor genes are involved in tumor progression, namely, the p53 gene on 17p and the DCC gene on 18q. The latter mutations are observed most often in the progression from a large adenoma to an in situ or invasive carcinoma.

With the above gene losses or mutations, additional chromosome changes occur that appear to be random and not associated with any particular stage or degree of malignancy (29,36,59–61). The chromosome changes in colon cancer tend to fall into two groups, one with a near-diploid chromosome number (41–49 chromosomes) and a second with near triploidy to hypertetraploidy. In an extreme case of an ascites from one patient (5), the number of chromosomes per cell was in the hundreds. For the near-diploid group, chromosome abnormalities tend to be structural rather than numerical and the near-diploid chromosome number seems to be maintained throughout all stages of malignancy (59–61). In general, the more complex the abnormalities, the more malignant is the cell clone.

D. Microsatellites and HNPCC Predisposition

In 1989, de la Chapelle and Vogelstein studied families with heritable nonpolyposis colon cancer (HNPCC), which accounts for about 15% of colon cancer, compared to FAP, which accounts for approximately 1%. Families with HNPCC are prone to other forms of cancer, including uterine, ovarian, and renal cancers. Although genes known to be involved in FAP could not be linked to cancer susceptibility, HNPCC was linked to a repeated microsatellite sequence on chromosome 2 (62,63). However, no deletions or mutations of genes on chromosome 2 were found. Instead, tumor-to-tumor variation in the length of the satellite sequence was observed. Furthermore, similar variations in microsatellite sequence lengths were found at several

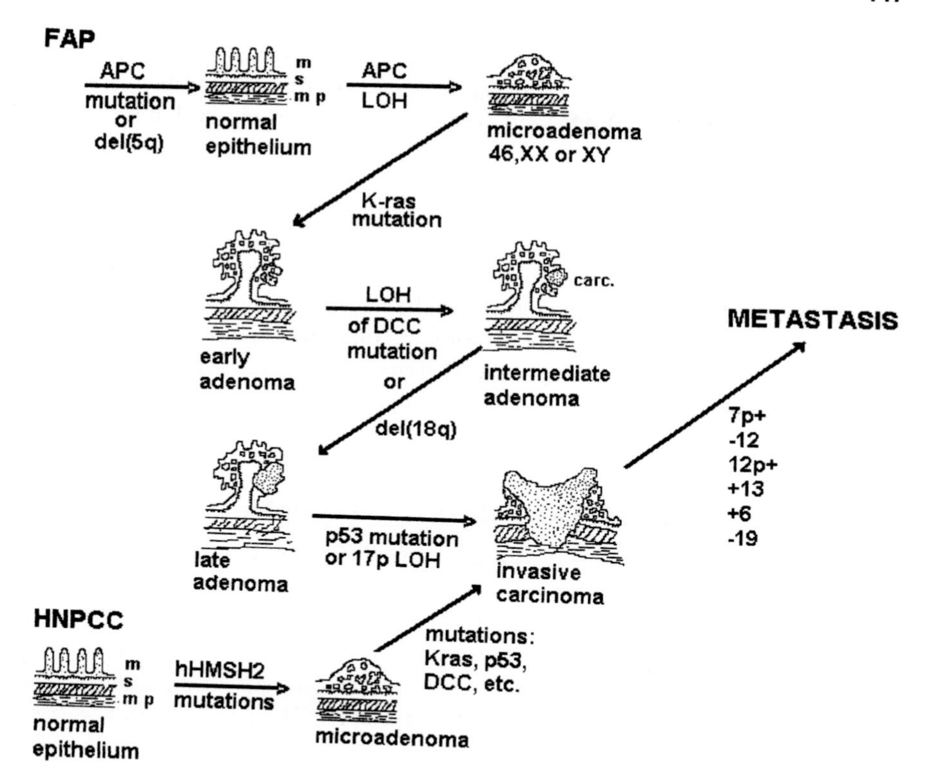

Figure 3 Schematic diagram showing possible disease progression in familial adenomatous polyposis coli (FAP) and hereditary nonpolyposis colon cancer (HNPCC) with accumulation of loss of heterozygosity (LOH) and/or protooncogene mutations at various stage of tumorigenesis. The exact order and relationship of mutations and LOH, following APC somatic or germline mutations, to progression and disease variability are speculative and probably vary in different tumors and with specific disease type and tissues involved. (Adapted from Ref. 10, p. 812.)

other chromosome sites in the same tumors. Independently, Thibodeau et al. (64) found variation in microsatellite size at still other chromosome sites in 13 of 90 colon tumors they studied. Microsatellite variations were not restricted to HNPCC but were found in 13% of sporadic colon cancers as well, particularly in cancers in the most proximal part of the colon. These replication error positive (RER+) tumors were subsequently studied by microdissection techniques to map DNA markers within 0.8 Mb of the HNPCC locus. Then a candidate gene was identified in this region that is homologous to a bacterial mismatch DNA repair gene, mutS, which maintains microsatellite stability and is also homologous to the yeast gene, MSH2 (65). The human homolog named hMSH2 was shown to be mutated in one copy in germline cells in two kindreds with typical HNPCC and in one kindred with atypical HNPCC. One somatic RER+ tumor was a compound heterozygote with a germline and a somatic mutation in different alleles. Parsons et al. (66) studied RER+ and RER− sporadic colorectal tumors as well as RER+ HNPCC tumors and found a 100-fold increase in the mutability of RER+ colorectal and HNPCC tumors over the more common RER− tumors. They showed that this increased mutability resulted from a deficiency in repair of slipped-strand mismatches. Furthermore, mismatched heteroduplexes prepared in vitro could not be repaired by extracts from the RER+ cells but could be repaired by extracts of RER− cells.

E. Familial Predisposition to Cancer and DNA Repair

The above examples of retinoblastoma, FAP, and HNPCC are representative of familial predispositions to a specific cancer, but as we have seen for HNPCC, predisposition is not necessarily restricted to one type of cancer. For example, colon and breast cancer families have increased risks for a variety of cancers. In the aforementioned studies, Leach et al. cloned the hMSH2 mismatch repair gene which cosegregates with the HNPCC gene on chromosome band 2p16. This gene acts on single-base mismatches and small displaced loops up to four or five base pairs long that result from slippage within repeated sequences during DNA replication. "DNA repair enzymes are now indispensable actors in any script for spontaneous and environmentally induced cancers," according to Cleaver (67). Patients with HNPCC inherit a mutation in the MSH2 gene and acquire additional gene mutations in tumors. Mismatch repair defects typically occur in polymorphic microsatellite regions, consisting of $(CA)_{10-30}$ repeats distributed throughout the genome where mismatch repair is required to maintain a faithful copy of the parental haplotype. Mismatch repair is initiated at methylated sites in the parent strand but requires the newly replicated DNA strand to be transiently unmethylated, and thus will repair any methylations that occur in the new strand. However, if mismatch repair is defective, methylations showing up in the new strand might not be removed and over time might occur in gene promoters that normally control tumorigenesis. This mechanism also explains how normal cells exposed to alkylating agents acquire aberrantly methylated genes due to misdirected mismatch repair, and die, while cells that are deficient in mismatch repair are resistant to killing by alkylating agents (67).

Defects in mismatch repair need to be distinguished from defects in excision repair that are well known in xeroderma pigmentosum. Excision repair normally deals with almost all other categories of damaged DNA bases, including oxidative lesions, ultraviolet photoproducts, and bulky chemical adducts.

More recently, defects in a third DNA repair mechanism that repairs double-strand breaks have been associated with X-ray hypersensitivity, immune deficiency, and familial predisposition to breast and ovarian cancers (68).

III. CHROMOSOME INSTABILITY SYNDROMES

A. Chromosome Instability (Breakage Syndromes) Associated with Neoplasia

The rare chromosome "instability" or "breakage" syndromes have genetic defects that appear to predispose to the accumulation of mutations leading to cancer or leukemia at a faster rate than in normal individuals. The most familiar of these disorders are ataxia telangiectasia (AT), Bloom syndrome (BS), Fanconi anemia (FA), and xeroderma pigmentosa (XP) (5). More recently, Werner syndrome (WS) has also come to be regarded as a chromosome breakage syndrome (69). Another, Nijmegen syndrome (NS), initially thought to be a variant of ataxia telangiectasia, is now known to be a different syndrome (70). All are autosomal recessive disorders (Table 4), distinguished by higher-than-normal rates of chromosome aberrations and/or neoplasias that are most evident in homozygotes. Each disorder is characterized by specific types of aberrations and associated cancers or leukemias.

Considerable progress has been made in mapping the genes. For AT, FA, and XP, more than one gene or mutation causes similar results, so that different complementation groups correct for deficiencies in one another. BS, on the other hand, has only one complementation group. Another typical feature of these disorders is that although the genes have been mapped to specific chromosomes, rearrangements in these chromosomes are not generally involved in the associated neoplasms.

1. Ataxia Telangiectasia

Ataxia telangiectasia (AT, also known as Louis-Bar syndrome) (Fig. 4) is characterized by progressive cerebellar ataxia in infancy, followed by conjunctival telangiectasia, sinopulmonary infection, and malignancies. Telangiectasias occur between 3 and 5 years of age (71). Oculomotor apraxia

Table 4 Chromosome Instability Syndromes

Disease	Abbreviation	Gene location	Gene identity	Proposed function	Neoplastic predisposition	Chromosome abnormalities
Ataxia telangiectasia	AT	11q22.3	Phosphatdyl-inositol 3′-kinase	Mitotic signal transformation, meiotic recombination, and possible cell cycle control of G1 to S checkpoint	T-cell leukemia (CLL) B-cell lymphoma Breast cancer (heterozygotes)	Increased chromosome breaks or clonal rearrangements; 25% have normal chromosomes
Nijmegen syndrome	NS	8q	Not identified	Abnormal cell cycle regulation	Lymphoma	Spontaneous breaks, gaps, and rearrangements of chromosomes 7 and 14
Bloom syndrome	BS	15q26.1	RecQ-type helicase	RNA/DNA unwinding	Acute leukemia Pulmonary fibrosis Hodgkin lymphomas	High rate of sister chromatid exchanges; quadriradial formation
Fanconi anema:	FA			Defect in ability to excise UV-induced pyrimidine dimers (possibly a defect in the passage of DNA repair enzymes from the cytoplasm to the nucleus)	Pancytopenia Leukemias Cancers	Increased chromosome breakage; nonhomolog chromatid exchanges induced by mitomycin C and diepoxybutane (DEB); high frequency of −7 and dup(1q) in leukemias
Complement group A	FAA	16q24.3				
Complement group B	FAB	Not mapped				
Complement group C	FAC	9q22.3				
Complement group D	FAD	3p22-26				
Complement groups E–H	FAE-H	Not mapped				
Werner syndrome	WS	8p12-p21	Helicase	Loss of function	Malignant sarcoma Meningioma Carcinoma	No increased chromosome breakage; increased number of stable nonconstitutional rearrangements
Xeroderma pigmentosa (nine complementation groups and one variant group)	XPA	9q34.1	A single UV endonuclease involved in pyrimidine dimer excision appears to require cooperative function of at least 9 different gene products	Inability to insert new bases into DNA after UV-induced pyrimidine dimer formation	Skin cancer	No increased chromosome breakage
	B	2q21				
	C	13q25				
	D	19q13.2-13.3				
	E	Not mapped				
	F	Not mapped				
	G	Not mapped				
	H	Not mapped				
	I	Not mapped				
	var	Not mapped				
Roberts syndrome	RB	Not mapped			No malignancies	Premature centromeric division
ICF syndrome	ICFS	Not mapped			No malignancies	Centromeric heterochromatin elongation; centromere fragility and whole-arm multiplication and deletion

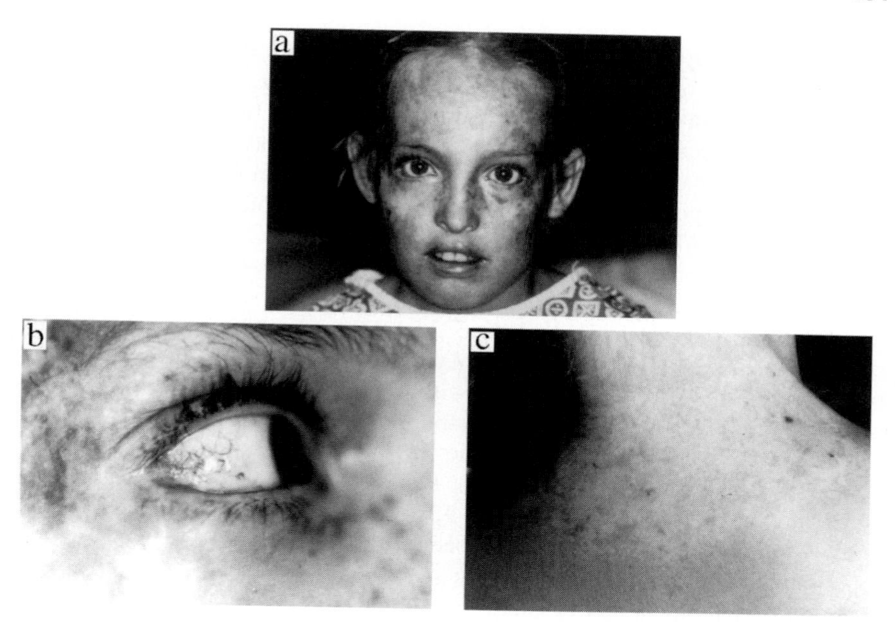

Figure 4 (a) Patient, age 13, with ataxia telangiectasia. (b) Conjunctival telangiectasia. (c) Cutaneous telangiectasia primarily in sun-exposed areas, atrophy and mottled patches on hyperpigmentation on back of neck. (Photographs courtesy of Farah Shah, M.D., Department of Pediatrics and Dermatology, Texas Tech University Health Sciences Center, Lubbock.)

often precedes the formation of telangiectasia. Truncal ataxia usually occurs by 3 years of age, while deep tendon reflexes become diminished or absent by age 8. Older patients, in their 20s or early 30s, develop progressive spinal muscular atrophy. Although mental retardation is not a feature of AT, some older patients develop severe short-term memory loss (72,73).

AT patients have a strong predisposition to develop lymphocytic leukemias and lymphomas. Lymphomas tend to be B cell in origin (74), whereas leukemias generally involve the T cells (i.e., chronic lymphocytic leukemia) (75–78). Other associated solid tumors include medulloblastomas and gliomas (72). Thymus hypoplasia and defects in the immune mechanism result in diminished or absent IgA and IgG2 levels in 60–80% of patients (79–81). The incidence of AT has been estimated at 1 in 89,000 in the U.S. Caucasian population and 1 in 300,000 in England (82,83). The estimated gene frequency based on a homogeneous genotype is 0.007. However complementation analysis has revealed genetic heterogeneity with a least five complementation groups (A–E), so some estimates suggest heterozygote frequencies as high as 0.08. Heterozygous carriers of the gene have been reported to have a three- to sixfold increased risk for breast cancer, over that of the general population. An estimated 7% of all breast cancers in the United States occur in women who are AT heterozygotes, a frequency that is higher than of women carrying the BRCA1 mutation (84–86).

Elevated spontaneous breakage occurs in fibroblasts and peripheral lymphocytes of AT patients: 50% of AT patients show high levels of chromosome breaks or clonal rearrangements, 25% have lower levels of breakage, and 25% have normal chromosomes. Impaired responsiveness to phytohemaglutinin and reduced numbers of circulating lymphocytes are typical, so chromosome studies of T lymphocytes are usually difficult (87,88). Fibroblast cultures are also sluggish, with a prolonged S phase. Rearrangements occur between chromosomes rather than between chromatids and tend to be unstable categories (i.e., acentric fragments, dicentrics, and rings) that disappear during the next mitosis (5). Patients with AT and associated cultured cells are especially sensitive to X-rays. Treatment of malignancy with conventional doses of radiation has been fatal (89).

A diffusible clastogenic peptide (MW = 500–1000) is present in the plasma of AT patients and in culture media of AT skin fibroblasts. However, clastogenic activity has not been demonstrated in extracts of AT cultured fibroblasts, and is not found in long-term lymphoblastoid cells. Intrachromosomal recombination rates are 30 to 200 times higher in AT fibroblasts than interchromosomal recombination rates, which are near normal. Hyperrecombination seems to be a specific feature of the AT phenotype, rather than a genetic consequence of defective DNA repair, and may be the primary contributor to increased cancer risk because of increased loss of heterozygosity rather than increased breakage per se (89,90).

Clonal abnormalities associated with neoplasias in AT occur most often in older patients and involve abnormalities of chromosome 14. The chromosomes in bone marrow are usually normal. The neoplasm-associated rearrangements are usually balanced and commonly involve chromosomes 2, 7, 8, 14, 22, and X. A common clone with a long D-group chromosome (90) was shown by G banding (Fig. 5) to be a tandem rearrangement of two chromosome 14s (91). The break in 14 always involves the 14q11–q12 region. Hecht and McCaw, along with others, suggested that such rearrangements were more than fortuitous (87,91). Sparkes et al. (77) provided evidence that a patient with a tandem duplication of 14q was associated with a T-lymphocyte form of CLL. The same patient was shown by Saxon et al. (76) to have the rearrangement in E-rosetting T lymphocytes with both IgM and IgG Fc receptors. Wake et al. (78) reported an atypical clonal abnormality, t(9;6), in a patient with an atypical T-cell subset. Common breakpoints in AT-related leukemia and tumors include 7p14, 7q35, 14q12, 14qter,

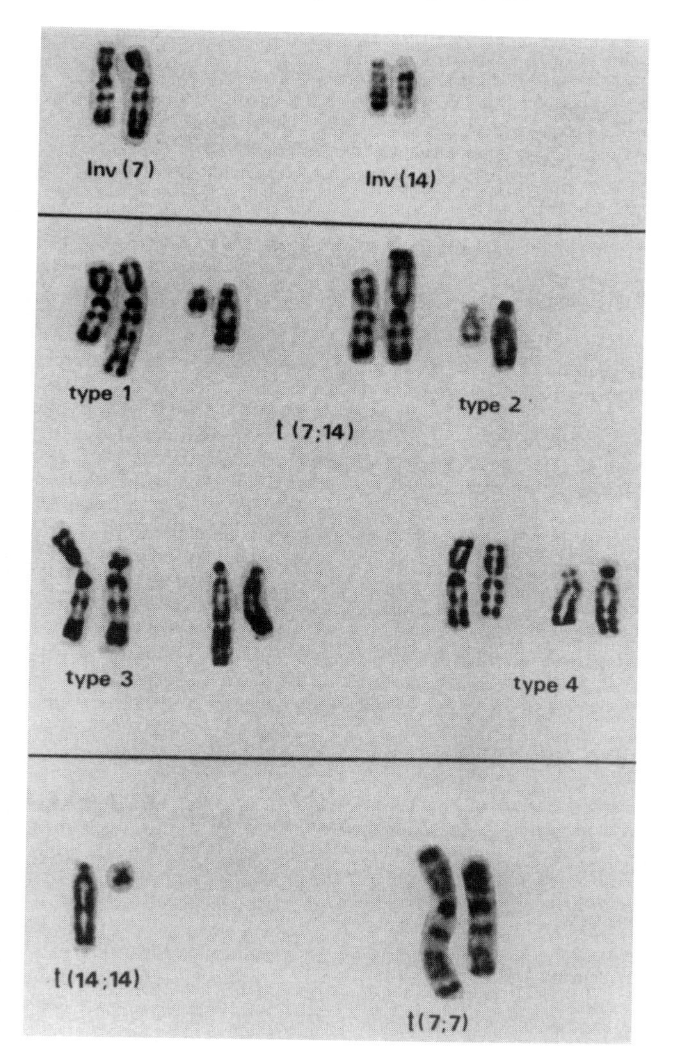

Figure 5 Representative rearrangements involving chromosome 14 associated with CLL and B-cell lymphoma. (From Ref. 5.)

2p11, 2p12, and 22q11–q12. These are all close to immunoglobulin superfamily gene locations which include IgK, IgH, IgL, TCRA, TCRB, and TCRG (93). Croce (94) suggested that the alpha subunit of the T-cell antigen receptor TCRA at 14q11.2, for example, is brought into juxtaposition with the oncogene TLC1, at 14q32.3, resulting in reduced immunoglobulin expression.

The molecular nature of the defect in AT has recently been revealed by several studies that mapped the AT gene to a 4-cM region on chromosome 11 at 11q22.3. Although at least five complementation groups have been described, these all appear to be clustered within the 11q22.3 region (95–97). At least two checkpoints in the cell cycle, G1-S and G2-M, are known to respond to DNA damage. These are points at which cells delay DNA synthesis in order to repair DNA damage before entering either S or M (98). The G1-S checkpoint is eliminated in AT patients. Furthermore, there is strong evidence that p53 is necessary for the G1-S checkpoint. AT genes upstream of p53 activate the G1-S checkpoint such that p53 levels increase three- to fivefold after gamma-irradiation, coincident with a delay in the G1-S transition (99). The AT gene with a transcript of 12 kb is mutated in AT patients from all complementation groups, suggesting that this is probably the sole gene responsible. The AT gene is homologous to genes with similar functions in a number of organisms, including mei41 in *Drosophila* and rad3 and mec1p in yeast. The specific protein product is a phosphatidylinositol-3-kinase (100).

2. Nijmegen Syndrome

Nijmegen syndrome (NS), suggested to be an ataxia telangiectasia "variant," is characterized by microcephaly, short stature, immunodeficiency, and a high incidence of cancer. The syndrome was first described by Weemaes et al. in 1981 (101). Approximately 42 patients have since been reported (102). Pre- or postnatal growth retardation, severe microcephaly after the first months of life, receding forehead, prominent midface with long nose and receding mandible, and profound immunodeficiency of both cellular and humoral systems are characteristic. Respiratory infections occur in almost all patients. Urinary and gastrointestinal infections repeatedly recur in 15% of patients. Twelve patients have developed lymphoma (102).

Mitogen-stimulated T cells show spontaneous chromosomal breaks, gaps, and rearrangements that, as in AT, preferentially occur in chromosomes 7 and 14. Increased sensitivity to radiation-induced cell killing and abnormal cell cycle regulation (i.e., radioresistant DNA synthesis) are also similar to AT. Overall, cell biology findings resemble AT, suggesting that the same pathway is impaired in both syndromes. Clinical features, however, are different in that NS patients do not show AFP elevation, cerebellar ataxia, or telangiectasia. A recent study mapped NS to the long arm of chromosome 8, refuting previous suggestions that this was a variant of AT. The study also excluded the existence of more than one complementation group for NS (103).

3. Bloom Syndrome

Bloom syndrome (BS) is a rare autosomal recessive disorder involving both pre- and postnatal growth deficiencies, mild microcephaly with dolichocephaly, and telangiectatic erythema in a butterfly pattern over the midface (Fig. 6). Patients typically show slow growth from infancy and frequently have feeding problems. The facial erythema is seldom present at birth, usually appearing upon exposure to sunlight. Occasional abnormalities include mild mental retardation, telangiectatic erythema of the hands and forearm, colloid-body-like spots in the Bruch membrane of the eye, absence of the upper lacteral incisors, prominent ears, ichthyotic skin, hypertrichosis, pilonidal cyst, sacral dimple, syndactyly, polydactyly, clinodactyly, short lower extremity, talipes, and café-au-lait spots. There is also occasional IgA or IgM deficiency. Males are typically infertile, but females have occasionally given birth (104–107).

The majority of cases are found in Ashkenazic Jews, in whom the gene carrier frequency is estimated to be 1 per 110 (5). More than 132 patients with BS were recorded in the Bloom syndrome registry as of 1990 (108).

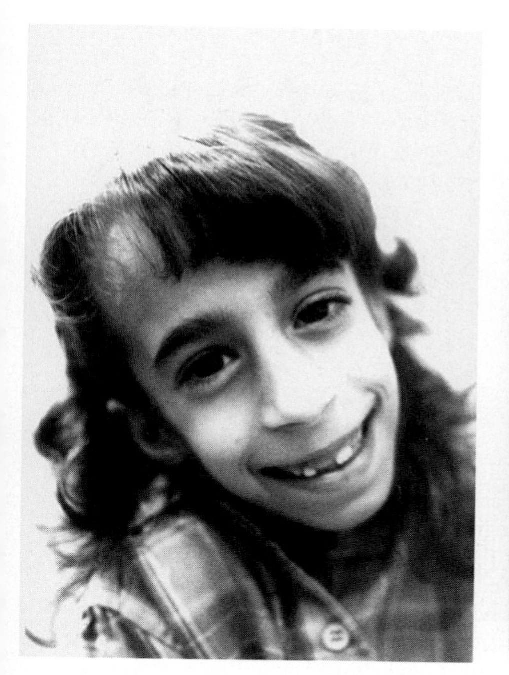

Figure 6 Patient, age 12, with Bloom syndrome.

Malignancy is the major cause of death. Thirty-one of these cases died of cancer at a mean age of 27.8. Ninety-three registry patients were still alive. The variety of cancers includes acute leukemia, pulmonary fibrosis and bronchiectasis, and Hodgkin's lymphoma (108). The high rate of cancers and leukemia, as in AT, has been causally related to the increased rate of chromosome breaks observed in BS (104,108–110). However, chromosome studies of the neoplasms themselves are conspicuously lacking (5). Some studies suggest that loss of heterozygosity due to somatic recombination may be a cause (5,111). In contrast to the other chromosome breakage disorders, most interchanges are between chromatids of homologous chromosomes resulting in symmetric quadriradial formations, or between sister chromatids. Sister chromatid exchange (SCE), in fact, represents the most distinctive cytologic diagnostic marker for BS (110), with SCE rates that are typically 10 times the rate in normal cells (Fig. 7).

A single complementation group has been found in Bloom syndrome for all ethnic groups (112). The gene for BS was mapped, to chromosome

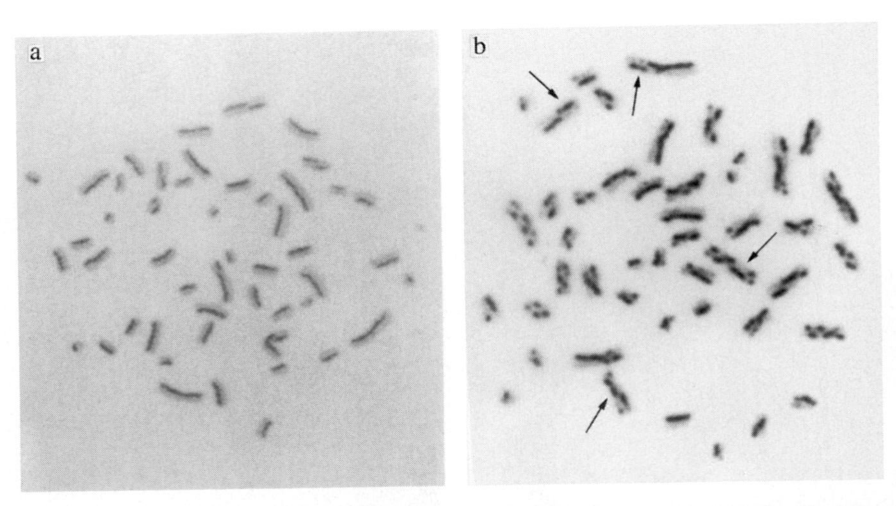

Figure 7 Metaphases from a normal cell culture (a) and from cultured cells from a patient with Bloom syndrome (b), comparing the sister chromatid exchange (SCE) rates. Arrows point to some exchanged segments in (b). Both cultures incorporated 5-bromodeoxyuridine (BrdU) in place of thymidine through two cell cycles (112). Metaphases are photoactivated and stained with Giemsa. Dark-staining segments represent monofiliarly substituted DNA and light-staining segments represent bifiliarly substituted DNA. If no SCE has occurred, one chromatid stains uniformly dark, the other light along the entire length. Checkerboard appearance represents multiple cross-overs between the first and second cycles of BrdU incorporation.

15q26.1 (114–116). The gene encodes a 1,417-amino acid peptide showing homology to the RecQ helicases (115), a subfamily of Dexh box-containing DNA and RNA helicases. The BS mutation seems to involve premature termination of enzyme translation with resultant loss of enzyme activity (115).

4. Fanconi Anemia

Fanconi anemia (FA) is a heterogeneous disorder characterized by congenital and skeletal malformations, progressive pancytopenia, and predisposition to malignancies (Fig. 8). As in AT and BS, increased chromosome breakage is observed. Rearrangements in FA tend to be between nonhomologous chro-

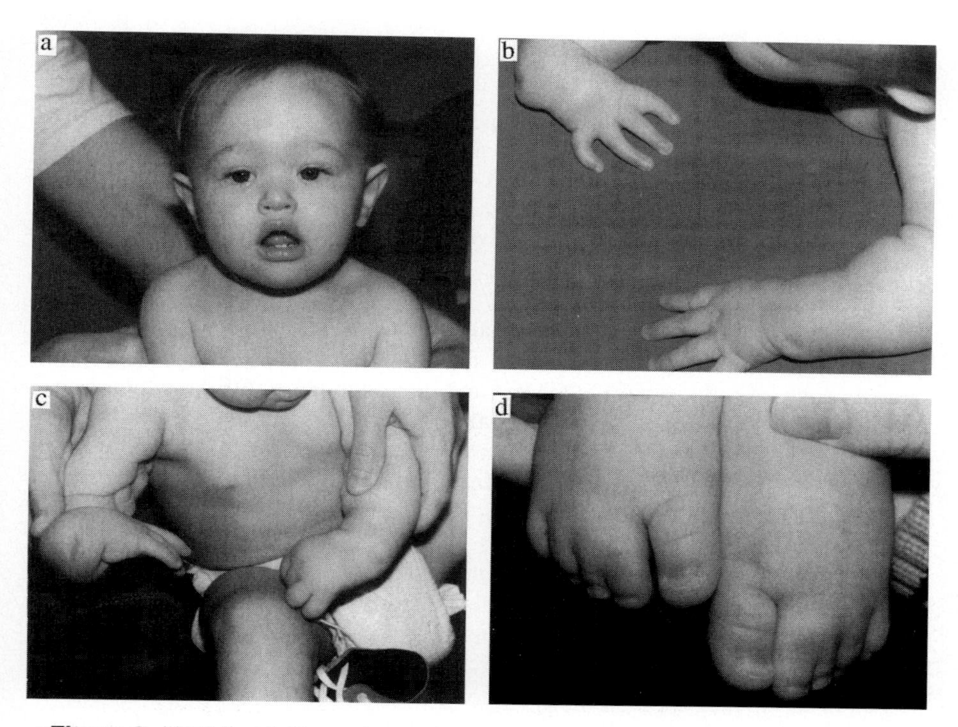

Figure 8 Child with Fanconi anemia. (a) Facial view, showing dysmorphic facies and low-set ears. (b) Upper trunk and arms, showing bilateral hypoplastic radii. (c) Hands and forearms, again showing hypoplastic radii, absent thumbs, and fifth-finger clinodachtyly. (d) Feet, showing 2–4 toe syndactyly. (Photographs courtesy of Larsen Haidle, genetic counselor; patient was confirmed to have FA in the Cytogenetics Laboratory of Shivanand R. Patil, Ph.D., University of Iowa Hospitals and Clinics, Iowa City.)

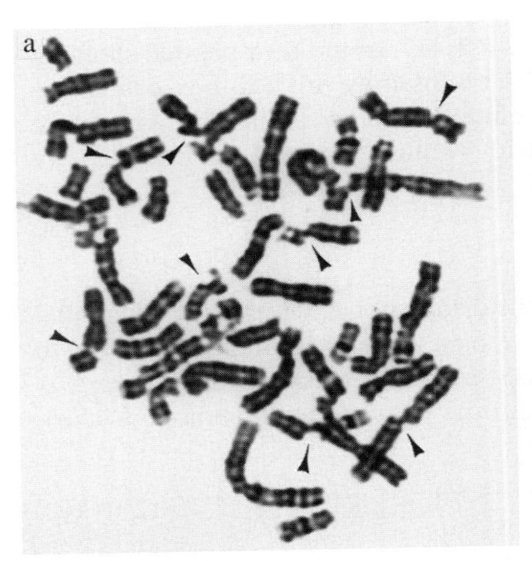

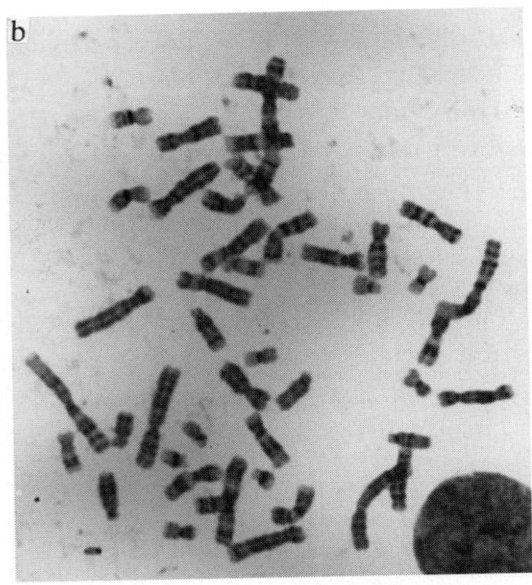

Figure 9 (a) Chromosome breakage (arrows) induced with diepoxybutane (DEB) in a metaphase from cell culture from a patient with Fanconi anemia is compared (b) to a metaphase from similarly treated cell culture from a normal patient. (Photograph courtesy of Jerome McCombs, Ph.D., Department of Laboratory Medicine, University of Texas Medical Branch, Galveston.)

matids, in contrast to AT where exchanges tend to be between homologous chromosomes. Chromosomes of FA patients are sensitive to cross-linking agents such as mitomycin C (MMC) and diepoxybutane (DEB), while their ability to repair radiation damage is almost normal. DEB-induced chromosomal breakage (Fig. 9) in particular has been used by Auerbach et al. (117,118) as a diagnostic technique. In 30 reported prenatal diagnoses from amniotic fluid cultures, 7 were diagnosed as affected. Two carried to term were affected, while 2 of 5 abortuses showed congenital abnormalities including abnormalities of the thumb and radius. There was no evidence of FA in the remaining 23 cases. Prenatal diagnosis of 10 first-trimester chorionic villus samples detected two cases with FA, but results were ambiguous in the remaining 8 cases.

FA is associated with a high rate of cancers and leukemias. Auerbach and Allen (119) found a high frequency of monosomy 7 and duplications of 1q in FA-associated leukemias, but no occurrence of t(8;21), t(15;17) or abnormalities of 11q.

Eight complementation groups have now been identified in FA (FAA-FAH) (Table 4) (120). Two of these, FAC and FAA, are cloned and mapped to chromosomes 9q22.3 (121,122) and 16q24.3 (123,124), respectively. A third (FAD) is mapped to chromosome 3p22-26 (125).

Poon et al. (126) showed that FA patients cannot excise UV-induced pyrimidine dimers from their DNA. They are capable, however, of single-strand break production and unscheduled DNA synthesis. From this, Poon et al. inferred that there might be a deficiency in an exonuclease that recognizes and excises distortions in the tertiary DNA structure. Because of an unusually high concentration of DNA topoisomerase I in the cytoplasm and a low level in nuclear sap, Wunder et al. (127) suggested the defect in FA might be in the transfer of DNA repair enzymes from the cytoplasm to the site of action of the nucleus, or the high cytoplasmic concentration of DNA topoisomerase I might be due to an impediment to enzyme entry into the nucleus (128). Auerbach suggested that the cellular defect in FA results in chromosomal instability, hypersensitivity to DNA damage, and hypermutability resulting in allele loss, thus predisposing to leukemia by a multistep process (129). She pointed to topoisomerase I and proliferating cell nuclear antigen (PCNA) as candidate genes for FAA on chromosome 20. Saito et al. (130), however, found no mutation of topoisomerase cDNA from an FAA patient. Foe et al. (131) identified a cDNA clone that corrected cross-linker hypersensitivity of FAA cells, but not of FAC cells. This cDNA clone mapped to the telomere of chromosome 16q. The cDNA sequence with an open reading frame of 4,368 bp encodes a 1,455-amino acid polypeptide predicted to contain two overlapping bipartite nuclear localization signals and a partial leucine zipper consensus sequence typical of a nuclear protein. Normal controls and FA patients from complementation groups other than FAA showed a 5.5-kb transcript that was virtually undetectable in FAA cell lines. The FA/Breast Cancer Consortium (132) also identified a candidate cDNA clone for FAA that encoded an identical polypeptide to that of Foe, except that it has an additional 13 bp of 5′ untranslated sequence and two amino acid substitutions. Reverse transcriptase PCR and direct sequencing of PCR products from FAA patients revealed four different mutations which led the FA/Breast Cancer Consortium to postulate that introns of the FAA gene may contain unstable repetitive sequences susceptible to deletion.

5. Werner Syndrome

Werner syndrome (WS) is a rare condition with multiple progeroid features, but it is an imitation of aging rather than accelerated or premature senescence. The condition was first described by Otto Werner in 1904 in four siblings with scleroderma in association with cataracts (133). The condition was reported in America in 1934 by Oppenheimer and Kugel (134). Thannhauser (135) reviewed all reported cases and noted three important characteristics: (a) skin changes different from those of classical scleroderma; (b) 12 principal symptoms (Table 5), and (c) the tendency of the condition to

Table 5 Principal Characteristics of Werner Syndrome of Thannhauser (135)

1. Short stature and characteristic habitus (thin extremities with stocky trunk)
2. Premature graying of hair
3. Premature baldness
4. Patches of apparently stiffened skin on face and lower extremities ("schleropoikiloderma")
5. Trophic ulcers of the legs
6. Juvenile cataracts
7. Hypogonadism
8. Tendency to diabetes
9. Calcification of blood vessels
10. Metastatic calcifications
11. Osteoporosis
12. Tendency to occur in siblings

Additional features added by others:

13. Increasing incidence of neoplasia (133)
14. High-pitched, hoarse voices and laryngeal abnormalities
15. Flat feet, hyperflexia, and excessive urinary excretion of hyaluronic acid (134)
16. Irregular dental development (136)

recur in siblings. Epstein and colleagues (136) reviewed 122 reported cases, defined the pathologic and histologic changes in the condition, and did extensive pedigree analysis which confirmed autosomal recessive inheritance. The average lifespan of WS patients is 47, the principal causes of death being malignancy and myocardial or cerebrovascular accidents. Among patients reviewed by Epstein et al., 10% had significant neoplasms that included malignant sarcomas, meningiomas, and carcinomas. Goto et al. (137) reviewed 195 patients in the Japanese literature and found 5.6% with malignant tumors. They noted an unusual number of neoplasms of mesenchymal origin but no cases of prostatic, pancreatic, gastric, or lung cancer.

Somatic chromosome aberrations are observed in vivo and in vitro in a variety of tissues and there is an increased incidence of neoplasia. Increased chromosome breakage is not observed in WS, but a variety of stable, de-novo chromosome rearrangements are observed in cultured skin fibroblasts (138–140).

A recent study has mapped the WS gene to chromosome 8p12-p21 (69). The gene product is a 1432-amino acid protein with a central domain consisting of seven motifs (I, Ia, and II–VI) which are a typical product of the superfamily of helicases. The presence of DEAH and ATP-binding sequences indicate that the gene is a functional helicase. Although there is significant homology with *Escherichia coli* RecQ, *Saccharomyces cerevisiae* Sgs1, human RECQL, and other helicases, the N- and C-terminal domains show no similarity to other helicases. A total of nine mutations were found in the WS gene. Four were in the C-terminal region and five were in the middle helicase region. All nine mutations are believed to result in complete loss of function.

6. Xeroderma Pigmentosum

Xeroderma pigmentosum (XP) is characterized by sensitivity to sunlight and a high risk for development of skin cancer at an early age (Fig. 10). Freckle-like lesions usually occur in the first years of life. Some patients with XP have neurologic abnormalities such as mental deterioration or sensorineural deafness associated with progressive and accelerated degeneration of the skin, eyes, and nervous system. Although the disease is autosomal recessive, blood relatives have a significantly higher frequency of nonmelanoma skin cancer than spouses, indicating that heterozygosity may predispose persons to skin cancer, especially with substantial exposure to sunlight. XP has been found in all races. The frequency in the European population is about 1/250,000, and in Japan it is about 1/40,000. The number of affected males and females is nearly equal. Consanguinity has been reported in 31% of cases (141).

Neoplasia associated with XP are mainly premalignant actinic keratoses and basal or squamous cell carcinomas, but also include melanomas, sarcomas, keratocanthomas, and angiomas. The squamous and basal cell carcinomas occur mainly on the face, head, and neck, regions of greatest exposure to sunlight. The median age of onset of skin cancer is 8 years. The rate of skin cancer in XP is approximately 2000 times higher than for the general population under age 20.

Generally, chromosomes in XP are normal. However, chromosomal changes have been described, primarily as balanced translocation clones in diploid fibroblasts, in vitro. Unusual rates of chromosome breakage, translocation figures, or dicentrics characteristic of the other chromosome instability syndromes do not occur with a high frequency in XP (8,141) but are increased after exposure to UV light and chemical carcinogens. The sister chromatid exchange rate is also normal but increases after UV and other chemical exposure. The proportion of cells able to form colonies after exposure to UV light and other chemicals has been used as a measure of sensitivity.

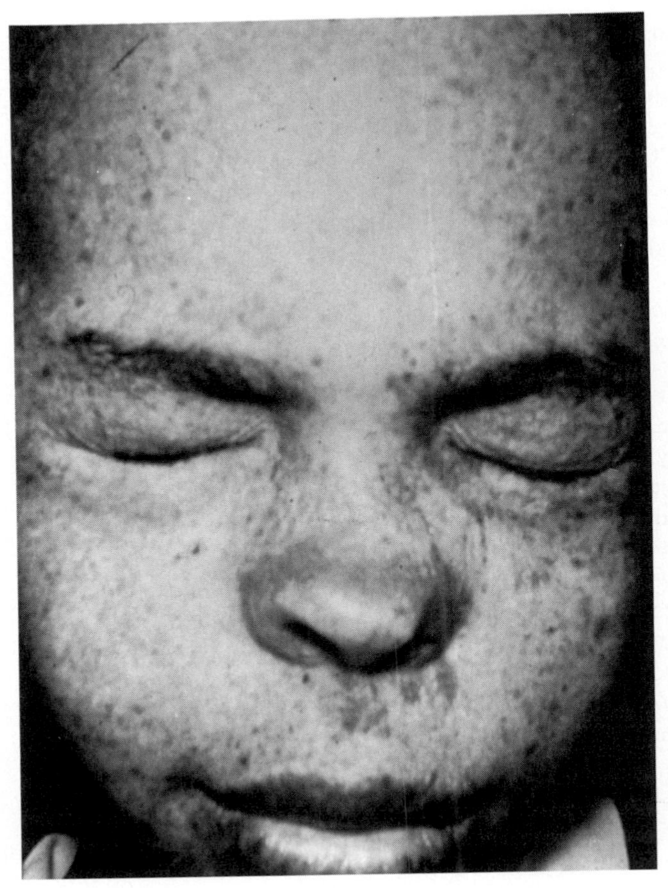

Figure 10 Patient with xeroderma pigmentosum, showing characteristic freckling and actinic keratosis in sun-exposed areas of face. (Photograph courtesy of Golder Wilson, M.D., Ph.D., Department of Pediatrics, Division of Genetics, University of Texas Southwestern Medical Center, Dallas.)

The defect in XP is mainly one of DNA repair after exposure to UV irradiation. Cleaver, in 1968 (142), first showed that, unlike normal cells, XP cells lacked the ability to insert new bases into DNA after UV-induced pyrimidine dimer formation. By 1974, complementation groups A–D were found, indicating at least four unliked loci leading to XP. Nine complementation groups (A–I) are now known (141). One XP patient with normal excision repair, described by Burk et al. in 1971 (143), was subsequently recognized as a variant form (144). Groups A, B, D, G, H, and I are associated with neurologic degeneration. Non-neurological forms fall into groups C, E, and F. Cultured cells from any one group will restore DNA repair capability when fused with cells from any other group. Groups A and C are the most common XP mutants worldwide. Three different pathways are involved in DNA repair: (a) photoreactivation, (b) excision repair, and (c) postreplicative repair. Groups A–I are deficient in gene products involved in the initial excision of damaged DNA. Patients with the variant form of XP are deficient in the postreplicative repair of damaged sites.

Viruses such as adenovirus, SV40, and herpes simplex use cellular enzymes for their DNA replication and repair. The ability of UV-damaged viruses to undergo repair and replicate when in infected XP cells has been used as an assay (145) for detecting XP, and is the only assay that detects the abnormality in every form of XP. The DNA repair of adenovirus is 20-fold less in XPA cells than in normal cells (141).

In 1989, Kaur and Athwal (146) corrected the DNA repair deficiency in XPA by transferring a normal human chromosome 9 into affected cells. Henning et al. (147) showed that microcell-mediated transfer of a piece of chromosome 9, 9q22.2→9q34.3, from a human–mouse somatic cell hybrid, also resulted in complementation of defective repair and UV sensitivity in XPA cells. Tanaka et al. (148) cloned a mouse gene that restored resistance to UV light upon transfection into either of two XPA cell lines. No correction was observed for cell lines from groups XPC, XPD, XPF, or XPG. These investigators subsequently localized the human and mouse XPA genes to human chromosome 9q34.1 and mouse chromosome 4C2, respectively. Almost all XPA cell lines showed an abnormality or absence of XPA messenger RNA. The XPA-correcting gene encodes a hydrophilic protein with a relative molecular mass of 31,000. It has a distinct zinc-finger motif that indicates it interacts directly with DNA, presumably as part of an enzyme complex that makes an incision near damaged sites. Most Japanese patients with type XPA have a splice-site mutation resulting in mRNA of reduced amount and size. Two patients had a complete absence of mRNA, and two patients with mild symptoms had normal mRNAs.

7. Cockayne Syndrome

Initially described by Cockayne in 1946 (149), Cockayne syndrome (CS) is characterized by dwarfism, a precociously senile appearance, pigmentary retinal degeneration, optic atrophy, deafness, marble epiphyses in some digits, mental retardation, and sensitivity to sunlight (Fig. 11). Some patients have had features of both CS and XP, and of 11 patients studied by Lehmann (150), one patient with clinical and biologic features of both syndromes represents the sole representative of the XP complementation group C.

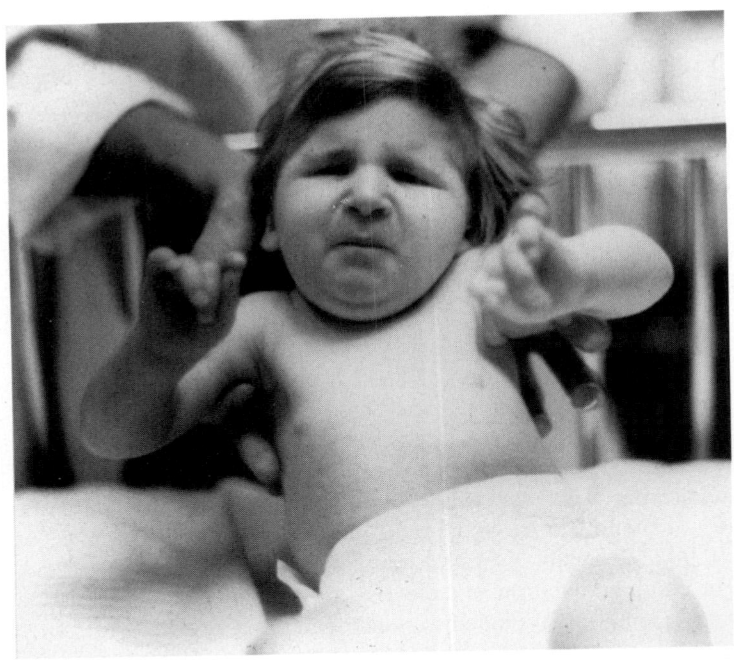

Figure 11 Patient, age 1 year, with Cockayne syndrome. (Photograph courtesy of Goldar Wilson, M.D., Ph.D.)

Schweiger (151) suggested that the DNA repair defect is located beyond incision, exonuclease reaction, and DNA synthesis and probably involves a DNA ligase. Jaeken et al. (152) studied three patients with unusually severe CS. Fibroblasts from two of these patients exhibited a complete defect in removal of UV-induced thymine dimers from their DNA. The third patient's cells had the classic in vitro repair characteristics of CS. These results supported the suggestion by Marshall et al. (153) of transitional forms between CS and XP. Although characterized by UV sensitivity and pyrimidine dimer-repair defect, neither an increase in cancer nor increased chromosome abnormalities have been reported in CS (154,155).

B. Chromosome Instability Syndromes Not Associated with Neoplasia

1. Roberts Syndrome

Initially described by Roberts (156) and by Appelt et al. (157), Roberts syndrome (RBS) is characterized by pre- and postnatal growth retardation, mild to severe mental retardation, and limb and craniofacial abnormalities (Fig. 12) (158). Limb abnormalities involve varying degrees of symmetric absence (phocomelia) or reduction (hypomelia) of bones in the upper extremities, including hypoplasia or absence of radii, first metacarpels, thumbs, fibulae, ulnae, carpels, and tibiae. There is also syndactyly, fifth finger clinodactyly, talipes deformity, and incomplete dermal ridge development. Craniofacial abnormalities include microbrachycephaly, cleft lip and/or palate, microgenathia, prominent premaxilla, hypertelorism, midfacial capillary hemangioma, thin nares, shallow orbits, and prominent eyes with bluish sclerae. Ears are malformed, with hypoplastic lobules. Hair is sparse and may be silvery blond. Cryptorchidism and a large phallus relative to body size have also been noted. Satar et al. (159) described additional abnormalities in a male infant which include atrial septal defect, rudimentary gall bladder, and accessory spleen. Masserati et al. (160) emphasized the wide variability in phenotype within families.

Also referred to as "Roberts-SC phocomelia syndrome," cases reported by Herrmann et al. (161) as "pseudothalidomide or SC syndrome" and a case reported by Hall and Greenberg (162) as "hypomelia-hypotrichosis-facial hemangioma syndrome" are possible examples of this disorder. Uncertainty exists as to whether or not RBS and SC syndrome are the same entity (163–166). Increased risk for neoplasias has not been described for any of these syndromes.

The chromosome observations have been similar for syndromes at both ends of the phenotypic spectrum. Namely, the heterochromatic regions detected by C banding appear to be puffy or "in repulsion" (167) around the centromeres of most chromosomes, including the long arm of the Y chromosome, which is often splayed out in metaphase spreads (Fig. 13) (168). Masserati et al. (160) studied the centromeres directly by immunofluorescence with serum antibodies from patients with CREST and concluded that the centromeres themselves are normal and that the phenomenon is one of centromeric heterochromatin repulsion. Keppen et al. (169) describe an infant with a diagnosis of RBS but without premature centromere division (PCD). Allingham-Hawkins and Tomkins (170) point out that some RBS patients have an abnormality of their constitutive heterochromatin (RS+) and are hypersensitive to mitomycin C and other DNA-damaging agents, while others (RS−) are not. Somatic cell hybrids of RS+ cases did not correct each other, whereas RS+/RS− fusions corrected the defect in the hybrids, suggesting different mutations for RS+ and RS− cases. Jabs et al. (171) presented evidence that RBS is a mitotic mutant. They pointed out additional characteristic features of aneuploidy with random chromosome loss and micronuclei and/or nuclear lobulation in interphase. Stioui et al.

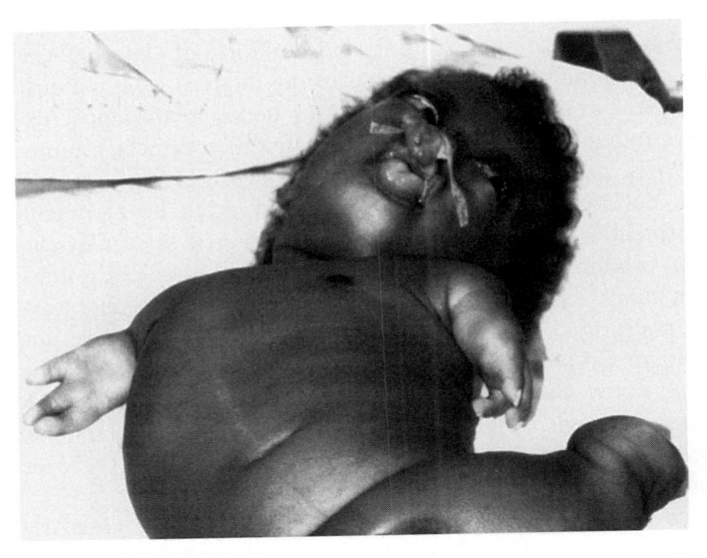

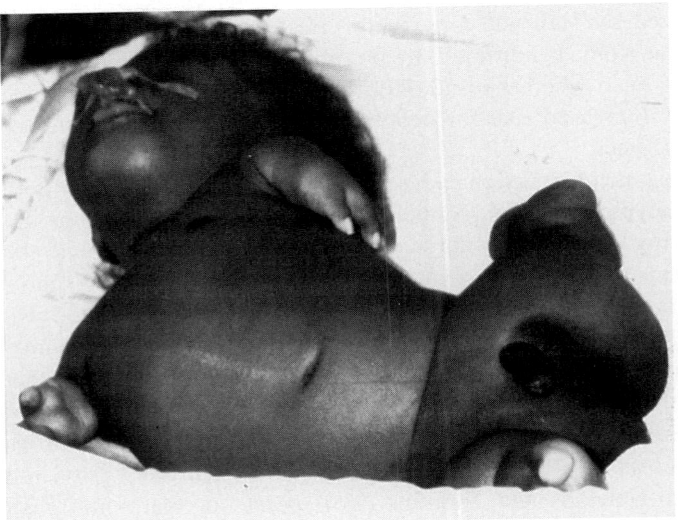

Figure 12 Patient, age 6 months, with Roberts syndrome. (Photograph courtesy of Goldar Wilson, M.D., Ph.D., Department of Pediatrics, Division of Genetics, University of Texas Southwestern Medical Center, Dallas.)

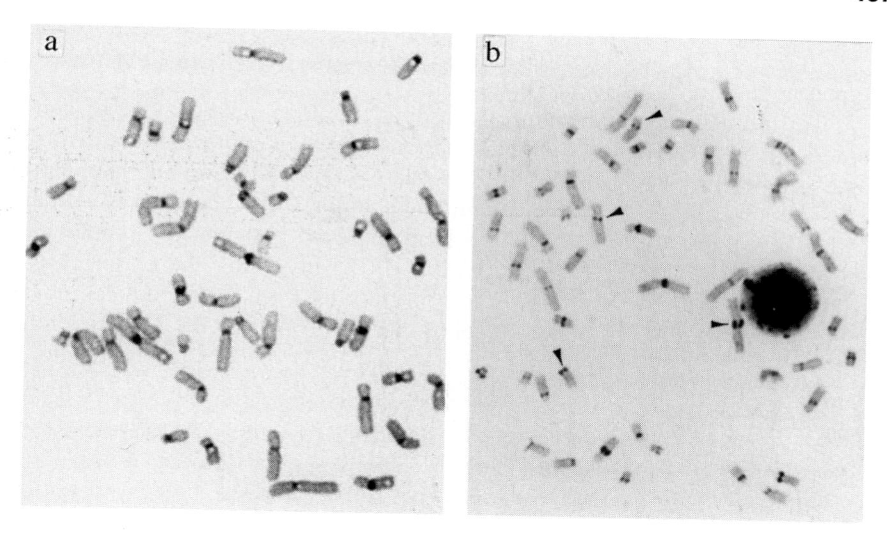

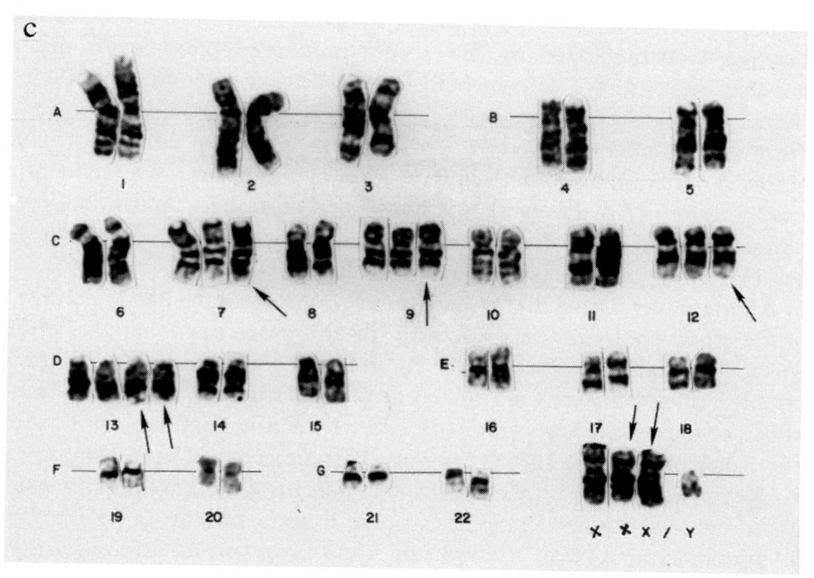

Figure 13 C-banded metaphases (a) from normal subject and (b) from patient with Roberts syndrome. Latter shows premature centromere division (PCD), revealed by puffing or repulsion of chromatids in the centromeric regions (C-banded regions), compared to lack of PCD in metaphases from a normal subject. (c) Karyotype from the patient showing a hyperdiploid clone, a less known chromosomal feature of the syndrome. (Photographs courtesy of Roger Schultz, Ph.D., McDermott Center for Human Growth and Development, University of Texas Southwestern Medical Center, Dallas.)

(172) found PCD in a chorionic villus sample at 8-weeks gestation in a woman at risk for RBS recurrence.

The published nomenclature regarding the apparent behavior of centromeric regions in RBS and other situations has been somewhat confusing. Some perspective was provided by Chamla (173), who describes three separate phenomena involving centromeric behavior: (a) C-anaphases which involve resistance to colchicine treatment in some patients; (b) premature centromere division (PCD) or separation (PCS) involving the entire chromatids of the inactive X chromosome in cultures from older females; and (c) the generalized phenomenon of centromeric puffing in RBS. Clarification of the subject does not appear imminent, and PCD or PCS are used interchangeably for all three phenomena even though the mechanisms for their respective occurrences may be different.

2. ICF Syndrome

Immunodeficiency, centromeric heterochromatin instability, and facial anomalies (ICF) syndrome was first reported by Hulten (174). The immunologic phenotype is characterized by low serum immunoglobulin levels that are higher than seen in X-linked or autosomal recessive agammaglobulinemia, by higher numbers of B cells with a surface Ig expression, and by normal function and numbers of T cells. The findings are more consistent with a common variable immunodeficiency. Autoimmune disease is a frequent additional finding (175). Eleven cases have been described. Autosomal recessive inheritance is suggested by recurrence in sibs and by lack of chromosome aberrations in parents.

Although not associated with neoplasia, chromosome abnormalities involve dramatic configurations primarily of chromosomes 1 and 16, present in affected individuals in the form of centromeric heterochromatin stretching, persistence of somatic pairing and centromeric fragility along with multibranched chromosome arms (Fig. 14), and whole arm deletions, duplications, and/or isochromosomes (176). Similar aberrations have been reported involving chromosomes 1, 2, 9, and 10, with chromosome 1 found universally, in all patients. Abnormal metaphases range in frequency from 45% to 94%, with variable expression in chromosomes and tissues from different patients, but with consistent expression in lymphocytes.

Howell and Taylor (176) did in situ hybridization with fluorescence-labeled painting probes to show the loss of chromosomes 1 and 16 in interphase and the presence of complex configurations in nuclear projections and micronuclei. Approximately 10% of the interphase cells had nuclear projections or micronuclei that hybridized with the chromosome 1 probe. Chromosome instability can be explained by both undermethylation (177)

Figure 14 Rosette configuration of multiple-replicated q arms of chromosome 1 seen in metaphases from a patient with ICF syndrome. (Photograph courtesy of J. R. Sawyer, Ph.D., Cytogenetics Laboratory, Arkansas Childrens Hospital, Little Rock.)

and viral infection (175). Jeanpierre et al. (177) observed an embryonic-like methylation pattern of the DNA from ICF patients. They also found that the hypomethylating agent, 5-azacytidine, caused similar failure of decondensation of heterochromatin in normal cells.

C. Telomere Associations and Chromosome Instability

The capping of the ends of chromosome arms by theoretical structures called telomeres was proposed by Muller (178) to explain why broken ends of chromosomes were sticky and nonbroken chromosome ends were not. Molecular studies of prokaryotes revealed specific repeated sequences at the ends of chromosome arms (179,180) which are recognized by the enzyme, telomerase, that completes replication of the lagging strand at its terminal end (180). Without this mechanism, the ends of chromosomes would become progressively shorter with each DNA replication. A similar mechanism was recently demonstrated in human chromosomes. The number of telomeric repeats on the ends of human chromosomes varies from 500 to 2000 copies (181). The number of copies is highest in sperm and appears to decrease in senescing cells (182). The number of telomeric repeats also appears to be lower in tumor cells, except for HELA cells which have a higher number of repeats than normal sperm cells (183). These observations have led some investigators to suggest that chromosome instability in some types of tumors is due to a gradual diminution of telomeric sequences, which results in an increase in the number of dicentric chromosomes and consequent chromosome loss or rearrangement leading to cancer (184–186). However, the majority of tumor cell lines have not been found to have a different number of telomere repeats than normal cells (187). On the other hand, telomerase activity, which is not usually demonstrable in normal somatic cells, is found in germline tissue, in mitotically active hematopoietic cells, and in tumor and immortal cell lines (188). Ohyashiki and Ohyashiki (188) have recently presented a model for the role of telomerase in human hematologic neoplasias, based on observations that telomere shortening is most pronounced in actively dividing tissues, such as germ cells, hematopoietic cells, and immortalized lymphoblastoid or tumor cells. However, the number of telomere repeats does not appear to decrease below a certain "critical length." Therefore, they suggest that telomerase activity is increased in these circumstances in order to maintain this critical length. An elevated level of telomerase was initially found in HELA cells (189) and reportedly occurs in 90% of primary tumors, in most established cancer cell lines, in some stem cells, in hair follicle cells, and in cryptic intestinal cells, but not in cells in G0. The RNA moiety of the ribonucleoprotein enzyme telomerase from *Euplotes crassus* has been identified, sequenced, and shown to prime telomere synthesis in vitro. It was described as a specialized reverse transcriptase by Shippen-Lentz and Blackburn (179). A model for the function of telomerase in completing replication of the terminal lagging DNA strand was proposed by Blackburn (180). The precise mechanism of how telomerase works is still being characterized (189–191).

IV. FRAGILE SITES ON HUMAN CHROMOSOMES

Fragile sites on human chromosomes represent lesions or breaks that recur at specific chromosome sites, either spontaneously or in response to a variety of treatments or culture conditions, some of which are known to cause mutations. The specificity of such breaks or lesions was not appreciated until chromosome banding techniques were developed which allowed localization to specific bands.

Even before chromosome banding techniques, however, some lesions were recognized as being heritable features of chromosomes that could serve as useful markers to map genes on chromosomes by linkage analysis. The first such site, used in this way in 1970, was a heritable site on the long arm of chromosome 16, later determined to be on band 16q22, which was used to confirm the location of the gene for alpha haptoglobin (α-Hp) on chromosome 16 by Magenis et al. (192). Breaks were consistently observed in the middle to distal third of the long arm of a chromosome 16, as either single or isochromatid breaks (Fig. 15) in 5–24% of metaphases from cultured peripheral blood lymphocytes or uncultured preparations from bone

(a)

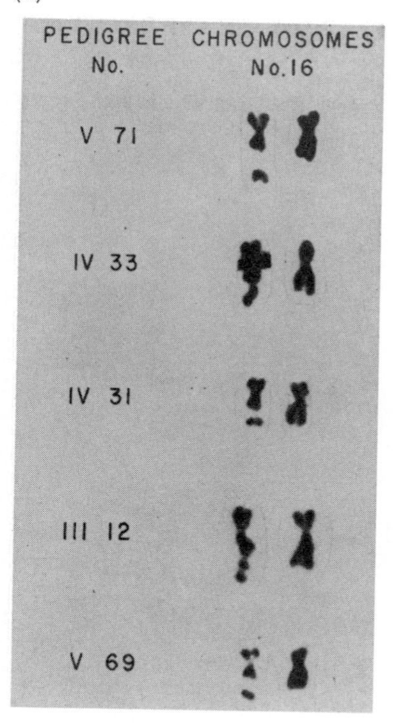

Figure 15 (a) Fragile site on chromosome 16 from several relatives in a large family. (b) Pedigree of large family. (Reprinted with permission from Ref. 192.)

marrow. Multiple endoreduplicated copies of the distal fragment were observed in some cells. Robson et al. (193) had previously mapped α-Hp to chromosome 16, using heritable translocations of chromosome 16. However, the study of a very large pedigree of 238 family members revealed that 50 people had a parent with a fragile 16 who transmitted the fragile site to approximately half of them. Linkage analysis of α-Hp and a variety of blood group polymorphisms localized the gene to the fragile site itself.

The term "fragile site" is attributed to Frederick Hecht, who, in describing the lesion on chromosome 16 in 1968 (194), wanted to convey the concept of transmissible points of chromosome fragility in the human genome. At that time, one other site on chromosome 2 was recognized as a possible heritable fragile site (195,196). Neither site was associated with genetic disease.

(b)

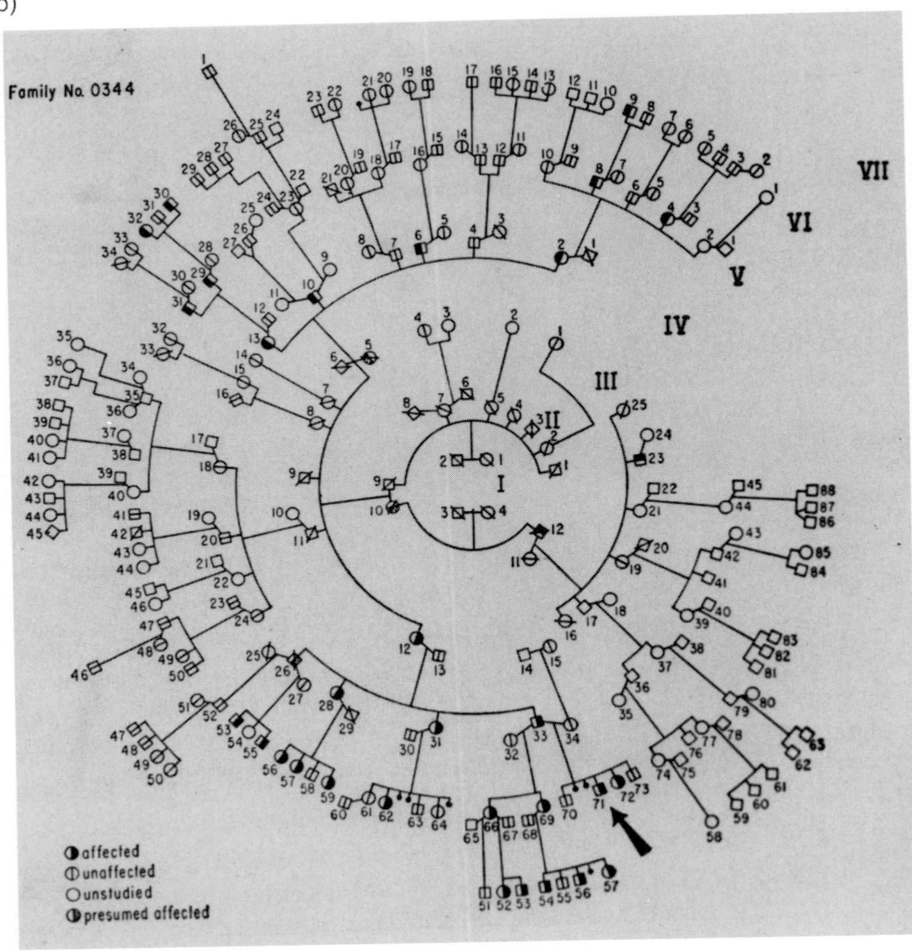

Figure 15 Continued

A. Fragile X Syndrome

A third "fragile site," described as a "marker X" by Lubs in 1969 (197), was associated with mental retardation in four retarded males and one of two obligate female carriers, in a family with X-linked mental retardation. Escalante et al. (1971), also described a marker X chromosome transmitted from a mother to her son with severe sex-linked mental retardation. This observation was included in a report by Giraud in 1976 (198) describing "constitutional chromosome breakage" in 18 patients (which included six unrelated patients, five males and one female) associated with mental retardation. Harvey et al. in 1977 (199) described eight families with mental retardation and an abnormal X. Although Lubs described the first family with a "marker X," it was not until the reports of Giraud and of Harvey et al. that widespread attention was focused on fragile X as a major cause of X-linked mental retardation. The term "fragile X" associated with X-linked mental retardation was proposed by Kaiser-McCaw et al. (200).

The fragile X syndrome is regarded as the second most common mental retardation syndrome after Down syndrome. The frequency of occurrence is approximately 1/1200 males and 1/2000 females (201). The disorder in males is characterized by moderate to severe mental retardation, macroorchidism, and characteristic facial appearance including a long narrow face, prominent forehead, prognathism, and protruding ears. Common behavioral features include hyperactivity, autism, and characteristic speech patterns. Other reported features include joint laxity and mitral valve prolapse, suggestive of an underlying connective tissue disorder (201). A significant proportion of female carriers also may have mild mental impairment, and a few isolated cases have the morphologic features of long face, prominent forehead, prognathia, and protruding ears. It should be emphasized that more than 50 X-linked disorders lead to mental retardation, approximately 15 of which have been associated with fragile X. The classical form of fragile X syndrome was first described by Martin-Bell (202).

The turning point for the recognition of so-called "fragile X syndrome" came with the demonstration that specific cell culture conditions were necessary for the optimal expression of the fragile site (Fig. 16). As for fragile 16, expression varies from 0 to usually less than 50% of cells. The detection of the marker X by Lubs and others was not widely observed for families with X-linked mental retardation and so remained for almost a decade until the discovery by Sutherland (203) that several fragile sites were expressed only when cells were cultured in Eagle's medium that was deficient in folic acid and thymidine (204). Subsequently various chemical antagonists of folic acid synthesis were also shown to increase the frequency of expression of fragile X and other fragile sites in cell culture (205).

Even with improved conditions for expressing the fragile X chromosome, frequencies varied greatly in different individuals, in different tissues, in affected males compared to obligate female carriers, and even at different times within the same individual. Generally, fragile X expression is higher in hemizygous affected males than in heterozygous carrier females. Female carriers may also be affected, but expression is still usually lower than in affected males. Preferential inactivation of the mutant fragile X gene was

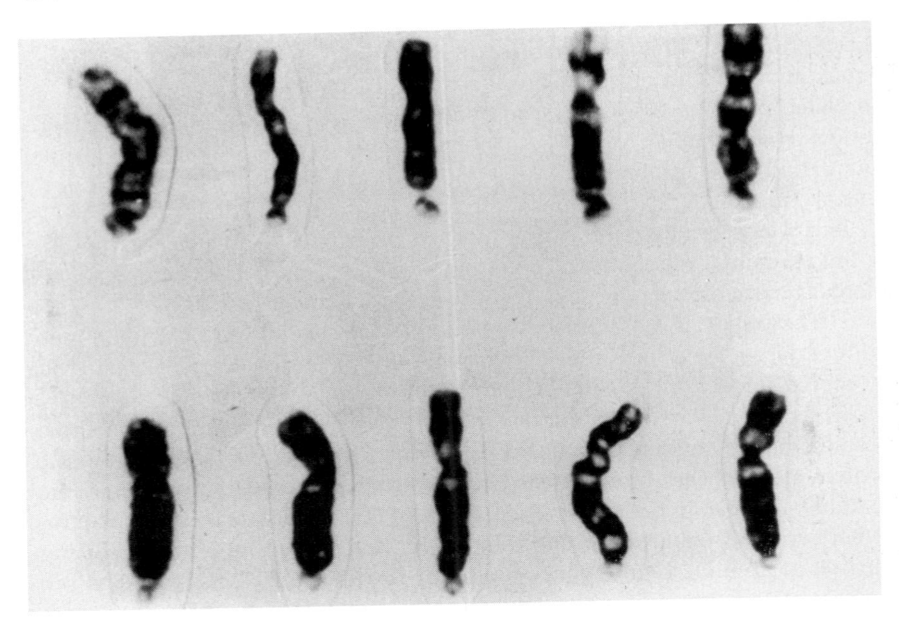

Figure 16 Fragile sites on X chromosome. Upper row with fragile site at Xq27.2 represents a common site that can be confused with lower row showing fragile site at Xq27.3. Lower row is associated with fragile X syndrome; upper row is a common fragile site, not associated with disease.

initially offered as an explanation of the difference in disease severity between affected males and affected female carriers. However, the relative contribution of X-inactivation in determining the degree of mental retardation in affected females is uncertain. In studies by Uchida et al. (206) and others the percentage of cells in which the fragile X chromosome was active, as determined by BrdU labeling, was higher in affected females than in intellectually normal carriers. Other studies maintained that X-inactivation played only a small role in determining mental retardation in females (207). This difference in conclusions is not easy to resolve without testing the relative proportion of fragile X alleles on the late-replicating X chromosome in the brain cells of carrier females.

Until recently, prenatal diagnosis and the cytogenetic diagnosis of carrier females, in particular, were continuing problems. Peripheral blood lymphocytes were the usual cells used in detecting the fragile X in children and adults. In contrast, long-term fibroblasts and amniocytes require especially stringent conditions of folate depletion for expression (208). This was problematic with regard to prenatal diagnosis of affected males. Prenatal detection of carrier females by cytogenetics simply was not feasible.

Normal transmitting males (NMTs) presented another problem in counseling and diagnosing fragile X syndrome. Initially described by Sherman et al. (209), this conundrum in clinical genetics became known as "Sherman's paradox." Sherman observed that less than 10% of male sibs of NTMs had fragile X syndrome, whereas 37% of grandsons of NMT were affected. Granddaughters of NTMs who were obligate carriers showed nearly 50% of their sons to be affected (Fig. 17). Furthermore, a significant proportion of their daughters were also affected. There was no precedent for such markedly different penetrance in different generations with known genetic models. Laird (210) proposed a genetic imprinting model which sug-

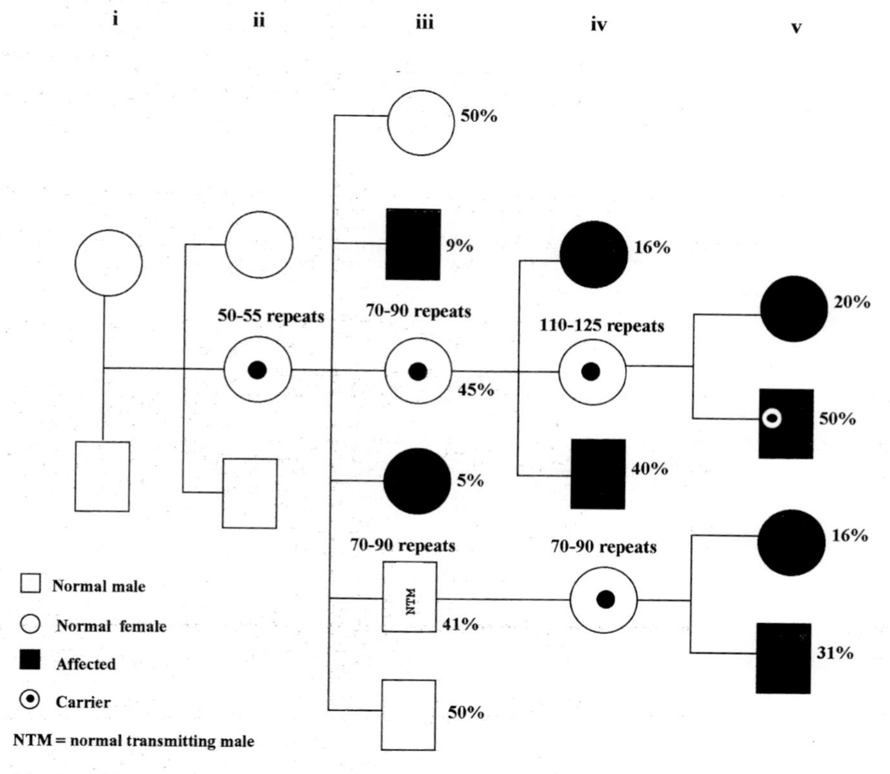

Figure 17 Idealized pedigree showing pattern and frequency of inheritance of trinucleotide repeats and the risk for being affected. Where appropriate, the symbols in generations III, IV, and V are labeled with the percent of offspring with fragile X syndrome. (Adapted from Ref. 201.)

gested that the fragile X gene when normally passed through female oogenesis would be inactivated and then reactivated in the female germ line. In the case of a fragile X mutation, however, he proposed a failure of the gene to be reactivated. He further resolved the mathematical predictions of such a model that would explain Sherman's paradox (211). Molecular characterizations have since coincided with many aspects of this model's predictions.

In 1991 the gene for fragile X was independently identified by four different groups of investigators (212–215). Their results explained anticipation and the complex paradox of fragile X syndrome and provided a possible explanation of why the fragile X appears "fragile." The fragile X gene (FMR1) contains a tandemly repeated trinucleotide CGG sequence near its 5′ end. In the normal population the number of repeats of this sequence is from 6 to 49. Males and females who are unaffected but have intermediate-length alleles of 50–200 repeats have a premutation. Premutation males who are the "normal transmitting males" pass the gene, generally unchanged in size, to all of their daughters, who are also unaffected. Although the mechanism is still not understood, the premutation undergoes expansion when it passes through female meiosis, so that these daughters are at risk for having affected children. Neither the normal FMR1 gene nor the premutation is methylated in males or females, whereas the full mutation is methylated and remains so in both affected males and females (216–218).

The FMR1 repeat has been correlated with a delay in replication timing in male lymphoblasts (219), based on BrdU labeling and isolation of specific DNA regions by flow cytometry. Delayed timing was found to extend at least 150 kb in the 5′ direction from the gene, and does not seem to result from a delay in polymerase progression through the expanded CGG element. Selig et al. (220) developed a different method of examining DNA replication in interphase nuclei based on the detection of the proportion of singlet to doublet signals for different loci along the length of a chromosome labeled by in-situ hybridization. Late-replicating loci show a high proportion of cells with one signal compared to early-replicating loci, which predominantly show a double signal in the G2 phase of the cell cycle. This method, applied by Boggs and Chinault (221) to the FMR1 locus, was consistent with other methods of measuring DNA replication. Subramanian et al. (222) used this technique to assess the DNA replication patterns of male and female fragile X syndrome lymphoblasts with a full CGG-repeat expansion.

Expansion of FRAXE associated with a milder form of mental retardation was also examined. Relative to normal male and female controls, it was found that replication timing was delayed over a 1–1.2 Mb region except in the premutation state (222). The proximal border maps from 300–450 kb upstream of FMR1 to 150–300 kb downstream, through the FRAXE locus. The FRAXE delay begins approximately 50 kb proximal to the CGG repeats, extends through the locus, and ends in the same region as for the FMR1 delay. These results suggest that long-range disruption is a normal feature of Fra(X) syndrome and of FRAXE disease.

The number of CGG repeats is correlated with FMR1 gene activity. The fragile X phenotype results when the trinucleotide repeats exceed 200 copies and the promoter region is methylated to inhibit the expression of the FMR1 gene (223). Males with approximately 180 CGG repeats and only a small proportion of methylated allelic repeats are nearly normal (224), while males with longer, fully methylated repeats express the full fragile X phenotype. Thus, inactivation of the single FMR1 gene in a hemizygous male results in fragile X syndrome (225).

In females, one X chromosome is randomly inactivated in the blastocyst stage of the early embryo. The clinical expression of the same size FMR1 allelic repeat in carrier females is dependent on the proportion of female cells with active X chromosomes carrying the normal FMR1 allele in each affected tissue. Nevertheless, females nearly always exhibit less severe symptoms than their affected male counterparts.

PCR reliably amplifies allelic gene regions with up to 100 trinucleotide repeats but not alleles with more than 100 repeats. Thus, Southern blotting is required to rule out longer repeats in female patients with only one amplified trinucleotide repeat length (Figs. 18–20).

B. Rare and Common Fragile Sites in the Human Genome

With the discovery of fragile X syndrome, considerable effort was made to locate other heritable fragile sites on human chromosomes that might be associated with disease. Although a large number of sites have been observed, no other microscopically visible site has been associated directly with a specific disease. At the majority of these sites chromosome breakage occurs spontaneously, but the frequency is enhanced in cell culture by various induction techniques. Some of these sites, observed only in selected families, are regarded as "heritable" or "rare" fragile sites. Others, which can be induced or occur spontaneously in almost all cell lines, are referred to as "constitutive" or "common" fragile sites. The definition of a heritable fragile site was initially quite restrictive and required not only that it segregate in a pedigree in a Mendelian fashion, but that deletions, fragments, and multiple copies of the broken segment were also found (226). Gradually, as many sites have been shown to exhibit variable frequency under different conditions, even within an individual, the criteria for heritable fragile sites as well as the distinction between common and rare fragile sites has become less clear. The definition of a "common fragile site" has come to mean any

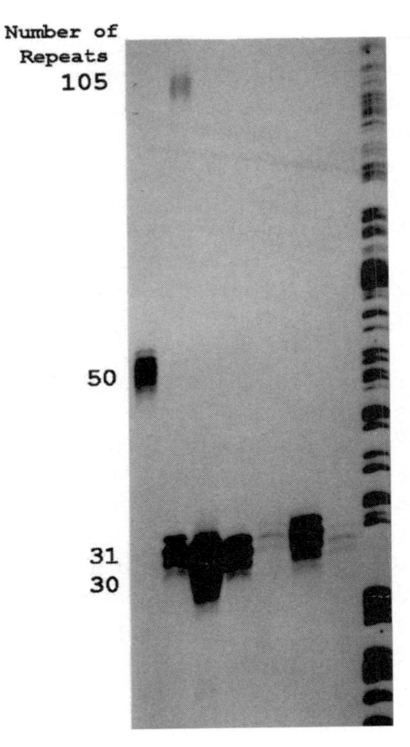

Figure 18 PCR amplification and acrylamide gel electrophoresis of the CGG trinucleotide repeat region 5′ to the untranslated FMR1 gene sequence characterizes allelic lengths up to about 100 repeats. PCR-amplified DNA in the first lane (at left) is from a premutation male with 50 repeats. The second lane characterizes both alleles of a premutation female carrier with 105 repeats in the abnormal allele and 31 repeats in the normal allele. The next five lanes are from normal male subjects, each with one allele of either 30 or 31 repeats. Size marker at the far right is the G nucleotide lane of an M13 sequencing ladder.

chromosomal gap or lesion that involves both chromatids and is demonstrated to recur at a precise location on a chromosome with a frequency that distinguishes it statistically from random breakage. However, the specific statistical criteria applied by different laboratories have varied (227–230).

Common and rare fragile sites are induced or enhanced by thymidine depletion or by the addition of antagonists of folic acid metabolism. In 1984, Glover et al. (231) found that the inhibitor of alpha DNA polymerase, aphidicolin, combined with low folate levels, reliably induced many common fragile sites. Sutherland et al. (232), thereafter, reported two new classes of

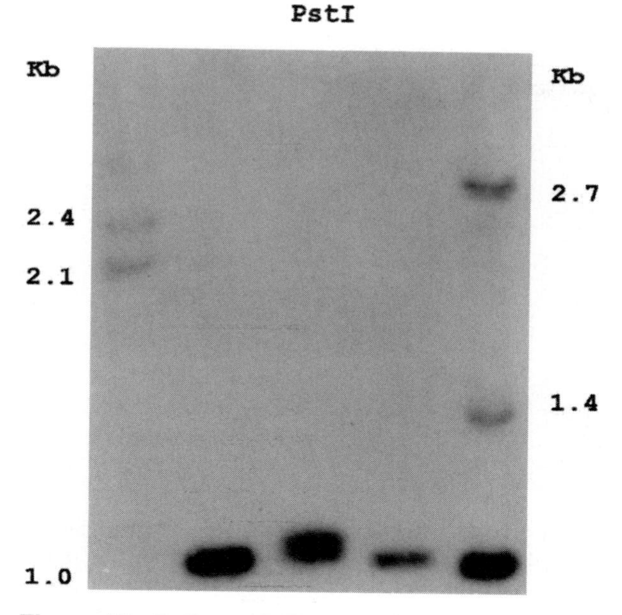

Figure 19 PstI restriction enzyme analysis of DNAs from five different patients hybridized to PX6 probes characterizes alleles with longer trinucleotide repeats. DNA in the left lane is from an affected male with approximately 350 repeats in the 2.1-kb fragment and 450 repeats in the 2.4-kb fragment. The next two samples, with 1.0-kb fragments, are from females with normal-length trinucleotide repeats in the FMR1 gene. The fourth sample is from a normal male with a normal number of trinucleotide repeats. The sample at the far right is from a carrier female with mosaic expansions of approximately 550 repeats at 2.7 kb in cells with the affected allele, 150 repeats at 1.4 kb in cells with a premutation, and 1.0-kb normal-length allele in all cells.

fragile sites induced by 5-azacytidine and 5-bromodexoyuridine (BrdU). Other chemicals, such as methotrexate, FUdr, and caffein, have also been shown to enhance fragile site expression (233). The characterization and listing of 22 rare fragile sites and some 100 common fragile sites (Table 6) has since been derived at several international human gene mapping workshops (234,235). In some cases rare and common sites apparently involve the same breakpoint with different induction systems.

While no fragile site has been reported to be coinherited with a genetic disease other than fragile X, correlations have been found between fragile sites and breakpoints in constitutional chromosome rearrangements (236,237) and chromosome breakpoints associated with cancer. Correlation

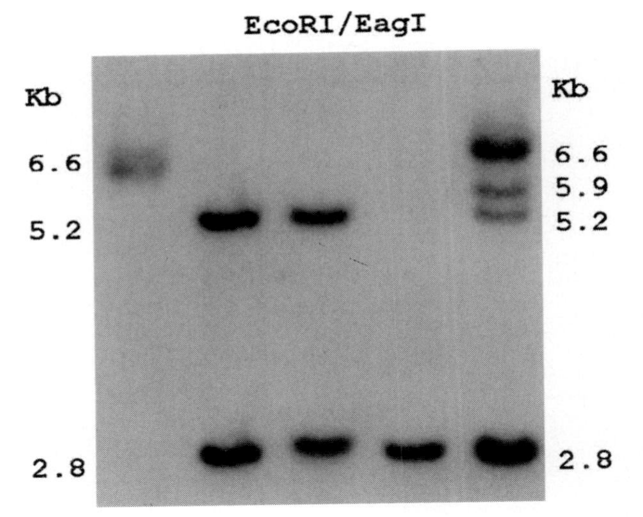

Figure 20 Dual EcoRI/Eagl restriction enzyme digestion with the methylation-sensitive Eagl enzyme, hybridized to the same PX6 probe, characterizes the methylation status of the same DNAs tested by Pst I in Fig. 19. The left lane illustrates a 6.6-kb methylated expanded male allele that is different in size from the 2.8-kb normal allelic fragments at the bottom of the adjacent lanes. Lanes 2 and 3 illustrate the female samples with a 2.8-kb normal active allele at the bottom and a 5.2-kb normal methylated inactive X allele. The fourth lane is the 2.8-kb normal, fully active, unmethylated allele of a normal male. The DNA of an affected female at the far right shows a normal 2.8-kb active allele, a normal methylated 5.2-kb inactive allele, and two different-length expanded alleles of 5.9 kb and 6.6 kb. The latter are also methylated and probably inactive based on the size difference between the two top bands and the normal fragments.

was initially noted between breakpoints in cancer rearrangements and oncogene sites (233,238,239). Yunis and Soreng (240) studied the distribution of fragile sites induced by thymidine deprivation and caffeine in lymphocyte metaphases analyzed by high-resolution banding from 10 human subjects. Twenty of 51 constitutive and 6 of 16 rare fragile sites mapped at or close to 26 of 31 breakpoints commonly seen in human malignancies. The most recent map of recurrent breakpoints in balanced rearrangements in human neoplasias compiled by Mitelman et al. (241) reveals that approximately one-half occur in bands identified as being fragile sites in the Human Genome database (235). From these and other studies (242), it is evident that there are ''hotspots'' for so-called fragile sites. Wenger et al. (243) recently suggested that such hotspots may represent unstable areas of the genome

Table 6 Common and Rare Fragile Sites and Oncogene Locations

Common fragile sites[a]				Rare fragile sites[a]		Oncogene locations[b]	
1p36	3p14.2	7p14.2	12q21.32	2q11.2	11p15.1	1p34-36	11p15.5
1p32	3q25	7p13	12q24	2q13	11q13.3	1p32	11q23-24
1p31	3q27	7q11	13q13.2	2q22.3	11q23.3	1p22 and/or 1p11-12	12pter-q14
1p22	4p16.1	7q22	13q21	6p23	12q13.1	1q22-qter	12p12.1
1p21.3	4p15	7q31.2	13q21.2	7p11.2	16p12.3	2p23-24	14q21-31
1p21.2	4q12	7q32.3	14q21.2	8q22.3	16q22.2	3p24-25	14q32
1q12	4q27	7q36	14q23	8q24.1	17p12	5q34	15q25-26
1q21	4q31.1	8q13	14q24.11	9q21.1	19p13	6p23-q12	17q11-21
1q25.1	5p14	8q22.1	15q22	9q32	20p11.23	6q15-24	17q21-22
1q31	5p13	8q24.3	16q32.2	10q23.3	22q13	7p12-14	18q21.3
1q42	5q15	9p21	17q23.1	10q25.2	Xq27.3	8q11-22	20q12-13
1q44	5q21	9q12	18q12.2			8q24	22q12.3-13.1
2p24.2	5q31.1	9q22.1	18q21.3			9q34	Xpter-q28
2p16.2	5q35	10q21	19q13				
2p13	6p25.1	10q22.1	20p12.2				
2q21.3	6p22.2	10q26.13	22q12.2				
2q31	6q13	11p15.1	Xp22.31				
2q32.1	6q15	11p14.2	Xq22.1				
2q33	6q26	11p13	Xq27.2				
2q37.3	7p22	11q14.2	Xq28				
3p24.2	7p21.2	11q23.3	Yq12				

[a] *Source*: Ref. 235.
[b] *Source*: Ref. 241.

which are prone to breaks and unequal crossing over, resulting in minute deletions and duplications. In studies of balanced translocations in phenotypically normal parents and affected offspring who appeared to have the same balanced translocations, 83% of breakpoints were in known fragile sites. This was two to four times higher than breaks in fragile sites in de novo rearrangements or in unbalanced rearrangements of offspring who had parents with balanced karyotypes.

Does detection of a fragile site in an individual indicate a higher risk for rearrangements that might be heritable or predispose to cancer? This has been a difficult question to answer, particularly for the common fragile sites that can be induced in almost any cultured cells and vary in frequency under different conditions. However, Jordan et al. (244) noted a twofold increase in frequency of expression of the 3p14 fragile site in patients with von Hippel-Lindau disease (a disease which predisposes to diverse tumors including renal cell carcinomas) over unaffected subjects. More recently, Egeli et al. (245) compared the expression of the 3p14 and other common fragile sites in patients with squamous cell lung cancer, in first-degree relatives, and in unrelated healthy controls. Expression of 3p14 was highest in patients and was significantly higher in patients and their relatives than in controls. Rates of expression of 2p21, 2q37, and 16q23 were also higher in patients and their relatives than in controls. They also compared rates of expression of 3p14 in smokers in the control group with patients (80% of whom were smokers) and with relatives who smoked. Again, patients had the highest rate of expression, but the difference between relatives and controls who smoked was not significant.

To date, no rare autosomal fragile site has been shown to predispose to any specific cancer, disease, or heritable chromosome abnormality. It is of interest that several known rare fragile sites are near sites of trinucleotide repeats expansions associated with diseases such as Huntington, Friedreick ataxia, and myotonic dystrophy (246,247). Nevertheless, no one has reported increased fragility of 4p16.1, 9q13-q21.1, or 19q13.3 in affected individuals with these respective diseases. One explanation may be that the sizes of expansions are less in these disorders than in fragile X. Repeat sequence stability apparently decreases dramatically when certain trinucleotide repeats reach a critical size, proposed by Chung et al. (248) to be larger than an Okazaki fragment, above which size rapid amplification results in disease. To date, both CGG repeats in fragile X and CAG repeats in the neurologic disorders mentioned have shown this property, but only the fragile X amplification appears to result in a microscopically visible fragile site (246).

In fact, the phenomena of mutations, deletions, rearrangements, fragile sites, sister chromatid exchange, recombination, and transposition of heritable elements in human and eukaryotic chromosomes have only just begun

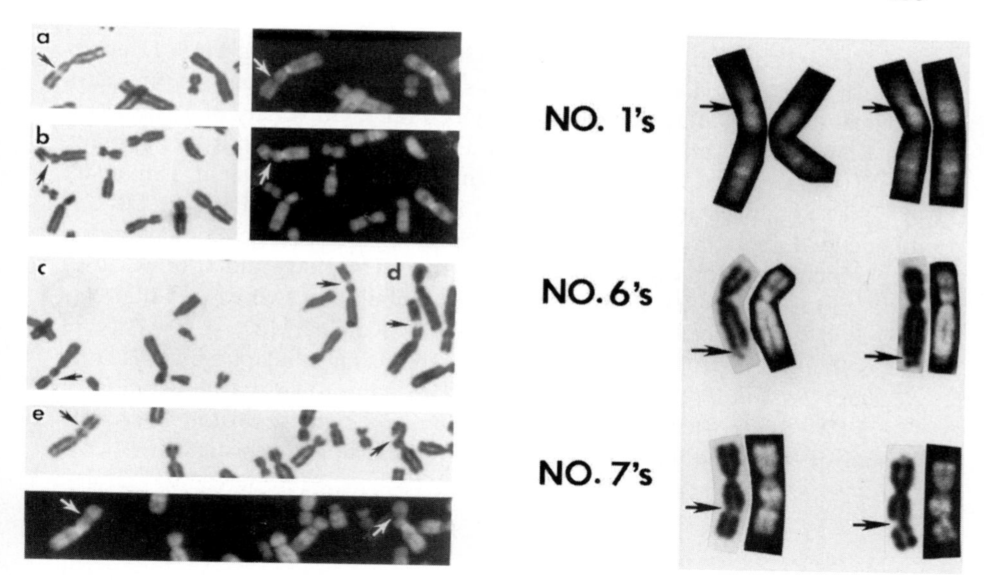

Figure 21 Examples of some common fragile sites frequently seen in cultures which have been treated with methotrexate or other antagonists of folic acid metabolism. (a–e) Common fragile site at 3p14 stained by plain Giemsa (unbanded) and by Q banding showing (a) gap in both chromatids, (b) cell with single chromatid break, (c) cell with break in one chromatid and gap in the other, (d) cell with gap involving both chromatids in both homologs, and (e) cell with gap in chromatids of one homolog and chromatid break in the other homolog. Pairs of chromosome at right show common fragile sites (arrows) at 1p31.2, 6q26, and 7q22.

to be understood. One of the projected outcomes of the Human Genome Project will be that as more of the 3 billion base pairs of the human genome are sequenced, certain recurrent themes will become evident and specific classes of DNA sequences and their physical locations will undoubtedly throw light on each of these processes, with some having a common sequence or structure and others having unique features. In the meantime, the stability of various genomes has taken a long time to evolve, and the heterogeneity in chromosome structure, visualized cytologically and revealed at the molecular level, is unlikely to be trivial or inconsequential.

ACKNOWLEDGMENTS

We would like to thank the following colleagues for providing photographs and other illustrative material: Farah Shah, M.D., Department of Dermatology, Texas Tech University, Lubbock; Golder Wilson, M.D., Ph.D., Department of Pediatrics, Division of Genetics, University of Texas Southwestern Medical Center, Dallas; Jerome McCombs, Ph.D., University of Texas Medical Branch, Department of Laboratory Medicine, Galvaston; Roger Schultz, Ph.D., McDermott Center for Human Growth and Development, University of Texas Southwestern Medical Center, Dallas; Shivanand Patil, Ph.D., Cytogenetics Laboratory, and Joy Larsen Haidle, MS, CGC genetic counselor, Department of Pediatrics, University of Iowa Hospitals and Clinics, Iowa City; and J. R. Sawyer, Ph.D., Cytogenetic Laboratory, Arkansas Children's Hospital, Little Rock. We also thank Vinay Kumar, M.D., University of Texas Southwestern Medical Center, for permission to adapt materials from *Robbin's Pathological Basis of Disease* (5th edition) in the preparation of Fig. 1. We thank Alex Dow, Center for Human Genetics, Boston University School of Medicine, for his photographic expertise and assistance.

REFERENCES

1. Poland BJ, Miller JR, Harris M, Livingston J. Spontaneous abortion. A study of 1,961 women and their conceptuses. Acta Obstet Gynecol Scand 1981; 102(supl):5–32.
2. Warburton D, Kline J, Stein Z, Strobino B. Cytogenetic abnormalities in spontaneous abortions of recognized conceptions. In Porter IH, Hatcher NH, Wiley A, eds. Perinatal Genetics: Diagnosis and Treatment. New York, Academic Press, 1986:23–40.
3. Porter IH, Hook EB. Population Cytogenetics. Studies in Humans. New York: Academic Press, 1976.
4. Radinsky R. Modulation of tumor cell gene expression and phenotype by the organ-specific metastatic environment (rev). Cancer Metastasis Rev 1995; 14: 323–338.
5. Sandberg A. The Chromosomes in Human Cancers and Leukemia. 2d ed. New York: Elsevier, 1992.
6. Sandberg A. Concurrent Workshop I: Applications of molecular cytogenetic technology in cancer (HFL Mark, Chair.), Fourth Joint Clinical Genetics Meeting (ACMG and March of Dimes), Fort Lauderdale, FL, Feb. 28–Mar. 2, 1997.
7. Varmus HE. Oncogenes and transcriptional control. Science 1987; 238:1337.
8. Yamamoto T. Molecular basis of cancer: oncogenes and tumor suppressor genes. Microbiol Immunol 1993; 37:11–22.
9. Sutherland GR, Richards RI. Simple tandem repeats and human genetic disease. Proc Natl Acad Sci USA 1995; 92:3636–3641.
10. Cotran RS, Kumar V, Robbins SL. Robbins Pathological Basis of Disease, 5th ed. Philadelphia: Saunders, 1994:241–303.

11. Knudson AG. Mutation and cancer: statistical study of retinoblastoma. Proc Natl Acad Sci USA 1971; 68:820–823.
12. Cavenee WK, Dryja TP, Phillips RA, Benedict WF, Godbout R, Gallie BL, Murphree AL, Strong LC and White RL. Expresson of recessive alleles by chromosomal mechanisms in retinoblastoma. Nature 1983; 305:779–784.
13. Cowell JK, Hogg A. Genetics and cytogenetics of retinoblastoma. Cancer Genet Cytogenet 1992; 64:1–11.
14. Horsthemke B. Genetics and cytogenetics of retinoblastoma. Cancer Genet Cytogenet 1992; 63:1–7.
15. Gessler M, Poustka A, Cavenee W, Neve RL, Orkin SH, Bruns GAP. Homozygous deletion in Wilms' tumors of a zinc-finger gene identified by chromosome jumping. Nature 1990; 343:774–778.
16. Lynch M, Ruiz A, Puig S, Lazaro C, Castel T, Estivill X. p16 (CDKN2) is commonly mutated in melanoma predisposed patients, but rarely in melanoma tumors. Am J Hum Genet 1995; 57(suppl):A71.
17. Fountain JW, Wallace MR, Brereton AM, O'Connell P, White RL, Rich DC, Ledbetter DH, Leach RJ, Fournier RE, Menon AG, Gusella JF, Barker D, Stephens K, Collins FS. Physical mapping of the von Recklinhausen neurofibromatosis region on chromosome 17. Am J Hum Genet 1989; 44:58–67.
18. Latiff F, Tory K, Gnarra J, Yao F, Orcutt ML, Stackhouse T, Kuzmin I, Modi W, Geil L, Schmidt L, Zhou F, Li H, Wei MH, Chen F, Glenn G, Choyke P, Walther MM, Weng Y, Duan D, Dean M, Glavac D, Richards FM, Crossey PA, Ferguson-Smith MA, Le Paslier D, Chumakov I, Cohen D, Chinault AC, Maher ER, Lineman WM, Zbar B, Lerman MI. Identification of the von Hippel-Lindau disease tumor suppresor gene. Science 1993; 260:1317–1320.
19. Herrera L, Kakati S, Gibas L, Pietrzak E, Sandberg AA. Brief clinical report: Gardner syndrome in a man with an interstitial deletion of 5q. Am J Med Genet 1986; 25:473–476.
20. Johnson RL, Rothman AI, Xie H, Goodrich LV, Bare JW, Bonifas JM, Quinn AG, Myers RM, Cox DR, Epstein EH, Scott MP. Human homolog of patched, a candidate gene for the basal cell nevus syndrome. Science 1996; 272:1668–1671.
21. Povey S, Burley MW, Attwood J, Benham F, Hunt D, Jeremiah SJ, Franklin D, Gillett G, Malas D, Robson EB. Two loci for tuberosclerosis: one on 9q34 and one on 16p13. Ann Hum Genet 1994; 58:107–127.
22. Li J, Yen C, Liaw D, Podsypanina K, Bose S, Wang SI, Puc J, Miliaresis C, Rodgers L, McCombie R, Bigner SH, Giovanella BC, Ittmann M, Tycko B, Hibshoosh H, Wigler MH, Parson R. PTEN, a putative protein tyrosine phophatase gene mutated in human brain, breast, and prostrate cancer. Science 1997; 275:1943–1947.
23. Stickens D, Clines G, Burbee D, Ramos P, Thomas S, Hogue D, Hecht JT, Lovett M, Evans GA. The EXT2 multiple exostoses gene defines a family of putative tumor suppressor genes. Nature Genet 1996; 14:25–32.
24. Tavtigian SV, Simard J, Rommens J, Couch F, Shattuck-Eidens D, Neuhausen S, Merajver S, Thorlacius S, Offit K, Stoppa-Lyonnet D, Belanger C, Bell R, Berry S, Bogden R, Chen Q, Davis T, Dumont M, Frye C, Hattier T, Jam-

mulapati S, Janecki T, Jiang P, Kehrer R, Leblanc J, Mitchell JT, McArthur-Morrison J, Nguyen K, Peng Y, Samson C, Schroeder M, Snyder SC, Steele L, Stringfellow M, Stroup C, Swedlund B, Swensen J, Teng D, Thomas A, Tran T, Tranchant M, Weaver-Feldhaus J, Wong AKC, Shizuya H, Eyfjord JE, Cannon-Albright L, Labrie F, Skolnick MH, Weber B, Kamb A, Goldgar DE. The complete BCRA2 gene and mutations in chromosome 13q-linked kindreds. Nature Genet 1996; 12:333–337.

25. Lee WH, Bookstein R, Hong F, Young LJ, Shew JY, Lee EY-HP. Human retinoblastoma susceptibility gene: cloning, identification and sequence. Science 1987; 235:1394–1399.
26. Sepp T, Yates JRW, Green AJ. Loss of heterozygosity in tuberous schlerosis hamartomas. J Med Genet 1996; 33:926–964.
27. Finlay CA, Hinds PW, Levine AJ. The p53 proto-oncogene can act as a suppressor of transformation. Cell 1989; 57:1083–1093.
28. Miki Y, Swensen J, Shattuck-Eidens D, Futreal PA, Harshman K, Tavtigian S, Liu Q, Cochran C, Bennett LM, Ding W, Bell R, Rosenthal J, Hussey C, Tran T, McClure M, Frye C, Hattier T, Phelps R, Haugen-Strano A, Katcher H, Yakumo K, Gholami Z, Shafer D, Stone S, Bayer S, Wray C, Bogden R, Dayananth D, Ward J, Tonin P, Narod S, Bristow PK, Norris FH, Helvering L, Morrison P, Rosteck P, Lei M, Barrett JC, Lewis C, Neuhausen S, Cannon-Albright L, Goldgar D, Wiseman R, Kamb R, Skolnick MH. A strong candidate for the breast and ovarian cancer susceptibility gene BCRA1. Science 1994; 266:66–71.
29. Fearon ER, Cho KR, Nigro JM, Kern SE, Simons JW, Ruppert JM, Hamilton SR, Preisinger AC, Thomas G, Kinzler KW, Vogelstein B. Identification of a chromosome 18q gene which is altered in colorectal cancers. Science 1990; 247:49–56.
30. Trofatter JA, MacCollin MM, Rutter JL, Murrell JR, Duyao MP, Parry MD, Eldridge R, Kley N, Menon AG, Pulaski K, Haase VH, Ambrose CM, Munroe D, Bove C, Haines JL, Martuza RL, MacDonald ME, Seizinger BR, Short MP, Buckler AJ, Gusella JF. A novel moesin-, ezrin-, radixin-like gene is a candidate for the neurofibromatosis 2 tumor suppressor. Cell 1993; 72:791–800.
31. Abeln EC, Kuipers-Dijkshoorn NJ, Berns EM, Henzen-Logmans SC, Fleuren GJ, Cornelisse CJ. Molecular genetic evidence for unifocal origin of advanced epithelial ovarian cancer and for minor clonal divergence. Br J Cancer 1995; 72:1330–1336.
32. Tatebe S, Ishida M, Kasagi N, Tsujitani S, Kaibara N, Ito H. Apoptosis occurs more frequently in metastatic foci than in primary lesions of human colorectal carcinomas: analysis by terminal-deoxynucleotydly-transferase-mediated dUTP-biotin nick and labeling. Int J Cancer 1996; 65:173–177.
33. Kerbel RS. Impact of multicellular resistance on the survival of solid tumors including micrometastases (rev.). Invasion and Metastasis 1994–1995; 14: 50–60.
34. Adeyinka A, Pandis N, Nilsson J, Idvall I, Mertens F, Petersson C, Heim S, Mitelmen F. Different cytogenetic patterns in skeletal breast cancer metastases. Genes Chromosom Cancer 1996; 16:72–74.

35. Simpson JF, Qian DE, Ho JP, Slovak ML. Genetic heterogeneity of primary and metastatic breast carcinoma defined by fluorescence in situ hybridization. Am J Pathol 1996; 149:751–758.
36. Pirc-Danoewinata H, Bull JP, Okamoto I, Karner J, Breiteneder S, Liebhardt A, Budinsky A, Marosi C. Cytogenetic findings in colorectal cancer mirror multistep evolution of colorectal cancer. Wien Klin Wochenschr 1996; 108: 752–758.
37. Schwendal A, Langreck H, Reichel M, Schrock E, Ried T, Dietel M, Petersen I. Primary small-cell lung carcinomas and their metastases are characterized by a recurrent pattern of genetic alterations. Int J Cancer 1997; 74:86–93.
38. Shiseki M, Kohno T, Adachi J, Okazaki T, Mizoguchi H, Noguchi M, Hirohashi S, Yokota J. Comparative allelotype of early and advanced stage non-small cell lung carcinomas. Genes Chromosom Cancer 1996; 17:71–77.
39. Perry A, Jenkins RB, Dahl RJ, Moertel CA, Scheithauer BW. Cytogenetic analysis of aggressive meningiomas: possible diagnostic and prognostic implications. Cancer 1996; 77:2567–2573.
40. Healy E, Belgaid CE, Takate M, Vahlquist A, Rehman I, Rigby H, Rees JL. Allelotypes of primary cutaneous melanoma and benign melanocytic nevi. Cancer Res 1996; 56:589–593.
41. Miele ME, Robertson G, Lee JH, Coleman A, McGary CT, Fisher PB, Lugo TG, Welch DR. Metastasis suppressed, but tumorgenicity and local invasiveness unaffected, in human melanoma cell line MelJuso after introduction of human chromosomes. Mol Carcinogen 1996; 15:289–299.
42. Morse HG, Moore GE, Ortiz LM, Gonzalez R, Robinson WA. Malignant melanoma from subcutaneous nodule to brain metastasis. Cancer Genet Cytogenet 1994; 72:16–23.
43. Ichikawa T, Nihei N, Kuramochi H, Kawana Y, Killary AM, Rinker-Schaeffer CW, Barrett JC, Isaacs JT, Kugoh H, Oshimura M, Shimazaki J. Metastasis suppressor genes for prostate cancer. Prostate 1996; suppl 6:31–35.
44. Jenkins RB, Qian J, Lieber MM, Bostwick DG. Detection of c-myc oncogene amplification and chromosomal anomalies in metastatic prostatic carcinoma by fluorescence in situ hybridization. Cancer Res 1997; 57:524–531.
45. Geburek BM, Kollmorgen TA, Qian J, D'Souza-Gburek SM, Lieber MM, Jenkins RB. Chromosome anomalies in stage D1 prostate adenocarcinoma primary tumors and lympyh node metastases detected by fluorescence in situ hybridization. J Urol 1997; 157:223–227.
46. Soder AI, Hopman AH, Ramaekers FC, Conradt C, Bosch FX. Distinct nonrandom patterns of chromosomal aberrations in the progression of squamous cell carcinomas of the head and neck. Cancer Res 1995; 55:5030–5037.
47. Fletcher JA, Gebhardt MC, Kozakewich HP. Cytogenetic aberrations in osteosarcomas. Nonrandom deletions, rings and double minute chromosomes. Cancer Genet Cytogenet 1994; 77:81–88.
48. Bussey HJR. Familial Polyposis Coli: Family Studies, Histopathology, Differential Diagnosis, and Results of Treatment. Baltimore: Johns Hopkins University Press, 1975.

49. Burt RW, Samowitz WS. The adenomatous polyp and the hereditary polyposis syndromes. Gastroenterol Clin N Am 1988; 17:657–678.
50. Fearon ER. Molecular genetic studies of adenoma-carcinoma sequence. Adv Intern Med 1994; 39:123–145.
51. Utsunomiya J, Lynch HT, eds. Hereditary Colorectal Cancer. New York: Springer-Verlag, 1990.
52. Nagase H, Myoshi Y, Horii A, Aoki T, Ogawa M, Utsunomiya J, Baba S, Sasazuki T, Nakamura Y. Correlation between the location of germline mutations in the APC gene and the number of colorectal polyps in familial adenomatous polyposis patients. Cancer Res 1992; 52:4055–4057.
53. Bodmer WF, Bailey C, Bodmer J, Bussey HJR, Gorman P, Lucibello FC, Murday VA, Rider SH, Scambler P, Sheer D, Solomon E, Spurr NK. Localization of the gene for familial polyposis on chromosome 5. Nature 1987; 328:614–616.
54. Leppert M, Dobbs M, Scambler P, O'Connell P, Nakamura V, Stauffer D, Woodward S, Burt R, Hughs J, Gardner E, Lathrop M, Wasmuth J, Lalouel J-M, White R. The gene for familial polyposis coli maps to the long arm of chromosome 5. Science 1987; 238:1411–1413.
55. Kinzler KW, Nilbert MC, Su L-K, Vogelstein B, Bryan T, Levy D, Smith K, Preisinger AS, Hedge P, McKechnie D, Finniear R, Markham A, Groffer J, Bogaski MS, Alfschul SF, Horii A, Ando H, Myoshi Y, Miki Y, Nishisho I, Nakamura Y. Identification of FAP locus genes from chromosome 5q21. Science 1991; 253:661–665.
56. Groden J, Thlivers A, Samowitz W, Carlson M, Gelbert L, Albertson H, Joslyn G, Stevens J, Spirio L, Robertson M, et al. Identification and characterization of the familial adenomatous polyposis gene. Cell 1991; 66:589–600.
57. Nishisho I, Nakamura Y, Miyoshi Y, Miki Y, Ando A, Horii A, Koyama K, Utsunomiya J, Baba S, Hedge P. Mutations of chromosome 5q21 genes in FAP and colorectal cancer patients. Science 1991; 253:661–665.
58. Vogelstein B, Fearon ER, Hamilton SR, Kern SE, Preisinger AC, Leppert M, Nakamura Y, White R, Smits AMM, Bos JL. Genetic alterations during colorectal tumor development. N Engl J Med 1988; 319:525–532.
59. Delhanty JD, Davis MB, Wood J. Chromosome instability in lymphocytes, fibroblasts and colon epithelial-like cells from patients with familial polyposis coli. Cancer Genet Cytogenet 8:27–50.
60. Muleris M, Salmon JR, Dutrillaux B. Existence of two distinct processes of chromosomal evolution in near-diploid colorectal tumors. Cancer Genet Cytogenet 1987; 32:43–50.
61. Muleris M, Salmon JR, Dutrillaux B. Cytogenetics of colorectal adenocarcinomas. Cancer Genet Cytogenet 1990; 46:143–156.
62. Marx J. Research News. New colon cancer gene discovered. Science 1993; 260:751–752.
63. Aaltonen LA, Peltomaki P, Leach F, Sistonen P, Plykkanen L, Mecklin J-P, Jarvinen H, Powell SM, Jen J, Hamilton SR, Petersen GM, Kinzler KW, Voegelstein B, de la Chapelle A. Clues to the pathogenesis of familial colorectal cancer. Science 1993; 260:812–815.

64. Thibodeau SN, Bren G, Schaid D. Microsatellite instability in cancer of proximal colon. Science 1993; 260:816–819.
65. Leach FS, Nicolaides NC, Papadopoulos N, Liu B, Jen J, Parsons R, Peltomaki P, Sistonen P, Aaltonen A, Nystrom-Lahti M, Guan X-Y, Zhang J, Meltzer PS, Yu J-W, Kao F-T, Chen DJ, Cerosaletti KM, Fournier REK, Todd S, Lewis T, Leach R, Naylor S, Weissenbach J, Mecklin J-P, Jarvinen H, Petersen GM, Hamilton SR, Green J, Jass J, Watson P, Lynch HT, Trent JM, de la Chapelle A, Kinzler KW, Vogelstein B. Mutations of mutS homolog in hereditary nonpolyposis colorectal cancer. Cell 1993; 75:1215–1225.
66. Parsons R, Li G-M, Longley MJ, Fang W-H, Papadopoulos N, Jen J, de la Chapelle A, Kinzler KW, Vogelstein B, Modrich P. Hypermutability and mismatch repair deficiency in RER+ tumor cells. Cell 1993; 75:1215–1225.
67. Cleaver JE. It was a very good year for DNA repair (minireview). Cell 1994; 76:1–4.
68. Hendrickson EA. Insights from model systems. Cell-cycle regulation of mammalian DNA double-strand-break repair. Am J Hum Genet 1997; 61:795–800.
69. Yu C-H, Oshima J, Wijsman EM, Nakara J, Tetsuro M, Piussan C, Matthews S, Fu Y-H, Mulligan J, Martin GM, Schellenberg GD, Werner's Syndrome Collaborative Group. Mutations in the consensus helicase domains of Werner syndrome gene. Am J Hum Genet 1997; 60:330–341.
70. Matsura S, Weemaes C, Smeets D, Takami H, Kondo N, Sakamoto S, Yano N, Nakamura A, Tauchi H, Endo S, Oshimura M, Komatsu K. Genetic mapping using microcell-mediated chromosome transfer suggests a locus for Nijmegen breakage syndrome at chromosome 8q21-24. Am J Hum Genet 1997; 60:1487–1494.
71. Louis-Bar D. Sur un syndrome progressif comprenant des telangiectasies capillaires cutanees et conjonctivales, a disposition naevoide et des troubles cerebelleux. Confin Neurol 1941; 4:32.
72. Gatti RA, Boder E, Vinters H, Sparkes RS, Norman A, Lange K. Ataxia-telangiectasia: an interdisciplinary approach to pathogenesis. Medicine 1991; 70:99–117.
73. Woods CG, Taylor AMR. Ataxia telangiectasia in the British Isles: the clinical and laboratory features of 70 affected individuals. Quart J Med 1992; 82:169–179.
74. Rosen SF, Harris NL. Case record of the Massachusetts General Hospital: a 30-year-old man with ataxia-telangiectasia and dysphagia. N Engl J Med 1987; 316:91–100.
75. Hecht F, Koler RD, Rigas DA, Dahnke GS, Case MP, Tisdale V, Miller RW. Leukemia and lymphocytes in ataxia-telangiectasia. Lancet 1966; II: 1193.
76. Saxon A, Stevens RH, Golde DW. Ataxia telangiectasia and chronic lymphocytic leukemia: evaluation of a T-lymphocyte clone with differentiation to both helper and suppressor T lymphocytes. N Engl J Med 1979; 300:700–704.

77. Sparkes RS, Como R, Golde DW. Cytogenetic abnormalities in ataxia telangiectasia with T-cell chronic lymphocytic leukemia. Cancer Genet Cytogenet 1980; 1:329–336.
78. Wake N, Minowanda J, Park B, Sandberg AA. Chromosomes and causation of human cancer and leukemia. XLVIII. T-cell acute leukemia in ataxia telangiectasia. Cancer Genet Cytogenet 1982; 6:345–357.
79. Petersen RD, Kelly WD, Good RA. Ataxia-telangiectasia: its association with a defective thymus, immunological-deficiency disease and malignancy. Lancet 1964; I:1189–1193.
80. Boder E. Ataxia-telangiectasia: an overview. In: Gatti RA, and Swift M, eds. Ataxia-telangiectasia: Genetics, Neuropathology and Immunology of a Degenerative Disease of Childhood. New York: Alan R Liss, 1985; 1–63.
81. Ammann AJ, Cain WA, Ishizaka K, Hong R, Good RA. Immunoglobulin deficiency in ataxia-telangiectasia. N Engl J Med 1969; 281:469–472.
82. Swift M, Morrell D, Cromartie E, Chamberlin AR, Skolnick MH, Bishop DT. The incidence and gene frequency of ataxia telangiectasia in the United States. Am J Hum Genet 1986; 39:573–583.
83. Swift M, Morrell D, Massey RB, Chase CL. Incidence of cancer in 161 families affected by ataxia-telangiectasia. N Engl J Med 1991; 325:1831–1836.
84. Athma P, Rappaport R, Swift M. Molecular genotyping show that ataxia-telangiectasia heterozygotes are predisposed to breast cancer. Cancer Genet Cytogenet 1996; 92:130–134.
85. Swift M, Reitnauer PJ, Morrell D, Chase CL. Breast and other cancers in families with ataxia-telangiectasia. N Engl J Med 1987; 316:1289–1294.
86. FitzGerald MG, Bean JM, Hedge SR, Unsal H, MacDonald DJ, Harkin DP, Finkelstein DM, Isselbacher KJ, Haber DA. Heterozygote ATM mutations do not contribute to early onset of breast cancer. Nature Genet 1997; 15:307–310.
87. Hecht F, McCaw BK. Chromosome instability syndromes: In: JJ Mulvihill, RW Miller, JF Fraumeni, eds. Genetics of Human Cancer: Cancer Research and Therapy. Vol. 3. New York: Raven Press, 1977:105–113.
88. Kohn PH, Whang-Peng J, Levis WR. Chromosomal instability in ataxia telangiectasia. Cancer Genet Cytogenet 1982; 6:289–302.
89. Cohen MM, Simpson SJ. Absence of a clastogenic factor in ataxia telangiectasia lymphoblastoid cells. Cancer Genet Cytogenet 1980; 2:327–334.
90. Shaham M, Becker Y. The ataxia telangictasia clastogenic factor is a low molecular weight peptide. Hum Genet 1981; 58:422–424.
91. Hecht F, McCaw BK, Koler RD. Ataxia-telangiectasia—clonal growth of translocation lymphocytes. N Engl J Med 1973; 289:286–291.
92. Cohen MM, Shaham M, Dagan J, Shmueli E, Kohn G. Cytogenetic investigations in families with ataxia-telangiectasia. Cytogenet Cell Genet 1975; 15: 338–356.
93. Aurias A, Dutrillaux B. Probable involvement of immunoglobulin superfamily genes in most recurrent chromosomal rearrangements from ataxia telangiectasia. Hum Genet 1986; 72:210–214.

94. Croce CM, Isobe M, Palumbo A, Puck J, Ming J, Tweardy D, Erikson J, Davis M, Rovera G. Gene for alpha-chain of human T-cell receptor: location on chromosome 14 region involved in T-cell neoplasms. Science 1985; 227: 1044–1047.
95. Gatti RA, Berkel I, Boder E, Braedt G, Charmley P, Concannon P, Ersoy R, Foroud T, Jaspers NG, Lange K, Lathrop GM, Leppert M, Nakamura Y, O'Connell P, Patterson M, Salser W, Sanal O, Silver J, Sparkes RS, Susi E, Weeks DE, Wei S, White R, Yoder F. Localization of an ataxia-telangiectasia gene to chromosome 11q22-23. Nature 1988; 336:577–580.
96. Foroud T, Wei S, Ziv Y, Sobel E, Lange E, Chao A, Goradia T, Huo Y, Tolun A, Chessa L, Charmley P, Sanal O, Salman N, Julier C, Concannon P, McConville C, Taylor AMR, Shiloh Y, Lange K, Gatti RA. Localization of an ataxia-telangiectasia locus to a 3-cM interval on chromosome 11q23: linkage analysis of 111 families by an international consortium. Am J Hum Genet 1991; 49:1263–1279.
97. Gatti RA, Peterson KL, Novak J, Chen X, Yan-Chen L, Liang T, Lange E, Lange K. Prenatal genotyping of ataxia-telangiectasia. Lancet 1993; 342:376.
98. Hartwell L. Defects in a cell cycle checkpoint may be responsible for the genomic instability of cancer cells. Cell 1992; 71:543–546.
99. Kastan MB, Zhan Q, El-Deiry WS, Carrier F, Jacks T, Walsh WV, Plunkett BS, Vogelstein B, Fornace AJ Jr. A mammalian cell cycle checkpoint pathway utilizing p53 and GADD45 is defective in ataxia-telangiectasia. Cell 1992; 71:587–597.
100. Savitsky K, Sfez S, Tagle DA, Ziv Y, Sartiel A, Collins FS, Shiloh Y, Rotman G. The complete sequence of the coding region of the ATM gene reveals similarity to cell cycle regulators in different species. Hum Mol Genet 1995; 4:2025–2032.
101. Weemaes CMR, Hustinx TWJ, van Munster PJJ, Bakkeren JAJM, Taalman RDFM. A new chromosomal instability disorder: the Nijmegen breakage syndrome. Acta Paedatr Scand 1981; 70:557–564.
102. van der Burgt I, Chrzanowska KH, Smeets D, Weemaes C. Nijmegen breakage syndrome. J Med Genet 1996; 33:153–156.
103. Matsura S, Weemaes C, Smeets D, Takami H, Kondo N, Sakamoto S, Yano N, Nakamura A, Tauchi H, Endo S, Oshimura M, Komatsu K. Genetic mapping using microcell-mediated chromosome transfer suggests a locus for Nijmegen syndrome at chromosome 8q21-24. Am J Hum Genet 1997; 60:1487–1494.
104. Sawitsky A, Bloom D, German J. Chromosomal breakage and acute leukemia in congenital erythema and stunted growth. Ann Intern Med 1966; 65:487–495.
105. Bloom D. The syndrome of congenital telangiectatic erythema and stunted growth. J Pediat 1966; 68:103–113.
106. German J, Archibald R, Bloom D. Chromosomal breakage in a rare and probably genetically determined syndrome of man. Science 1965; 148:506–507.
107. German J. Bloom's syndrome. I. Genetical and clinical observations in the first twenty-seven patients. Am J Hum Genet 1969; 21:196–227.

108. German J. Bloom's syndrome: incidence, age of onset, and types of leukemia in the Bloom's Syndrome Registry. In: Bartsocas CS, Loukopoulos D, eds. Genetics of Hematological Disorders. Washington DC: Hemisphere, 1992: 241–258.
109. German J, Crippa LP, Bloom D. Bloom's syndrome. III. Analysis of the chromosome aberrations characteristic of this disorder. Chromosoma 1974; 48: 361–366.
110. German J, Schonberg S, Louie E, Chaganti RSK. Bloom's syndrome. IV. Sister chromatid exchanges in lymphocytes. Am J Hum Genet 1977; 29: 248–255.
111. German J, Takabe H. Bloom's syndrome. XIV. The disorder in Japan. Clin Genet 1989; 35:93–110.
112. Latt SA. Localization of sister chromatid exchanges in human chromosomes. Science 1974; 185:74–77.
113. Weksberg R, Smith C, Anson-Cartwright L, Maloney K. Bloom sydrome: a single complementation group defines patients of diverse ethnic origin. Am J Hum Genet 1988; 42:816–824.
114. Woodage T, Prasad M, Dixon JW, Selby RE, Romain DR, Columbano-Green LM, Graham D, Rogan PK, Seip JR, Smith A, Trent RJ. Bloom syndrome and maternal uniparental disomy for chromosome 15. Am J Hum Genet 1994; 55:74–80.
115. Ellis NA, Groden J, Ye TZ, Straughen J, Lennon DJ, Ciocci S, Proytcheva M, German J. The Bloom's syndrome gene product is homologous to RecQ helicases. Cell 1995; 83:655–666.
116. Straughen J, Ciocci S, Ye TZ, Lennon DN, Proytcheva M, Alhadeff B, Goodfellow P, German J, Ellis NA, Groden J. Physical mapping of the Bloom syndrome region by the identification of YAC and P1 clones from human chromosome 15 band q26.1. Genomics 1996; 35:118–128.
117. Auerbach AD, Sagi M, Adler B. Franconi anemia: prenatal diagnosis in 30 fetuses at risk. Pediatrics 1985; 76:794–800.
118. Auerbach AD. Franconi anemia diagnosis and the diepoxybutane (DEB) test. Exp Hematol 1993; 21:731–733.
119. Auerbach AD, Allen RG. Leukemia and preleukemia in Fanconi anemia patients: a review of the literature and report of the International Fanconi Anemia Registry. Cancer Genet Cytogenet 1991; 51:1–12.
120. Joenje H, Oostra AB, Wijker M, di Summa FM, van Berkel CGM, Rooimans MA, Ebell W, van Weel M, Pronk JC, Buchwald M, Arwert F. Evidence of at least eight Fanconi anemia genes. Am J Hum Genet 1997; 61:940–944.
121. Strathdee CA, Duncan AMV, Buchwald M. Evidence for at least four Fanconi's anemia genes including FACC on chromosome 9. Nature Genet 1992; 1:196–198.
122. Savoia A, Centra M, Ianzano L, Pio de Cillis G, Buchwald M, Zelante L. Molecular characterization of Fanconi anemia group C (FAC) gene polymorphisms. Mol Cell Probes 1996; 10(3):213–218.
123. Gschwend M, Levran O, Kruglyak L, Ranade K, Verlander PC, Shen S, Faure

S, Weissenbach J, Altay C, Lander ES, Auerbach AD, Botstein D. A locus for Fanconi anemia on 16q determined by homozygosity mapping. Am J Hum Genet 1996; 59(2):377–384.

124. Pronk JC, Gibson RA, Savoia A, Wijker M, Morgan NV, Melchionda S, Ford D, Temtamy S, Ortega JJ, Jansen S, Havenga C, Cohn RJ, deRavel TJ, Roberts I, Westerveld A, Easton DF, Joenje H, Mathew CG, Arwert F. Localisation of the Fanconi anemia complementation group A gene to chromosome 16q24.3. Nature Genet 1995; 11(3):338–340.
125. Whitney MA, Theyer M, Reifsteck C, Olson S, Smith L, Jakobs PM, Leach R, Naylor S, Joenje H, Grompe M. Microcell mediated chromosome transfer maps the Fanconi anemia group D gene to chromosome 3p. Nature Genet 1995; 11:341–343.
126. Poon PK, O'Brien RL, Parker JW. Defective DNA repair in Fanconi anemia. Nature 1974; 250:223–225.
127. Wunder E, Burghardt U, Lang B, Hamilton L. Fanconi's anemia; anomaly of enzyme passage through the nuclear membrane? Anomalous intracellular distribution of topoisomerase activity in placental extracts in a case of Fanconi's anemia. Hum Genet 1981; 58:149–155.
128. Wunder E. Further studies on compartmentalisation of DNA-topoisomerase I in Fanconi anemia and xeroderma pigmentosa families. Am J Hum Genet 1984; 34:781–793.
129. Auerbach AD. Fanconi anemia and leukemia: tracking the genes. Leukemia 1992; 6(suppl. 1):1–4.
130. Saito H, Grompe M, Neeley TL, Jakobs PM, Moses RE. Fanconi anemia cells have a normal gene structure for topoisomerae I. Hum Genet 1994; 93:583–586.
131. Lo Ten Foe JR, Rooimans MA, Bosnoyan-Collins L, Alon N, Wijker M, Parker L, Lightfoot J, Carreau M, Callen DF, Savoia A, Cheng NC, van Berkel CGM, Strunk MHP, Gille JJP, Pals G, Kruyt FAE, Pronk JC, Arwert F, Buchwald M, Joenje H. Expression cloning of a cDNA for the major Fanconi anaemia gene, FAA. Nature Genet 1996; 14:320–323.
132. Fanconi Anaemia/Breast Cancer Consortium. Positional cloning of the Fanconi anaemia group A gene. Nature Genet 1996; 14:324–328.
133. Werner O. Uber Kataraki in Verbindung mit Sklerodermie. Doctoral diss., Zkiel Univ. Kiel: Schmidt and Klaunig, West Germany, 1904.
134. Oppenheimer BS, Kugel VH. Werner's syndrome—a hereditary familial disorder with schleroderma, bilateral juvenile cataract, precocious graying of hair and endocrine stigmatization. Trans Assoc Am Physicians 1941; 49:358.
135. Thannhauser SJ. Werner's syndrome (progeria of the adult) and Rothmund's syndrome: two types of closely related heredo-familial atrophic dermatosis with juvenile cataracts and endocrine features. A critical study with five new cases. Ann Intern Med 1945; 23:559.
136. Epstein CJ, Martin GM, Schultz AL, Motulsky AG. Werner's syndrome: a review of its symptomatology, natural history, pathologic features, genetics and relationship to natural aging process. Medicine 1966; 45:177–221.

137. Goto M, Tanimoto K, Horiuchi Y, Sasazuji T. Family analysis of Werner's syndrome: a survey of 42 Japanese families with a review of the literature. Clin Genet 1981; 19:8–15.
138. Hoehn H, Bryant EM, Au K, Norwood TH, Boman H, Martin GM. Variegated translation mosaicism in human skin fibroblast cultures. Cytogenet Cell Genet 1975; 15:282–298.
139. Salk D. Werner's syndrome: a review of recent research with an analysis of connective tissue metabolism, growth control of cultured cells and chromosomal aberrations. Hum Genet 1982; 62:1–15.
140. Scappaticci S, Cerimele D, Fraccaro M. Clonal structural chromosomal rearrangements in primary fibroblast cultures and in lymphocytes of patients with Werner's syndrome. Hum Genet 1982; 62:16–24.
141. Cleaver JE, Kraemer KH. Xeroderma pigmentosum. In: Stanbury JB, Wyngaarden JB, Frederickson DS, eds. Metabolic Basis of Inherited Disease. 4th ed. New York: McGraw-Hill, 1991:2949–2971.
142. Cleaver JE. Defective repair replication of DNA in xeroderma pigmentosum. Nature 1968; 218:652–656.
143. Burk PG, Lutner MA, Clarke DD, Robbins JH. Ultraviolet light-stimulated thymidine incorporation in xeroderma pigmentosum in lymphocytes. J Lab Clin Med 1971; 77:759.
144. Cleaver JE. Xeroderma pigmentosum: variants with normal DNA repair and normal sensitivity to ultraviolet light. J Invest Dermatol 1972; 58:124–128.
145. Day RS III. Xeroderma pigmentosum variants have decreased repair of UV-damaged DNA. Nature 1975; 253:748–749.
146. Kaur GP, Athwal RS. Complementation of a DNA repair defect in xeroderma pigmentosum cells by transfer of human chromosome 9. Proc Natl Acad Sci USA 1989; 86:8872–8876.
147. Henning KA, Schultz RA, Sekhon GS, Frieberg EC. Gene complementing xeroderma pigmentosum group A cells maps to distal human chromosome 9q. Somat Cell Mole Genet 1990; 16:395–400.
148. Tanaka K, Miura N, Satokata I, Miyamoto I, Yoshida MC, Satoh Y, Kondo S, Yasui A, Okayama H, Okada Y. Analysis of a human DNA excision repair gene involved in group A xeroderma pigmentosum and containing a zinc-finger domain. Nature 1990; 348:73–76.
149. Cockayne EA. Dwarfism with retinal atrophy and deafness. Arch Dis Child 1946; 21:52.
150. Lehmann AR. Three complementation groups in Cockayne syndrome. Mutat Res 1982; 106:347–356.
151. Schweiger M, Auer B, Burtscher HJ, Hirsch-Kauffmann M, Klocker H, Schneider R. DNA repair in human cells: biochemistry of hereditary diseases Fanconi's anaemia and Cockayne syndrome. Eur J Biochem 1987; 165:235–242.
152. Jaeken J, Klocker H, Schwaiger H, Bellman R, Hirsch-Kaufmann M, Schweiger M. Clinical and biochemical studies in three patients with severe early infantile Cockayne syndrome. Hum Genet 1989; 83:339–346.

153. Marshall RR, Arlett CF, Harcourt SA, Broughton BA. Increased sensitivity of cell strains from Cockayne's syndrome to sister-chromatid-exchange induction and cell killing by UV light. Mutat Res 1980; 69:107–112.
154. Lehmann AR, Thompson AF, Harcourt SA, Stefanini M, Norris PG. Cockayne's syndrome: correlation of clinical features with cellular sensitivity of RNA synthesis to UV irradiation. J Med Genet 1993; 30:679–682.
155. Schmickel RD, Chu EHY, Trosko JE, Chang CC. Cockayne syndrome: a cellular sensitivity to ultraviolet light. Pediatrics 1977; 60:135–139.
156. Roberts JB. A child with double cleft of lip and palate, protrusion of the intermaxillary portion of the upper jaw and imperfect development of the bones of the extremities. Ann Surg 1919; 70:252–253.
157. Appelt J, Gerken H, Lenz W. Tetraphokomelie mit Lippen-Keifer-Gaumenspalte und Klitorishhypertrophie—ein Syndrom. Paediat Padeol 1966; 2:119–124.
158. Van Den Berg DJ, Francke U. Roberts syndrome: a review of 100 cases and a new rating system for severity. Am J Med Genet 1993; 47:1104–1123.
159. Satar M, Atici A, Bisak U, Tunali N. Roberts-SC phocomelia syndrome: a case with additional anomalies. Clin Genet 1994; 45:107–108.
160. Maserati E, Pasquali F, Zuffardi O, Buttita P, Cuoco C, Defant G, Gimelli G, Fraccaro M. Roberts syndrome: phenotypic variation, cytogenetic definition, and heterozygote detection. Ann Genet 1991; 34:239–246.
161. Hermann J, Feingold M, Tuffli GA, Opitz JM. A familial dysmorphogenetic syndrome of limb deformities, characteristic facial appearance and associated anomalies: the "pseudothalidomide" or "SC-syndrome." Birth Defects: Original Article Series 1969; 5:81–89.
162. Hall BD, Greenberg MH. Hypomelia-hypotrichosis-facial hemangioma syndrome. Am J Dis Child 1972; 123:602–604.
163. Tomkins DJ, Hunter A, Roberts M. Cytogenetic findings in Roberts-SC phocomelia syndrome(s). Am J Med Genet 1979; 4:17–26.
164. Fryns J, Goddeeris P, Moerman F, Herman F, van den Berghe H. The tetraphocomelia-cleft palate syndrome in identical twins. Hum Genet 1980; 53: 279–281.
165. Zergollern L, Hitrec V. Four siblings with Robert's syndrome. Clin Genet 1982; 21:1–6.
166. Romke C, Froster-Iskenius U, Heyne K, Hohn W, Hof M, Grzejszczyk G, Rausskolb R, Rehder H, Schwinger E. Roberts syndrome and SC phocomelia: a single genetic entity. Clin Genet 1987; 31:170–177.
167. German J. Roberts' syndrome. I. Cytological evidence for a disturbance in chromatid pairing. Clin Genet 1979; 16:441–447.
168. Louie E and German J. Robert's syndrome. II. Aberrant Y-chromosome behavior. Clin Genet 1981; 19:71–74.
169. Keppen LD, Gollin SM, Seibert JJ, Sisken JE. Roberts syndrome with normal cell division. Am J Med Genet 1991; 38:21–24.
170. Allingham-Hawkins DJ, Tomkins DJ. Heterogeneity in Roberts syndrome. Am J Med Genet 1995; 55:188–194.

171. Jabs EW, Tuck-Miller CM, Cusano R, Rattner JB. Studies of mitotic and centromeric abnormalities in Roberts syndrome: implications for a defect in the mitotic mechanism. Chromosoma 1991; 100:251–261.
172. Stioui S, Privitera O, Brambati B, Zuliani G, Lalatta F, Simoni G. First-trimester prenatal diagnosis of Roberts syndrome. Prenat Diag 1992; 12: 145–149.
173. Chamla Y. C-anaphases in lymphocyte cultures versus permature centromere division syndromes. Hum Genet 1988; 78:111–114.
174. Hulten M. Selective somatic pairing and fragility at 1q12 in a boy with common variable immunodeficiency. Clin Genet 1978; 14:294.
175. Sawyer JR, Swanson CM, Wheeler G, Cunniff C. Chromosome instability in ICF syndrome: formation of micronuclei from multibranched chromosomes 1 demonstrated by fluorescence in situ hybridization. Am J Med Genet 1995; 56:203–209.
176. Howell RT, Taylor AMR. Chromosome instability syndromes. In: Rooney DE, Czepulkowski BH, eds. Human Cytogenetics: A Practical Approach. Vol. II. Oxford: IRL Press, 1992.
177. Jeanpierre M, Turleau C, Aurias A, et al. An embryonic-like methylation pattern of classical satellite DNA is observed in ICF syndrome: implication of a defect in the mitotic mechanism. Chromosoma 1993; 100:251–261.
178. Muller HJ. The remaking of chromosomes. The Collecting Net—Woods Hole 1939; 13:181–195.
179. Shippen-Lenz D, Blackburn EH. Functional evidence for an RNA template in telomerase. Science 1990; 247;546–552.
180. Blackburn EH. Structure and function of telomeres. Nature 1991; 350:569–573.
181. Moyzis RK, Buckingham JM, Cram LS, Dani M, Deaven LL, Jones MD, Meyne J, Ratliff RL, Wu J-R. A highly conserved repetitive sequence, $(TTAGGG)_n$, present at the telomeres of human chromosomes. Proc Natl Acad Sci USA 1988; 85:6622–6626.
182. deLange T, Shiue L, Meyers R, Cox DR, Naylor SL, Killery AM, Varmus HE. Structure and variability of human chromosome ends. Mol Cell Biol 1990; 10:518–527.
183. Greider CW. Telomeres, telomerase and senescence. Bioessays 1990; 12(8): 363–369.
184. Morgan R, Jarzavbek V, Jaffe JP, Hecht BK, Hecht F, Sandberg A. Telomeric fusion in pre-T-cell acute lymphoblastic leukemia. Hum Genet 1986; 73: 260–263.
185. Riboni R, Casata A, Nardo T, Zaccaro E, Ferretti L, Nuzzo F, Mondello C. Telomeric fusion in cultured human fibroblasts as a source of genomic instability. Cancer Genet Cytogenet 1997; 95:130–136.
186. Wan TSK, Chan LC, Ngan HYS, Tsao S-W. High frequency of telomeric associations in human ovarian surface epithelial cells transformed by papilloma viral oncogenes. Cancer Genet Cytogenet 1997; 95:166–172.
187. Schmitt H, Blin N, Zanki H, Scherthan H. Telomere length variation in normal and malignant human tissues. Genes Chrom Cancer 1994; 11:171–177.

188. Oyashiki K, Oyashiki JH. Telomere dynamic and cytogenetic changes in human hematologic neoplasias: a working hypothesis. Cancer Genet Cytogenet 1997; 94:67–72.
189. Morin GB. The human telomere terminal transferase enzyme is a ribonucleoprotein that synthesizes TTAGGG repeats. Cell 1989; 59:521–529.
190. Harrington LA, Greider CW. Telomere primer specificity and chromosome healing. Nature 1991; 353:451–456.
191. Lingner J, Hughes TR, Shevchenko A, Mann M, Lundblad V, Cech TR. Reverse transcriptase motifs in the catalytic subunit of telomerase. Science 1997; 276:561–567.
192. Magenis RE, Hecht F, Lovrien EW. Heritable fragile site on chromsome 16: probable localization of haptoglobin locus in man. Science 1970; 170:85–87.
193. Robson EB, Polani PE, Dart SJ, Jacobs PA, Renwick JH. Probable assignment of the alpha locus of haptoglobin to chromosome 16 in man. Nature 1969; 223:1163–1165.
194. Sutherland GR, Hecht F. Fragile Sites on Human Chromosomes. New York: Oxford University Press, 1985.
195. Lejeune J, Dutrillaux B, Lafourcade J, Berger R, Abonyi D, Rethore MO. Endoreduduplication selective du bras long du chromosome 2 chez une femme et sa fille. C R Acad Sci Paris, 1968; 266:24–26.
196. Shaw MW. In: CB Lozzio, AI Chernoff, eds. Symposium on Human Cytogenetics. Birth Defects Evaluation Center, University of Tennessee, Knoxville, 1968.
197. Lubs HA. A marker X chromosome. Am J Hum Genet 1969; 21:231–244.
198. Giraud F, Ayme S, Mattei JF, Mattei MG. Constitutional chromosome breakage. Hum Genet 1976; 34:125–136.
199. Harvey J, Judge C, Wiener S. Familial X-linked mental retardation with an X chromosome abnormality. J Med Genet 1977; 14:46–50.
200. Kaiser-McCaw B, Hech F, Cadian JD, Moore BC. Fragile X-linked mental retardation. Am J Med Genet 1980; 7:503–505.
201. Hagerman RJ, Silverman AC. Fragile X syndrome: Diagnosis, Treatment and Research. Baltimore: Johns Hopkins University Press, 1991.
202. Martin JP, Bell J. A pedigree of mental defect showing sex-linkage. J Neurol Psychiatr 1943; 6:154–157.
203. Sutherland GR. Fragile sites on human chromosomes: demonstration of their dependence on the type of tissue culture medium. Science 1977; 197:265–266.
204. Sutherland GR. Heritable fragile sites on human chromosomes. I. Factors affecting expression in lymphocyte culture. Am J Hum Genet 1979; 31: 125–135.
205. Glover TW. FUdR induction of the X chromosome fragile site: evidence for the mechanism of folic acid and thymidine inhibition. Am J Hum Genet 1981; 33:234–242.
206. Uchida IA, Freeman VCP, Jamro H, Partington MW, Soltan HC. Additional evidence for fragile X activity in heterozygous carriers. Am J Hum Genet 1983; 35:861–868.

207. Fryns JP, Kleczkowska A, Kubien E, Petit P, Van den Berghe H. Inactivation patterns of the fragile X in heterozygous carriers. Hum Genet 1984; 65: 401–405.
208. Jacky PB, Dill FJ. Expression in fibroblast culture of satellited-X chromosome associated with familial sex-linked mental retardation. Hum Genet 1980; 53: 267–269.
209. Sherman SL, Jacobs PA, Morton NE, Froster-Iskenius U, Howard-Peeble PN, Nielsen KB, Partington MW, Sutherland GR, Turner G, Watson M. Further segregation analysis of the fragile X syndrome with special reference to transmitting males. Hum Genet 1985; 69:289–299.
210. Laird CD. Proposed mechanism of inheritance and expression of the human fragile X syndrome of mental retardation. Genetics 1987; 117:587–599.
211. Laird CD, Lamb MM, Thorne JL. Two progenitor cells for human oogonia inferred form pedigree data and the X-inactivation imprinting model of the fragile-X syndrome. Am J Hum Genet 1990; 46:696–719.
212. Hirst MC, Nakahori Y, Knight SJL, et al. Genotype prediction in the fragile X syndrome. J Med Genet 1991; 28:824–829.
213. Verkerk AJMH, Pierretti M, Sutcliffe JS, Fu Y-H, Kuhl DPA, Pizzuti A, Reiner O, Richards S, Victoria MF, Zhang F, Eussen BE, van Ommen G-JB, Blonden LA-J, Riggins GJ, Chastain JL, Kunst CB, Galjaard H, Caskey CT, Nelsen DL, Oostr BA, Warren ST. Identification of a gene (FMR-1) containing a CGG repeat coincident with a breakpoint cluster region exhibiting length variation in fragile X syndrome. Cell 1991; 65:905–914.
214. Oberle I, Rousseau F, Heitz D, Kretz C, Devys D, Hanauer A, Boue J, Bertheas MF, Mandel JL. Instability of a 550-base pair DNA segment and abnormal methylation in fragile X syndrome. Science 1991; 252:1097–1102.
215. Yu S, Pritchard M, Kremer E, Lynch M, Nancarrow J, Baker E, Holman K. Mulley JC, Warren ST, Schlessinger D, Sutherland G, Richards RI. Fragile X genotype characterized by an unstable region of DNA. Science 1991; 252: 1179–1181.
216. Webb T. Molecular genetics of fragile X: a cytogenetics viewpoint. Report of the Fifth International Symposium on X-Linked Mental Retardation, Strausbourg, France, 12–16 August 1991 (organizer Dr J-L Mandel). J Med Genet 1991; 28:814–817.
217. Warren ST. The expanding world of trinucleotide repeats. Science 1996; 271: 1374–1375.
218. Connor JM. Cloning the gene for fragile X syndrome: implications for the clinical geneticist. J Med Genet 1991; 28:811–813.
219. Hansen RS, Canfield TK, Lamb MM, Gartler SM, Laird CD. Association of fragile X syndrome with delayed replication of the FMR1 gene. Cell 1993; 73:1403–1409.
220. Selig S, Okumura K, Ward DC, Cedar H. Delineation of DNA replication time zones by fluorescence in situ hybridization. EMBO J 1992; 11:1217–1225.
221. Boggs BA, Chinault AC. Analysis of replication timing properties of human

X-chromosomal loci by fluorescence *in situ* hybridization. Proc Natl Acad Sci USA 1994; 91:6083–6087.

222. Subramanian PS, Nelson DL, Chinault CA. Large domains of apparent delayed replication timing associated with triplet repeat expansion at FRAXA and FRAXE. Am J Hum Genet 1996; 59:407–416.
223. Smeets HJM, Smits APT, Verheij CE, Theelen JPG, Willemsen R, van de Burgt I, Hoogeveen AT, Oosterwijk, Oostra BA. Normal phenotype in two brothers with a full FMR1 mutation. Hum Mol Genet 1995; 4:2103–2108.
224. Zhong N, Yang W, Dobkin C, Brown WT. Fragile X gene instability: anchoring AGGs and linked microsatellites. Am J Hum Genet 1995; 57:351–361.
225. Lugenbeel KA, Peier AM, Carson NL, Chudley AE, Nelson DL. Intragenic loss of function mutations demonstrate the primary role of FMR1 in fragile X syndrome. Nature Genet 1995; 10:483–485.
226. Sutherland GR. Heritable fragile sites on human chromosomes II. Distribution, phenotypic effects, and cytogenetics. Am J Hum Genet 1979; 31:136–148.
227. Jordan DK, Burns TL, Divelbiss JE, Woolson RF, Patil SR. Variability in expression of common fragile sites: in search of a new criterion. Hum Genet 1990; 85:462–466.
228. Michels VV. Fragile sites on human chromosomes: description and clinical significance. Mayo Clin Proc 1985; 60:690–696.
229. Craig-Holms A, Strong L, Goodacre A, Pathak S. Variation in the expression of aphidicolin-induced fragile sites in human lymphocyte cultures. Hum Genet 1987; 76:134–137.
230. Krawczun MS, Jenkins EC, Duncan CJ, Stark-Houck SL, Kunaporn S, Schwartz-Richstein C, Gu H, Brown T. Distribution of autosomal fragile sites in specimens cultured for prenatal fragile X diagnosis. Am J Med Genet 1991; 38:456–463.
231. Glover TW, Berger C, Coyle J, Echo B. DNA polymerase a inhibition by aphidicolin induces gaps and breaks at common fragile sites in human chromosomes. Hum Genet 1984; 67:136–142.
232. Sutherland GR, Parslow MI, Baker E. New classes of common fragile sites induced by 5 azacytidine and bromodeoxyuridine. Hum Genet 1985; 69: 233–237.
233. Yunis JJ, Soreng AL, Bowe AE. Fragile sites are targets of diverse mutagens and carcinogens. Oncogene 1987; 1(1):59–69.
234. Human Gene Mapping 11 (London Conference). Eleventh International Workshop on Human Gene Mapping. Cytogenet Cell Genet 1991; 58(1–2): 82–84.
235. Genome Data Base (GDB). http://www.gdb.org/gdb-bin/genera/genera/hgd/ Fragile Site? Nov. 1997.
236. Hecht F, Hecht BK. Fragile sites and chromosome breakpoints in constitutional rearrangements. I. Amniocentesis. Clin Genet 1984; 26:169–173.

237. Hecht F, Hecht BK. Fragile sites and chromosome breakpoints in constitutional rearrangements. II. Spontaneous abortions, stillbirths and newborns. Clin Genet 1984; 26:174–177.
238. Yunis JJ. The chromosomal basis of neoplasia. Science 1983; 221:227–236.
239. Rowley JD. Human oncogene locations and chromosome aberrations. Nature 1983; 301:290–291.
240. Yunis JJ, Soreng AL. Constitutive fragile sites and cancer. Science 1984; 226: 1199–1204.
241. Mitelman F, Mertens F, Johanson B. A breakpoint map of recurrent chromosomal rearrangements in human neoplasia. Nature Genet 1997 (special issue):417–474.
242. DeBraekeleer M, Smith B, Lin CC. Fragile sites and structural rearrangements in cancer. Hum Genet 1985; 69:112–116.
243. Wenger SL, Steele, MW, Boone LY, Lenkey SG, Cummins JH, Chen X-Q. Balanced karyotypes in six abnormal offspring of balanced reciprocal translocation normal carrier parents. Am J Med Genet 1995; 55:47–52.
244. Jordan DK, Patil SR, Divelbiss JE, Vemuganti S, Headley C, Waziri MH, Guyrll NJ. Cytogenetic abnormalities in tumors of patients with von Hippel-Lindau disease. Cancer Genet Cell Genet 1989; 42:227–241.
245. Egeli U, Karadag M, Tunca B, Ozyardimci N. The expression of common fragile sites and genetic predisposition to squamous cell lung cancers. Cancer Genet Cell Genet 1997; 95:153–158.
246. Sutherland GR, Richards RI. Simple tandem DNA repeats and human genetic disease. Proc Natl Acad Sci USA 1995; 92:3636–3641.
247. Warren ST. The expanding world of trinucleotide repeats. Science 1996; 271: 1374–1375.
248. Chung M-Y, Ranum LPW, Duvick LA, Servadio A, Zoghbi HY, Orr HT. Evidence for a mechanism predisposing to intergenerational repeat instability in spinocerebellar ataxia type I. Nature Genet 1993; 5:254–258.

APPENDIX: GLOSSARY OF TERMS

acentric fragment
Chromosomal fragment lacking a centromere.

adenovirus
Common virus in upper and lower respiratory infections in humans, determined to produce E1A proteins that are active transforming agents in vitro.

allele (=allelomorph)
Alternate forms of a gene found at the same locus on homologous chromosomes.

anaphase
Stage in mitosis at which the centromeres have divided and chromatids have begun migrating toward opposite poles of the dividing cell.

anticipation
increased risk for an affected offspring, for disease of increased severity, and/or earlier disease onset due to trinucleotide amplification in maternal meiosis. First discovered in fragile X syndrome, amplification occurs when the number of repeats of the trinucleotide sequence CGG is in the premutation or mutation range in one allele.

APC
Tumor-repressor gene on human chromosome 5 in adenomatous polyposis coli.

autosome
One of 22 pairs of chromosomes in humans or one of any pair of chromosomes in higher organisms that is not involved in sex determination.

autosomal dominant
Refers to an allele or trait on an autosome that is preferentially expressed in an individual who is heterozygous at that locus.

balanced rearrangement
Rearrangement within a chromosome or between chromosomes with no loss or gain of genetic material.

bp
Basepairs.

BrdU
The base analog, 5-bromodeoxyuridine.

C-terminal
The carboxyl end of a polypeptide, or protein.

cDNA
Single-stranded DNA sequence that is complementary to a messenger RNA (usually produced by reverse transcriptase).

centromere
Site on the chromosome to which spindle fibers attach during cell division.

checkpoint
Theoretical point in the cell cycle at which a lag in transition from one stage in the cell cycle to the next allows the cell to correct errors in DNA replication.

chromatid
Replicated DNA strand comprising one-half of a metaphase chromosome. Chromatids remain attached at the centromere until the centromere divides, at which time each chromatid becomes a chromosome, with one migrating to each of two daughter cells.

clastogenic
Causing chromosome breakage.

clonal
Of or pertaining to a clone. By definition, all cells in a clone arise from a single cell and have the same genetic constitution.

cM
CentiMorgan, a genetic unit of measurement equal to a frequency of 1% recombination between two genetic loci in meiosis.

common fragile site
Site on a chromosome which is intrinsically more fragile than other sites, manifest as a gap or break in the chromosome, usually requiring some form of induction to be expressed.

complementation group
Group of recessive mutations in different genes, each of which, if homozygous, results in loss of the same cellular function, but each of which, if combined by somatic cell fusion with any of the other mutant genes in the group, will restore normal cellular function.

consanguinity
Mating between two individuals who share one or more ancestors by descent, usually between first- or second-degree relatives such as cousins.

constitutional
Referring to a chromosome abnormality that is congenital, i.e., is present in some or all cells from birth, as opposed to arising in a somatic cell and proliferating as a neoplastic cell line.

contiguous gene syndrome
Syndrome that is recognizable as a collection of clinical abnormalities, all of which may not be manifested in the same individual, usually due to a small chromosomal deletion that may or may not be detectable microscopically, but which involves one or more genes that are adjacent to each other.

cross-linked
The two strands in a DNA double helix are cross-linked when attached to each other by one or more chemical bonds.

cytogenetic
Of or pertaining to chromosomes or the study of chromosomes.

deletion
Cytogenetically, refers to a microscopically visible loss of a chromosome segment.

dicentric
Cytogenetically, describes a chromosome with two functional centromeres.

diploid (=2n)
The *n* refers to one complete haploid set of chromosomes, the number present in spermatozoa or ova which have completed both meiotic cell divisions. The combination of maternal and paternal haploid sets results in a diploid cell with two complete sets of chromosomes, the number present in most somatic cells. In humans the diploid number is 46.

DNA ligase
Enzyme that joins two DNA molecules or nucleotides.

domain
Refers to the space occupied by a single chromosome in the interphase nucleus.

duplication
Cytogenetically, refers to a microscopically visible chromosome segment that is present in two copies.

early replicating
Chromosomal regions that are replicated in the first half of the S period.

Endoreduplicated
Refers to the process by which chromosomes go through two or more DNA replication cycles but fail to undergo cell division.

eukaryote
Refers to any organism that has a distinct separation of nuclear and cytoplasmic functions.

excision repair
Mechanism of DNA repair by which aberrant nucleotide pairs are removed or excised and replaced by the correct pair.

exonuclease
Enzyme which catalyzes the removal of nucleotides from the ends of a DNA molecule.

familial
Refers to a disease or trait that is more common in relatives than in the general population.

FMR1
The most common characterized mental retardation gene linked to fragile site on the X chromosome. Although the precise mechanism is not known, retardation presumably occurs when the gene is abnormally methylated and thus inactivated due to expansion of the trinucleotide sequence adjacent to the gene.

fragile site
Site on a chromosome that shows an increased tendency to break.

full mutation
Refers to the degree of expansion (number of trinucleotide repeats) that inevitably results in an abnormal phenotype such as mental retardation in fragile X syndrome present in hemizygous males. When the full mutation is present in a female fragile X carrier, one-half of her sons are likely to be affected.

G banding
The banding pattern obtained on chromosomes following mild treatment with trypsin and staining with any one of a number of Giemsa blood stains.

G1
Stage in the cell cycle in which the cell contains a $2n$ quantity of DNA. Although often referred to as the resting stage, in fact many chemical and physiologic events are occurring in preparation for the cell entering S phase.

G2
Stage in the cell cycle following S phase and preceding mitosis.

gene carrier
One who has the allele for a particular trait on at least one chromosome, but does not express it.

genome
The total cellular DNA carrying the genetic blueprint of an organism.

haploidy
In humans, a set of 23 chromosomes each contributed by the egg and sperm to make up a diploid chromosome complement of 46 chromosomes.

Hela cell
Refers to the immortal cell line established from a malignant cervical carcinoma in 1952.

helicase
DNA-dependent ATPase involved in the unwinding of the DNA double helix by the breaking of hydrogen bonding between complementary base pairs.

heterochromatin
Genetically inert chromatin corresponding to condensed dark-staining nuclear regions in interphase as well as to chromosome bands determined to contain highly repetitive DNA sequences by molecular techniques.

heteroduplex
Hybrid double-stranded DNA molecule formed by pairing between two different DNA molecules that are not completely complementary.

heterozygote
An individual who carries two different alleles at one particular locus on homologous chromosomes.

homogeneously staining region
A form of gene amplification in cancer cells where a light-staining chromosome band or bands become(s) visibly longer and uniformly staining and is usually associated with increased gene expression, because a larger number of gene copies are present.

homologous chromosomes (=homologs)
Pairs of chromosomes in a diploid cell, with one of each pair being contributed by each parent.

hypertriploid
Form of aneuploidy in which the basic chromosome complement is composed of three haploid sets of chromosomes (i.e., triploid) plus four copies of some chromosomes. In humans, a hypertriploid cell would be a cell with more than 69 chromosomes but less than 81 chromosomes.

in situ hybridization

Technique by which a fluorescence- or radionucleotide-labeled, single-stranded DNA sequence is hybridized with the complementary DNA sequence in intact cells or otherwise immobilized DNA on slides.

isochromosome

Duplication of an entire chromosome arm, thought to be due to a misdivision through the centromere. The two arms of an isochromosome are presumed to be genetically identical.

karyotype

Standardized arrangement of the chromosomes of a cell, by pairs and in a particular numerical and size order, usually starting with the largest autosome and ending with the sex chromosome complement.

kb

Kilobase or kilobase pairs (= 1000 nucleotides or nucleotide base pairs).

late replicating

DNA which is replicated in the second half of the S period of the cell cycle.

linked

Refers to genes that are on the same chromosome and undergo recombination less than 50% of the time. Genes that can be demonstrated to be on the same chromosome but undergo recombination 50% or more of the time are said to be syntenic.

loss of heterozygosity

Loss of genetic function of one or a pair of alleles by any combination of the following: loss or deletion of the chromosome, gene mutation, somatic recombination, or uniparental disomy.

lymphoblastoid

Pertaining to or having the characteristics of a transformed lymphocyte cell line.

M

Referring to mitosis or stage of centromere division in the cell cycle.

marker

Cytogenetically, refers to a chromosome, no part of which can be identified with regard to its origin.

Mb

Megabase: equals 1000 kilobases or 1,000,000 base pairs.

methylation

Covalent addition of a methyl ($—CH_3$) group to a nucleotide sequence of DNA.

microcell-mediated transfer

Transfer of a fraction of a genome from a micronucleus to a host cell with a complete diploid chromosome complement by fusion of the cell membranes of the microcell and host cell.

microsatellite
Category of satellite DNA that consists of small tandem repeat units that are usually two, three, or four basepairs long.

mosaicism
Cytogenetically, refers to two or more cell lines within the same individual that have different chromosome constitutions. Mosaicism usually is present in more than one tissue and may occur in all tissues, as opposed to clonal abnormalities which usually have a single focal origin.

motif
A three-dimensional structure that represents a common theme in multiple, different molecules.

N-terminal
Refers to the amino end of a polypeptide or protein.

near-diploid
In humans, a cell with more or less than the normal diploid number of 46 chromosomes (35–57 chromosomes).

obligate carrier
In genetics, one who, by elimination of any other possibilities, must be a carrier of a particular allele that is present in offspring and/or other relatives.

oncogene
A gene that usually serves a normal cell function, but when mutated may result in unrestrained cell division resulting in cancer.

p53
An oncogene on the short arm of human chromosome 17 that, when mutated and/or deleted in both homologs, depending on the tissue, can contribute to a variety of cancers including colon, breast, and other cancers.

painting probe
Labeled DNA library that is complementary to a particular portion or segment of a chromosome that can be used, by in situ hybridization, to determine the locations of all homologous chromosome sequences in metaphase chromosomes.

partial leucine zipper
Leucine zipper refers to a class of DNA-binding regulating proteins that have a helical structure with a leucine at regular intervals projecting in register on one side of the helix. Basic amino acids (arginine, glutamine, and aspartic acid) project uniformly on the opposite side. Protein binds to DNA in the form of a dimer. Dimer formation requires both leucine zipper and basic amino acid components.

PCR (=polymerase chain reaction)
Equisitely specific logarithmic amplification (10^6- to 10^9-fold) of short DNA sequences targeted by homologous DNA primers less than 50 (usually 20–24) base pairs long.

pedigree
Schematic representation of the transmission of a genetic trait or disorder through multiple generations.

premature centromere division
Separation or division of the centromere of a chromosome or chromosomes before the onset of anaphase.

phenotype
The physical and/or clinical manifestations of a particular genotype.

photoreactivation
Direct enzyme repair (removal) of cyclobutyl linkage between two adjacent thymidine residues requiring activation by ordinary visible light.

polyploidy
Multiples of a haploid set of chromosomes (see haploidy).

postreplicative repair
Repair of errors in newly synthesized nucleotide pairs detected by DNA repair mechanisms in the cell after the DNA replication has been completed.

premutation
Refers to the number of trinucleotide repeats (usually 50–200 in the fragile X gene locus, for example) that has a high risk of being amplified to a full mutation when passing through one or more parental meioses.

pyrimidine dimer
Caused by absorption of ultraviolet radiation by DNA, most frequently a fusion between two adjacent thymidine residues in the form of a cyclobutyl linkage.

rad3
Radiation-sensitive DNA repair gene in yeast that is related to helicases.

rare fragile site (=heritable fragile site)
Consistently inducible breaks at a particular locus on a chromosome that cannot be induced in the homologous chromosome by the same induction system unless the site is homozygous (i.e., inherited like any other hemizygous allele from both parents).

repetitive sequences
Polynucleotide DNA sequences that occur more than once within a genome.

retrovirus
A type of RNA virus that can reverse-transcribe its RNA into DNA which then inserts into the DNA of the host cell, where it encodes more retroviral RNA.

reverse transcriptase
Enzyme that synthesizes a DNA sequence using messenger RNA as the template.

ring chromosome
Cytogenetically, refers to a chromosome in the form of a ring, considered to be formed by a break at each end of a chromosome, followed by fusion of the two broken ends.

S
Refers to the period most of the DNA is synthesized in the cell cycle.

satellite
Cytogenetically, refers to a small accessory body composed of genetically inert material seen in human acrocentric chromosome attached to the short arm by a secondary constriction or stalk region that is typically nucleolar organizing in function. Satellite sequences also refer to highly repetitive nucleotide sequences that were originally isolated from the bulk of genomic DNA by virtue of their property of AT- or GC-richness and consequent difference in buoyant density when centrifuged in a cesium chloride gradient.

sister chromatid exchange
Exchange during G2 between a newly replicated DNA strand and the old (native strand) of sister chromatids such that when replication is complete, part of one daughter strand has both halves of the double helix composed of the old (native) strands, and the homologous segment of the other daughter strand has both halves of the double helix composed of the two newly replicated strands. Classic autoradiographic experiments showing switching of label from one chromatid to the other in a harlequin pattern due to such exchange was the first evidence of the semiconservative nature of DNA replication.

slipped-strand mismatch
Misalignment of complementary nucleotide sequences in the two halves of a DNA double helix due to a difference in the number of copies of a repeated sequence between the two strands.

somatic
Pertaining to cells which give rise to differentiated tissue, as opposed to germline cells which give rise to gametes.

somatic cell hybridization
The forming of a genetic hybrid by the fusion of somatic cell nuclei, frequently involving fusion of rodent and human cells, where the rodent cell line is deficient in some cellular function and the human cell line in another so that both characteristics can be used to select for hybrid cells. Since the nonessential human chromosomes are lost over time in such a hybrid, somatic cell hybridization has been a useful method for mapping genes to specific human chromosomes or chromosome segments.

somatic pairing
Pairing of chromosomes in somatic cells, proposed as a mechanism for certain phenomenon involving the exchange or cross-linking between homologous chromosomes in disorders such as Bloom syndrome.

Southern blot
Technique for transferring denatured DNA fragments from an agarose gel to a nitrocellulose filter on which they can be hybridized to a complementary labeled DNA probe.

splice site mutation
Mutation resulting in the aberrant removal of introns and joining of exons during RNA transcription.

tandem duplication
Duplication of a piece of chromosome or DNA in which the two copies are immediately adjacent to each other (end to end) and the order of genes is in the same direction in both copies.

telomerase
A complex of RNA primer and enzyme which specifically recognizes the $(TTAGGG)_n$ repeated sequence at the 5′ ends of a linear DNA strand or chromosome arm and extends the number of repeats, which are then thought to form a hairpin loop by special bonding between adjacent G strings. The addition of extra telomeric repeats is thought to facilitate complete replication of the lagging strand, cap the ends of the chromosome, and prevent a gradual diminution in length of the DNA molecule.

telomere
Theoretical structure at the end of a chromosome arm, proposed by Muller to prevent one chromosome arm from fusing with another. We now know the telomere to be a specific hexameric repeat that is recognized as a terminal sequence by the enzyme telomerase.

telomeric repeat
Specific repeated nucleotide sequence $(TTAGGG)_n$, located at the ends of chromosome arms.

tertiary DNA structure
Supercoiling and folding of the DNA, possibly related to its availability or unavailability for transcription to messenger RNA.

Topoisomerase
Class of enzymes involved in uncoiling of DNA during replication. Two types of topoisomerase are known. Type 1 nicks the DNA, removes a coil, and reseals it. Type II causes superuncoiling by producing two breaks in the same strand, which allows the double helix in between to uncoil around itself.

transcript
Messenger RNA formed as a complementary copy of a specific gene sequence.

transfection
Incorporation of new or foreign genetic material into an existing eukaryotic genome.

translation
The process of translating the one-dimensional messenger RNA code into a three-dimensional amino acid sequence comprising a polypeptide or protein product.

trinucleotide repeat
Sequence of three nucleotide pairs repeated in a tandem array that when too long may result in abnormal gene expression in genetic disorders such as fragile X syndrome and Huntington disease.

triploid
A cell with three times the haploid number of chromosomes.

tumor-suppressor
Gene product that prevents the overexpression of another gene product that promotes cell proliferation.

upstream
Refers to sequences 5′ to the location under discussion in a DNA sequence.

X-inactivation
Process by which X chromosomes in excess of one per diploid set of chromosomes are genetically turned off in human and mammalian cells.

X-linked
Genes carried on the X chromosome. A father passes his X chromosome only to his daughters, while a mother passes one of her two X chromosomes to each child.

zinc finger
DNA-binding motif of a class of regulatory protein having a characteristic repeated domain with histidine and cysteine at regular intervals as the common ligands for the binding of zinc. Phenylalanine or tyrosine and leucine are also required at specific positions for DNA binding.

15
Cytogenetics of Breast Cancer

Sandra R. Wolman
Uniformed Services University of the Health Sciences, Bethesda, Maryland

I. REVIEW OF EXISTING DATA

A. Introduction

It is evident in other contributions to this volume that cytogenetic analysis has been critical to the localization, identification, and cloning of genes that define tumor development in many differentiated tissues. In contrast, breast cancer-specific cytogenetic associations have been elusive, hampered by the inherent heterogeneity, slow growth, and other features characteristic of breast cancer biology. Although almost 400 breast carcinomas were included in the last *Catalog of Chromosome Aberrations in Cancer* (1), many still were derived from effusion fluids and solid metastatic lesions, and many were incompletely karyotyped, reflecting the difficulties in analysis of primary breast carcinomas. The most prevalent aberrations reported have been gains of chromosomes 7, 8, 18, and 20; loss of chromosome 17; and structural rearrangements of 1q, 3p, and 6q. However, many other alterations, singly and combined, have been reported, and there is little agreement on identification of the earliest relevant events. At least some of the heterogeneity of karyotypic findings in breast cancer may be attributable to differing pathways of breast carcinogenesis. Until recently, little attention has been paid by cytogeneticists to the differences in disease stage and pathologic classification that may contribute to the marked variations observed in existing cytogenetic data. The existing complex and often confusing accumulated data have been reviewed recently (2–4). This chapter will focus largely on new findings and alternative cytogenetic approaches.

B. Accessibility of Samples

The overall incidence of breast cancer continues to rise. In addition, increasing awareness of the value of screening and increasing utilization of mammographic techniques contribute to the numbers of new neoplastic lesions of the breast, either invasive or in situ, that are detected. For the same reasons, the average size of resected breast lesions is declining. Sufficient tumor must be available for the primary purpose of histologic evaluation, and the next marker studies generally required are hormone receptor studies and ploidy determinations. Thus, relatively few tumors are sufficiently large that, after these determinations, remaining tumor is available for culture and for attempts at cytogenetic evaluation. In the same time frame, the development of in situ techniques has greatly increased access to cytologic samples from needle aspirates, to archival tissues, and to lesions of small size, thus enabling studies of preneoplastic diseases. The in situ methods, however, targeted to specific sites within the genome, do not provide the broad perspective that can be gained from complete karyotyping made possible with conventional cytogenetic assays.

C. Problems in Interpretation

A major advantage of the cytogenetic approach is the ability to derive information from individual cells without a preconceived commitment to one or more specific genetic lesions. Thus, unexpected aberrations are readily detectable, and cell-to-cell differences permit assessments of heterogeneity. Inherent limitations of conventional cytogenetics include a relatively low level of resolution, the requirement for mitotic cells, and problems deriving from selection during in vitro culture. Because of their dependence on cultured cells, the results are neither fully nor stably representative of the originating tumors in vivo. Difficulties in obtaining successful harvests from a high proportion of the tumors persist, and even the successes are tempered by small numbers of metaphases of variable quality. Nevertheless, it is the only method that provides an overview of the complexity of genetic alteration at the level of individual tumor cells, and it is therefore essential for observations of intratumor heterogeneity and clonal evolution.

Over the past few years, several laboratories have achieved limited success in obtaining clonally aberrant populations from relatively large numbers of breast cancers (5–7), either from direct harvests or, more often, from relatively brief culture periods (i.e., 4–13 days). Attempts to distinguish epithelial or malignant cell types by morphology in culture, and to support preferential growth of epithelial cells by minor modifications of medium, growth on specialized substrates, and varying modes of disaggregation (8,9) account for an increasing body of data on primary breast cancers. Nevertheless, harvests from significant numbers of cases fail to yield metaphase cells suitable for complete karyotyping, e.g., 30% in (7), although this baseline figure is not reported in many studies. Fewer than half of some series, 21% in (6) and 39% in (7), demonstrated clonal chromosome aberrations, and the remaining majority of cases showed nonclonal or diploid karyotypes. On the other hand, this pattern was reversed in the series of 125 breast cancers reviewed by Pandis et al. (5); 83% of their cases were clonally aberrant and only 17% were diploid. The differences in these studies may reflect the culture modifications used in each, but appear more likely to be influenced by time in culture. Direct harvests, although less often successful, provided larger numbers of metaphases with complex aberrations and more frequent evidence of polyploidy.

D. Data Summary and Update

The above-cited studies as well as other small series, despite certain similarities in their results, reveal a complex picture that is difficult to summarize. The proportion of cases that show clonal chromosomal aberrations varies widely, as does the frequency of breast cancers yielding metaphase cells suitable for analysis. Some cases and many cells in individual cases continue to demonstrate neither clonality nor nondiploidy. There is disagreement as to whether the earliest events are simple trisomies or specific rearrangements, such as t(1;16) or del(3p). The most commonly reported findings varied with the study. Longer average time in culture resulted in trisomies of chromosomes 7, 8, and 18 (6). Direct harvests and shorter cultures focused on recurrent structural anomalies, including der(1;16), i(1q), del6(q21), and del(1)(p22), with breakpoints clustered in the centromeric regions of chromosomes 1, 3, 11, 15, and 16, and in the short (p) arms of 7, 17, and 19 (7). Several of the major recurrent aberrations—der(1;16), i(1q), del(6q), +7, +18—reported in the third series (5) overlap with those from both of these studies. Additional recurrent aberrations, some associated clinical findings, and relevant gene localizations are listed in Table 1. Many investigators have also reported a great increase in the number of nonclonal aberrations among cells from breast cancer cultures in comparison to control cultures. Heterogeneity within, as well as between, tumors is the rule rather than the exception.

At least one recent study has added a clinical context to preexisting cytogenetic information. The review of Pandis and colleagues (5) amplified earlier work from the same group with the added benefit of relatively extensive correlation with clinical and histologic findings. They summarized data on a large series of primary breast cancers that included 114 invasive

Table 1 Genetic Alterations in Breast Cancer

Chromosome region	Associations	References
del(1p)	Poor prognosis	(47–50)
del(1)(p35-36)	Lymph node metasases; large tumor size; nondiploid tumor; common in DCIS; ?*PK58*	(49,51,52)
gain of 1 or 1q	36% of breast tumors; observed in both primary and metastatic disease	(24,53–57)
del(1)(q21-31)	?Tumor-suppressor gene	(57)
gain 1q41-44	?Oncogene	(57)
del(3p)	Early event; three LOH regions identified: 3p11-14, 3p14-23, 3p24-26	(42,58–60)
del(6)(q13-27)	No relationship with ER status	(61,62)
+7	Early event	(9,45,63)
del(7)(q31)	Early event, poor prognosis; metastatic potential, *TP53* alterations	(64–67)
+8/gain of 8q abn of 8q24	33% breast tumors; observed in both primary and metastatic disease; tumor proliferative anomaly (*CMYC*); invasive disease	(1,6,37,45, 55,68–70)
del(8)(p21-22)	Early event; invasive/metastatic potential	(71,72)
del(11)(p15.5)	Observed in 25% of breast tumors; familial association; independent of 11q loss	(73–75)
del(11)(q13)	Early prevent in sporadic tumors; ?MEN-1 TSG, invasion/metastatic potential	(76)
amplif 11q13	4–17% of breast tumors; poor prognosis; invasive; recurrence in node negative tumors; ?*CCND* and *EMS1*	(77–82)
del(11)(q23-24)	No correlation with PR status; aggressive postmetastatic course; poor prognosis; familial predisposition association; LOH of 17p13.1 with metastatic disease	(56,73,83–86)
LOH 13q12-13	Breast cancer susceptibility gene (*BRCA2*); autosomal dominant; early onset; male breast cancer	(72,87)
del(16)(q12.1) del(16)(q24.2) del(16)(q21)	Early, preinvasive event	(72,84)
del(16)(q22-23) del(16)(q24.3)	Invasive disease; distant metastases	(45,58,89–91)
del(17p)	Two sites identified; 17p13.3 and *TP53*; commonly early event; invasive disease; LOH with 17q and 13q; LOH 1p or amplification of 8p and progression	(30,72,92–98)

Table 1 Continued

Chromosome region	Associations	References
abn 17p13.1/*TP53*	Early event; more common in medullary, ductal invasive disease; mutations frequent in ER/PR-negative tumors; germline mutations in Li-Fraumeni syndrome; poor prognosis; increased chromosomal instability	(99–106)
amplif 17q12	HER-2/*neu* (*erb B-2*); 20% of invasive breast tumors; observed in all stages (initiation and early progression); high-grade tumors; ER/PR-negative tumors; short survival	(78,107–110)
LOH 17q21	Breast cancer susceptibility gene *BRCA1*; early onset; autosomal dominant; mutations observed in sporadic tumors; increased risk of ovarian cancer; possible second site associated with aberrant expression of plakoglobin	(111–113) (114)
amplif 17q21-24)	Distal to HER-2/*neu*	(24,25)
LOH 17q25	?Unknown tumor-suppressor gene	(115)
amplif 20q13.2	Aggressive tumors	(24,25,116)
Xq11-13	Germline mutation of androgen receptor gene; male breast tumors; Reifenstein syndrome	(117,118)

Partially reproduced (columns 1 and 2) with permission of the author and publisher from Slovak ML. Breast tumor cytogenic markers. (From Ref. 120.)

tumors and 11 carcinomas in situ (CIS). A number of potentially important clinical correlations emerged from this analysis. Higher modal chromosome numbers were associated with a younger age of onset, increasing tumor grade, and mitotic activity. Lobular carcinomas were invariably diploid or near-diploid. Similarly, increasing tumor grade, histologic classification, and mitotic activity were correlated with increased numbers of chromosomal aberrations. Lobular carcinomas and grade I or II ductal carcinomas were more likely to show normal karyotypes or simple alterations, findings that parallel their generally less aggressive clinical course. In contrast, poorly differentiated ductal tumors often had more than three aberrations, many in the form of unbalanced rearrangements. Polyclonality (the presence of two or more populations containing unrelated chromosome aberrations), which was found more often in invasive than in CIS lesions, was also associated with increasing tumor grade. None of the aberrations that are considered to be common, or "recurrent," in breast cancer [der(1;16), i(1q), del(1q), del(3p), del(6q), +7, +18, +20] was associated with specific clinicopathologic subgroups. An unexpected result was the observation that tumors with associated CIS showed more structural aberrations than those without a CIS component.

As these recurrent patterns of aberration begin to emerge, they raise the expectation that candidate loci for oncogenes and tumor-suppressor genes relevant to breast cancer will be identified. Some genes, including p53 (17p13.1), *BRCA1* (17q21), *BRCA2* (13q12-3), and *ATM* (11q22.3), are already recognized as important determinants in breast cancer predisposition for certain individuals. It is interesting that chromosomal aberrations identified in prophylactic mastectomies from women bearing such mutations have shown breaks in 1q and 3p (10), well-known regions of rearrangement in breast malignancy. The chromosomal findings indicate that other, as yet unrecognized, genes may be critical in the formation of sporadic cancers.

Difficulties pertaining to sufficient tissue retrieval, initial tissue processing, tissue culture methodology, and cytogenetic interpretation indicate that conventional karyotypic analysis of breast carcinomas and associated lesions is likely to remain an investigative rather than a diagnostic or prognostic approach. The newer methods of fluorescent in situ hybridization (FISH) and comparative genomic hybridization (GCH) appear at present to have greater potential for clinical utility.

II. FLUORESCENT IN SITU HYBRIDIZATION (FISH) AND COMPARATIVE GENOMIC HYBRIDIZATION (CGH)

A. The Methods; Advantages and Disadvantages

1. FISH

FISH and other in situ assays are more adaptable than conventional chromosome analyses to small sample sizes, do not require tissue culture facilities or expertise, and can yield results in a time frame acceptable for clinical determinations. These methods are applicable to single-cell isolates in smears, "touch preparations" (imprints) from fresh or frozen tissue, and find needle aspirates (FNA). Solid tissues may be studied in thin sections from paraffin-embedded or frozen tissue blocks, or after disaggregation to single-cell suspension by enzyme treatments. Studies using these methods are of particular interest to pathologists because of their applicability to cytologic samples and histologic sections, and because direct correlations can be made with morphology so that tissue localization of genetic events is feasible. Further correlations are possible by applying these methods in combination with markers of proliferation or with immunocytochemical markers. An extensive discussion of the types of probes, methodology, and other considerations can be found in recent review by Hopman et al. (11).

2. CGH

In contrast to the targeted approach of FISH, CGH permits a view of changes over the entire genome, as does conventional karyotyping. Unlike karyotyping, it is based on changes averaged over a mixture of tumor cells (and usually the tumor cell population is diluted by the presence of normal cells as well). CGH is therefore not a good method for detecting heterogeneity within a single tumor. It appears to provide a rough measure of the extent of genetic alteration within individual lesions, which could serve as a basis for comparison among tumors of similar origin or morphology. CGH at present is still a research tool, dependent on computerized, expensive instrumentation (e.g., cooled and computerized camera systems, complex software) and requiring considerable expertise. Essentially, the DNA extracted from a tumor is probe-labeled and mixed with equimolar amounts of control DNA (e.g., from normal lymphocytes) labeled with a different probe. The mixture of DNAs is then hybridized to normal metaphase spreads, which have been karyotyped by DAPI (4′,6-diamidino-2-phenylindole-dihydrochloride) banding. The tumor and normal DNA probes are distinguished by reactions with different fluorochromes. The nature and intensities (ratios) of fluroescent color reactions result from the competition of both DNA sources for the same target sequences in the normal metaphase spreads. The end result is a copy number karyotype representing the tumor DNA as compared with DNA from normal cells. It depicts the averaged genetic imbalances in all cells of the source sample of tumor DNA. The principles of CGH are depicted in Fig. 1. The origins of double minute chromosomes (dmin) or homogeneously stained regions (HSR) can be targeted with ease by CGH (12), as can imbalances of whole chromosomes or chromosome arms. At present, the detection of small regions (less than 10 Mb) of imbalance is difficult unless the tumor DNA differs by many copies, but technical improvements are expected to increase the sensitivity of the method.

In summary, CGH requires a baseline preparation of normal metaphase cells against which the two labeled DNA samples must be hybridized, with the karyotypic spreads located and individual chromosomes identified. The hybridization methods (and their technical difficulties) are essentially those described for FISH. Instrumentation and automation are needed for identification of the longitudinal fluorescent-labeled patterns and intensities, and for reading, recording, and interpreting the results. With sufficient stimulus, eventually the analytic tools could be developed to make CGH measurements of large numbers of samples feasible and practical in terms of relia-

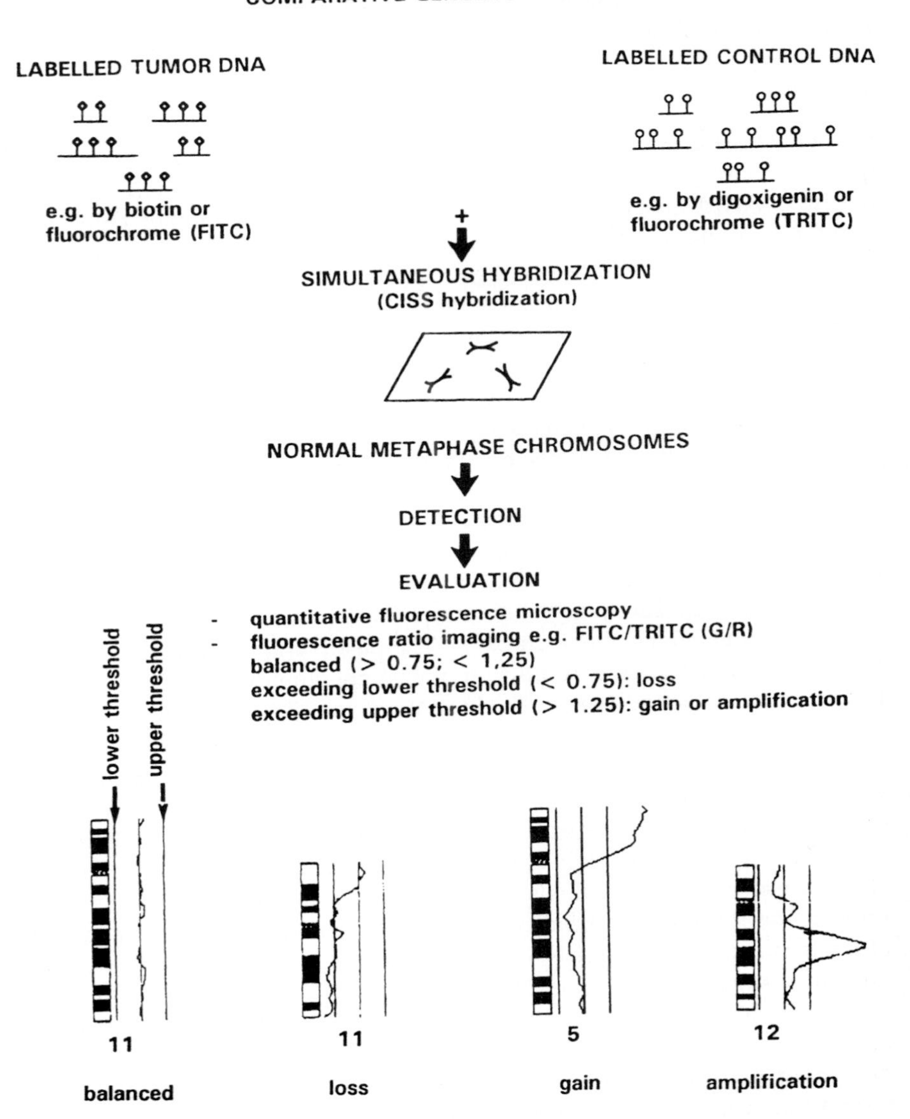

Figure 1 Diagrammatic representation of CGH. (Reproduced from Ref. 11 with permission of the authors and publisher.)

bility and cost. Such development will depend heavily on whether CGH results will prove to be uniquely informative in the diagnosis or, more likely, in the prognosis of some relatively common tumors such as breast cancers.

B. Emerging Clinical Applications

1. FISH in Different Stages of Malignancy

FISH has been an effective tool for the identification of tumor cells in pleural or peritoneal effusions, allowing for the distinction between tumor cells and reactive mesothelial cells by the identification of aneuploidy. One study using centromeric probes for chromosomes 7, 11, 12, 17, and 18 indicated that the detection of malignant cells was much improved and that in breast cancer, probes for 7 and 11 were the most successful combination for detection of aneuoploidy (13). The authors concluded that FISH was a valuable adjunct to routine cytologic examination in this context.

FISH has also been used to verify the metastatic spread of breast cancer to bone marrow (14). It was possible to detect changes in copy number in clusters of cytokeratin-positive cells in marrow aspirates after exposure to probes for centromere 17 or for the HER-*2/neu* oncogene (referred to below as *erb B-2*). Relatively few patients were studied and the results were consistent with known clinical data for some of them. The potential prognostic significance of this approach to the detection of residual micrometastatic disease deserves further study.

A broader attempt to establish the utility of FISH in primary breast cancer diagnosis utilized three centromeric probes applied simultaneously to FNA samples from 98 tumors (15). Material sufficient for analysis was obtained from 22 of 26 benign tumors (85%), from 2 of 2 phyllodes tumors, and 74 of 78 breast cancers (95%). None of the benign or phyllodes tumors was aneusomic. In striking contrast, all but 3 of the malignant tumors showed aneusomy for one or more of the probes. It should be noted that the criteria for establishing monosomy and polysomy in this study differed substantially from those of many of the other studies reviewed herein; the cutoff for monosomy was more than 15% of nuclei with one signal, and for polysomy it was more than 20% of nuclei with more than 3 signals per cell. If these criteria were stated correctly, then trisomic cells would not have been tabulated. The cutoff for polysomy is usually less stringent, and rarely excludes trisomies. The use of three simultaneous probes might be expected to result in higher frequencies of signal overlap, justifying the monosomic cutoff. Aneusomy for chromosome 1 was observed in 88% of cancers and for chromosome 11 in 66%. Approximately half of the tumors were aneusomic for chromosome 17. Polysomy was more common than monosomy in all cases. The frequent aberrations of chromosome 1 are concordant with karyotypic data, but are in disagreement with several other FISH studies. FISH was recommended as a tool in preoperative (FNA) diagnosis (15).

A similar study, which employed far more stringent cutoffs for aberration, was limited to invasive and metastatic tumors and examined aneusomy for chromosomes 11 and 17 (16). These investigators found abnormalities in essentially all of the 30 cases analyzed, and observed both inter- and intratumor heterogeneity. In several effusion fluids, FISH permitted an unambiguous identification of tumor cells.

Another group explored the use of FISH as a measure of aneuploidy, a parameter whose value in the prognostication for breast cancer has been hotly debated. Examination of 84 primary cancers and 20 metastatic lesions, using 8 different centromeric probes on disaggregated cells from fresh surgical samples, revealed polysomy for chromosomes 1, 3, and 17, less often for 7, 11, and 16, and rarely for 10 and 20 (17). Using a conservative threshold, 9 of 104 samples evaluated were classified as diploid, a lower figure than is usually obtained by conventional cytogenetics; in 8 cases there was aneusomy for only a single probe. The selective retention of multiple copies of specific chromosomes was suggested as a means to magnify oncogene amplification and, thus, to point out the potential value of FISH in prognosis.

2. FISH in Preinvasive Disorders

Perhaps the most important application of FISH has been in the studies of preinvasive breast lesions (Color Plate 15.2). Many such lesions are small and cannot be detected with great accuracy prior to surgery. Relatively few have been available for conventional cytogenetic analysis. An in situ approach permits a genetic analysis of small lesions that can be tightly correlated with cellular morphology. This has proved to be informative in studies of carcinoma in situ (CIS). Evidence of chromosome 17 copy-number alteration was prominent in both lobular and ductal CIS (LCIS, DCIS) (18) with less frequent losses and gains of the X chromosome and chromosomes 16 in DCIS and 8 and 7 in LCIS. Only 1 of 7 preneoplastic proliferative lesions showed numerical chromosome change. Considering the high proportion of CIS lesions that showed chromosome-specific losses or gains, it would have been of interest to study the concurrent invasive lesions in these cases. These authors found clear evidence that many of the alterations involved only subsets of intralesional cells, although no comment was made as to the geographic distribution of the aberrations. There was heterogeneity in the patterns of alterations among otherwise similar lesions: predominant patterns of losses of X and 17 in DCIS were associated with invasive disease and with (more pervasive) gains of other chromosomes.

A more extensive examination confined to DCIS employed centromeric probes for 12 chromosomes, in addition to probes for gene amplification (*erb B-2*) and 1p36 as a locus-specific check on the validity of numerical alterations detected by centromeric probes (19). Only 1 of 24 tumors failed to show chromosomal imbalance. The most common change was a gain of chromosome 17, but gains in chromosome 3 or 10 were found in approximately half of the cases. Losses were less frequent, the most common loss involving chromosome 18. Amplification of *erb B-2* was found in 4 of 15 tumors. Losses of centromeric 1 in three cases mirrored losses of the distal 1p locus. In several cases with tissues representing different areas within a tumor, or different tumors, the patterns of loss or gain supported an interpretation of clonal origin followed by divergent evolution. Thus, heterogeneity and clonal evolution were evident within preinvasive disease. In contrast, another study also reported polysomy for chromosome 17 with FISH; the gains appeared to correspond to increases in ploidy where data were available (20). However, no internal probe controls were used. One surprising observation was that gains found in DCIS were not noted in the concurrent invasive lesion. Both of these studies (19,20) noted the anomalous loss of loci by microsatellite analysis, concurrent with chromosomal gain.

Another study included data on some lesions even earlier in the postulated pathway to malignancy. Aneusomy for chromosomes 7 to 12, 17, 18, and X was examined in smears scraped from the cut surfaces of nonproliferative lesions (NPL) and atypical hyperplasia (AH) as well as carcinomas (CA) (21). The cutoffs for definition of true loss (15%) or gain (3%) were relaxed, perhaps contributing to the high frequency of positive findings. Numerical aberrations were found in 2 of 5 NPL, and in all of the 9 AH and 11 CA. Gains were more common than losses, and chromosome 17 was most often involved. The mean numbers of polysomic cells was 1.1 for NPL, 8.5 for AH, and 20.2 for CA, clearly indicating an increase of aberrations with disease progression. Gains of the X chromosome appeared to distinguish the different lesion groups reliably. The authors suggested that screening of needle aspirates by FISH may help to identify patients at increased risk for development of malignancy (21).

The greatest problem with the assessment of the FISH results for clinical utility is the disparities among the probes used, the standards (cutoff values) defined to evaluate aberration, the tissues examined, and the results obtained (see Table 2).

3. CGH

The clinical role for this new method is still uncertain. It is already clear that it is a powerful tool for detecting previously unknown genes that may

Table 2 Chromosome Alterations Detected by FISH

First author (reference)	Chromosomes assayed	Most common finding	Other common findings	Tissue(s) examined
Simpson (29)	1, 3, 7, 8, 16	+1, −16	All others	CIS, Ca, mets
Chen (20)	17	+17		DCIS
Murphy (19)	12 chromosomes	+17, −18	+3, +10, −1, −16	DCIS
Ichikawa (15)	1, 11, 17	+1	+11, +17	Ca
Sneige (21)	7–12, 17, 18, X	−17		Benign
		+9	−17, −18	AH
		+7	+8, +17	Ca (4 CIS)
Fiegl (16)	11, 17	+11	+17, −17	Ca, mets
Visscher (18)	7, 8, 16, 17, X	−17, −X	Several losses and gains	CIS
Afify (119)	8, 12	+8		5/11 ductal 1/5 DCIS No other Ca or benign
Shackney (17)	1, 3, 7, 10, 11, 16, 17, 20	+1	+3, +17	Ca, mets

play a role in breast cancer. An amplified region in 20q that was originally identified by this means now appears to contain several smaller regions of independent amplification (22,23). More important, amplification at 20q13.2 appears to correlate with poor prognosis (23).

Another new region of amplification at 17q22-24, distal to the *erb B-2* locus, has been identified (24,25), but the expected identification of known oncogenes by this technique has not been forthcoming. An important potential clinical application is implied by recent studies reporting that increases in the cumulative aberration frequency in primary node-negative breast carcinomas are correlated with poorer prognosis (26). Aberrations demonstrating losses rather than gains were significantly associated with tumor recurrence and poor survival. In this context it is interesting that cumulative aberration frequencies have not differed substantially between primary and metastatic tumors from the same individuals (27).

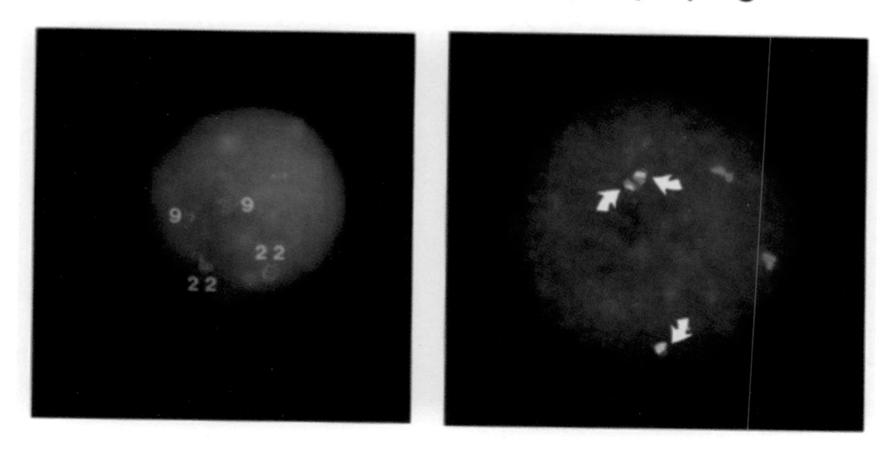

Figure 12.4 FISH with *bcr/abl* probes. Left, normal signals; right, abnormal fused signals.

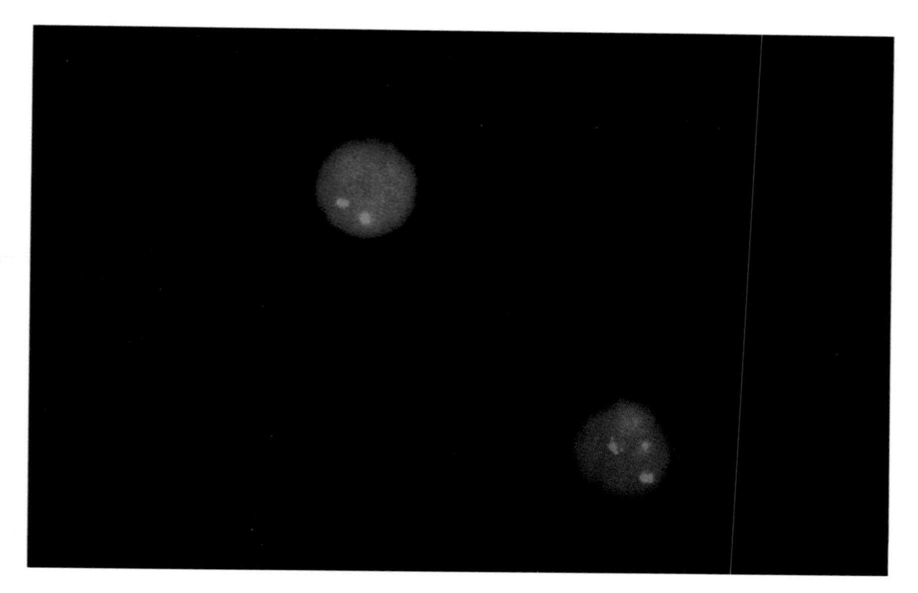

Figure 12.6 Case 2: FISH with MLL probe.

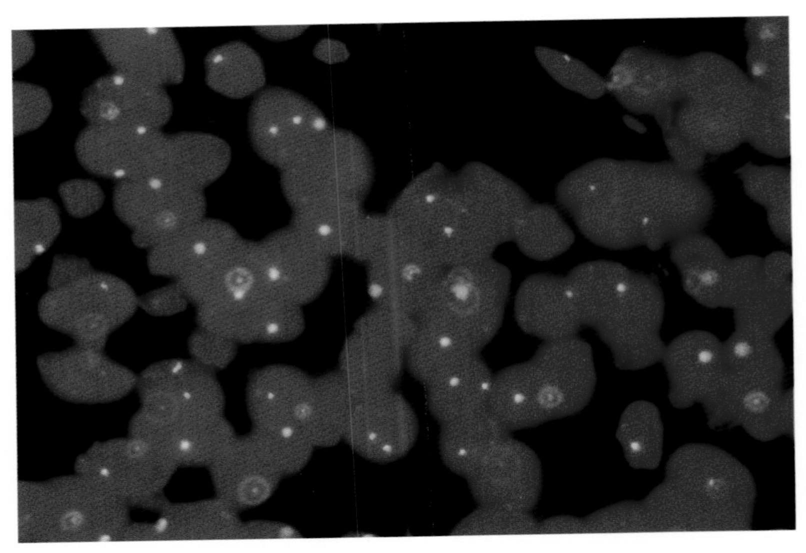

Figure 15.2 Tissue section (4 mm) of a typical hyperplasia of the breast probed with an α-centromeric probe for chromosome 17. Most cells in the field show only a single signal. In a few cells there is a halo around the signal, indicating that it is slightly out of the plane of section.

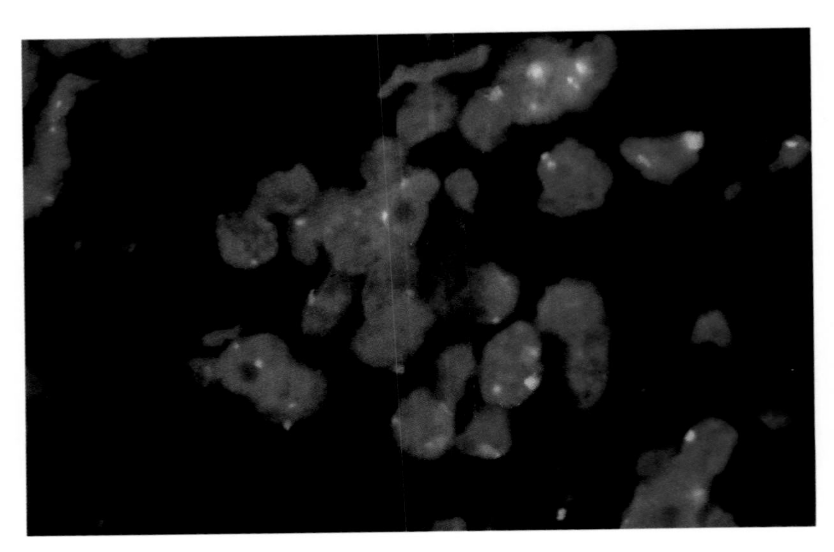

Figure 15.3 Tissue section (4 mm) from a male breast cancer probed for the X and Y centromeres; several cells contain two green (X) and a single red (Y) signals, suggesting preferential increases in X chromosome copy number within the tumor, although constitutional Klinefelter syndrome could be excluded.

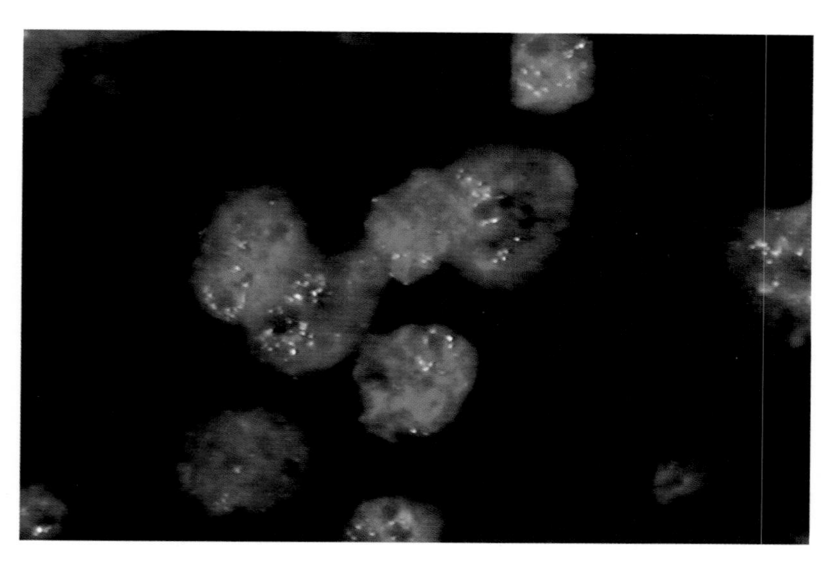

Figure 15.4 Imprint or "touch prep" of intact breast cancer cells showing approximately 10–20 green signals per cell; using a biotin-labeled probe for *erb B-2* to detect gene amplification.

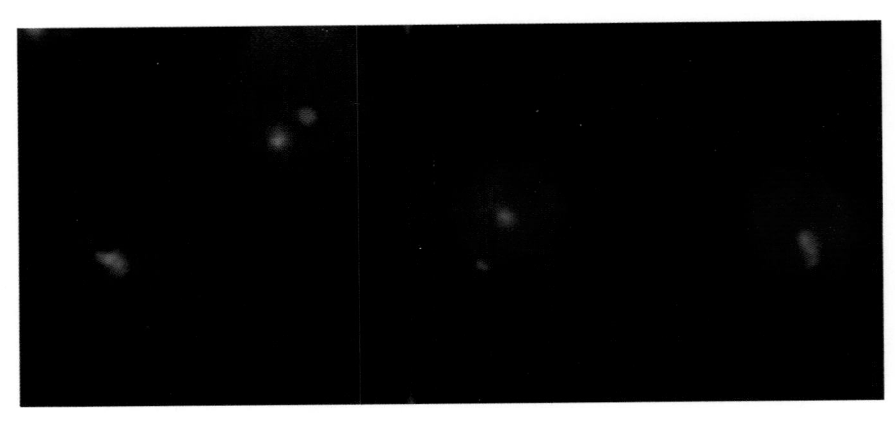

Figure 16.1 The Vysis dual-color X/Y probe for distinguishing host and donor cells in sex-mismatched bone marrow transplants.

III. CLINICAL UTILITY

A. Diagnosis

1. Comparison of Data Obtained by Different Methods

A direct comparison of data derived from different methods of analysis has been applied to the same tumors (28). Comparisons of molecular genetic with cytogenetic analyses emphasized that the two approaches can yield substantially different and complementary information. Cytogenetic results obtained from karyotyping (both from direct harvests and short-term cultures) and by FISH were similar; these were compared with loss of heterozygosity (LOH) for polymorphic markers on four chromosome arms. Studies of 25 breast carcinomas with clonal aberration revealed concordance in 70%, with significantly better matching from direct harvests than from the cultured cell karyotypes. The discrepancies were attributed to (a) balanced translocations without DNA loss, (b) sampling so that one or the other technique was not fully representative of the tumor, and (c) incomplete analysis of some clones.

On the other hand, direct comparisons among conventional cytogenetic analysis, FISH on the same preparations, and FISH on disaggregated cells from the paraffin-embedded tumor revealed significant similarities in results (29). The study included 15 breast tumors that included examples of DCIS, primary cancers, and metastatic lesions. From standard cytogenetic analyses, the most common gain was of chromosome 1, followed by gains of chromosomes 8, 7, 3, and 16. Loss of chromosome 16 was also observed in 4 cases. Concordant results were obtained with FISH on freshly harvested cells; an additional example of loss of chromosome 16 was detected. FISH on the deparaffinized cells was generally the same, but failed to detect some of the abnormal populations found in the fresh samples. One anomalous observation was that some low-grade or noninvasive lesions showed more numerical aberration than did some higher-grade lesions, although the numbers of lesions examined was small and the study was limited to five centromeric probes. Thus, similar to the larger study of Fiegl et al. (16), no consistent relationship was found between grade or stage and the extent of aneusomy. In 10 cases, a comparison with metastases revealed identical or a few additional clones. The authors stressed the value of FISH in permitting investigations of relatively early genetic events and in identifying dominant clones and their relation to metastases (29).

Concordance between FISH and flow cytometric data was found in 78% of a large series of primary breast cancers and metastases (17). Discordant results were attributed to a greater sensitivity of FISH in detecting near-diploid alteration, possible sampling errors, the high threshold adhered to in the study, and technical limitations of the FISH procedure. An increasingly adverse outcome was correlated with aneuploidy. The combined data were used to support the hypothesis that aneuploidy arises from a sequence of doubling of the chromosome complement followed by selective chromosome loss.

Dissociation between loss or imbalance and p53 mutation was observed in an investigation that used microdissection and PCR for mutation detection, and ISH (nonfluorescent labels) combined with immunohistochemical assessment of p53 and *erb B-2* expression on adjacent sections (20). Discordance between chromosome 17 copy number and p53 LOH was also noted. The cutoff values for ISH differed from those of other studies in paraffin sections, and the section thickness was not identified.

One recurrent observation, often presented anecdotally, is that LOH at a locus may be accompanied by polysomy for that chromosome (20,30). We suggested some years ago that "if a gene is truly significant in tumorigenesis, chromosomal loss should represent an extremely costly and, therefore infrequent, route to achieve gene loss because of the concomitant loss of other information" (31), and that poorly regulated chromosome replication may follow or even be driven by LOH (32). Uniparental disomy, or trisomy, may be the cell's attempt at compensation for such a loss.

2. Unusual Entities

The numbers of male breast cancers analyzed by conventional cytogenetics remains small. Various monosomies and trisomies have been reported. As with cases in females, the structural changes show no emerging dominant patterns. One recent case was trisomic for chromosomes 8 and 9 and showed rearrangement in 17q25 distal to genes known to be associated with female breast cancer (33). We demonstrated relative increases in X chromosome copy numbers in male breast cancers by FISH in tumors from males without evidence of Klinefelter syndrome (Color Plate 15.3).

Phyllodes tumors are an interesting subset of breast tumors because of their possibly dual cellular origin, containing both epithelial and mesodermal elements. There has been a continuing debate over whether both elements are neoplastic (35,36). Recently, there has been hope that separation techniques (differential sedimentation) or in situ methods will be able to resolve whether one or both cell types are chromosomally aberrant and therefore represent different stemlines, and whether both are potentially neoplastic. Similarly, there has been speculation on the relationship between stromal fibrous proliferation and the epithelial tumor component, based on similar findings of trisomies of chromosomes 8 and 20 in mesenchymal lesions and in breast epithelial malignancies (6,37,38). However, another form of fibrous proliferation, nodular fasciitis, revealed very different and mainly structural aberrations (39). A far more common mixed tumor, the fibroadenoma, is sometimes characterized by different clonal aberrations in the epithelial and connective tissue elements (40).

B. Cytogenetic Complexity and Prognosis

The correlation between increased complexity of chromosomal aberrations and poor clinical prognosis in breast cancer complements the observations made for other malignancies. It provides the theoretical justification for attempts to use CGH as a relative measure of quantitative cytogenetic aberration in order to distinguish different prognoses among tumors of like phenotype. Some FISH studies (e.g., 21) have also demonstrated increased complexity of numerical aberrations, which in the context of preinvasive disease could possibly serve as a marker of risk for the development of malignancy.

The frequency and significance of dmin or HSRs (both evidence of gene amplification) remain controversial (3). However, the identification of both forms of aberration and localization of amplified genes with a tumor cell population are ideally performed with FISH when suitable probes are available. Amplification of the *erb B-2* oncogene (Color Plate 15.4) has been associated with poor prognosis in breast cancer. New genes, whose significance in diagnosis or prognosis is as yet undetermined, have been partially localized by CGH and may soon be accessible for further examination.

IV. TUMOR BIOLOGY

A. Heterogeneity

Heterogeneity among and within breast cancers is evident from differences in proliferation rates, cellular morphology and structural organization, antibody reactivity, and protein synthesis. It is abundantly clear that heterogeneity of chromosomal aberrations is also a prominent feature of breast cancers, when data can be gathered from in situ approaches, by regional assessment, or by microdissection techniques. Much of the data from FISH studies reveal clear evidence of regional heterogeneity within individual tumors. Although CGH studies do not permit examination of intratumor heterogeneity, the interpretation that relative increases in intratumor aberration may have arisen from genetic instability and heterogeneity is widely accepted.

B. Definition of Aberrations in Premalignant Lesions

One group of investigators has successfully studied several proliferative breast lesions whose role in the pathway to malignancy is obscure and where the relative risk of emergency of cancer is minor or not increased (40,41). They have reported karyotypic abnormalities in fibroadenomas and other forms of proliferative breast disease. Fibroadenomas are of particular interest because of a reported two- to threefold increased risk. Nine of 50 cases were clonally aberrant—one with multiple, cytogenetically unrelated clones, and others with single abnormal clones (4 balanced translocations, 1 complex rearrangement, 1 numerical, 1 ring chromosome, and 2 with both numerical and structural changes). In all cases, the aberrant cells were mosaic with a population of normal diploid cells. After review of an additional 24 published aberrant cases, the recurrent aberrations that the authors felt could be attributed to epithelial rather than mesenchymal proliferations were del(1)(q12), der (1;16)(10;p10), and interstitial 3p deletions (41).

C. Chromosomal Specificity

Specificity of aberrations and the detection of the earliest chromosomal events in breast cancer have remained elusive. As is evident from Table 1, there are many candidates—perhaps too many. It can be argued that the 3p aberrations (37,42) found in a subset of breast carcinomas are relevant because they have also been noted in several benign lesions, including epithelial hyperplasias (40), fibroadenomas (41), and even a hamartoma (43). However, these lesions have not been clearly identified as precursors for breast cancer. Data from the more generally recognized precursor lesions of AH and DCIS are derived largely from FISH studies and focused on chromosome enumeration. With increasing use of locus-specific probes, the appropriate investigations may be launched, but to safeguard against uniparental disomy or polysomy, these studies will need to be controlled by concomitant molecular assays.

D. Multiclonality

The finding of multiple cytogenetically aberrant clones in individual cases of breast cancer has been emphasized in the work of Pandis and collaborators (37,42,44). Earlier they reported that 43% of cases with abnormality are multiclonal, some with related clones and others with unrelated, apparently independent chromosomal aberrations. This work was extended to demonstrate geographic heterogeneity in different tumor quadrants. Chromosomally aberrant bimodality has comprised up to 25% of some series of cultured breast cancers (3) and was absent from others (45). However, when cases are characterized by extensive complex rearrangements, the identification of modality may be difficult. More recently, the investigators who have provided the most extensive documentation of polyclonality summarized their data and concluded that approximately half of all breast cancers are polyclonal (46). Similar results have been found for benign breast tumors, and the specificity of their aberrations is related to those of the malignant tumors. The preferred interpretation favors a multicellular origin for breast cancer, thus distinguishing it from many other tumor types (46).

The resultant questions are whether such polyclonality is truly bound to tumor origins or results from more complex and extensive evolutionary divergence, and whether other common cancers arise similarly but show a greater convergence of evolution. Which of these or other interpretations is correct is the subject of future investigations, but it is already clear that the problem of heterogeneity in breast cancer is substantially different from that in other common cancers.

ACKNOWLEDGMENTS

Partial support from PHS grants CA61217 and CA70923 from the National Cancer Institute, NIH, DHHS, and EDT-116 from the American Cancer Society is gratefully acknowledged.

REFERENCES

1. Mitelman F. Catalog of Chromosome Aberrations in Cancer. 5th ed. New York: Wiley-Liss, 1994.
2. Bieche I, Lidereau R. Genetic alterations in breast cancer. Genes Chromosom Cancer 1995; 14:227–251.
3. Slovak ML, Wolman SR. Breast cancer cytogenetics: clues to genetic complexity of the disease. Breast J 1996; 2:124–140.
4. Slovak ML. Breast tumor cytogenetic markers. In: Wolman SR and Sell S, eds. Human Cytogenetic Cancer Markers. Totowa, NJ: Humana Press, 1997: 111–149.
5. Pandis N, Idvall I, Bardi G, et al. Correlation between karyotypic pattern and clinicopathologic features in 125 breast cancer cases. Int J Cancer 1996; 66: 191–196.
6. Rohen C, Meyer-Bolte K, Bonk U, et al. Trisomy 8 and 18 as frequent clonal and single-cell aberrations in 185 primary breast carcinomas. Cancer Genet Cytogenet 1995; 80:33–39.

7. Steinarsdottir M, Petursdottir I, Snorradottir S, Eyfjörd JE, Ögmundsdottir HM. Cytogenetic studies of breast carcinomas: different karyotypic profiles detected by direct harvesting and short-term culture. Genes Chromosom Cancer 1995; 13:239–248.
8. Pandis N, Heim S, Bardi G, Limon J, Mandahl N, Mitelman F. Improved technique for cytogenetic analysis of human breast cancer. Genes Chromosom Cancer 1992; 5:14–20.
9. Pandis N, Heim S, Bardi G, Idvall I, Mandahl N, Mitelman F. Chromosome analysis of 20 breast carcinomas: cytogenetic multiclonality and karyotypic-pathologic correlations. Genes Chromosom Cancer 1993; 6:51–57.
10. Petersson C, Pandis N, Mertens F, et al. Chromosome aberrations in prophylactic mastectomies from women belonging to breast cancer families. Genes Chromosom Cancer 1996; 16:185–188.
11. Hopman AHN, Voorter CEM, Speel EJM, Ramaekers FCS. In situ hybridization and comparative genomic hybridization. In: Wolman SR, Sell S, eds. Human Cytogenetic Cancer Markers. Totowa, NJ: Humana Press, 1996:45–69.
12. Mohamed AN, Macoska JA, Kallioniemi O-P, et al. Extrachromosomal gene amplification in acute nonlymphocytic leukemia; definition by metaphase analysis and comparative genomic hybridization. Genes Chromosom Cancer 1993; 8:185–189.
13. Zojer N, Fiegl M, Angeler J, et al. Interphase fluorescence in situ hybridization improves the detection of malignant cells in effusions from breast cancer patients. Br J Cancer 1997; 75:403–407.
14. Müller P, Weckermann D, Reithmüller G, Schlimok, G. Detection of genetic alterations in micrometastatic cells in bone marrow of cancer patients by fluorescence in situ hybridization. Cancer Genet Cytogenet 1996; 88:5–16.
15. Ichikawa D, Hashimoto N, Hoshima M, et al. Analysis of numerical aberrations in specific chromosomes by fluorescent in situ hybridization as a diagnostic tool in breast cancer. Cancer 1996; 77:2064–2069.
16. Fiegl M, Tueni C, Schenk T, et al. Interphase cytogenetics reveals a high incidence of aneuploidy and intra-tumour heterogeneity in breast cancer. Br J Cancer 1995; 72:51–55.
17. Shackney SE, Singh SG, Yakulis R, et al. Aneuploidy in breast cancer: a fluorescence in situ hybridization study. Cytometry 1995; 22:282–291.
18. Visscher DW, Wallis TL, Crissman JD. Evaluation of chromosome aneuploidy in tissue sections of preinvasive breast carcinomas using interphase cytogenetics. Cancer 1996; 77:315–320.
19. Murphy DS, Hoare SF, Going JJ, et al. Characterization of extensive genetic alterations in ductal carcinoma in situ by fluorescence in situ hybridization and molecular analysis. J Natl Cancer Inst 1995; 87:1694–1704.
20. Chen T, Dhingra K, Sahin A, Sneige N, Hortobagyi G, Aldaz CM. Technical approach for the study of the genetic evolution of breast cancer from paraffin-embedded tissue sections. Breast Cancer Res Treat 1996; 39:177–185.
21. Sneige N, Sahin A, Ba MD, El-Naggar A. Interphase cytogenetics in mammographically detected breast lesions. Hum Pathol 1996; 27:330–335.

22. Tanner MM, Tirkkonen M, Kallioniemi A, et al. Increased copy number at 20q13 in breast cancer: defining the critical region and exclusion of candidate genes. Cancer Res 1994; 54:4257–4260.
23. Tanner MM, Tirkkonen M, Kallioniemi A, et al. Independent amplification and frequent co-amplification of three nonsyntenic regions on the long arm of chromosome 20 in human breast cancer. Cancer Res 1996; 56:3441–3445.
24. Kallioniemi A, Kallioniemi O, Piper J, et al. Detection and mapping of amplified DNA sequences in breast cancer by comparative genomic hybridization. Proc Natl Acad Sci USA 1994; 91:2156–2160.
25. Muleris M, Almeida A, Gerbault-Seureau M, Malfoy B, Dutrillaux B. Detection of DNA amplification in 17 primary breast carcinomas with homogeneously staining regions by a modified comparative genomic hybridization technique. Genes Chromosom Cancer 1994; 10:160–170.
26. Isola JJ, Kalioniemi OP, Chu LW, et al. Genetic aberrations detected by comparative genomic hybridization predict outcome in node-negative breast cancer. Am J Pathol 1995; 147:905–911.
27. Kuukasjarvi T, Karhu R, Tanner M, et al. Genetic heterogeneity and clonal evolution underlying development of asynchronous metastasis in human breast cancer. Cancer Res 1997; 57:1597–164.
28. Valgardsdottir R, Steinarsdottir M, Anamthawat-Jonsson K, Petursdottir I, Ögmundsdottir HM, Eyfjörd JE. Molecular genetics and cytogenetics of breast carcinomas: comparison of the two methods. Cancer Genet Cytogenet 1996; 92:37–42.
29. Simpson JF, Quan DE, Ho JP, Slovak ML. Genetic heterogeneity of primary and metastatic breast carcinoma defined by fluorescence *in situ* hybridization. Am J Pathol 1996; 149:751–758.
30. Matsumura K, Kallioniemi A, Kallioniemi O, et al. Deletion of chromosome 17p loci in breast cancer cells detected by fluorescence *in situ* hybridization. Cancer Res 1992; 52:3474–3477.
31. Wolman SR, Macoska JA, Micale MA, Sakr WA. An approach to definition of genetic alterations in prostate cancer. Diagn Mol Pathol 1992; 1:192–199.
32. Macoska JA, Micale MA, Sakr WA, Benson P, Wolman SR. Extensive genetic alterations in prostate cancer revealed by dual PCR and FISH analysis. Genes Chromosom Cancer 1993; 8:88–97.
33. Cavalli RC, Rogatto SR, Rainho CA, dos Santos MJ, Cavalli IJ, Grimaldi DM. Cytogenetic report of a male breast cancer. Cancer Genet Cytogenet 1995; 81:66–71.
34. Wolman SR, Sanford J, Ratner S, Dawson PJ. Breast cancer in males, DNA content and sex chromosome constitution. Modern Pathol 1995; 8:239–243.
35. Dietrich CU, Pandis N, Bardi G, et al. Karyotypic changes in phyllodes tumors of the breast. Cancer Genet Cytogenet 1994; 78:200–206.
36. Dal Cin P, Moreman P, DeWever I, Van Den Berghe H. Is l(1)(q10) a chromosome marker in phyllodes tumor of the breast? Cancer Genet Cytogenet 1995; 83:174–175.

37. Pandis N, Jin Y, Gorunova L, et al. Chromosome analysis of 97 primary breast carcinomas: identification of eight karyotypic subgroups. Genes Chromosom Cancer 1995; 12:173–185.
38. Rohen C, Bartnitzke S, Bullerdiek J, Bonk U. Trisomy 8 and 20 in desmoid tumors and breast cancer. Cancer Genet Cytogenet 1996; 86:92.
39. Birdsall SH, Shipley JM, Summersgill BM, et al. Cytogenetic findings in a case of nodular fasciitis of the breast. Cancer Genet Cytogenet 1995; 81:156–158.
40. Dietrich CU, Pandis N, Teixera MR, et al. Chromosomal abnormalities in benign hyperproliferative disorders of epithelial and stromal breast tissue. Int J Cancer 1995; 60:49–53.
41. Petersson C, Pandis N, Rizou H, et al. Karyotypic abnormalities in fibroadenomas of the breast. Int J Cancer 1997; 70:282–286.
42. Pandis N, Jin Y, Limon J, et al. Interstitial deletion of the short arm of chromosome 3 as a primary chromosome abnormality in carcinomas of the breast. Genes Chromosom Cancer 1993; 6:151–155.
43. Dietrich CU, Pandis N, Bardi G, et al. Rearrangement of chromosomal bands 3p13-14 in two hamartomas of the breast. Int J Oncol 1995; 6:559–561.
44. Teixeira MR, Pandis N, Bardi G, Andersen JA, Mitelman F, Heim S. Clonal heterogeneity in breast cancer: karyotypic comparison of multiple intra- and extra-tumorous samples from 3 patients. Int J Cancer 63:63–68.
45. Thompson F, Emerson J, Dalton W, et al. Clonal chromosome abnormalities in human breast carcinomas. I. Twenty-eight cases with primary disease. Genes Chromosom Cancer 1993; 7:185–193.
46. Heim S, Teixera MR, Dietrich CU, Pandis N. Cytogenetic polyclonality in tumors of the breast. Cancer Genet Cytogenet 1997; 95:16–19.
47. Hainsworth PJ, Raphael KL, Stillwell RG, Bennett RC, Garson OM. Rearrangement of chromosome 1p in breast cancer correlates with poor prognostic features. Br J Cancer 1992; 66:131–135.
48. Bieche I, Champeme M, Matifas F, Cropp CS, Callahan R, Lidereau R. Two distinct regions involved in 1p deletion in human primary breast cancer. Cancer Res 1993; 53:1990–1994.
49. Genuardi M, Tshira H, Anderson DE, Saunders GF. Distal deletion of chromosome 1p in ductal carcinoma of the breast. Am J Hum Genet 1989; 45: 73–82.
50. Bieche I, Champeme M, Merlo G, Larsen C, Callahan R, Lidereau R. Loss of heterozygosity of the *L-myc* oncogene in human breast tumors. Hum Genet 1990; 85:101–105.
51. Borg A, Zhang Q, Olsson H, Wenngren E. Chromosome 1 alterations in breast cancer: allelic loss on 1p and 1q is related to lymphogenic metastases and poor prognosis. Genes Chromosom Cancer 1992; 5:311–320.
52. Eipers PG, Barnoski BL, Han J, Carroll AJ, Kidd VJ. Localization of the expressed human p58 protein kinase chromosomal gene to chromosome 1p36 and a highly related sequence to chromosome 15. Genomics 1991; 11:621–629.

53. Micale MA, Visscher DW, Gulino SE, Wolman SR. Chromosomal aneuploidy in proliferative breast disease. Hum Pathol 1994; 25:29–35.
54. Pandis N, Heim S, Bardi G, Idvall I, Mandahl N, Mitelman F. Whole-arm t(1;16) and i(1q) as sole anomalies identify gain of 1q as a primary chromosomal abnormality in breast cancer. Genes Chromosom Cancer 1992; 5: 235–238.
55. Dutrillaux B, Gerbault-Seureau M, Zafrani B. Characterization of chromosomal anomalies in human breast cancer: a comparison of 30 paradiploid cases with few chromosome changes. Cancer Genet Cytogenet 1990; 49:203–217.
56. Trent J, Yang J, Emerson J, et al. Clonal chromosome abnormalities in human breast carcinomas II. Thirty-four cases with metastatic disease. Genes Chromosom Cancer 1993; 7:194–203.
57. Bieche I, Champeme M, Lidereau R. Loss and gain of distinct regions of chromosome 1q in primary breast cancer. Clin Cancer Res 1995; 1:123–127.
58. Hainsworth PJ, Raphael KL, Stillwell RG, Bennett RC, Garson OM. Cytogenetic features of twenty-six primary breast cancers. Cancer Genet Cytogenet 1991; 52:205–218.
59. Chen L, Matsumura K, Deng G, et al. Deletion of two separate regions on chromosome 3p in breast cancers. Cancer Res 1994; 54:3021–3024.
60. Sato T, Akiyama F, Sakamoto G, Kasumi F, Nakamura Y. Accumulation of genetic alterations and progression of primary breast cancer. Cancer Res 1991; 51:5794–5799.
61. Magdelenat H, Gerbault-Seureau M, Dutrillaux B. Relationship between loss of estrogen and progesterone receptor expression and of 6q and 11q chromosome arms in breast cancer. Int J Cancer 1994; 57:63–66.
62. Iwase H, Greenman JM, Barnes DM, Bobrow L, Hodgson S, Mathew CG. Loss of heterozygosity of the oestrogen receptor gene in breast cancer. Br J Cancer 1995; 71:448–450.
63. Pandis N, Teixeira MR, Gerdes A, et al. Chromosome abnormalities in bilateral breast carcinomas. Cancer 1995; 76:250–258.
64. Deng G, Chen L, Schott DR, et al. Loss of heterozygosity and *p53* gene mutations in breast cancer. Cancer Res 1994; 54:499–505.
65. Bieche I, Champeme M, Matifas F, Hacene K, Callahan R, Lidereau R. Loss of heterozygosity on chromosome 7q and aggressive primary breast cancer. Lancet 1992; 339:139–143.
66. Zenklusen JC, Bieche I, Lidereau R, Conti CJ. $(C\text{-}A)_n$ microsatellite repeat D7S522 in the most commonly deleted region in human primary breast cancer. Proc Natl Acad Sci USA 1994; 91:12155–12158.
67. Champeme M, Bieche I, Beuzelin M, Lidereau R. Loss of heterozygosity on 7q31 occurs early during breast tumorigenesis. Genes Chromosom Cancer 1995; 12:304–306.
68. Atkin NB, Baker MC. Are human cancers ever diploid—or often trisomic? Conflicting evidence from direct preparations and cultures. Cytogenet Cell Genet 1990; 53:58–60.

69. Theillet C, Adelaide J, Louason G, et al. *FGFR1* and *PLAT* genes and DNA amplification at 8p12 in breast and ovarian cancers. Genes Chromosom Cancer 1993; 7:219–226.
70. Zafrani B, Gerbault-Seureau M, Mosseri V, Dutrillaux B. Cytogenetic study of breast cancer: clinicopathologic significance of homogeneously staining regions in 84 patients. Hum Pathol 1992; 23:542–547.
71. Yaremko ML, Recant WM, Westbrook CA. Loss of heterozygosity from the short arm of chromosome 8 is an early event in breast cancers. Genes Chromosom Cancer 1995; 13:186–191.
72. Radford DM, Fair KL, Phillips NJ, et al. Allelotyping of ductal carcinoma *in situ* of the breast: deletion of loci on 8p, 13q 16q, 17p and 17q. Cancer Res 1995; 55:3399–3405.
73. Carter SL, Negrini M, Baffa R, et al. Loss of heterozygosity at 11q22-q23 in breast cancer. Cancer Res 1994; 54:6270–6274.
74. Negrini M, Rasio D, Hampton GM, et al. Definition and refinement of chromosome 11 regions of loss of heterozygosity in breast cancer: identification of a new region at 11q23.3. Cancer Res 1995; 55:3003–3007.
75. Winqvist R, Mannermaa A, Alavaikko M, et al. Refinement of regional loss of heterozygosity for chromosome 11p15.5 in human breast tumors. Cancer Res 1993; 53:4486–4488.
76. Zhuang Z, Merino MJ, Chuaqui R, Liotta LA, Emmert-Buck MR. Identical allelic loss on chromosome 11q13 in microdissected *in situ* and invasive human breast cancer. Cancer Res 1995; 55:467–471.
77. Schuuring E, Verhoeven E, vanTinteren H, et al. Amplification of genes within the chromosome 11q13 region is indicative of poor prognosis in patients with operable breast cancer. Cancer Res 1992; 52:5229–5234.
78. Gaffey MJ, Frierson HF, Williams ME. Chromosome 11q13, *c-erb*B-2, and *c-myc* amplification in invasive breast carcinoma: clinicopathologic correlations. Modern Pathol 1993; 6:654–659.
79. Tsuda HS, Hirohashi Y, Shimosato T, et al. Correlation between long-term survival in breast cancer patients and amplification of two putative oncogene-coamplification units: hst-1/int-2 and *c-erb*B2/ear-1. Cancer Res 1989; 49: 3104–3108.
80. Champeme M, Bieche I, Lizard S, Lidereau R. 11q13 amplification in local recurrence of human primary breast cancer. Genes Chromosom Cancer 1995; 12:128–133.
81. Borg A, Sigurdsson H, Clark GM, et al. Association of int-2/hst-1 coamplification in primary breast cancer with hormone-dependent phenotype and poor prognosis. Br J Cancer 1991; 63:136–142.
82. Brookes S, Lammie GA, Schuuring E, et al. Amplified region of chromosome band 11q13 in breast and squamous cell carcinomas encompasses three CpG island telomeric of *FGF3*, including the expressed gene *EMS1*. Genes Chromosom Cancer 1993; 6:222–231.
83. Hampton GM, Mannermaa A, Winquist R, et al. Loss of heterozygosity in sporadic human breast carcinoma: a common region between 11q22 and 11q23.3. Cancer Res 1994; 54:4586–4589.

84. Singh S, Simon M, Meybohm I, et al. Human breast cancer: frequent p53 allele loss and protein overexpression. Hum Genet 1993; 90:635–640.
85. Winqvist R, Hampton GM, Mannermaa A, et al. Loss of heterozygosity for chromosome 11 in primary human breast tumors is associated with poor survival after metastasis. Cancer Res 1995; 55:2660–2664.
86. Lindblom A, Sandelin K, Iselius L, et al. Predisposition for breast cancer in carriers of constitutional translocation 11q; 22q. Am J Hum Genet 1994; 54: 871–876.
87. Wooster R, Neuhausen SL, Mangion J, et al. Localization of a breast cancer susceptibility gene, *BRCA2*, to chromosome 13q12-13. Science 1994; 265: 2088–2090.
88. Cleton-Jansen A, Moerland EW, Kuipers-Dijkshoorn JJ, et al. At least two different regions are involved in allelic imbalanced on chromosome arm 16q in breast cancer. Genes Chromosom Cancer 1994; 9:101–107.
89. Tsuda H, Callen DF, Fukutomi T, Nakamura Y, Hirohashi S. Allele loss on chromosome 16q24.2-qter occurs frequently in breast cancers irrespectively of differences in phenotype and extent of spread. Cancer Res 1994; 54:513–517.
90. Lindblom A, Rotstein S, Skoog L, Nordenskjold M, Larsson C. Deletions on chromosome 16 in primary familial breast carcinomas are associated with development of distant metastases. Cancer Res 1993; 53:3707–3711.
91. Harada Y, Katagiri T, Ito I, et al. Genetic studies of 457 breast cancers. Clinicopathologic parameters compared with genetic alterations. Cancer 1994; 74:2281–2286.
92. Mackay J, Elder PA, Steel CM, Forrest APM, Evans HJ. Allele loss on short arm of chromosome 17 in breast cancers. Lancet 1988; ii:1384–1385.
93. Rosenberg C, Andersen TI, Nesland JM, Lier ME, Brogger A, Borresen A. Genetic alterations of chromosome 17 in human breast carcinoma studied by fluorescence *in situ* hybridization and molecular DNA techniques. Cancer Genet Cytogenet 1994; 75:1–5.
94. Merlo GR, Venesio T, Bernardi A, et al. Evidence for a second tumor suppressor gene on 17p linked to high S-phase index in primary human breast carcinomas. Cancer Genet Cytogenet 1994; 76:106–111.
95. Coles C, Thompson AM, Elder PA, et al. Evidence implicating at least two genes on chromosome 17p in breast carcinogenesis. Lancet 1990; 336:761–763.
96. Cornelis RS, Van Vliet M, Vos CBJ, et al. Evidence for a gene on 17p13.3, distal to *TP53*, as a target for allele loss in breast tumors without *p53* mutations. Cancer Res 1994; 54:4200–4206.
97. Radford DM, Fair K, Thompson AM, et al. Allelic loss on chromosome 17 in ductal carcinoma *in situ* of the breast. Cancer Res 1993; 53:2947–2950.
98. Cheickh MB, Rouanet P, Louason G, Jeanteur P, Theillet C. An attempt to define sets of cooperating genetic alterations in human breast cancer. Int J Cancer 1992; 51:542–547.
99. Borresen A, Andersen TI, Eyfjord JE, et al. *TP53* mutations and breast cancer

prognosis: particularly poor survival rates for cases with mutations in the zinc-binding domains. Genes Chromosom Cancer 1995; 14:71–75.
100. Marchetti A, Buttitta F, Pellegrini S, et al. *p53* mutations and histological type of invasive breast carcinoma. Cancer Res 1993; 53:4665–4669.
101. Andersen TI, Holm R, Nesland JM, Heimdal KR, Ottestad L, Borresen AL. Prognostic significance of *TP53* alterations in breast carcinoma. Br J Cancer 1993; 68:540–548.
102. Cunningham JM, Ingle JN, Jung SH, et al. p53 gene expression in node-positive breast cancer: relationship to DNA ploidy and prognosis. J Natl Cancer Inst 1994; 86:1871–1873.
103. Davidoff AM, Kerns BM, Iglehart JD, Marks JR. Maintenance of *p53* alterations throughout breast cancer progression. Cancer Res 1991; 51:2605–2610.
104. Hartwell L. Defects in a cell cycle checkpoint may be responsible for the genomic instability of cancer cells. Cell 1992; 71:543–546.
105. Yin Y, Tainsky MA, Bischoff FZ, Strong LC, Wahl GM. Wild-type *p53* restores cell cycle control and inhibits gene amplification in cells with mutant *p53* alleles. Cell 1992; 70:937–948.
106. Li FP, Fraumeni JF, Mulvihill JJ, et al. A cancer family syndrome in twenty-four kindreds. Cancer Res 1988; 48:5358–5362.
107. Slamon DJ, Gogolphin W, Jones LA, et al. Studies of the HER-2/*neu* proto-oncogene in human breast cancer and ovarian cancer. Science 1989; 244:707–712.
108. Seshadri R, Firgaira FA, Horsfall DJ, McCaul K, Setlur V, Kitchen P, for the South Australian Breast Cancer Study Group. Clinical significance of HER-*2/neu* oncogene amplification in primary breast cancer. J Clin Oncol 1993; 11:1936–1942.
109. Lonn U, Lonn S, Nilsson B, Stenkvist B. Prognostic value of *erb*-B2 and *myc* amplification in breast cancer imprints. Cancer 1995; 75:2681–2687.
110. Iglehart JD, Kraus MH, Langton BC, Huper G, Kerns BJ, Marks JR. Increased *erb*B-2 gene copies and expression in multiple stages of breast cancer. Cancer Res 1990; 50:6701–6707.
111. Futreal PA, Liu Q, Shattuck-Eidens D, et al. *BRCA1* mutations in primary breast and ovarian carcinomas. Science 1994; 266:120–122.
112. Merajver SD, Pham TM, Caduff RF, et al. Somatic mutations in the *BRCA1* gene in sporadic ovarian tumours. Nature Genet 1995; 9:439–443.
113. Miki Y, Swensen J, Shattuck-Eidens D, et al. A strong candidate for the breast and ovarian cancer susceptibility gene *BRCA1*. Science 1994; 266:67–71.
114. Aberle H, Bierkeamp C, Torchard D, et al. The human plakoglobin gene localizes on chromosome 17q21 and is subjected to loss of heterozygosity in breast and ovarian cancers. Proc Natl Acad Sci USA 1995; 92:6384–6388.
115. Kirchweger R, Zeillinger R, Schneeberger C, Speiser P, Louason G, Theillet C. Patterns of allele losses suggest the existence of five distinct regions of LOH on chromosome 17 in breast cancer. Int J Cancer 1994; 56:193–199.
116. Berns EMJJ, Klijn JGM, van Staveren IL, Portengen H, Foekens JA. Sporadic amplification of the insulin-like growth factor I receptor gene in human breast tumors. Cancer Res 1992; 52:1036–1039.

117. Wooster R, Mangion J, Eeles R, et al. A germline mutation in the androgen receptor gene in two brothers with breast cancer and Reifenstein syndrome. Nature Genet 1992; 2:132–134.
118. Lobaccaro J, Lumbroso S, Belon C. Male breast cancer and the androgen receptor gene. Nature Genet 1993; 5:109–110.
119. Afify A, Bland KI, Mark HFL. Fluorescent *in situ* hybridization assessment of chromosome 8 copy number in breast cancer. Breast Cancer Res Treat 1996; 38:201–208.
120. Wolman SR, Sell S, eds. Human Cytogenetic Cancer Markers. Totowa, NJ: Humana Press, 1997:111–149.

16
Cytogenetics in Transfusion Medicine

Carolyn Te Young
Rhode Island Blood Center and Brown University, Providence, Rhode Island

Hon Fong L. Mark
Brown University School of Medicine and Rhode Island Department of Health, Providence, and KRAM Corporation, Barrington, Rhode Island

I. INTRODUCTION

Transfusion medicine is a broad medical specialty that encompasses blood utilization, blood donation and collection, preparation of blood components, immunologic and genetic principles, regulatory compliance issues, and vast clinical considerations. These clinical considerations include obstetric and perinatal transfusion practices, neonatal and pediatric problems, hematopoietic stem and progenitor cell transplantation, including umbilical cord blood and peripheral blood stem cell transplantation, tissue banking and organ transplantation, surgical procedures, bleeding disorders, and both potentially noninfectious and infectious complications of blood transfusions (1). While there are various excellent texts devoted to the subject of transfusion medicine, in this chapter we will focus our discussion on transplantation, diseases that may benefit from transplantation, and the role of cytogenetics in assessing clinical status and monitoring engraftment.

II. BONE MARROW TRANSPLANTATION, PERIPHERAL BLOOD STEM CELL TRANSPLANTATION, AND UMBILICAL CORD BLOOD TRANSPLANTATION

The cells composing the hematopoietic system have a relatively fast turnover rate. Erythrocytes circulate for about 120 days, platelets circulate for about

10 days, and granulocytes survive for less than 10 h. Approximately 1×10^{11} blood cells are removed and replaced each day in a process called hematopoiesis. It is important that the human body possess the means by which its entire volume of blood can be replenished efficaciously throughout life. Stem cells are defined by their function (2). Hematopoietic stem cells are pluripotential cells capable of differentiating into all cellular components of the hematopoietic system. The erythrocytic, leukocytic, and megakaryocytic cell lines are all derived from the hematopoietic stem cell as shown in Table 1. This cell is the "mother of all blood cells" and has unlimited reproductive potential. Although stem cell markers have not been clearly identified in human cells, markers are known for progenitor cells, i.e., cells

Table 1 Hematopoietic System

"Mother" of all cells:	Pluripotent stem cells
A. Produce:	Multilineage progenitor cells
Myeloid:	Granulocytic/monocytic/erythrocytic/megakaryocytic
Which produce	Unilineage progenitor cells
Granulocytic	
Monocytic	
Erythrocytic	
Megakaryocytic	
Eosinophilic	
Basophilic	
Which become:	Mature circulating cells
Neutrophils	
Monocytes	
Erythrocytes	
Platelets	
Eosinophils	
Basophils	
B. Produce:	Multilineage progenitor cells
Lymphoid	
Which produce:	Unilineage progenitor cells
B-cell line	
T-cell line	
Which become:	Mature cells
B cells	
T cells	

that have differentiated to pursue a committed cell line. For example, the earliest recognizable cells pursuing B-cell development have been characterized by the presence of class II human leukocyte antigen (HLA-DR), Tdt, CD 34, and CD 19. HLA-DR is seen in 10–40% of cases of T-cell acute lymphoblastic leukemia (T-ALL), Tdt in virtually all cases, and CD 34 in 10–20% of the cases. The antigens CD 7, CD 5, and CD 2 are present on the blast cells of T-ALL. Because the antigens on their surface can be identified, stem cells and progenitor cells are in turn suspected to be present in bone marrow, peripheral blood, and umbilical cord blood repopulating the hematologic system. Hence, all three sources may be used for hematopoietic stem cell and progenitor cell collection and transplantation.

In bone marrow, stem cells comprise between 0.1 and 1.0 cell per 1 million nucleated bone marrow cells (3). They also produce other stem cells and generate additional mature cells that differentiate into committed cell lines. As shown in Table 1, hematopoiesis is a complex process. It involves nonhematopoietic cells called stromal cells, such as fibroblasts, endothelial cells, and adipocytes. These cells supply hematopoietic growth factors and membrane-bound attachment molecules (4). Granulocyte colony-stimulating factor (G-CSF) has been used successfully to mobilize stem cells and progenitor cells. These cells may then be collected by bone marrow aspiration with subsequent separation of the hematopoietic cells by apheresis. This product is then cryopreserved until infusion into the patient. Research on a new drug, Flt3 ligand, has resulted in dramatic increases in both dendritic cells and stem cells in peripheral blood (5).

Stem/progenitor cell processing, whether obtained by umbilical cord blood, peripheral blood, or bone marrow, may be manual or semiautomated. Cryopreservation requires a cryoprotectant, usually dimethyl sulfoxide (DMSO), and controlled-rate freezing with liquid or vapor phase nitrogen. Immunomodulation such as T-cell removal or tumor-cell purging may be performed by pharmacologic means, radioisotopes, cytokines, photoactivated dyes, immunologic methods, or various combinations of these methods (6,7).

Bone marrow transplantation (BMT) is one of the fastest-growing and most important forms of medical therapy. The first successful bone marrow transplant in the United States in 1968 marked the beginning of a new course of treatment for leukemia, aplastic anemia, and many other life-threatening illnesses. This treatment provided an option that never existed previously. The early BMTs were from related donors because the patient's best chance of locating a match was within his or her own family. In fact, the first transplant in the United States between unrelated individuals was performed only in 1973. The most recent estimate is that, on average, an individual has approximately a 30% chance of finding a match within his or her own family. This is due to the downsizing of the average American family. An individual has a 25% probability of having a match with each sibling. However, because the average American family has 2.1 children, the probability of matching a family member has been reduced.

The National Marrow Donor Program (NMDP) was established in 1987 to make marrow transplants from unrelated donors available to patients as a life-saving therapy. The NMDP network includes 102 donor centers, 109 collection centers, 102 transplant centers, 11 recruitment groups, and a coordinating center in Minneapolis, Minnesota. Based on available statistics as of November 1996, there were 2,453,678 donors registered in the NMDP. Since the inception of the NMDP, 4912 transplants have been performed. There are now over 2000 formal ongoing searches for patient/donor matches (8,9).

However, fewer than 13% of the nearly 12,000 patients who have initiated an NMDP search have actually received a bone marrow transplant (6). The vast majority of participants in blood, solid organ, and bone marrow donation are Caucasian. The NMDP's distribution of donor groups consists of 74.4% Caucasian, 9.1% African American, 8.2% Hispanic, 6.7% Asian, and 1.6% Native American (statistics courtesy of NMDP, 1996). This fact is important because HLA is known to differ in frequency among various racial groups (10–15). Due to this lack of minority representation in the registry, many non-Caucasian patients may not find a match through the registry. Because of these statistics, recruitment of minorities has become an important focus of the NMDP.

Autologous peripheral blood stem/progenitor cell (PBSC) transplant is an alternative stem/progenitor cell source that may be considered when a patient has a malignancy with marrow involvement. This option may also be considered when an autologous bone marrow transplantation is contraindicated due to damage to the posterior iliac crests after irradiation, e.g., fibrosis, or when the patient is unable to undergo general anesthesia. In fact, the frequency in the utilization of autologous PBSC collection has already surpassed the traditional BMT procedure. The advantages of PBSC include reduced treatment toxicity, reduced duration of aplasia, reduced collection risks to the patient or donor, reduced costs (i.e., no operating room time), an ability to coordinate collections with treatment, and the possibility of outpatient procedures (6). There are numerous PBSC collection devices (e.g., Fenwal CS-3000, COBE Spectra, Haemonetics V50+) which are based on the same principle of separation by centrifugation. The PBSCs exist in the mononuclear cell (MNC) fraction of the leukocytes collected by apheresis. The duration and volume of collection are determined based on the patient's leukocyte count, weight, height, and hematocrit. In general, and based on our experience, $1-2 \times 10^8$ MNC/kg body weight is collected per apheresis, with an average dose of $5-7 \times 10^8$ MNC/kg (6) and an average of 5×10^6 CD 34+ cells/kg. Allogeneic peripheral blood stem cell collection is performed in a manner similar to autologous stem/progenitor cell collection.

Utilization of cord blood stem/progenitor cells (CBSC) as a source for transplantable hematopoietic stem cells was first suggested by Edward Boyse in the early 1980s. The idea was supported by animal studies which showed that lethally irradiated mice could be fully rescued and their hematopoietic cells could be reconstituted by receiving fetal blood (16,17). The first cord blood transplant was performed in 1988 for a patient with Fanconi's anemia; the cord blood was provided by his newborn sibling's cord blood (6,17,18). More recently, transplantation of cord blood stem/progenitor cells has been encouraged due to its abundance, ease of collection, immediate availability, representation of ethnic diversity, and insurance for future needs (i.e., in the case of autologous use). Characteristics of CBSC include a high degree of proliferative potential, high concentration, an earlier stage of development in cord blood, immunologic immaturity or increased immunologic tolerance, which may allow engraftment despite slight tissue mismatches and may reduce the occurrence of graft-versus-host disease (GVHD) while maintaining a graft-versus-leukemia activity. The New York Blood Center (NYBC) established the first Placental Blood Program (PBP) in 1992. The NYBC approached the National Heart, Lung and Blood Institute with a proposal to test the feasibility of using placental and umbilical cord blood for unrelated stem/progenitor cell transplantation in 1990. The PBP has provided cord blood for 34 transplants (19). Canada began a program to make CBSCs available to expectant parents in 1997.

Placental blood is collected through the umbilical vein by cannulating the vein and draining the blood into a sterile transfusion bag, or by draining blood from the severed end of the umbilical cord into a beaker with an anticoagulant. Blood volumes from the umbilical cord after separation from the placenta are between 40 and 200 ml. The dose of cord blood is $1-3 \times 10^8$ MNC/kg of the recipient's body weight. Hydroxyethyl starch (HES) may be used to separate mononuclear cells (95% recovery) and granulocytes (80% recovery) and to reduce red cell volume and plasma (19). A 10% DMSO solution may be used as a cryoprotectant with a cooling rate of 2°C per minute. Storage of the hematopoietic progenitor cells should be at a temperature below −120°C (20). Currently, the storage viability for CBSCs appears to be indefinite. The recovery of hematopoietic stem/precursor cells from frozen blood after 5 years of storage was excellent in comparison to recovery of precursor cells with the same samples tested before freezing (21).

III. CLINICAL INDICATIONS

Clinical indications for hematopoietic stem/progenitor cell transplantation may be divided into allogeneic and autologous transplantation categories, with some overlap. Preparation, collection/processing, transplant, and supportive care differ somewhat between the allogeneic transplant and the autologous transplant. The donor in an allogeneic transplantation must be selected through screening for histocompatibility, general health, and infectious disease. The patient in both allogeneic and autologous transplants receives chemotherapy and/or radiation therapy. A conditioning regimen may include iodorubicin 14 $mg/m^2/d \times 3$ days, Ara-C 100 $mg/m^2/d \times 2$ days, and cyclophosphamide 60 mg/kg/d $\times$ 2 days in addition to total body irradiation 160 cGy bid $\times$ 2 days and single-fraction 160 cGy for a total dose of 800 cGy (D. Oblon, personal communication, 1997). Priming donor leukocytes may be accomplished using recombinant human granulocyte colony-stimulating factor (G-CSF) with optimal collection of normal donor PBSCs 5 days after mobilization with high-dose G-CSF (22,23). After the collection of hematopoietic cells, the product may be manipulated or left unmanipulated and cryopreserved directly.

Manipulation of hematopoietic cells includes purging or enriching a certain cell population (24). Infusion of stem/progenitor cells is performed as a "rescue" after high-dose chemotherapy in autologous transplantations. Supportive care following transplantation involves transfusion support and infection prevention/treatment in both allogeneic and autologous transplant patients. However, allogeneic transplant patients also bear the risk of GVHD, which must be balanced by immunosuppression. This risk of GVHD is partly offset by an apparent lower frequency of relapse, which may be attributed to a graft-versus-leukemia effect. Studies suggest that transplantation early in the course of certain diseases yield a greater chance for success than later in the disease course. Furthermore, when patients are stratified by disease and its stage, unrelated donor transplants have results similar to transplants from related donors. A list of diseases which may be treated with transplantation (25) is given in Table 2.

IV. THE ROLE OF CYTOGENETICS AND MOLECULAR BIOLOGY IN ASSESSING CLINICAL STATUS

The number of recurrent karyotypic abnormalities found in neoplasms of hematopoietic and lymphoid origin continues to grow rapidly (refer to Chapters 11 and 12). However, centers providing cytogenetic analyses of solid tissues are still relatively few. The cytogenetic study of solid tissues has

Table 2 Clinical Indications for Bone Marrow Transplantation

A. Allogeneic transplantation	
Malignant diseases	
Acute nonlymphocytic leukemia (ANLL)	
Acute lymphocytic leukemia (ALL)	
Chronic myelogenous leukemia (CML)	
Myelodysplastic and myeloproliferative syndromes	
Non-Hodgkin's lymphoma	
Multiple myeloma	
Aplastic conditions	
Aplastic anemia	
Fanconi anemia	
Congenital pure red cell aplasia	
Paroxysmal nocturnal hemoglobinuria	
Genetic diseases	
Red cell disorders	Thalassemia major
	Sickle cell anemia
Immune deficiencies	Severe combined immunodeficiency
	Wiskott-Aldrich syndrome
Storage disorders	Gaucher disease
	Mucopolysaccharidoses
	Leukodystrophies
Other	Osteoporosis
B. Autologous transplantation	
Malignant diseases	
Acute leukemias (ANLL, ALL)	
Non-Hodgkin's lymphoma	
Multiple myeloma	
Breast cancer	
Germ cell cancer	
Ovarian cancer	
Neuroblastoma	
Sarcomas: soft tissue and osteosarcomas	

been hampered by a number of factors (26–30). Among these factors are the lack of an adequate number of dividing cells, the suboptimal quality of the metaphase spreads, and poor banding results. The first consistent karyotypic abnormality, Philadelphia chromosome, was reported by Rowley using a new banding technique which identified a reciprocal translocation between chromosomes 9 and 22 (31). Mitelman alone has cataloged thousands of cases of aberrant chromosomal abnormalities associated with numerous diseases (32). Chromosomal abnormalities have been reported in nearly

100% successfully studied non-Hodgkin's lymphomas, with most clones having multiple abnormalities (33,34). Many translocations have major disease associations such as the t(8;14)(q24.1;q32) association with malignant lymphoma (small noncleaved cell type) and the t(14;18)(q32;q21.3) association with malignant follicular lymphoma, especially of the small cleaved type. Many of these recurrent chromosomal abnormalities have been identified using conventional cytogenetics such as GTG banding, CBG staining, and QFQ banding. Further details are discussed elsewhere in this volume.

The most direct cytogenetic method of gene localization has been the use of fluorescent in situ hybridization (FISH). This method allows direct visualization of the location of a probe on a chromosome. Chromosome enumeration probes are designed for the detection of numerical chromosomal abnormalities. Painting probes are designed for the detection of structural chromosomal abnormalities such as translocations. Probes may be biotin-labeled, conjugated with avidin, and then visualized by the use of anti-avidin antibodies and immunofluorescent staining with fluorescein on counterstained chromosomes. Two separate fluorescent dyes may be used to visualize two different probes (see Color Plate 16.1). Combinations of probes may be used to identify multiple translocations (35–37). The principles and techniques of FISH are discussed in Chapter 6. The practical applications of FISH are discussed in Chapter 17. However, newer molecular cytogenetic techniques, such as spectral karyotyping, have become available in certain specialized laboratories, and the need to perform several separate experiments may soon be obviated. These emerging molecular techniques are discussed in Chapter 18.

The hallmark of a malignant hematologic disorder is a neoplastic clone. The standard cytogenetic definition of a clone requires the detection of at least two metaphases with the same extra chromosome or the same structural abnormality and at least three metaphases that are missing the same chromosome (38–40). Gene rearrangement analysis can be used as an aid to diagnose and classify hematopoietic diseases, and it has the sensitivity to detect clones even when they represent as little as 1–2% of the cell population under study using deoxyribonucleic acid (DNA) hybridization (e.g., Southern blot) (41).

Restriction fragment length polymorphisms (RFLPs) and variable number tandem repeats (VNTRs) are segments of DNA of varying lengths among individuals. Since they are inherited from one's parents, they are very useful in establishing parentage and in monitoring BMT or PBSC transplant engraftment (Fig. 2). The RFLPs used in parentage testing are introns rather than exons, and thus are not coding "functional" genes. These RFLPs may also be used in engraftment studies to detect complete chimerism (100% donor DNA).

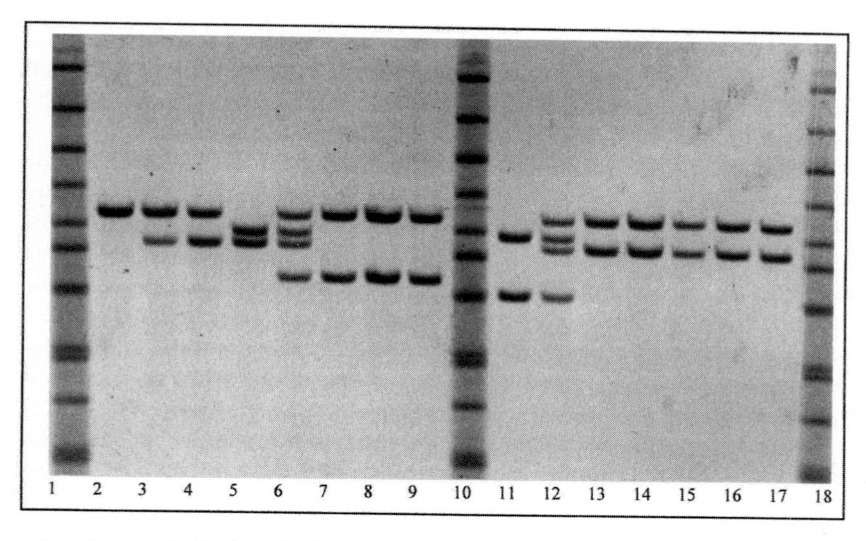

Lane 1- molecular weight ladder, lane 2- child, lane3- child and alleged father, lane 4- alleged father, lane 5- patient 1, pre-transplant, lane 6- patient 1, pre-transplant, and donor 1, lane 7- patient 1, post-transplant, lane 8- patient 1, post-transplant, lane 9- donor 1, lane 10- moleular weight ladder, lane 11- patient 2, pre-transplant, lane 12- patient 2, pre-transplant, and donor 2, lane 13- patient 2, post-transplant, lane 14- patient 2, post-transplant, lane 15- patient 2, post-transplant, mononuclear fraction post-Ficoll treatment, lane 16- patient 2, post-transplant, granulocyte fraction post-Ficoll treatment, lane 17- donor 2, lane 18- molecular weight ladder.

Figure 2 A representative RFLP membrane demonstrating a case of paternity inclusion and two cases of engraftment following peripheral blood stem/progenitor cell transplantation.

To detect gene rearrangements, DNA is extracted from clinical samples. The DNA is cut into fragments using restriction endonucleases that identify the sites flanking a portion of the DNA of interest. The use of these restriction endonucleases led to the term RFLP. These fragments of DNA are electrophoresed on an agarose gel. Probes designed to hybridize with complementary strands of single-stranded DNA are allowed to hybridize with the fragments. Detection may be performed with colorimetric, chemiluminescent, or radioactive systems. This same technique may also be used to monitor engraftment after transplants.

To monitor engraftment optimally following allogeneic transplantation, both the patient and the donor should possess nonidentical heterozygous RFLPs which can be detected prior to transplantation. These RFLPs may then be monitored to determine engraftment following transplantation. For example, if the patient's RFLPs following transplant are identical to the donor's RFLPs, engraftment has occurred. Figures 2 and 3 contain examples of patients who engrafted following peripheral blood stem/progenitor cell transplantation. Conversely, if the patient's RFLPs remain identical to the patient's pretransplant sample, engraftment has not occurred.

Polymerase chain reaction (PCR) is an ingenious molecular technique whereby segments of DNA may be replicated selectively (42). The concept of PCR was created by Kary Mullis during a moonlit drive in the spring of 1983, then refined and practiced months later. He was awarded the 1993 Nobel Prize in chemistry for his invention. Beginning with a single molecule of DNA, PCR can generate 100 billion similar molecules (43). Basically, PCR consists of a sequence of three steps: (a) denaturation of double-stranded DNA into single-stranded DNA, (b) annealing of two oligonucleotide primers or probes which match their target sequences perfectly, and (c) extension of the primer or probe using Taq polymerase, a thermostable DNA polymerase, and a mix of the four deoxyribonucleotide triphosphates. The paired primers are designed so that the nucleotide sequence of one primer is complementary to the sequence flanking one side of the target DNA while the nucleotide sequence of the other primer is complementary to the sequence flanking the other side of the target DNA. When the extension product of one primer extends beyond the other primer's annealing site, the newly synthesized DNA strand can be used as a new template for the next cycle because the target DNA is flanked by nucleotide sequences which are complementary to the paired primers. Steps (a), (b), and (c) repeated in sequence about 25 to 30 times or cycles with the number of target DNA copies doubling with each cycle. Due to the exponential yield of replicated

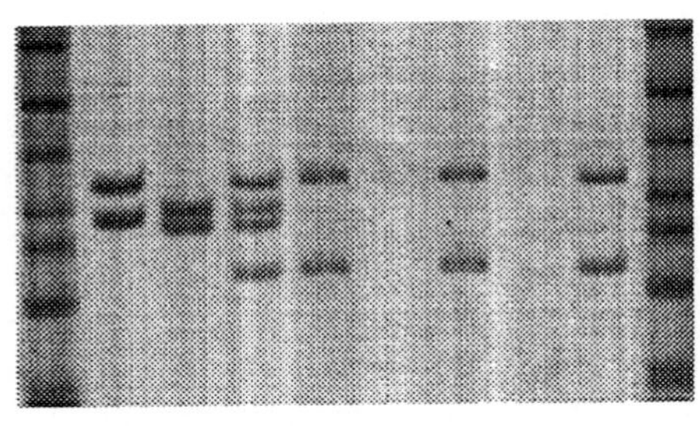

Lane 1- molecular weight ladder, lane 2- known control, lane 3- patient, pre-transplant, lane 4- patient, pre-transplant, and donor, lane 5- patient, post-transplant, lane 6- blank, lane 7- patient, post-transplant, lane 8- blank, lane 9- donor, lane 10- molecular weight ladder.

Figure 3 A representative RFLP membrane demonstrating engraftment following peripheral blood stem/progenitor cell transplantation.

DNA in the final product, one can understand how only minute amounts of DNA may be sufficient. Detection of the final product may be performed with chemiluminescence, colorimetry, fluorescence, enzyme-linked immunoabsorbant assays, or radioactive methods (42).

The application of PCR may detect infectious agents, aid in clinical diagnosis, and monitor therapy for numerous conditions. In transfusion medicine, PCR techniques have recently been applied to evaluate ABO, Rh, Kell, and Duffy systems (44). With respect to monitoring engraftment following hematopoietic cell transplant, both endonuclease restriction and PCR techniques may be used. A comparison of RFLPs by Southern hybridization with VNTRs using PCR reported by Sreenan et al. revealed 80% concordance between the two methodologies, and 20% discordance: 10% due to a failure to detect mixed chimerism by VNTR, 8.75% due to a failure to detect mixed chimerism by RFLP, and 1.25% due to insufficient DNA for RFLP (45). The major theoretical advantage of PCR is its increased sensitivity, which allows the use of minute quantities of DNA. The use of PCR has the possibility of detecting as few as one abnormal cell in 1 million (46–48). The major disadvantage of PCR is the possibility of contamination using a system that is based on amplification.

V. THE ROLE OF CYTOGENETICS IN MONITORING ENGRAFTMENT

After transplantation, cytogenetics and other molecular studies may be beneficial in monitoring the clinical status of patients, since molecular relapse is thought to precede clinical relapse. Karyotyping a female patient whose donor was a male may demonstrate the Y chromosome, which signifies chimerism (Fig. 4). The molecular technique of RFLP was instrumental in detecting an abnormality in a subcutaneous hematopoietic lesion arising from the patient's own DNA about 1 year following allogeneic peripheral blood stem cell transplant (Young et al., unpublished data). Use of FISH with the bcr/abl probe (36) and the PML/RARα probe (49) to monitor the success of a transplantation protocol and to detect the presence of minimal residual disease is discussed in Chapter 17. Figures illustrating various probes are also given in the same chapter.

VI. CONCLUDING COMMENTS

The future for molecular studies is bright. Day by day, more and more discoveries are made unlocking the mysteries of the human genome. Some

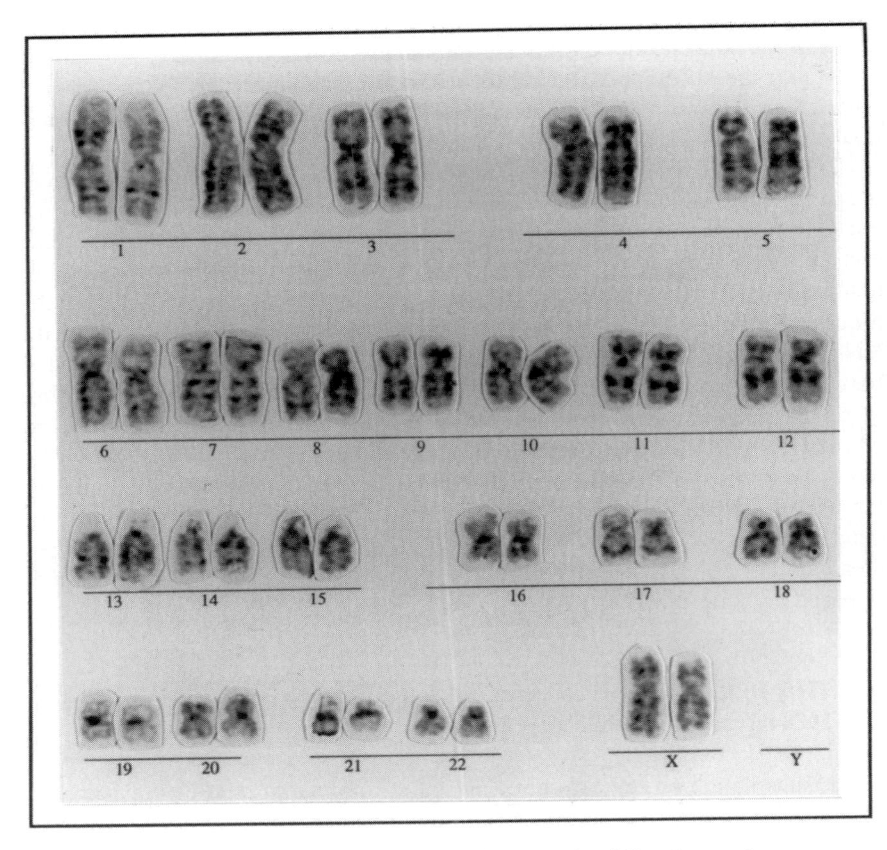

Figure 4 A representative 46,XX cell in a male following a bone marrow transplant.

of the goals of the human genome project are being realized. An increasing number of genes associated with diseases have been identified. With the tremendous progress being made in the human genome project, many molecular probes are now available. Today, genetic tests exist for approximately 450 disorders, according to the National Human Genome Research Institute (50). It is highly likely that in the near future, every cancer with a genetic alteration may be monitored using the techniques of cytogenetics, molecular biology, or combinations of both. Continued advances will no doubt lead to better tools to aid in genetic counseling. Let us hope that the U.S. government continues to promote these advances of science without penalizing those individuals who could benefit most from these genetic tests (50), so that disease prevention may be possible in the next millennium.

ACKNOWLEDGMENTS

We thank Dr. Roger Mark, Ms. Sandy Chang, Ms. Peggy Kenney, and the scientific reviewers for their comments. The technical support of the former members of the Lifespan Academic Medical Center at the Cytogenetics Laboratory at Rhode Island Hospital and the DNA Laboratory at the Rhode Island Blood Center are acknowledged.

REFERENCES

1. Vengelen-Tyler V. Technical Manual. 12th ed. Bethesda, MD: American Association of Blood Banks, 1996.
2. Metcalf D, Moore MAS. Hematopoietic cells. In: Neugberger A, Tatum EL, eds. Frontiers of Biology. Amsterdam: North-Holland, 1971:25.
3. Emerson SG, Farnen JP. Hematopoiesis and hematopoietic growth factors. In: McClatchey KD, ed. Clinical Laboratory Medicine. Baltimore: Williams & Wilkins, 1994:817–825.
4. Tsai S, Sieff CA, Nathan DG. Stromal cell associated erythropoiesis. Blood 1986; 67:1418.
5. Phillips P. 3rd International Society of Hematotherapy and Graft Engineering. Bordeaux, France, 1997.
6. Dracker RA. Clinical apheresis: new application and technology evolution. ASCP/CAP 1994 Fall Meeting, Washington, DC, Oct. 22–28, 1994.
7. 58 Federal Register 53247, FDA, Oct. 14, 1993.
8. National Marrow Donor Program Minority Fact Sheet, National Marrow Donor Program, #892, Oct. 1996.
9. Chance of a Lifetime: Questions and Answers About Unrelated Marrow Transplants, National Marrow Donor Program, #410, March 1996.
10. National Blood Resource Education Program: Profile of Minority Blood Donors, US. Bethesda, MD: U.S. Department of Health and Human Services, Public Health Service, National Institutes of Health, National Heart, Lung and Blood Institute, May 1993.
11. Oswalt R, Gordon J. Blood donor motivation: a survey of minority college students. Psychology Report 1993; 72:785–786.
12. Lee TD. Distribution of HLA antigens. In: Lee J, ed. The HLA System. A New Approach. New York: Springer-Verlag, 1990:141–178.
13. Baur MP, Danilors JA. Population analysis of HLA-A, B, C, DR and other genetic markers. In: Terasaki PA, ed. Histocompatibility Testing 1980: Report of the Eighth International Histocompatibility Workshop. Los Angeles: UCLA Tissue Typing Laboratory, 1980:955–993.

14. Hurley CK. HLA: a legacy of human evolution. Transplant Proc 1989; 6:3876–3877.
15. King KE, Ness PM, Braine HG, Armstrong KS. Racial differences in the availability of human leukocyte antigen-matched platelets. J Clin Apheresis 1996; 11:71–77.
16. Broxmeyer HE, Kurtzberg J, Gluckman E, et al. Human umbilical cord blood: a clinically useful source of transplantable hematopoietic stem/progenitor cells. Int J Cell Cloning 1990 (suppl 1); 8:76.
17. Wagner JE. Umbilical cord blood stem cell transplantation. Am J Ped Hematol Oncol 1993; 15(2):169–174.
18. Stephenson J. Terms of engraftment: umbilical cord blood transplants arouse enthusiasm. JAMA 1995; 273(23):1813–1815.
19. Brecher ME, Lasky LC, Sacher RA, Issitt LA. Hematopoietic Progenitor Cells: Processing, Standards and Practice. Bethesda, MD: American Association of Blood Banks, 1995.
20. Standards Committee of the American Associaton of Blood Banks. Standards for Blood Banks and Transfusion Services. 17th ed. Bethesda, MD: American Association of Blood Banks, 1996.
21. Broxmeyer HE, Hangoc G, Cooper S, et al. Growth characteristics and expansion of human umbilical cord blood and estimation of its potential for transplantation in adults. Proc Natl Acad Sci USA 1992; 89:4109.
22. Weinthal JA, Nemunaitis JJ, Aston S, Magsamen MJ, Rosenfeld CS. Analysis of apheresis content of progenitor cell collections from normal donors to whom granulocyte-colony stimulating factor is administered. Transfusion 1996; 36: 943–947.
23. Tabilio A, Falzetti F, Giannoni C, et al. Stem cell mobilization in normal donors. J Hematother 1997; 6:227–234.
24. Draft document concerning the regulation of peripheral blood hematopoietic stem cell products intended for transplantation or further manufacture into injectable products. FDA, Feb. 11, 1996.
25. Read EJ, AuBuchon JP. To tell the truth, I've got a secret . . . about hematopoietic progenitor cells. AABB Technical/Scientific Workshop, New Orleans, LA, Nov. 10, 1995.
26. Mark HFL. Fluorescent in situ hybridization as an adjunct to conventional cytogenetics. Ann Clin Lab Sci 1994; 24(2):153–163.
27. Mark HFL. Recent advances in molecular cytogenetics: fluorescent in situ hybridization. Rhode Island Med 1994; 77:377–381.
28. Afify A, Bland KI, Mark HFL. Fluorescent in situ hybridization assessment of chromosome 8 copy number in breast cancer. Breast Cancer Res Treat 1996; 38:201–208.
29. Afify A, Mark HFL. FISH assessment of chromosome 8 copy number in stage I and stage II infiltrating ductal carcinoma of the breast. Cancer Genet Cytogenet. 1997; 97:101–105.
30. Mark HFL, Jenkins R, Miller WA. Current applications of molecular cytogenetic technologies. Ann Clin Lab Sci 1996; 27(1):47–56.

31. Rowley JD. A new consistent chromosomal abnormality in chronic myelogenous leukemia identified by quinacrine fluorescence and Giemsa staining. Nature 1973; 234:290–293.
32. Mitelman F. Catalog of Chromosome Aberrations in Cancer. 5th ed. New York: Wiley-Liss, 1994.
33. Schouten HC, Sanger WG, Weisenburger DD, Anderson J, Armitage JO (Nebraska Lymphoma Study Group). Chromosomal abnormalities in untreated patients with non-Hodgkin's lymphoma: associations with histology, clinical characteristics, and treatment outcome. Blood 1990; 75:1841–1847.
34. Yunis JJ, Oken MM, Kaplan ME, Ensrud KM, Howe RR, Theologides A. Distinctive chromosomal abnormalities in histologic subtypes of non-Hodgkin's lymphoma. N Engl J Med 1982; 307:1231–1236.
35. Barch MJ, Knutsen T, Spurbeck JL, eds. The Association of Cytogenetics Technologists. The AGT Cytogenetics Laboratory Manual. 3d ed. Philadelphia: Lippincott-Raven, 1997.
36. Young CT, Di Benedetto J Jr, Glasser L, Mark HFL. A Philadelphia chromosome positive CML patient with a unique translocation studied via GTG-banding and fluorescence in situ hybridization. Cancer Genet Cytogenet 1996; 89: 157–162.
37. Pinkel D, Strawne R, Gray J. Cytogenetic analysis using quantitative high sensitivity, fluorescence hybridization. Proc Natl Acad Sci USA 1986; 83: 2934–2938.
38. Mitelman F, ed. An International System for Human Cytogenetic Nomenclature Basal: Karger, 1995.
39. Dewald GW, Morris MA, Lilla VC. Chromosome studies in neoplastic hematologic disorders. In: McClatchey KD. Clinical Laboratory Medicine. Baltimore, MD: Williams & Wilkins, 1994:703–739.
40. Mark HFL. Bone marrow cytogenetics and hematologic malignancies. In: Kaplan BJ, Dale KS, eds. The Cytogenetics Symposia. Burbank, CA: The Association of Cytogenetics Technologists, 1994; chapt 8:1–7.
41. Cossman JC, Uppenkamp M, Sundeen J, Couplcand R, Raffeld M. Molecular genetics and the diagnosis of lymphoma. Arch Pathol Lab Med 1988; 112: 117–127.
42. Erlich HA, Gelfand D, Sninsky JJ. Recent advances in the polymerase chain reaction. Science 1991; 252:1643–1651.
43. Mullis K. The unusual origin of the polymerase chain reaction. Sci Am 1990; 262:56–65.
44. Ni H, Blajchman MA. Understanding the polymerase chain reaction. Transfusion Med Rev 1994; 8(4):242–252.
45. Sreenan JJ, Pettay JD, Tbakhi A, et al. The use of amplified variable number tandem repeats (VNTR) in the detection of chimerism following bone marrow transplantation: a comparison with restriction fragment length polymorphism (RFLP) by Southern blotting. Am J Clin Pathol 1997; 107:292–298.
46. O'Leary T, Stetler-Stevenson M. Diagnosis of t(14;18) by polymerase chain reaction: the natural evolution of a laboratory test. Arch Pathol Lab Med 1994; 118:789–790.

47. Segal GH, Scott M, Jorgenson T, Braylan RC. Standard polymerase chain reaction analysis does not detect t(14;18) in reactive lymphoid hyperplasia. Arch Pathol Lab Med. 1994; 118:791–794.
48. Negrin RS. Use of the polymerase chain reaction for the detection of tumor cell involvement of bone marrow and peripheral blood: implications for purging. J Hematother 1992; 1:361–368.
49. Mark HFL, Dong H, Glasser L. A multimodal approach for diagnosing patients with acute promyelocytic leukemia (APL). Cytobios 1996; 87:137–149.
50. Aston G. Preventing genetic discrimination. Am Med News 1997; 40(29):3, 22.

APPENDIX: GLOSSARY OF TERMS

bone marrow transplant (BMT)
The process of collecting bone marrow from an individual and reinfusing that bone marrow into the same individual, i.e., autologous transplant, or into another individual, i.e., heterologous transplant.

engraftment monitoring
Testing transplanted patients by varying methods including cytogenetics, restriction fragment length polymorphisms by Southern blot, or variable-number tandem repeats by polymerase chain reaction, to determine the presence of donor DNA compared to the presence of patient (recipient) DNA.

peripheral blood stem cells (PBSC)
Multipotential cells present in peripheral blood, bone marrow, and cord blood, which may be specified by location.

restriction fragment length polymorphisms (RFLP)
Varying lengths of segments of DNA determined by the sites of splicing by restriction enzymes called endonucleases.

transfusion medicine
A broad medical specialty that encompasses blood utilization, blood donation and collection, preparation of blood components, immunologic and genetic principles, regulatory compliance issues, and vast clinical considerations.

variable number tandem repeats (VNTR)
Varying numbers of a specific sequence of DNA in consecutive order without intervening sequences.

17

Fluorescent In Situ Hybridization (FISH): Applications for Clinical Cytogenetics Laboratories

Hon Fong L. Mark
Brown University School of Medicine and Rhode Island Department of Health, Providence, and KRAM Corporation, Barrington, Rhode Island

I. INTRODUCTION

Although conventional cytogenetics via various banding techniques is a highly informative and established diagnostic procedure which enables one to inspect the entire human genome at a glance, its application is restricted to dividing cells. Interpretation is also difficult when one encounters a suboptimal specimen with an inadequate number of high-quality metaphases. The advent of fluorescent in situ hybridization (FISH), made possible by the explosion in recombinant DNA technology, offers an unprecedented opportunity for the analysis of a large number of nondividing cells in the interphase stage. Thus, FISH is sometimes called interphase cytogenetics.

The principles, methodology, rationale, and development of FISH have been discussed in the chapter by Blancato and Haddad. The present chapter focuses on the various potential applications of FISH for diagnostic and research purposes in the average clinical cytogenetics laboratory, usually located within the department of pathology. As FISH is an evolving area of research, there may potentially be other areas in which FISH or FISH-based technology may be applicable as well.

II. FISH FOR BETTER DELINEATION OF NUMERICAL AND STRUCTURAL CHROMOSOMAL ABNORMALITIES

An accurate identification of chromosomal abnormalities and mosaicism detection is important for genetic counseling, for prenatal diagnosis, and as part of the workup for patients with congenital malformations, developmental delay, or mental retardation. The various indications for performing routine cytogenetic analysis have been discussed in previous chapters. FISH has been used with increasing frequency to augment the power of conventional cytogenetics. Numerical chromosomal abnormalities can be detected readily using chromosome enumeration probes. For example, we have detected mosaicism for trisomy 8 using an α-satellite probe for chromosome 8 (1). The identification of structurally rearranged chromosomes can be achieved by using either "painting" probes or α-satellite probes. For example, using the latter we have identified an individual with i(18p), a well-defined syndrome described by Callen et al. (2), Callen (3), and Phelan (4). According to Callen (3), patients possessing the i(18p) marker demonstrate low birth weight, typical faces, neonatal hypotonia with subsequent limb spasticity, short stature, microcephaly, mental retardation, and seizure disorders. Chromosome painting for the detection of structural rearrangements, such as translocations (including cryptic translocations) as illustrated in Color Plate 17.1, rings, and marker chromosomes, are important uses of FISH.

III. FISH FOR MARKER CHROMOSOME IDENTIFICATION

During routine cytogenetic analysis in a clinical cytogenetics laboratory, an unidentifiable chromosome called a marker chromosome is occasionally encountered. An International System for Human Cytogenetic Nomenclature (5) defines a marker chromosome as a structurally abnormal chromosome in which no part can be identified. They are also called extra structurally abnormal chromosomes (ESACs) or supernumerary chromosomes.

Marker chromosomes can be classified into three groups for convenience, although each of the three groups contains markers of quite diverse origins. Rooney and Czepulkowski (6) have provided an excellent overall description of marker chromosomes, which is summarized below.

Bi-satellited markers, as the name implies, have satellites on both arms of the chromosome. They are assumed to be derived from the acrocentric chromosomes, which are chromosomes 13, 14, 15, 21, and 22. Chromosome 15, in particular, is often implicated in the origins of these markers. Bi-

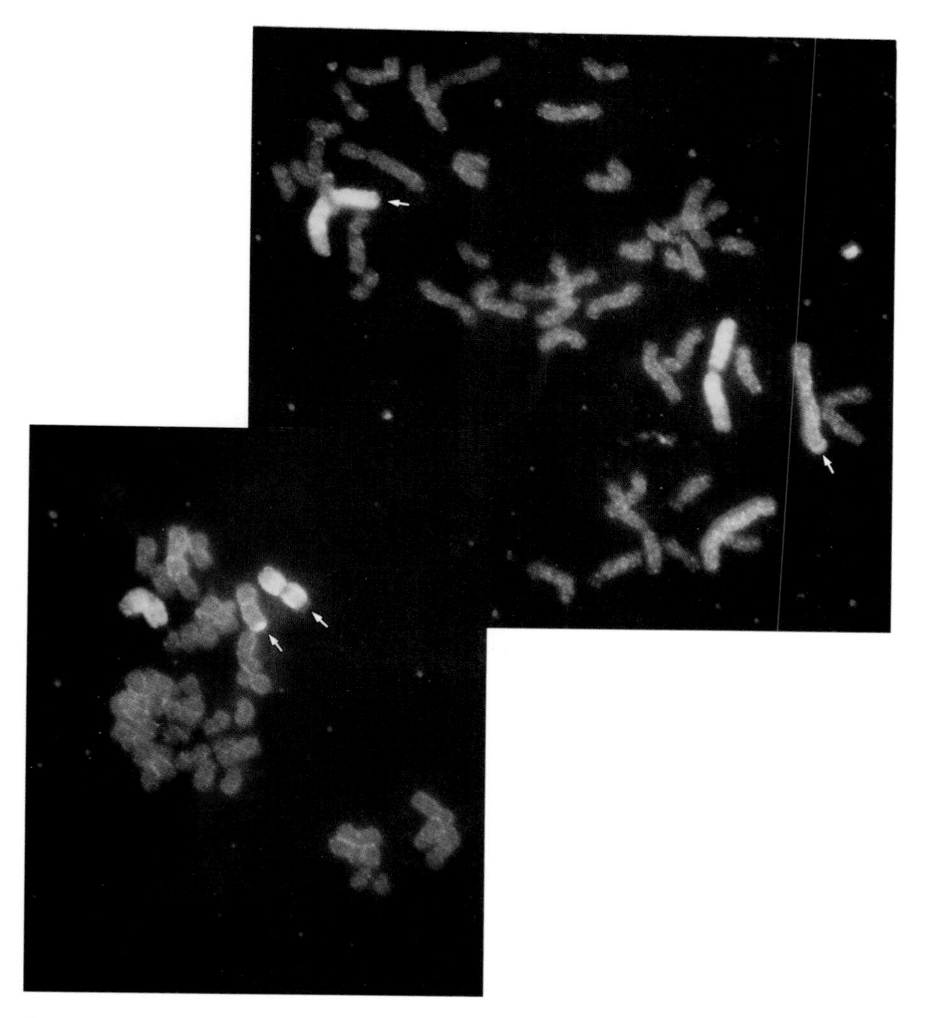

Figure 17.1 FISH using dual-color chromosome painting probes illustrating a subtle reciprocal translocation between chromosomes 1 and 2 that can easily be missed by conventional cytogenetics.

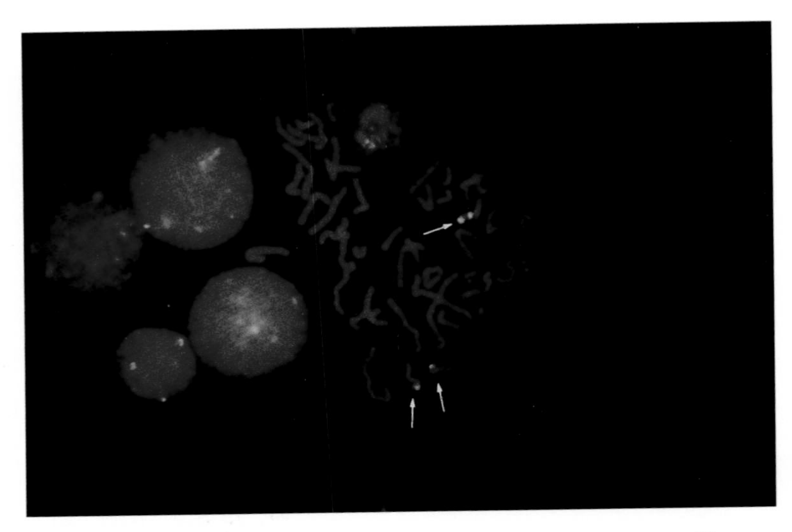

Figure 17.2 Fluorescent in situ hybridization using a chromosome 15 specific α-satellite probe; arrows point to marker and two normal chromosome 15s. Karyotype interpreted as 47,XY,+inv dup(15). (Reprinted with permission. Clinical Pediatrics, 1995.)

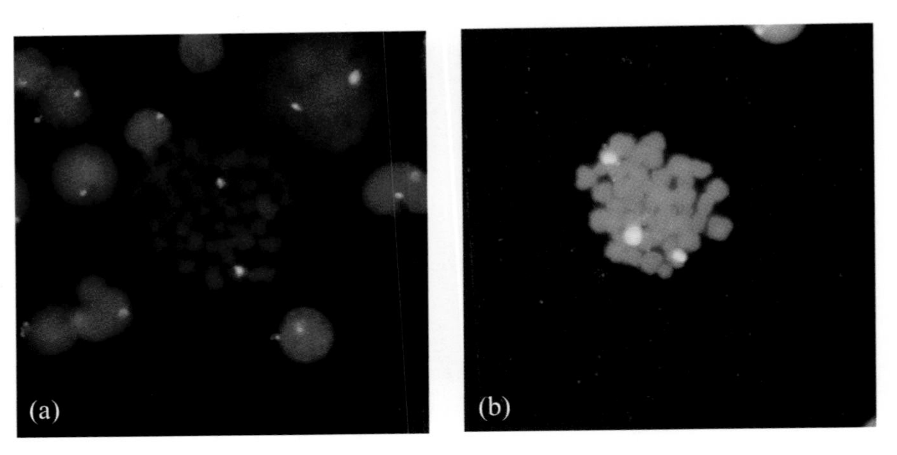

Figure 17.3 a,b Demonstration of disomy (a) and trisomy (b) for chromosome 8 in a sub-optimal bone marrow preparation using FISH.

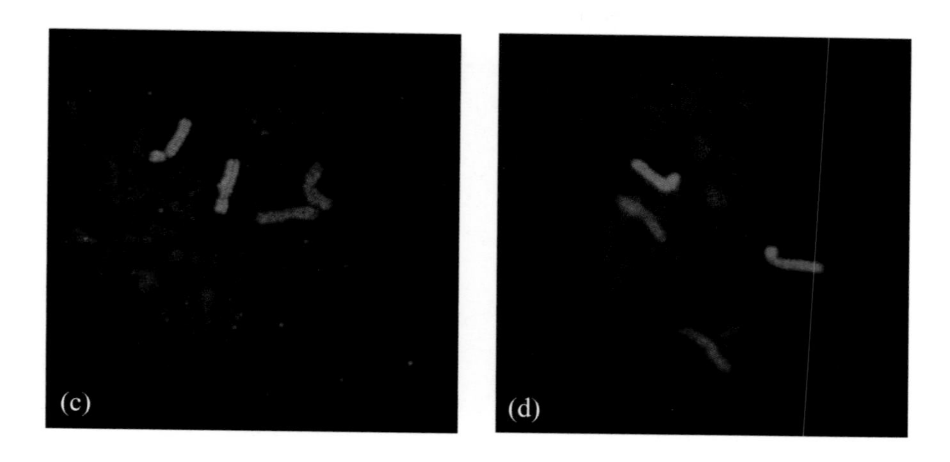

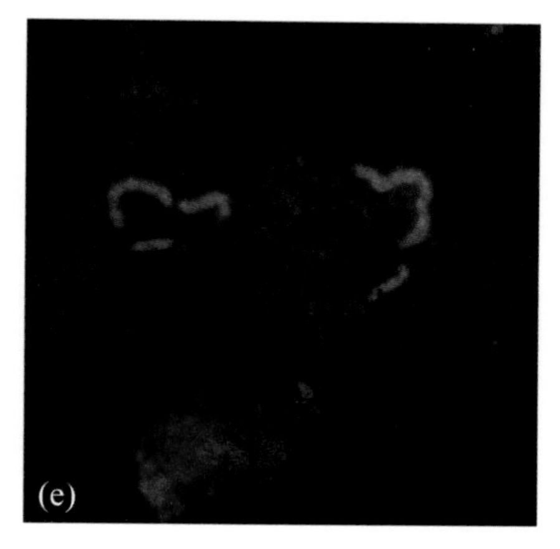

Figure 17.3 c,d,e (c,d) Fluorescent in situ hybridization of a suboptimal metaphase using whole chromosome painting probes for chromosomes 3 (orange) and chromosome 4 (green). (e) Fluorescent in situ hybridization of a suboptimal metaphase using whole chromosome painting probes for chromosome 2 (orange) and chromosome 18 (green)

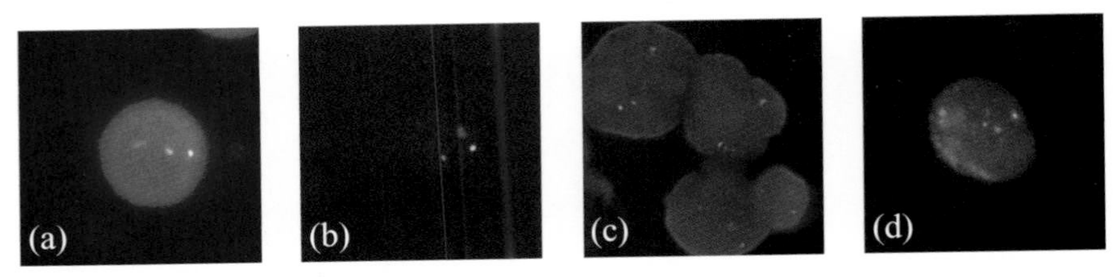

Figure 17.4 a,b,c,d (a,b,c) Three panels showing *bcr/abl* probe for detecting the Philadelphia translocation. (d) PML/RARα probe for detecting t(15;17) in APL.

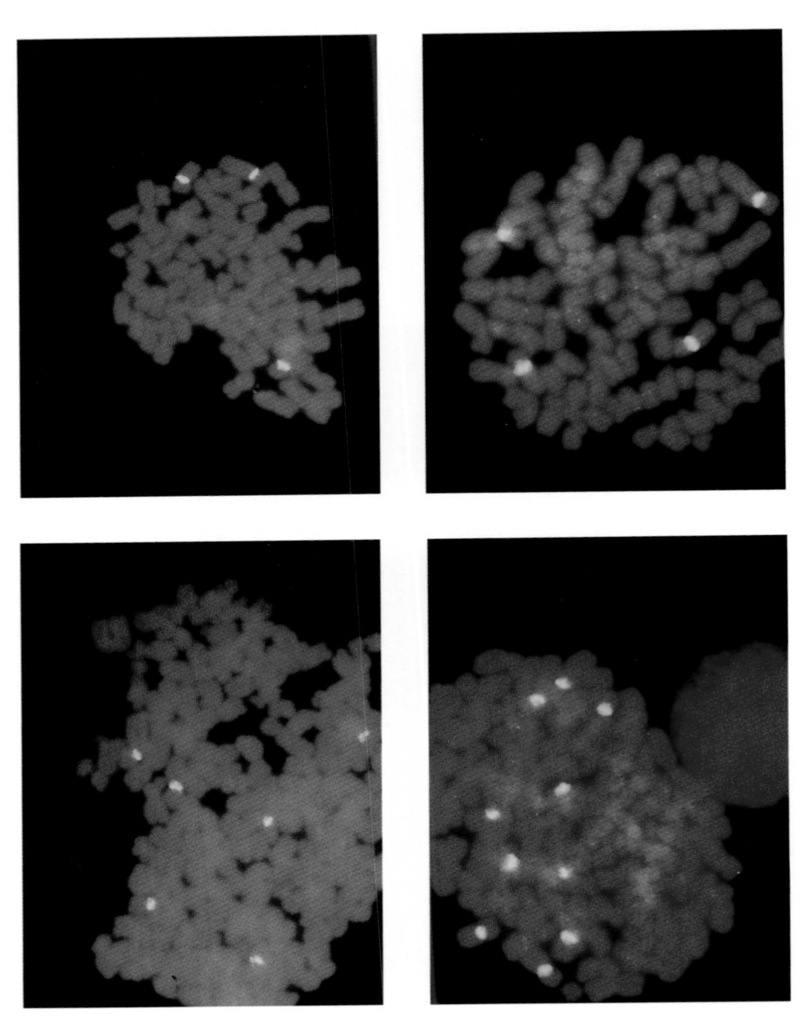

Figure 17.5 FISH using a chromosome 7 probe in the study of chromosome 7 copy number in cervical cancer cells.

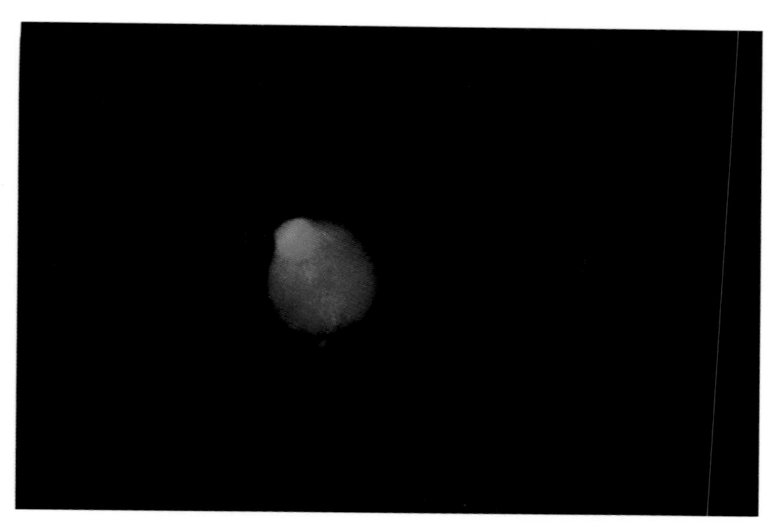

Figure 17.6 FISH on buccal smear using the dual-color X (red) and Y (green) probes. (Reprinted with permission from Ref. 35.)

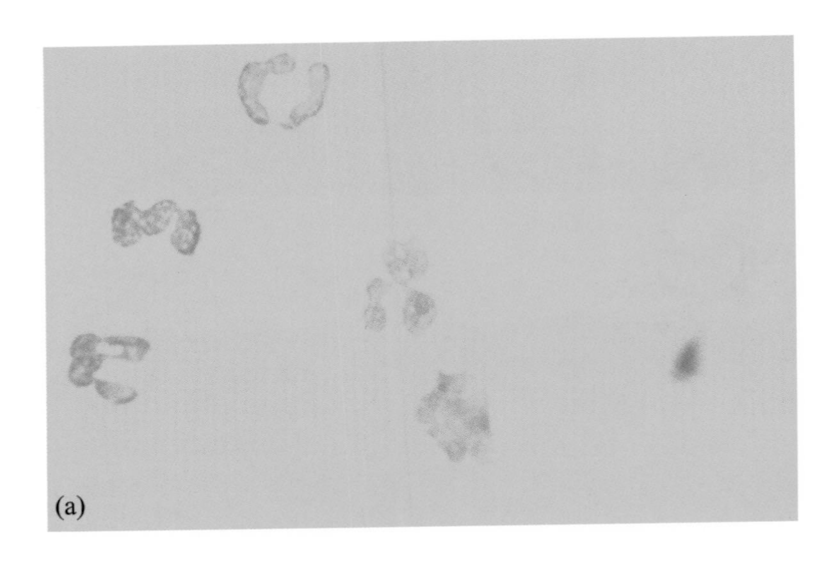

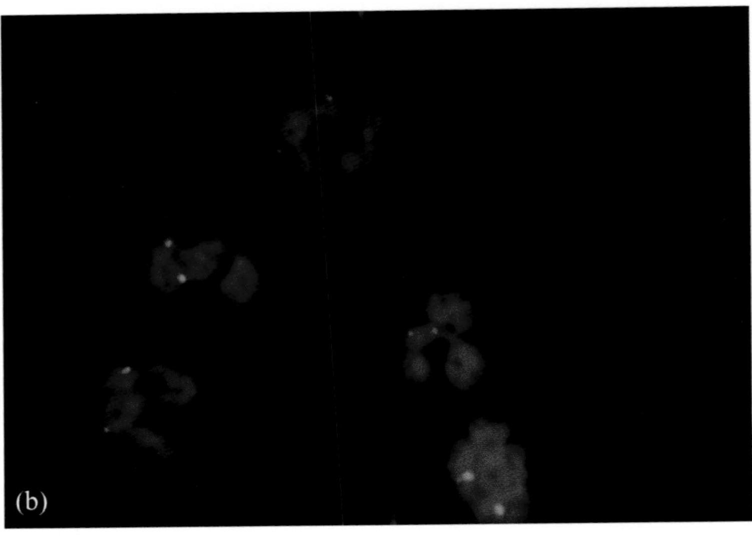

Figure 17.7 a,b Sequential Wright-staining and FISH. (Courtesy of Dr. Kathryn Lin.)

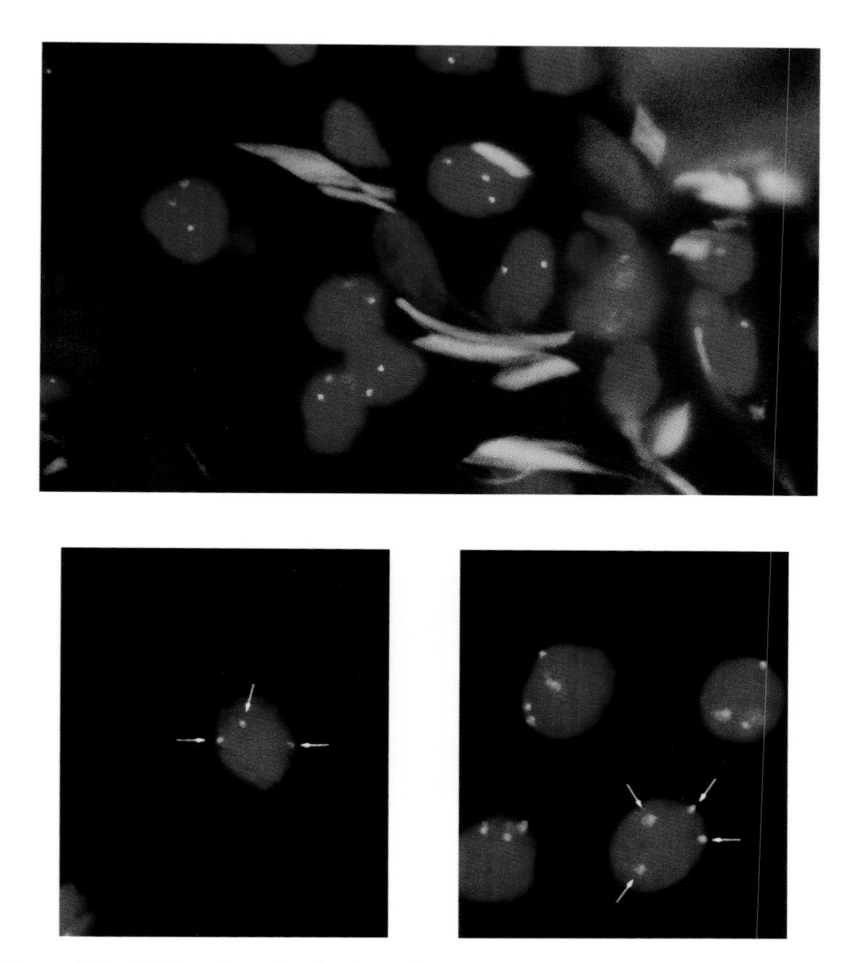

Figure 17.8 FISH on formalin-fixed paraffin-embedded breast cancer tissue illustrating three and four signals for the chromosome 8 α-satellite probe.

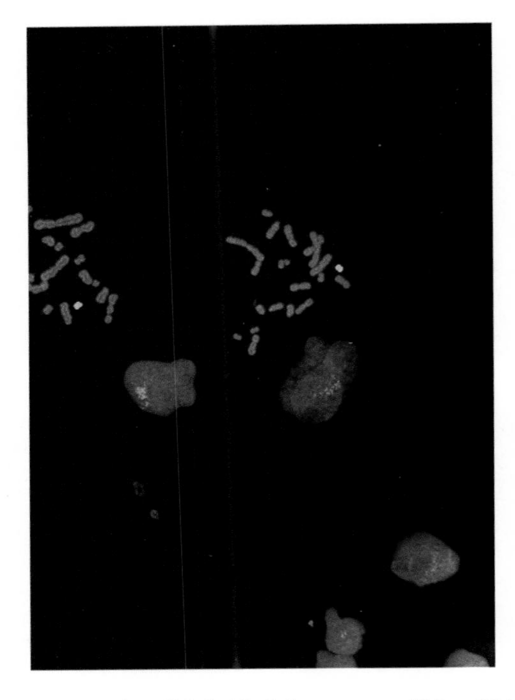

Figure 17.9 FISH on a somatic cell hybrid of chromosome 19 in a Chinese hamster background. (Reprinted with permission from Ref. 42.)

satellited markers may contain a single centromere or two centromeres. The region between the two centromeres may or may not contain G or R bands. The two centromeres may be adjacent to the breakpoints very close to the centromeric region; they contain little or no euchromatin. Many variants are possible. Alternatively, the two centromeres may be separated by euchromatin. In addition, one of the centromeres may be inactive, making the marker a pseudo-dicentric.

Markers can have satellites on one end only. This is the second class of markers. An individual marker may show a terminal deletion, an interstitial deletion, or a translocation product, for example, from a 3:1 segregation of a balanced rearrangement such as t(11;22)(q23;q11). This class of markers, with satellites on one end of the chromosome only, can be derived from three different sources: terminal or interstitial deletions of acrocentrics, familial rearrangements of acrocentrics, and *de novo* rearrangements of acrocentrics.

A third class of markers is nonsatellited markers. These include everything other than the above two classes of markers. Specific groups that have been identified include: (a) very small rings (usually heterochromatic) derived from the centromeric regions of chromosomes 1, 9, or 16; (b) isochromosomes derived from the short arms of chromosomes 12, 18, and the Y chromosome; and (c) deletions of the long arm of the Y chromosome.

Miscellaneous structurally rearranged chromosomes also exist. The derivative of chromosome 22 from t(11;22)(q23;q11), the isochromosome of the short arms of chromosome 18, the pseudo-isodicentric of chromosome 15, and the isochromosome of the short arm of chromosome 12 (mosaic in skin) are markers known to be associated with distinct phenotypes. One of the best-known examples of a marker is an additional isodicentric of chromosome 22 found in cat-eye syndrome.

A case report by Abuelo et al. (7) described how FISH was utilized to identify the origin of a marker in a child who presented with developmental delay. The marker was demonstrated to be an inv dup (15), more properly called pseudo-dicentric 15 in this case (Color Plate 17.2). Marker chromosomes occur in approximately 0.05% of live-born infants and in approximately 0.06% of amniocenteses. About one-half of the cases are familial and carry a low risk for an abnormal phenotype. However, nonfamilial cases may have cognitive, behavioral, and physical abnormalities. In many previously reported cases, conventional chromosome staining techniques could not determine the origin of the extra chromosomes. Only with recent advances in molecular cytogenetic technology could many supernumerary chromosomes be identifiable (7).

IV. MICRODELETION DETECTION USING FISH AND LOCUS-SPECIFIC PROBES

A microdeletion is a small deletion of chromosomal material. One of the most powerful methods for detecting microdeletions and microduplications is FISH. Indeed, Ledbetter et al. (8) attributed the "death of high resolution cytogenetics" to this particular application of FISH. Some of the specific applications of FISH include the detection of deletions and cryptic translocations involving 17p13.3 in Miller-Dieker syndrome (9), the detection of deletions of chromosome 15q11-13 in Prader-Willi and Angelman syndromes (10), and the detection of 22q11.2 deletions in DiGeorge and Velocardiofacial syndromes (11,12). For a growing number of microdeletion syndromes, a combination of routine cytogenetics and FISH analysis may be the optimal protocol. Included in this list of syndromes are deletion of 17p11.2 in Smith-Magenis syndrome, deletion of 7q11.23 in Williams syndrome, deletion of 4p16.3 in Wolf-Hirschhorn syndrome, deletion of 5p15.2 in cri-du-Chat syndrome, and deletion of Xp22.3 in steroid sulfatase deficiency/Kallman syndrome (13).

Molecular cytogenetics can also be applied to the detection of microduplications such as the microduplication at 17p11.2 which leads to Charcot-Marie-Tooth disease (type 1A) (13).

V. SEX CHROMOSOME IDENTIFICATION AND DETECTION OF MOSAICISM

An immediate consequence of the ability to assess chromosome copy number in interphase as well as metaphase cells is a potentially large sample of cells for mosaicism study. The utility of FISH using an X chromosome and a Y chromosome probe was illustrated by a case report by Mark et al. (14). The patient was an 18-year-old woman referred for delayed puberty, short stature, hypergonadotropic hypogonadism, primary amenorrhea, and growth retardation. Her past medical history was unremarkable, with the exception of a learning disability. Cytogenetic analysis was ordered to rule out a chromosomal basis for her clinical findings. GTG-banding analysis of peripheral blood lymphocytes revealed the presence of predominantly 46,XY cells. Using dual-color X-chromosome-specific and Y-chromosome-specific probes, a FISH analysis undertaken to further assess the contribution of a minor cell line yielded frequencies of 87% cells with the 46,XY constitution and 9% with the 45,X constitution. To establish the presence of mosaicism unequivocally, a skin biopsy was obtained for fibroblast culture, which further corroborated the results of the peripheral blood study. The FISH analysis revealed that 74% of the cells were 46,XY and 12% were 45,X. The unequivocal presence of XY cells puts the patient at risk for neoplastic transformation of the gonads. Laparoscopy and surgical removal of the patient's presumptive streak gonads were therefore undertaken. Cytogenetic results derived from the gonadal tissues further confirmed findings of the previous cytogenetic analyses. A picture of the dual-color X/Y probe is shown elsewhere in this volume.

In another study, a combined cytogenetic and molecular approach was employed to detect low-percentage mosaicism for a minor cell line in a 15½-year-old patient with delayed puberty and short stature. The use of DNA analysis further helped delineate the nature of a marker present in a low percentage of cells initially detected using cytogenetics. Adoption of a combined cytogenetic and molecular approach can be useful for the diagnosis and management of patients with specific sex chromosome aneuploidy. It has been estimated that monosomy X with a normal male cell line (45,X/46,XY) accounts for only 4% of cases of Turner syndrome. However, the occurrence of neoplasia in dysgenetic gonads associated with karyotypes containing the Y chromosome is well known. Gonadoblastoma is the most common tumor arising in dysgenetic gonads, although dysgerminoma and other histologic types may also occur. Because the risk of a gonadoblastoma may be as high as 30% in Turner subjects with Y-chromosomal material, removal of the streak gonads is recommended. Thus, determining with certainty the presence or absence of Y-chromosomal material is of critical importance, because a recommendation for a surgical procedure, gonadectomy, is based on this information [15].

VI. PRENATAL DIAGNOSTIC APPLICATIONS OF FISH FOR RAPID ANEUPLOIDY DETECTION

Rapid aneuploidy detection in prenatal diagnosis using FISH technology usually involves the analysis of interphase nuclei rather than metaphase chromosome spreads. The test is often used to detect the most common aneuploidies, such as trisomy 13 (Patau syndrome), trisomy 18 (Edwards syndrome), trisomy 21 (Down syndrome), and the sex chromosome aneuploidies leading to such syndromes as Klinefelter syndrome (47,XXY) and Turner syndrome (45,X). The major advantage of this technique is that certain information about fetal chromosome status is obtained within 48 h rather than in 7–10 days without the requirements of *in vitro* culture. However, concerns have been expressed about the sensitivity and specificity of this technique. Accordingly, enthusiasm for using this technique in clinical practice has been considerably dampened (13).

According to Mark et al. (13), the main utility of FISH for prenatal applications is for select situations such as very late gestational age pregnancies with ultrasound-identified abnormalities that are relatively specific for FISH-diagnosable aneuploids. These cases include cystic hygroma (monosomy X, trisomy 18, and trisomy 21), holoprosencephaly with cleft lip (trisomy 13), clenched fists with overlapping fingers and a two-vessel cord (trisomy 18, etc.), and in biochemical screen-positive populations. For those cases, FISH testing may be clinically applicable and may prove to be cost-effective (13).

FISH also holds promise for use in preimplantation genetics. It may also have potential applications in the detection of fetal cells in maternal blood, a new and exciting area of investigation which may be feasible as a routine test in the future. FISH may be useful for the detection of certain chromosomal abnormalities after fetal cell sorting. Male fetal cells can be detected using classical and α-satellite probes derived from the Y chromosome. Amniocentesis and chorionic villus sampling (CVS) are invasive techniques that pose small but significant risks. The applications of FISH for confined placental mosaicism have been briefly discussed by Blancato and Haddad in another chapter.

Issues relating to prenatal diagnosis are further explained in the chapter by Lamb and Miller.

VII. "STAT CHROMOSOMES"

"STAT chromosomes" can be performed using slides of fixed cells from the bone marrow or peripheral blood. This technique has been utilized only in cases where an answer must be obtained rather quickly, such as in cases where there is a question of a trisomy in a newborn. FISH can be used for "STAT chromosomes" with the appropriate probes by using slides prepared from fixed cells from bone marrow or even bone marrow smears.

VIII. DETECTION OF TRISOMIES IN CANCER USING FISH

One of the most important uses of FISH is for chromosome enumeration in the detection of aneuploidies such as trisomy 8, monosomy 7, or trisomy 13 (in acute leukemia). Mark et al. (16) used FISH to assess the proportion of cells with trisomy 4 in a patient with acute nonlymphoblastic leukemia, and recognized FISH as a potentially useful tool for monitoring the minimal residual disease in this patient. Another study described the concurrence of congenital trisomy 8 mosaicism and gestational trophoblastic disease in a

42-year-old woman (17). Subsequently, archival materials from this patient were retrieved, and FISH was used to assess the chromosome 8 copy number in the cells of various tissues. Mark et al. (18) compared the results of the FISH studies with each other and with the original cytogenetic studies. Subsequent studies of other available archival material raised the question of whether there is a subset of GTD characterized by this trisomy (19).

Recently, Vysis, Inc. (Downer's Grove, IL) received 510(k) clearance from the U.S. Food and Drug Administration to market the company's CEP 12 SpectrumOrange™ DNA probe kit. The products were approved for use as an adjunct to GTG banding for the identification and enumeration of chromosome 12 in interphase cells of patients with B-cell chronic lymphocytic leukemia. As of May 1997, approvals were also obtained for chromosome 8 and the X and Y chromosome probe kits.

IX. DETECTION OF NUMERICAL AND STRUCTURAL CHROMOSOMAL ABNORMALITIES IN CANCER USING FISH: LOW MITOTIC INDEX, SUBOPTIMAL PREPARATIONS, NONDIVIDING, AND TERMINALLY DIFFERENTIATED CELLS

Conventional cytogenetics has been hampered by the requirement that there be sufficient dividing cells with adequate quality for karyotypic analysis. FISH is a useful adjunct to conventional cytogenetics in cases where dividing cells are either inadequate or lacking in number (low mitotic index), or where the quality of the metaphase preparation is suboptimal (Color Plates 17.3a, b, c, d, and e). Interphase cells are excluded from conventional cytogenetic analysis. Equally inaccessible for karyotyping are terminally differentiated cells. FISH is clearly superior to conventional cytogenetics in its ability to screen a large number of cells in a relatively short period of time. Furthermore, FISH does not require a new specimen. It can be performed on the original specimen submitted for routine cytogenetic analysis or even hematologic evaluation. Both air-dried slides and flame-dried slides prepared from fixed cells are suitable for FISH. FISH can be performed on either unstained or previously stained cells. We have even performed FISH on archival material nearly 20 years old (18). Thus, FISH greatly expands the capabilities of the clinical cytogenetics laboratory in the study of cancer cells.

X. FISH USING TRANSLOCATION PROBES

The Philadelphia (Ph) chromosome was the first consistent chromosome abnormality associated with a single cancer type, chronic myelogenous leukemia (CML). Using a "new" banding technique in the 1970s, Rowley (20) identified the Ph chromosome as a reciprocal translocation between chromosomes 9 and 22. The Ph translocation can be detected even in suboptimal metaphase cells and in interphase nuclei through the use of molecular probes. Young et al. (21) used FISH with paired painting probes in various combinations and the Oncor bcr/abl probe to identify a unique complex translocation involving chromosomes 9, 13, 15, 17, and 22 associated with Ph chromosome-positive CML. With the increasing availability of newer clinical cytogenetic technologies, many steps in the cumbersome process as described will become unnecessary.

XI. APPLICATIONS OF FISH IN BONE MARROW TRANSPLANTS

One of the most important applications of FISH in cancer is the monitoring of engraftment in sex-mismatched bone marrow transplants (BMT). Conventional cytogenetics has the capability of distinguishing female (XX) cells from male (XY) cells in metaphase spreads. FISH can distinguish between host and donor cells in both interphase and metaphase cells. By detecting the ratio of XX to XY cells using FISH with the X chromosome and Y chromosome probes in a post-BMT sample, the degree of success for engraftment can be assessed and the recurrence rate can be predicted.

XII. DETECTION OF NUMERICAL AND STRUCTURAL CHROMOSOMAL ABNORMALITIES IN CANCER USING FISH: DETECTION OF MINIMAL RESIDUAL DISEASE

The ability to detect very low levels of disease (i.e., minimal residual disease) by the polymerase chain reaction (PCR) has assisted in designing the appropriate therapy, in predicting the outcome of treatment, and in creating novel molecular therapies. However, as Cotter (22) noted, molecular analysis has its limitations. The analysis is primarily performed on a "soup" of DNA or RNA from both normal and diseased cells. Thus, using PCR, it is impossible to identify the individual cell bearing the abnormality or to determine its phenotypic characteristics. In addition, only a small focused area of genetic change can be assessed. Other relevant genetic alterations that are outside the region or in other parts of the genome may not be assessed. As Cotter noted, "Cytogenetics took a quantum leap in detection methodology with FISH and was able to fill some of the shortcomings of PCR-based technology. FISH has strongly enhanced gene and genome analysis, rapidly mapping large segments of DNA, defining chromosomal alterations and permitting whole chromosome analysis with individual chromosome paints'' (22).

Using FISH, useful information on remission and relapse can be obtained and applied prognostically to determine treatment options and to monitor the response to therapy. Shown in Color Plates 17.4a, b, and c is the bcr/abl translocation probe used to detect cells carrying the Philadelphia translocation in CML, and in Color Plate 17.4d, the PML/RARα probe used to detect cells carrying the 15;17 translocation in acute promyelocytic leukemia (APL) (23). The presence of even a very low level of leukemic cells can be detected by FISH.

XIII. FISH ANALYSIS OF SINGLE-CELL TRISOMIES FOR DETERMINATION OF CLONALITY

In addition to its use for the identification of clonal trisomies associated with specific hematopoietic disorders, FISH can be used to study cases of single-cell trisomies identified in standard 20-cell GTG-banded analyses.

An International System for Human Cytogenetic Nomenclature (5) defines a clone as a cell population derived from a single progenitor cell. It is common practice to infer a clonal origin when a number of cells have the same or closely related abnormal chromosome complements. It is generally accepted that numerical abnormalities are reported only if two or more cells with the same abnormalities are found in the sample (see, for example, Ref. 24). One-cell numerical abnormalities in standard 20-cell cytogenetic analyses are by convention considered random or nonclonal.

Nonclonal chromosomal abnormalities were the subject of study in two prior reports. Chen et al. (25) used interphase FISH to analyze hematologic (bone marrow or peripheral blood) samples and demonstrated clonality of single-trisomic cells in six cases of myelodysplastic syndromes (MDS), two cases of acute myeloid leukemia (AML), and one case each of a myeloproliferative disorder and chronic lymphocytic leukemia. Kasprzyk et al. (26) asked whether single-cell trisomies in bone marrow samples were truly random events or whether they represent cells from a mitotically inactive or minor clone. After using FISH to study a total of seven patients with single abnormal cells at diagnosis, these investigators concluded that single-cell abnormalities may not be clonal and ''do not represent the tip of an iceberg.'' They thus contradicted somewhat the conclusions of Chen et al. (25). In view of these conflicting data, and in view of their important implications for subsequent patient follow-up, we (27) decided to conduct a retrospective study of single-cell trisomies in patients previously studied by routine GTG banding at the cytogenetics laboratory at Rhode Island Hospital from July 1, 1990, to November 30, 1996. FISH was performed using chromosome enumeration probes on previously fixed bone marrow and, in some cases, peripheral blood slides from patients identified to have single-cell trisomies of selected chromosomes in 20-cell routine GTG-banded analyses. The results of the study of our cohort of 16 patients with available and analyzable slides indicated that a single unifying answer to the above question does not seem to exist. While some cases are apparently random events as predicted by chance, other cases appear to represent "the tip of an iceberg." It is therefore important for cancer cytogeneticists to interpret the results of each patient on a case-by-case basis and to formulate the most optimal strategy for follow-up in the particular case under study.

XIV. GENE AMPLIFICATION STUDIES, ANALYSIS OF SOLID TUMORS AND MISCELLANEOUS ONCOLOGIC APPLICATIONS

The value of FISH for improving the diagnosis in effusions from various malignancies was reported by Chen et al. (28). FISH is also useful as a diagnostic tool to detect premalignant lesions or secondary primary tumors in bladder washes (29).

Gene amplification analysis (such as HER-2/neu, c-myc, and N-myc), gene deletion (such as p53), as well as translocation analysis in cancer are also discussed in the chapter by Blancato and Haddad. Solid tumor analysis of breast is further discussed in the chapter by Wolman.

XV. FISH AS AN ADJUNCT TO CONVENTIONAL CYTOGENETICS IN THE CHARACTERIZATION OF CELL LINES DERIVED FROM PRIMARY TUMORS

Mark (30,31) noted that FISH is a useful adjunct technique for the analysis of tumor cells in cases in which there are insufficient metaphases for karyotypic analysis, where the quality of the preparation is suboptimal, or when there is extensive aneuploidy. An illustration of the utility of FISH in the cytogenetic characterization of cervical cancer cells with multiple numerical and structural chromosomal abnormalities is shown in Color Plate 17.5 (32).

XVI. ASSESSMENT OF CHROMOSOME COPY NUMBER IN INTERPHASE CELLS FOUND IN CYTOLOGICAL SPECIMENS, SUCH AS BUCCAL SMEARS, CERVICAL SMEARS, SEMEN SMEARS, AND EXTRACELLULAR FLUIDS

Sex chromatin (Barr body) analysis of buccal mucosal cells has been recognized for many years as an inexpensive, noninvasive, and rapid means of sex determination. Conventional Barr body analysis using Papanicolaou stain was discontinued as a routine test because of its lack of reliability and its inability to detect mosaicism and other chromosomal abnormalities. With the advent of recombinant DNA technology and the availability of molecular probes, however, the value of this simple, albeit obsolete, test was reevaluated. We reported the results of our laboratory's experience in optimizing a FISH assay on buccal mucosal cells (33). Color Plate 17.6 shows a representative FISH image of a buccal smear from a male, using the dual-color X (red) and Y (green) probes, illustrating one copy each of the X and Y chromosomes. In contrast, no Barr body would be expected in normal males or in females with Turner syndrome (45,X). Only one Barr body per cell would be expected in normal females (46,XX) and in males with Klinefelter syndrome (47,XXY). A picture of the Barr body (sex chromatin) is given in the chapter by Mark and Mark.

FISH is useful when other sources for conventional cytogenetic analysis, such as peripheral blood or bone marrow, are not available for a variety of reasons, as discussed by Mark et al. (33). Buccal smears can be utilized in the absence of a more suitable specimen. Obtaining a buccal smear specimen is a relatively simple, noninvasive procedure that requires only minimally trained personnel. Part of the specimen can also be sent for DNA analysis.

Even in cases in which other tissues are available, buccal smear results can be used to corroborate data derived from these tissues or to assess the proportions of different cell lines in cases of mosaicism. Because the FISH procedure is rapid, it may be useful as an adjunct to routine 3-day stimulated peripheral blood cytogenetic study for diagnosing newborns with ambiguous genitalia.

Similarly, FISH can be performed on cervical smears. The Papanicolaou smear has been established as a useful cytologic screening tool that has contributed greatly to the reduction of cervical cancer-related mortalities. This test, however, cannot reveal underlying genetic damage, e.g., numerical abnormalities, that predispose the individual to a potentially life-threatening neoplasm. The interphase FISH assay is feasible as a potential future screening tool for cervical cancer (34).

Other body tissues that are amenable to interphase cytogenetics analysis using α-satellite probes include, but are not limited to, cerebrospinal fluid, frozen and fixed suspensions of blood, bone marrow, and tumor, hair follicles, urine sediments in bladder and cervical cancer, and other exfoliated cells.

XVII. FISH-BASED ASSAYS FOR ANEUPLOIDY DETECTION IN SPERM

It is estimated that in humans a minimum of 10% of all recognized conceptions are chromosomally abnormal, and that 1–2% are the result of fertilization by spermatozoa with chromosomal abnormalities (35). The potential applications of FISH for analysis of sperm for infertility as well as genotoxicology are discussed in the chapter by Sigman and Mark.

XVIII. SEQUENTIAL FLOW CYTOMETRY AND FISH FOR THE STUDY OF FORMALIN-FIXED, PARAFFIN-EMBEDDED BREAST CANCER CELLS/ SEQUENTIAL HEMATOPATHOLOGY AND FISH

Both flow cytometry and FISH are useful techniques for the analysis of cancer tissues. When the two methods are used to study the same specimens, they are usually performed separately, in parallel. This protocol is problematic in cases where there is a scarcity of material, making completion of both studies impossible. FISH procedures that utilize excess material discarded from flow cytometry would be advantageous. Thus, Mark et al. (36) described a protocol for performing sequential flow cytometry and FISH using formalin-fixed, paraffin-embedded archival material of breast cancer tissues. Analogously, sequential hematopathology and FISH study can also be performed, as illustrated in Color Plates 17.7a and b.

XIX. ASSESSMENT OF CHROMOSOME COPY NUMBER IN FORMALIN-FIXED, PARAFFIN-EMBEDDED ARCHIVAL MATERIALS

FISH is especially suited for retrospective studies in which a cytogenetic study has not been part of the patient's initial workup. FISH analysis allows for the correct cytogenetic results to be coordinated with previous studies. Probes such as bcr/abl, p53, HER-2/neu, c-myc, and N-myc can be used.

FISH analysis can be performed on archival materials of a wide variety of tumors, including breast cancer, prostate cancer, bladder cancer, ovarian cancer, neuroectodermal tumors, testicular carcinoma, endometrial carcinoma, and stomach cancer. We have analyzed 20-year-old archival materials of gestational trophoblastic disease using FISH (18). Here, we will illustrate this approach with our studies of breast cancer.

Breast cancer affects approximately 1 in 8 women in the United States. The current criteria for breast cancer staging include: pathologic parameters such as tumor size, histologic classification, and nodal involvement; cytologic parameters such as nuclear grade; and biologic parameters such as patient age, hormone receptor status, and oncogene amplification. Although there is evidence that some of these factors may be of prognostic significance, information currently available is still fragmented and incomplete. Clearly, more biomarkers are needed. We began our FISH studies in an attempt to address this need, as molecular cytogenetics has until recently been an underdeveloped area of breast cancer research.

The tumors were examined morphologically by H&E staining. Tumors were then analyzed cytogenetically using FISH and a chromosome 8-specific, α-satellite probe as the test probe and a chromosome 12 probe as a control probe, in some cases, to distinguish disomy from diploidy and trisomy from triploidy. Various approaches for specimen preparations were tested to compare their adequacies according to defined criteria. The approaches tested included modifications in specimen preparation such as cell disaggregation from formalin-fixed, paraffin-embedded tissue sections prior to FISH, FISH on thin sections from paraffin blocks, FISH on touch preparations, and FISH from suspensions of cells from fresh tumors. A total of 34 specimens was analyzed in the first study (37). The second and third studies (38,39), using FISH on archival breast tissue (Color Plate 17.8), focused on stage I and stage II disease and employed sample sizes of 34 and 36 cases, respectively. Our data thus far indicate that chromosome 8 trisomy in infiltrating ductal carcinoma is associated with aggressive biologic behavior and poor prognosis. While these results need to be further confirmed and extended in other centers, the data that currently exist suggest that chromosome 8 copy number may be utilized as a possible biomarker to predict patient prognosis.

XX. FISH FOR LOCALIZING DNA SEQUENCES ONTO METAPHASE CHROMOSOMES AND FOR MAPPING VIRAL INTEGRATION SITES

FISH is the most direct method to map a DNA sequence onto metaphase chromosomes. However, mapping can be done only if the cloned DNA sequence is available. Using this approach, Mark and Chow (40), for example, localized the gene encoding the secretin receptor, SCTR, on human chromosome 2q14.1 by FISH and chromosome morphometry. Similarly, the site of integration of human papillomavirus sequences in a cervical tumor cell line was mapped by the technique of FISH (41).

XXI. FISH AS AN INDISPENSABLE TOOL IN THE HUMAN GENOME PROJECT

Somatic cell hybrids (Color Plate 17-9) containing one or more human chromosomes in a rodent background have been used extensively in gene mapping. For mapping of subchromosomal fragments, radiation hybrids are employed. The utility of the irradiation-fusion approach, however, depends on the purity of the hybrid cell lines. Initial characterization and periodic verification are therefore of paramount importance. Both conventional techniques (e.g., GTG banding and G-11 staining) and molecular cytogenetic techniques are useful, but FISH is more sensitive in detecting human chromosomes or chromosomal fragments in a rodent background (42).

XXII. CONCLUDING COMMENTS

In this chapter, we illustrate the delineation of numerical and structural chromosomal abnormalities using FISH with either α-satellite or chromosome painting probes. FISH is a powerful adjunct technique which has revolutionized the field of cytogenetics. Perhaps one of the most important aspects of the FISH technology is that it is within the grasp of most diagnostic cytogenetics laboratories. Because of its accessibility, the impact of FISH in the 1990s rivals that of banding technology in the 1970s.

Although FISH does not have the sensitivity of PCR, it does have the ability to combine with immunophenotyping. Furthermore, both metaphase and interphase cells can be analyzed and can be applied to a range of tissues. Besides FISH developments, computer-enhanced digital imaging has come of age and has expanded the role of cytogenetics, especially for the more complex applications. With comparative genomic hybridization (CGH) (43) and the appropriate computer software, it is possible to compare DNA from normal and diseased cells even in the absence of tumor metaphases and to detect chromosome losses or gains provided they are no less than 10 megabases. The principles and applications of CGH have been discussed in the chapter by Blancato and Haddad.

Moreover, the next generation of cytogenetic techniques based on the ability of the digital image analysis system to designate a specific color to a mixture of fluorochromes has arrived (44). By altering the ratio of four different fluorescence dyes, it is possible to create a different color for each chromosome paint to cover the whole genome. The whole chromosome content of a cell can be analyzed in one hybridization in a process known as multicolor spectral karyotyping (SKY). These newer and more sophisticated techniques are discussed in greater details in the chapter by Blancato and Haddad and the chapter by Green et al.

Because of the rapid advances in the area of molecular cytogenetic technologies (13) and the fact that, for the most part, many FISH probes are not yet FDA-approved for diagnostic purposes, a number of organizations have issued statements urging caution in interpreting FISH-based data in diagnostic settings. For example, the New England Regional Genetics Group (NERGG) (45) recommended that issues of sensitivity, accuracy, and applicability need to be addressed more fully before accepting this technology in routine practice. This is also consistent with the recommendations of the American College of Medical Genetics (ACMG) (46), which state that investigational studies must be distinguished from accepted clinical tests and that this information be clearly given in a qualifying statement.

Important recommendations made by the ACMG (47), the College of American Pathologists (CAP), the U.S. Health Care Financing Administration (HCFA), and the U.S. Food and Drug Administration (FDA), as they relate to genetic testing, have been previously summarized and discussed elsewhere (48,49). Because genetic testing is an area with implications extending far beyond that of the patient, it is only appropriate that this is an area that is under increased scrutiny. The ACMG, founded in 1991, has assumed responsibility for developing standards of practice for the medical genetics community to ensure that high quality is maintained in genetic testing using interphase FISH. The recently published *Standards and Guidelines: Clinical Genetics Laboratories* is expected to be an evolving document that reflects the consensus of participating ACMG members (47–49). Strict adherence to quality assurance/quality control measures is critical, because the rapid pace of new developments with potential applications in the field of medical genetics is placing great pressures on clinical laboratories to implement new procedures quickly and to adopt new test systems for diagnostic purposes.

Summarizing, with the ready access of commercially available recombinant DNA probes, FISH has become an integral part of genetic testing laboratories, even with increased selectivity resulting from the advent of managed care. Although other molecular cytogenetic technologies such as CGH and SKY exist, discussion in this chapter has focused on FISH using commercially available probes in practical applications that can be utilized readily by the average clinical cytogenetic laboratory. The stark reality is that in this age of corporate downsizing and increased emphasis on cost containment, the most advanced technology may not be within the reach of most small clinical cytogenetic laboratories many of which are located in small departments of pathology throughout the country.

ACKNOWLEDGMENTS

Thanks are extended to Dr. Roger Mark for continued support and to Dr. Yvonne Mark for computer and editorial assistance. The support of the former staff of the Lifespan Academic Medical Center Cytogenetics Laboratory, which celebrated its third decade of service, is also acknowledged.

REFERENCES

1. Mark HFL, Bier JB. Disappearing trisomy 8 mosaicism in a patient initially referred to rule out Fragile X Syndrome. Ann Clin Lab Sci 1997; 27:293–298.
2. Callen D, Freemantle C, Ringenbergs M, et al. The isochromosome 18p syndrome: confirmation of cytogenetic diagnosis in nine cases by *in situ* hybridization. Am J Hum Genet 1990; 47:493–498.
3. Callen DF. The classification of marker chromosomes by *in situ* hybridization. Mol Cytogenet 1991; 1:1–2.
4. Phelan M, Anderson K, Martin K, et al. Isochromosome 18p in two unrelated males. Proc Greenwood Genet Center 1991; 10:23–27.
5. Mitelman F. ISCN, 1995: An International System for Human Cytogenetic Nomenclature. Basel: Karger, 1995.
6. Rooney DE, Czepulkowski BH. Human Cytogenetics: A Practical Approach. Volume I: Constitutional Analysis. Oxford, England: Oxford University Press, 1992.
7. Abuelo D, Mark HFL, Bier J. Developmental delay caused by a supernumerary chromosome, inv dup (15), identified by fluorescent *in situ* hybridization. Clin Ped 1995; 34:223–226.
8. Ledbetter DH. Microdeletion syndromes: FISH is responsible for the death of high resolution cytogenetics. The Genetic Viewpoint 1995; 2:5–6.
9. Kuwanao A, Ledbetter SA, Dobyns WB, Emanuel BS, Ledbetter DH. Detection of deletions and cryptic translocations in Miller-Dieker syndrome by *in situ* hybridization. Am J Hum Genet 1991; 49:707–714.
10. Kuwano A, Mutirangura A, Dittrich B, et al. Molecular dissection of the Prader-Willi/Angelman syndrome region (15q11-13) by YAC cloning and FISH analysis. Hum Molec Genet 1992; 1:417–425.

11. Driscoll DA, Salvin J, Sellinger B, et al. Prevalence of 22q11 microdeletions in DiGeorge and velocardiofacial syndromes: implications for genetic counseling and prenatal diagnosis. J Med Genet 1993; 30:813–817.
12. Desmaze C, Scambler P, Prieur M, et al. Routine diagnosis of DiGeorge syndrome by fluorescence *in situ* hybridization. Hum Genet 1993; 90:663–665.
13. Mark HFL, Jenkins R, Miller WA. Current applications of molecular cytogenetic technologies. Ann Clin Lab Sci 1996; 27(1):47–56.
14. Mark HFL, Meyers-Seifer CH, Seifer DB, DeMoranville BM, Jackson IMD. Assessment of sex chromosome composition using fluorescent *in situ* hybridization as an adjunct to GTG-banding. Ann Clin Lab Sci 1995; 25(5):402–408.
15. Mark HFL, Bayleran JK, Seifer DB, Meyers-Seifer CH. A combined cytogenetic and molecular approach for diagnosing delayed puberty. Clin Ped 1996; 35(2):62–66.
16. Mark HFL, Sikov W, Safran H, King TC, Griffith RC. Fluorescent *in situ* hybridization for assessing the proportion of cells with trisomy 4 in a patient with acute non-lymphoblastic leukemia. Ann Clin Lab Sci 1995; 25:330–335.
17. Mark HFL, Ahearn J, Lathrop JC. Constitutional trisomy 8 mosaicism and gestational trophoblastic disease. Cancer Genet Cytogenet 1995; 80:150–154.
18. Mark HFL, Grollino MG, Sulaiman RA, Lathrop JC. Fluorescent *in situ* hybridization (FISH) assessment of chromosome 8 copy number in gestational trophoblastic disease. Ann Clin Lab Sci 1995; 25:291–296.
19. Mark HFL, Afify A, Taylor W, Santoro K, Lathrop JC. A subset of gestational trophoblastic disease (GTD) characterized by abnormal chromosome 8 copy number detected by FISH? Cancer Genet Cytogenet 1997; 96:1–16.
20. Rowley JD. A new consistent chromosomal abnormality in chronic myelogenous leukemia identified by quinacrine fluorescence and Giemsa staining. Nature 1973; 234:290–293.
21. Young C, DiBenedetto J, Glasser L, Mark HFL. A Philadelphia chromosome positive CML patient with a unique translocation studied via GTG-banding and fluorescence *in situ* hybridization. Cancer Genet Cytogenet 1996; 89:157–162.
22. Cotter FE. The emergence of FISH and Chip technology in molecular pathology. VYSIONS 1997; 2:1–7.
23. Mark HFL, Dong H, Glasser L. A multimodal approach for diagnosing patients with acute promyelocytic leukemia (APL). Cytobios 1996; 87:137–149.
24. Mark HFL. Bone marrow cytogenetics and hematologic malignancies. In: Kaplan B, Dale K, eds. The Cytogenetics Symposia. Burbank, CA: The Association of Cytogenetic Technologists, 1994.
25. Chen Z, Morgan R, Stone JF, Sandberg AA. FISH: a useful technique in the verification of clonality of random chromosome abnormalities. Cancer Genet Cytogenet 1993; 66:73–74.
26. Kasprzyk A, Mehta AB, Secker-Walker LM. Single-cell trisomy in hematologic malignancy. Random change or tip of the iceberg? Cancer Genet Cytogenet 1995; 85:37–42.

27. Mark HFL, Rehan J, Mark S, Santoro K, Zolnierz K. FISH analysis of single-cell trisomies for determination of clonality. Cancer Genet Cytogenet 1998; 102:1–5.
28. Chen Z, Wang DD, Peier A, Stone JF, Sandberg AA. FISH in the evaluation of pleural and ascitic fluid. Cancer Genet Cytogenet 1995; 84:116–119.
29. Cajulis RS, Haines GK III, Frias-Hidvegi D, McVary K, Bacus JW. Cytology, flow cytometry, image analysis, and interphase cytogenetics by fluorescence *in situ* hybridization in the diagnosis of transitional cell carcinoma in bladder washes: a comparative study. Diagn Cytopathol 1995; 13:214–223.
30. Mark HFL. Fluorescent *in situ* hybridization as an adjunct to conventional cytogenetics. Ann Clin Lab Sci 1994; 24(2):153–163.
31. Mark HFL. Recent advances in molecular cytogenetics: fluorescent *in situ* hybridization. Rhode Island Med 1994; 77:377–381.
32. Mark HFL, Hann E, Mikumo R, Lauchlan S, Beauregard L, Braun L. Cytogenetic characterization of three cell lines derived from primary cervical tumors. Ann Clin Lab Sci 1995; 25:193–207.
33. Mark HFL, Mills DR, Kim E, Santoro K, Quddus MR, Lathrop JC. Fluorescent *in situ* hybridization assessment of chromosome copy number in buccal mucosal cells. Cytobios 1996; 87:117–126.
34. Mark HFL, Mills DR, Santoro K, Quddus MR, Lathrop JC. Fluorescent *in situ* hybridization (FISH) analysis of cervical smears: a pilot study of 20 cases. Ann Clin Lab Sci 1997; 27:224–229.
35. Mark HFL, Sheu M, Lopes L, et al. Towards the development of a fluorescent *in situ* hybridization based genotoxicologic assay for aneuploidy detection in sperm. Cytobios 1995; 82:171–180.
36. Mark HFL, Rausch ME, Taylor WM, et al. Sequential flow cytometry and fluorescent *in situ* hybridization (FISH) for the study of formalin-fixed paraffin-embedded breast cancer cells. Cancer Genet Cytogenet. 1998; 100:129–133.
37. Afify A, Bland KI, Mark HFL. Fluorescent *in situ* hybridization assessment of chromosome 8 copy number in breast cancer. Breast Cancer Res Treat 1996; 38:201–208.
38. Afify A, Mark HFL. FISH assessment of chromosome 8 copy number in stage I and stage II infiltrating ductal carcinoma of the breast. Cancer Genet Cytogenet 1997; 97:101–105.
39. Mark HFL, Afify A, Taylor W, et al. Stage I and stage II infiltrating ductal carcinoma of the breast analyzed for chromosome 8 copy number using FISH. Pathobiology 1997; 65:184–189.
40. Mark HFL, Chow BK-C. Localization of the gene encoding the secretin receptor, SCTR, on human chromosome 2q14.1 by fluorescence *in situ* hybridization and chromosome morphometry. Genomics 1995; 29:817–818.
41. Mark HFL, Santoro K, Hann E, Mikumo R, Lauchlan S, Braun L. Integration of human papillomavirus sequences in cervical tumor cell lines. Ann Clin Lab Sci 1996; 26:147–153.
42. Mark HFL, Santoro K, Jackson C. A comparison of approaches for the cytogenetic characterization of somatic cell and radiation hybrids. Appl Cytogenet 1992; 18:149–156.

43. Kallioniemi O-P, Kallioniemi A, Sudar D, et al. Comparative genomic hybridization: a rapid new method for detecting and mapping DNA amplification in tumors. Semin Cancer Biol 1993; 4:41–46.
44. Schrock E, duManoir S, Veldman T, et al. Multicolor spectral karyotyping of human chromosomes. Science 1996; 273:494–497.
45. New England Regional Genetics Group: Fluorescent *in situ* hybridization policy statement. The Genetics Resource 1992; 6:56.
46. American College of Medical Genetics: Prenatal interphase fluorescence *in situ* hybridization (FISH) policy statement. Am J Hum Genet 1993; 53:526–527.
47. American College of Medical Genetics (ACMG). Standards and Guidelines: Clinical Genetics Laboratories. ACMG Laboratory Practice Committee, 1994; 1–34.
48. Mark HFL, Kelly T, Watson MS, Hoeltge G, Miller WA, Beauregard L. Current issues of personnel and laboratory practices in genetic testing. J Med Gene 1995; 32:780–786.
49. Mark HFL, Watson MS. Evolving standards of practice for clinical cytogenetics. Rhode Island Med 1994; 77:375–376.

APPENDIX: GLOSSARY OF TERMS

ACMG

American College of Medical Genetics. Founded in 1991, the ACMG is the clinical standards and practice arm of the medical genetics community.

acrocentric chromosome

A chromosome with the centromere positioned near one end of the chromosome. Acrocentric chromosomes possess satellites that are connected to the main chromosomes by stalks.

α-satellite probe

A labeled nucleic acid fragment used in fluorescent *in situ* hybridization that hybridizes to its respective target sequences at the centromeric locations of specific chromosomes.

AML

Acute myeloid leukemia. A type of acute nonlymphocytic leukemia in which there is an accumulation of leukemic cells in the bone marrow, which end up circulating into the bloodstream and infiltrating other tissues such as the lymph nodes, the liver, the spleen, the skin, other viscera, and the central nervous system.

amniocentesis

A prenatal diagnostic procedure which involves the removal of amniotic fluid for cytogenetic and/or biochemical analysis.

Angelman syndrome
A syndrome associated with a maternally derived deletion of 15q11-13. Some common characteristics include short stature, characteristic facies, obesity, small extremities, poor muscle tone, and mental retardation.

Barr body
An inactivated X chromosome that, upon Papanicolau staining, is manifested as a darkly stained mass in the somatic cells of the mammalian female.

CAP
College of American Pathologists. The organization that develops standards for accreditation of clinical pathology laboratories, and defines parameters for physical resources and environment, organizational structure, personnel, safety, QA/QC, and record keeping.

centromere
The central location of the chromosome to which spindle fibers attach, also called the primary constriction.

CGH
Comparative Genomic Hybridization. A molecular cytogenetic method for analyzing a tumor for genetic changes, such as gain or loss of chromatin material.

Charcot-Marie-Tooth disease
A demyelinating disorder; some symptoms include distal weakness in the legs, decreased reflexes, and hypertrophic neuropathy.

chromosome painting probe
A FISH reagent that allows for studying the presence or absence of particular chromosomes or chromosomal fragments in a variety of settings, including detection of species-specific material in somatic cell hybrids. Chromosome painting is an especially effective technique for delineating structural rearrangements such as translocations.

clone
A subpopulation of genetically identical cells.

Cri-du-Chat syndrome
Deletion of 5p15.2. Some characteristic manifestations of the disease are microencephaly, a moonlike face, and distinct mewlike cries of a cat during infancy.

de novo rearrangements
Structural chromosomal changes that apparently are not inherited from either parent.

DiGeorge syndrome
Deletion of 22q11.2 leading to a T-cell deficiency from the maldevelopment of thymic epithelial components. Some characteristics include abnormal ears, micrognathia, malfunctioning antibody responses, and a tiny thymus.

diploid
Having two sets of chromosomes.

disomy
Two copies of a particular chromosome.

Down syndrome
Usually trisomy 21; the most prevalent kind of human aneuploidy, commonly caused by nondisjunction of chromosome 21 in a normal parent, especially if the mother gives birth at a later age. Some characteristics include mental retardation, eyes with an epicanthic fold, short stature, and a wide, flat face. A small percentage of Down syndrome patients have a translocation, and some are mosaics.

Edwards syndrome
Trisomy 18; some common characteristics of those affected with the syndrome include "faunlike" ears, rockerbottom feet, a small jaw, a narrow pelvis, and death within a few weeks of birth.

ESAC
Extra structurally abnormal chromosome. A chromosome that cannot be identified, also called a marker chromosome or supernumerary chromosome.

euchromatin
Chromatin that is universally perceived as gene-bearing and which also stains normally.

FDA
Food and Drug Administration. The U.S. governmental agency that regulates the distribution of food and drugs to the general public.

FISH
Fluorescent *in situ* hybridization. A technique that employs fluorescent probes, which hybridize to homologous sequences on the chromosomes.

GTD
Gestational trophoblastic disease. A term that includes a number of conditions such as choriocarcinoma, hydatidiform mole, and invasive mole, all disorders that are due to the epithelial malignancy of the trophoblasts.

GTG banding
A technique that uses trypsin and Giemsa to induce light- and dark-stained regions along the lengths of the chromosomes.

HCFA
Health Care Financing Administration. A government organization that oversees the activities of U.S. clinical laboratories.

heterochromatin
Chromatin that is tightly coiled and devoid of genes, usually stained positive with C-banding, which can be constitutive or facultative.

holoproencephaly
The condition of having underdeveloped facial features (e.g., cleft palate and lip and low-set ears) due to the failure of the proencephalon to cleave into separate hemispheres.

hypergonadotropic
Related to or caused by a surfeit of gonadotropins, stimulatory hormones of the gonads.

hypotonia
Reduced toning of the skeletal muscles.

i(18p)
A marker comprised of two short arms of chromosome 18, which results in symptoms of low birth weight and distinct facial features.

interstitial deletion
Chromosomal loss of a segment between the two ends of the chromosome that have rejoined.

inverted duplication
A segment of the chromosome that is duplicated and is found in the chromosome in the reverse orientation.

ISCN
An International System for Human Cytogenetic Nomenclature. The standard classification and naming system for numeral and structural chromosomal abnormalities as well as a system establishing rules of nomenclature.

isochromosome
A chromosome with two identical arms.

isodicentric chromosome
A chromosome with two identical arms and two centromeres.

Klinefelter syndrome
A condition that is usually associated with a XXY chromosome constitution. Some characteristics of those affected include gynecomastia, slight mental retardation, tall stature, underdeveloped testes, and sterility.

marker chromosome
A structurally abnormal chromosome that cannot be identified with currently available technology.

MDS
Myelodysplastic syndrome. A condition caused by a disorder in the maturation of erythroid, myeloid, and megakaryocytic cells. Some common cytogenetic abnormalities found in MDS include a deletion of the long arm of chromosome 5 and trisomy for chromosome 8.

microdeletion
A very small deletion that is usually not detectable by routine chromosome analysis.

Miller-Dieker syndrome

Deletion of 17p13.3. Some characteristics include lissencephaly, microencephaly, epilepsy, and micrognathia.

mitotic index

The percentage of cells that are in metaphase among the entire sample of cells (in interphase and in metaphase).

monosomy

A missing homolog of a specific chromosome pair (e.g., in monosomy 7, a chromosome 7 is missing).

mosaicism

The state of having tissues that are composed of two or more genetically distinct cell types, derived from the same zygote.

NERGG

New England Regional Genetics Group. A genetics organization that is comprised of medical geneticists, genetic counselors, and other health care professionals.

Papanicolaou smear test

A cytologic procedure that has been established as a useful screening tool for cervical cancer.

Patau syndrome

Trisomy 13; some common characteristics of those affected with the syndrome include a harelip, a tiny deformed head, and death within a few days of birth.

Prader-Willi syndrome

A syndrome usually associated with a paternally derived deletion of 15q11-13. Some common characteristics include short stature, obesity, small extremities, poor muscle tone, and mental retardation.

pseudodicentric chromosome

A dicentric chromosome with only one active centromere.

radiation hybrids

Human–rodent hybrid cells that are produced from a lethal irradiation of a somatic hybrid cell line and subsequent fusion with a rodent cell line to produce hybrids containing small pieces of human chromosomes in a rodent background for gene mapping studies.

ring chromosome

A circular-like chromosome usually resulting from the loss of its telomeric chromosome sequences and the rejoining of its chromosome ends.

satellite

A terminal area of the chromosome that extends beyond the centromere and is connected to the main chromosome by a stalk.

sex chromatin

The manifestation of the inactivated X chromosome in female mammals, also called the Barr body.

SKY
Spectral karyotyping. The labeling of chromosomes using a spectrum of different colors, one of the newest molecular cytogenetic techniques.

Smith-Magenis syndrome
A syndrome that is usually associated with a deletion of 17p11.2.

somatic cell hybrids
Cells formed by a fusion between cells from two different species (e.g., human–mouse, human–hamster or mouse–hamster combinations) which are utilized extensively by researchers participating in the Human Genome Project.

Steroid sulfatase deficiency/Kallman syndrome
Microdeletion of Xp22.3, which results in olfactory bulb defects and hypogonadotropic hypogonadism.

supernumerary chromosome
An extra marker chromosome that cannot be identified by current cytogenetic technology.

tandem duplication
A repeated segment on the chromosome that lies immediately adjacent to the original in the same orientation.

terminal deletion
Loss of a segment of a chromosome from one of its ends.

3:1 segregation
An aberrant mode of segregation in a translocation heterozygote in which three chromosomes migrate to one pole and one chromosome migrates to the opposite pole.

translation
The process by which proteins are synthesized by the aid of ribosomal units that assemble the polypeptide, whose amino acid sequence is determined by the mRNA codon.

triploidy
The state of having three sets of chromosomes.

trisomy
An extra copy of a specific chromosome, usually resulting from nondisjunction.

Turner syndrome
Usually manifested in a female with a single X chromosome. Some characteristics of those affected include short stature, poor breast development, underdeveloped reproductive organs, characteristic facial features, a constricted aorta, and sterility.

Velocardiofacial syndrome
A condition associated with a deletion of the 22q11.2 region.

Williams syndrome
A condition associated with a deletion of the 7q11.23 region. A few outstanding characteristics include excessive talkativeness, mental retardation, a hoarse voice, stellate iris, congenital heart disease, periorbital fullness, thick lips, and a lack of visuospatial coordination.

Wolf-Hirschhorn syndrome
A condition associated with a deletion of the 4p16.3 region.

18
Evolving Molecular Cytogenetic Technologies

Gabriela Adelt Green, Evelin Schröck, and Tim Veldman
National Human Genome Research Institute, National Institutes of Health, Bethesda, Maryland

Kerstin Heselmeyer-Haddad, Hesed M. Padilla-Nash, and Thomas Ried
National Cancer Institute, National Institutes of Health, Bethesda, Maryland

I. INTRODUCTION

Conventional cytogenetics and molecular biology have each played important roles in the identification of chromosomal aberrations associated with human disease, particularly malignancies. Nonrandom chromosomal changes often reflect events occurring at the molecular level within the cell and provide important clues about the location of genes involved in these events (1,2). Although traditional chromosomal banding techniques are critical in the assessment of karyotype changes, these techniques have certain inherent limitations that complicate accurate characterization of tumor genomes (3). These limitations apply particularly to solid tumors and include: (a) difficulty in culturing tumor cells, which typically have a low mitotic index and produce chromosomes of poor quality; (b) specimens contaminated by necrotic tissue, making subsequent analyses problematic; (c) the selective growth, in culture, of subclones that may not be representative of the in vivo tumor; and (d) the presence of complex karyotypes, which often precludes reliable, comprehensive identification and characterization of chromosomal abnormalities.

Molecular cytogenetic techniques utilizing fluorescent in situ hybridization (FISH) with chromosome-specific DNA probes have facilitated the identification of chromosomal aberrations and can be performed on nondividing cells (4,5). The use of FISH is limited to targeted chromosomes or subchromosomal regions for confirmation of the presence of a previously detected aberration (6,7). Therefore, in contrast to traditional cytogenetics, FISH cannot serve as an initial screening tool for identifying unknown chromosomal abnormalities (8). Recent advances in molecular cytogenetics have led to the development of techniques that combine the sensitivity and specificity of FISH with the global screening ability of conventional cytogenetics. Comparative genomic hybridization and spectral karyotyping are two important new technologies that permit surveillance of entire tumor genomes for chromosomal aberrations in a single experiment. Each of these methods offers distinct advantages for karyotypic evaluation of tumors.

II. COMPARATIVE GENOMIC HYBRIDIZATION

Comparative genomic hybridization (CGH) is a molecular cytogenetic assay that screens for DNA copy number changes globally within a complete genomic complement and maps these changes onto normal chromosomes (9,10). The technique is based on the simultaneous hybridization of differentially labeled total genomic tumor DNA and normal reference DNA onto normal human metaphase chromosomes. CGH does not require culturing of tumor cells (only tumor DNA is needed) or prior knowledge of the aberrations in the genome to be studied. The results of such analyses are depicted in karyograms, which indicate regions of tumor-specific DNA gains and losses.

A. General Methodology of CGH

CGH utilizes quantitative two-color fluorescent in situ suppression hybridization (Fig. 1). Total genomic DNA is isolated from a tumor specimen (test DNA) and from an individual with a normal karyotype (reference DNA). The two DNA samples are then differentially labeled with fluorochrome-tagged nucleotides (e.g., the test DNA labeled with ''green'' fluorescein-dUTP and the reference DNA labeled with ''red'' rhodamine-dUTP) by standard nick translation procedures (Fig. 1a). The samples are pooled (using equal amounts of each) and hybridized to normal human metaphase chromosome spreads (Fig. 1b). During the hybridization, there is a ''competition'' between homologous segments of tumor and normal DNA for the ability to anneal to the same chromosomal targets on the metaphase chromosomes. The ubiquitous presence of repetitive sequences in human DNA requires that steps be taken to suppress the cross-hybridization of the repeats within the samples. This is accomplished by including a large excess of unlabeled human DNA, fractionated to enrich for repetitive sequences (e.g., Cot1 DNA). Use of such a ''suppression hybridization'' strategy prevents the repetitive sequences from interfering with the detection of single-copy-

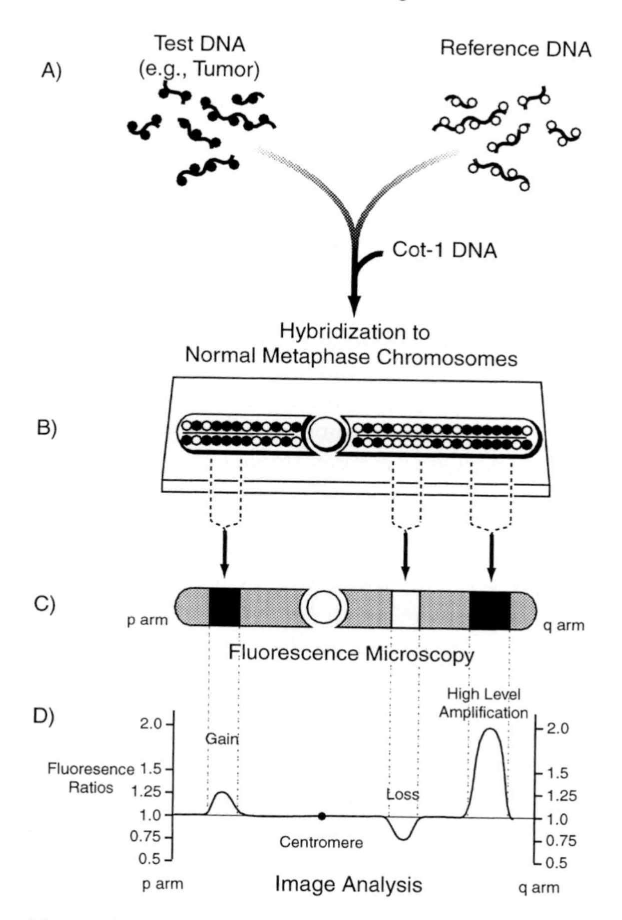

Figure 1 Schematic diagram outlining the CGH technique. (a) Test and reference DNAs are differentially labeled with fluorochromes [e.g. test DNA in green (closed circles) and reference DNA in red (open circles)] and (b) hybridized to normal metaphase chromosome spreads. (c) Fluorescence microscopy is used to generate a ratio image, which displays the differences in fluorescence intensities along the length of each chromosome, with green staining (depicted in black) indicative of tumor gain(s) and red staining (depicted in white) indicative of tumor loss(es). The areas depicted in gray represent regions of the chromosome that do not have copy number changes (''normal'' regions). (d) Fluorescence ratio profiles are calculated along the axis of each chromosome, indicative of copy number changes in the test DNA relative to the reference DNA. A ratio of 1.0 indicates that there are no copy number changes (e.g. a ''normal'' region of the chromosome), while ratios of 0.75 or less indicate a loss of DNA and ratios of 1.25 or greater indicate DNA gains at the corresponding location along the chromosome axis. Ratios of 2.0 or greater indicate the presence of a high level DNA amplification.

number changes in the tumor genome. The metaphase chromosomes are counterstained with DAPI for subsequent classification by banding pattern recognition.

Following hybridization, differences in fluorescence intensities along individual metaphase chromosomes are measured by fluorescence microscopy (Fig. 1c). These differences reflect copy number changes in the corresponding tumor DNA. If the tumor karyotype is normal, the observed fluorescence reflects equal hybridization of tumor (green) and reference (red) DNA. If there is gain of tumor DNA (involving either an entire chromosome or subchromosomal region), there is an increase in fluorescence signal from the tumor DNA. This is visualized as green staining of the corresponding region on the chromosome. Conversely, loss of tumor DNA is visualized as red staining of the lost region on the chromosome. For example, amplification of a tumor oncogene would appear as a green signal at the cytogenetic position corresponding to the location of the oncogene on normal metaphase chromosomes.

Quantitative measurements of fluorescence intensities are made by digital image analysis using fluorescence microscopy with a cooled charge-coupled device (CCD) camera, specific optical filters, and specialized computer software (11,12). Following image acquisition, fluorescence ratio profiles are calculated along the axis of each chromosome (Fig. 1d). Average ratio profile calculations are based on the analysis of at least five metaphase spreads per tumor. The ratios represent tumor genome (green) to reference genome (red) fluorescence hybridization signal intensities and are indicative of changes in copy number. The "normal" ratio of tumor genome to reference genome is 1.0, indicating that no copy number changes have occurred in the tumor. Ratios of 0.75 or less indicate loss or deletion of a whole chromosome or portion of a chromosome, and ratios of 1.25 or more indicate chromosomal gains. A ratio of greater than 2.0 is indicative of a high-level amplification. The end result of the image analysis is the generation of a "copy number karyotype" for the tumor (Color Plate 18.2), indicating relative gains and losses of segments along the length of each chromosome.

B. CGH Applications

In a single experiment, CGH provides a global view of the chromosomal aberrations present within a tumor. It only requires genomic tumor DNA, without demanding the often difficult task of culturing tumor cells. In addition, and most important, DNA extracted from formalin-fixed, paraffin-embedded tissue can be utilized (13). The ability to gain insight about the karyotype of a tumor derived from archived tissue has several advantages, including the ability: (a) to use large numbers of specimens for study; (b) to evaluate tumors that are difficult to culture; (c) to retrospectively identify chromosomal aberrations within a previously characterized tumor (thereby facilitating the correlation of cytogenetic findings with histologic/histochemical information, clinical course, and prognosis); and (d) to analyze small targeted regions of a histologically defined lesion (e.g., foci of carcinoma microdissected from surrounding normal tissue) (14,15).

Another distinct advantage of CGH is the need for small amounts of tumor DNA (less than 1 μg). Modified methods for extracting and labeling DNA for CGH analysis of formalin-fixed, paraffin-embedded tissues are available (16). Using a universal DNA amplification technique, termed degenerate oligonucleotide primed PCR (DOP-PCR) (17,18), as few as 2000 cells from a formalin-fixed, paraffin-embedded tumor are required, representing a tumor size of less than 1 mm^3 (13). Such a small amount of material can be easily obtained from almost any routine tissue section. Concordance between CGH analyses performed on matched fresh and formalin-fixed tissue samples is high (95%) (16). In addition, 70–90% of all archived, formalin-fixed material (including very old samples and tissues derived from autopsies) has been found to be suitable for CGH analysis (16), reiterating the great utility of this technique for retrospective genome analysis.

CGH has been used to evaluate a wide variety of tumor specimens. It has been used to map chromosomal copy number changes in common neoplasms, such as lung (19,20), breast (21–23), colon (14,24), brain (25–28), head and neck (29,30), prostate (31–34), hematologic malignancies (35–38), kidney (39–41), ovary (42), bladder (43), and uterine cervix (15,44,45). Analyses of these different tumor types have revealed general patterns of chromosomal aberrations, with both the number and genomic location of such aberrations defining certain tumor types. For example, squamous cell carcinomas of the head and neck typically acquire extra copies of chromosome arm 3q accompanied by loss of 3p (29,30), while glioblastomas characteristically demonstrate a gain of chromosome 7 and loss of chromosome 10 (25,28).

Chromosomal mapping by CGH may direct gene identification efforts to specific regions of the genome. The analysis of breast cancers, for instance, has revealed a number of previously undetected amplification sites on chromosomes 17q and 20q (21). Investigation of these areas is ongoing in an attempt to identify the gene(s) whose overexpression may be involved in breast carcinogenesis (46,47).

CGH has also been important in elucidating some of the chromosomal changes associated with tumor progression, allowing for a better correlation between genotype and phenotype. A good example is found in studies performed on neoplastic lesions of the uterine cervix (15,45). A gain of chromosome 3q was identified in 9 of 10 cases of invasive carcinoma, but observed only sporadically in severe dysplasias. The gain of chromosome 3q may be required for the transformation of cervical epithelium from an in situ to an invasive lesion, possibly making this a useful clinical marker for predicting the potential of cervical dysplasias to progress.

The study of other tumors by CGH has improved our understanding of the correlation between tumor behavior and genetic changes. A recent study compared chromosomal aberrations in primary prostate tumors to recurrent prostate tumors that had become resistant to pharmaceutical castration (androgen ablation therapy) (33,34). Amplification of a well-defined region on the short arm of the X chromosome was identified in the therapy-resistant tumors, corresponding to the androgen receptor gene. Thus, these recurrent tumors selectively amplified this receptor, permitting utilization of minute amounts of residual androgen to sustain growth.

The use of CGH for the identification of chromosomal aberrations in tumors has contributed to our understanding of the genetic changes associated with carcinogenesis. This technology will continue to guide efforts to isolate the gene(s) responsible for tumor formation and progression. CGH should also prove to be useful in the clinical arena, with the information it yields being translated into improved diagnostic and prognostic tools.

C. CGH Limitations

Comparisons of CGH with chromosome banding analyses (10) as well as confirmatory FISH experiments (25) have demonstrated the accuracy of this technique in identifying gains and losses in tumor DNA. The ability of CGH to detect small regions of DNA gain or loss is limited primarily by the degree of condensation of the metaphase chromosomes (the more condensed the chromosomes, the lower the resolution) and the magnitude of the copy number changes. The resolution limit for low-level copy number changes (loss or gain of a single copy) is estimated to be in the range of 10–20 million base pairs (48). If the copy number changes are high (as in high-level amplifications seen with some oncogenes), amplicons as small as a few hundred kilobases may be detected by CGH (48). The smallest detectable deletion is thought to be in the range of approximately 3–5 million base pairs (48). In addition, CGH only detects copy number changes relative to the average copy number in the entire tumor (e.g., diploid tumors cannot be distinguished from triploid or tetraploid tumors).

Another important limitation of CGH is its inability to detect balanced genetic aberrations that fail to produce copy number changes (such as inversions and balanced translocations). Similarly, it does not provide any information about the way in which chromosomal segments involved in gains and losses are arranged in aberrant chromosomes. Furthermore, CGH results reflect an averaging of the most common aberrations found in a tumor population and can reliably detect a chromosomal gain or loss only if present in greater than 50% of the cells analyzed (13). This is an important point when analyzing tumors that are intermixed with an abundant population of normal reactive and/or stromal cells or those having a great deal of clonal heterogeneity or genetic diversity.

III. SPECTRAL KARYOTYPING

Another molecular cytogenetic technique, termed spectral karyotyping (SKY), has recently been developed to complement CGH, FISH, and conventional cytogenetics in karyotype analysis (49,50). SKY permits the simultaneous visualization of all human chromosomes, with each chromosome appearing in a different color. This method combines the karyotype screening ability of CGH with the ability of FISH to identify structural and numerical aberrations that may not affect copy number. SKY generates a readily interpretable "color karyotype" of each metaphase analyzed.

A. General Methodology of SKY

SKY is based on the simultaneous hybridization of a pool of probes (specific for each of the 24 human chromosomes) to test (tumor) metaphase spreads (Fig. 3). A specialized imaging technique (termed spectral imaging) is used to produce a karyotype in which each chromosome is represented by a unique color. The chromosome-specific probes are generated from flow-sorted (51) or microdissected (52) chromosomes (Fig. 3a). Each chromosome-specific probe is labeled with either a single fluorochrome or a unique combination of two to three fluorochromes (Fig. 3b) (49). By using a total of five fluorochromes, a pool of 24 (representing 22 autosomes and 2 sex chromosomes) uniquely labeled DNA probes can be generated, each producing a distinct color. This probe pool is subsequently hybridized to tumor metaphase chromosome spreads (Fig. 3c).

The hybridization is visualized using spectral imaging, which combines epifluorescence microscopy, CCD imaging, and Fourier spectroscopy to measure the entire fluorescence spectrum emitted by each chromosome (Fig. 3d) (53,54). Due to the unique fluorochrome combinations used for labeling, each chromosome emits a unique fluorescence spectrum, which is detected when viewing the metaphase through a specialized optical filter. These unique spectra are converted into and visualized as distinct colors. Thus, each chromosome appears to be "painted" a different color, depending on the fluorochrome or multifluorochrome combination with which it was labeled. The imaging technique allows for the simultaneous detection

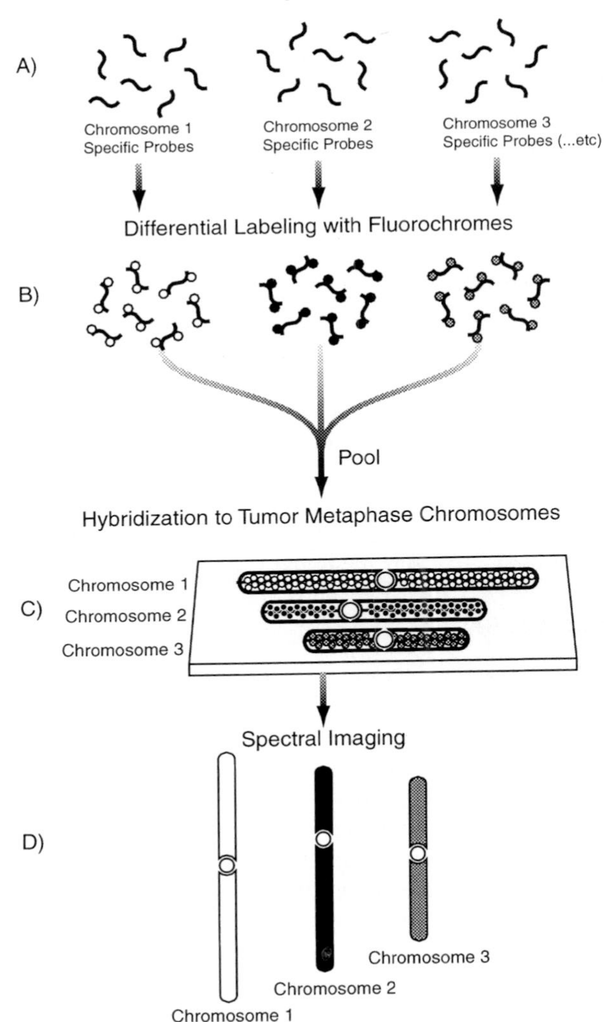

Figure 3 Schematic diagram outlining the SKY technique. (a) Chromosome-specific probes are generated by flow sorting or microdissection of chromosomes and (b) differentially labeled with fluorochromes, to produce a pool of 24 uniquely labeled probes (representing each human chromosome). (c) The probe pool is hybridized to tumor metaphase chromosomes. (d) Spectral imaging permits the simultaneous visualization of all of the chromosomes, with each chromosome "painted" in a different color. In this example, chromosome 1 is labeled with probe represented by open circles, visualized as a chromosome "painted" in white, while chromosome 2 is labeled with a probe represented by closed circles, visualized as a "black" chromosome. Likewise, chromosome 3 is labeled with a probe represented by gray circles, visualized as a chromosome "painted" in gray.

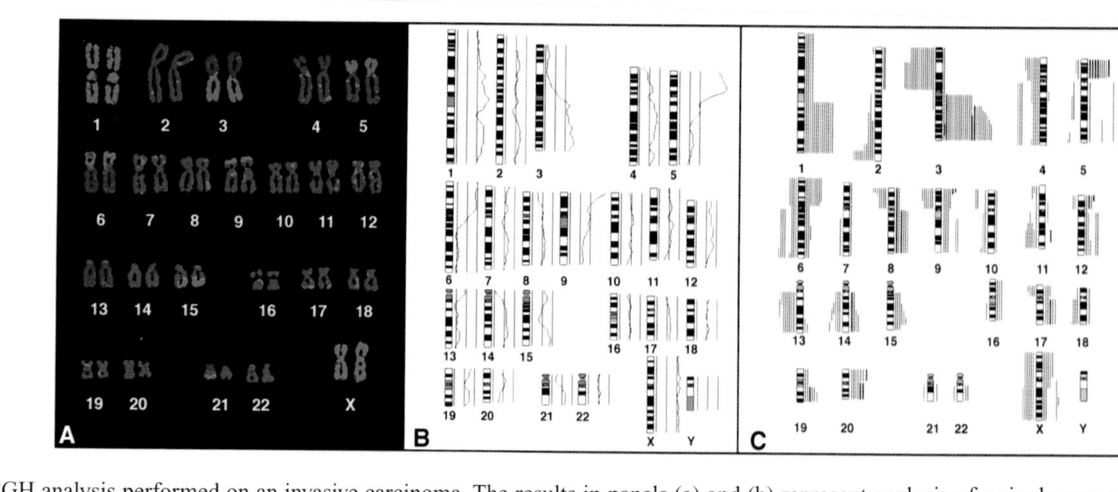

Figure 18.2 CGH analysis performed on an invasive carcinoma. The results in panels (a) and (b) represent analysis of a single case of cervical carcinoma, stage IIB. (a) The ratio image karyotype of one metaphase analyzed from the tumor. The green regions represent areas of tumor gain and the red regions represent areas of tumor loss. The blue regions represent areas without copy number changes. (b) The average fluorescence ratio profile computed on the basis of 10 ratio images (each ratio image is derived from the analysis of a single metaphase). Three lines are present to the right of the chromosome ideograms (the ideograms depict the chromosomes with the "p" or short arm oriented up). The center line represents a ratio of 1.0 (regions of the chromosome without losses or gains in DNA). The line to the right of center represents a ratio of 1.25. Deviations along or to the right of this line represent areas of chromosomal gain. The line to the left of center (nearest the ideogram) represents a ratio of 0.75. Deviations along or to the left of this line represent areas of chromosomal loss. For example, the short arm of chromosome 3 is lost (-3p), while the long arm of chromosome 3 is gained (+3q). This is depicted in the ratio image karyotype as a red 3p and a green 3q arm. In contrast, no copy number changes are detected for chromosome 2, whose ratio image does not deviate from the center line and whose karyotype color is blue. Copy number increases (gains) are present on chromosomes 1, 3q, 5p, 6p, 9p, and X, while copy number decreases (losses) are found on chromosomes 3p, 6q, and 13. (c) Summary karyogram of net gains and losses in 30 cases of advanced-stage cervical carcinoma (including the single case example used in panels a and b). Lines to the right of the ideograms represent regions of chromosomal gains and lines to the left of the ideograms represent losses. Each line represents the results from an individual case. Heavy lines indicate regions with high-level amplifications. For example, the karyogram of chromosome 3 can be interpreted as follows, from left to right: 3 cases with partial loss of 3p, 11 cases with complete loss of 3p, 1 case with complete loss of 3p extending through the centromere into proximal 3q, 1 case with gain of chromosome 3 (with a high-level amplification detected distally on 3q), 13 cases with gains of 3q (including 2 with high-level amplifications), and 9 cases with partial gains of 3q (including 1 with high-level amplification). (From Ref. 45.)

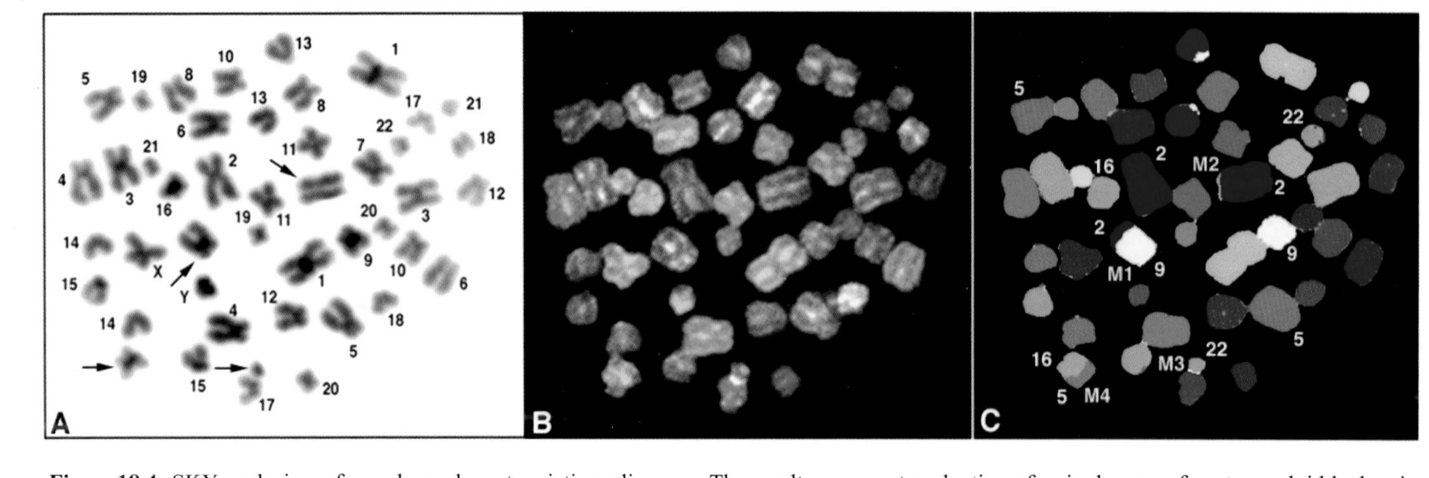

Figure 18.4 SKY analysis performed on a hematopoietic malignancy. The results represent evaluation of a single case of acute myeloid leukemia (subclassified as an M1 acute myeloid leukemia). (a) The tumor spread stained with DAPI to allow visualization of conventional chromosomal banding patterns. The arrows indicate aberrations that are further characterized in panel (c). (b) The raw or "display" computer image of the chromosomes after hybridization to the differentially labeled probe pool. (c) The computer-enhanced "classification" image in which each of the 24 human chromosomes is represented by a different color, generating a "color karyotype" of the tumor. The aberrations present in this single tumor are noted in panels (a) and (c). M1 (or marker chromosome 1) represents a translocation of chromosomes 9 and 2 (with chromosome 9 depicted in white and chromosome 2 depicted in brown). M2 represents a translocation of chromosomes 2 and 22 (with chromosome 2 in brown and 22 in pink). M3 is designated the Philadelphia chromosome, representing the remainder of the translocated chromosome 22 (in pink) and a portion of 9 (too small to be visualized). Similarly, M4 represents a translocation of chromosomes 5 and 16 (with chromosome 5 depicted in rust brown and chromosome 16 depicted in orange). Also, there is a monosomy 7 or loss of one of the homologs of chromosome 7 (chromosome 7 is depicted in light pink). The karyotype of this tumor would therefore be reported as 45,XY,-7,t(2;9;22),der(5)t(5;16).

and visualization of all of the chromosomes on the metaphase spread. The resulting spectral karyotype represents each chromosome in a different color and is therefore easily interpreted (Color Plate 18.4).

It is worth noting that another technique has recently been reported for visually discerning all human chromosomes in distinct colors (55). In contrast to SKY, this technique utilizes conventional filter technology and requires sequential exposures of the hybridized chromosomes with six optical filters to obtain a "color karyotype." Whereas SKY allows for a one-step image acquisition, this technique requires multiple images to be acquired for visualization of all chromosomes.

B. SKY Applications

The potential utility of SKY for the study of chromosomal aberrations involved in human disease is enormous. The low mitotic index of many cultured tumor cells necessitates the comprehensive analysis of each of the few available metaphases. SKY permits the identification of abnormalities that are too subtle to detect by traditional chromosome banding techniques. Tumors often possess numerous chromosomal abnormalities that are difficult to characterize with traditional techniques (such aberrant chromosomes are termed "marker" or "derivative" chromosomes). SKY has been shown to identify even complex rearrangements unambiguously, and allows for a determination of the origins of virtually all aberrant chromosomes (56). SKY can also be used in conjunction with conventional banding techniques to characterize other chromosomal structures (such as double minutes and homogeneously staining regions), as well as to determine breakpoint regions more accurately (49,56).

To date, SKY has been used to analyze hematologic malignancies (myelodysplastic syndrome, acute lymphoblastic leukemia, acute myelogenous leukemia, chronic myelogenous leukemia, and non-Hodgkin's lymphoma) which have chromosomal aberrations that could not be fully characterized by conventional cytogenetic techniques (56). In these cases, SKY provided additional information about the tumor karyotypes, including the classification of marker or derivative chromosomal rearrangements. The results obtained by SKY were confirmed by FISH analyses, thereby validating the accuracy of this form of analysis (56). Ongoing studies include evaluation of multiple myeloma, pancreatic malignancies, breast and cerebral neoplasms, and other tumor types as well as constitutional chromosome abnormalities.

Our increasing ability to correlate cytogenetic information with prognosis and treatment makes accurate tumor karyotyping necessary. By enhancing our ability to identify and classify chromosomal aberrations, SKY will allow us to expand our cytogenetic database for specific tumor types, facilitating the study of these tumors. Furthermore, the ease of interpreting the SKY results makes it a strong candidate for future use as a tumor screening method for specimens in the clinical laboratory.

C. SKY Limitations

While SKY is effective at detecting aberrations that change the "color" of a chromosome (e.g., translocations, insertions) and characterizing overall chromosomal structure, it does not reliably detect small chromosomal gains and deletions or pericentric and paracentric inversions. Therefore, while SKY is a powerful adjuvant to conventional cytogenetics and CGH, it will not replace these techniques. Rather, it can be viewed as another important form of karyotype analysis that should be used in conjunction with banding-based analyses and CGH to enhance information about chromosomal structure and aberrations (49,56). Although the painting probes utilized for SKY are commercially available, the technique is relatively expensive and requires specialized equipment. As with CGH, the instrumentation requirement for SKY may hinder its utility for the average clinical cytogenetic laboratory in the present climate of managed care. Despite these obstacles, however, the power of the technology will continue to promote its use for comprehensive karyotype analysis.

IV. CONCLUSION

The application of molecular approaches to karyotype analysis has revolutionized the field of cytogenetics. Until recently, global cytogenetic screening was limited to conventional chromosomal banding techniques. The development of CGH and SKY now permits genome-wide screening to be performed using molecular techniques. These technologies, used in conjunction with classic banding methods, markedly improve the sensitivity and accuracy of karyotyping. The information obtained from such a "multifaceted" approach to whole genome analysis will increase detection of otherwise unidentified chromosomal aberrations, advance the understanding of mechanisms involved in the genesis of chromosomal aberrations, and guide positional cloning efforts to identify the gene(s) involved in the development of particular tumor types. The field of molecular cytogenetics will continue to evolve, contributing to an ever-increasing understanding of human disease and facilitating the development of improved diagnostics, prognostics, and ultimately perhaps, gene therapeutics.

ACKNOWLEDGMENTS

We would like to thank Applied Spectral Imaging for contribution of the color page, as well as Darryl Leja for graphics assistance. We would also like to thank Drs. Eric Green, Elaine Jaffe, and Shimareet Kumar for critical review of this chapter, and Drs. Merryn Macville, Zoë Weaver, Danny Wangsa, Joan Packenham, Turid Knutsen, Marek Liyanage, and Stan du Manoir for general advice and support.

REFERENCES

1. Rabbitts TH. Chromosomal translocations in human cancer. Nature 1994; 372: 143–149.
2. DeVita VT, Hellman S, Rosenberg SA, eds. Cancer Principles and Practice of Oncology. 3rd ed. Philadelphia, New York: Lippincott-Raven, 1997.
3. Heim S, Mitelman F, eds. Cancer Cytogenetics. 2d ed. New York: Wiley-Liss, 1995.
4. Gray JW, Pinkel D, Brown JM. Fluorescence in situ hybridization in cancer and radiation biology. Radiat Res 1994; 137:275–289.
5. Cremer T, Landegent JE, Brueckner A, et al. Detection of chromosome aberrations in the human interphase nucleus by visualization of specific target DNAs with radioactive and nonradioactive in situ hybridization techniques. Hum Genet 1986; 74:346–352.
6. Pinkel D, Landegent J, Collins C, et al. Fluorescence in situ hybridization with human chromosome specific libraries: detection of trisomy 21 and translocation of chromosome 4. Proc Natl Acad Sci USA 1988; 85:9138–9142.
7. Cremer T, Lichter P, Borden J, Ward DC, Manuelidis L. Detection of chromosome aberrations in metaphase and interphase tumor cells by in situ hybridization using chromosome specific library probes. Hum Genet 1988; 80:235–246.
8. LeBeau M. Detecting genetic changes in human tumor cells: have scientists "gone fishing?" Blood 1993; 81:1979–1983.
9. Kallioniemi A, Kallioniemi O-P, Sudar D, et al. Comparative genomic hybridization for molecular cytogenetic analysis of solid tumors. Science 1992; 258: 818–821.
10. DuManoir S, Speicher MR, Joos S, et al. Detection of complete and partial chromosome gains and losses by comparative genomic in situ hybridization. Hum Genet 1993; 90:590–610.
11. Du Manoir S, Schröck E, Bentz M, et al. Quantification of comparative genomic hybridization. Cytometry 1995; 19:27–41.
12. Piper J, Rutovitz D, Sudar D, et al. Computer image analysis of comparative genomic hybridization. Cytometry 1995; 19:10–26.
13. Speicher MR, duManoir S, Schröck E, et al. Molecular cytogenetic analysis of formalin fixed, paraffin embedded solid tumors by comparative genomic hybridization after universal DNA amplification. Hum Mol Genet 1993; 2:1907–1914.

14. Ried T, Knutzen R, Steinbeck R, et al. Comparative genomic hybridization reveals a specific pattern of chromosomal gains and losses during the genesis of colorectal tumors. Genes Chromosom Cancer 1996; 15:234–245.
15. Heselmeyer K. Schröck E, duManoir S, et al. Gain of chromosome 3q defines the transition from severe dysplasia to invasive carcinoma of the uterine cervix. Proc Natl Acad Sci USA 1996; 93:497–484.
16. Isola J, DeVries S, Chu L, Ghazrini S, Waldman F. Analysis of changes in DNA sequence copy number by comparative genomic hybridization in archival paraffin-embedded tumor samples. Am J Pathol 1994; 145:1301–1308.
17. Telenius H, Carter NP, Bebb CE, Nordenskjold M, Ponder BAJ, Tunnacliffe A. Degenerate oligonucleotide-primed PCR: general amplification of target DNA by a single degenerate primer. Genomics 1992; 13:718–725.
18. Telenuis H, Pelmear AH, Tunnacliffe A, et al. Cytogenetic analysis by chromosome painting using DOP-PCR amplified flow-sorted chromosomes. Genes Chromosom Cancer 1992; 4:257–263.
19. Ried T, Petersen I, Holtgreve-Grez H, et al. Mapping of multiple DNA gains and losses in primary small cell lung carcinomas by comparative genomic hybridization. Cancer Res 1994; 54:1801–1806.
20. Levin NA, Brzoska PM, Warnock ML, Gray JW, Christman MF. Identification of novel regions of altered DNA copy number in small cell lung tumors. Genes Chromosom Cancer 1995; 13:175–185.
21. Kallioniemi A, Kallioniemi O-P, Piper J, et al. Detection and mapping of amplified DNA sequences in breast cancer by comparative genomic hybridization. Proc Natl Acad Sci USA 1994; 91:2156–2160.
22. Ried T, Just KE, Holtgreve-Grez H, et al. Comparative genomic hybridization of formalin fixed, paraffin embedded breast carcinomas reveals different patterns of chromosomal gains and losses in fibroadenomas and diploid and aneuploid carcinomas. Cancer Res 1995; 55:5415–5423.
23. Isola JJ, Kallioniemi O-P, Chu LW, et al. Genetic aberrations detected by comparative genomic hybridization predict outcome in node-negative breast cancer. Am J Pathol 1995; 147:905–911.
24. Schlegel J, Stumm G, Scherthan H, et al. Comparative genomic *in situ* hybridization of colon carcinomas with replication error. Cancer Res 1995; 55:6002–6005.
25. Schröck E, Thiel G, Lozanova T, et al. Comparative genomic hybridization of human glioma reveals consistent genetic imbalances and multiple amplification sites. Am J Pathol 1994; 144:1203–1218.
26. Kim DH, Mohapatra G, Bollen A, Waldman FM, Feuerstein BG. Chromosomal abnormalities in glioblastoma multiforme tumor and glioma cell lines detected by comparative genomic hybridization. Int J Cancer 1995; 60:812–819.
27. Mohapatra G, Kim DH, Feuerstein BG. Detection of multiple gains and losses of genetic material in ten glioma cell lines by comparative genomic hybridization. Genes Chromosom Cancer 1995; 13:86–93.
28. Schröck E, Blume C, Meffert MC, et al. Recurrent gain of chromosome 7q in low grade astrocytic tumors studied by comparative genomic hybridization. Genes Chromosom Cancer 1996; 15:199–205.

29. Brzoska PM, Levin NA, Fu KF, et al. Frequent novel DNA copy number increases in squamous cell head and neck tumors. Cancer Res 1995; 55:3055–3059.
30. Speicher MR, Howe C, Crotty P, duManoir S, Costa J, Ward DC. Comparative genomic hybridization detects novel deletions and amplifications in head and neck squamous cell carcinomas. Cancer Res 1995; 55:1010–1013.
31. Cher ML, MacGrogan D, Bookstein R, Brown JA, Jenkins RB, Jensen RH. Comparative genomic hybridization, allelic imbalance, and fluorescence in situ hybridization on chromosome 8 in prostate cancer. Genes Chromosom Cancer 1994; 11:153–162.
32. Joos S, Bergerheim USR, Pan Y, et al. Mapping of chromosomal gains and losses in prostate cancer by comparative genomic hybridization. Genes Chromosom Cancer 1995; 14:267–276.
33. Visakorpi T, Hyytinen E, Koivisto P, et al. In vivo amplification of the androgen receptor gene and progression of human prostate cancer. Nature Genet 1995; 9: 401–406.
34. Visakorpi T, Kallioniemi AH, Yvanen A-C, et al. Genetic changes in primary and recurrent prostate cancer by comparative genomic hybridization. Cancer Res 1995; 55:342–347.
35. Bentz M, Dohner H, Huck K, et al. Comparative genomic hybridization in the investigation of myeloid leukemias. Genes Chromosom Cancer 1995; 12:193–200.
36. Bentz M, Huck K, duManoir S, et al. Comparative genomic hybridization in chronic B-cell leukemias shows a high incidence of chromosomal gains and losses. Blood 1995; 85:3610–3618.
37. Joos S, Otano-Joos MI, Ziegler S, et al. Primary mediastinal (thymic) B-cell lymphoma is characterized by gains of chromosomal material including 9p and amplification of the REL gene. Blood 1996; 87:1571–1578.
38. Monni O, Joensuu H, Franssila K, Knuutila S. DNA copy number changes in diffuse large B-cell lymphoma—comparative genomic hybridization study. Blood 1996; 87:5269–5278.
39. Presti JC Jr, Moch H, Reuter VE, Huynh D, Waldman FM. Comparative genomic hybridization for genetic analysis of renal oncocytomas. Genes Chromosom Cancer 1996; 17:199–204.
40. Moch H, Presti JC Jr, Sauter G, et al. Genetic aberrations detected by comparative genomic hybridization are associated with clinical outcome in renal cell carcinoma. Can Res 1996; 56:27–30.
41. Bentz M, Bergerheim US, Li C, et al. Chromosome imbalances in papillary renal cell carcinoma and first cytogenetic data of familial cases analyzed by comparative genomic hybridization. Cytogenet Cell Genet 1996; 75:17–21.
42. Iwabuchi H, Sakamotot M, Sakunaga H, et al. Genetic analysis of benign, low-grade, and high-grade ovarian tumors. Cancer Res 1995; 55:6172–6180.
43. Kallioniemi A, Kallioniemi O-P, Citro G, et al. Identification of gains and losses of DNA sequences in primary bladder cancer by comparative genomic hybridization. Genes Chromosom Cancer 1995; 12:213–219.

44. Atkin NB, Baker MC, Fox MF. Chromosome changes in 43 carcinomas of the cervix uteri. Cancer Genet Cytogenet 1990; 44:229–241.
45. Heselmeyer K, Macville M, Schröck E, et al. Advanced stage cervical carcinomas are defined by a recurrent pattern of chromosomal aberrations revealing high genetic instability and a consistent gain of chromosome 3q. Genes Chromosom Cancer 1997; 19:233–240.
46. Tanner MM, Tirkkonen M, Kallioniemi A, et al. Independent amplification and frequent co-amplification of three nonsyntenic regions on the long arm of chromosome 20 in human breast cancer. Cancer Res 1996; 56:3441–3445.
47. Tanner M, Tirkkonen M, Kallioniemi A, et al. Increased copy number at 20q13 in breast cancer; defining the critical region and exclusion of candidate genes. Cancer Res 1994; 54:4257–4260.
48. Kallioniemi O-P, Kallioniemi A, Piper J, et al. Optimizing comparative genomic hybridization for analysis of DNA sequence copy number changes in solid tumors. Genes Chromosom Cancer 1994; 10:231–243.
49. Schröck E, duManoir S, Veldman T, et al. Multicolor spectral karyotyping of human chromosomes. Science 1996; 273:494–497.
50. Garini Y, Macville M, duManoir S, et al. Spectral karyotyping. Bioimaging 1996; 4:65–72.
51. Telenius H, Pelear AH, Tunnacliffe A, et al. Cytogenetic analysis by chromosome painting using DOP-PCR amplified flow sorted chromosomes. Genes Chromosom Cancer 1992; 4:257–263.
52. Guan X-Y, Trent JM, Meltzer PS. Generation of band-specific painting probes from a single microdissected chromosome. Hum Mol Genet 1993; 2:1117–1121.
53. Malik Z, Cabib D, Buckwald RA, Talmi A, Garini Y, Lipson SG. Fourier transform multipixel spectroscopy for quantitative cytology. J Microscopy 1996; 182: 133–140.
54. Garini Y, Katzir N, Cabib D, Buckwald RA, Soenksen D, Malik Z. Spectral bioimaging. In: Wang XF, Herman B, eds. Fluorescence Imaging Spectroscopy and Microscopy. New York: Wiley, 1996.
55. Speicher M, Ballard SG, Ward DC. Karyotyping human chromosomes by combinatorial multi-fluor FISH. Nature Genet 1996; 12:368–375.
56. Veldman T, Vignon C, Schröck E, Rowley JS, Ried T. Hidden chromosome abnormalities in hematological malignancies detected by multicolor spectral karyotyping. Nature Genet 1997; 15:406–410.

APPENDIX: GLOSSARY OF TERMS

Comparative genomic hybridization
A molecular cytogenetic assay that screens for DNA copy number changes (both losses/deletions and gains/amplifications) within a complete genomic complement.

Spectral karyotyping
A molecular cytogenetic *in-situ* hybridization technique which permits the simultaneous visualization of all human chromosomes, with each chromosome appearing in a different color.

19

Preanalytical Variables in Cytogenetic Studies

Jila Khorsand
Boston University, Boston, Massachusetts, and Roger Williams Medical Center, Providence, Rhode Island

I. INTRODUCTION

Cytogenetic analysis of neoplastic disease has become a significant part of diagnostic pathology. Chromosomal evaluation of hematologic specimens (i.e., blood, bone marrow, lymph nodes) and solid neoplasms plays a crucial role in diagnostic surgical pathology. Such studies have been shown to be particularly valuable in the diagnosis and prognosis of hematologic disorders (1–5). Chromosomal evaluations have also improved the diagnostic accuracy of solid neoplasms, especially bone and soft tissue tumors (6–8).

In a study using a combination of conventional or ''metaphase cytogenetics'' and fluorescent in-situ hybridization (FISH) or ''interphase cytogenetics'' on soft tissue sarcomas, Sozzi et al. (8) demonstrated the feasibility of these combined approaches in determining tumor-specific chromosomal aberrations, such as t(X;18) in synovial sarcoma, t(12;16) in myxoid liposarcoma (9), t(11;22) in peripheral primitive neuroectodermal tumors (pPNETs) (10), t(2;13) in alveolar rhabdomyosarcoma (11), and ring chromosomes in dermatofibrosarcoma protuberans. The determination of chromosome-specific alterations was crucial in reaching definitive diagnoses in multiple clinical situations. Throughout the years, pathologists have developed morphologic indicators, such as the degree of differentiation, nuclear grade, mitotic index, and presence or absence of necrosis, to predict the biological behavior of a particular neoplasm. Although such information is a significant component of a surgical pathology report that provides important prognostic variables, the genetic analysis of some neoplastic disorders may provide additional information that is invaluable for credible therapeutic decisions. Chromosomal abnormalities such as deletions and translocations, detected in some leukemic and preleukemic conditions, have also greatly improved the understanding of the role of genetic information in malignant transformation (12). With the availability of emerging molecular cytogenetic technologies, cytogenetics will no doubt play an ever-increasing role in diagnostic pathology. That is why the proper collection of surgical specimens to ensure optimized cytogenetic results is so important. This chapter is devoted to quality assurance and quality control issues that optimize the interaction between surgical pathology and the cytogenetics laboratory.

II. SURGICAL SPECIMEN COLLECTION

Direct communication between the clinician and the pathologist is key to a successful cytogenetic study. The surgical pathologist and/or hematopathologist must be consulted in all complex cases where additional ancillary studies may be necessary to augment routine histologic studies.

Pertinent patient history, clinical diagnosis, differential diagnosis, and other preanalytic information must be provided at the time of specimen collection to avoid errors in tissue handling and processing. Complete written instructions outlining the different steps of this process must be available to all staff members, including residents and fellows. Establishing an efficient tissue procurement section staffed by qualified personnel will assure an appropriate and expeditious delivery of tissue to its designated site. Such a facility requires adequate space, ideally adjacent to the surgical grossing room, as well as specific tools and equipment, including special fixatives and transport media.

III. HEMATOLOGIC SPECIMEN COLLECTION

Conventional cytogenetic testing which evaluates metaphase cells requires fresh, dividing tissue. Bone marrow is the preferred specimen for standard cytogenetic evaluation of hematologic diseases. Bone marrow (1–5 ml) is collected by aspiration into a sodium-heparin tube, under aseptic conditions (13). (Details are provided elsewhere in this volume.) There are, however, clinical situations in which bone marrow aspiration may be unsuccessful (13) or collected specimens may be of poor quality. Myelofibrosis is a condition typically associated with poor specimen collection (14). Bone marrow fibrosis occurs either as a primary idiopathic disorder, or, more commonly, as a secondary phenomenon related to a wide variety of diseases such as metastatic carcinoma, malignant lymphoma, acute leukemia, and chronic myeloproliferative diseases (15–17). In cases of myelofibrosis or other conditions associated with difficulties in bone marrow aspiration, bone marrow biopsy and/or peripheral blood collection may be preferred. Five to 10 ml of venous blood from adults and 2–5 ml of blood from children, collected in sodium-heparin tubes, is adequate for genetic evaluation. Additional details can be found elsewhere in this volume.

IV. COLLECTION OF SOLID TISSUE SPECIMENS

Standard cytogenetic testing of lymphoid or other solid neoplasms requires fresh tissue. Specimens must be submitted to the laboratory in an expeditious fashion to maximize cell viability. If the cytogenetic laboratory is far away from the original facility, the tissue should be submitted within 24–48 h after removal for optimal results (12). A block of tissue (0.5–1.0 g or 1 cm^3) cut under sterile conditions is generally adequate. If tissue of this magnitude is unavailable, smaller tissue fragments may also be submitted. Since morphologic heterogeneity is frequently seen in soft tissue tumors and in several other types of solid neoplasms, special attention must be paid to adequate sampling of large neoplasms. Necrotic tissue must be avoided. A common problem in handling solid tissue is drying artifacts and autolysis. Upon arrival of fresh tissue to the surgical pathology grossing room, the pathologist is faced with the difficult task of submitting tissue for routine morphologic evaluation and other ancillary studies. This complex process should be performed in an expeditious and organized way to prevent tissue degeneration (See Table 1). In the case of intraoperative consultation, special attention must be paid to the remaining tissue to assure viability while the frozen section process is in progress. It is generally good practice to keep the tissue covered with a saline-soaked gauze and to avoid direct heat from an overhead light.

V. TRANSPORT OF TISSUE

Expeditious delivery of tissue to the cytogenetics laboratory is of paramount significance for achieving satisfactory results. In general, it is recommended that tissue be submitted within 24 h after collection, for optimal results. In facilities where the cytogenetic laboratory is in close proximity to the surgical grossing room, the tissue may be transported in 5–10 ml of sterile

Table 1 Tissue Allocations for Ancillary Studies

Fresh tissue	Snap frozen tissue	Alcohol fixed tissue	Formalin fixed tissue	Air dried imprints or cytocentrifuged cell preparation
Cytogenetics (sterile)	Immunohistochemistry	Molecular genetics	Routine Hematoxylin-Eosin stain	Giemsa or Wright-Giemsa stain
Flow cytometry	Cytochemistry		Special stains	Cytochemical stains
Molecular genetics	Molecular genetics		Immunohistochemistry	Immunofluorescent stains
Microbiology (sterile)	Immunofluorescence		Fluorescent in situ hybridization (FISH)	Fluorescent in situ hybridization (FISH)
Electron microscopy (after fixation in glutaraldehyde)			Molecular genetics	Molecular genetics
				Storage at −70°C

saline or phosphate-buffered saline (PBS) and delivered to the laboratory as soon as possible. If longer transport time is anticipated (5–48 h), the tissue must be collected in sterile transport media to prevent autolysis (12). One of the most widely used media is RPMI 1640 (18) (Gibco, Grand Island, NY). Upon arrival of the specimen at the laboratory a cell suspension is prepared (19). To prepare cell suspensions, peripheral blood samples are diluted with sterile phosphate-buffered saline (PBS, Gibco, Grand Island, NY). Peripheral blood samples with a normal white cell count, (5,000–10,000/mm^3) are diluted 1:4. For specimens with a higher number of white cells, one volume is added for each additional 10,000 cells (e.g., if the white blood cell count is 30,000/mm^3, it is diluted 1:6). Cell suspensions of aspirated bone marrow samples are prepared by mechanical grinding of the spicules with a sterile instrument (e.g., the flat rubber bottom of a plunger from a 10-cm^3 syringe). The specimen is then diluted by sterile phosphate-buffered saline, 1:5 to 1:10. Highly cellular bone marrow specimens require further dilution to 1:20 or greater. Solid tissue cell suspensions (e.g., lymph nodes) require tissue disruption by a combination of cutting, teasing, and scraping manipulations in the presence of tissue culture media, RPMI 1640. These manipulations which result in cell isolation are carried out until all tissues have been disrupted. The highly cellular specimen is then assessed for cell count and cell viability. A simple way of cell counting and viability assessment is through manual technique using a hemocytometer. This procedure is described in Table 2 (19).

VI. FLUORESCENT IN SITU HYBRIDIZATION (FISH)

The application of in situ hybridization (ISH) has expanded in the past decade (20–22). Once the sole perview of the research laboratory, this technique has evolved into a method with invaluable clinical utility (23–26). Fluorescent in situ hybridization is a modified ISH technique used for the cytogenetic analysis of interphase or metaphase cells. Conventional cytogenetics involves obtaining fresh cytologic material, short-term cell cultures, inducing mitosis, followed by obligate metaphase arrest (27). Despite the proven utility of standard metaphase cytogenetics in the diagnosis and management of human disease, this technique has limitations. Since conventional cytogenetic testing is based on the microscopic evaluation of metaphase cells, the availability of fresh tissue is a prerequisite for this type of study. The procedure is also limited to an inherent bias toward dividing cells, hence eliminating nondividing populations from analysis (28). This is particularly troublesome in cell populations with a low mitotic index. It has been reported that 5–20% of bone marrow and 40% or more solid tumor specimens

Table 2 Assessment of Cell Viability

1. Clean a hemocytometer with alcohol.
2. Place a clean coverslip over the counting chamber.
3. Combine a 50-μl representative aliquot of the cell suspension with 50 μl 0.4% trypan blue dye (Gibco, Grand Island, NY) in sterile, physiologic phosphate-buffered saline (PBS) and allow to sit for 30–60 s.
4. Load one side of the hemocytometer with a 50-μl representative aliquot of the cell suspension to determine cells per milliliter.
5. Load the other side of the hemocytometer with 50 μl of the cell suspension treated with trypan blue dye to determine percent viability.
6. Allow 30–60 s prior to counting.
7. Count the number of cells that are not treated by trypan blue dye.
8. Proceed to count the cells that are treated with trypan blue dye.

The clear cells are viable and the blue cells are dead. Count a total of 200 consecutive cells (clear and blue) and calculate the viability as follows:

$$\frac{\text{Live cells}}{\text{Total live cells and dead cells}} \times 100 = \%\ \text{viability}$$

The number of viable cells per milliliter is calculated by multiplying the cell count per milliliter by the percent viability.

fail to provide a successful cytogenetic analysis due to low mitotic yield or poor specimen quality (29,30).

Finally, standard genetic testing is a time-consuming procedure and is labor intensive. Some of these issues are discussed in Chapter 17 in this volume on the applications of FISH. The term "interphase cytogenetics," which was first introduced by Cremer (27,31), refers to the cytogenetic evaluation of interphase nuclei. This is a highly sensitive method and is applicable to a wide variety of specimens such as cytospin preparation of cell suspensions, tissue sections, frozen sections, and smears. The feasibility of applying FISH technology to fixed tissue has created an enormous resource for cytogenetic evaluation and has enabled the study of a variety of disease entities with little or no prior cytogenetic information. Recent advances in ISH technology coupled with an ever-increasing number of available probes has broadened the scope of clinical applications of this technique. The sensitivity of the analysis has also been improved to the extent that specific DNA sequences of single-copy genes (27,32–34), viral genomes (35,36), mRNA molecules in single cells (37), or tissue sections (38–41) can now be detected. Studies comparing FISH with conventional cytogenetics demonstrate the validity of this method in detecting structural and numerical chromosomal aberrations (42–44). FISH is particularly sensitive in detecting minimal residual disease when there is a known genetic marker such as in chronic myelogenous leukemia (44). The success of this method is greatly influenced by the specificity and sensitivity of the probes. Several types of DNA probes are currently available for detecting chromosomal anomalies. They are discussed in the chapter by Blancato and Haddad.

VII. SPECIMEN COLLECTION FOR FISH

Fresh specimens are collected in RPMI 1640 (Flow Laboratories, Irvine, CA) which contains 17% fetal calf serum (Flow Laboratories), 50 g gentamycin, 50 g penicillin, and 50 g streptomycin/ml (45).

Cell suspensions derived from peripheral blood, fresh tissue, fine needle aspirations (27,44,46), and cell cultures are the most suitable specimens for interphase cytogenetics (27,44). In the case of solid tumors, a single cell suspension is prepared by mechanical disaggregation achieved by scraping and cutting the tissue in a petri dish. The tissue is then filtered through a 100-μm nylon filter (Ortho Diagnostic System, Breese, Belgium) (47). Filtered cells are then fixed in 70% ethanol at −20°C and stored at −30°C. Cell suspensions prepared in this manner can be stored for years without any complications (45).

In order to preserve the morphology and also provide optimal conditions for probes to access the target DNA sequences, specimens need to be fixed. Most fixatives used in pathology, such as ethanol or formaldehyde, are suitable for ISH (44), with ethanol being the most widely used (46). The fixation process is followed by digestion, which is achieved by proteolytic enzymes. Proteinase K (48–50), pepsin (36,46), and collagenase (51) are some of the more widely used enzymes. Although technically easier to perform, FISH on cell suspensions has some disadvantages. One important disadvantage is the loss of histopathologic features and inability to correlate the genetic and morphologic findings (44). Another limitation is the inability to discriminate among target cells, stromal cells, and inflammatory cells (44). Finally, the process of disaggregation may create a bias toward cells that disaggregate easily (27).

Fixed and paraffin-embedded tissue application to FISH is technically more challenging because there are tremendous variations among fixatives, the duration of fixation, and tissue types (27). There is, however, a great deal of interest in improving this methodology, because an invaluable amount of genetic information lies in archival material stored in basements or in remote locations.

There are two ways to handle fixed tissue for FISH. One method is to retrieve nuclei from fixed, paraffin-embedded tissue by disaggregation, similar to that used in flow cytometry (52–54). Thick sections (50 μm) are cut, dewaxed, and then dehydrated. The tissue is then treated with 0.5% pepsin (P-7000, Sigma) for 30 min at 37°C in 0.01 or 0.2 M HCl and centrifuged. The pellet is washed and resuspended in 30 ml of distilled water. After sedimentation for a few minutes at $1 \times g$, the supernatant containing single nuclei is transferred to another tube and fixed in ethanol. To make slides from the cell suspension, a few drops of fixed nuclei are added to 1 ml of 70% acetic acid and centrifuged to clean, poly L-lysine-coated slides, dried overnight, and washed in PBS and distilled water. The slides are then dehydrated and stored at room temperature (52,55). The advantage of this method using 50-μm thick sections is that since the loss of nuclear material is minimal in thick sections, the copy number of chromosome is more accurate (54). It has been demonstrated that as the thickness of sections increases, so does the number of signals apparent from each nucleus (56). The disadvantage, however, is that in cell suspensions, the histomorphologic and architectural features of tumor cells are completely lost.

The second method of applying interphase cytogenetics to paraffin-embedded tissues is to use intact 4- to 6-μm-thick sections without the disaggregation process. This method is technically more difficult and requires optimization of the different steps of the process, including tissue adhesion to glass slides, proteolysis, denaturation, hybridization, and signal detection (57). Each step needs to be optimized for every tissue block (46). Another technical problem is loss of the nuclear material due to thinness of the sections (27). This problem can be controlled by analyzing inflammatory and stromal cells as internal controls and by developing ratios (56–58).

VIII. SAFETY ISSUES IN THE CYTOGENETICS LABORATORY

Laboratories are hazardous environments which necessitate the implementation and enforcement of strict safety guidelines for the health and well-being of all workers. Some general rules are mandated by government-appointed agencies and must all be strictly adhered to in order to avoid penalties and other serious consequences. The Occupational Safety and Health Administration (OSHA) is such an agency, charged with the task of overseeing the safety of the workplace and limiting occupational hazards. Accreditation by the College of American Pathologists (CAP) also requires compliance with laboratory safety practices. Discussion of general safety measures is beyond the scope of this chapter. However, there are rules pertaining to handling specimens which will be briefly reviewed here. It is imperative that all laboratory personnel, pathologists, pathology residents, and fellows comply with the Universal Precautions as described by the Centers for Disease Control (59). All personnel who come into contact with fresh specimens, blood, or body fluids must wear gloves and appropriate protective clothing. Specimens must be received in covered containers. It is highly desirable to examine the specimens in a laminar-flow hood to prevent contamination and possible aerosolization (19). For conventional cytogenetics, specimens must be received in sterile condition and handled in a sterile laminar-flow tissue culture hood (Sterilegard Hood, Baker, Sandford, ME). This practice allows thorough evaluation of specimens with little or no risk to the operator or other laboratory personnel.

IX. SUMMARY

In closing, it should be emphasized that with continued technological advances in ancillary techniques and their increased utility in diagnostic pathology, the surgical pathologist is expected to play a more interactive role in patient management from initial diagnosis to therapeutic decision making. To face the new challenges and increasing professional responsibilities, pathologists need to familiarize themselves with the basic principles of these methods and their much-varied clinical applicability. One way to succeed is to develop well-structured pathology training programs that provide a balanced knowledge of basic morphologic skills and adequate exposure to complementary techniques such as cytogenetics and molecular genetics. With dedication and commitment, new trainees will be sufficiently prepared for the challenges of the next century.

REFERENCES

1. Sandberg AA. The chromosomes in human cancer and leukemia. 2nd ed. New York: Elsevier, 1990.
2. Heim S, Mitelman M. Cancer Cytogenetics. New York: Liss, 1987.
3. Sandberg AA. Cytogenetics of human neoplasia. Advances and perspectives. In: Willey AM, Murphy PD, eds. Fragile X/Cancer Cytogenetics: Progress in Clinical and Biological Research. New York: Willey-Liss, 1991.
4. Sandberg AA. Chromosome abnormalities in human cancer and leukemia. Mutat Res 1991; 247:231–240.
5. Trent JM, Meltzer PS. The last shall be first. Nature Genet 1993; 3:101–102.
6. Sandberg AA. Chromosomal lesions and solid tumors. Hosp Prac 1988; 15: 93–106.

7. Heim S, Mitelman F. Cytogenetics of solid tumors. Recent Adv Histopathol 1992; 15(36):37–66.
8. Sozzi G, Minoletti F, Mizzo M, et al. Relevance of cytogenetic and fluorescent in situ hybridization analyses in the clinical assessment of soft tissue sarcoma. Hum Pathol 1997; 28:134–142.
9. Aman P, Ron D, Mandahle N, et al. Rearrangement of the transcription factor gene CHOP in myxoid liposarcomas with t(12;16) (q13;p11). Genes Chromosomes Cancer 1992; 5:278–285.
10. Delattre O, Zucman J, Plougastel B, et al. Gene fusion with an ETS DNA-binding domain caused by chromosome translocation in human tumours. Nature 1992; 359:162–165.
11. Barr FG, Galili N, Holick J, Biegel JA, Rovera G, Emanuel BS. Rearrangement of the PAX3 paired box gene in the pediatric solid tumours alveolar rhabdomyosarcoma. Nature Genet 1993; 3:113–117.
12. Sandberg AA, Bridge JA. Introduction, cytogenetic methodologies. In: Sandberg AA, Bridge JA, eds. The Cytogenetics of Bone and Soft Tissue Tumors. Austin, TX: R.G. Landes, 1994:1–13.
13. LeBeau MM. The role of cytogenetics in the diagnosis and classification of hematopoietic neoplasms. In: Knowles DM, ed. Neoplastic Hematopathology. Baltimore: Williams & Wilkins, 1992.
14. McCarthy DM. Fibrosis of the bone marrow, content and causes. Br J Hematol 1985; 59:1–7.
15. Brunning RD. Bone marrow. In: Rosai J, ed. Ackerman's Surgical Pathology. Vol. 2. 8th ed. New York: Mosby-Year Book, 1996.
16. Burston J, Pinniger JL. The reticulin content of bone marrow in hematological disorders. Br J Haematol 1963; 9:172–184.
17. Sanerkin NG. Stromal changes in leukemia and related bone marrow proliferations. J Clin Pathol 1964; 17:541–547.
18. Gibas LM, Gibas Z, Sandberg AA. Technical aspects of cytogenetic analysis of human solid tumors. Karyogram 1984; 10:25–27.
19. Knowles DM. Organization and operation of Hematopathology Laboratory. In: Knowles DM, ed. Neoplastic Hematology. Baltimore: Williams & Wilkins, 1992:323–366.
20. DeLellis RA, Wolfe HJ. New techniques in gene product analysis. Arch Pathol Lab Med 1987; 111:620–627.
21. Fenoglio-Preiser C, Willman C. Molecular biology and the pathologist: general principles and applications. Arch Pathol Lab Med 1987; 111:601–619.
22. Wolfe HJ. DNA probes in diagnostic pathology. Am J Clin Pathol 1988; 90: 340–344.
23. Bentz M, Dohner H, Cabot G, Lichter P. Fluorescence in situ hybridization in leukemias: the FISH are spawning!! Leukemia 1994; 8:1447–1452.
24. LeBeau MM. Detecting genetic changes in human tumor cells: have scientists "gone fishing." Blood 1993; 81:1979–1983.
25. Ledbetter DH. The "colorizing" of cytogenetics: is it ready for prime time? Hum Mol Genet 1992; 5:297–299.
26. Price CM. Fluorescence in situ hybridization. Blood Rev 1993; 7:127–134.

27. Herrington CS. Interphase cytogenetics: principles and applications. J Histotechnol 1994; 17:219–234.
28. Anastasia J, LeBeau MM, Wardiman JW, Westbrook CA. Detection of numerical chromosomal abnormalities in neoplastic hematopoietic cells by in situ hybridization with a chromosome-specific probe. Am J Pathol 1990; 136:131–139.
29. Najfeld V. Fishing among myeloproliferative disorders. Semin Hematol 1997; 34:55–63.
30. Limon J, Dalcin P, Sandberg AA. Application of long-term collagenase disaggregation for the cytogenetic analysis of human solid tumors. Cancer Genet Cytogenet 1986; 23:305–313.
31. Cremer T, Landegent J, Bruckner A, et al. Detection of chromosome aberrations in the human interphase nucleus by visualization of specific target DNAs with radioactive and non-radioactive in situ hybridization techniques: diagnosis of trisomy 18 with probe L1.84. Hum Genet 1986; 74:346–352.
32. Lichter P, Boyle AL, Cremer T, Ward DC. Analysis of genes and chromosomes by non isotopic in situ hybridization. Genet Anal Technol Appl 1991; 8:24–35.
33. Yurov YB, Soloviev IV, Vorsanova SG, Marcais B, Roizes G, Lewis R. High resolution multicolor fluorescence in situ hybridization using cyanine and fluorescein dyes: rapid chromosome identification by directly fluorescently labeled alphoid DNA probes. Hum Genet 1996; 97:390–398.
34. Harper ME, Ullrich A, Saunders GF. Localization of human insulin gene to the distal end of the short arm of chromosome 11. Chromosoma 1981; 83: 431–439.
35. Brigati DJ, Myerson D, Leary JJ, et al. Detection of viral genomes in cultured cells and paraffin-embedded tissue sections using biotin-labeled hybridization probes. Virology 1982; 126:32–50.
36. Burns J, Redfern DRM, Esiri MM, McGee Jo'D. Human and viral gene detection in routine paraffin-embedded tissue by in situ hybridization with biotinylated probes: viral localization in herpes encephalitis. J Clin Pathol 1986; 39:1066–1073.
37. Coghlan JP, Aldred P, Haralambridis J, Niall HD, Penschow JD, Tregear GW. Hybridization histochemistry. Anal Biochem 1985; 149:1–28.
38. Angerer RC, Cox KH, Angerer LM. In situ hybridization to cellular RNAS. Genet Eng 1985; 7:43–65.
39. Denijn M, DeWeger RA, Berends MJ, et al. Detection of calcitonin encoding mRNA by radioactive in situ hybridization: improved colorimetric detection and cellular localization of mRNA in thyroid sections. J Histochem Cytochem 1990; 38:351–358.
40. Burns J, Graham AK, Frank C, Fleming KA, Evans MF, McGee Jo'D. Detection of low copy human papilloma virus DNA and mRNA in routine paraffin sections of cervix by non-isotopic in situ hybridization. J Clin Pathol 1987; 410:858–864.
41. McGee Jo'D, Burns J, Fleming KA. New molecular techniques in pathological diagnosis: visualization of genes and mRNA in human tissue. In: Bleehen NM,

ed. Investigational Techniques in Oncology. London: Springer-Verlag, 1987: 21–34.

42. Poddighe PJ, Moesker O, Smeets D, Awwad BH, Ramaekers FCS, Hopman AHN. Interphase cytogenetics of hematological cancer: comparison of classical karyotyping and in situ hybridization using a panel of 11 chromosome specific DNA probes. Cancer Res 1991; 51:1959–1967.
43. Nederlof PM, Van der Flier S, Raap AK, et al. Detection of chromosome aberrations in interphase tumor nuclei by non-radioactive in situ hybridization. Cancer Genet Cytogenet 1989; 42:87–97.
44. Poddighe PJ, Ramaekers FCS, Hopman AHN. Interphase cytogenetics of tumours. J Pathol 1992; 166:215–224.
45. Hopman AHN, Poddighe PJ, Smeets AWGB, et al. Detection of numerical chromosome aberrations in bladder cancer by in situ hybridization. Am J Pathol 1989; 135:1105–1117.
46. Hopman AHN, Poddighe PJ, Moesker O, Ramaekers FSC. Interphase cytogenetics: an approach to the detection of genetic aberrations in tumors. In: Diagnostic Molecular Pathology: A Practical Approach. Vol. 1. Oxford: Oxford University Press, 1992:141–167.
47. Smeets AWGB, Pauwels RPE, Beck HLM, et al. Comparison of tissue disaggregation techniques of transitional cell bladder carcinomas for flow cytometry and chromosomal analysis. Cytometry 1987; 8:14–19.
48. Hopman AHN, Raap AK, Landegent JE, Wiegant J, Boerman RH, Van der ploeg M. Non-radioactive in situ hybridization. In: Van Leeuwen FW, Buijs RM, Pool CW, Pach O, eds. Molecular Neuroanatomy. Amsterdam: Elsevier, 1988:43–68.
49. Raap AK, Arnoldus EPJ, Nederlof PM, Smit VTHBM, Cornelisse CJ, Van der ploeg M. Detection of numerical and structural chromosome aberrations in interphase nuclei by fluorescence in situ hybridization. In: Elder Hy, ed. Transactions of the Royal Microscopic Society. London: MICRO 90, 1990:661–667.
50. Hopman AHN, Ramaekers FCS, Raap AK, et al. In situ hybridization as a tool to study numerical chromosome aberrations in solid bladder tumors. Histochemistry 1988; 89:307–316.
51. Cajulis RS, Frias-Hidvegi D. Detection of numerical chromosomal abnormalities in malignant cells on body fluids by fluorescence in situ hybridization of interphase cell nuclei with chromosome-specific probes. Diagn Cytopathol 1992; 8:627–631.
52. Hedley DW, Friedlander ML, Taylor IW, Rugg CA, Musgrove EA. Method for analysis of cell DNA content of paraffin-embedded pathological material using flow cytometry. J Histochem Cytochem. 1983; 31:1333–1335.
53. Arnoldus EP, Peters AC, Bots GT, Raap AK, Van der ploeg M. Somatic pairing of chromosome 1 centromeres in interphase nuclei of human cerebellum. Hum Genet 1989; 83:231–234.
54. Arnoldus EP, Dreef EJ, Noordermeer IA, et al. Feasibility of in situ hybridization with chromosome specific DNA probes on paraffin wax embedded tissue. J Clin Pathol 1991; 44:900–904.

55. Hedley DW. Flow cytometry using paraffin-embedded tissue: five years on. Cytometry 1989; 10:229–241.
56. Hopman AHN, Van Hooren E, Van de Kaa CA, Vooijs GP, Ramaekers FCS. Detection of numerical chromosome aberrations using in situ hybridization in paraffin sections of routinely processed bladder cancer. Mod Pathol 1991; 4: 503–513.
57. Kim SY, Lee JS, Ro JY, Gay ML, Hong WK, Hittelman WN. Interphase cytogenetics in paraffin sections of lung tumors by non-isotopic in situ hybridization. Mapping genotype/phenotype heterogeneity. Am J Pathol 1993; 142: 307–317.
58. Herrington CS, McGee Jo'D. Interphase cytogenetics. Neurochem Res 1990; 15:467–474.
59. Department of Labor, Occupational Safety and Health Administration. Occupational exposure to blood borne pathogens; final rule, part II (excerpt). Fed Reg 1994; 64175–64182.

APPENDIX: GLOSSARY OF TERMS

autolysis
The spontaneous disintegration of tissues or of cells by the action of their own autogenous enzymes.

chronic myelogenous leukemia (CML)
A clonal disorder associated with the t(9;22) chromosome, known as the Philadelphia (PH) chromosome, and with the molecular translocation of the *abl* oncogene from chromosome 9 to the breakpoint cluster region (*bcr*) on chromosome 22.

chronic myeloproliferative disease
A group of closely related disorders characterized by proliferation of one or more myeloid cell lines.

fine-needle aspiration (*FNA*)
The method of aspiration biopsy using thin needles (18–22 gauge).

myelofibrosis
Fibroblastic proliferation of bone marrow, which can be primary or secondary in nature.

20
Setting the Standards for Cytogenetics Laboratories

Gerald A. Hoeltge
The Cleveland Clinic Foundation, Cleveland, Ohio

I. INTRODUCTION

Clinical cytogenetics is a medical specialty. Physicians integrate the testing for chromosomal abnormalities into the care of children with congenital abnormalities (Chapter 7), expectant mothers (Chapter 8), and young couples hoping to become parents (Chapter 10). Cytogenetic diagnosis has become the standard of care in the diagnosis and management of leukemia (Chapter 11), for many types of lymphoma (Chapter 12), and may even be essential for some solid tumors (Chapters 14 and 15). A karyotypic analysis of a patient's genetic constitution carries lifelong implications for that individual. The analysis must be accurate, reproducible, and communicated with clarity. The referring physician must have full confidence in the testing facility chosen.

The best way for a laboratory to demonstrate excellence to patients and their physicians is by submitting to external, independent review. The value of this external audit is a function of the expertise of the reviewer and the benchmarks that are chosen for comparison. A thorough understanding of the agencies that set the standards and review or accreditation of the laboratory facilitates an evaluation of the laboratory's clinical performance.

In this chapter we will look at those organizations, both voluntary and regulatory, that define the standards of care. Our task is made easier because of the partnership between the American College of Medical Genetics (ACMG) (1) and the College of American Pathologists (CAP) (2). The CAP is the world's largest accrediting organization for clinical laboratories. Since 1962, the CAP's Laboratory Accreditation Program (LAP) has offered medical laboratories the opportunity for external peer review. More than 300 of the 5200 laboratories accredited by the CAP provide clinical cytogenetic services. Most accredited laboratories are located in the United States or Canada, but CAP-accredited facilities may be found all over the world. The CAP also offers the world's largest program for proficiency testing, with more than 100 different surveys currently available. Among these are tests of karyotypic diagnosis, in-situ DNA hybridization, molecular genetics, and biochemical genetics.

From where does the CAP draw its expertise? The CAP commissions recognized experts within each specialty to define standards for accreditation and to develop credible and realistic proficiency testing challenges. In cytogenetics these goals are achieved by a partnership with the ACMG. Collectively, these two professional organizations determine the process for accreditation. The ACMG and the CAP specify how laboratories can best demonstrate their proficiency in the practice of laboratory genetics. Each organization has promulgated written standards. The CAP's standards (3) are broadly phrased principles of laboratory excellence which codify the expectations for any laboratory seeking CAP accreditation, regardless of its specialty. The ACMG's standards target genetic laboratories, and they include specific standards and guidelines for cytogenetics. Figure 1 illustrates the cooperative relationship between these two professional organizations. The accreditation requirements incorporate the standards of organizations such as the National Committee for Clinical Laboratory Standards (4) and the National Fire Protection Association (5), as well as specific regulatory requirements.

II. PERSONNEL

Cytogenetic laboratory directors should be certified in cytogenetics by the American Board of Medical Genetics (ABMG; see Chapter 1) (6). Since 1991, the ABMG has been a member of the American Board of Medical Specialties (7). Training programs for clinical cytogenetics must be accredited by the Accreditation Council of Graduate Medical Education (8). In order to be board-certified, a cytogeneticist must have defended her or his skill in the diagnosis and interpretation of a wide range of cytogenetic problems. As in any other medical discipline, certification by a recognized specialty board is an important means by which a referring physician can seek and choose a consultant, and ABMG's certification is recognized as the preferred credential when seeking a consultant for a patient with a genetic

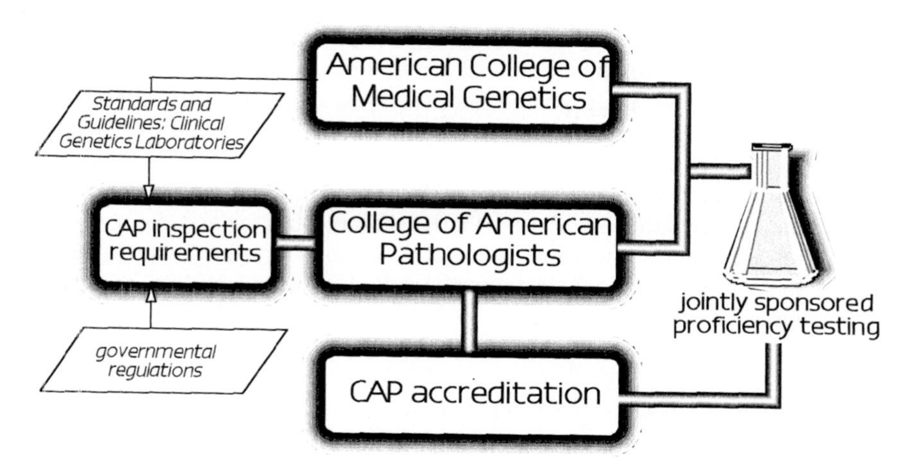

Figure 1 The relationship between the American College of Medical Genetics and the College of American Pathologists. The ACMG provides the professional expertise for the CAP's inspection requirements. The two organizations together produce proficiency testing programs in laboratory genetics.

problem. Just as certification by the American Board of Pathology is required for fellowship in the College of American Pathologists, ABMG certification is a prerequisite for Fellows of the ACMG.

A discussion of the responsibilities of the "laboratory director" is in order here. The title is often used in different ways and often leads to confusion. To the LAP, the laboratory director is the person in whom all laboratory responsibilities are vested. The laboratory director selects the technical staff, approves all procedures, and stands behind all results. She provides patient-specific interpretations of karyotypic results. She sets the standard of performance and sees to it that a quality improvement program is initiated. She ensures that training programs and developmental work are appropriate for the laboratory's scope of care. She sees to it that the laboratory is a safe environment for personnel, patients, and visitors. A director has administrative, management, and strategic planning responsibilities. So pivotal is the laboratory directorship that a laboratory's CAP accreditation does not automatically survive a change in directorship.

But what about those multispecialty laboratories within which the cytogenetics laboratory is one part? The overall laboratory director is likely to be a general pathologist without special expertise in cytogenetics. In this setting, the director must ensure that the head of the cytogenetics area is an individual who is competent in the field (as demonstrated, for example, by certification by the ABMG in clinical cytogenetics). The regulations of the Clinical Laboratories Improvement Act (CLIA), as amended in 1988, refer to the cytogeneticist in a multispecialty laboratory as a ''technical supervisor.'' A technical supervisor must have at least 4 years of training or experience in genetics, 2 of which have been in clinical cytogenetics. In a standalone cytogenetics laboratory, the laboratory director and the technical supervisor are usually the same individual. In such a setting, the ''general supervisor,'' the individual customarily called ''supervisor'' in a clinical cytogenetics laboratory, must have at least a bachelor's degree in science and at least 2 years' experience in cytogenetic methods under a qualified director.

Analytic personnel must understand the idiosyncrasies and special needs of cytogenetic testing. Specimen acceptability issues, culture techniques, slide preparation, and photography occur in combinations different from the rest of the clinical laboratory. The visual acuity and cognitive skills necessary to ''read'' a metaphase spread demand a substantial knowledge base, a depth of experience, and a flair for imagery and pattern recognition. The National Certification Agency for Medical Laboratory Personnel (NCAMLP) (9) offers a certification in clinical cytogenetics. Clinical Laboratory Specialists in Cytogenetics [CLS(CG)], as certified by NCAMLP, have demonstrated that they have mastered a set of relevant core competencies, which have been defined by professionals in the field. NCAMLP is sponsored by the American Society for Clinical Laboratory Science (10), among other professional organizations. While it is not required that all testing personnel be NCAMLP certified, it is highly appropriate. A certified individual at the baccalaureate level will meet the criteria for general supervisor of a cytogenetics laboratory as defined by the CLIA. The professional organization that represents cytogenetic analysts is the Association of Genetic Technologists (AGT; formerly the Association of Cytogenetic Technologists) (11). The AGT publishes the standard reference volume of technical procedures for chromosome analysis (12). The AGT is also a professional sponsor of the NCAMLP.

III. STANDARDS

The ACMG, through its *Standards and Guidelines: Clinical Genetics Laboratories* (13), has defined the minimal standard of care for diagnostic testing in genetics. This document covers personnel requirements, general principles of analytic technique, quality control, and record keeping (14). The ACMG *Standards* cover clinical cytogenetics, clinical biochemical genetics, and clinical molecular genetics. Specific and useful information on the choice of methodology, statistically defensible sample sizes, and benchmarks for success are provided in the document. The LAP references the appropriate portions of the *Standards* in the checklist used by its on-site inspectors. Like CAP accreditation, use of the *Standards* is voluntary. They have been written by experts in the field and reviewed by the laboratorian's own peers.

IV. REGULATION

All laboratories in the United States that test human samples for the assessment of health or disease must be licensed by the Health Care Financing Administration (HCFA). This includes cytogenetic laboratories. The CLIA mandates specific personnel requirements for cytogenetic laboratories through its published regulations (15). Although there are no CLIA regulations specific to cytogenetic laboratories, all general requirements apply. Appropriate management of the testing process is expected, from test requisitioning through specimen labeling and transport, throughout analytic testing, and all the way to result reporting (16). Records must be complete and retained for at least 2 years (17). The laboratory must pursue a quality assurance program that covers training, competency testing of personnel, and the integration of the testing process within the total patient context (preanalytic and postanalytic testing) (18).

Most states require some form of laboratory licensure. Accreditation by a voluntary program such as the CAP's LAP usually qualifies the laboratory for state licensure in such cases. There are exceptions, and a cytogenetic laboratory that is located in a state with a licensure requirement (or even one that intends to care for the residents of such a state) should request from that state the details of its particular regulations. HCFA does not require cytogenetic laboratories to enroll in formal proficiency testing, but there are states that require laboratories that practice therein to do so.

Cytogenetic laboratories must also follow the regulations of the U.S. Food and Drug Administration (FDA). The FDA's Center for Devices and Radiological Health defines the standards for diagnostic "devices" (19), which include reagent kits and image analysis systems (20). An example of the former would be a packaged set of materials for fluorescent in situ hybridization (FISH). The DNA probe in a FISH kit is a regulated reagent. Licensed kits must be used *as directed by the manufacturer*. Any other use requires the user to document performance characteristics (see below). An example of the latter is the microcomputer and its software used in the cytogenetics laboratory to generate a digitized karyotype and file the results for later retrieval. If the computer program in a licensed system is changed beyond the bounds defined by the vendor's instructions, the user bears the responsibility for documenting the clinical reliability of this modified system.

The laboratory that chooses to prepare its own reagents, to develop its own methods, or to write its own computer software—or one that modifies a licensed kit or program that it has purchased—has assumed the burden of establishing the validity of its process. This means that the accuracy and precision of the laboratory's local method must be determined, and a reference range must be defined (21). For qualitative results, such as "47,XY,+21," the term *precision* does not apply, but accuracy is confirmed through proficiency testing and regular quality assurance activities. (For example, one can document the correlation between a karyotypic and a clinical diagnosis or may replicate a prenatal diagnosis that was originally established from abortus tissue.) For standard human cytogenetic analysis the reference values are well understood to be 46,XX and 46,XY.

But what about FISH or chromosomal fragility studies? In analytic methods such as these, the diagnosis is based on the observation of a population of cells. The set of observations must be large because artifacts can deceive an observer. In the case of FISH, background fluorescence might resemble a true hybridization signal, or two independent signals might coincide and appear as one. Fragility studies are likewise interpretable only on the basis of a statistical occurrence of morphologic events within a population of cells. In any quantitative analysis, the accuracy, precision, and reference ranges of the method may be measured using standard statistical techniques. There is no difference in this regard between a test developed in a cytogenetics laboratory and one developed in a chemistry or hematology laboratory. These statistics must be calculated to meet the regulatory requirement. Indeed, the laboratory director would want to calculate them were no regulations to exist.

V. PROFICIENCY TESTING

A proficiency test (PT) is any external challenge by which a laboratory measures its expertise in a discipline by comparing its performance with others performing the same test. A PT challenge in clinical cytogenetics may be based on any medium that is familiar to the laboratory. One might be offered a blood sample for culture and analysis or a prepared karyotype ready for diagnosis. As a general principle, the more the specimen simulates a real clinical situation, the better is the challenge. A good system minimizes those artifacts and variables that are peculiar to the survey process.

Some PT programs provide the laboratory with prepared glass slides to be examined microscopically and diagnosed. Such a medium simulates an important component of the routine process, but a program of this type suffers from two serious flaws. First, the participants in the program necessarily examine different slides. It is impossible to guarantee uniformity in the level of difficulty in a glass-slide challenge. Second, the acceptable response(s) for any particular slide is based on a small set of reference observations (usually determined by a panel of experts). It is extremely difficult to demonstrate that a referee panel reasonably represents the prevailing standard of care.

The Interlaboratory Comparison Program in cytogenetics developed and produced by the CAP and the ACMG is based on four principles: (a) The PT program must first and foremost be a tool for laboratory improvement. (b) All participants must study identical survey materials. (c) All responses must be phrased in an objective, short-answer format. (d) The "correct" response will be defined on the basis of group consensus (22). The ACMG/CAP program uses photographed metaphases as its primary medium for challenge. Multiple cells are offered from each case. The graded response is a karyotypic diagnosis written in the format familiar to a cytogeneticist. Whole blood samples supplement the photographs. To determine the correct diagnosis from a fresh, living blood sample ensures that the entire analytic process is intact. The range of abnormalities that can be surveyed in a fresh blood sample is limited, however, because the donor must be an adult in good health and able to consent to the donation. Paper challenges are not only less expensive to distribute but can depict a far wider array of abnormalities (Fig. 2). Over 320 laboratories now subscribe to the ACMG/CAP cytogenetics PT survey, far more than any other cytogenetics PT survey. The ACMG and the CAP have completed pilot studies for extending the range of cytogenetic PT challenges to include FISH (23).

Proficiency testing requirements for cytogenetics are undefined within the CLIA regulations, but some mechanism for comparison of laboratory results is required (24). External comparisons such as split-sample testing (contemporaneous analysis of the same sample by two or more laboratories) or other shared arrangements whereby cytogenetic laboratories may challenge each other are acceptable under the current regulations.* Similar strategies are necessary for those genetic tests for which there are no available PT programs. Accreditation programs' requirements may of course be more stringent than the regulatory requirements.

---

*Split-sample testing is an acceptable substitute for proficiency testing when there is no formal alternative. To refer a PT sample or a portion of a PT sample to another laboratory is fraudulent and unacceptable.

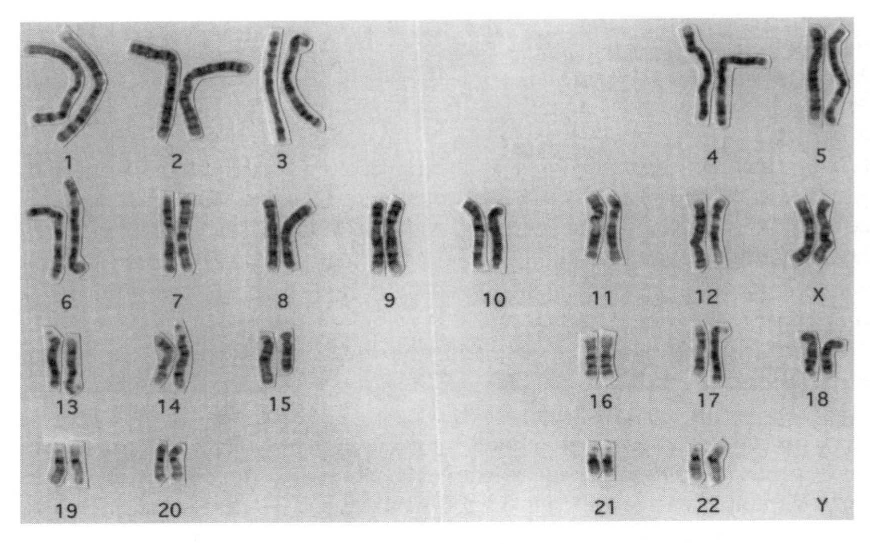

Figure 2 A disarranged karyotype used as a CAP/ACMG proficiency testing challenge. It contains the following errors:

The right chromosome 2 is incorrectly oriented.
The left chromosome depicted as a 3 is a chromosome 5.
The right chromosome depicted as a 4 is a chromosome 3.
The right chromosome depicted as a 5 is a chromosome 6.
The left chromosome depicted as a 6 is a chromosome 7.
The left chromosome depicted as a 7 is a chromosome 11.
The right chromosome depicted as an 8 is an X chromosome.
The right chromosome depicted as a 10 is an abnormal chromosome 4 [del(4)(p12)].
The right chromosome depicted as an 11 is a chromosome 12.
The left chromosome depicted as a 12 is a chromosome 8.
The right chromosome depicted as a 13 is a chromosome 14.
The left chromosome depicted as a 14 is a chromosome 17.
The right chromosome depicted as a 15 is a chromosome 18.
The right chromosome depicted as a 17 is a chromosome 13.
The left chromosome depicted as an 18 is a chromosome 15.
The left chromosome depicted as a 19 is a chromosome 22.
The right chromosome depicted as a 22 is a chromosome 19.
The left chromosome depicted as an X is a chromosome 10.
(Reproduced with permission of the College of American Pathologists.)

VI. INSPECTION

The LAP of the College of American Pathologists is a voluntary program for laboratory improvement based on peer review, on-site inspection, and proficiency testing. Accreditation by the CAP may be used by a laboratory to meet federal and most state regulatory requirements. A cytogenetic laboratory may elect the CAP to be its accrediting agency under CLIA. Alternative accrediting agents for laboratories include the Joint Commission on Accreditation of Health Care Organizations (JCAHO) and the Commission on Office Laboratory Accreditation (COLA). Of the available private accrediting agencies, only the CAP offers an inspection and PT program for clinical cytogenetics. A clinical cytogenetics laboratory may ask HCFA to inspect it directly rather than to opt for a nongovernmental program, in which case it will likely be surveyed by its local state department of health, which will use an inspection instrument (25) developed directly from the CLIA regulations.

CAP accreditation is awarded when a laboratory demonstrates that it meets the *Standards for Laboratory Accreditation*. To do so means that a qualified inspector has visited the laboratory once every 2 years, that the laboratory has inspected itself using the same set of checklist criteria in the intervening years, and that the laboratory continues to perform satisfactorily on a CAP-approved PT program (of which the ACMG/CAP survey is the only current example). Any serious deficiencies that are identified during an inspection must be corrected before a favorable accreditation decision is granted. Unsatisfactory performance in PT may precipitate an interim on-site inspection and could lead to loss of CAP accreditation. Sanctions requiring corrective actions or more punitive measures may be imposed by HCFA if the public health is endangered or results pose an immediate risk to patients. CAP accreditation documents that the laboratory has a comprehensive program for laboratory improvement, one that has detailed procedures for the cardinal elements of laboratory performance: people, procedures, equipment, and its environment (Fig. 3).

CAP accreditation is an economic necessity for many American laboratories. JCAHO is the primary accrediting body for health-care providers seeking Medicare reimbursement. CAP accreditation is recognized by the JCAHO, which qualifies a cytogenetics laboratory within a hospital for Medicare reimbursement. Laboratories bidding on provider contracts may find CAP accreditation a competitive advantage. Such marketplace realities increase the value of CAP accreditation even as they may detract from its voluntary intent.

Eight essentials of CAP accreditation should be noted because they go beyond the requirements of CLIA.

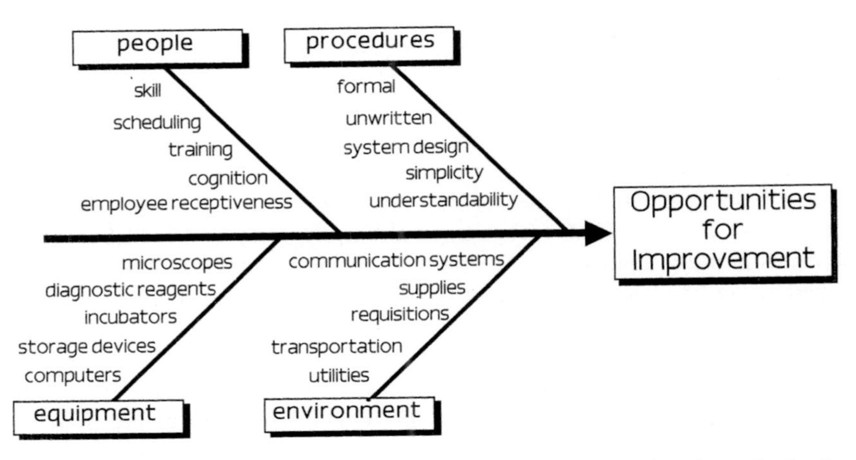

Figure 3 The cardinal elements of laboratory performance that form the basis of a quality improvement program.

1. Participation in the CAP/ACMG survey is expected. PT samples must be handled in the same manner as patient samples whenever possible. For example, multiple analysts may participate in the diagnosis of a PT sample only if routine samples are studied by a team approach. Each "unacceptable" evaluation must precipitate a prompt and appropriate corrective action.
2. The quality improvement (QI) program and the director's role in it must be defined. As a minimum, the QI program must include an analysis of failed or suboptimal cultures, and the laboratory should try to follow up on each abnormal prenatal diagnosis.
3. *An International System for Human Cytogenetic Nomenclature* (ISCN) must be used correctly in the final report (26).
4. Minimal requirements for turnaround time must have been defined. At least 90% of final reports must be reported in 21 days (for prenatal studies), 28 days (for blood and bone marrow studies), and 6 weeks (for non-neoplastic fibroblast analyses). The target for urgent neonatal studies is 72 h.
5. Redundancy of certain systems is expected. All cultures must be set up in duplicate (or established independently). Duplicate amniotic fluid and chorionic villi cultures must be harvested independently to distinguish true mosaicism from pseudo-mosaicism. Prenatal samples must be split between two incubators that have independent power, gas, and alarm systems.

6. Sufficient numbers of cells must be studied to achieve a statistically significant sample. As a general rule, at least 20 cells must be counted, 5 cells analyzed, and 2 cells karyotyped per case.
7. The level of band resolution must be sufficiently adequate to answer the clinical question (for example, ≥550 bands per haploid set when studying birth defects, mental retardation, or spontaneous abortion).
8. FISH questions are now included in the CAP inspection of a cytogenetics laboratory. For example, each probe used should have been validated by the laboratory (whether commercial or prepared in-house), reference ranges should have been established, and controls must be run with each case.

VII. CONCLUSION

Every standard, guideline, and regulation has been written to ensure that laboratory services provided to the public are of high quality. Familiarity with source documents facilitates compliance and helps the pathologist or cytogeneticist interpret their intent. Professional organizations such as the ACMG and the CAP, which have developed the most detailed standards, assist their members to meet or exceed these standards. By submitting to external review, a laboratory can determine how it compares with its peers. External review includes the credentialing of laboratory personnel, regulatory oversight, voluntary inspection and accreditation, and proficiency testing.

An internal quality assurance program that focuses on all critical laboratory processes is crucial to optimize the value of external review. Unfortunately, the pace of change taking place in the field of cytogenetics too often exceeds the external agencies' ability to refine the published standards. The wise cytogenetics professional will examine and reexamine each step in the testing process from sample collection to the interpretability of the final report to discover new opportunities for improvement. External standards, which define the expectations for quality service and are the basis for acceptable practice, are very important, but they should not be viewed as the complete definition of excellence.

REFERENCES

1. American College of Medical Genetics, 9650 Rockville Pike, Bethesda, MD 20814.

2. College of American Pathologists, 325 Waukegan Road, Northfield, IL 60093.
3. Laboratory Accreditation Program. Standards for Laboratory Accreditation. Northfield, IL: College of American Pathologists, 1996.
4. National Committee for Clinical Laboratory Standards, 940 West Valley Road, Suite 1400, Wayne, PA 19087.
5. National Fire Protection Association, 1 Batterymarch Park, Quincy, MA 02269.
6. American Board of Medical Genetics, 9650 Rockville Pike, Bethesda, MD 20814.
7. American Board of Medical Specialties, 1007 Church Street, Suite 404, Evanston, IL 60201.
8. Accreditation Council of Graduate Medical Education, 515 North Street, Suite 2000, Chicago, IL 60610.
9. National Certification Agency for Medical Laboratory Personnel, P.O. Box 15945-289, Lenexa, KS 66285.
10. American Society for Clinical Laboratory Science, 7910 Woodmont Avenue, Suite #530, Bethesda, MD 20814.
11. Association of Genetic Technologists, P.O. Box 15945-288, Lenexa, KS 66285.
12. Association of Cytogenetic Technologists. In: Knutsen T, Spurbeck TL, eds. The AGT Cytogenetics Laboratory Manual, 3rd ed. Philadelphia: Lippincott-Raven Press, 1997.
13. The American College of Medical Genetics. Standards and Guidelines: Clinical Genetics Laboratories. Bethesda, MD: ACMG, 1993.
14. Mark HFL, Watson MS. Evolving standards of practice for clinical cytogenetics. Rhode Island Med 1994; 77:373–374.
15. Title 42 CFR, Part 493, §493.1449(p).
16. Title 42 CFR, Part 493, §493.1103 and 493.1109.
17. Title 42 CFR, Part 493, §493.1107.
18. Title 42 CFR, Part 493, Subpart P.
19. Title 21 CFR, Part 820.
20. Mark HFL, Kelly T, Watson MS, Hoeltge G, Miller WA, Beauregard L. Current issues of personnel and laboratory practices in genetic testing. J Med Genet 1995; 32:780–786.
21. Title 42 CFR, Part 493, §493.1213.
22. Hoeltge GA, Dewald G, Palmer CD, et al. Proficiency testing in clinical cytogenetics: a 6-year experience with photographs, fixed cells, and fresh blood. Arch Pathol Lab Med 1993; 117:776–779.
23. Dewald GW, Brothman AR, Butler MG, et al. Pilot studies for proficiency testing using fluorescence in situ hybridization with chromosome-specific DNA probes. Arch Pathol Lab Med 1997; 121:359–367.
24. Title 42 CFR, Part 493, §493.801(a)(2)(ii).
25. Department of Health and Human Services, Health Care Financing Administration. State Operations Manual; Provider Certification, Appendix C. Item PB 92-146-174, National Technical Information Service, U.S. Department of Commerce, Springfield, VA.
26. Mitelman F, ed. ISCN (1995): An International System for Human Cytogenetic Nomenclature. Basel: S. Karger, 1995.

APPENDIX: GLOSSARY OF TERMS

accreditation

Organizational recognition that a laboratory maintains standards of professional practice as determined by voluntary peer review, inspection, and proficiency testing.

accuracy

Degree of conformity of a laboratory measure to a standard or a true value.

certification

A written statement that an individual has met all requirements of the granting agency; "board certification" refers to the demonstrated qualifications of medical specialists such as clinical cytogeneticists.

director of laboratory

The individual in whom is vested the authority for laboratory administration and the responsibility for maintaining the laboratory's professional expertise.

licensure

Permission granted by governmental authority to engage in the business of laboratory practice.

precision

The degree of reproducibility with which an analytic operation may be performed.

proficiency testing

An external challenge by which a laboratory measures its expertise in a discipline by comparing its performance with others performing the same test.

reference range

The span or set of analytic values that is normative for a particular patient population.

supervisor, general

A CLIA term for the laboratory technologist responsible for test performance.

supervisor, technical

A CLIA term for the laboratory employee with clinical cytogenetics expertise and in whom the responsibility for test selection and quality is vested.

Afterword

Avery A. Sandberg
St. Joseph's Hospital and Medical Center, Phoenix, Arizona

I presume that I have been asked to write the Afterword with an emphasis on the future of cytogenetics mainly because I am the oldest (by way of age) of all contributors to the present volume. After all, the elderly are supposedly the best seers. It is also true that I have been involved with human cytogenetics for a period longer than any of the authors in this volume (some of whom were born just after my entry into the field of cytogenetics in the late 1950s). Thus, I will take advantage of the privileges offered to me by my age and experience. In order to predict or provide you with a plausible glimpse of the future of our field, I feel that the present must be critically evaluated.

Over the course of more than four decades of active personal participation (in the areas of constitutional and primarily cancer cytogenetics) and as witness to the development of the field of human cytogenetics, remarkable strides have led not only to methodological breakthroughs but also to an explosion of the body of knowledge called molecular genetics. The methodological advances include the application of various banding techniques for the identification of chromosomes and their changes, the use of fluorescent in situ hybridization (FISH) techniques (centromeric, cosmid, and whole painting probes) and recently spectral karyotyping (SKY). All of these have led to a precise recognition of a myriad of chromosomal changes in all areas of human anomalies and diseases.

The following comments relate to the field of cancer cytogenetics, since my experience and expertise reside primarily in this area. Despite the ready availability of the methodologies mentioned above, a significant percentage of physicians (including hematologists, oncologists, and pathologists) have not availed themselves of the cytogenetic approaches to the diagnosis and follow-up of patients who have hematological disorders and cancers of various tissues and organs. This situation presents a challenge to cytogeneticists undertaking and expanding informational and educational efforts to acquaint, if not educate, those colleagues who have not kept up-to-date so that they can understand and reap the full benefits of cytogenetics for diagnostic purposes in a wide range of neoplastic diseases. Considering the specificity with which cytogenetics can define a disease, it is unfortunate that a significant segment of patients, having disorders that can be unequivocally defined through cytogenetics, are not given the advantage of this approach. This is particularly true of solid tumors, which seem confusing or complex to the pathologist. Hopefully, this challenge will be met in the future through presentations in lecture rooms, publications, and computerized information at appropriate levels. This educational effort must encompass discussions of the true value of cytogenetic (chromosomal, karyotypic) analysis, FISH examinations, and molecular approaches when they are appropriate for the case being studied. We must somehow overcome the almost inchoate resistance (behind which there is usually a thick element of misinformation or ignorance) from a segment of the medical profession to new developments and approaches.

The ongoing identification of which genes play a role in the development and progression of cancer and leukemia will lead to the availability of an increasing number of probes that can be applied for diagnostic and follow-up purposes. Incidentally, the identification of such genes, particularly their functions, may also lead to the planning of therapeutic approaches related to the functions of such genes.

A good example is the introduction of successful vitamin therapy in the treatment of acute promyelocytic leukemia (APL). Its effect (although not curative) probably operates through changes affecting a vitamin receptor. Specifically, t(15;17)(q22;q11), the translocation that characterizes APL, involves a gene encoding the retinoic acid receptor-alpha (RARα) mapped to the chromosomal breakpoint 17q21 and has been found to be rearranged in all cases of APL. The breakpoints on chromosome 15 are in a region termed PML. The PML/RARα fusion protein is present in all patients with APL; the PML/RARα fusion gene is found in about two-thirds of APL cases. Thus, the molecular detection of the PML/RARα rearrangement by PCR, besides previously available cytogenetic or FISH analyses, can be used as a diagnostic assay for the disease.

In about 85% of patients with APL treated with all-*trans*-retinoic acid (ATRA), complete remission is achieved even though the length of each remission may vary. While it may be surprising that a defect causing a maturation in a receptor can result in a remarkable sensitivity of that receptor to one of its ligands, the initial clinical response to ATRA correlates precisely with the presence of the t(15;17). Through the use of ATRA and conventional chemotherapy it has been possible to achieve 80% five-year survival with the prospect of decreasing the relapse rate further by ablating the residual disease with monoclonal antibody–based protocols and by the possible use of bone marrow transplantation.

I present the situation in APL in somewhat great detail because its essential meaning and application may ultimately be extended to many other conditions, particularly leukemias and bone and soft tissue tumors characterized by a translocation as the sole karyotypic abnormality. Thus, one can envision that in the future the genes (and their functions) affected by translocations or other genetic changes will be defined and appropriate therapy (including gene therapy) devised.

Although FISH (and SKY) methodologies have added an important avenue in the cytogenetic analysis of leukemia and cancer, routine or classical cytogenetic analyses will continue to offer the most panoramic view of the cytogenetic changes in the above-mentioned conditions. The presence of single cells with abnormalities reflecting a clone with such changes is best established through routine cytogenetics (and a follow-up with appropriate FISH probes), as is the variety of changes that may be present, particularly in adenocarcinomas. Thus, for the foreseeable future, routine cytogenetics will remain the most informative approach to the study of karyotypic changes in neoplastic states.

From a realistic point of view, the economic aspects of cancer cytogenetics, particularly in the area of solid tumors, will require a more aggressive approach to justify the extension of third-party payments to cover the tests required. The lack of such payments at present continues to thwart the extension of this area of cytogenetics, despite its intrinsic diagnostic value. It should be an integral part of the care of patients. Why such payments are extended to cover some cytogenetic tests and not others remains a nagging problem and one that all interested parties must address in the near future. An effective way to address this is through professional genetic organizations such as the American College of Medical Genetics.

Considerable automation has already been incorporated into cytogenetic methodologies. This is likely to be expanded in the future as new approaches to establishing karyotypic changes, particularly in interphase nuclei and archival tissues, are developed. Thus, computerized analysis of karyograms, as used in SKY analysis, is likely to be an approach that will see further exploration and application in the future. The use of "cocktail" probes of various types will afford a more rapid cytogenetic diagnosis. Federal grant support, such as the National Institutes of Health's Small Business Innovative Research (SBIR) and Small Business Technology Transfer (STTR) programs that fund small commercial companies, will continue to accelerate the research and development efforts in this and other areas of conventional and molecular cytogenetics.

In the future we will also most likely see continued transmission and dissemination of cytogenetic and molecular genetic data together with their meaning through electronic media (e.g., Internet) reducing the time consumed in transmitting results to physicians. Also, the karyotypes, FISH results, and other data can be readily transmitted to physicians through a number of systems that are and will become available in the near future. Thus, what has been called bioinformatics—or, more properly, computational biology and medicine—will likely play an important role in the transmission of information to those depending on cytogenetics and molecular genetics for the diagnosis and care of their patients.

More intensive education of physicians who are likely to utilize cytogenetic and molecular genetic approaches should lead not only to a keener appreciation and understanding of results but also to cogent physician inputs relating to anticipated and likely findings of their patients. This will not only help the personnel performing these tests but also result in more efficient (labor- and time-saving) and meaningful yields of the studies.

In summary, more active involvement of physicians in the utilization of cytogenetic and related approaches in the diagnosis of tumors of uncertain or confusing histology (or in the confirmation of diagnosis) is key to the future expansion of the field. I envision future cancer cytogenetics to include more automated, and hence more expeditious, analyses encompassing cytogenetics with an appropriate application of FISH methodologies, particularly for the analysis of interphase cells and archival tissues. I also foresee more readily accessible bioinformatics on the Internet (or related outlets) to help clinicians with the interpretation and application of cytogenetic data. The increasing utilization of molecular techniques as an adjunct to conventional cytogenetics in medicine, electronic transmission, and interpretation of cytogenetic data will become the wave of the future.

Questions

CHAPTER 1

1. When was the correct number of chromosomes in human beings discovered, and by whom?
2. What does Mendel's principle of segregation predict?
3. What does Mendel's principle of independent assortment predict?
4. What was Mendel's recognized profession?
5. What did Theodor Boveri hypothesize?
6. Who discovered phytohemagglutinin (PHA), and what is its purpose?
7. Name some other mitogens employed by cytogeneticists. (Answers to this question can be found in another chapter of this book.)
8. What is the name of the marker chromosome that was consistently found to be associated with chronic myelogenous leukemia (CML)? Describe its composition according to ISCN (International System for Human Cytogenetic Nomenclature).
9. What was the first banding technique that was developed, and by whom?
10. List four major developments during the 1960s in cytogenetics and medical genetics.
11. List four major developments during the 1970s in cytogenetics and medical genetics.
12. List four major developments during the 1980s for medical genetics.
13. Name two of the newest techniques that have been developed in molecular cytogenetics.
14. What became the 24th primary specialty board of the American Board of Medical Specialties?
15. What is the ACMG?

CHAPTER 2

1. What are chromosomes?
2. Describe how a single molecule of eukaryotic DNA is organized.

3. How are the DNA/histone complexes released? Describe one chemical process.
4. Name the two types of classification for chromosomal material and describe the major distinction between the two.
5. How is heterochromatin subclassified? Distinguish between the different types.
6. What is the number of chromosomes in the somatic cells of an organism called?
7. What is the number of chromosomes in the gametes of an organism called?
8. What is the relationship between the diploid number and the haploid number?
9. What is the true diploid number of humans? How many autosomes are there? How many sex chromosomes are there?
10. What types of cells undergo mitosis? Meiosis?
11. Briefly describe what happens in each stage of mitosis.
12. What is interphase?
13. What is the G_0 stage?
14. What must one take into account when counting fluorescent signals in interphase nuclei by FISH (fluorescent in-situ hybridization)?
15. Compare the processes of meiosis and mitosis.
16. What is gametogenesis?
17. When does male meiosis begin? Female meiosis?
18. What is the sex-determining factor in humans?
19. What is the primary constriction, and what is its purpose?
20. What are some likely fates of chromosomes without telomeres?
21. How does the International System for Human Cytogenetic Nomenclature (ISCN) define random loss?

CHAPTER 3

1. A patient was diagnosed with classical Turner syndrome soon after birth. Several years later a bone marrow chromosome analysis showed a modal number of 47 chromosomes with two X chromosomes and three copies of chromosome 8. Designate the bone marrow karyotype of this patient.
2. A Klinefelter patient had a constitutional chromosome analysis performed to confirm the clinical diagnosis. A 24-cell analysis on stimulated blood cells showed three different cell lines, 46,XY, 47,XXY, and 48,XXXY, all in equal numbers (8 each). Designate the karyotype of the patient.
3. A Down syndrome female patient with nondisjunctional trisomy 21 was found to have 46 chromosomes in the bone marrow cells with only two copies of chromosome 21. Write the karyotype designation.
4. Bone marrow chromosome analysis on a Down syndrome female patient has a modal number of 48 chromosomes with four copies of chromosome 21. How would you designate the karyotype?
5. A normal female has an extra X chromosome in her bone marrow cells. What is the karyotype?
6. Chromosome analysis on a couple referred for multiple miscarriage showed that the male partner had a karyotype with a paracentric inversion of the long

arm of chromosome 3. The breakpoints were at bands 3q21 and 3q26. Designate the karyotype.

7. A baby boy was found to have an unbalanced karyotype with one copy of chromosome 2 showing extra chromosome material at band q32. However, from the band morphology, the origin of this extra chromosome segment was unclear. What is the most accurate way to write the karyotype?
8. Insertion of a chromosome segment can occur within a single chromosome. Insertion can also be direct or inverted. Describe the following rearrangements in words.
   (a) 46,XY,ins(2)(p13q21q31)
   (b) 46,XY,ins(2)(p13q31q21)
   (c) 46,XY,ins(5;2)(p14;q22q32)
9. A direct duplication of a chromosome 3q21−q25 was seen in one chromosome 3 of a male patient. Write the karyotype.
10. A balanced Robertsonian translocation involving chromosomes 14 and 21 was seen in a woman. Her first pregnancy resulted in the birth of a baby boy with Down syndrome. Write (a) the mother's karyotype; (b) the child's karyotype.
11. An unbalanced translocation karyotype was designated as 47,XX,+der(1;3)(p10;q10). Describe in words the chromosome regions that have trisomy, monosomy, or both.
12. An isochromosome for the long arm of an X chromosome was seen in a female with 46 chromosomes, and the nature of the centromere is not known. What is the karyotype?
13. In a female chronic myelogenous leukemia (CML) patient, a classical t(9;22)(q34;q11.2) was seen along with a second copy of der(22) with 47 chromosomes. What is the karyotype?
14. Fluorescent in-situ hybridization was done on the case in question 13 and it showed the same result. How would you designate the karyotype?
15. Nuclear in-situ hybridization was performed on the interphase nuclei of bone marrow cells from a CML patient. The study showed two fusion signals for the *bcr/abl* probe and one signal each for *bcr* and *abl* separately. Write the *ISH* karyotype.
16. There were 3 signals of a centromeric (α satellite) sequence for chromosome 18 seen on interphase nuclei. Designate the *ish* karyotype.
17. 46,XX,ish inv(16)(p13q22)(pcp16q sp). Describe this karyotype in words.
18. 46,X,?(Y)(p10).ish idic(Y)(DYZ3++,DYZ1−). Describe this karyotype in words.

CHAPTER 4

1. If conventional cytogenetic analysis is based on the examination of dividing cells, how does one perform such an analysis on cells with a low division rate?
2. What is a short-term culture?
3. What is a long-term culture?

4. What are the best kind of cells that sustain very well in a long-term culture?
5. What is the most common source of cells for cytogenetic analysis?
6. Which type of chromosomal abnormality is characteristic of tumor cell populations and is localized to tumor cells?
7. What is the tissue of choice for the cytogenetic study of most hematologic disorders?
8. Why would a cytogenetic analysis of bone marrow cells more closely indicate what is happening in vivo?
9. How can a constitutional chromosomal abnormality be ruled out?
10. What are two common procedures used for the prenatal diagnosis of genetic disorders?
11. What are the four major elements required in culturing and harvesting a peripheral blood culture?
12. Briefly describe a typical procedure to set up skin tissue for fibroblast culture.
13. Briefly describe a typical procedure for harvesting a fibroblast culture.
14. Name some enzymes used for tissue disaggregation.
15. What is an effective way to increase the mitotic activity of myeloid and undifferentiated leukemia cultures?
16. What kind of improvements can be made in the cytogenetic analysis of acute lymphocytic leukemias, given that the chromosomes of those affected often result in poor morphology, indistinct bands, and spreading resistance?

CHAPTER 5

1. What is a chromosome band?
2. How were chromosomes grouped prior to the advent of banding techniques?
3. How was each unbanded chromosome identified unequivocally in the pre-banding era?
4. Describe the first banding technique reported in the literature.
5. What does the three-letter code for banding signify?
6. Evaluate the utility of solid staining.
7. Name an example of a Romanovsky-type stain.
8. What is the most commonly used method for inducing G-banding?
9. It has been hypothesized that G- and Q-bands reflect the inherent genetic composition and underlying structure of the chromosome, since the banding patterns are distinct and consistent from experiment to experiment. Please elaborate.
10. For what application is high-resolution banding most useful?
11. Match the following:

(a) Deletions/cryptic translocations involving 17p13.3	(i) Steroid sulfatase deficiency/ Kallman syndrome
(b) Deletions of chromosome 15q12	(ii) DiGeorge syndrome and velocardiofacial syndrome

(c) 22q11.2 deletions — (iii) Prader-Willi and Angelman syndromes
(d) Deletions of 17p11.2 — (iv) Williams syndrome
(e) Deletions of 7q11.23 — (v) Cri-du-chat syndrome
(f) Deletions of 4p16.3 — (vi) Wolf-Hirschhorn syndrome
(g) Deletions of 5p15.2 — (vii) Charcot-Marie-Tooth disease
(h) Deletions of Xp22.3 — (viii) Miller-Dieker syndrome
(i) Microduplication at 17p11.2 — (ix) Smith-Magenis syndrome

12. What reagents are involved in Q-banding? Say a few words about one of them.
13. To what do the Q-bright bands correspond? The Q-dull bands?
14. What is the most useful feature of Q-banding?
15. Discuss the drawbacks of Q-banding.
16. What is constitutive heterochromatin?
17. What region of the human genome does C-banding stain?
18. Why is C-banding useful?
19. What does C-banding entail?
20. What is R-banding?
21. What does R-banding entail?
22. Why do some cytogeneticists think NOR staining is a valuable technique?
23. Name two uses of 5-bromodeoxyuridine.
24. What is DA-DAPI staining?
25. What special staining technique may be useful for the study of somatic hybrids?
26. What research and clinical application does kinetochore (Cd) staining have?
27. What does the Barr body represent?
28. Discuss the phenotypes of the individuals whose karyotypes are shown in (a) Fig. 5.9 and (b) Fig. 5.10 of the text.

CHAPTER 6

1. What category of DNA sequences is located at the centromere of human chromosomes?
2. Which nuclear counterstains are most commonly used for FISH studies?
3. What is meant by direct probe labeling versus indirect probe labeling?
4. Can whole chromosome probes or "paints" be used on interphase cells?
5. Which oncogenes are analyzed with FISH on tumor tissues?
6. What is the advantage of using FISH for translocation analysis?
7. What are marker chromosomes?
8. Fluorescence microscopy requires regular equipment upkeep. What are two important factors in maintenance?
9. How does comparative genomic hybridization assist in the determination of chromosomal regions of interest in tumor cells?
10. How are both comparative genomic hybridization and microFISH examples of reverse hybridization?

CHAPTER 7

1. What are the major indications for performing routine chromosome analysis?
2. Which are more common, autosomal trisomies or monosomies, and why?
3. What are the various chromosomal causes of Down syndrome?
4. What is the clinical importance of identifying a mosaicism for a structurally rearranged Y chromosome in an individual with Turner syndrome?
5. How can a balanced chromosome rearrangement in an individual lead to recurrent miscarriage?
6. How can chromosome painting be useful in the identification of structurally abnormal chromosomes?
7. How does uniparental disomy lead to a clinical phenotype?
8. What is the major known risk factor for nondisjunction?
9. How can chromosome inversion lead to production of germ cells with unbalanced chromosomes?
10. Why is it important to karyotype both parents if a supernumary marker chromosome is found on amniocentesis?

CHAPTER 8

1. Outline the historical development of prenatal chromosome diagnosis.
2. What are the major indications for prenatal chromosome studies?
3. What concern(s) are there for offspring of carriers of Robertsonian translocations, other than unbalanced rearrangements?
4. Is there a difference in risk for unbalanced offspring for carriers of pericentric inversions in comparison to carriers of paracentric inversions?
5. Briefly compare and contrast the two major techniques for amniotic fluid culture.
6. What causes should be considered if there is a discrepancy between ultrasound-identified fetal sex and the karyotype?
7. Cite the major approaches to evaluating the significance of supernumerary marker chromosomes.
8. What are the five basic steps of genetic counseling for prenatal chromosome studies?
9. What options should be presented to a couple if an abnormal fetal chromosome constitution is identified.
10. Identify the steps involved in evaluating the significance of chromosome variants.

CHAPTER 9

1. How can aneuploidy in human sperm be assessed?
2. How can preimplantation diagnosis of aneuploidy be accomplished?

3. Which is more common among sperm, aneuploidy or chromosome rearrangements? Which is more common among oocytes?
4. What are two possible mechanisms of origin for diandric triploid conceptions? For each mechanism, what proportions of triploids with XXX, XXY, and XYY sex chromosome complements are expected?
5. What is the most common mechanism of origin for: (a) trisomy; (b) monosomy X; (c) tetraploidy?
6. What is the most common trisomy in recognized human conceptions? Which trisomies are the least common?
7. Which human trisomies are compatible with survival to term?
8. Which karyotype group of spontaneous abortions is associated with the highest risk for recurrent abortion?
9. What pathological presentation is most often found in tetraploid spontaneous abortions?
10. What chromosome constitution is most often associated with: (a) hydatidiform mole; (b) partial mole; (c) ovarian teratoma?
11. What is the single most likely cause of a spontaneous abortion in a woman aged 40 or over?
12. What is the frequency of unbalanced chromosomal rearrangements among spontaneous abortions? What is the frequency of balanced rearrangement carriers among couples with two or three spontaneous abortions?
13. What further testing is required for a couple whose first pregnancy ends in a spontaneous abortion with a 45,X karyotype?

CHAPTER 10

1. Male infertility is an unknown quandary. What possible causes may lead to male infertility?
2. Why does spermatogenesis require that the hypothalamic neurosecretory cells produce GnRH (gonadotropin-releasing hormone)?
3. What is the stem cell of the male germ line? What are the structural changes that occur in spermiogenesis when round spermatids mature into elongated mature spermatozoa?
4. How do spermatogonia multiply and increase in number?
5. Describe male meiosis or spermatogenesis.
6. What is the difference between male and female meiosis?
7. What are the important elements of the patient's history a physician needs to look for in the evaluation of a patient who is suspected to be infertile?
8. When does gonadal differentiation occur?
9. What is the default pathway for gonadal differentiation when a functional Y chromosome and proper genetic factors are lacking?
10. What evidence exists that suggests that spermatogenesis is the result of a polygenic effect?

11. What is the general trend in chromosomal abnormalities observed in infertile men?
12. What is the most common chromosomal abnormality in infertile men?
13. What have recent FISH studies shown about the spermatozoa of men with an XYY karyotype?
14. Karyotype analysis of XX males shows what would appear to be a normal karyotype for 46,XX females. How could this be possible? List some of the phenotypes of an XX male.
15. What is one hypothesis as to why reciprocal translocations of autosomes interfere with spermatogenesis?
16. What is a Robertsonian translocation?
17. Why have sperm nuclei been difficult to study? Describe one way this problem can be circumvented.

CHAPTER 11

1. What is "Ph-negative CML," and how can it be identified?
2. What are "atypical CML" and "Juvenile CML"?
3. What secondary changes are most frequent in CML, and what is their significance if they are present at diagnosis, or if they evolve during the course of disease?
4. What are the three most common structural chromosomal changes in AML?
5. What is the core binding factor (CBF), and how does it relate to AML?
6. What is the "5q− syndrome," and how does it differ from other cases of MDS with del(5q)?
7. What is the significance of i(12p) in a hematopoietic malignancy?
8. What morphologic features in AML show some correlation with the following:
   (a) t(6;9)(p23;q34)
   (b) t(8;16)(p11;p13)
   (c) inv(3)(q21q26)
   (d) inv(16)(p13q22)
   (e) t(8;21)(q22;q22)
9. What cytogenetic findings might be expected in a case of acute leukemia showing lineage infidelity (i.e., co-expression of myeloid and lymphoid markers, lineage switch from lymphoid to myeloid, or two distinct clones with myeloid and lymphoid phenotype)?
10. What cytogenetic abnormalities are in the "favorable group" and what abnormalities are in the "unfavorable group" for AML?
11. What are the two classes of drugs associated with therapy-related MDS/AML, and with what cytogenetic findings are they commonly seen?
12. What would you think of a case submitted as "essential thrombocythemia" in which you found the Ph chromosome?
13. What is the significance of a positive RT-PCR reaction for AML1/ETO mRNA in a patient in the fifth year of remission from AML associated with t(8;21)?

14. Which leukemias have a particular association with the following:
    (a) Extramedullary disease
    (b) Central nervous system (CNS) involvement
    (c) Disseminated intraveascular coagulation (DIC)

CHAPTER 12

1. In ALL, ploidy is
    (a) Recognized as a distinct feature
    (b) Predictive of clinical outcome
    (c) Used in the stratification scheme for risk-specific therapy
    (d) All of the above
2. Regarding hyperdiploidy in ALL:
    (a) It is divided into two subgroups.
    (b) The DNA index of the leukemic cell is approximately correlated with the number of chromosomes.
    (c) The proportion of hyperdiploid cases is similar for children and adults.
    (d) (a) and (b) are true.
3. Trisomy of chromosome 12 is
    (a) Frequently observed in cases of hyperdiploidy 47–50
    (b) Typically observed in cases of hyperdiploidy 51+
    (c) Highly associated with lymphomas
    (d) All of the above
4. t(12;21)(p13;q22) is
    (a) Observed in 20–25% of children with B-lineage ALL
    (b) Rarely observed by conventional cytogenetic analysis
    (c) Associated with a good prognosis
    (d) All of the above
5. Regarding hypodiploidy in ALL:
    (a) Most hypodiploid cases have a modal number <40.
    (b) Dicentric chromosomes are among the structural abnormalities most frequently identified in hypodiploid cases.
    (c) The most frequent abnormality is t(9;22).
    (d) All of the above.
6. Regarding near-haploidy in ALL:
    (a) It is associated with an excellent prognosis.
    (b) It is frequently seen in pediatric cases.
    (c) Occasionally, the duplication of the near-haploid clone mimics an hyperdipoid line.
    (d) All of the above.
7. t(1;19)(q23;p13.3)
    (a) Is the most common translocation detected by conventional cytogenetics in adult ALL
    (b) Is typically found in 20–25% of early pre-B ALL cases

(c) Results in expression of the chimeric transcription factor E2A-PBX1
(d) All of the above

8. t(9;22)(q34;q11.2) is
(a) The most frequent translocation in adult ALL
(b) Associated with a good prognosis
(c) Associated with low white blood counts
(d) (b) and (c) are true

9. t(4;11)(q21;q23) is
(a) The most frequent translocation involving the 11q23 region
(b) Observed in a large proportion of infants
(c) Associated with poor prognosis
(d) All of the above

10. In cases of Burkitt's lymphoma:
(a) t(8;14)(q24.1;q32) is the most frequent structural abnormality.
(b) The t(2;8)(q12;q24.1) and t(8;22)(q24.1;q22) are less frequent variants.
(c) *MYC* is dysregulated occasionally.
(d) All of the above.

11. In T-cell leukemia:
(a) Conventional cytogenetics detects an abnormal clone in 60–70% of cases.
(b) The breakpoints are nonrandom in 30–40% of cases with abnormal karyotypes.
(c) t(11;14)(p13;q11.2), t(11;14)(p15;q11.2), and t(10;14)(q24;11.2) are among the most frequently observed aberrations.
(d) (a) and (b) are correct.
(e) All of the above.

12. t(2;5)(p23;q35) is
(a) Typically identified in cases of anaplastic large cell lymphoma
(b) Seen only in adults
(c) Usually associated with lack of expression of CD30
(d) Typically associated with B-lineage cases

13. t(11;14)(q13;q32)
(a) Is strongly associated with mantle cell lymphoma
(b) Is frequently seen in B-CLL
(c) Leads to expression of *BCL6*
(d) All of the above

14. The test(s) that contribute(s) to definitive diagnosis of mantle cell lymphoma is/are
(a) Identification of IGH-*BCL2* by PCR
(b) Positive staining for cyclin D1 antibody
(c) Identification of *BCL2* rearrangements by Southern analysis
(d) (a) and (c)

15. t(14;18)(q32;q21.3)
(a) Is observed in 40% of children with NHL
(b) Is typically associated with Burkitt lymphomas
(c) Leads to deregulation of *BCL2* by IGH
(d) None of the above

CHAPTER 13

1. Wilms tumor is associated with which of the following chromosome deletions?
   (a) 3p−
   (b) 11p−
   (c) 13q−
   (d) 20q−
2. Tumor specimens for cytogenetic analysis
   (a) May be kept at room temperature overnight
   (b) May be refrigerated overnight
   (c) May be transported in a small amount of sterile saline solution or phosphate-buffered saline (PBS)
   (d) All of the above
3. Which of the following statements is correct?
   (a) The presence of karyotypic changes is always indicative of the existence of malignancy within a tumor.
   (b) Benign tumors are not associated with recurrent chromosome anomalies.
   (c) Chromosomal anomalies have never been observed in normal tissues.
   (d) Karyotypic analysis can be used to delineate cytogenetic subgroups in tumors of similar histology.
4. Which of the following statements applies to chromosome changes in malignant tumors?
   (a) Most mesenchymal tumors demonstrate relatively simple abnormalities.
   (b) Most cancers of epithelial origin show complex karyotypic changes.
   (c) The diverse changes in malignant tumors have been related to a stepwise process of genetic changes thought to characterize these cancers.
   (d) All of the above.
5. Which of the following observations is correct?
   (a) Amplification of the *MYCN* gene is often associated with a poor outcome for the patient with neuroblastoma.
   (b) Overexpression of the *NM23* gene is often correlated with low metastatic potential in patients with breast cancer.
   (c) Specific changes may be present in histologic subtypes, such as t(12;16) in myxoid liposarcoma, t(2;13) in alveolar rhabdomyosarcoma, and t(X; 18) in synovial sarcoma.
   (d) All of the above.
6. Which of the following may be associated with only one cytogenetic change?
   (a) Adenocarcinoma of the prostate
   (b) Synovial sarcoma
   (c) Cancer of the breast
7. For optimal cytogenetic results on surgically obtained specimen(s), which of the following should be utilized?
   (a) Freezing
   (b) Fixation
   (c) Frozen tissue

(d) Fixed tissue
(e) Fresh specimen

8. A translocation (11;22) (q24;q12) is characteristic of which tumor?
   (a) Rhabdomyosarcoma
   (b) Ewing sarcoma
   (c) Osteosarcoma
   (d) Liposarcoma
9. In differentiating a lipoma from a myxoid liposarcoma, the presence of which of the following is crucial?
   (a) Double minute chromosomes
   (b) Extra chromosomes
   (c) t(12;16)
   (d) Missing chromosomes
10. The cytogenetic changes in tumors can be found in the following:
   (a) Blood cells
   (b) Marrow cells
   (c) Tumor tissues
   (d) All of the above

CHAPTER 14

1. What is the relationship between/among the following?
   (a) Tumor suppressor gene
   (b) Loss of heterozygosity
   (c) Retinoblastoma
   (d) Familial adenomatous polyposis
2. What is the significance of a somatic versus a germline mutation in:
   (a) Retinoblastoma?
   (b) Adenomatous polyposis coli?
3. What is/are the primary gene(s) involved in:
   (a) Familial adenomatous polyposis coli?
   (b) Hereditary nonpolyposis colon cancer?

   What is the distinction between these two types of colon cancer?
4. Which of the phenomena that follow are associated with (a) ataxia telangiectasia, (b) Bloom's syndrome, (c) Fanconi's anemia, (d) xeroderma pigmentosum?
   (i) Increased chromosome breakage
   (ii) Increased frequency of sister chromatid exchange
   (iii) Formation of acrocentric, dicentric, and ring chromosomes
   (iv) Decreased response to phytohemagglutinin
   (v) Formation of quadriradials
   (vi) Rearrangements between nonhomologous chromosomes
   (vii) Sensitivity to diepoxybutane

(viii) Telomeric associations
(ix) Sensitivity to radiation

5. Match gene loci at the following chromosome locations (a–e, left column) with the diseases (i–v, right column):

(a) 16q24.3	(i) Ataxia telangiectasia
(b) 9q34.1	(ii) Werner's syndrome
(c) 11q22.3	(iii) Xeroderma pigmentosum, type A
(d) 8p12-21	(iv) Xeroderma pigmentosum, type D
(e) 19q13.2-13.3	(v) Fanconi's anemia

6. Match the disease (a–c, left column) with the neoplastic predisposition (i–v, right column) with which it is most likely to be associated:

(a) Fanconi's anemia	(i) Pulmonary fibrosis
(b) Ataxia telangiectasia	(ii) Hodgkins lymphoma
(c) Bloom's syndrome	(iii) Pancytopenia
(d) Xeroderma pigmentosum	(iv) T-cell leukemia
(e) ICF syndrome	(v) No neoplastic disorder

7. How do each of the following relate to chromosome instability or to cancer?
   (a) Telomerase
   (b) Mismatch repair
   (c) Excision repair
   (d) Fragile site
8. How does each of the following relate to fragile X syndrome?
   (a) CGG trinucleotide repeat
   (b) Anticipation
   (c) Methylation
   (d) Folate-deficient medium
9. What is the risk for each of the following for a woman who carries a 110–125 CGG repeat fragment:
   (a) That she is affected with fragile X syndrome?
   (b) That she will have sons who are affected?
   (c) That she will have sisters who are affected?
10. Which was the most likely the source of the allele that gave rise to the 110–125 CGG repeat in the woman in Question 9?
   (a) A new mutation
   (b) Her mother
   (c) Her father
   (d) Her mother or her father
   Why?
11. If the woman in Question 9 had a CGG fragment of 70–90 repeats, which would be the most likely source of her allele?
   (a) A new mutation
   (b) Her mother
   (c) Her father
   (d) Her mother or her father

12. If the woman in Question 11 had an affected sister, which parent most likely had an expanded premutation?
    Why?

CHAPTER 15

1. What primary problems contribute to difficulties in obtaining data on breast cancers from conventional cytogenetic studies?
2. Loss of chromosome 17 from breast cancer has been reported frequently. What genes on chromosome 17 are thought to be involved in breast carcinogenesis?
3. Is clonal cytogenetic aberration definitive evidence of malignancy?
4. What is the evidence that cytogenetic aberrations are biologically significant in breast cancer?
5. Why are the new methods of FISH and CGH desirable for breast cancer studies?
6. What kinds of breast samples are suitable for FISH analyses?
7. What are the major benefits of the FISH approach?
8. In the FISH data reviewed, what is the consensus of its utility?
9. How could cytogenetic analyses contribute to assessment of prognosis?
10. What are the most common sites of cytogenetic aberration in breast cancer?
11. Have the relevant genes of these common sites been clearly identified?
12. What cytogenetic evidence supports the apparent heterogeneity of breast cancer?

CHAPTER 16

1. What are hematopoietic stem cells?
2. Where are hematopoietic stem cells located?
3. Approximately how many blood cells are removed and replaced from the circulation each day?
4. What is the ratio of stem cells to nucleated bone marrow cells?
5. What cells are involved in hematopoiesis?
6. How are stem/progenitor cells processed after collection?
7. What is the probability that an individual will find a match within his/her family?
8. What are some of the advantages of peripheral blood stem/progenitor cell collection?
9. What are some of the advantages of cord blood collection?
10. What are restriction fragment length polymorphisms?
11. Can cytogenetics monitor engraftment with Y chromosome analysis if the recipient is a female and the donor is a male?
12. Can cytogenetics monitor engraftment with Y chromosome analysis if the recipient is a male and the donor is a female?

13. How are restriction fragment length polymorphisms used to monitor engraftment following stem/progenitor cell transplantation?

CHAPTER 17

1. How does FISH allow for the identification of marker chromosomes? Discuss the different types of markers.
2. For the following microdeletions, name the corresponding syndrome(s) that can result from each.
   (a) 17p13.3
   (b) 15q12
   (c) 22q11.2
   (d) 17p11.2
   (e) 7q11.23
   (f) 4p16.3
   (g) 5p15.2
   (h) Xp22.3
3. What is the major advantage of using rapid aneuploidy detection by FISH?
4. How might FISH on fetal cells detected in maternal blood be a better method for prenatal diagnosis than conventional cytogenetic analyses of amniocentesis and chorionic villus sampling?
5. What are STAT chromosomes? How are they prepared?
6. List the differences between using FISH and flow cytometry for cell analysis.
7. What is the Philadelphia chromosome (Ph), and with what cancer type is it associated?
8. How can FISH be applied in bone marrow transplants?
9. How does FISH fill a gap created by PCR-based technology in the detection of cancer?
10. How can FISH be used in rapid sex determination?
11. FISH is the most direct way to map DNA sequences onto metaphase chromosomes. Why, then, is FISH not used all the time?

CHAPTER 18

1. Comparative genomic hybridization (CGH) requires:
   (a) Normal metaphase chromosomes
   (b) Normal DNA
   (c) Tumor DNA
   (d) All of the above
   (e) Only (a) and (c)
2. The results of comparative genomic hybridization (CGH) allow one to determine:
   (a) The presence of amplifications
   (b) The presence of translocations

(c) The presence of deletions
(d) All of the above
(e) Only (a) and (c)

3. Comparative genomic hybridization (CGH) can be performed on:
   (a) Cultured cells
   (b) Formalin-fixed tissue
   (c) Fresh tissue
   (d) Paraffin-embedded tissue
   (e) All of the above
4. Spectral karyotyping (SKY) can be performed on:
   (a) Cultured cells
   (b) Formalin-fixed tissue
   (c) Fresh tissue
   (d) Paraffin-embedded tissue
   (e) Only (a) and (c)
5. Prior knowledge of the presence of a chromosomal aberration is required for the use of:
   (a) Comparative genomic hybridization (CGH)
   (b) Fluorescent in-situ hybridization (FISH)
   (c) Spectral karyotyping (SKY)
   (d) All of the above
   (e) Only (a) and (c)

CHAPTER 19

1. What is the most common karyotypic abnormality in synovial sarcoma?
2. Name a hematologic disease which is typically associated with poor specimen collection.
3. What is the safest way to transport tissue to a remote cytogenetics laboratory?
4. Name two advantages of FISH in comparison to standard cytogenetics.
5. What is the relationship between proliferative activity of a neoplasm and the success rate of standard cytogenetics?
6. Name a clinical situation in which FISH is utilized to detect minimal residual disease.
7. What kind of cell or tissue preparation is most suitable for interphase cytogenetics?
8. Name some of the most widely used enzymes for the digestion process in FISH.
9. Name two important disadvantages of using cell suspension in FISH.
10. Name some variables that influence the success of FISH on fixed tissue.

CHAPTER 20

1. What is the most objective way a laboratory can demonstrate excellence to its referring physicians?
2. Name the two professional organizations that partner to provide proficiency testing for cytogenetics laboratories.
3. Name the organization that accredits training programs in clinical cytogenetics.
4. List five functions of the cytogenetics laboratory director.
5. What is the CLIA term for a laboratory's clinical cytogeneticist?
6. What professional organization publishes the standard reference book on cytogenetic methods?
7. What professional organization publishes *Standards and Guidelines: Clinical Genetics Laboratories*?
8. Identify the U.S. federal agency that writes the regulations under CLIA.
9. Name two examples of cytogenetic "devices" regulated by the U.S. Food and Drug Administration.
10. Why is group consensus the preferred mechanism for determining the "correct" response to a PT challenge?
11. How does split-sample testing differ from proficiency testing?
12. What are the cardinal elements of laboratory performance?
13. Of the following, which is not required for CAP accreditation?
    (a) Participation in the ACMG/CAP proficiency test
    (b) Investigation of failed and suboptimal culture attempts
    (c) Use of FISH in prenatal analysis
14. List the four elements of external laboratory review.

Index